Translational Interventional Radiology

Handbook for designing and conducting clinical and translational research

Series Editor
Adam E. M. Eltorai, MD, PhD

Handbook for Designing and Conducting Clinical and Translational Research

Translational Interventional Radiology

Edited by

Adam E. M. Eltorai
Brigham & Women's Hospital, Harvard Medical School, Boston, MA, United States

Tao Liu
Department of Biostatistics, Brown University, Providence, RI, United States

Rajat Chand
Department of Radiology, University of North Carolina Medical Center, University of North Carolina School of Medicine, Chapel Hill, NC, United States

Sanjeeva P. Kalva
Department of Radiology, Massachusetts General Hospital, Harvard Medical School, Boston, MA, United States

ELSEVIER

ACADEMIC PRESS
An imprint of Elsevier

Academic Press is an imprint of Elsevier
125 London Wall, London EC2Y 5AS, United Kingdom
525 B Street, Suite 1650, San Diego, CA 92101, United States
50 Hampshire Street, 5th Floor, Cambridge, MA 02139, United States
The Boulevard, Langford Lane, Kidlington, Oxford OX5 1GB, United Kingdom

ISBN: 978-0-12-823026-8

For information on all Academic Press publications visit our website at https://www.elsevier.com/books-and-journals

Publisher: Mica H. Haley
Acquisitions Editor: Stacy Masucci
Editorial Project Manager: Barbara L. Makinster
Production Project Manager: Omer Mukthar
Cover Designer: Mark Rogers

Typeset by TNQ Technologies

Contents

* **Capt. Tej Ishaan Mehta, MD -** Interventional Radiology, The Johns Hopkins Hospital, Baltimore, MD, United States & The United States Air Force Medical Corps

List of contributors

Rayan Abboud, Wake Forest School of Medicine, Winston-Salem, NC, United States

Akanksha Acharya, W.L. Gore and Associates, Inc, Elkton, MD, United States

Ragheed Al-Dulaimi, Department of Radiology and Imaging Sciences at the University of Utah- School of Medicine, Salt Lake City, UT, United States

Sayf Al-Katib, Department of Diagnostic Radiology and Molecular Imaging, Beaumont Health, Royal Oak, MI, United States

Ahmed Al-Nowfal, University College Hospital, London, United Kingdom

Omair Ali, Department of Radiology, Section of Interventional Radiology, University of Illinois at Chicago College of Medicine, Chicago, IL, United States

Tarek Almsaddi, Oakland University William Beaumont School of Medicine, Rochester, MI, United States

Joao Amaral, Division of Interventional Radiology, Diagnostic Imaging Department, The Hospital for Sick Children, Toronto, ON, Canada

Alvin Anene, School of Medicine, Texas Christian University, Fort Worth, TX, United States; Radiology Associates of North Texas, Fort Worth, TX, United States

Christopher Awad, Emory University School of Medicine, Atlanta, GA, United States

Zain Badar, Interventional Radiology Department, University Health Network, University of Toronto, Toronto, ON, Canada

Samyuktha Balabhadra, Department of Diagnostic and Interventional Radiology, University of Texas-Houston, Houston, TX, United States

Amanda Bronte Balon, Wake Forest School of Medicine, Winston-Salem, NC, United States

Shelly Bhanot, Rush University, Chicago, IL, United States

Walter G. Bircher, Yale University, New Haven, CT, United States

Ryan Bitar, University of Texas Health Science Center San Antonio, San Antonio, TX, United States

Scott Bittle, Diagnostic Radiology Residency, Baylor University Medical Center, Dallas, TX, United States

William Borror, Department of Diagnostic and Interventional Imaging, University of Texas, Houston, TX, United States

Maryam Boumezrag, Medstar Georgetown University Hospital, Washington, DC, United States

Daniel Braga, Department of Vascular and Interventional Radiology, Mount Sinai Medical Center, Miami Beach, FL, United States

Sakshum Chadha, Rutgers New Jersey Medical School, Newark, NJ, United States

Kristina M. Chapple, Trauma and Acute Care Surgery Department, St. Joseph's Hospital and Medical Center, Phoenix, AZ, United States

George Koshy Chiramel, Department of Diagnostic Imaging, Hospital for Sick Children, Toronto, ON, Canada

R. Chitra, Department of Anatomy, Guntur Medical College, Guntur, Andhra Pradesh, India

Arun Chockalingam, Department of Radiology, Mount Auburn Hospital/Harvard Medical School, Cambridge, MA, United States

Priyam Choudhury, Department of Radiology, Mayo Clinic, Jacksonville, FL, United States

Geoffrey D. Clarke, University of Texas Health Science Center San Antonio, San Antonio, TX, United States

Kimberly Coffman, School of Public Health, The University of Texas Health Science Center, Houston, TX, United States

Sydney Cooper, Michigan State University College of Osteopathic Medicine, MI, United States

Kyle Cooper, Department of Interventional Radiology, Loma Linda University, Loma Linda, CA, United States; Department of Radiology, University of Michigan Health Systems, Ann Arbor, MI, United States

Santiago M. Cornejo, Department of Radiology, St. Joseph's Hospital and Medical Center, Phoenix, AZ, United States

Joshua Cornman-Homonoff, Department of Radiology and Biomedical Imaging, Yale University School of Medicine, New Haven, CT, United States

Brian Covello, Department of Interventional Radiology, HCA Florida Aventura Hospital, Aventura, FL, United States

Chaitu Dandu, Wayne State University School of Medicine, Detroit, MI, United States

Jennifer J. Dennison, Vascular and Interventional Radiology, Medical College of Wisconsin, Milwaukee, WI, United States

Ashwin Deshmukh, Department of Radiology, Yale New Haven Health Bridgeport Hospital, Bridgeport, CT, United States

Purushottam K. Dixit, Department of Radiology, Division of Interventional Radiology, Beaumont Health, Royal Oak, MI, United States

Rachel Drummey, The Russell H. Morgan Department of Radiology and Radiological Science, The Johns Hopkins Hospital, Baltimore, MD, United States

Sean Duguay, Department of Radiological Sciences, University of Oklahoma College of Medicine, University of Oklahoma Health Sciences Center, Oklahoma City, OK, United States

Stephanie S. Dybicz, Department of Radiology, University of Minnesota, Minneapolis, MN, United States

N. El Sehemawi, Faculty of Medicine, University of Montreal, Montreal, QC, Canada

Marvee Espiritu, Oakland University William Beaumont School of Medicine, Rochester, MI, United States

Brandon Ewing, Department of Radiology, University at Buffalo Jacobs School of Medicine and Biomedical Sciences, Buffalo, NY, United States

Ahmed Farag, Interventional Radiology Fellowship, Baylor University Medical Center, Dallas, TX, United States

Jason A. Fisher, Hartford Hospital Radiology, Hartford, CT, United States

Wylie T. Foss, University of Texas Houston MD Anderson Cancer Center, Houston, TX, United States

Roberto Fourzali, Department of Interventional Radiology, HCA Florida Aventura Hospital, Aventura, FL, United States

Katherine Shin-Ying Fu, University of Florida College of Medicine, Gainesville, FL, United States

Ron C. Gaba, Department of Radiology, College of Medicine University of Illinois at Chicago, Chicago, IL, United States

Gaurav Gadodia, Division of Vascular and Interventional Radiology, Department of Radiology, Medical College of Wisconsin, Milwaukee, WI, United States

Tushar Garg, Division of Vascular and Interventional Radiology, Russell H. Morgan Department of Radiology and Radiological Science, The Johns Hopkins University School of Medicine, Baltimore, MD, United States

Sona Ghorashi, University of Maryland School of Pharmacy, Baltimore, MD, United States

Brianna L. Gibney, Oakland University William Beaumont School of Medicine, Rochester, MI, United States

Alexandra Gilbert, Leeds Institute of Medical Research at St. James's, University of Leeds, Leeds, United Kingdom

Andrew C. Gordon, Department of Radiology, Section of Interventional Radiology, Northwestern University, Feinberg School of Medicine, Chicago, IL, United States

Aakash N. Gupta, Department of Radiology, Section of Interventional Radiology, Northwestern University, Feinberg School of Medicine, Chicago, IL, United States

Matthew Henry, Wayne State University School of Medicine, Detroit, MI, United States; Rackham Graduate School, University of Michigan, Ann Arbor, MI, United States; Department of Radiology, Medical College of Wisconsin, Milwaukee, WI, United States

Mauricio Hernandez, Diagnostic Radiology, UF College of Medicine, Jacksonville, FL, United States

Amber Hood, University of Oklahoma College of Medicine, University of Oklahoma Health Sciences Center, Oklahoma City, OK, United States

Neil K. Jain, MedStar Georgetown University Hospital, Washington, DC, United States

Jalil Kalantari, Department of Radiology, Section of Interventional Radiology, Loma Linda School of Medicine, Loma Linda, CA, United States

Arun Kamireddy, Department of Interventional Radiology, Medical College of Wisconsin, Milwaukee, WI, United States

Joshua Katz, Rutgers Robert Wood Johnson Medical School, New Brunswick, NJ, United States

Nathan D. Kauffman, Michigan State University College of Human Medicine, Grand Rapids, MI, United States

Charissa Kim, Department of Radiology, Beth Israel Deaconess Medical Center, Boston, MA, United States

Daniel Kirkpatrick, Department of Radiology, Michigan Medicine, University of Michigan, Ann Arbor, MI, United States

Joshua Kogan, Department of Radiology, Vanderbilt University Medical Center, Nashville, TN, United States

Menelaos Konstantinidis, Institute of Health Policy, Management and Evaluation, University of Toronto, Toronto, ON, Canada; Division of Biostatistics, Dalla Lana School of Public Health, University of Toronto, Toronto, ON, Canada; Child Health Evaluative Sciences, Hospital for Sick Children, Toronto, ON, Canada

Stefan Kovac, Heersink School of Medicine, The University of Alabama at Birmingham, Birmingham, AL, United States

Adam Christopher Krajewski, Interventional Radiology Department, University Health Network, University of Toronto, Toronto, ON, Canada

Jonas Kruse, School of Medicine, Texas Christian University, Fort Worth, TX, United States

E.A. Lalla, Centre for Research in Earth and Space Science, Lassonde School of Engineering, York University, Toronto, ON, Canada

Rebecca Tuan Le, Diagnostic Radiology, Rochester General Hospital, Rochester, NY, United States

Robert J. Lewandowski, Department of Radiology, Section of Interventional Radiology, Northwestern University, Feinberg School of Medicine, Chicago, IL, United States; Department of Medicine, Division of Hematology and Oncology, Northwestern University, Feinberg School of Medicine, Chicago, IL, United States; Department of Surgery, Division of Transplant Surgery, Northwestern University, Feinberg School of Medicine, Chicago, IL, United States

Millie Liao, Department of Interventional Radiology, Loma Linda University, Loma Linda, CA, United States

Tao Liu, Department of Biostatistics, Brown University, Providence, RI, United States

Abraham Liu, Department of Biostatistics, Brown University, Providence, RI, United States

G. Lopez-Reyes, Universidad de Valladolid, Valladolid, Spain

Thaddeus Maguire, Vascular and Interventional Radiology, Medical College of Wisconsin, Milwaukee, WI, United States

Gregory C. Makris, Department of Endovascular and Interventional Radiology, Guys and St. Thomas' Hospitals NHS Foundation Trust, London, United Kingdom

Alexander Martinek, Morehouse School of Medicine, Atlanta, GA, United States

Travis William McCain Pebror, Interventional and Diagnostic Radiology, Department of Radiology and Imaging Sciences, Indiana University School of Medicine Indianapolis, IN, United States

Delaney McGuirt, University of Maryland School of Pharmacy, Baltimore, MD, United States

Capt. Tej Ishaan Mehta, Department of Interventional Radiology, The Johns Hopkins Hospital, Baltimore, MD, United States; United States Air Force Medical Corps, Falls Church, VA, United States

Travis E. Meyer, Interventional Radiology, UF College of Medicine, Jacksonville, FL, United States

Syamak Moattari, Health Sciences Department, Worcester State University, Worcester, MA, United States

Ashutosh Mohapatra, Department of Orthopaedics, Mohapatra Hospital, Ulhasnagar, Maharashtra, India; Department of Radiology, Mayo Clinic, Jacksonville, FL, United States

Babak Mohit, Sleep Disorders Center, Division of Pulmonary and Critical Care Medicine, Department of Medicine, University of Maryland School of Medicine, Baltimore, MD, United States

David Bradley Money, University of Southern Florida Morsani College of Medicine, Tampa, FL, United States

John T. Moon, Division of Interventional Radiology and Image-Guided Medicine, Department of Radiology and Imaging Sciences, Emory University School of Medicine, Atlanta, GA, United States

Satya K. Morar, Integrated Medical Student, Case Western Reserve University School of Medicine, Cleveland, OH, United States

Yechiel Mor, Apogee Physicians, Scottsdale, Arizona, United States

Pranav Moudgil, Oakland University William Beaumont School of Medicine, Rochester, MI, United States

Sandeep Murthy, University of Virginia School of Medicine, Charlottesville, VA, United States

Shashidhara Murthy, Mount Sinai Health System, New York, NY, United States; Department of Interventional Radiology, Icahn School of Medicine at Mount Sinai Hospital, New York, NY, United States

Prakash Muthusami, Department of Diagnostic Imaging, Hospital for Sick Children, Toronto, ON, Canada

Girish B. Nair, Division of Pulmonary and Critical Care Medicine, Beaumont Health, Royal Oak, MI, United States

Mark Nassar, Interventional Radiology Department, University Health Network, University of Toronto, Toronto, ON, Canada

Nariman Nezami, Division of Vascular and Interventional Radiology, Department of Diagnostic Radiology and Nuclear Medicine, University of Maryland School of Medicine, Baltimore, MD, United States; Experimental Therapeutics Program, University of Maryland Marlene and Stewart Greenebaum Comprehensive Cancer Center, Baltimore, MD, United States

Han G. Ngo, Oakland University William Beaumont School of Medicine, Rochester, MI, United States

S. Nourouzpour, Toronto General Hospital, University Health Network, Toronto, ON, Canada

Brandon Olivieri, Department of Vascular and Interventional Radiology, Mount Sinai Medical Center, Miami Beach, FL, United States

Christopher Ovanez, Russell H. Morgan Department of Radiology and Radiological Science, Johns Hopkins Hospital, Baltimore, MD, United States

Merve Ozen, University of Kentucky, Lexington, KY, United States

Matt Parker, University of Texas Health Science Center San Antonio, San Antonio, TX, United States

Shrey Patel, Interventional Radiologist, Baylor Radiologists, Houston, TX, United States

Neal Patel, Nova Southeastern University Dr. Kiran C. Patel College of Osteopathic Medicine, Ft. Lauderdale, FL, United States

Mounica Paturu, Department of Neurosurgery, Duke University Medical Center, Durham, NC, United States

Eric Pham, University of Oklahoma College of Medicine, University of Oklahoma Health Sciences Center, Oklahoma City, OK, United States

Zahi Qamhawi, Department of Radiology, Oxford University Hospitals, Oxford, United Kingdom

Ranjan Ragulojan, University of Minnesota, Minneapolis, MN, United States

Ishmael Raheem, Wake Forest School of Medicine, Winston-Salem, NC, United States

Shakthi Kumaran Ramasamy, Department of Radiology, Stanford University School of Medicine, Stanford, CA, United States

Husayn F. Ramji, University of Oklahoma College of Medicine, University of Oklahoma Health Sciences Center, Oklahoma City, OK, United States; Hudson College of Public Health, University of Oklahoma Health Sciences Center, Oklahoma City, OK, United States

Daniel Reyes, Department of Vascular and Interventional Radiology, Mount Sinai Medical Center, Miami Beach, FL, United States

Conner D. Reynolds, Transitional Year Residency, John Peter Smith Health Network, Fort Worth, TX, United States; Diagnostic Radiology Residency, Department of Medical Imaging, University of Arizona College of Medicine, Tucson, AZ, United States

Tony H. Rizk, Medical University of South Carolina, Charleston, SC, United States

Karishma Shah, Botnar Research Centre, Nuffield Department of Orthopaedics, Rheumatology and Musculoskeletal Sciences, University of Oxford, Oxford, United Kingdom

Raj Shah, Department of Radiology, Section of Interventional Radiology, University of Illinois at Chicago College of Medicine, Chicago, IL, United States

Jatin Sharma, Oakland University William Beaumont School of Medicine, Rochester, MI, United States

Rahul A. Sheth, Department of Interventional Radiology, The University of Texas MD Anderson Cancer Center, Houston, TX, United States

Li Ka Shing, Knowledge Institute, St. Michael's Hospital, Unity Health Toronto, Toronto, ON, Canada

Apurva Shrigiriwar, Division of Gastroenterology and Hepatology, The Johns Hopkins University School of Medicine, Baltimore, MD, United States

Zachary T. Smith, Robert Larner, M.D. College of Medicine at the University of Vermont, Burlington, VT, United States

Alex J. Solomon, The Johns Hopkins University School of Medicine, Baltimore, MD, United States

Kilari Sreenivasulu, Departments of Radiology Mayo Clinic, Vascular and Interventional Translational Laboratory, Rochester, MN, United States

Mehdi Taghipour, Einstein Healthcare Network, Philadelphia, PA, United States

Tulasi Talluri, Oakland University William Beaumont School of Medicine, Rochester, MI, United States

Benedict Thomson, Department of Interventional Radiology, Guy's and St. Thomas' NHS Foundation Trust, London, United Kingdom

Siddhant Thukral, Division of Vascular and Interventional Radiology, Department of Radiology, Duke University Medical Center, Durham, NC, United States

Ravi Tyagi, Department of Radiology, Section of Interventional Radiology, Rush University Medical Center, Chicago, IL, United States

M.Veneranda, Universidad de Valladolid, Valladolid, Spain

Siddharth Venkatraman, The Johns Hopkins University School of Medicine, Baltimore, MD, United States

Nicholas Vollano, College of Osteopathic Medicine, Kansas City University, Joplin, MO, United States

William Wagstaff, The University of Chicago Pritzker School of Medicine, Chicago, IL, United States

John Walker, University of Texas Health Science Center San Antonio, San Antonio, TX, United States

Linzi Arndt Webster, Emory University School of Medicine Department of Radiology, Division of Interventional Radiology and Image Guided Medicine, Atlanta, GA, United States

Austin-Marley Windham-Herman, Yale University, New Haven, CT, United States

Gregory J. Woodhead, Department of Medical Imaging, University of Arizona College of Medicine, Tucson, AZ, United States

Jim Zhong, Leeds Institute of Medical Research at St. James's, University of Leeds, Leeds, United Kingdom

Preface

Translational research is essential to the advancement of Interventional Radiology (IR). Written for clinicians, scientists, students, and biotech/medtech entrepreneurs, this IR-specific guide to translational research serves as a practical, step-by-step roadmap for designing and conducting basic, translational, and clinical research, including basic regulatory considerations for new technology. *Translational Interventional Radiology* aims to help bridge the gap between clinical problems, research, and practice. Comprehensively spanning from preclinical research, clinical research, clinical implementation, to public health, the book provides a clear process for understanding, designing, executing, and analyzing clinical and translational research.

The reader will learn how to more critically evaluate the quality of published studies with respect to measuring outcomes and to make effective use of all types of evidence in patient care. This book covers the principles of evidence-based medicine and applies these principles to the design of translational investigations.

The book's straightforward approach aims to help the aspiring investigator navigate challenging considerations in study design and implementation. Researchers will benefit from greater confidence in their ability to initiate and execute investigations along the translational spectrum, gain valuable insights on funding, avoid common pitfalls, and know what is needed in collaborators.

Whether you are a basic scientist with a promising therapeutic agent interested in moving from laboratory to clinical experimentation, an engineer needing to complete a pilot study for a new biomedical device, a student learning the principles of statistics or epidemiology, a clinician orchestrating and preparing grants for a randomized clinical trial, or an entrepreneur seeking insight on regulatory considerations, *Translational Interventional Radiology* is a practical guide for effectively advancing and translating an IR research question.

Adam E. M. Eltorai

April 2023

Part I

Introduction

Chapter 1

Translational process

Priyam Choudhury and Ashutosh Mohapatra
Department of Radiology, Mayo Clinic, Jacksonville, FL, United States

Key points

- Translational Research (TR) aims to build on basic scientific research to create new therapies, medical procedures, and diagnostics for the betterment of public health.
- Collaboration between multidisciplinary investigators is crucial for TR success.
- TR results in public policy changes and is also beneficial for the community's economy.

Translational research (TR) aims to improve public health by applying the outcomes of multidirectional and multidisciplinary research based on basic, clinical, patient, population, and policy-based science. It promotes collaboration among various disciplines, translating knowledge from one science to another (for example, from basic science to clinical science)[1] to mitigate health concerns and generate novel ideas. TR provides public health benefits, clinical benefits, and economic benefits by adopting new tools and procedures into practice settings and amending the policy-making process formally.

While the definition of TR has been less clear and was first associated with cancer research in the 1990s, today, various fields have attempted to define the term. The NIH defines TR, while applying for the CTSA (Clinical and Translational Science Award) program, as encompassing two areas of translation: applying discoveries generated through research in the laboratory and preclinical studies to the development of trials and studies in humans and enhancing and adopting best practices in the community through research. Treatment strategies and cost-effectiveness in preventing diseases are also essential parts of translational science.[1]

In the continuum, T1 refers to the first part of research where knowledge is transferred from basic research to clinical research, and T2 refers to the second part of research that transfers findings from clinical research to practice, aiding in improving the community's health.

Successful TR examples include cancer therapy, requiring collaboration between basic researchers, clinicians, and industry to generate target compounds with reduced toxicity and enhanced efficacy,[2] the introduction of insulin, and the advent of antibiotics.

The National Center for Advancing Translational Sciences (NCATS) describes translational research broadly as the process in which observations in the laboratory, clinic, and community are translated into interventions to improve the people's health in the community. It involves implementing diagnostics and therapeutics in medical procedures and behavioral changes in the community.[3]

The translational science is not unidirectional, but rather, each stage is built on information acquired from other stages. It consists of stages starting from the basics of health and disease to interventions in improving public health. Patient involvement in all stages is crucial to develop new approaches, demonstrate their usefulness, and disseminate those findings in treatment (Fig. 1.1).

In June 2013, the Institute of Medicine (IOM), which supplies major investment to **University of Wisconsin Institute for Clinical and Translational Research (ICTR)** and more than 60 CTSA sites nationwide, made recommendations to change research classification definitions. The T0–T4 classification replaced the T1 and T2 categories.[4]

T0 Research includes basic biomedical research, including preclinical and animal studies, not including interventions on human subjects.

T1 Research encompasses translation to humans, including proof of concept studies, Phase 1 clinical trials,

Translational Interventional Radiology. https://doi.org/10.1016/B978-0-12-823026-8.00019-5

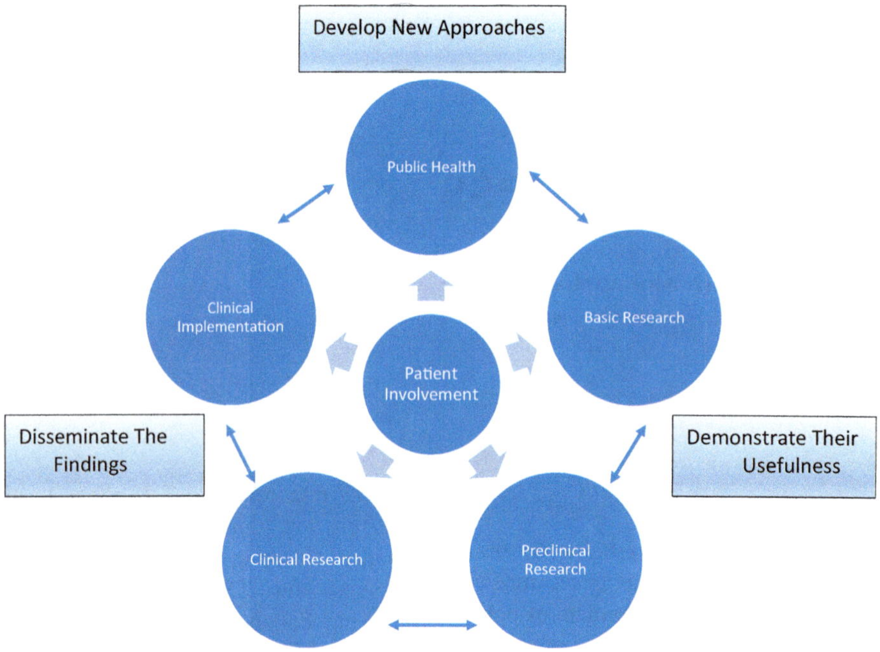

FIGURE 1.1 Stages in translational research.

and focus on new methods of diagnosis, treatment, and prevention in highly controlled settings.

T2 Research entails translation to patients, including Phase 2 and 3 clinical trials, and controlled studies leading to clinical application and evidence-based guidelines.

T3 Research includes translation to practice, including comparative effectiveness research, postmarketing studies, clinical outcomes research, as well as health services, and dissemination and implementation research; and

T4 Research translates to communities, including population level outcomes research, monitoring of morbidity, mortality, benefits and risks, and impacts of policy and change.

TR programs are trying to design an effective training program where trainees are expected to develop certain skills and knowledge that are not traditionally taught in training. Although it's challenging to track outcomes as every individual program and trainee has different needs and expectations, it can be documented whether the pre-established goals were met or not. The Institute for Clinical and Translational Sciences (ICTS) Tracking and Evaluation Team developed the Translational Science Benefits Model (TSBM), which can be used as a tool to assess and evaluate the outcome to measure community health benefits and clinical benefits from TR.[5] They plan to form a resource library which can be freely accessed by academicians, evaluators, and scientists who wish to evaluate effort.

TR does not come without barriers despite aiming at improving patient's health. Cultural and training differences between groups of investigators, lack of communication between the groups, lack of resources and the complex regulatory issues that encompass tissue banking, FDA, ethics involved in human research, and Medicines and Healthcare Regulatory Agency (MHRA) approvals—all these factors can be intimidating to the researchers.[6] These issues can be overcome with support from regulatory bodies, as well as by the building of and collaboration between interdisciplinary forums. In the United States of America, the CTSA scheme in 2006 and the NIH Roadmap in 2003 were designed to support TR.[7]

References

1. Rubio DM, Schoenbaum EE, Lee LS, et al. Defining translational research: implications for training. *Acad Med.* 2010;85(3):470–475. https://doi.org/10.1097/ACM.0b013e3181ccd618.
2. Goldblatt EM, Lee WH. From bench to bedside: the growing use of translational research in cancer medicine. *Am J Transl Res.* 2010;2(1):1–18. Published 2010 January 1.

3. *Ncats.nih.gov*; 2021 [online] Available at: https://ncats.nih.gov/files/NCATS_Factsheet_508.pdf [Accessed 17 March 2021].
4. *What Are the T0 to T4 Research Classifications?* ICTR; 2021. https://ictr.wisc.edu/what-are-the-t0-to-t4-research-classifications/ [Accessed 17 March 2021].
5. Luke DA, Sarli CC, Suiter AM, et al. The translational science benefits Model: a new framework for assessing the health and societal benefits of clinical and translational sciences. *Clin Transl Sci.* 2018;11(1):77−84. https://doi.org/10.1111/cts.12495.
6. Homer-Vanniasinkam S, Tsui J. The continuing challenges of translational research: clinician-scientists' perspective. *Cardiol Res Pract.* 2012;2012:246710. https://doi.org/10.1155/2012/246710.
7. Zerhouni E. Medicine. The NIH Roadmap. *Science*. 2003;302(5642):63−72. https://doi.org/10.1126/science.1091867.

Chapter 2

Scientific method

Joshua Katz[1] and Zachary T. Smith[2]

[1]Rutgers Robert Wood Johnson Medical School, New Brunswick, NJ, United States; [2]Robert Larner, M.D. College of Medicine at the University of Vermont, Burlington, VT, United States

Key points

- The scientific method is an organized process of reproducible discovery.
- It consists of Observation, Question, Hypothesis, Experiment, Analysis, and Conclusion.
- Proper understanding of these steps is important to produce reliable research results.

Why it matters?

The scientific method is the fundamental means by which humanity makes reliable discoveries. It consists of six core steps, as illustrated in Fig. 2.1. Understanding these components and their limitations is crucial to successfully and accurately uncover, report, and apply knowledge.

In medicine, the rigorous process of inquiry enabled by the scientific method allows researchers to identify pathologies and develop new technologies in an organized and verifiable manner. This process ensures that the knowledge gained is reliable and can be used to make informed decisions.

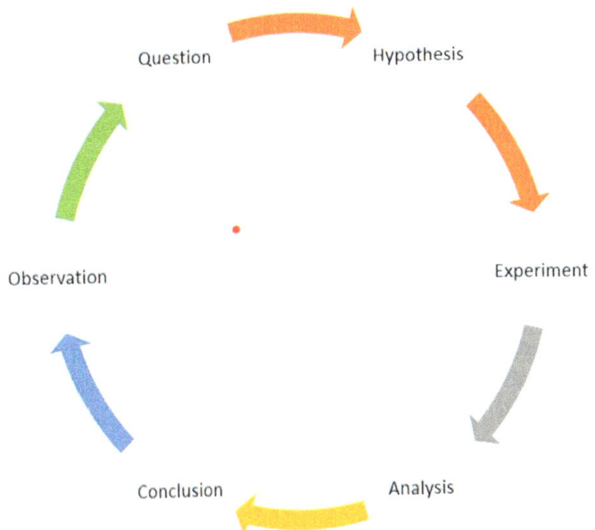

FIGURE 2.1 Steps of the scientific method

Observation

The scientific method begins with observation, which kickstarts the entire process. While this step may seem simple relative to the ones to come, its success is critical to the entire process. Only by paying attention to the world around them can researchers recognize the surprising or unknown and find inspiration to ask novel questions or solve interesting problems.

For instance, a clinician may notice that a certain subset of her patients experiences fewer complications than another or may observe that patients on a particular medication regimen have decreased mortality. Observation is fundamental to question generation and is the initial step in making important discoveries.

Question

The research question forms the basis for the remainder of the scientific process (as shown in Fig. 2.2). A quality research question can be a novel question that has not been

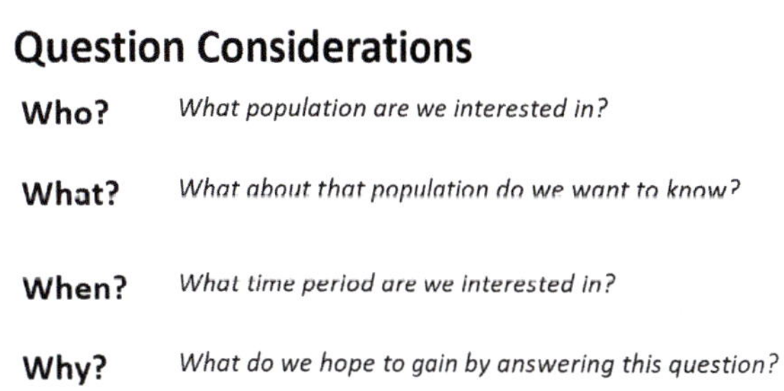

FIGURE 2.2 Considerations when asking a research question.

Translational Interventional Radiology. https://doi.org/10.1016/B978-0-12-823026-8.00043-2

asked before, or a question that has been previously investigated, but the current literature does not provide a definitive answer. By asking novel or unanswered questions, the investigator can ensure that the research is relevant and worthwhile.

It is essential to note that research questions are always evolving, and discovering answers to previously asked questions leads to the development of new and novel questions, allowing for a better understanding of the subject matter. For example, Breen et al. decided to investigate the efficacy of percutaneous cryoablation for renal tumors.[1] Following the advent of percutaneous ablation for renal cell carcinoma, Lourenco et al. questioned and compared outcomes of percutaneous ablation versus nephrectomy in patients with T1 renal cell carcinoma.[2] New discoveries always lead to the development of new and novel research questions.

Before asking the research question, it is crucial to be well-versed in the current published literature. Developing a thorough understanding of the literature allows for a more thoughtful and meaningful research question to be asked. Additionally, current NIH-funded projects underway can be searched online using the NIH Research Portfolio Online Reporting Tools at report.nih.gov.

It is common for studies to ask more than one research question. Having secondary research questions can add complexity to the study by introducing multiple variables and complicating data analysis, but this will be discussed in later chapters on study design.

Hypothesis

Once good research questions have been formulated, it is crucial for the researcher to consider the possible answers they may expect. This is important as generating hypotheses forces a more in-depth consideration of the exact questions to be asked and clarifies the means by which answers will be found during experimentation.

For example, when a clinician generates a differential diagnosis list for a given patient's concern, such as new-onset calf pain and swelling, they are generating a set of hypotheses to answer the question, "What is going on here?" Potentially, there was trauma to the limb, or the patient may have a deep vein thrombosis (DVT), or there could be a superficial infection. These hypotheses then guide the clinician's next steps: to ask about trauma and look for signs of injury, to inquire about risk factors for DVT and perform a doppler ultrasound, and to observe for signs of infection.

Hypothesis generation, therefore, helps determine the specifics of the next step, experimentation. It allows researchers to identify the variables to be tested, the methods for data collection, and the statistical analyses to be performed. Overall, hypothesis generation is a critical component of the scientific process that ensures the validity and reliability of research findings.

Experiment

Experimentation can take many forms depending on the research question, ranging from observation of risk factors or disease incidence in a specified population to randomized manipulation of a single variable between two groups (as shown in Fig. 2.3). In this step, it is essential to consider extraneous factors that may affect the collected data. The most powerful studies examine the effect of changing one variable between large sample groups and control for other differences between the groups through study design adjustments.

An illustrative example of good experimental design is seen in Granger et al.'s trial, which compared apixaban versus warfarin for patients with atrial fibrillation. The study found that apixaban is superior to warfarin in preventing stroke or systemic embolism, caused less bleeding, and resulted in lower mortality.[3] This was an effective trial as it was randomized, double-blinded, had a large sample of 18,201 patients, and was focused on assessing one variable's effect on outcomes.

Overall, experimentation is a critical step in the scientific method as it allows researchers to test their hypotheses and generate new knowledge. However, it is essential to ensure that the experimental design is rigorous and well-controlled to minimize bias and ensure the validity and reliability of the findings.

Analysis

Once the experiment has been conducted and the data has been collected, the next step is to perform an analysis of the results. This is generally done through a series of statistical tests that allow the investigator to objectively understand the results.

The analysis process is highly variable and depends heavily on the design of the study, as certain statistical tests are preferred for different experimental designs. For instance, parametric tests, such as t-tests and ANOVA, are commonly used for normally distributed continuous data, while non-parametric tests, such as Wilcoxon rank-sum and Kruskal-Wallis tests, are used for non-normally distributed or ordinal data.

In addition to selecting the appropriate statistical test, it is crucial to consider potential confounding variables, control for them through statistical adjustments, and interpret the results with caution. Various statistical methods of analyzing results of a study will be discussed in detail in later chapters, including the advantages and limitations of different statistical tests and strategies to minimize bias and improve the accuracy and generalizability of the findings.

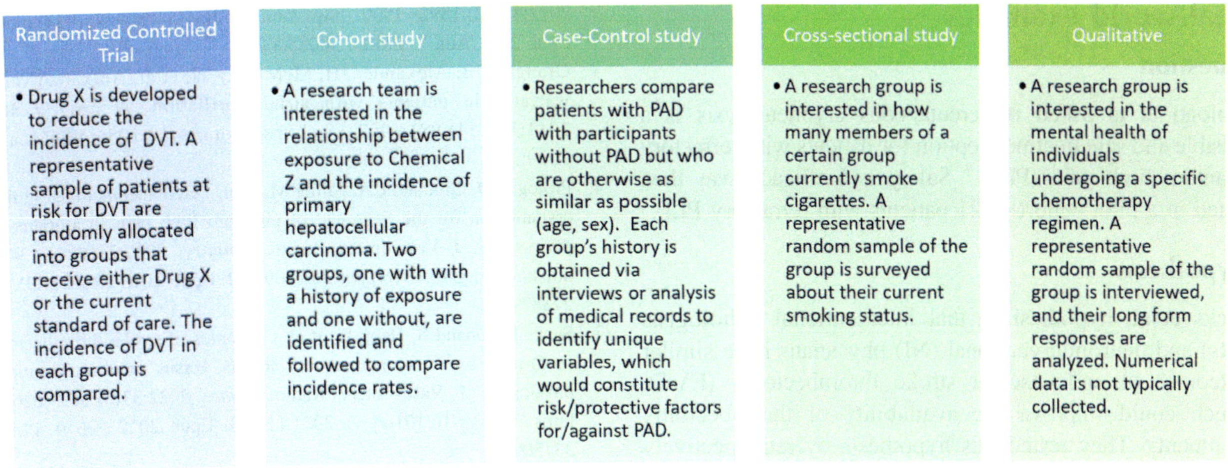

FIGURE 2.3 Types of experiments with examples. *DVT*, deep vein thrombosis. *PAD*, peripheral artery disease.

Conclusion

After the data has been analyzed, and the results are available for interpretation, the investigator needs to return to the initial hypotheses made before implementing the experiment. Based on the findings of the study, the researcher makes an informed decision on how the study's results relate to the initial hypotheses. If the study's results support the hypotheses, the researcher can consider them validated. However, if the results contradict the hypotheses, the researcher needs to re-evaluate their assumptions and develop new hypotheses that can explain the findings.

Interesting findings may serve as the observations that spark continued iteration of the scientific method. The researcher may use the findings to develop new research questions or to refine existing hypotheses, leading to further studies that build upon the previous work. The scientific method is, therefore, an iterative process that relies on the constant re-evaluation of assumptions and the refinement of research questions to generate new knowledge and improve our understanding of the natural world.

Get started

Now that the fundamentals of the scientific method have been reviewed, it is time to begin refining the research question and establish the study population. Please reference the following chapters for further discussion regarding these next steps:

Chapter 19 – "The question: Types of research questions and how to develop them"

Chapter 20 – "Study population: Who and why them?"

Potential pitfalls

- Poorly defined variables
  - Before jumping to experimental design, it is crucial to clearly define the variables under investigation and the metrics by which they will be measured. For instance, a researcher may wonder, "Is percutaneous biliary drainage safer than cholecystectomy?" However, they must define the safety metrics, such as all-cause mortality, hospital readmission rate, complication rate, and others, before beginning the experiment. Failure to do so may result in data being collected or analyzed incorrectly, leading to inaccurate conclusions.
- Not controlling for variation between study groups
  - During experimentation, it is essential to control for as many variables as possible to avoid confounding factors. For example, a researcher planning to compare the safety of two procedures must ensure that the patient populations receiving each procedure are similar in terms of illness severity and risk factors. Failure to do so may lead to inaccurate conclusions and flawed study design.
- Understanding limitations
  - Researchers must accurately understand the generalizability of the results of any study. It is crucial to recruit a large, representative sample of the population under investigation to extrapolate the findings to the larger population of interest. Additionally, researchers must manage expectations and accurately report the impact of any study. For example, a study examining the life expectancy of patients diagnosed with pancreatic cancer in a single urban tertiary care center in the United States may not be applicable to rural patients in another country, or even to rural patients in the same state. Understanding the study's limitations and reporting them accurately is crucial for scientific integrity.

Real-world examples

Question

Prologo et al asked if percutaneous cryoneurolysis is a feasible and safe treatment option for patients with refractory phantom limb pain (PLP).[4] Safety and efficacy was then tested in a pilot cohort of 21 patients with refractory PLP.

Hypothesis

Sacks et al. hypothesized that interventional radiologists (IRs) and neurointerventional (NI) physicians have similar outcomes of endovascular stroke thrombectomy (EVT), which could improve the availability of thrombectomy treatment.[5] They tested this hypothesis by retrospectively analyzing outcomes between IRs and NI physicians across eight institutions.

Experiment

In an effort to explore the potential applications of histotripsy, Swietlik et al designed an experiment to evaluate the feasibility and safety of in-vivo histrotripsy of subcutaneous fat in a porcine model.[6] Subcutaneous fat volume reduction was measured through serial MRI at 7, 28, and 56 days post-treatment.

References

1. Breen DJ, Bryant TJ, Abbas A, et al. Percutaneous cryoablation of renal tumours: outcomes from 171 tumours in 147 patients. *BJU Int.* October 2013;112(6):758–765. https://doi.org/10.1111/bju.12122. Epub 2013 Apr 12. PMID: 23581293.
2. Lourenco P, Bilbey N, Gong B, Bahrabadi A, Halkier B. Percutaneous ablation versus nephrectomy for small renal masses: clinical outcomes in a single-center cohort. *Cardiovasc Intervent Radiol.* December 2018;41(12):1892–1900. https://doi.org/10.1007/s00270-018-2050-9. Epub 2018 Aug 7. PMID: 30088062.
3. Granger CB, Alexander JH, McMurray JJ, et al. Apixaban versus warfarin in patients with atrial fibrillation. *N Engl J Med.* 2011;365(11):981–992. https://doi.org/10.1056/NEJMoa1107039.
4. Prologo JD, Gilliland CA, Miller M, et al. Percutaneous image-guided cryoablation for the treatment of phantom limb pain in amputees: a pilot study. J Vasc Interv Radiol. *January.* 2017;28(1):24–34.e4. https://doi.org/10.1016/j.jvir.2016.09.020. Epub 2016 Nov 23. PMID: 27887967.
5. Sacks D, Dhand S, Hegg R, et al. Outcomes of stroke thrombectomy performed by interventional radiologists versus neurointerventional physicians. J Vasc Interv Radiol. *June.* 2022;33(6):619–626.e1. https://doi.org/10.1016/j.jvir.2021.11.018. Epub 2022 Feb 9. PMID: 35150837.
6. Swietlik JF, Knott EA, Longo KC, et al. Histotripsy of subcutaneous fat in a live porcine model. Cardiovasc Intervent Radiol. *September.* 2022 12. https://doi.org/10.1007/s00270-022-03262-4. Epub ahead of print. PMID: 36097074.

Additional reading

Reiff-Cox R. Exchanging the myth of a step-by-step scientific method for a more authentic description of inquiry in practice. In: McComas W, ed. *Nature of Science in Science Instruction. Science: Philosophy, History and Education.* Cham: Springer; 2020. https://doi-org.proxy.libraries.rutgers.edu/10.1007/978-3-030-57239-6_6.

Gauch JHG. *Scientific Method in Brief.* ProQuest Ebook Central; 2013. https://ebookcentral-proquest-com.proxy.libraries.rutgers.edu.

Further reading

1. *InformedHealth.org [Internet].* Cologne, Germany: Institute for Quality and Efficiency in Health Care (IQWiG); 2006. What types of studies are there? 2016 Jun 15 [Updated 2016 Sep 8]. Available from: https://www.ncbi.nlm.nih.gov/books/NBK390304/.

Chapter 3

Basic science research

Amanda Bronte Balon
Wake Forest School of Medicine, Winston-Salem, NC, United States

Introduction to basic science research

Basic science research is a fundamental study incorporating the principles of biological sciences with the central aim of discovering, understanding, and refining scientific theories and mechanisms.[1] This type of research encompasses all forms of experimentation that are directed toward increasing the fundamental knowledge of observable scientific phenomena. This research includes a multitude of subjects, including biochemistry, microbiology, physiology, pharmacology, and pathology extending to social and behavioral sciences. The rationale for basic research in medical and information science is best summarized by Horst Kunge: "Learning to master theoretically and in practical application, the ground rules of research creates the best foundation for continuing growth in a profession."[2] This rationale is especially applicable within the field of medicine, a field in which the structure of the profession relies on providing advice to others derived from a body of generalized and systemic knowledge that comprises its theoretical core.

The scientific method of inquiry

The process of research involves systematic collection and analysis of data to facilitate organized exploration of scientific principles. Basic science research can be categorized into quantitative and qualitative methods, which are highly structured and rely on purposeful measurement, analysis, and evaluation.[3] The observation of scientific phenomena leads to inductive reasoning, contributing to the development of the scientific method of inquiry.[4] This method of scientific discovery and research gained support in the 16th century and has remained a valid method for answering questions and solving problems in modern research.[4] The scientific method of inquiry generally starts with identifying a problem or question. The scientific method initially involved identifying a problem or interest as a lead point to theory construction, resulting in the derivation of theoretical hypotheses.[3] Frankfort-Nachmias and Nachmias describe the research process as consisting of seven principal stages: problem, hypothesis, research design, measurement, data collection, data analysis, and generalization.[5] These stages outline modern scientific exploration and experimentation.

Hypothesis development is a critical stage in this process. Farrugia et al. state that a well-formulated hypothesis is based on a good research question and is developed before the start of the study, serving as the driving focus behind data collection for the study. The authors also highlight the importance of the hypothesis as the central element of the study, serving as the basis for testing and establishing the statistical and clinical significance of the research.[6]

Translating basic science research

An important component of creating impactful basic science research is translating the discoveries of that research into clinical trials. Eventually, the findings from these clinical trials can be applied to diverse populations in an effort to develop tools and knowledge for implementing well-developed scientific changes into healthcare delivery systems.[7] This process of applying data from a clinical trial to large populations is known as translational research and is the ultimate goal of clinically relevant basic science research. Surgeons and clinicians rely on updated studies founded in basic science research to guide their clinical practice. The process of translating research into actual patient care guidelines is heavily prioritized, and the National Institutes of Health (NIH) works towards minimizing natural barriers to the utilization of updated data in daily clinical practice. These barriers include improving the applicability of research, motivating clinicians to increase adherence, and improving the availability of data to a broader group of healthcare workers.[8] The importance of minimizing barriers to the utilization of basic science research as a means of informing healthcare cannot be

Translational Interventional Radiology. https://doi.org/10.1016/B978-0-12-823026-8.00032-8

overstated, as it contributes to the mission of medicine by improving patient health outcomes and the scientific basis for the practice of medicine.

Barriers to research

Although basic science research is fundamental to progress healthcare and improve patient health outcomes, numerous barriers exist to designing research, accessing research opportunities, resources, or equipment, funding research, motivating scientists toward specific disciplines, and sharing knowledge globally, as well as limitations within studies. A study by Okoduwa et al. revealed that at a particular institution many more students were motivated or interested in research but had extreme barriers impeding their ability to participate.[9] This phenomenon has been described across multiple studies at many other institutions.[10–12] Even within the medical field, AlGhamdi et al. found that although majority of participating medical students believe that research is an important component of the progression of medicine; however, only around half of the students participated in research during medical school due to barriers to conducting research.[13]

Once accessing and designing research there are still limitations within the study. Limitations are defined as any potential weakness of the study that is often out of the researcher's control and are usually associated with the research design, statistical model constraints, funding constraints, or other predictable or unpredictable factors.[14] Limitations can ultimately affect the conclusions and results of the study. A common example of a basic science limitation is a study in which the researcher only has access to a small number of participants' experience as data points. This is a limitation of the study because it does not give proper representation of the average person within a larger, more diverse population. All research unavoidably has some limits. Therefore, it is extremely important for researchers to acknowledge and discuss the limitations to their study within their final publishing to create space for less limited research on the topic in the future.

Finding ways to overcome these obstacles to and within research is an ongoing challenge and a topic that is itself studied heavily. The improvement in these barriers is imperative at the individual and institutional level to guarantee enduring progression in health research.

Conclusion

Basic science research serves as the foundation of healthcare advancements across a multitude of disciplines and requires a methodical and organized approach to data collection and analysis. Conducting research with these fundamentals in mind creates a pathway to clinical relevancy and application. The translation of basic science research to clinical trials and eventually generalizing to large populations changes the way clinicians practice medicine. Despite the ongoing acknowledgement of the importance of basic science research, there are still many avoidable barriers to these types of studies, which continue to pose ongoing challenges for future researchers. Finding ways to overcome these barriers is imperative at both the individual and institutional level to ensure enduring progression in health research.

References

1. La Caze A. The role of basic science in evidence-based medicine. *Biol Philos*. 2011;26:81–98. https://doi.org/10.1007/s10539-010-9231-5.
2. Mouly, George J. *Educational Research: The Art and Science of Investigation*. Boston: Allyn and Bacon; 1978 (Print).
3. Guthrie G. *Basic Research Methods: An Entry to Social Science Research*. New Delhi: SAGE Publications India Pvt Ltd; 2010. SAGE Knowledge, 26 September 2021 http://www.doi.org/10.4135/9788132105961.
4. Connaway LS, Powell RR. *Basic Research Methods for Librarians*. Santa Barbara, Calif: Libraries Unlimited; 2010 (Internet resource.
5. Andrews J. Book review – research methods in the social sciences, by Chava Frankfort-Nachmias, David Nachimas and Jack De waard (8th edition). *Popul Ageing*. 2019;12:195–198. https://doi.org/10.1007/s12062-017-9191-5.
6. Farrugia P, Petrisor B, Farrokhyar F. Practical tips for surgical research: research questions, hypotheses and objectives". *J Can Chir*. 2010;53(4):278–281.
7. Hulley SB, Cummings SR, Browner WS, Grady DG, Newman TB. *Designing Clinical Research*. 4th ed. Lippincott Williams and Wilkins; 2013.
8. Kahn K, Ryan G, Beckett M, et al. Bridging the gap between basic science and clinical practice: a role for community clinicians. *Implement Sci*. 2011;6:34. https://doi.org/10.1186/1748-5908-6-34.
9. Okoduwa SIR, Abe JO, Samuel BI, Chris A. Attitudes, perceptions, and barriers to research and publishing among research and teaching staff in a Nigerian research Institute. *Front Res Metr*. 2018;3:26. https://doi.org/10.3389/frma.2018.00026.
10. Baro EE, Bosah GE, Obi IC. Research funding opportunities and challenges: a survey of academic staff members in Nigerian tertiary institutions. *Bottom Line*. 2017;30:447–644. https://doi.org/10.1108/BL-07-2016-0027.
11. Aslam F, Shakir M, Ahad-Qayyum M. Why medical students are crucial to the future of Research in South Asia. *PLoS Med*. 2005;2:e322. https://doi.org/10.1371/journal.pmed.0020322.
12. Dickson-swift V, James EL, Kippen S, Liamputtong P. Blurring the boundaries in qualitative health research on sensitive topics. *Qual Res*. 2006;16:853–871. https://doi.org/10.1177/1049732306287526.
13. AlGhamdi KM, Moussa NA, Alessa DS, Alothimeen N, Al-Saud AS. Perceptions, attitudes and practices toward research among senior medical students. *Saudi Pharmaceut J*. 2014;22(2):113–117. https://doi.org/10.1016/j.jsps.2013.02.006.
14. Theofanidis D, Fountouki A. Limitations and delimitations in the researsch process. *J Perioper Nurs*. 2019;7(3):155–162. https://doi.org/10.5281/zenodo.2552022. E-ISSN:2241-3634.

Part II

Pre-clinical: discovery and development

Chapter 4

Overview of preclinical research

Siddharth Venkatraman[1] and Capt. Tej Ishaan Mehta[2,3]

[1]The Johns Hopkins University School of Medicine, Baltimore, MD, United States; [2]Department of Interventional Radiology, The Johns Hopkins Hospital, Baltimore, MD, United States; [3]United States Air Force Medical Corps, Falls Church, VA, United States

Key points

- Preclinical research is an essential starting point for the development and evaluation of nearly all translational, clinically relevant medical interventions.
- Understanding the underlying components of preclinical research is essential for designing robust experimental frameworks.
- Early consideration of experimental biases in preclinical research may help improve experimental design.

What is preclinical research?

While clinical research refers to studies in humans, preclinical research encompasses the body of research that lays the groundwork for ethical justification of human trials. In this chapter, we will focus on hypothesis-testing in preclinical studies, explain general concepts related to types of experiments and the design of in vivo experiments, and discuss steps to mitigate biases in preclinical research.[1]

Why it matters?

Preclinical research is an essential component of nearly all medical interventions. Preclinical research tests for safety and efficacy of therapeutics prior to human trials, ultimately reducing the risk of potential treatment-related sequelae and refining treatment delivery parameters.

Preclinical research models

There are three broad categories of preclinical research: in silico, in vitro, and in vivo.

In silico ("within the silicon") studies broadly cover studies performed using computer simulation. In silico studies allow for rapid modeling of greatly varied biological systems including but not limited to the following: novel pharmaceutical applications with virtual drug screening, creation of new pharmaceutical compounds using machine learning, developing mechanistic models of action for drug–protein interactions, and establishing structure–activity relationships, radiotherapy comparative dosimetric studies, etc. In silico imaging has even recently emerged as a field, potentially allowing for rapid development of novel imaging techniques and optimization of current techniques.[2] Although still in its infancy, in silico modeling has been applied within interventional radiology, such as patient-specific simulation of yttrium-90 dosimetry.[3] Results from in silico models may help guide study design for future in vitro and in vivo studies.

In vitro ("within the glass") studies provide a generally well-controlled environment for hypothesis testing, often using biological molecules or cell cultures outside the body. Cell lines or biological molecules are often derived from either humans or nonhuman animals and can be introduced to a therapy in question to monitor a response. As compared to in vivo studies, in vitro studies are generally cheaper, higher throughput and do not require animal husbandry. Historically, results from in vitro studies guided in vivo studies, but in vitro results poorly extrapolated directly to in vivo results as in vitro models do not fully model the complex cellular environment and interactions in vivo. In vitro studies may be viewed as reducing the "degrees of freedom" compared to in vivo studies, allowing for more rapid and simplified modeling, but sacrificing many of the complexities of independent biological systems. To help bridge the gap between in vitro and in vivo systems, modern three-dimensional cell cultures are being implemented, which theoretically maintain cellular heterogeneity and tissue architecture.[4] If there is evidence of

Translational Interventional Radiology. https://doi.org/10.1016/B978-0-12-823026-8.00066-3

efficacy in in vitro experiments in vivo trials may be attempted to gather additional information in more complex biological systems.

In vivo ("within the living") studies cover a broad range of topics but, generally speaking, encompass trials within independent, complex biological systems. In medical pre-clinical research, in vivo studies almost always occur in nonhuman animals. These studies are able to better model efficacy and sequelae of therapies in complex biological environments, which can improve its predictions of safety, https://www.news-medical.net/life-sciences/What-is-Toxicity.aspx toxicity, and efficacy as compared to in vitro studies. However, in vivo studies often require significantly more money, time, and attention to comply with ethical policies and regulatory laws governing animal testing. Often, scientists are required to demonstrate that no alternative methodology can be used to conduct the experiment and that the benefits of the study outweigh the suffering caused to the animals. This field is also expanding, as technologies like CRISPR allow scientists to generate animal models of specific pathologies cheaper and faster than before.[4]

Within the groups of in silico, in vitro, and in vivo studies, there are several common preclinical study models that are conducted and worth describing:

1. Screening test: Simple and rapid initial screen to determine the presence or absence of a pharmacological property in the putative drug. Avenues to investigate this may include (but are not limited to) molecular modeling of drug binding sites to specific proteins, application of pharmaceuticals to cell cultures, or applications of pharmaceuticals to living animals.

   Ex: Testing if an antibiotic has efficacy on a specific bacterial strain.
2. General observation test: Observation of the effects from a therapy, generally focused on identifying sequelae associated with the therapy. One experimental design may include a drug of interest being injected in triplicate into a series of animals at varying doses and observing the animals for sequelae.

   Ex: Assess if a novel contrast medium causes renal damage and, if so, to describe the relationship between contrast dose and degree of renal damage.
3. Mechanism of action: Experiments are conducted to determine the underlying mechanism of action for the drug. Avenues of investigation may include (but are not limited to) molecular simulations of drug-protein interactions and a litany of biochemical investigative techniques.

   Ex: Determine if a drug with an antihypertensive effect is acting on α-receptors or β-receptors.
4. Systemic pharmacology experiment: Determining effects on individual and major organ systems (cardiovascular, respiratory, neurological). This requires in vivo experiments in most cases. At this stage, scientists determine the absorption, distribution, metabolism, excretion, and toxicity profile of the therapy in question.

   Ex: The drug cimetidine is used to reduce stomach acid excretion, but has various other GI, neurological, and genitourinary side effects that require in vivo tests to describe and cannot be adequately determined in silico or in vitro.
5. Quantitative tests and pharmacokinetics: examining dose-response relationship, maximal effects, and therapeutic range. This also includes the movement of the drug across body systems (absorption, distribution, metabolism, localization in tissues, and excretion). These parameters provide data to compare the new therapy to an existing standard of care, as well as a preferred safe dose and route of administration.

Get started

Different aspects of preclinical research generally have different levels of entry beyond technical knowledge. In silico modeling has relatively low barriers to entry, requiring only a standard computer, though increased processing capabilities, such as with advanced graphics processing units, generally improve processing times. Online access to third-party graphics processing units and tensor processing units, such as through the Google Colaboratory, can also provide access to increased processing capabilities without the need for local advanced processing capabilities.

In vitro models may (in principle) be conducted with simple equipment such as cell-culture media and a microscope. In modern practice, a wide-range of equipment is often required such as labeling tools (I.e., histopathology media), purification/extraction tools (I.e., ultracentrifuges) and safety equipment (I.e., fume hoods). Evaluation of specific molecules further requires advanced biochemical technologies such as nuclear magnetic resonance spectroscopy, high-performance liquid chromatography or mass spectrometry.

Akin to in vitro and in silico models, in vivo models may be greatly varied in size, scope and complexity. Accordingly, establishing models for in vivo testing vary greatly in complexity. In general, there is an inverse relation between phylogenetic "distance" from humans and complexity of starting in vivo testing. While plant and fungi models require horticultural skills, animal models require animal husbandry with associated testing, approval and funding above and beyond that of the therapeutics being tested. Additionally, matters of equipoise become increasingly important with reduced phylogenetic distance from humans, further escalating the complexity of in vivo models.

Designing in vivo experiments

Designing an in vivo experiment requires meticulous planning beforehand to create a robust experimental plan. Typically, the framework is divided into the following components:

Hypothesis and effect size

By convention in experimental design, there are generally two hypotheses for any given experiment, a null hypothesis and an alternative hypothesis. The *null hypothesis* states that the intervention or the drug had no statistically significant effect or change on the outcome. The *alternative hypothesis* states that the intervention produced a statistically significant effect on the outcome. In statistics, the *P*-value reports the probability of observing an effect as extreme or more extreme than what would be observed if the null hypothesis was true. In other words, the smaller the *P*-value, the more likely the null hypothesis may be rejected.

Crucial to this process is determining an effect of interest, which is the variable that is measured as an outcome of the intervention. Then, the scientist must decide on a *minimum effect size* or the smallest effect the experiment is designed to be able to detect (often the minimum difference that would be of biological relevance). This can then be used to determine an adequate sample size to ensure that there is enough experimental power to detect changes that are biologically relevant, not just statistically significant. While there is no easy way to ensure that results will translate to clinical relevance, it is suggested that the minimum effect size fulfill the following criteria:[5]

- Cause a beneficial effect for individuals rather than large cohorts
- The difference must be experimentally testable and reasonable to achieve
- There should be a rationale for translation into patients in the long run

Experimental unit and sample size

In a comparative in vivo experiment, the animals are separated into groups and exposed to different interventions (drugs, dosages, delivery mechanisms, surgical procedures, etc.). The *sample size* is the number of experimental units per group. The *experimental unit* is the entity subjected to an intervention independently of all other units. Under this definition, a unit could refer (for example) to individual mice, or to an entire cage of mice. Once identified, the experimental units are separated and subjected to different interventions. It is generally good practice to randomize the distribution of experimental units to reduce the effects of random error and selection bias. It is important to note that one of these groups should be used as a *control group* that minimizes the effect of confounding variables. There are several types of control groups, including negative control, vehicle control, positive control, sham control, comparative control ,and naïve control, and it is important to know when to use each of these.[6] We refer readers elsewhere for full descriptions of these types of control groups.

Outcome data and measurements

Outcome measurements are made throughout the course of the experiment. These are also known as dependent variables. While multiple outcome measurements can be made, there should be a set of outcome measures that are of greatest importance to the experiment. In statistical analysis, these outcomes are classified as either continuous or categorical. *Continuous variables* are quantitative data that can include any real number, as well as any real value in between (ex. bodyweight, body temperature). *Categorical variables* can take on a fixed and limited number of possible values, separating outcomes into categories that are *nominal* (no intrinsic order to them; ex. male/female/other) or *ordinal* (intrinsic order present; ex. mild/moderate/severe).

Reducing biases in preclinical studies

Experimental biases can cause significant weakness in the design, conduct, and analysis of in vivo animal studies, producing misleading results and wasting valuable resources. A list of experimental biases have been summarized in Table 4.1 (adapted from Cochrane Handbook for Systematic Reviews of Interventions)[7]:

Real-world examples

Scientific and layman literature abounds with examples of pre-clinical research. As a brief recap of the pre-clinical research examples described in this chapter, recall the use of in silico models for patient-specific simulation of yttrium-90 dosimetry, the use of in vitro models to develop three-dimensional cell cultures, and the use of CRISPR technology to generate animal models to mimic specific pathologies. [3,4]

Resources

Exploration of institutional and local resources is generally the recommended starting point for establishing and developing pre-clinical research programs. The National Institutes of Health preclinical research toolbox provides a wide-range of publicly available tools and repositories for preclinical research. Pending institutional and local

TABLE 4.1 Common experimental biases that may affect the design and analysis of preclinical research studies, as well as potential ways to mitigate the effects of these biases.

Name of bias	Definition	Mitigation strategies
Selection bias	The biased allocation of experimental units such that they differ systematically from the population of interest. Ex. only sick mice are given the intervention while healthy mice are given placebo, which artificially increases mortality from the drug due to preexisting comorbidities in the treatment group.	Randomization software should be used to avoid systemic differences between animals allocated to different treatment groups.
Performance bias	Systematic differences between groups in the care that is provided, or in exposure to factors other than the interventions of interest. Ex. Applying different housing conditions to experimental and control animals in an experiment such that the control animals live in worse conditions when that is not an intrinsic feature of the study.	Randomization and blinding of the researchers can help ensure equal care for all groups. Researchers may also collect some data to evaluate the conditions of each group.
Detection bias	Systematic differences between groups in how outcomes are determined. Ex. Control groups in a study are not as closely monitored for side effects as experimental groups.	Complete blinding of the experimenters, including those who are analyzing outcomes data so that they are not aware which animals belong to which treatment group.
Attrition bias	Systematic differences between groups in withdrawals from a study. Withdrawals from the study lead to incomplete outcome data and can be caused by exclusion or attrition. **Exclusions** refer to situations in which some participants are omitted from reports of analyses, despite outcome data being available to the trialists. **Attrition** refers to situations in which outcome data are not available (ex. Patient dies).	Experimenters should report attrition information for each experimental group and attempt to include outcomes that will not be affected by attrition.
Reporting bias	Reporting bias occurs when statistically significant, "positive" results that indicate that an intervention works are "more likely to be published, more likely to be published rapidly, more likely to be published in English, more likely to be published more than once, more likely to be published in high impact journals, and more likely to be cited by others."	May be reduced by journals by publishing high-quality studies regardless of novelty or unexciting results and by publishing protocols or full-study data sets.

policies, support from private industry may also be considered.

References

1. Cressman EN, Newton I, Larson AC, et al. State of the research enterprise in IR and recommendations for the future: proceedings from the Society of Interventional Radiology Foundation Investigator Development Task Force. *J Vasc Intervent Radiol.* 2018;29(6):751–757.
2. Badano A. In silico imaging clinical trials: cheaper, faster, better, safer, and more scalable. *Trials.* 2021;22(1):1–7.
3. Roncali E, Taebi A, Foster C, Vu CT. Personalized dosimetry for liver cancer Y-90 radioembolization using computational fluid dynamics and Monte Carlo simulation. *Ann Biomed Eng.* 2020;48(5):1499–1510.
4. Lorian V. Differences between in vitro and in vivo studies. *Antimicrob Agents Chemother.* 1988;32(10):1600–1601.
5. Huang W, du Sert NP, Vollert J, Rice AS. General principles of preclinical study design. *Handb Exp Pharmacol.* 2020;257:55.
6. Bate ST, Clark RA. *The Design and Statistical Analysis of Animal Experiments.* Cambridge University Press; 2014.
7. Higgins J, Thomas J, Chandler J, Cumpston M, Li T, Page M. *Cochrane Handbook for Systematic Reviews of Interventions Version 6.2.* Cochrane; 2021. Available from: www.training.cochrane.org/handbook.

Additional reading

1. Good research practice in non-clinical pharmacology and biomedicine. In: *Handbook of Experimental Pharmacology.* Springer International Publishing; 2020:55–69. https://doi.org/10.1007/164_2019_277 (Chapters: Minimum Information and Quality Standards for Conducting, Reporting, and Organizing In Vitro Research; Minimum Information in In Vivo Research).

Chapter 5

Defining the problem to solve

Samyuktha Balabhadra
Department of Diagnostic and Interventional Radiology, University of Texas-Houston, Houston, TX, United States

Key points

- Create an initial project plan.
- Begin evaluating the impact of your research in the real world.
- Assess the benefit of investing resources into this project.

Why it matters?

Translational research projects, while impactful and imperative to progress in medicine and interventional radiology, can be time and resource consuming. It is, therefore, essential to create a plan of action and assess project impact before initiation.

Table 1. Five-Step Process to Translational Research Planning
Step 1: Ask broad questions
• Prepare a list of questions on broad topics that warrant investigation • Examples can be derived from: o Work environment o Academic discussions o Scientific/medical literature
Step 2: Literature review
• Review current literature regarding chosen topics • While reviewing, it is important to: o Actively modify the question list from Step 1 o Review disease pathophysiology o Understand pervasiveness of the issue
Step 3: Impact evaluation
• Detailed assessment of project impact and goals • Consider: o Target population o Predicted outcomes o Sustainability o Comparison to current or pipeline therapies
Step 4: Study design
• Create design outlines • Consider necessary resources: o Temporal o Financial o Human • Discuss with other collaborators and modify plan
Step 5: Initiate project
• Commence project

Translational Interventional Radiology. https://doi.org/10.1016/B978-0-12-823026-8.00027-4

Step 1

Ask broad questions. What are some broad, overarching topics or issues that you have observed in your work environment? What areas or disease processes could benefit from innovation? Examples may be derived from patients you have seen, cases you have read about in medical literature, or discussions with colleagues. Prepare a list of ideas and questions on topics that you are interested in pursuing.

Case 1: Prostate artery embolization

Problem: Benign prostatic hyperplasia (BPH) causes lower urinary tract symptoms in males; a major quality of life issue (Fig. 5.1).[1]

What happens to patients who do not respond to medication? What happens to patients who are not surgical candidates? What about patients who cannot stop anticoagulants for surgery? How can we create a beneficial, novel therapy for enlarged prostates using IR tools?

Case 2: Inferior vena cava (IVC) filters

Problem: Deep vein thrombosis (DVT) and pulmonary embolism (PE) are a major cause of morbidity and mortality.[2]

What happens to patients who do not qualify for anticoagulation therapy? What happens to patients who are not surgical candidates? How can we create a beneficial, novel therapy for PE using IR tools?

Case 3: Arterial gene transfer

Problem: Peripheral arterial disease (PAD) is a major cause of morbidity and mortality.[3] The progression of PAD is such that it requires multiple occasions of medical, surgical, and/or endovascular care.

How can we create a more sustainable therapy? Is there a way to alter the course of this disease at a molecular level? Is there scope for a molecular/genetic transcatheter therapy?

Step 2

Literature review. Gather and read preexisting articles that are relevant to your chosen topic. What have others already found? As you read relevant articles, you will find answers to some of your questions, but you will also continue to build on your list. Continue to narrow your questions and make them more specific. The literature review is also an opportunity to understand the basic pathophysiology of the problem and to understand how pervasive it is within the medical field.

Case 1: Prostate artery embolization

The accepted treatments for BPH are medication, transurethral resection of the prostate (TURP), and open prostatectomy.[4] Initially, internal iliac artery branch

Flowchart outlining a five-step process in project planning and development.

embolization was performed for massive hemorrhage.[5] Few case reports have discussed a benefit in this method reducing the size of prostates overall.

Is it possible to use internal iliac artery branch embolization for BPH? Would reducing blood supply to the prostate be beneficial, possibly reducing its size? If this method is successful in reducing prostate size, it can be high impact, given that this is a condition that affects a majority of middle-aged and elderly men.

Case 2: IVC filters

For many years, the widely accepted treatment for managing DVT has been anticoagulation or open thrombectomy, if possible. For patients who could not receive anticoagulant medication, inferior vena cava (IVC) ligation was practiced. The initial intent for developing the IVC filter was to create an alternative method to IVC ligation for preventing recurrent pulmonary embolism.[6]

Can an endovascularly delivered mechanical barrier to prevent pulmonary embolism be developed? How can such a barrier be made temporary for the period when anticoagulation is contraindicated? If this method is successful, it may reduce mortality and morbidity rates secondary to pulmonary embolism and its high-risk treatments.

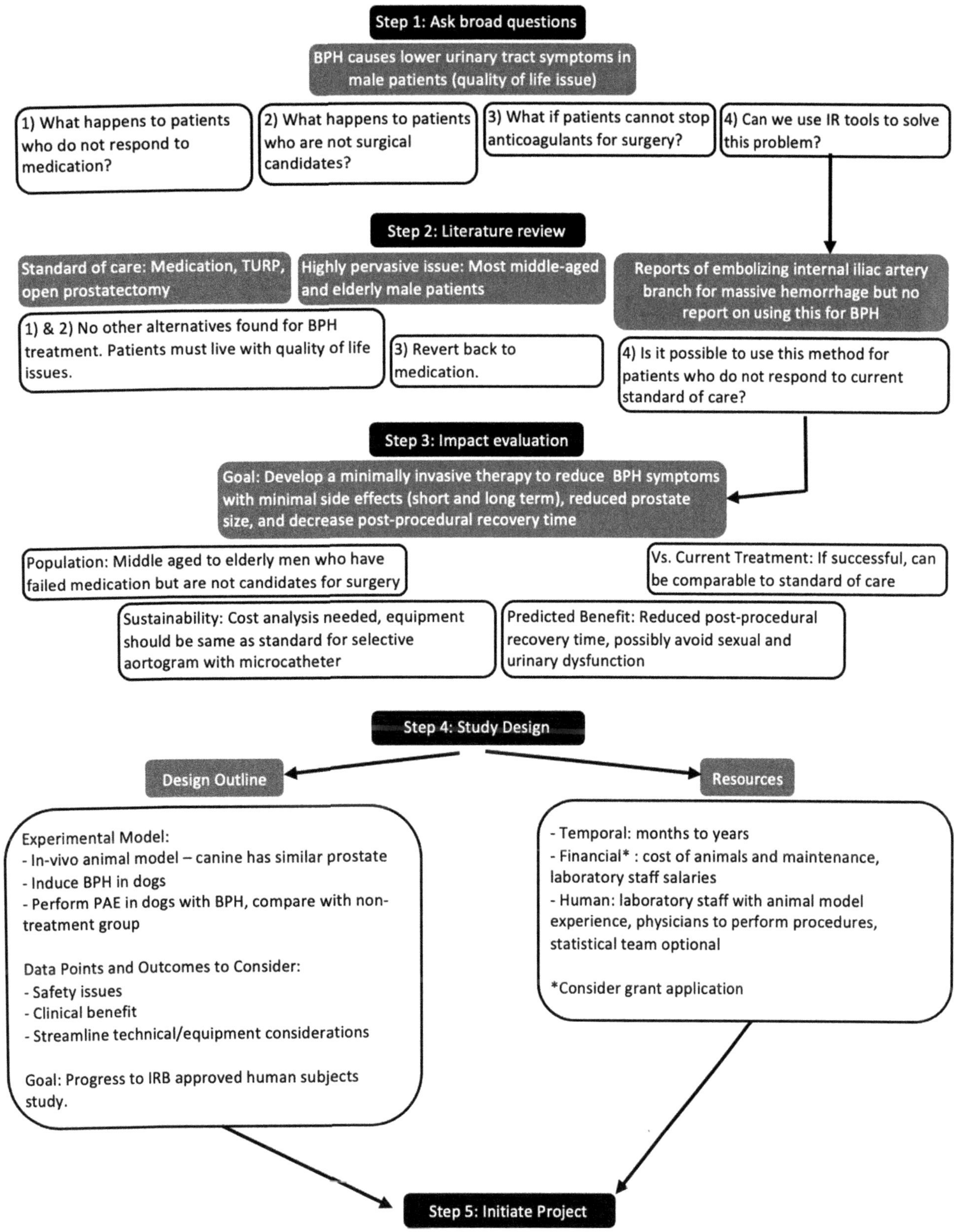

FIGURE 5.1 Flowchart demonstrating a five-step process using Case 1: Prostate artery embolization.

Case 3: Arterial gene transfer

Current therapy for these patients involves a combination of medication, endovascular therapy, and open surgery.[7] These are often in a graded process because PAD is a progressive disease, and it is difficult to develop a sustainable therapy in this context. Several molecular pathways have been identified as being involved with angiogenesis.[8] Vascular endothelial growth factor (VEGF) has been identified as one key signaling protein that promotes angiogenesis. Catheter-directed gene therapy is emerging as an arena with great potential.

Would it be possible to transfect the VEGF gene into vulnerable arteries using an endovascular approach? Could this method provide a long-term solution?

Step 3

Impact evaluation. This phase requires a more detailed assessment of the impact of pursuing this project. Establish the goals of the project. What patients or organizations will benefit? How much will they benefit? Is this therapy a sustainable option as compared to current therapies or those already in the pipeline?

Case 1: Prostate artery embolization

The goal of this project would be to develop a minimally invasive therapy to reduce symptoms associated with enlarged prostate with minimal side effects, both in the short term and long term, reduce prostate size, and decrease postprocedural recovery time. The outcomes would have to be comparable to the accepted standard of care; in this case medical management and prostatectomy. If the technique is successful, patients can benefit from a procedure that reduces immediate postprocedural recovery time. In addition, the procedure could preserve sexual and urinary function, which are reported postsurgical effects.[9]

Case 2: IVC filters

IVC filters were initially developed to prevent recurrent pulmonary embolism in patients; a significant population with high morbidity and mortality. It was reasoned that a mechanical, endovascularly delivered barrier would reduce recovery time as compared to mechanical thrombectomy.

Over time, IVC filters underwent evolution in terms of design and indications. Landmark studies demonstrated that these devices were most beneficial in patients who cannot undergo currently accepted treatments.[10] Even with the current specific indications, a large subset of patients is receiving benefits from this treatment.

Case 3: Arterial gene transfer

The goal of this project would be to use catheter-directed gene therapy to promote angiogenesis. This technique could potentially benefit thousands of PAD patients and at least maintain, if not improve, their disease state.

Step 4

Study design. For each of these ideas, come up with a rough study design. In the design, it is important to understand the feasibility of the project. What resources are needed? How long will it take? At this stage, it is wise to discuss the project with other collaborators.

Case 1: Prostate artery embolization

One method of approaching a new technique is to attempt it in an in vivo animal model. This can help identify safety issues, estimate clinical benefits, and streamline technical/equipment considerations. For example, one group utilized a canine model, which demonstrated promising outcomes.[11] Of note, this type of model requires adequate laboratory facilities and an institutional review board (IRB) proposal. It is after this step that human clinical trials can be commenced.

Case 2: IVC filters

Device development requires collaboration with biotechnologists and materials scientists. Sponsorship or a financial grant will be needed. Several prototypes must be developed and tested in vitro to assess the ease of delivery, design integrity, and retrievability. Subsequently, in vivo experiments are needed to assess safety. This phase will likely require IRB approval.

In the case of the development and evolution of IVC filters, scientific collaboration was needed to create a device that considered all of the aforementioned factors as well as more specific areas such as clot trapping and flow dynamics. In vitro studies demonstrated that certain designs were more effective than others.[12] In vivo studies in animal models further demonstrated design integrity.[13] It is after this step that human clinical trials can be commenced.

Case 3: Gene therapy

Collaboration with a biomedical scientist will be needed to extract and create recombinant VEGF. In conjunction with a team of interventionalists, multiple delivery methods may need to be tested. Which catheters, wires, embolic agents, and polymers will be most effective? Initially, ex vivo studies, on living tissues in a laboratory setting, will need to be performed to assess the true viability of the extracted gene and method of delivery.[14] Subsequently, in vivo studies, in an intact organism, can be performed to further assess the integrity of the new technique. The final step is to commence human trials. Many of the steps in this method of translational research will require IRB approval.

Step 5

Initiate project. Now you are ready to commence your project, according to your plan.

Case 1: Prostate artery embolization

Prostate artery embolization was examined in a prospective randomized trial, which demonstrated that this technique is safe and effective for improving BPH symptoms.[15]

Case 2: IVC filters

Since the development of the first IVC filter, this device has been examined in several large-scale studies, subsequently undergoing many changes in both design and indications. The most recent studies agree that, if indicated, IVC filters can provide a safe and effective mechanical method of preventing PE.[16]

Case 3: Gene therapy

In the case of VEGF transfection for PAD, VEGF recombinant DNA was successfully transfected into the internal iliac artery in an animal model, using an angioplasty balloon.[17] This team later demonstrated successful angiogenesis in a human patient following a similar technique.[18]

Get started

Begin the five-step plan to initiate a research project with the first phase—creating a list of broad questions. If you are in training, it may be helpful to find a mentor who can help you navigate the process.

References

1. Members of the Florey Adelaide Male Ageing Study, Martin SA, Haren MT, Marshall VR, Lange K, Wittert GA. Prevalence and factors associated with uncomplicated storage and voiding lower urinary tract symptoms in community-dwelling Australian men. *World J Urol.* 2011;29(2):179–184. https://doi.org/10.1007/s00345-010-0605-8.
2. Horlander KT, Mannino DM, Leeper KV. Pulmonary embolism mortality in the United States, 1979–1998: an analysis using multiple-cause mortality data. *Arch Intern Med.* 2003;163(14):1711–1717. https://doi.org/10.1001/archinte.163.14.1711.
3. Golomb BA, Dang TT, Criqui MH. Peripheral arterial disease: morbidity and mortality implications. *Circulation.* 2006;114(7):688–699. https://doi.org/10.1161/CIRCULATIONAHA.105.593442.
4. Foster HE, Dahm P, Kohler TS, et al. Surgical management of lower urinary tract symptoms attributed to Benign prostatic hyperplasia: AUA guideline amendment 2019. *J Urol.* 2019;202(3):592–598. https://doi.org/10.1097/JU.0000000000000319.
5. Appleton DS, Sibley GN, Doyle PT. Internal iliac artery embolisation for the control of severe bladder and prostate haemorrhage. *Br J Urol.* 1988;61(1):45–47. https://doi.org/10.1111/j.1464-410x.1988.tb09160.x.
6. Mobin-Uddin K, Utley JR, Bryant LR. The inferior vena cava umbrella filter. *Prog Cardiovasc Dis.* 1975;17(5):391–399. https://doi.org/10.1016/s0033-0620(75)80014-4.
7. Gerhard-Herman MD, Gornik HL, Barrett C, et al. 2016 AHA/ACC guideline on the management of patients with lower extremity peripheral artery disease: a report of the American College of Cardiology/American Heart Association Task Force on Clinical Practice Guidelines. *Circulation.* 2017;135(12):e726–e779. https://doi.org/10.1161/CIR.0000000000000471.
8. Risau W. Mechanisms of angiogenesis. *Nature.* 1997;386(6626):671–674. https://doi.org/10.1038/386671a0.
9. Reich O, Gratzke C, Stief CG. Techniques and long-term results of surgical procedures for BPH. *Eur Urol.* 2006;49(6):970–978. https://doi.org/10.1016/j.eururo.2005.12.072. Discussion 978.
10. Jones TK, Barnes RW, Greenfield LJ. Greenfield vena caval filter: rationale and current indications. *Ann Thorac Surg.* 1986;42(6 Suppl):S48–S55. https://doi.org/10.1016/s0003-4975(10)64643-7.
11. Sun F, Sánchez FM, Crisóstomo V, et al. Transarterial prostatic embolization: initial experience in a canine model. *AJR Am J Roentgenol.* 2011;197(2):495–501. https://doi.org/10.2214/AJR.10.5947.
12. Katsamouris AA, Waltman AC, Delichatsios MA, Athanasoulis CA. Inferior vena cava filters: in vitro comparison of clot trapping and flow dynamics. *Radiology.* 1988;166(2):361–366. https://doi.org/10.1148/radiology.166.2.3336712.
13. Irie T, Yamauchi T, Makita K, Kusano S. Retrievable IVC filter: preliminary in vitro and in vivo evaluation. *J Vasc Interv Radiol JVIR.* 1995;6(3):449–454. https://doi.org/10.1016/s1051-0443(95)72840-8.
14. Thomas JW, Kuo MD, Chawla M, et al. Vascular gene therapy. *Radiogr Rev Publ Radiol Soc N Am Inc.* 1998;18(6):1373–1394. https://doi.org/10.1148/radiographics.18.6.9821188.
15. Pisco JM, Bilhim T, Pinheiro LC, et al. Medium- and long-term outcome of prostate artery embolization for patients with Benign prostatic hyperplasia: results in 630 patients. *J Vasc Interv Radiol JVIR.* 2016;27(8):1115–1122. https://doi.org/10.1016/j.jvir.2016.04.001.
16. DeYoung E, Minocha J. Inferior vena cava filters: guidelines, best practice, and expanding indications. *Semin Interv Radiol.* 2016;33(2):65–70. https://doi.org/10.1055/s-0036-1581088.
17. Isner JM, Walsh K, Symes J, et al. Arterial gene transfer for therapeutic angiogenesis in patients with peripheral artery disease. *Hum Gene Ther.* 1996;7(8):959–988. https://doi.org/10.1089/hum.1996.7.8-959.
18. Isner JM, Pieczek A, Schainfeld R, et al. Clinical evidence of angiogenesis after arterial gene transfer of phVEGF165 in patient with ischaemic limb. *Lancet Lond Engl.* 1996;348(9024):370–374. https://doi.org/10.1016/s0140-6736(96)03361-2.

Chapter 6

Types of problems

Joshua Kogan[1] and Charissa Kim[2]

[1]*Department of Radiology, Vanderbilt University Medical Center, Nashville, TN, United States;* [2]*Department of Radiology, Beth Israel Deaconess Medical Center, Boston, MA, United States*

Key points

- Translational research in interventional radiology can fill the gaps in knowledge that cannot be readily addressed by existing medical or surgical approaches, because interventional radiology presents the unique advantages of minimally invasive techniques.
- Interventional radiology can improve patient care by harnessing antitumor immunity with interventional oncology and by drug repurposing across multiple disease systems.
- Technology and device development research in interventional radiology can enhance the effectiveness and safety of existing procedures and also provide innovative foundations for new procedures.

Introduction (why it matters)?

Translational research, as defined by the National Institute of Health, incorporates two areas of translation:

> *One is the process of applying discoveries generated during research in the laboratory, and in preclinical studies, to the development of trials and studies in humans. The second area of translation concerns research aimed at enhancing the adoption of best practices in the community.[1]*

The ultimate objective of translational research is to transfer basic science research to the clinic in a way that is applicable to a broad patient population.[1] Although translational research in interventional radiology (IR) has resulted in positive advances across multiple disease systems, it has yet to reach its full potential due to the relatively new nature of the field. The Society of Interventional Radiology acknowledges that there has been a "[d]eficiency of durable, long-term, and/or longitudinal research projects undertaken across broad range of investigators and institutions"[2] and has proposed actionable solutions to support IR innovation and research. These efforts, which include mentorship networks and mechanisms to parlay institutional grants into extramural funding,[2] alluding to the fact that IR as a field depends on research for growth. Indeed, within the 13 body systems in the human body, there exist roughly 70,000 separately diagnosable conditions, according to the World Health Organization's 10th revision of the International Classification of Diseases and Related Health Problems.[3] However, many of these ailments lack cures or adequate treatments, leaving much room for medical advancements. IR, with its noninvasive procedures, decreased time in hospital, and lower complication rate,[4] can improve patient outcomes across many diseases. In this chapter, we discuss three major ongoing and upcoming topics of translational research in IR, with an emphasis on areas where IR can fill current gaps in knowledge that cannot be addressed with existing medical or surgical approaches: (1) intersections of interventional oncology and immuno-oncology, (2) technology and device development, and (3) drug repurposing (Fig. 6.1).

Intersection of interventional oncology and immuno-oncology

Interventional oncology (IO) is one of the branches of IR that has been emerging as the fourth pillar of oncology treatment next to medical, surgical, and radiation oncology.[5] While the other pillars of oncology focus on chemotherapeutic agents, targeted agents, radiotherapy, and surgery, IO involves percutaneous and transcatheter endovascular therapies that are unique and specific to IR, in addition to biopsies, venous access techniques, and port placement.[6,7] Increasingly, translational research endeavors in IO have combined the capabilities of IR with advances in medical, surgical, and radiation oncology.[6,7] However, one of the most promising intersections of IO is with translational immunology, which has been gaining

Translational Interventional Radiology. https://doi.org/10.1016/B978-0-12-823026-8.00083-3

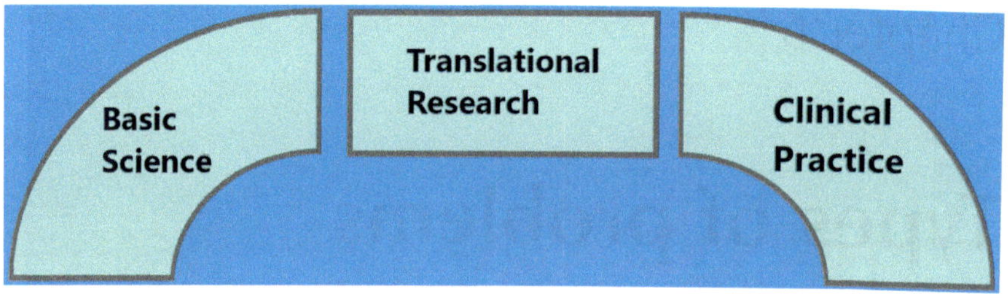

FIGURE 6.1 Translational research is what bridges the basic-science research discoveries occurring at the bench-top to the patient-oriented clinical research occurring at the bedside. This allows better application and utilization of the most cutting-edge scientific discoveries of all types, to specifically improve patient care.

traction due to recent breakthroughs in immune check point blockade therapy[8] and chimeric antigen receptor T-cell therapy in cancer.[9] Immune checkpoint receptors on cell surfaces can inhibit or activate antitumor immunity, and antibodies against the inhibitory checkpoints CTLA-4 and PD-1 have been shown to up-regulate antitumor immune responses across several tumor types.[8] Conse quently, IO technologies that enhance immune activation can complement cancer immunotherapy and improve patient care. These techniques are being effectively combined with percutaneous tumor ablation, to reduce tumor burden and stimulate systemic immunologic response. The localized tumor necrosis, induced by radiofrequency, microwave, or cryo-ablation, releases tumor antigens and induces a systemic immunologic response, known as the abscopal effect (Fig. 6.2).[10]

In 2017, a panel of 18 interventional oncologists and immuno-oncologists convened to discuss the translation of immuno-oncology to IO patient practice, ranging from IO-based stimulation of the immune system to combinations of IO with immunotherapy.[11] The white paper that resulted from this discussion conceded that while there were too few studies to reach definitive conclusions, considerable current data supported the synergy between IO and immuno-oncology. For instance, cryoablation of breast tumors in combination with immune checkpoint blockade was found to promote antitumor immunity by increasing intratumor T-cell clone number and diversity.[12] Moreover, transarterial chemoembolization (TACE) of hepatocellular carcinoma (HCC), increased tumor-associated antigens that could be recognized by T-cells, possibly potentiating the effects of T-cell therapy.[13] Combinations of IO and immunotherapy,

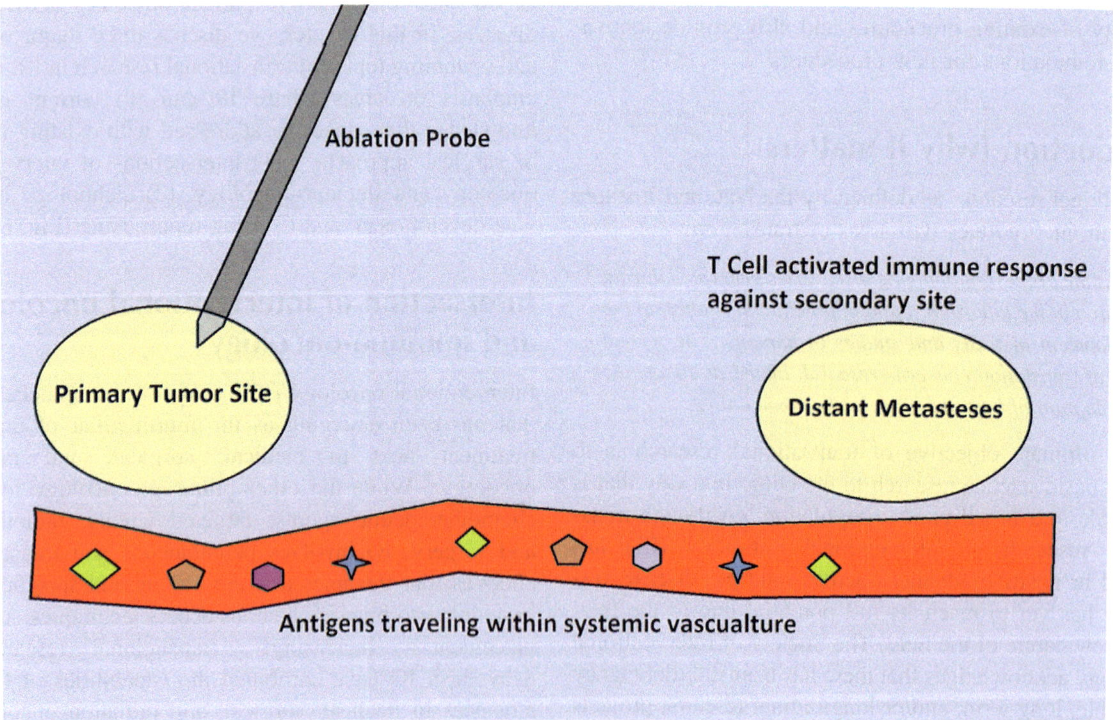

FIGURE 6.2 Treating tumors locally at their primary site results in tissue destruction and the release of many tumorigenic antigens. The travel of these antigens throughout systemic circulation can result in activation of a T-cell-mediated immune response and destruction of distant metastases at a secondary tumor site.

such as in TACE or radiofrequency ablation (RFA) with cytokine-induced killers, increased overall survival in HCC patients compared to TACE/RFA alone.[14]

Early data show that IO can also create an abscopal effect, wherein immune-mediated control of one tumor site leads to responses in distant metastatic sites.[15] In a published case of an 80-year-old male with squamous cell carcinoma and bilobar hepatic metastases, targeted IO yttrium-90 (Y90) radioembolization of the right hepatic lobe lesion resulted in complete regression of the left lesion.[16] Another case report in metastatic breast cancer demonstrated that in a patient with a right breast mass and lymph node metastases, cryoablation of the right mass yielded an abscopal effect on the metastatic sites.[17] These clinical examples of the abscopal effect indicate that there is value in conducting translational research in the immuno-oncological abscopal benefits of IO.[18]

Drug repurposing

The fundamental arsenal of interventional radiologists consists of needles, wires, catheters, and coils. However, adjunctive pharmaceuticals significantly increase the effectiveness and applicability of this equipment and the minimally invasive procedures they are used to perform.[19] A variety of medication types are currently being used for therapeutic purposes in multiple organ systems, but most diseases demonstrate treatment limitations or resistance. These treatment limitations can potentially be overcome by utilizing new types of medications or delivery methods that have not been previously attempted. Repurposing existing medications in this way opens the door for future innovations within IR, with the intention of amplifying treatment efficacy and the quality of care for patients.

To best understand the future of drug repurposing, it is worth noting examples currently in use. Thrombin, which normally cleaves fibrinogen into fibrin and contributes to the platelet plug formation during hemostasis, has been increasingly used for the treatment of pseudoaneurysms.[20] Commonly injected under ultrasound guidance, thrombin initiates clot formation of the blood trapped within the pseudoaneurysm, eliminating the channel that feeds the collection of blood, until the clot is naturally resorbed.[21]

The rising popularity of radial artery access during angiographic procedures, as opposed to the classically used femoral artery, has allowed patients to experience lower rates of vascular complications and has increased post-procedural mobilization rates.[22] Radial artery access also led to the use of verapamil, an antihypertensive calcium channel blocker, as a primary component of the spasmolytic cocktail, which prevents intraprocedural radial artery vasospasm.[23]

Drug repurposing can be both cost-effective and time efficient, as commonly used medical compounds are broadly available and have well-documented safety profiles for human use. With increased attention on some commonly prescribed medications for drug repurposing, researchers are starting to investigate previously unknown antineoplastic properties.[24] With the increased growth of interventional oncology as a mainstay of cancer treatment, a number of promising repurposed have the potential to be combined with interventional techniques, to provide IR patients with more effective chemotherapy options. At its core, interventional oncology has redesigned the way chemotherapy is delivered. Medications classically administered systemically by medical oncologists are now commonly injected intra arterially, to treat tumors locally.[25] TACE allows tumors deemed unresectable to have antineoplastic medications, such as doxorubicin, mitomycin, fluorouracil, and cisplatin, injected directly into their feeding vessels. The recent advent of the embolization of polyethylene glycols microspheres has opened the door to the drug-eluting bead—TACE, in which the beads are soaked in antineoplastic medication, before they are injected (Table 6.1).[26]

The importance of applying the principles of chemotherapy treatment repurposing is particularly important in illnesses like pancreatic adenocarcinoma, a malignancy with the highest mortality rate that carries a 5-year survival rate of 7%.[27] At the time of diagnosis, an estimated 85% of patients with pancreatic adenocarcinoma are at an advanced stage and only 15%–20% are surgical candidates.[28] These dismal statistics signify the urgent need for alternative treatment options, a gap in knowledge that IR is helping to fill. Currently, there are phase III clinical trials being performed to compare the treatment of locally advanced pancreatic cancer with systemic chemotherapy versus localized intraarterial injection of gemcitabine.[29] A 2017 study of this technique on 20 patients found 58% of patients have a reduction in their CA 19-9 tumor markers and display an overall 12-month survival rate of 60% after local delivery of gemcitabine.[29]

There are additional studies being conducted of IR-enhanced drug repurposing, which adds to our arsenal of treatment options for advanced pancreatic cancer. For instance, irreversible electroporation (IRE), with the NanoKnife device (Angiodynamics), slows the growth of pancreatic adenocarcinoma by disrupting cellular membranes via nonthermal, localized ablation. A 2018 meta-analysis of 15 studies articles involving 535 patients that underwent IRE found a survival rate of 54.5% at 12 months.[30] As new properties of existing medications are discovered and technologies to deliver them evolve, there will be increased potential to develop novel approaches to treatment. Reimagined solutions to some of our patient's problems, allow us to increase treatment efficacy and safety while also keeping costs down. It is no doubt that there will soon be many more advancements in the field of drug repurposing and delivery.

TABLE 6.1 A number of commonly used medications are being repurposed in a variety of ways that increase the quality of care for interventional radiology patients.

Drug	Group	Original indication	Repurposed indication
Thrombin	Cleaves fibrinogen to fibrin	Naturally occurring within the coagulation cascade	Pseudoaneurysm resorption
Verapamil	Calcium channel blocker	Hypertension	Antispasmodic during endovascular procedures
Aspirin	NSAID/COX 1 and 2 inhibitor	Analgesia and antithrombotic	Antineoplastic
Propranolol	Beta adrenergic receptor antagonist	Antihypertensive and antiarrhythmic	Antineoplastic
ACEi and ARBs	Inhibit formation of/block receptor of Ang II	Antihypertensive	Antineoplastic
Metformin	Activation of AMPK and inhibition of mGPD	Antihyperglycemic	Antineoplastic

ACE, angiotensin-converting enzyme inhibitors; *AMPK*, AMP-activated protein kinase; *Ang II*, angiotensin II; *ARBs*, angiotensin II receptor blockers; *COX*, Cyclooxygenase; *mGPD*, mitochondrial glycerophosphate dehydrogenase.

Technology and device development

Translational research has expanded the landscape of IR by increasing the effectiveness and range of procedures that can be safely conducted by IR physicians. Moreover, advances in biotechnology and device development have permitted growth in IR via collaborations with industry partners. For example, percutaneous devices for dialysis have enabled arteriovenous fistula creation for vascular access in end-stage renal disease patients at a lower infection rate and at a better cost-benefit ratio.[10] Optical coherence tomography has produced high-resolution intravascular images for enhanced visualization during IR and vascular procedures,[31,32] and systems like the Echo-Navigator have merged real-time echocardiographic and fluoroscopic images to guide interventional procedures.[33] Future research along the same vein in device development and imaging is projected to further extend the breadth and depth of IR in clinical practice.

Advances in simulated technologies have applied extended reality to both IR education and practice. Virtual reality (VR) is a simulated experience with a completely virtual background created by an immersive display while augmented reality (AR) utilizes the true background with a see-through display. These two modalities are considered to be opposite ends of the extended reality spectrum, with the highest medical utility found somewhere in between, with merged (MeR) reality and mixed reality (MxR). The immersive display of MeR and the see-through display of MxR similarly simulate interaction with digital objects, such as projected holograms, while allowing the user to remain in contact with their actual environment.[34] Future uses for merged and mixed reality include visualization of patient data during procedures and virtual reconstruction of patient anatomy for remote patient assessments and preoperative planning or training.[10]

Get started

Interventional radiologists are encouraged to actively evaluate ongoing preclinical and basic science studies with the intention of translating novel concepts to IR practice. Research in IO immuno-oncology, drug repurposing, and technology and device development have significant potential in not only enhancing existing IR procedures but also creating new approaches for IR treatments. IR physicians should be more involved in the translational research process, lending their unique clinical perspectives to the clinical applications of basic science. Interventional radiology began through the innovation of clinicians who decided to challenge the status quo and approach clinical dilemmas with novel approaches. The specialty still has a tremendous capacity for growth, to increase outcomes and lower burdens for patients. We as interventional radiologists owe it to the founders of this specialty and to our patients to continually seek ways to improve existing treatments and develop novel ones.

References

1. Rubio DM, Schoenbaum EE, Lee LS, et al. Defining translational research: implications for training. *Acad Med.* 2010;85(3):470–475.
2. Cressman ENK, Newton I, Larson AC, et al. State of the research enterprise in IR and recommendations for the future: proceedings from the Society of Interventional Radiology Foundation Investigator Development Task Force. *J Vasc Intervent Radiol.* 2018;29(6): 751–757.

3. World Health Organization W. *ICD-10: International Statistical Classification of Diseases and Related Health Problems: Tenth Revision*. 2nd ed. 2004.
4. Baum RA, Baum S. Interventional radiology: a half century of innovation. *Radiology*. 2014;273(suppl 2):S75–S91.
5. Kim HS, Chapiro J, Geschwind JH. From the guest editor: interventional oncology: the fourth pillar of oncology. *Cancer J*. 2016;22(6):363–364.
6. Hickey R, Vouche M, Sze DY, et al. Cancer concepts and principles: primer for the interventional oncologist-part II. *J Vasc Intervent Radiol*. 2013;24(8):1167–1188.
7. Schoenberg SO, Attenberger UI, Solomon SB, Weissleder R. Developing a roadmap for interventional oncology. *Oncol*. 2018;23(10):1162–1170.
8. Esfahani K, Roudaia L, Buhlaiga N, Del Rincon Sv, Papneja N, Miller Jr WH. A review of cancer immunotherapy: from the past, to the present, to the future. *Curr Oncol*. 2020;27(suppl 2):S87–S97.
9. Miliotou AN, Papadopoulou LC. CAR T-cell therapy: a new era in cancer immunotherapy. *Curr Pharmaceut Biotechnol*. 2018;19(1):5–18.
10. Midulla M, Pescatori L, Chevallier O, et al. Future of IR: emerging techniques, looking to the Future...and learning from the past. *J Belg Soc Radiol*. 2019;103(1):12.
11. Erinjeri JP, Fine GC, Adema GJ, et al. Immunotherapy and the interventional oncologist: challenges and opportunities-A Society of Interventional Oncology white paper. *Radiology*. 2019;292(1):25–34.
12. Page DB, Yuan J, Redmond D, et al. Deep sequencing of T-cell receptor DNA as a biomarker of clonally expanded TILs in breast cancer after immunotherapy. *Cancer Immunol Res*. 2016;4(10):835–844.
13. Mizukoshi E, Nakamoto Y, Arai K, et al. Comparative analysis of various tumor- associated antigen-specific t-cell responses in patients with hepatocellular carcinoma. *Hepatology*. 2011;53(4):1206–1216.
14. Huang ZM, Li W, Li S, et al. Cytokine-induced killer cells in combination with transcatheter arterial chemoembolization and radiofrequency ablation for hepatocellular carcinoma patients. *J Immunother*. 2013;36(5):287–293.
15. Hickey RM, Kulik LM, Nimeiri H, et al. Immuno-oncology and its opportunities for interventional radiologists: immune checkpoint inhibition and potential synergies with interventional oncology procedures. *J Vasc Intervent Radiol*. 2017;28(11):1487–1494.
16. Ghodadra A, Bhatt S, Camacho JC, Kim HS. Abscopal effects and yttrium-90 radioembolization. *Cardiovasc Intervent Radiol*. 2016;39(7):1076–1080.
17. Wehrenberg-Klee E, Goyal L, Dugan M, Zhu AX, Ganguli S. Y-90 radioembolization combined with a PD-1 inhibitor for advanced hepatocellular carcinoma. *Cardiovasc Intervent Radiol*. 2018;41(11):1799–1802.
18. Mehta A, Oklu R, Sheth RA. Thermal ablative therapies and immune checkpoint modulation: can locoregional approaches effect a systemic response? *Gastroenterol Res Pract*. 2016;2016:9251375.
19. Oppenheimer J, Ray Jr CE, Kondo KL. Miscellaneous pharmaceutical agents in interventional radiology. *Semin Intervent Radiol*. 2010;27(4):422–430.
20. Lundblad RL, Bradshaw RA, Gabriel D, Ortel TL, Lawson J, Mann KG. A review of the therapeutic uses of thrombin. *Thromb Haemostasis*. 2004;91(5):851–860.
21. La Perna L, Olin JW, Goines D, Childs MB, Ouriel K. Ultrasound-guided thrombin injection for the treatment of postcatheterization pseudoaneurysms. *Circulation*. 2000;102(19):2391–2395.
22. Kiemeneij F, Vajifdar BU, Eccleshall SC, Laarman G, Slagboom T, van der Wieken R. Evaluation of a spasmolytic cocktail to prevent radial artery spasm during coronary procedures. *Cathet Cardiovasc Interv*. 2003;58(3):281–284.
23. Harmon CS S, Falluji N. Spasmolytic cocktail of nitroglycerin and verapamil versus verapamil alone for the prevention of radial artery spasm (RAS). *J Cardiol Study Res*. 2016.
24. Ishida J, Konishi M, Ebner N, Springer J. Repurposing of approved cardiovascular drugs. *J Transl Med*. 2016;14:269.
25. Mihlon F, Ray Jr CE, Messersmith W. Chemotherapy agents: a primer for the interventional radiologist. *Semin Intervent Radiol*. 2010;27(4):384–390.
26. Lee IJ, Lee JH, Lee YB, et al. Effectiveness of drug-eluting bead transarterial chemoembolization versus conventional transarterial chemoembolization for small hepatocellular carcinoma in Child-Pugh class A patients. *Ther Adv Med Oncol*. 2019;11, 1758835919866072.
27. Ferrone CR, Ryan DP. Pancreatic cancer: a time to change. *Ann Surg*. 2020;271(6):1003–1004.
28. Liu X, Yang X, Zhou G, Chen Y, Li C, Wang X. Gemcitabine-based regional intra- arterial infusion chemotherapy in patients with advanced pancreatic adenocarcinoma. *Medicine (Baltim)*. 2016;95(11):e3098.
29. Rosemurgy AS, Ross SB, Vitulli PL, Malek R, Li J, Agah R. Safety study of targeted and localized intra-arterial delivery of gemcitabine in patients with locally advanced pancreatic adenocarcinoma. *J Pancreat Cancer*. 2017;3(1):58–65.
30. Tian G, Liu X, Zhao Q, Xu D, Jiang T. Irreversible electroporation in patients with pancreatic cancer: how important is the new weapon? *BioMed Res Int*. 2018;2018:5193067.
31. Karnabatidis D, Katsanos K, Paraskevopoulos I, Diamantopoulos A, Spiliopoulos S, Siablis D. Frequency-domain intravascular optical coherence tomography of the femoropopliteal artery. *Cardiovasc Intervent Radiol*. 2011;34(6):1172–1181.
32. Prati F, Guagliumi G, Mintz GS, et al. Expert review document part 2: methodology, terminology and clinical applications of optical coherence tomography for the assessment of interventional procedures. *Eur Heart J*. 2012;33(20):2513–2520.
33. Balzer J, Zeus T, Hellhammer K, et al. Initial clinical experience using the EchoNavigator((R))-system during structural heart disease interventions. *World J Cardiol*. 2015;7(9):562–570.
34. Uppot RN, Laguna B, McCarthy CJ, et al. Implementing virtual and augmented reality tools for radiology education and training, communication, and clinical care. *Radiology*. 2019;291(3):570–580.

Chapter 7

Types of interventions

Husayn F. Ramji[1,2], Amber Hood[1], Eric Pham[1] and Sean Duguay[3]

[1]University of Oklahoma College of Medicine, University of Oklahoma Health Sciences Center, Oklahoma City, OK, United States; [2]Hudson College of Public Health, University of Oklahoma Health Sciences Center, Oklahoma City, OK, United States; [3]Department of Radiological Sciences, University of Oklahoma College of Medicine, University of Oklahoma Health Sciences Center, Oklahoma City, OK, United States

Key points

- Interventional radiology procedures can be diagnostic, therapeutic, or palliative.
- Multiple types of interventions are often used in conjunction as part of patient-centered care.
- To provide comprehensive patient care, IR physicians must be capable of performing a wide range of procedures utilizing various techniques, drugs, devices, telehealth, behavioral modification, and palliative care.

Why it matters?

The field of interventional radiology (IR) encompasses a variety of interventions across all organ systems of the body. For any interventional radiologist, it is important to understand not only the types of procedures but also the intended outcomes. This understanding serves as a foundation to build upon when working with other medical fields in delivering patient-centered care and in formulating and conducting translational research. This chapter will provide a brief overview of some of the most common types of interventional procedures, as well as the drugs and devices utilized.

Types of interventions

Interventional radiology procedures can fall into one of three categories: diagnostic, therapeutic, and palliative.[1] The category of intervention for any individual patient is dependent on indication and acuity. Diagnostic procedures, such as biopsy and venous sampling, provide an opportunity to further direct the pathway for disease management. Therapeutic procedures, such as tumor ablation or active bleed embolization, provide a minimally invasive means to treat pathology of all acuities. Palliative procedures, such as tunneled peritoneal catheter placement or venting gastrostomy tube placement, provide patients enhanced comfort in otherwise difficult times. IR is a unique field that contains a plethora of interventions in its repertoire, many of which will be covered in this chapter.

Drugs

Pharmaceuticals are an essential element of modern medicine (Fig. 7.1). They are as integral to medicine as the X-ray is to radiology. Medications are by far the most common type of medical intervention, with one in two Americans being prescribed at least one medication and one in five being prescribed three or more, which is why the development process of each drug is highly regulated and coordinated.[2] The many steps and procedures involved in drug development and in the subsequent trials to test a proposed drug's safety and efficacy are discussed in detail in Chapters 8–9 and Chapters 83–95.

Interventional radiologists utilize drugs from various classes that have undergone this rigorous testing.[3] Some of the most common are illustrated in Table 7.1.[4] The drugs

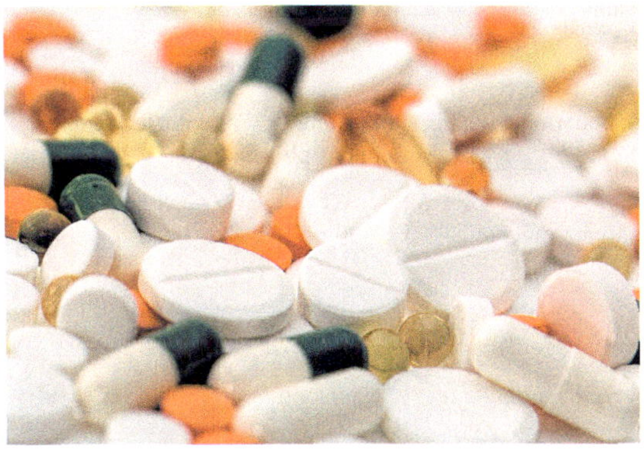

FIGURE 7.1 Oral pharmaceuticals. *From https://www.pexels.com/photo/bunch-of-white-oval-medication-tablets-and-white-medication-capsules-159211/*

Translational Interventional Radiology. https://doi.org/10.1016/B978-0-12-823026-8.00101-2

TABLE 7.1 Common medications used by interventional radiologists.

Drug class	Generic name	Brand name	Typical adult dosing, notes[4,7,8]
Anticoagulant	Apixaban	Eliquis (Bristol-Myers Squib Company, New York, NY, USA)	Prophylaxis: 2.5 mg PO BID Therapeutic: 5 mg PO BID
Anticoagulant	Enoxaparin sodium	Lovenox (Sanofi-Aventis, Bridgewater, NJ, USA)	Prophylaxis: 40 mg SC q12h Therapeutic: 1 mg/kg/dose q12h
Anticoagulant	Heparin		Intermittent IV: 8000–10,000 units IV initially, THEN 50–70 units/kg (5000–10,000 units) q4-6hr Continuous IV: 5000 units IV injection, followed by continuous IV infusion of 20,000–40,000 units/24 h
Antiplatelet	Aspirin		81–500 mg per day
Antiplatelet	Clopidogrel bisulfate	Plavix	Loading dose 150 mg 75 mg per day
Bowel antiperistalsis	Glucagon		Inhibit motility of stomach and small bowel: 0.2–0.5 mg IV over 1 min or 1 mg IM Inhibit motility of colon: 0.5–0.75 mg IV over 1 min or 1–2 mg IM
Contrast	Ethiodized oil	Lipiodol (Guerbet, Villepinte, France)	Typically, 0.25 mg/kg
Embolic	Ethylene vinyl alcohol (EVOH)	Onyx liquid embolic system (LES) (Medtronic, Minneapolis, MN, USA)	1.5 mL of 6% or 8% EVOH
Embolic	*N*-butyl cyanoacrylate (NBCA) "glue," enbucrilate	Histoacryl (B/Braun, Tuttilengen, Germany) Trufill (Cordis, Miami lakes, FL, USA)	Volume determined intraprocedurally
Local anesthetics	Lidocaine		Can come with epinephrine. If patient is allergic to amides, tetracaine, an ester, can be used.
Prothrombotic	Absorbable gelatin Compressed sponge	Gelfoam (Pharmacia and Upjohn Company, Kalamazoo, MI, USA)	Volume determined intraprocedurally
Prothrombotic	Thrombin topical	Recothrom (Baxter, Deerfield, IL, USA)	For complete thrombosis: 300–3000 IU/mL
Radial access sheath cocktail	Heparin, nitroglycerin, calcium channel blocker		Example doses (variable): Heparin 3000 μg Nitroglycerin 200 μg Verapamil 2.5 mg
Radiopharmaceutical embolic	Yttrium-90 (Y-90) coated resin microspheres	SIR-Spheres (Sirtex Medical, Woburn, MA, USA)	Dose specific to liver size, tumor size, tumor type, tumor location, lung shunt, etc.
Radiopharmaceutical embolic	Y-90 coated glass beads	TheraSphere (Boston Scientific, Marlborough, MA, USA)	Dose specific to liver size, tumor size, tumor type, tumor location, lung shunt, etc.
Reversal agents	Flumenazil		200 μg IV, repeat q1 min (max 1 g)
Reversal agents	Naloxone		40 μg IV, repeat q2 min (max 2 g)
Reversal agents	Protamine		10 mg for every presumed 1000u heparin still active

Continued

TABLE 7.1 Common medications used by interventional radiologists.—cont'd

Drug class	Generic name	Brand name	Typical adult dosing, notes[4,7,8]
Sclerosant	Alcohol/Ethanol, 98% dehydrated ethyl alcohol	ScleroGel (AbMedica, Milan, Italy)	Usually <100 mL in for a successful procedure
Sclerosant	Bleomycin		60u in 50–100 mL normal saline
Sclerosant	Doxycycline		500 mg powder mixed with normal saline
Sclerosant	Sodium tetradecyl sulfate (STS)	Sotradecol (Mylan Pharma, Viatris, Canonsburg, PA, USA)	0.5–2 mL, max 10 mL in a single session
Sclerosant	Talc		4 g, or 5 g if in slurry form
Sedative (benzodiazepine)	Midazolam	Versed	Dosed 0.5–1 mg IV PRN typically
Sedative (opiate)	Fentanyl		Dosed 25–50 μg IV PRN typically
TACE (conventional)	Combo of doxorubicin powder/ adriamycin liquid, and/or mitomycin C powder		Dose highly variable depending on disease process and treatment goal
TACE (drug-eluting bead)	Doxorubicin		Dose determined intraprocedurally, but typically 150 mg max in a single session
Thrombolytic	Tissue plasminogen activator (tPA), Alteplase		Dose variable depending on disease process and treatment goal. Example doses: 4–8 mg IV for AV graft de-clot 4–24 mg IA to provoke GI bleed
Vasoconstrictor	Epinephrine	Adrenalin (par pharmaceutical, Woodcliff lake, NJ, USA)	Asystole/PEA/VF/pulseless VT: 1 mg IV q3-5min prn Bradycardia: 2–10 μg/min IV Septic shock: 0.05–2 μg/kg/min IV
Vasodilator	Nitroglycerin		Between 50 and 300 μg of 10 μg/mL through a polycarbonate or glass syringe
Vasodilator	Verapamil hydrochloride	Verelan	10 ± 3 mg per arterial distribution in a dilution of 10 mL of saline

themselves can either be administered intravenously (IV), intraarterially (IA), percutaneously, or released intravascularly as part of a drug-eluting stent or drug-coated balloon, which are discussed in further detail in the procedural section of this chapter.[5]

One unique drug class used by interventional radiologists is radiopharmaceuticals. These are compound drugs that include radioactive isotopes, which contain emission and pharmaceutical components. The emission component can be captured and utilized to generate an image. The pharmaceutical component can affect and treat a disease process. An example of a radiopharmaceutical is Yttrium-90, used in treating hepatocellular carcinoma.[6]

Tools and devices

Interventional radiology relies heavily on medical devices, tools, and substances. The field as a whole was arguably pioneered with the Seldinger technique, followed by Dotter's percutaneous angioplasty, both of which set the stage for many techniques and devices seen in today's IR suites.[9] Similar to the process of drug approval, much goes into the process from device prototype to market.[10] Much of the product design takes inspiration from the "Biodesign Innovation Process," a philosophy pioneered at Stanford involving needs-based innovation to bring new medical devices to the table.[11,12] After the devices have faced rigorous testing by their developers, they must undergo another rigorous approval process by the FDA.[10,11] Biodesign is a focus of the Society of Interventional Radiology, which has held webinars and competitions to further discuss the biodesign process.[13] Biodesign, prototyping, and the FDA process are discussed further in Chapters 10–13.

The IR toolkit includes a myriad of tools and devices that you become familiar with during a career in IR. The following section briefly describes common IR tools you will see used in many IR procedures, adapted from various IR resources.[14–18]

Procedural imaging: IR procedural suites often contain a combination of fluoroscopic, sonographic, and computed tomographic (CT) equipment. Fluoroscopic equipment, such as c-arms, are important IR tools used during interventions. Fluoroscopy utilizes X-rays to produce real-time images, allowing for the evaluation of body anatomy and function. Many fluoroscopy suites are equipped with movable and easily adjustable c-arm X-ray machines that can provide real-time imaging, playing roles in visualizations of blood vessels, renal collecting systems, bile ducts, guidewires, and contrast. Some are able to perform CTs, producing three-dimensional and reformatted images. Ultrasound machines are utilized to provide real-time images to access vessels, perform biopsies, and place drains, without the use of ionizing radiation. Stand-alone CT scanners are also found in the IR department, where they are utilized in biopsy, ablation, and drain placement procedures, among others.

Contrast: One the most utilized substances in IR. The most common types of contrast are iodine or gadolinium-based, used in CT and magnetic resonance imaging (MRI) studies, respectively.[19] Other types include barium-based agents, often used in GI studies, and negative contrasts in the form of saline and gas, also called "microbubbles," which can be used to enhance ultrasound studies.[14,20] Due to the importance of renal insufficiency and its prohibitive effect on contrast use, labs should be drawn ahead of any imaging studies or procedures that require the use of contrast. However, complications can still occur due to contrast depending on the patient, including contrast-induced nephropathy from iodine-based contrast and nephrogenic systemic fibrosis from gadolinium-based contrast.[14,19] In patients like this, especially those with renal failure, CO_2 can be used as a contrast agent.[14]

Needles: There are three primary types of needles: puncture, biopsy, and ablation needles. Regardless of the type of needle, they are often described by their outer diameter, or "gauge." The higher the gauge, the smaller the needle.

Puncture needles: Puncture needles can be one-part, two-part, or sheathed. One-part needles have a cutting bevel and are the needles often used in blood draws or vascular access. Two-part needles contain an inner stylet that extends beyond the edges of the outer shaft and are often used in scenarios such as a lumbar puncture or renal collecting system initial access. Sheathed needles contain an inner stylet that extends beyond the edges of the outer sheath. The outer sheath is often used for the aspiration of fluid collections.

Biopsy needles: Biopsy needles come in aspiration and cutting variants, and both are used in specific scenarios, depending on the disease type and biopsy goal. Aspiration needles are typically smaller in size and used to remove a small amount of tissue cells for pathologic evaluation. Cutting needles can be solitary or coaxial and are often used to obtain larger specimen. Coaxial needles have an outer component used for guidance/stability and an inner component used to cut and extract tissue for sampling.

Microwave ablation needles: Microwave ablation involves using an applied electromagnetic field to heat up water molecules in target tissues to the point of death and denaturation.[21] The microwave ablation antenna, which may be in the shape of a needle, is placed into the lesion under image guidance.[21] Once this probe is in place, the electrical current is run through the probe. To prevent damage to tissues near the tip of the probe, the probe is often cooled with the use of chilled saline or gas.[14,21]

Radiofrequency ablation needles/probe: Radiofrequency ablation involves the application of an electrical current to heat up tissue to cytotoxic levels. The ablation probe, or the electrode, is placed in the tissue under image guidance and grounding pads are placed directly on the patient. An alternating current is passed through, heating up the tissue and causing coagulative necrosis.[16] The size of the ablation zone depends on the type of needle or probe utilized. See "Thermal Ablation" in the "IR Procedures" section of this chapter.

Cryoablation needles/probe: Cryoprobes, or cryoablation needles, are similar to microwave ablation antennae but instead of heating the tissue, it cools the tissue down to cytotoxic temperatures via argon gas. See "Cryoablation" in the "IR Procedures" section of this chapter.

Guidewires: Guidewires are used in conjunction with catheters to guide and support other devices. There are multiple functional categories of guidewires: nonsteerable, steerable, hydrophilic, stiff/support, and hybrid guidewires.[22] The nonsteerable tips are often "J-shaped," ensuring a relatively atraumatic passage into and through a vessel. Once the guidewire is in place, it provides a relatively safe passage of catheters, sheaths, and other devices to slide further within the vessel into the desired position.

Sheaths: A sheath is a hollow tube that allows the insertion and controlled passage of another instrument such as a guidewire, catheter, or dilator.[17] Sheaths are vitally important in interventional procedures requiring the introduction of an indwelling device, exchange of multiple interventional instruments, protect from mechanical injury, and/or prevent bleeding at the access site as they may have kink-resistant walls, varying column strength, and may come braided for improved structural integrity.[14] Two things to keep in mind when utilizing sheaths is the size used depends on the goal of the procedure and the sheath French size.[14] The outer diameter of a sheath is typically 2 Fr larger than its inner diameter, although there are thinner walled sheaths available for radial artery access.[22]

There are two main types of sheaths: vascular and peel away. As the name suggests, vascular sheaths are primarily used for vascular access that also requires the exchange of multiple instruments during a particular procedure. Due to the nature of the intervention, these sheaths are typically more mechanically resistant compared to simple access sheaths and/or peel away sheaths.[14] Peel away sheaths on the other hand serve the purpose of introducing indwelling devices that may stay with the patient for a prolonged amount of time. These types of sheaths are significantly thinner compared to the vascular access sheaths and are designed to split along their midline into halves and removed after the intervention is complete.[14] However, removal of the peel-away sheath is a bit nuanced as the rest of the sheath must be split without removing the indwelling device as well. Thus, the sheath must be slowly and carefully retracted while keeping the indwelling device in place.[14]

Catheters: Catheters are flexible tubes used in multiple specialties in varying degrees, from draining fluid to delivering medications. They come in many varieties and can be described in the size of their outer diameters (French or "Fr," where 3 Fr = 1 mm in diameter) and categorized into either flush/nonselective or selective catheters. Catheters are often advanced through a sheath. Flush, or nonselective, catheters allow high-flow injections into the vasculature and often have multiple sideholes along with an endhole. Selective catheters, alternatively, only have a single endhole, and come in various shapes, including the "cobra" and "sidewinder," that help improve access to different types of vessels. Catheters may also be braided or coated in various substances that improve visibility on imaging or alter how hydrophilic the catheter is, which can modify the maneuverability of the catheter after insertion. Microcatheters are another variant that allows access to even smaller vessels in the body.

Balloons: Balloons catheters contain an inflatable device at the distal end of a straight catheter that is deflated on vascular entry, advanced into position over a guidewire, then is expanded to dilate stenotic or occluded vessels through outward radial pressure.[14] The structure of a balloon-catheter system is comprised of a catheter with one to two channels: one for the guidewire and one for inflation/deflation of the balloon. Some constructions allow a single channel to serve both purposes. The two primary constructions are over-the-wire balloons and monorail balloons.[14] Over-the-wire balloons are the most commonly used. Its general construction is shown in Fig. 7.2. The primary difference between these two constructions is the guidewire: with over-the-wire balloons, the guidewire enters at the hub of the catheter and runs the entire length of the catheter to the tip; with monorail-rapid-exchange balloons, the guidewire enters the catheter at the tip and exits at approximately 15 cm distal to the end of catheter. The construction of a monorail angioplasty balloon is shown in Fig. 7.3. The monorail construction allows for shorter guidewires and allows the catheter shaft to exist as only the inflate/deflate channel. Balloons also come with measurements of nominal and burst pressure, or the pressure it can reach without bursting.[17] Standard balloons will dilate most stenoses, but in the event of a resistant stenosis, high-pressure balloons and cutting angioplasty balloons can be used. High-pressure balloons are made of a thicker material that allows for higher inflation pressures.[17] Cutting

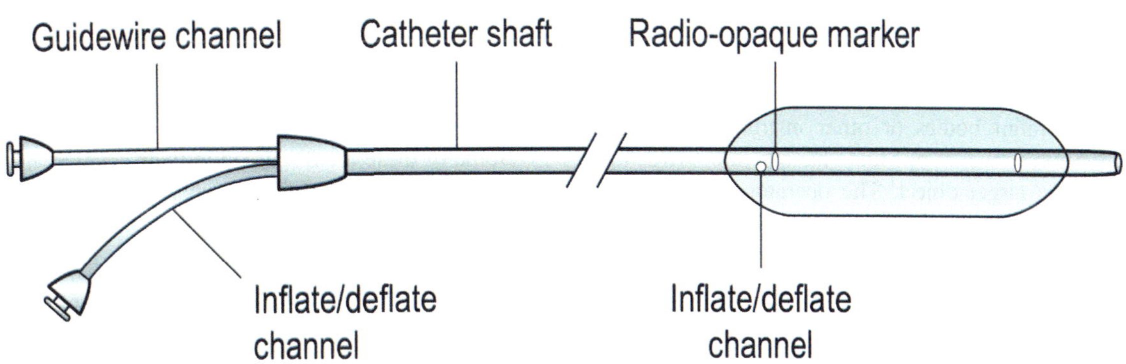

FIGURE 7.2 Typical angioplasty balloon construction. *Reprinted from "18: Essential Equipment: Angioplasty Balloons", Kessel D, Robertson I, Interventional Radiology: A Survival Guide, p.77, Copyright Elsevier 2017.*

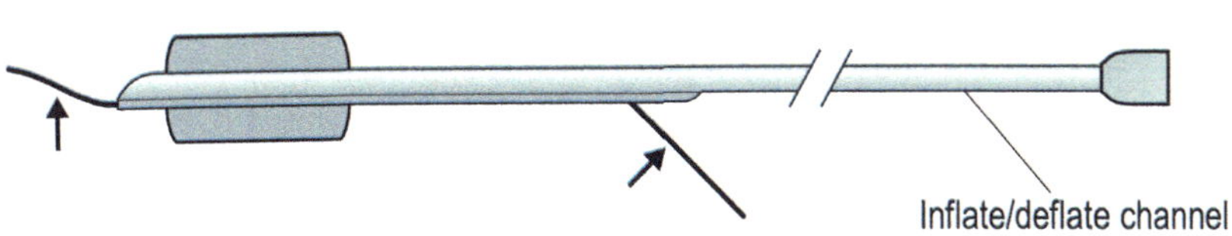

FIGURE 7.3 Schematic of a monorail angioplasty balloon. *Reprinted from "18: Essential Equipment: Angioplasty Balloons", Kessel D, Robertson I, Interventional Radiology: A Survival Guide, p.77, Copyright Elsevier 2017.*

balloons have razor blades in the balloon that cut into the stenosed tissue.[17] Both of these balloons typically require larger sheaths and are higher profile and higher cost than standard balloons.

Coils/vascular plugs: Coils and vascular plugs are types of permanent embolic devices used in during procedures. Although the coil itself provides minimal disruption of blood flow, it increases the chance of thrombus formation on its large surface area and/or by damaging the intima.[14] Often made from metal, coils can be pushable or detachable and intertwined with various kinds of fibers to increase thrombogenicity and the chance of full occlusion.[14,22] Some coils, such as the "MReye coil" (Cook Medical, Bloomington, NJ, USA) made from Inconel and those made from platinum are MRI compatible. Deployment techniques depend on the coil type and location of use. Vascular plugs disrupt blood flow and provide increased surface area for thrombus formation. Vascular plugs are detachable and can be repositioned if necessary.

Embolic protection devices (EPD): Devices utilized in patients who at risk of emboli due to underlying conditions, interventions, or contraindications to anticoagulation.[23] One type of EPD is a venous filter, most commonly positioned within the inferior vena cava (IVC).[14] We discuss the IVC placement procedure later in this chapter. The IVC filters themselves may be retrievable or permanent.[24] Other embolic protection devices may be utilized in procedures with a risk of embolization, such as in revascularization procedures. These types of embolic protection devices include other types of filters and distal or proximal occlusion balloons.[16] They can differ in pore size and whether or not they let blood flow through. These are often placed before the revascularization procedure and removed afterward.

Snares: Snares, in general, are tools utilized to capture an item or object. Similarly, in interventional radiology, snares are wire loops designed to capture, manipulate, and/or remove foreign bodies or other instruments within a vessel.[14] Snares work by first having the wire loop encompass the target object. The operator then decreases the diameter of the loop by advancing a catheter until the target object is secure.[14] The snare, catheter, and object are then removed via the access site. Snares come in either a single- or multiloop design, but both function similarly. However, multiloop snares do have an advantage as they allow more maneuverability when in a confined space compared to conventional single-loop snares.[14] This is primarily due to the shape of the wire as single-loop GooseNeck snares, for example, are circular, and multiloop snares are pear shaped.

Stents/stent grafts: Stents are metal structures placed into vessels or ducts to maintain a patent lumen in systems with stenosis or occlusion.[14] Alternatively, they can also be used to trap a thrombus against a vessel wall to prevent embolization during angioplasty. The composition of stents is a lattice-like structure made of metal that can be compressed for insertion and advancement and expanded at its destination.[14] Stents expand primarily in two ways: with a balloon or through intrinsic radial force. Balloon-expandable stents are preconstructed on an angioplasty balloon, which is then inflated, causing the stent to expand with the balloon.[17] Self-expanding stents are preconstructed on a delivery catheter and expand due to intrinsic radial force; however, balloon dilation is often required for full stent expansion. Stents are usually composed of steel, nickel, or titanium alloys and may have drug-eluting or radioactive coatings that can further reduce the likelihood of restenosis or thrombosis.[14] Stent grafts are another type of stent covered in the material used in grafts, including polytetrafluorethylene, polyethylene tetraphthalate, and silicon.[25] They are often used in covering damaged vessels or recreating tracts of ruptured vessels.[14] They are also used in creating the tract through the liver in transjugular intrahepatic portosystemic shunt (TIPS) cases.[14] See the section on "Shunt Placement" later in this chapter.

Thrombectomy and thrombolysis tools: Thrombectomy and thrombolysis are utilized to restore perfusion to tissues in acute, subacute, and chronic scenarios.[14] Thrombolysis is usually achieved via the use of pharmaceutical agents, such as tPA mentioned in Table 7.1. Once the clot is crossed with a guidewire, a catheter with several sideholes is placed into the thrombus, and the thrombolytic is slowly deployed.[17] See "Thrombolysis" in the "IR Procedures" of this section. Mechanical thrombectomy, in contrast to thrombolysis techniques, involves devices with the goal of physically breaking the thrombus or removing the thrombus entirely via "Thrombosuction."[14,26] Mechanical thrombectomy often utilizes rheolytic catheters and fragmentation baskets while suction thrombectomy involves aspiration of the thrombus with a syringe or vacuum.[17] See "Thrombectomy" in the "IR Procedures" section of this chapter.

Some of these are discussed in their practical use in the "Procedural" section of this chapter and in Chapters 10–13.

Laboratory analysis and IR

Laboratory tests are paramount for preoperative assessment. The results of certain laboratory tests, like coagulation studies and renal function tests, may drive the timing of and type of interventional procedure to be performed. For example, a transjugular approach to liver biopsy may be chosen over a percutaneous approach if the patient is coagulopathic or CO_2 contrast may be chosen over iodinated contrast if acute renal failure is present. However, IRs also utilize the laboratory to evaluate specimens obtained during a procedure, such as fluid collection aspiration and adrenal venous sampling. Fluid collection aspirate can be evaluated to determine if the collection represents an abscess, hematoma, seroma, urinoma, pancreatic fluid collection, bile leak, or lymphocele, to name a few. Adrenal vein sampling is performed in patients with unexplained hypertension and concern for primary hyperaldosteronism. IR physicians obtain venous blood samples from both adrenal glands and a peripheral site for laboratory analysis to determine aldosterone and cortisol ratios.[27]

IR procedures

The procedures performed by interventional radiologists in the IR suite are performed using the imaging modalities described in the previous section, more specifically fluoroscopy or CT. Some common IR procedures include angiogram/angiography, angioplasty and stent insertion, chemoembolization, embolization, image-guided biopsy, IVC filter placement, thermal ablation, shunt placement (TIPS), thrombolysis, and others. In this section, we will discuss these procedures, their techniques, and their goals.

Ablation—thermal ablation

Microwave and radiofrequency ablations are procedures that use heat to damage a localized area of tissue.[28] Radiofrequency ablation uses heat generated from a high-frequency alternating current. Microwave ablation uses heat generated by electromagnetic waves. There are multiple indications for this procedure including tumor ablation, cardiac arrhythmias, pain management, and varicose veins. This procedure may be preferred over surgery if surgery is contraindicated due to the location of the area of interest or if the patient is at risk for operative complications. In this procedure, an image-guided ablation needle is directed toward the area of interest. The needle is then heated to the desired temperature to damage the surrounding tissue. The target temperature varies, depending on the type of procedure. For example, a study determined the optimal temperature for neurological ablation in the setting of back pain as 90°C.[29] However, it is generally optimal to keep temperatures below 100°C, as anything higher that can lead to gas production within the tissue.[16] Complications associated with this procedure include tissue damage to the surrounding areas and a common phenomenon called postradiofrequency ablation syndrome. This syndrome is associated with flu-like symptoms such as fever, pain, nausea, vomiting, fatigue, and muscle aches.[30]

Ablation—cryoablation

Cryoablation is a procedure similar to thermal ablation, but uses argon gas to cool tissues to cytotoxic temperatures, followed by a thawing cycle using helium.[31] The change in air pressure leads to a decrease in temperature, also known as

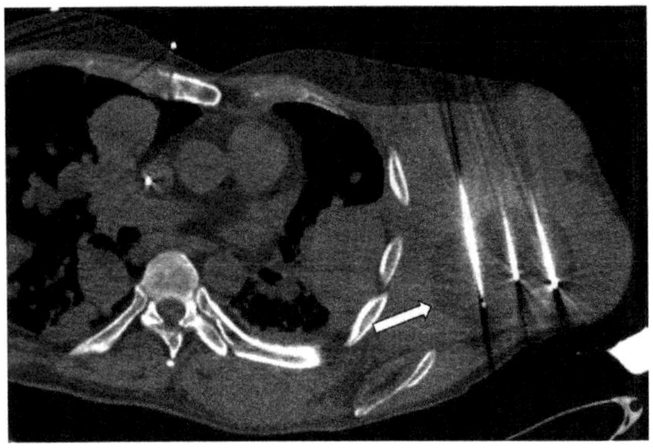

FIGURE 7.4 CT image demonstrating a cryoablation and a cryoablation zone (arrow). *Reprinted from Cam I, Gencturk M, Young S. Cryoablation of a Large Metastatic Chest Wall Mass, 1888–89. Journal of Vascular and Interventional Radiology 2020/11/02, Copyright 2020, with permission from Elsevier.*

the Joule–Thomson effect.[31] At least two freeze–thaw cycles are performed, during which ice crystals form extracellular or intracellularly leading to cell death.[14] The cryoprobe, or cryoablation needle, is placed within the target under image guidance. When the gas is turned on, an "iceball" forms at the probe tip. This is advantageous in certain procedures as the "iceball" and subsequent ablation zone can actually be visualized on CT, MRI, and US, as seen in Fig. 7.4.[14] Multiple probes can be placed near each other to increase the size of the "iceball." Another advantage of cryoablation is that it is relatively less destructive to nearby tissues, as opposed to thermal ablation techniques. This principle can be utilized in treating bone metastases, where published literature reports cryoablation of where the tumor meets the bone can lead to reduced pain, improved quality of life, and even reduced use of opioids for pain control.[32]

Angiogram/angiography

An angiogram is used to visualize and study the structure and anatomy of blood vessels.[33,34] This procedure is performed by inserting a catheter into an anatomical vessel of choice and maneuvering the catheter to an area of interest under the guidance of fluoroscopy. At the site of interest, contrast dye is injected into the blood vessel, as shown in Fig. 7.5. Indications for this procedure include evaluations of aneurysms, blood vessel malformations, tumors, stenotic vessels, and coronary thromboembolisms. In some procedures, CT angiography is used as the diagnostic gold standard to evaluate the effectiveness of treatment, including aneurysm patency after embolization, which may necessitate a return to the IR suite in the near future.[35]

Angioplasty and stent insertion

Indication for this procedure is primarily for the treatment of narrowed blood vessels (e.g., carotid, renal) to improve blood flow to poststenotic regions.[36] Diseases benefitting

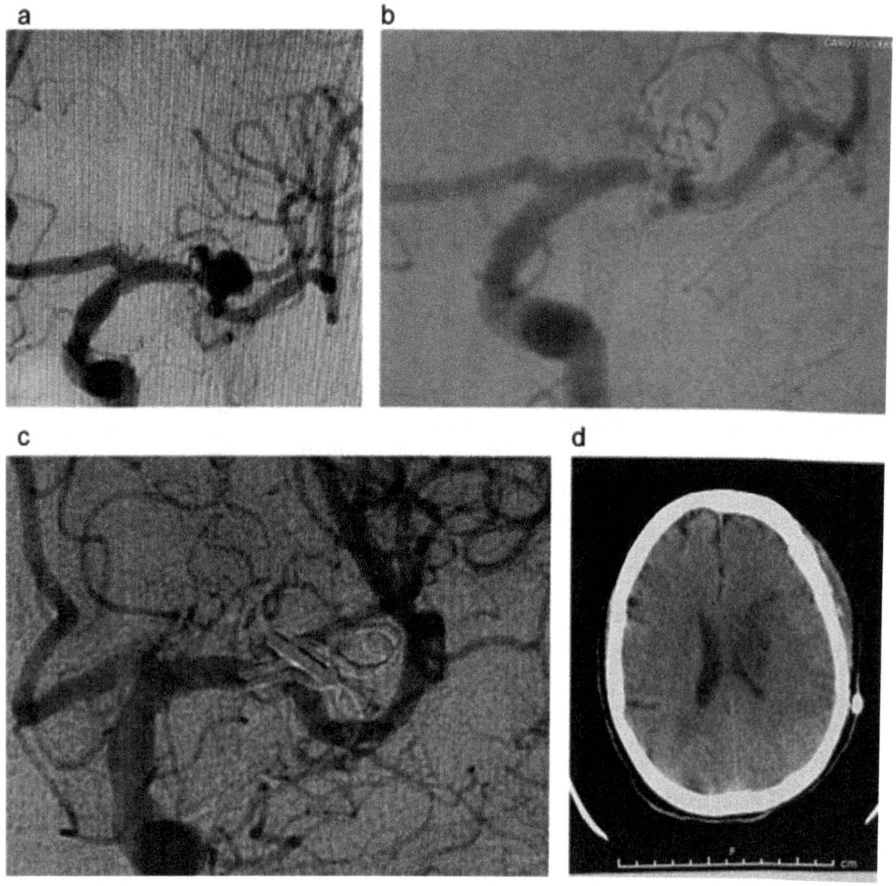

FIGURE 7.5 (A) Angiogram showing a Middle Cerebral Artery Aneurysm. (B) An MCA vasospasm from an angiogram taken intraoperatively. (C) Post-op angiogram showing resolution of the vasospasm. (D) CT showing infarct in L MCA territory. *Reprinted from Interdisciplinary Neurosurgery, 25, Moufarrij N, Stroke due to middle cerebral artery aneurysm clipping when the intraoperative angiogram was normal, Copyright 2021, with permission from Elsevier.*

from this treatment modality include coronary artery disease, peripheral artery disease, chronic deep vein thrombosis and other causes of vessel narrowing. Angioplasty involves the insertion of a cylindrical balloon along a catheter that is maneuvered to the stenotic area of interest and inflated. This inflation stretches the vessel open, as seen in Fig. 7.6 with the multichannel approach, and in Fig. 7.7 in the Superficial femoral artery with a single channel approach. A stent is then placed to ensure the vessel remains open. This entire procedure is commonly done under angiography with fluoroscopy.[33]

Fluid drainage

This procedure is indicated for the drainage of fluid from a body cavity, such as the thorax (thoracentesis) or abdomen (paracentesis).[37] Fluid accumulation in a body cavity is usually a sequela of a pathological process, such as congestive heart failure, cirrhosis, or malignancy. This procedure involves the insertion of a needle into the desired cavity under the guidance of ultrasound, as seen in Fig. 7.8, or CT.[37] The cavity is aspirated and the fluid can be analyzed for further diagnostic testing. As mentioned earlier, aspirate can be evaluated to determine the type of the collection, such as abscess, hematoma, seroma, urinoma, pancreatic fluid collection, bile leak, or lymphocele.

Chemoembolization

Chemoembolization is used to treat tumors that originate predominantly within the liver or have spread to the liver from a distant site (metastasis).[38,39] This procedure is performed through the insertion of a catheter into the femoral or radial artery, which is then guided through the arterial system to the liver. When the catheter is within the hepatic artery, an angiogram locates the arteries feeding the tumor. The catheter is then maneuvered to the arteries of interest and chemotherapeutic medications are injected. These medications may be injected in combination with or without particles to provide additional blocking of the arteries feeding the tumor, causing the tumor to lose its supply of nutrients. See Table 7.1 for common medications used during this Transarterial Chemoembolization (TACE).

Embolization

Catheter embolization is similar to chemoembolization in that an embolic medium is delivered to a blood vessel to cause cessation of blood flow, similar to the method

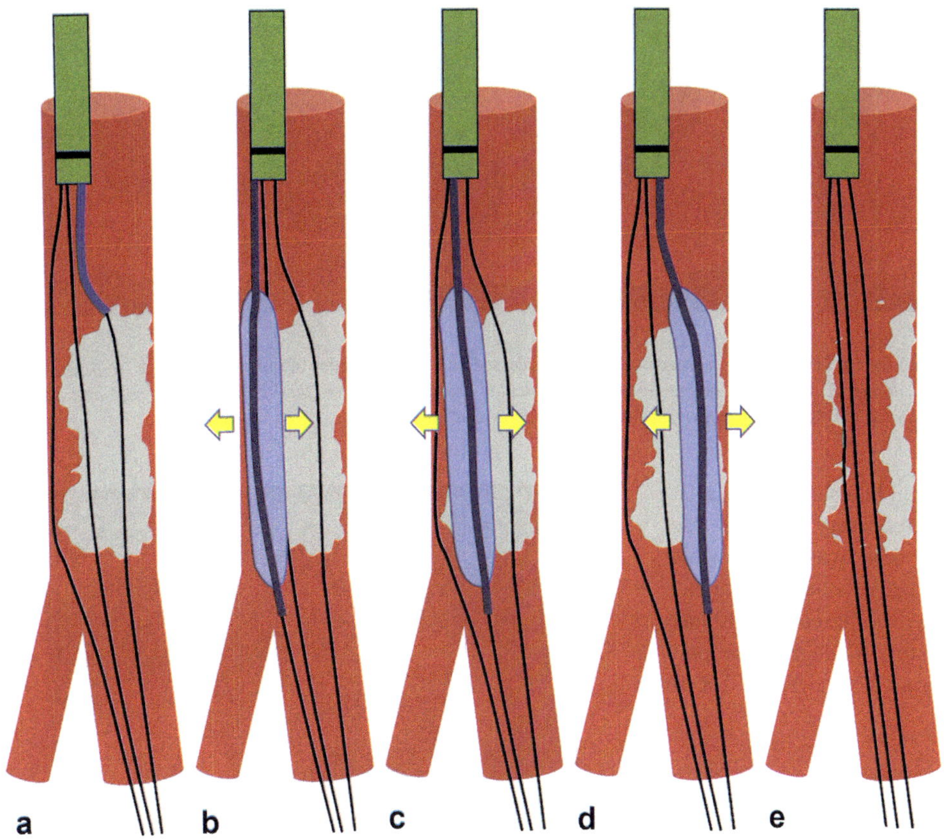

FIGURE 7.6 A–E. Multichannel angioplasty, with balloon angioplasty performed along each channel. *Reprinted from Nagatomi S, Ichihashi S, Yamamoto H et al. Efficacy of multichannel balloon angioplasty in calcified common femoral artery lesions. Journal of Vascular and Interventional Radiology, 32(2), 181–186, Copyright 2020, with permission from Elsevier.*

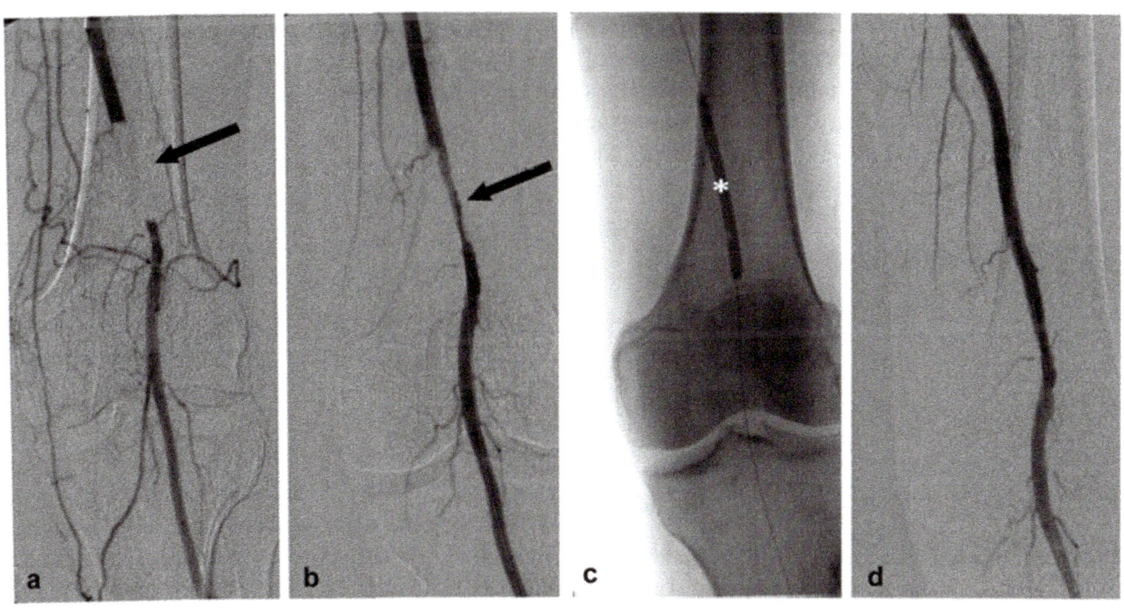

FIGURE 7.7 Total occlusion of superficial femoral artery treated successfully with mechanical thrombectomy and drug-coated balloon angioplasty (asterisk in C). *Reprinted from Wang Q, Zhu RM, Ren HL, Leng R, Zhang WD, Li CM. Combination of Percutaneous Rotational Thrombectomy and Drug-Coated Balloon for Treatment of Femoropopliteal Artery Nonembolic Occlusion: 12-Month Follow-up. J Vasc Intervent Radiol, 31(10), 1661–67. Copyright 2020, with permission from Elsevier.*

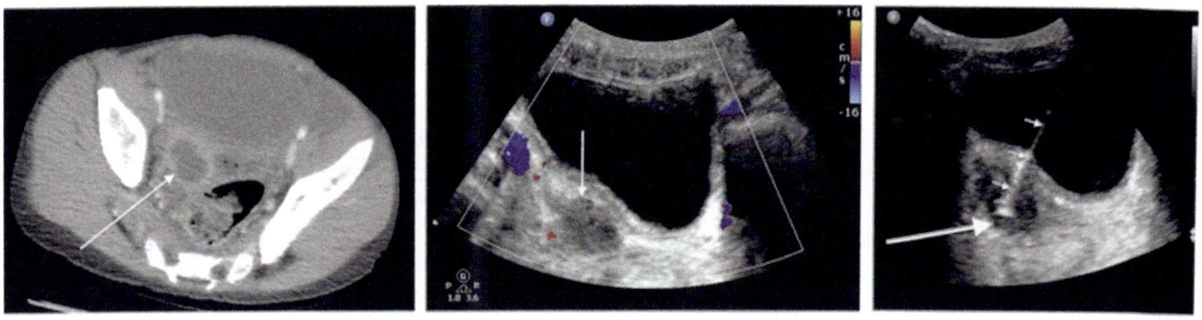

FIGURE 7.8 Image of an appendicitis and right pelvic abscess in a pediatric patient accessed and drained under ultrasound guidance. *Reprinted from Ayyagari RR, Yeh C, Arici M, Mojibian H, Reiner E, Pollak JS. Image-guided transvesicular drainage of pelvic fluid collections: a safe and effective alternative approach. J Vasc Intervent Radiol 27(5), 689–693, Copyright 2016, with permission from Elsevier.*

described above. This technique can be used to stop active bleeding from a traumatically severed artery or the GI tract, treat tumors, or divert flow, among other reasons. However, a wide array of embolic agents instead of chemotherapeutic medications is available, such as metallic coils, calibrated microspheres, shape-memory foams, biodegradable polymers, and in situ gelling solutions.[40] See Table 7.1 for common embolics.

Image-guided biopsy

An image-guided biopsy is a minimally invasive procedure used to access difficult-to-reach areas of the body to sample tissue for diagnostic testing, as shown in Figs. 7.9 and 7.10 under ultrasound and CT guidance, respectively.[41] There are two primary methods of biopsy, core needle biopsy (CNB) and fine-needle aspiration (FNA). The indications for each depend on the goals of the biopsy. FNA allows for cytological visualizations assisted by immunohistochemistry. CNB maintains the structural integrity of the tissue being sampled. Thus, the diagnostic value varies depending on tissue types, such as CNB for head and neck malignancies and FNA for thyroid malignancies.[42,43]

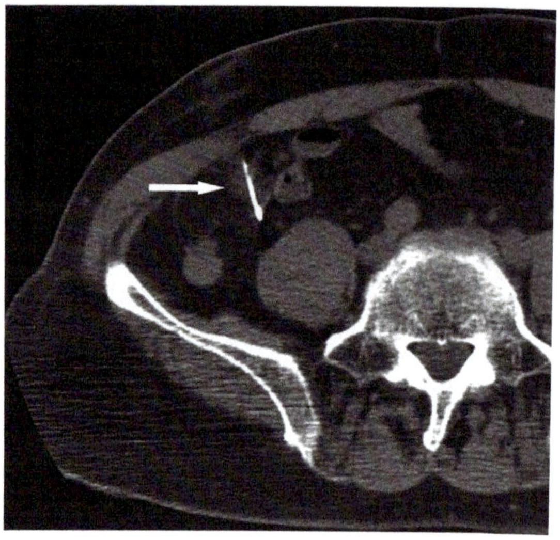

FIGURE 7.10 An omental nodule biopsy conducted with CT guidance. Arrow indicates omental nodule. *Reprinted from Journal of Vascular and Interventional Radiology, 28(11), Vadvala HV, Furtado VF, Kambadakone A, Frenk NE, Mueller PR, Arellano RS, Image-Guided Percutaneous Omental and Mesenteric Biopsy: Assessment of Technical Success Rate and Diagnostic Yield,1569–1576, Copyright 2017, with permission from Elsevier.*

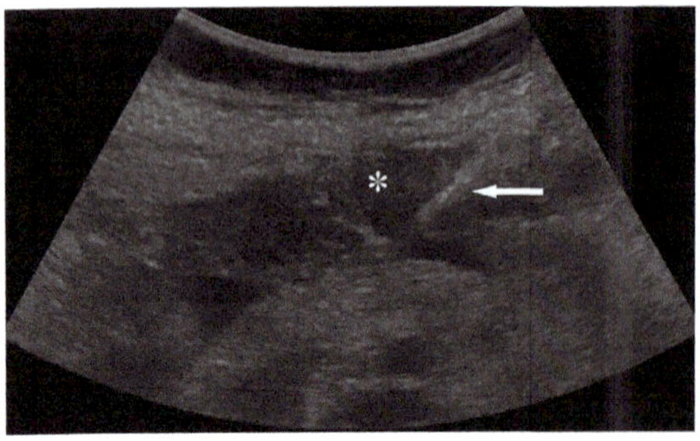

FIGURE 7.9 Omental mass biopsy conducted with ultrasound guidance. Arrow indicates biopsy needle. Asterisk indicates omental mass. *Reprinted from Journal of Vascular and Interventional Radiology, 28(11), Vadvala HV, Furtado VF, Kambadakone A, Frenk NE, Mueller PR, Arellano RS, Image-Guided Percutaneous Omental and Mesenteric Biopsy: Assessment of Technical Success Rate and Diagnostic Yield, 1569–1576, Copyright 2017, with permission from Elsevier.*

Inferior vena cava filters

The purpose of IVC filters is to prevent pulmonary emboli by blocking the propagation of an embolus from the lower extremities at the inferior vena cava in patients at high risk for fatal pulmonary embolism (PE) and contraindications to anticoagulation.[44] The filters can be temporary or permanent; however, permanent filters have been associated with a significant increase in deep vein thrombosis risk as well as filter thrombosis and IVC penetration.[45] Thus, permanent filters or delayed retrieval of filters are not

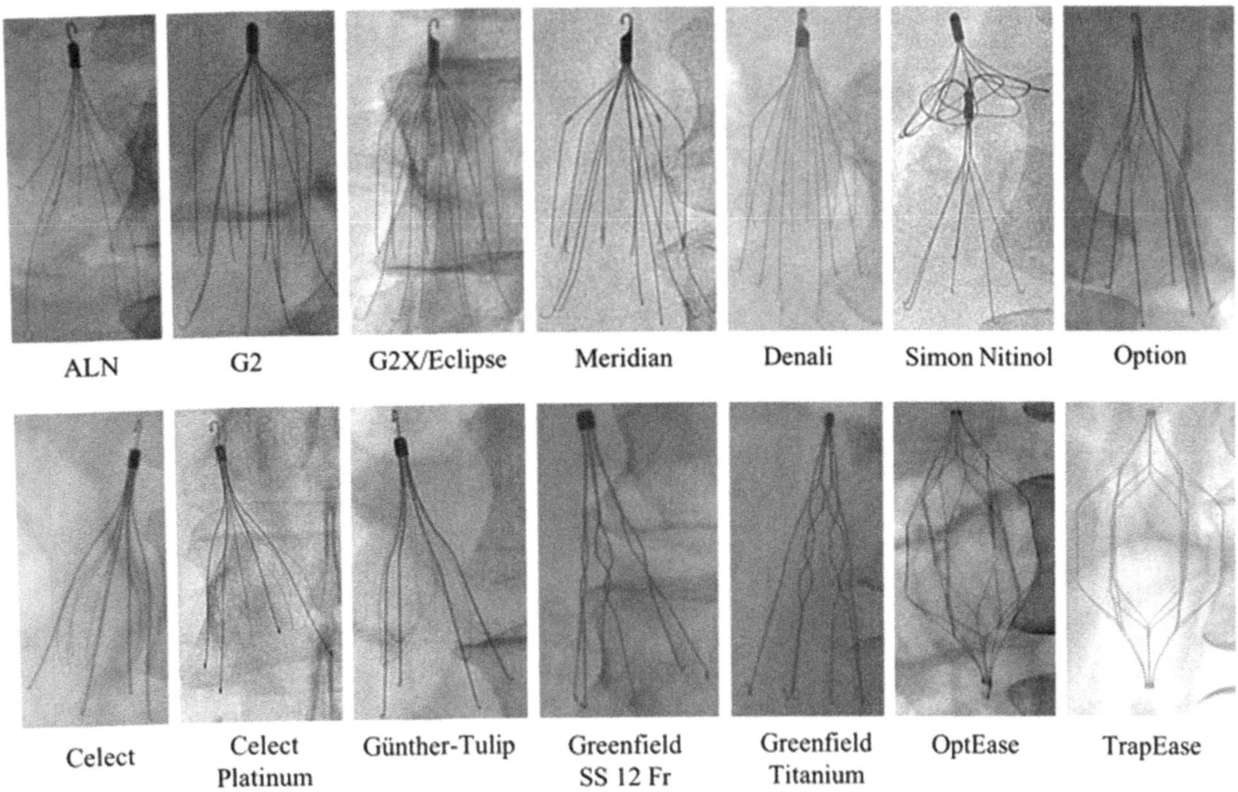

FIGURE 7.11 Examples of various types of IVC filter. "SS" is stainless steel. *Reprinted from Journal of Vascular and Interventional Radiology, 31(1), Ni JC, Shpanskaya K, Han M et al. Deep Learning for Automated Classification of Inferior Vena Cava Filter Types on Radiographs, 66–73. Copyright 2020, with permission from Elsevier.*

recommended.[46] Insertion of an IVC filter is most commonly endovascular with femoral or internal jugular access. The filter is compressed within a catheter, which is guided fluoroscopically from the femoral or internal jugular vein into the IVC, where the filter is deployed, usually below the junction of the IVC and the renal veins. Various types are shown in Fig. 7.11.[47]

Shunt placement (TIPS)

TIPS is a procedure used to quickly and effectively treat portal hypertension and its complications, including uncontrolled variceal hemorrhage (esophageal, gastric, and intestinal) and ascites.[48] In this procedure, interventional radiologists create a vascular communication using a stent between the portal vein and hepatic vein. This essentially diverts backed-up portal blood and reroutes it into the systemic circulation. Thus, venous drainage from the intestines and other organs that normally drain to the portal system bypass the congested and poorly functioning liver to directly drain into the systemic venous system that returns blood to the heart. This reduces pressure in the portal system, which decreases engorged veins from the esophagus, for example, decreasing rupture and bleeding risk. One complication associated with this procedure is the accumulation of ammonia due to a lack of hepatic metabolism to break down the toxin, leading to shunt-induced hepatic encephalopathy.[49] A successful TIPS placement is shown in Fig. 7.12.

Thrombolysis

This procedure, like many other procedures in interventional radiology, requires the use of imaging to direct a catheter into an area of interest. In catheter-directed thrombolysis, a catheter is introduced to the blood vessel of interest, and thrombolytic medication, such as tPA, is delivered to dissolve a blood clot, freeing an obstructed vessel. The thrombolytic is often delivered with heparin to prevent rethrombosis.[16] It is essential a clot be broken up in the timeliest manner, as tissues rely on a constant stream of nutrients and oxygen provided by blood. Thus, if initial systemic anticoagulation does not provide a satisfactory improvement, interventional treatment has been shown to improve morbidity and mortality. One group that benefits from this procedure is patients with acute portal vein thrombosis or iliocaval thrombosis.[50] Thrombolysis techniques may also involve devices used during mechanical thrombectomy in a technique called pharmacomechanical thrombolysis.

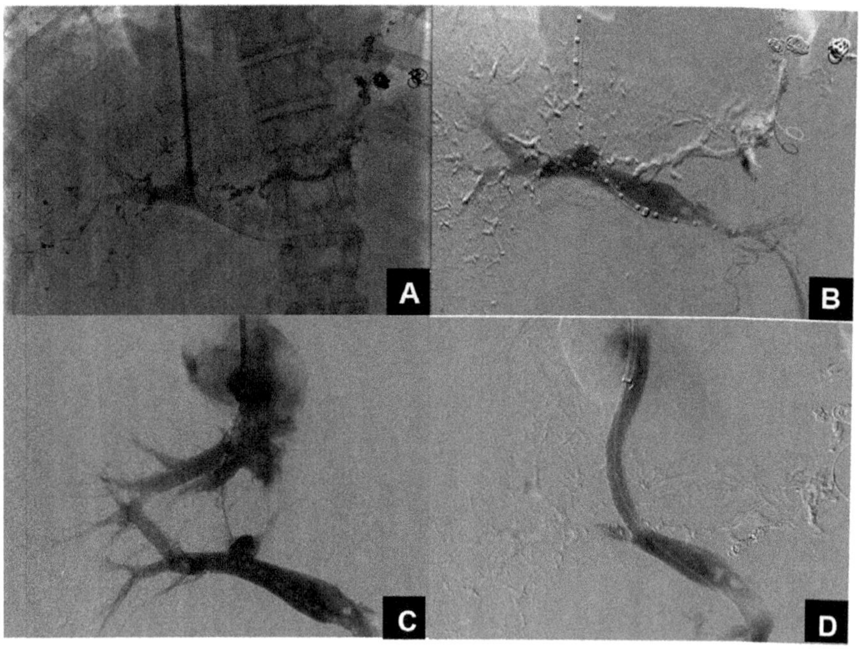

FIGURE 7.12 Images demonstrating Pre- (A) and Post TIPS (D) placements, with resultant flow through the shunt. Image (B) shows the alignment of the right hepatic and main portal vein, and (C) shows the shunt being placed. *Reprinted from Yan J, Browne W, Kesselman A. Transjugular intrahepatic portosystemic shunt (TIPS) as rescue therapy for endoscopic glue migration and bleeding gastric varices, Radiology Case Reports, 16(8), 2035–37, Copyright 2021, with permission from Elsevier.*

Thrombectomy

Thrombectomy, as mentioned earlier in the chapter, can be achieved via mechanically breaking down the clot or aspirating it. In its simplest form, thrombectomy can be performed by applying a negative pressure to a catheter or sheath positioning near a clot, effectively aspirating out the clot. Another common technique for mechanical thrombectomy involves placing a rheolytic catheter near the thrombus. The rheolytic catheter then utilizes a high-velocity saline jet that creates a vortex of negative pressure, sucking the thrombus through one of the holes at the catheter tip.[14,17] Mechanical thrombectomy may also involve a rotating basket that breaks up the clot and aspirates it through a catheter.[16] Many of these techniques may also utilize thrombolytics, which is referred to as pharmacomechanical thrombectomy.

Percutaneous drainage

Abscesses form throughout the body via various processes, including bowel perforation and pancreatitis. These abscesses are poorly penetrated by antibiotics and, thus, usually require drainage. An added benefit of percutaneous drainage, as opposed to surgical drainage, is that percutaneous drainage can usually be performed under minimal-moderate sedation. The procedure involves preprocedure image-guided mapping, usually CT or ultrasound, and ideal positioning of the patient. A catheter is placed using the Seldinger technique and drainage after confirmation of appropriate placement. Catheters are removed after resolution of clinical infection or when drainage volume decreases to <10–20 mL/day.[51]

Telehealth and telesurgery

Telehealth might not be an intervention in the conventional sense, but it is slowly becoming a convenient and socially-distanced way to communicate with, diagnose, and treat patients.[52,53] It is not a new innovation, with origins dating back to 1905 with Dr. Willem Einthoven interpreting electrocardiograms remotely and to the 1920s with the "Radio Flying Doctor" in Australia.[52,54] It progressed further with innovations by NASA and their need to remotely care for astronauts. With increased access to the internet, the reaches of telehealth have also increased, even to medical specialties that would seem unreachable to telehealth, including radiology.[52]

Radiology is a unique specialty in that it does not necessitate the physical presence of the radiologist in a hospital or clinic environment at all times. Due to the degree of "digitalization," radiology work can be done remotely, so long as the radiologist has access to the radiology information system (RIS) and picture archive and communication system (PACS).[55] This is essentially the basis of what is termed "teleradiology."[56] Teleradiology involves the interpretation of images outside of the facility at which the images were acquired.[56] Teleradiology and advances in digitization have also paved the way to incorporating artificial intelligence into radiology, especially in terms of pattern and image recognition.[57,58]

FIGURE 7.13 Example of telehealth. *From Edward Jenner, https://www.pexels.com/photo/woman-having-a-video-call-4031818/*

The adoption of telehealth by both patients and providers was slow until the arrival of the SARS-Cov-2, or COVID-19, pandemic, which led to an upsurge in the use of telehealth and its applications, with many telehealth visits looking like Fig. 7.13.[52] It is not only becoming a more common way to conduct traditional outpatient interactions but is also used in emergency situations and even radiology and interventional radiology. Teleradiology is playing a large role in the COVID-19 pandemic, allowing interventional radiologists to provide necessary services while minimizing the risk of virus spread.[55] However, radiology technicians and associated staff still need to be present at institutions to conduct the imaging itself. So, while teleradiology provides benefits, it does not eliminate the need for in-person radiology personnel at institutions and additionally requires a reliant IT and telecommunication service in place.

Interventional radiology is a specialty whose outpatient clinic may benefit from telehealth sessions, as advancements in technology may permit interventional radiologists to perform procedures remotely.[59] This unique type of intervention is called "telesurgery" and, in contrast to telehealth, is only about 20 years old.[60] Telesurgery involves a physician performing a procedure when he or she is not in the immediate vicinity of the patient.[61] The first documented telesurgery was conducted in 2001 where surgeons in New York conducted a laparoscopic cholecystectomy on a patient in France.[62] Since then, numerous specialties have experimented with telesurgery, made more accessible with the arrival of robotic surgery and 5G wireless technology.[60,63] While this technology requires further advancements, it is only a matter of time before the first telesurgical IR procedure takes place, perhaps within the next decade.

Behavioral change

Incentivizing behavior change is an important part of treating patients. It can play a role in preventing disease, in addition to improving the prognosis of existing disease. In the United States, it is estimated that around 45% of the population, or 133 million Americans, have one or more chronic conditions.[64] Some of the more common conditions include obesity, nicotine addiction, and alcoholism. These chronic conditions promote complications that may eventually require the interventions of IR physicians. For example, peripheral arterial disease or various cancers may result in complications that require stent placement or chemoembolization, procedures described in an earlier section.[14,65] While prevention of a disease is preferable to treatment of an existing disease due to improved outcomes, IR is usually not involved in a patient's care until after preventative behavioral measures would be of any benefit. By the time IR sees a patient, they are often already facing the long-term symptoms of their disease, whether the disease is a result of ongoing behavior or the accumulation of small decisions and exposures over the years. At that point, IR can intervene with minimally invasive procedures that can not only improve the patient's condition, but in a way also promote behavior change, such as gastric artery embolization.[66] Embolization of the gastric artery leads to appetite suppression and can therefore promote weight loss.[66] Beyond this, however, as part of patient-centered care, interventional radiologists can encourage patients to adhere to behavioral interventions to prevent additional future complications.

Behavioral intervention can include assisting a patient with smoking cessation or in better medication compliance.[67] These behavioral interventions are often utilized by interventional radiologists in conjunction with the treatment of peripheral arterial disease. In that sense, behavior intervention is the human component of drug interventions that were described earlier in the chapter. Changing behavior is a difficult practice, as it involves many aspects of a healthcare team, and a patient's lifestyle and personality, but behavioral interventions can play a large role in addressing preventable diseases and even racial disparities in health.[67]

Palliative

The goal of palliative medicine is prolonged survival or improved quality of life, rather than cure. Interventional radiology allows for a minimally invasive approach to palliation, which allows for more rapid recovery with improved functioning with reduced toxicity and risk when compared to surgical approaches. Examples of palliative procedures in interventional radiology include thermal ablation and vertebroplasty for pain, placing tunneled pleural/peritoneal catheters for fluid drainage, embolization of symptomatic tumors, venting gastrostomy and duodenal stents for nausea/vomiting, and percutaneous nephrostomy or percutaneous biliary drainage for delirium secondary to urinary tract infection or biliary infection, respectively.[68] Because of the minimally invasive approach of these procedures, they can be performed on patients who have advanced illness stages or who are poor surgical candidates. The benefit of these procedures can be twofold, providing symptom relief as well as possibly changing the course of a patient's disease. For example, percutaneous nephrostomies can provide symptomatic relief while also preventing ultimate organ failure.

While IR procedures can safely provide relief of symptoms as well as possibly change the course of the disease, it is imperative to remember the central tenant to palliative care: what matters most to the patient. Necessary factors to consider before approaching a patient with a possible IR intervention include risks of the intervention, likelihood of benefit to the specific patient, and alignment of the procedure with the patient's goals and values.

Real-world examples

Many real-world examples were discussed in the "Procedural" section. For further interest though, we recommend looking into the following case reports.

- Interventional Treatment of Acute Portal Vein Thrombosis
  - https://pubmed.ncbi.nlm.nih.gov/30045398/
  - Wolter K, Decker G, Kuetting D, Trebicka J, Manekeller S, Meyer C, Schild H, Thomas D. Interventional Treatment of Acute Portal Vein Thrombosis. Rofo. 2018 Aug; 190(8):740–746. English. https://doi.org/10.1055/a-0631-9265. Epub 2018 Jul 25. PMID: 30045398.
- Interventional Radiology Approaches for Intra-Hepatic Cholangiocarcinoma
  - https://pubmed.ncbi.nlm.nih.gov/32008330/
  - Mosconi C, Calandri M, Javle M, Odisio BC. Interventional radiology approaches for intra-hepatic cholangiocarcinoma. Chin Clin Oncol. 2020 Feb; 9(1):8. https://doi.org/10.21037/cco.2019.12.15. Epub 2020 Jan 18. PMID: 32008330.
- Interventional Radiology Techniques for Pain Reduction and Mobility Improvement in Patients with Knee Osteoarthritis
  - https://pubmed.ncbi.nlm.nih.gov/30935863/
  - Filippiadis D, Charalampopoulos G, Mazioti A, Alexopoulou E, Vrachliotis T, Brountzos E, Kelekis N, Kelekis A. Interventional radiology techniques for pain reduction and mobility improvement in patients with knee osteoarthritis. Diagn Interv Imaging. 2019 Jul–Aug; 100(7–8):391–400. https://doi.org/10.1016/j.diii.2019.02.011. Epub 2019 Mar 29. PMID: 30935863.

Potential pitfalls

While there are many types of medical interventions, they do not all involve interventional radiology. Therefore, it is important as an IR to know where and whom to contact to involve other specialties in patient care, as part of the patient-centered care model.

It may be difficult to know which types of interventions are necessary for any specific patient. It is best to gauge this on a patient-to-patient basis and do only what is necessary. It is important to not order any extraneous studies, especially if the study will not change the treatment course. Everything must be goal-directed and patient-centered.

With any procedure, both IR-related and otherwise, there are inherent associated risks. The standard risks of bleeding, infection, damage to surrounding structures, and procedure failure apply to IR procedures as they do all others. The benefit of IR procedures, however, is that the risk of these complications is lower than with traditional open procedures. In addition to these standard risks, a unique risk of IR procedures is radiation exposure in CT and fluoro-guided interventions. While complications associated with radiation from procedures are uncommon, they are a necessary consideration, especially in children, who can develop cancers later in life from increased radiation exposure.

As in any other medical specialty, there may be complications associated with interventional radiology due to different types of errors. These errors affect a patient's quality of care as well as their clinical outcome. Thankfully, the majority (54.6%) of errors are preventable and are associated with ineffective planning, communication error, and equipment difficulties as opposed to a physician's lack of knowledge or technical skills.[69] Therefore it is always important to remember medical and IR interventions, patient-centered care, research, and even innovation, never exists in a vacuum. They rely upon and exist alongside one another. Familiarity with these fields and interventions and collaboration between them is as essential to the workings of IR as it is to research, patient outcomes, and the field of medicine.

Get started

- Identify the different types of interventions involved in interventional radiology, including drugs, devices, diagnostic imaging, and labs
- Understand the importance of other types of medical intervention and how they relate to interventional radiology
- Identify pitfalls of the various medical interventions and of specific interventional radiology procedures

Additional resources

- American College of Radiology (ACR): https://www.acr.org
- Cardiovascular and Interventional Radiology Society of Europe (CIRSE): https://www.cirse.org
- *Handbook of Interventional Radiologic Procedures*, second ed. Available for purchase at https://shop.lww.com/Handbook-of-Interventional-Radiologic-Procedures/p/9781496302076
- *Interventional Radiology: A Practical Guide*: Available for purchase at https://www.amazon.com/Interventional-Radiology-Practical-Andreas-Adam/dp/1857750314
- IR Quarterly: https://connect.sirweb.org/e-irq/home
- Journal of Vascular and Interventional Radiology (JVIR): https://www.jvir.org
- Radiological Society of North America (RSNA): https://www.rsna.org
- Society of Interventional Radiology (SIR) Residents, Fellows, Students (RFS) section: http://rfs.sirweb.org
- Society of Interventional Radiology (SIR): https://www.sirweb.org
- US Food and Drug Administration: https://www.fda.gov
- US Centers for the Disease Control and Prevention: https://www.cdc.gov

References

1. Chapter 2: types of interventions and their development. In: Smith PG, Morrow RH, Ross DA, eds. *Field Trials of Health Interventions: A Toolbox*. OUP Oxford London School of Hygiene and Tropical Medicine; 2015, 2015.
2. Statisics NC.H. Therapeutic Drug Use Centers for Disease Control and Prevention. Updated 3/1/2021. Accessed 4/4/2021, https://www.cdc.gov/nchs/fastats/drug-use-therapeutic.htm.
3. Administration UFaD. Information by Drug Clas. US Food and Drug Adminsistraton. Accessed 04/02/2021, https://www.fda.gov/drugs/drug-safety-and-availability/information-drug-class.
4. Oppenheimer J, Ray Jr CE, Kondo KL. Miscellaneous pharmaceutical agents in interventional radiology. *Semin Intervent Radiol*. December 2010;27(4):422–430. https://doi.org/10.1055/s-0030-1267854.
5. Yerasi C, Case BC, Forrestal BJ, et al. Drug-coated balloon for De Novo coronary artery disease: JACC state-of-the-art review. *J Am Coll Cardiol*. Mar 10 2020;75(9):1061–1073. https://doi.org/10.1016/j.jacc.2019.12.046.
6. Herring W. *Learning Radiology: Recognizing the Basics*. 3rd ed. Elsevier; 2016.
7. Sehy JV, Holloway WE, Lin S-P, Cross DT, Derdeyn CP, Moran CJ. Improvement in angiographic cerebral vasospasm after intra-arterial verapamil administration. *Am J Neuroradiol*. 2010;31(10):1923–1928. https://doi.org/10.3174/ajnr.A2215.
8. Medscape Drugs and Disease. Medscape. Accessed 06/20/2021, https://reference.medscape.com.
9. Tang Z, Jia A, Li L, Li C. [Brief history of interventional radiology]. *Zhonghua Yi Shi Za Zhi*. May 2014;44(3):158–165.
10. Administration UFaD. Classification of Products as Drugs and Devices and Additional Product Classification Issues. US Food and Drug Administration. Updated 07/12/2018. Accessed 04/10/2021, https://www.fda.gov/regulatory-information/search-fda-guidance-documents/classification-products-drugs-and-devices-and-additional-product-classification-issues.
11. Steinberger JD, Denend L, Azagury DE, Brinton TJ, Makower J, Yock PG. Needs-based innovation in interventional radiology: the biodesign process. *Tech Vasc Intervent Radiol*. June 2017;20(2):84–89. https://doi.org/10.1053/j.tvir.2017.04.006.
12. Stanford Byers Center for Biodesign. Accessed 04/10/2021, https://biodesign.stanford.edu.
13. Biodesign Center. Society of Interventional Radiology. Accessed 03/28/21, http://rfs.sirweb.org/medical-student-section/biodesign-center/.
14. Kessel D, Robertson I. *Interventional Radiology: A Survival Guide*. 4th ed. Elsevier; 2017.
15. Playbook IR. *A Comprehensive Introduction to Interventional Radiology*. Springer; 2018.
16. Kaufman JA, Lee MJ. Vascular and Interventional Radiology. In: *The Requisites*. 2nd ed. Elsevier; 2014.
17. Warhadpande S, Lionberg A, Cooper KJ. *Pocketbook of Clinical IR. Thieme*. 2019.
18. Taslakian B, Ingber R, Aaltonen E, Horn J, Hickey R. Interventional radiology suite: a primer for trainees. *J Clin Med*. 2019;8(9):1347. https://doi.org/10.3390/jcm8091347.
19. Beckett KR, Moriarity AK, Langer JM. Safe use of contrast media: what the radiologist needs to know. *Radiographics*. 2015;35(6):1738–1750. https://doi.org/10.1148/rg.2015150033.
20. Bercovich E, Javitt MC. Medical imaging: from roentgen to the digital revolution, and beyond. *Rambam Maimonides Med J*. October 4, 2018;9(4). https://doi.org/10.5041/rmmj.10355.
21. Lubner MG, Brace CL, Hinshaw JL, Lee Jr FT. Microwave tumor ablation: mechanism of action, clinical results, and devices. *J Vasc Interv Radiol. Aug*. 2010;21(8 Suppl):S192–S203. https://doi.org/10.1016/j.jvir.2010.04.007.
22. Haug S. Tools of the trade. In: Keefe NA, Haskal ZJ, Park AW, Angle JF, eds. *IR Playbook*. Springer; 2018.
23. Radvany MG, Wholey M. Expanding use of embolic protection devices. *Semin Intervent Radiol*. March 2008;25(1):27–36. https://doi.org/10.1055/s-2008-1052303.
24. Grewal S, Chamarthy MR, Kalva SP. Complications of inferior vena cava filters. *Cardiovasc Diagn Ther*. December 2016;6(6):632–641. https://doi.org/10.21037/cdt.2016.09.08.
25. Uthamaraj S, Tefft BJ, Jana S, et al. Fabrication of small caliber stent-grafts using electrospinning and balloon expandable bare metal stents. *JoVE*. October 26, 2016:116. https://doi.org/10.3791/54731.
26. Rajan DK. Dialysis fistulae and grafts. In: Keefe NA, Haskal ZJ, Park AW, Angle JF, eds. *IR Playbook*. Springer; 2018.
27. Daunt N. Adrenal vein sampling: how to make it quick, easy, and successful. *Radiographics*. 2005;25(suppl_1):S143–S158. https://doi.org/10.1148/rg.25si055514.
28. Izzo F, Granata V, Grassi R, et al. Radiofrequency ablation and microwave ablation in liver tumors: an update. *Oncol*. October 2019;24(10):e990–e1005. https://doi.org/10.1634/theoncologist.2018-0337.

29. Costandi S, Garcia-Jacques M, Dews T, et al. Optimal temperature for radiofrequency ablation of lumbar medial branches for treatment of facet-mediated back pain. *Pain Pract.* November 2016;16(8):961–968. https://doi.org/10.1111/papr.12346.
30. Carrafiello G, Laganà D, Ianniello A, et al. Post-radiofrequency ablation syndrome after percutaneous radiofrequency of abdominal tumours: one centre experience and review of published works. *Australas Radiol.* December 2007;51(6):550–554. https://doi.org/10.1111/j.1440-1673.2007.01871.x.
31. Mauro DM. Lung, kidney, and bone ablation. In: Keefe NA, Haskal ZJ, Park AW, Angle JF, eds. *IR Playbook*. Springer; 2018.
32. Jennings JW, Prologo JD, Garnon J, et al. Cryoablation for palliation of painful bone metastases: the MOTION multicenter study. *Radiology: Imaging Cancer.* 2021;3(2):e200101. https://doi.org/10.1148/rycan.2021200101.
33. Angioplasty and Vascular Stenting. Radiological Society of North America, Inc. (RSNA). Updated 02/26/2019. Accessed 04/01/2021, https://www.radiologyinfo.org/en/info/angioplasty.
34. Yu J, Cockburn J. Angiography. The Royal Australian and New Zealand College of Radiologists. Updated 03/29/2017. Accessed 04/01/2021, https://www.insideradiology.com.au/angiography/.
35. Uricchio M, Gupta S, Jakowenko N, et al. Computed tomography angiography versus digital subtraction angiography for postclipping aneurysm obliteration Detection. *Stroke. Feb.* 2019;50(2):381–388. https://doi.org/10.1161/strokeaha.118.023614.
36. Clark W, Burnes J. Angioplasty and Stent Insertion. The Royal Australian and New Zealand College of Radiologists. Updated 03/29/2017. Accessed 04/01/2021, https://www.insideradiology.com.au/angioplasty-stent-insertion/.
37. Chew A, Lyon S. Pleural Aspiration. The Royal Australian and New Zealand College of Radiologists. Updated 07/26/2017. Accessed 04/01/2021, https://www.insideradiology.com.au/pleural-aspiration/.
38. Goh G, Joseph T. Transarterial Chemoembolisation (TACE). The Royal Australian and New Zealand College of Radiologists. Updated 08/30/2018. Accessed 04/01/2021, https://www.insideradiology.com.au/transarterial-chemoembolisation/.
39. Maeda N, Osuga K, Higashihara H, et al. Transarterial chemoembolization with cisplatin as second-line treatment for hepatocellular carcinoma unresponsive to chemoembolization with epirubicin-Lipiodol emulsion. *Cardiovasc Intervent Radiol.* February 2012;35(1):82–89. https://doi.org/10.1007/s00270-010-0086-6.
40. Hu J, Albadawi H, Chong BW, et al. Advances in biomaterials and technologies for vascular embolization. *Adv Mater.* 2019;31(33):1901071. https://doi.org/10.1002/adma.201901071.
41. Pathi R, Burnes J. Image Guided Liver Biopsy. The Royal Australian and New Zealand College of Radiologists. Updated 07/26/2017. Accessed 04/03/2021, https://www.insideradiology.com.au/image-guided-liver-biopsy/.
42. Soumya WA, Ooi EH, Lockwood C. Accuracy of core needle biopsy compared to fine needle biopsy for the diagnosis of neoplasm in patients with suspected head and neck cancers: a systematic review protocol of diagnostic test accuracy. *JBI Database System Rev Implement Rep.* January 8, 2020. https://doi.org/10.11124/jbisrir-d-19-00225.
43. Renshaw AA, Pinnar N. Comparison of thyroid fine-needle aspiration and core needle biopsy. *Am J Clin Pathol. Sep.* 2007;128(3):370–374. https://doi.org/10.1309/07tl3v58337txhmc.
44. Lyon S, Burnes J. Inferior Vena Cava Filters. The Royal Australian and New Zealand College of Radiologists. Updated 07/26/2017. Accessed 04/04/2021, https://www.insideradiology.com.au/inferior-vena-cava-filters/.
45. Duffett L, Carrier M. Inferior vena cava filters. *J Thromb Haemostasis.* January 2017;15(1):3–12. https://doi.org/10.1111/jth.13564.
46. Group PS. Eight-year follow-up of patients with permanent vena cava filters in the prevention of pulmonary embolism: the PREPIC (Prevention du Risque d'Embolie Pulmonaire par Interruption Cave) randomized study. *Circulation.* July 19, 2005;112(3):416–422. https://doi.org/10.1161/circulationaha.104.512834.
47. Kinney TB. Inferior vena cava filters. *Semin Intervent Radiol.* September 2006;23(3):230–239. https://doi.org/10.1055/s-2006-948760.
48. Strunk H, Marinova M. Transjugular intrahepatic portosystemic shunt (TIPS): pathophysiologic basics, actual indications and results with review of the literature. *Röfo.* August 2018;190(8):701–711. https://doi.org/10.1055/a-0628-7347. Transjugulärer intrahepatischer portosystemischer Stentshunt (TIPS): Pathophysiologische Grundlagen, gegenwärtige Indikationen und Ergebnisse mit Review der Literatur.
49. Schindler P, Heinzow H, Trebicka J, Wildgruber M. Shunt-induced hepatic encephalopathy in TIPS: current approaches and clinical challenges. *J Clin Med.* November 23, 2020;9(11). https://doi.org/10.3390/jcm9113784.
50. Wolter K, Decker G, Kuetting D, et al. Interventional treatment of acute portal vein thrombosis. *Röfo.* August 2018;190(8):740–746. https://doi.org/10.1055/a-0631-9265. Interventionelle Therapie der akuten Pfortaderthrombose.
51. Hynes D, Aghajafari P, Janne d'Othée B. Role of interventional radiology in the management of infection. *Semin Ultrasound CT MR.* February 2020;41(1):20–32. https://doi.org/10.1053/j.sult.2019.10.006.
52. Wijesooriya NR, Mishra V, Brand PLP, Rubin BK. COVID-19 and telehealth, education, and research adaptations. *Paediatr Respir Rev.* 2020;35:38–42.
53. Smith AC, Thomas E, Snoswell CL, et al. Telehealth for global emergencies: implications for coronavirus disease 2019 (COVID-19). *J Telemed Telecare.* June 2020;26(5):309–313. https://doi.org/10.1177/1357633x20916567.
54. Bertolini A, Capaccione K, Austin JHM, et al. Teleradiology: an opportunity to improve outcomes in pulmonary fibrosis. *Clin Imaging. Apr.* 2020;60(2):263–264. https://doi.org/10.1016/j.clinimag.2019.05.011.
55. Martín-Noguerol T, Lopez-Ortega R, Ros PR, Luna A. Teleworking beyond teleradiology: managing radiology departments during the COVID-19 outbreak. *Eur Radiol.* 2021;31(2):601–604. https://doi.org/10.1007/s00330-020-07205-w.
56. Silva E, Breslau J, Barr RM, et al. ACR white paper on teleradiology practice: a report from the task force on teleradiology practice. *J Am Coll Radiol.* 2013;10(8):575–585. https://doi.org/10.1016/j.jacr.2013.03.018.
57. Ni JC, Shpanskaya K, Han M, et al. Deep learning for automated classification of inferior vena cava filter types on Radiographs. In: FCrioIftSs steel, ed. *Journal of Vascular and Interventional Radiology*. Elsevier; 2020:66–73.
58. Hosny A, Parmar C, Quackenbush J, Schwartz LH, Aerts H. Artificial intelligence in radiology. *Nat Rev Cancer. Aug.* 2018;18(8):500–510. https://doi.org/10.1038/s41568-018-0016-5.
59. Bulman JC, Moussa M, Lewis TK, et al. Transitioning the IR clinic to telehealth: A Single-Center experience during the COVID-19 pandemic. *J Vasc Interv Radiol.* 2020;31(8):1315–1319.e4. https://doi.org/10.1016/j.jvir.2020.05.008.

60. Choi PJ, Oskouian RJ, Tubbs RS. Telesurgery: past, present, and future. *Cureus*. May 31, 2018;10(5):e2716. https://doi.org/10.7759/cureus.2716.
61. Evans CH, Schenarts KD. Evolving educational techniques in surgical training. *Surg Clin*. 2016;96(1):71–88. https://doi.org/10.1016/j.suc.2015.09.005.
62. Marescaux J, Leroy J, Rubino F, et al. Transcontinental robot-assisted remote telesurgery: feasibility and potential applications. *Ann Surg*. *Apr*. 2002;235(4):487–492. https://doi.org/10.1097/00000658-200204000-00005.
63. Dohler M. 5G Networks, haptic codecs, and the operating theatre. *Digital Surgery*. August 1, 2020:71–86.
64. Raghupathi W, Raghupathi V. An empirical study of chronic diseases in the United States: a visual analytics approach. *Int J Environ Res Public Health*. *Mar 1*. 2018;15(3). https://doi.org/10.3390/ijerph15030431.
65. Sturm R, An R, Maroba J, Patel D. The effects of obesity, smoking, and excessive alcohol intake on healthcare expenditure in a comprehensive medical scheme. *SAMJ (S Afr Med J): S Afr Med J*. 2013;103:840–844.
66. Weiss CR, Abiola GO, Fischman AM, et al. Bariatric embolization of arteries for the treatment of obesity (BEAT obesity) trial: results at 1 year. *Radiology*. 2019;291(3):792–800. https://doi.org/10.1148/radiol.2019182354.
67. Council NR. *Critical Perspectives on Racial and Ethnic Differences in Health in Late Life*. The National Academies Press; 2004:752.
68. Buss MK. The intersection of palliative care and interventional radiology: enhancing understanding and collaboration. *Semin Intervent Radiol*. Jun 2017;34(2):140–144. https://doi.org/10.1055/s-0037-1602756.
69. Morbi AH, Hamady MS, Riga CV, et al. Reducing error and improving efficiency during vascular interventional radiology: implementation of a preprocedural team rehearsal. *Radiology*. Aug 2012;264(2):473–483. https://doi.org/10.1148/radiol.12110530.

Chapter 8

Drug discovery

Ryan Bitar, Matt Parker and John Walker
University of Texas Health Science Center San Antonio, San Antonio, TX, United States

- Drug discovery serves as the foundation of modern medicine and provides the tools for IR translational research.
- The drug discovery process is a multistage progression, beginning with the selection and validation of a target, a high-throughput screening process of many compounds, and then the execution of several clinical trials.
- Modifications of established drugs can be applied to the field of IR translational research.

Drug discovery serves as the fundamental process of modern medicinal research. Conventionally, drug developers identify a dysfunctional protein linked to a disease process, which leads to the synthesis and screening of several compounds, one of which will ideally demonstrate high treatment efficacy and limited toxicity to a patient. Typically this drug discovery process occurs well before the time in which it undergoes translation research in an IR setting; after all, most drugs used in IR were previously investigated on a systemic scale in the field of Medicine. Particularly, IR may explore drug discovery research by (1) employing already-established drugs in a novel means, which aim to increase efficacy yet minimize systemic toxicity or (2) administering hazardous agents in a local environment, which may treat the disease but would otherwise be lethal systemically. Therefore, IR translational research really explores drug delivery systems or drug modification; however, all translational medicine researchers should have an understanding on the steps required to discover a novel drug, from the initial design to its eventual clinical use (as illustrated in Fig. 8.1).

Historically, Drug Discovery relied on the incidental finding of natural remedies that could potentially relieve ailments. Scientists determined this traditional method to be greatly problematic, given the difficulty of purifying high amounts of the active compound from the additional intrinsic toxic substances of a natural source. With the implementation of synthetic chemistry in the 19th century, scientists could generate more pure and potent remedies via the synthesis of new compounds or the production of analogs of preexisting compounds.[1] This more effective approach is first dependent on the selection of a "target," an existing biological structure involved in the disease pathway of interest, which may be modified by a drug to improve the patient's condition.[2]

Characterizing the pathology of a disease will inherently allow researchers to identify a potential target for a novel drug. The vast majority of targets are proteins; examples are four classic candidates: G-coupled protein receptors, enzymes, ion channels, and transporters.[3] Molecular biologists may employ various strategies to identify a target protein and subsequentially validate it; however, primary strategies exist: Forward genetics, in which a disease phenotype is selected and genotype mapping traces the underlying etiology; and the reciprocal, Reverse genetics, in which a normal genotype is modified and followed by observation of the resulting phenotype. Eventually, the collection of gene expression data, proteomics data, and transgenic phenotyping is key to profiling diseases, thus providing drug targets.[4] The discovery of EGFR as a potential valid target for the treatment of cancer is a classic example of a reverse and forward approach to target discovery. Exposure of genotypically normal cell lines to growth factors led to an unregulated proliferation similar to that of cancer cells; additionally, isolated cancer cell lines were found to demonstrate overexpression of the EGFR, establishing EGFR as an oncogene.[5] This validation process specifically involves the development of a molecular compound, typically an analog of the natural substrate or antagonist of the protein, which may then be assayed to assess target inhibition or activation.[6] In addition to using natural substrates as a starting scaffold, technology such as NMR and X-ray crystallography can rather reveal the structure of a target and even demonstrate a potential binding site for which to design a de novo low molecular weight modulator.[7]

Translational Interventional Radiology. https://doi.org/10.1016/B978-0-12-823026-8.00029-8

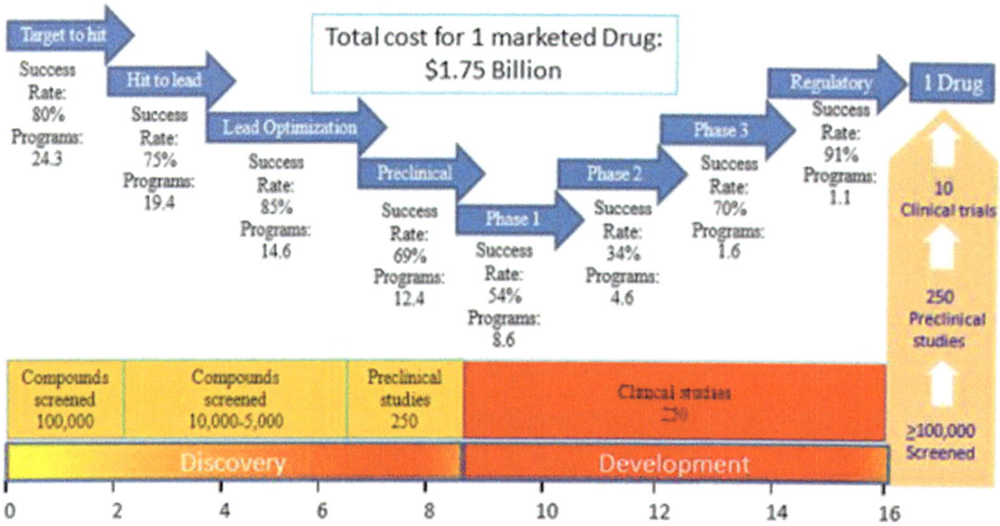

FIGURE 8.1 An overview of the drug discovery and development process, starting from the identification of a candidate target to the complete development of a marketable drug. *Figure borrowed from Blass B. Classical targets in drug discovery. In B. Blass. Basic Principles of Drug Discovery and Development (pp 87–142). Waltham, MA: Academic Press.*

Once the target is validated, it undergoes high-throughput screening (HTS). HTS, now the standard in modern drug discovery, refers to the development of a library of candidate molecular compounds, which may then be screened against the target protein for the purpose of selecting "hits," compounds that achieve the desired effect. Unsuitable compounds will be discarded while hits will undergo further evaluation in hopes of identifying a "lead compound." Leads will undergo further optimization for potency, selectivity, toxicity, and other pharmacokinetic properties. While this process is extremely lengthy and taxing, many recent innovations have significantly expedited the process. In silico methods have proven a massive impact on drug discovery by utilizing virtual screening to simulate docking and alignment of compounds to more rapidly identify a hit (as exemplified in Fig. 8.2). Additionally, high-pressure liquid chromatography has proven an extremely important addition to the field of drug discovery as a means of purifying desired compounds from synthetic reactions.[8] The identified hits will further be assayed to generate respective dose–response curves, which should ideally demonstrate a normal competitive and reversible binding pattern, important toward patient safety. A secondary assay (such as a second messenger assay or cell-based assay) should then be implemented to not only confirm the target's functional response but to also cluster hits into groups for lead series.[4]

As lead compounds emerge from HTS, the pharmacokinetic ADME profile of these compounds should be established: absorption, distribution, metabolism, and excretion. Traditionally, this process was completely dependent upon in vivo trials; however, due to recent innovation, this process may be conducted in an in vitro setting and even in silico via virtual screening. Examples include biological permeable membranes to assess cell absorbance or liver microsomes to predict Phase 1 and Phase 2 metabolism.[9] Additionally, predicting human toxicity before preclinical therapeutic trials has been explored via in silico, in vitro, and in vivo methods; these methods may uncover toxic metabolite predictions, drug–drug interactions, cell toxicity, and genotoxicity, thus establishing a maximum tolerated dose before therapeutic in vivo trials.[10]

Before use in humans, in vivo therapeutic studies must be executed. Translational research is predicated on identifying or creating an animal model that mimics human pathophysiology via genomic manipulation or exposure to an inducing agent. The lead compound may then be administered to the animal in hopes of improving outcome and identifying therapeutic dose; a historical example being the L1210 and P388 leukemia murine models. These oncogenic mutations would be induced via painting DBA/2 mice with methylcholanthrene, or the disease could be propagated in these mice via injection of ascitic fluid loaded with leukemia cells. Antineoplastic agents along with their doses and delivery modalities could then be evaluated for disease outcome based on qualities such as increase in life span, tumor cell death, and tumor growth, all while limiting toxicity. Further modification of the compound structure, delivery conditions, and dose may occur throughout this process.[11]

Once a lead has demonstrated therapeutic value with a favorable preclinical toxicity, the lead may enter a formal clinical evaluation as an investigational new drug under the FDA. Preclinical evaluations of the candidate drug up to this point may be considered a Phase 0 trial. The primary

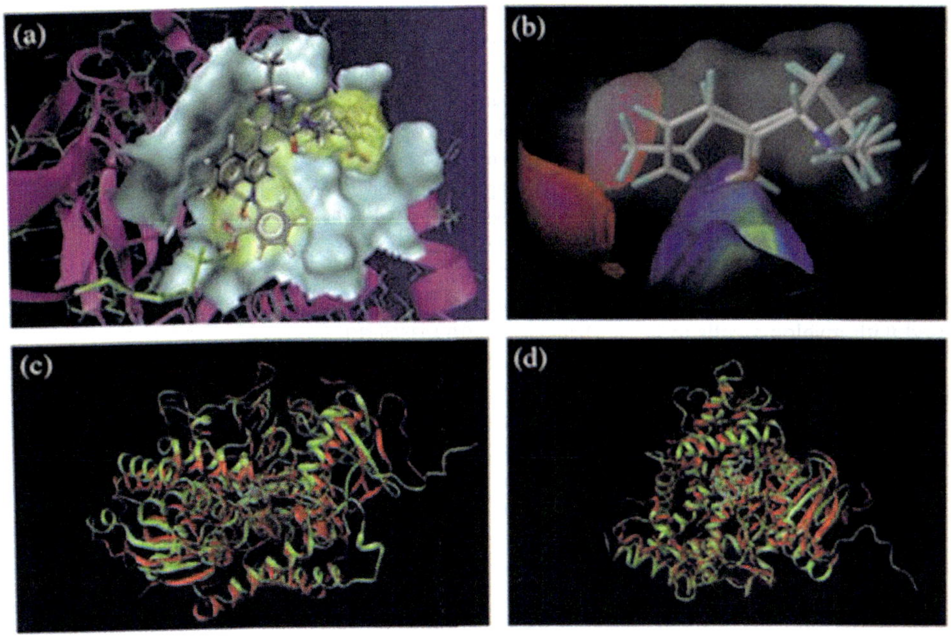

FIGURE 8.2 Virtual programs such as Benchware 3D explorer provide a platform for visualization of protein–ligand interactions, greatly expediting drug design and high throughput screening. (A–D) give examples of docking candidate compounds into virtual protein reconstructions. *Figure borrowed from Blass B. Classical targets in drug discovery. In B. Blass. Basic Principles of Drug Discovery and Development (pp 87–142). Waltham, MA: Academic Press.*

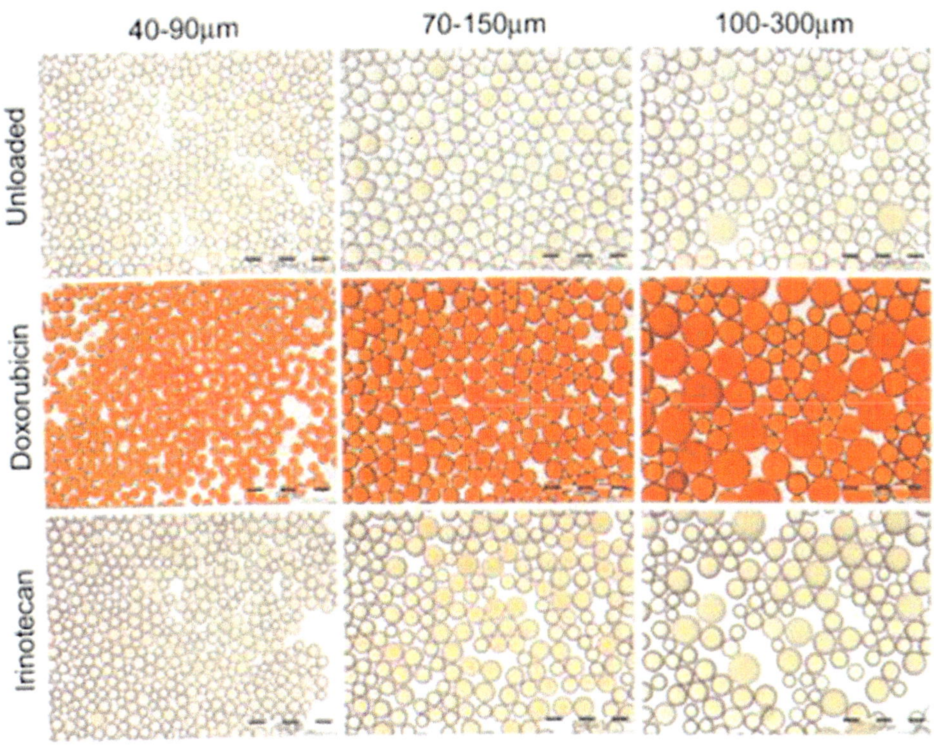

FIGURE 8.3 Microscopy images of drug eluting DC bead LUMI loaded with either doxorubicin or irinotecan for a delayed drug release effect, maximizing anti-tumor efficacy and minimizing systemic toxicity. *Figure borrowed from Ashrafi K et al Characterization of a Novel Intrinsically Radiopaque Drug-eluting Bead for Image-Guided Therapy: DC Bead LUMI.*

objective of the Phase 1 trial is to establish a maximum tolerated dose and route of administration while gathering a pharmacokinetic profile of the drug typically in health volunteers.[12] The subsequent Phase 2 trial provides preliminary data regarding the response rate of the experimental drug in a subset of patients in comparison to a similar population who had received the standard of care.[13] The final step for the experimental drug (the Phase 3 trial) is the large-scale, highly regulated comparison of overall clinical performance and superiority of an experimental drug in comparison to the standard of care, which may grant the drug eligibility to receive FDA approval, be added to the market, and potentially undergo a Phase 4 trial, a long-term postmarketing study.[14]

The fundamental guidelines behind the process outlined earlier remains greatly conserved in the field of Drug Discovery; however, as pharmacological advances are made, innovative methods begin to surface. The emergence of biological agents such as monoclonal antibodies (rituximab, bevacizumab, and trastuzumab) has ushered a new wave of antineoplastic agents via the hybridoma method, in which murine B cells are immunized against a foreign target antigen of interest and subsequentially fused with myeloma cells to rapidly produce high amounts of highly specific monoclonal antibodies.[15] In addition to new molecular compounds, new drug formulations usher in new drug delivery systems that may enhance therapeutic efficacy, improve pharmacokinetic profile, and reduce toxicity. Liposomal formulations of already-established antineoplastic agents (Doxil for doxorubicin, DuanoXome for daunorubicin, and Onyvide for Irinotecan) are employed due to their ability to maximize dose, increase solubility, improve tumor uptake, delay excretion, and minimize normal tissue toxicity.[16] Of the most significant novel drug delivery systems in IR, drug-eluting beads have become widely adopted for the treatment of nonresectable hepatocellular carcinoma. Fig. 8.3 demonstrates images of polyvinyl alcohol microspheres modified with anionic sulfonyl groups may be loaded with cationic Doxorubicin that permits the delivery of high amounts of chemotherapy while limiting systemic toxicity.[17] It can be anticipated that drug discovery will only continue to advance, providing more novel contributions toward Interventional Radiology Translational Research.

References

1. Sneader W. *Drug Discovery: A History*. Hoboken, NJ: John Wiley & Sons; 2006:3–5.
2. Li J, Corey EJ. *Drug Discovery: Practices, Processes, and Perspectives*. Hoboken, NJ: John Wiley & Sons; 2013:44.
3. Blass B. Classical targets in drug discovery. In: *B. Blass. Basic Principles of Drug Discovery and Development*. Waltham, MA: Academic Press; 2015:87–142.
4. Hughes JP, Rees S, Kalindjian SB, Philpott KL. Principles of early drug discovery. *Br J Pharmacol*. 2011;162(6):1239–1249. https://doi.org/10.1111/j.1476-5381.2010.01127.
5. Mendelsohn J, Baselga J. The EGF receptor family as targets for cancer therapy. *Oncogene*. 2000;19:6550–6565. https://doi.org/10.1038/sj.onc.1204082.
6. Wu T, Ding S. Target validation in chemogenomics. In: Metcalf B, Dillon S, eds. *Target Validation in Drug Discovery*. Waltham, MA: Academic Press; 2007:28–29.
7. Bleicher K, Böhm H, Müller K, et al. Hit and lead generation: beyond high-throughput screening. *Nat Rev Drug Discov*. 2003;2:369–378. https://doi.org/10.1038/nrd1086.
8. Szymański P, Markowicz M, Mikiciuk-Olasik E. Adaptation of high-throughput screening in drug discovery-toxicological screening tests. *Int J Mol Sci*. 2012;13(1):427–452. https://doi.org/10.3390/ijms13010427.
9. Blass B. In vitro ADME and in vivo pharmacokinetics. In: *B. Blass. Basic Principles of Drug Discovery and Development*. Waltham, MA: Academic Press; 2015:245–306.
10. Kerns E, Di L. Toxicity methods. In: Kerns E, Di L, eds. *Drug-like Properties: Concepts, Structure Design and Methods*. Waltham, MA: Academic Press; 2008:386–398.
11. Waud W. Murine 1210 and P388 leukemias. In: Teicher B, Andrews P, eds. *Anticancer Drug Development Guide*. Totowa, NJ: Humana Press Inc; 2004:79–97.
12. Acevedo P. Phase I trial design and methodology for anticancer drugs. In: Teicher B, Andrews P, eds. *Anticancer Drug Development Guide*. Totowa, NJ: Humana Press Inc; 2004:351–362.
13. Flaherty K. Phase II trials. In: Teicher B, Andrews P, eds. *Anticancer Drug Development Guide*. Totowa, NJ: Humana Press Inc; 2004:363–380.
14. Honig S. FDA role in cancer drug development and requirements for approval. In: Teicher B, Andrews P, eds. *Anticancer Drug Development Guide*. Totowa, NJ: Humana Press Inc; 2004:429–441.
15. Lu R, Hwang Y, Liu I, et al. Development of therapeutic antibodies for the treatment of diseases. *J Biomed Sci*. 2020;27:1. https://doi.org/10.1186/s12929-019-0592-z.
16. Bulbake U, Doppalapudi S, Kommineni N, Khan W. Liposomal formulations in clinical use: an updated review. *Pharmaceutics*. 2017;9(2):12. https://doi.org/10.3390/pharmaceutics9020012.
17. Wáng YX, De Baere T, Idée JM, Ballet S. Transcatheter embolization therapy in liver cancer: an update of clinical evidences. *Chinese J Canc Res Chung-kuo yen cheng yen chiu*. 2015;27(2):96–121. https://doi.org/10.3978/j.issn.1000-9604.2015.03.03.

Chapter 9

Drug safety in vitro, in vivo, harm, toxicity, dosing, good laboratory practices

Sona Ghorashi[1], Delaney McGuirt[1] and Nariman Nezami[2,3]

[1]University of Maryland School of Pharmacy, Baltimore, MD, United States; [2]Division of Vascular and Interventional Radiology, Department of Diagnostic Radiology and Nuclear Medicine, University of Maryland School of Medicine, Baltimore, MD, United States; [3]Experimental Therapeutics Program, University of Maryland Marlene and Stewart Greenebaum Comprehensive Cancer Center, Baltimore, MD, United States

Introduction

Until the early 20th century, drugs were not developed based on scientific research addressing safety (toxicology) and efficacy and were primarily extracted or developed from plants, with only a few substances of synthetic origin. Those drugs were empirically administered to patients without any safety workup or dosing assessment. The birth of the Food and Drug Administration (FDA) in 1906, as the first agency to regulate the quality of food and medicine, was a revolution in determining the safety of drugs.[1] Since then, drug safety approaches evolved to ensure patients' safety and health are guaranteed. The entire process of determining drug efficacy, dosing, toxicity, and safety would require an entire textbook; this chapter will briefly cover the different aspects of drug safety and efficiency from the experimental phase to clinical application.

Preclinical studies: in vitro and in vivo

Preclinical research is widely used to evaluate drug safety, maximum tolerated dose, pharmacodynamics and pharmacokinetics, and potential harm. The main goal of preclinical studies is to assess a drug's potential to cause harm in humans before advancing to the next phase, clinical research in humans, with the ultimate goal of making it to market. Two types of preclinical research are widely used in drug screening processes: in vitro and in vivo assays. The difference between these two types is explained in Table 9.1.

"In vitro," Latin for "within the glass," refers to experiments performed outside of a living organism. Compared to in vivo, these studies are typically faster and lower cost. However, they do not always translate to humans because they fail to simulate identical cellular conditions, thus in vivo tests and clinical trials are vital to determine safety.[2] An example of an in vitro study as it relates to interventional radiology: a study showed the use of 3D printed antibiotics and chemotherapeutic catheters could be utilized for personalized medicines in interventional radiology procedures.[3]

"In vivo," Latin for "within the living," refers to the experiments performed in live environments such as animals and humans. They are a critical step in the preclinical process. These experiments are more specific, reliable, and costly than in vitro tests. Due to their use in live organisms, in vivo studies are also subject to strict and precise regulations. In vivo tests are critical for the development of medical devices and therapeutics.[2] An example of an in vivo study as it relates to interventional radiology: a study done on mice showed electromagnetic heating of breast tumors in mice is an assuring technique for interventional radiologic treatments.[4]

Clinical studies: clinical trials

Although preclinical studies answer some basic questions about drug safety, they cannot replace human studies to assess drug interaction within the human body. Clinical assessment of drug safety and dosing is done through three phases of human-based clinical trials (Table 9.2). Phase 1 trials are performed on a small group (20−100) of healthy volunteers or people with the disease. Their purpose is to define the safety of a new treatment. If the safety of the

Translational Interventional Radiology. https://doi.org/10.1016/B978-0-12-823026-8.00022-5

TABLE 9.1 Comparison of in vivo and in vitro experiments.

	In vitro	In vivo
Accuracy	Less precise	More precise
Time	Quick results	More time consuming
Cost	Less expensive	More expensive
Limitation	Less restrictions	More restrictions
Example	Cell cultures in petri dishes, test tubes	Experiments performed on rabbits, pigs, mice, etc

TABLE 9.2 Clinical trials.[5]

Phase	Purpose	Number of participants	Length of study
1	Safety of a new treatment	20–100	Several months
2	Determination of efficacy and side effects	Hundreds of participants with the disease of interest	Several months to 2 years
3	Comparison of the new treatment with the current standard of care	300–3000	1–4 years

drug is approved, it moves to phase 2 trials. Phase 2 involves several hundreds of participants with the disease of interest, and the purpose is to determine efficacy and side effects. If the drug is deemed effective, it moves to phase 3 where 300–3000 participants are recruited and the new treatment is compared with the current standard of care. If investigators can prove the investigational drug is as safe and effective as other drugs on the market, the FDA will grant approval of the new medication.[5]

Dosing

Dose is defined as the specific amount of medication administered/taken at one time, which could be expressed in various forms such as weight of the drug, volume of drug solution, number of dosage forms, etc. The two main factors used to define the dosing of medications are effective dose (ED50) and toxic dose (TD50). The ED50 (median effective dose) is the specific dose of a medication that results in a particular effect in 50% of the population taking that dose.[6] Doses that are around ED50 are usually sufficient. Studies have shown that doses higher than the ED50 are only associated with a slight improvement in efficacy, while the risk of adverse effects increases.[6]

On the other hand, toxic dose (TD50) is a dose that results in toxicity in 50% of the population. The ED50 would ideally be significantly less than TD50, indicating an effective medication at a lower dose.[6] Doses below ED50 or above TD50 are either nonbeneficial or toxic, respectively.

To analyze the safety and efficacy of drugs, dose-ranging studies are conducted. Dose-ranging studies are phase 1 or early phase 2 clinical trials, and their ultimate goal is to evaluate the response to a specific dose to determine at which dose the drug is least harmful. The duration of action for any drug is also determined during a dose-ranging study, which allows the finalization of the dosage schedule. The maximum tolerable dose is an essential component for designing dose-ranging studies.

A recent hot topic in healthcare, personalized medicine has the potential to diminish adverse effects and risks linked to medicines by individualizing treatment selection for each patient. Personalized medicine began at the beginning of the 21st century and revolutionized treatment by increasing patient safety and decreasing medication-related adverse effects.[7]

Toxicity

Toxicity is considered any adverse effect of a drug administered therapeutically or during diagnostic tests/procedures. However, the description of drug toxicity is diverse and variable. Clinicians might evaluate toxicity differently based on the signs and symptoms of each patient. Over the years, there has been a long-lasting challenge to develop common terminology and criteria for adverse effects leading to multiple systems currently in use. The most commonly used US classification system is the "Common Terminology Criteria for Adverse Events (CTCAE)." The latest version of CTCAE was created in November 2017 and can be found at the following website[8]:

https://ctep.cancer.gov/protocolDevelopment/electronic_applications/ctc.htm.

There are different mechanisms for drug toxicity that are classified using five categories:

1. On-target adverse effects: exaggerated and adverse pharmacologic effects at the target of interest in the test system.[9]
2. Off-target adverse effects: adverse effects as a result of modulation of other targets; these may be related biologically or unrelated to the target of interest.[9]
3. Biological activation to toxic metabolites
4. Hypersensitivity and related immunological reactions
5. Idiosyncratic toxicities[10]

Medication harm

Medication harm is defined as any negative outcomes or injury to a patient as a result of medication administration, regardless of severity or preventability. Medication harm may lead to hospitalization, mortality, and/or morbidly.[11]

Clinicians must select a dose that balances a drug's efficacy with potential harm. This can be more difficult with some certain medications, such as warfarin, digoxin, and lithium.[12]

Adverse drug reaction (ADR), adverse drug event (ADE), and medication error (ME) are terms widely used in healthcare, and yet there has been significant confusion among healthcare professionals regarding these terms.[13] The World Health Organization defines adverse drug reactions as "a response to a drug that is noxious and unintended and that occurs at doses normally used in man for prophylaxis, diagnosis, or therapy of disease, or the modification of physiological function."[11] An example of an ADR is developing a rash in a patient who was administered penicillin at the correct dose.

Adverse drug events are defined by the Institute of Medicine as "an injury resulting from medical intervention related to a drug."[13] The National Coordinating Council for Medication Error Reporting and Prevention (NCC MERP) classifies ADEs as preventable and nonpreventable. Preventable ADEs are medication-related harm that occurs due to error. An example of a preventable ADE is developing respiratory distress in a patient who was prescribed hydromorphone by mistake. Nonpreventable ADEs are medication-related harm that is not due to an error. An example of a nonpreventable ADE is difficulty breathing and facial swelling in a patient following administration of amoxicillin and clavulanate potassium, in which case the patient had no history of previous adverse effects with this antibiotic.[13]

NCC MERP, defines medication error as "any preventable event that may cause or lead to inappropriate medication use or patient harm while the medication is in the control of the health care professional, patient, or consumer."[13] There are many different types of medication errors such as improper dose, wrong time, omission, and unauthorized drug. MEs usually occur during medication prescribing or ordering; however, many are preventable and typically do not result in harm.[14]

Good laboratory practice

Good laboratory practice (GLP) is a set of principles designed to guarantee the quality of nonclinical laboratory studies, ranging from in vitro studies to preclinical animal experiments. These principles ensure that nonclinical studies are planned and performed safely. GLP monitors the following aspects of the research process:

1. Resources: organization, personnel, facilities, and equipment
2. Characterization: test items and test systems
3. Rules: protocols and standard operating procedures (SOPs)
4. Results: raw data, final reports, and archives
5. Quality Assurance: independent monitoring of research processes[15]

Resources

Organization and personnel: GLP requires a clear definition of the tasks and duties of all members of the research team. An organization should keep an up-to-date version of the manuals used for laboratory functions and job descriptions.

Facilities and equipment: According to GLP, there should be enough people to perform the required tasks in a GLP-compliant way. Training, education, and qualifications of all staff members should be recorded.[15]

Characterization

Test item: GLP requires researchers have a comprehensive knowledge of all test items and pharmaceutical compounds used during the study, including identity, stability, and potency.

Test system: In the case in which the test system is an animal, it is necessary to have information regarding health status, biologics, and strain.[15]

Rules

Protocol or study plan: The GLP outlines the main sections that should be listed in a study protocol. The study plan or protocol shapes the design of the study and should be approved by the study director before the start of the study.

SOPs: SOPs are detailed, step-by-step instructions for all procedures conducted during the research process. The main goal of SOPs is to reduce bias and achieve high productivity and performance.[15]

Results

Raw data: Raw data or source data are the original data collected during the study protocol. They may be used in various ways, directly or statistically. In any case, results should be consistent with the raw data.

Study report: It is the study director's responsibility to ensure that the study report is accurate and follows GLP principles.[15]

Archives: It is important to properly store study data in a manner that is safe for a long duration of time and could be recovered quickly if needed.[15]

Quality assurance

The quality assurance team is charged with ensuring GLP compliance, both in individual studies and the research facility as a whole.[15]

Summary

In summary, drug safety assessment is crucial for patient safety. There are several factors that must be considered before administering a drug during an interventional radiology procedure. These include individualized dosing for every patient, drug toxicity, adverse effects, and medication harm. In vitro and in vivo studies are crucial steps in the preclinical phase of drug research utilized to determine drug safety and assess potential human harm. Clinical studies assess safety and efficacy of new medications through three phases of human-based clinical studies. Good laboratory practices assure that preclinical studies are performed safely. All of the above factors are critical for planning any interventional radiology research.

References

1. Andrade EL, Bento AF, Cavalli J, et al. Non-clinical studies required for new drug development - Part I: early in silico and in vitro studies, new target discovery and validation, proof of principles and robustness of animal studies. *Braz J Med Biol Res*. 2016;49(11). https://doi.org/10.1590/1414-431X20165644.
2. Differences between in Vitro, in Vivo, and in Silico Studies (MPKB). [Accessed 27 July 2020]. https://mpkb.org/home/patients/assessing_literature/in_vitro_studies.
3. Weisman JA, Ballard DH, Jammalamadaka U, et al. 3D printed antibiotic and chemotherapeutic eluting catheters for potential use in interventional radiology: in vitro proof of concept study. *Acad Radiol*. 2019;26(2):270–274. https://doi.org/10.1016/j.acra.2018.03.022.
4. Hilger I, Andrä W, Hergt R, Hiergeist R, Schubert H, Kaiser WA. Electromagnetic heating of breast tumors in interventional radiology: in vitro and in vivo studies in human cadavers and mice. *Radiology*. 2001;218(2):570–575. https://doi.org/10.1148/radiology.218.2.r01fe19570.
5. *Commissioner O of the. Step 3: Clinical Research*. FDA; April 18, 2019. Published online https://www.fda.gov/patients/drug-development-process/step-3-clinical-research [Accessed 27 July 2020].
6. Kenny BJ, McPhee AS. ED50. In: *StatPearls*. StatPearls Publishing; 2020. http://www.ncbi.nlm.nih.gov/books/NBK538269/ [Accessed 27 July 2020].
7. Vogenberg FR, Isaacson Barash C, Pursel M. Personalized medicine. *Pharm Ther*. 2010;35(10):560–576.
8. Common Terminology Criteria for Adverse Events (CTCAE). Protocol Development. CTEP. https://ctep.cancer.gov/protocolDevelopment/electronic_applications/ctc.htm#ctc_50. [Accessed 27 July 2020].
9. Rudmann DG. On-target and off-target-based toxicologic effects. *Toxicol Pathol*. 2013;41(2):310–314. https://doi.org/10.1177/0192623312464311.
10. Liebler DC, Guengerich FP. Elucidating mechanisms of drug-induced toxicity. *Nat Rev Drug Discov*. 2005;4(5):410–420. https://doi.org/10.1038/nrd1720.
11. Falconer N, Barras M, Martin J, Cottrell N. Defining and classifying terminology for medication harm: a call for consensus. *Eur J Clin Pharmacol*. 2019;75(2):137–145. https://doi.org/10.1007/s00228-018-2567-5.
12. Blix HS, Viktil KK, Moger TA, Reikvam A. Drugs with narrow therapeutic index as indicators in the risk management of hospitalised patients. *Pharm Pract*. 2010;8(1):50–55.
13. Becker SC. *Contemporary View of Medication– Related Harm. A New Paradigm*; March 2015. Published online https://www.nccmerp.org/sites/default/files/nccmerp_fact_sheet_2015-02-v91.pdf [Accessed 25 July 2020].
14. Tariq RA, Vashisht R, Scherbak Y. Medication errors. In: *StatPearls*. StatPearls Publishing; 2020. http://www.ncbi.nlm.nih.gov/books/NBK519065/ [Accessed 31 July 2020].
15. *Who. Handbook: Good Laboratory Practice (GLP): Quality Practices for Regulated Non-clinical Research and Development*. World Health Organization Stylus Publishing, LLC [distributor]; 2010. http://public.ebookcentral.proquest.com/choice/publicfullrecord.aspx?p=579101 [Accessed 27 July 2020].

Chapter 10

Device discovery

Neil K. Jain
MedStar Georgetown University Hospital, Washington, DC, United States

With a constant demand for providing patient-centered care with the most cutting-edge technology, interventional radiologists have had a long history of being at the forefront of designing innovative solutions. In 1964, Charles T. Dotter, often referred to as the father of interventional radiology, performed the first percutaneous transluminal angioplasty of the leg.[1] By introducing this new concept of arterial stenting percutaneously and fibrinolysis with streptokinase injections directly into a clot, he pioneered a new way of disease management: using catheters for diagnosis and treatment. This movement became the start of using minimally invasive procedures as a standard of care, a hopeful attempt to replace scalpels and open surgeries.

Dotter's inventions did not only mark the creation of interventional radiology but also established strong grassroots between the specialty and medical industry. The most difficult challenge with catheters was ironically the inherent design of the catheter. While Dotter trialed several household materials for his catheter prototypes such as guitar strings and speedometer cables, none compared with the industrialized materials and corporal mindset that Bill Cook brought to the table. Bill, who later became the chief executive officer of Cook Medical, was able to manufacture all the catheters Dotter designed and eventually was able to secure a corporate partnership with him. Dotter was undoubtedly the mastermind behind the products, but without Cook's commercial resources, Dotter's ideas would have never come to life.

The process of developing medical devices is a timely investment, but a simple conceptual algorithm can help advance your idea from inception to product. Using a modified version of the Stanford Biodesign Process, device discovery can be separated into three applicative stages: (1) identification, (2) creation, and (3) execution. These phases can also be easily remembered by the I—C—E acronym. It is important to note these phases are dynamic and interlinked as you move forward in the development of your product. It is common to revert from the creation or execution phases to the identification phase before proceeding forward. Fig. 10.1 provides an outline, critical points, and examples along each phase.

Identification. There are two ways to formulate ideas to jumpstart your device development: organic discussion and active brainstorming. Organic ideas emerge naturally in an unprompted setting, usually when confronted with a problem or challenge with a diagnosis or procedure. Active brainstorming requires a forced thought process where potential solutions are discussed for a list of problems. Regardless of the method, defining the clinical or technical problem is necessary to move forward. The identification of this "unmet need" can be best represented by the construction of a "problem statement." A comprehensive problem statement should emulate the following format: "How might we [*action*] in [*stakeholder*] in order to [*measurable criteria*]." Here is an example of a problem statement that addresses the aforementioned elements in tackling prostate artery embolization: "How might we improve the identification of the prostatic artery [action] in prostate artery embolizations [stakeholder] to increase operator confidence, ultimately reducing procedure time, radiation exposure, and anesthesia time [measurable criteria]." Notice that the problem statement does not mention the specific invention to improve embolization. It is critical to address the unmet need in a clear and comprehensive manner before proposing a solution.

Creation. This phase involves three major actionable items: concept generation, concept screening, and concept prototyping. After a problem statement is successfully developed, the generation of your idea can ensue. Explain and accurately design your solution to meet the unmet need. Before a viable product can be created, a thorough screening process of the concept needs to be accomplished. Patents of similar inventions need to be reviewed to prevent copyright infringement. Subsequently, prototyping and designing of your device can begin.

Translational Interventional Radiology. https://doi.org/10.1016/B978-0-12-823026-8.00063-8

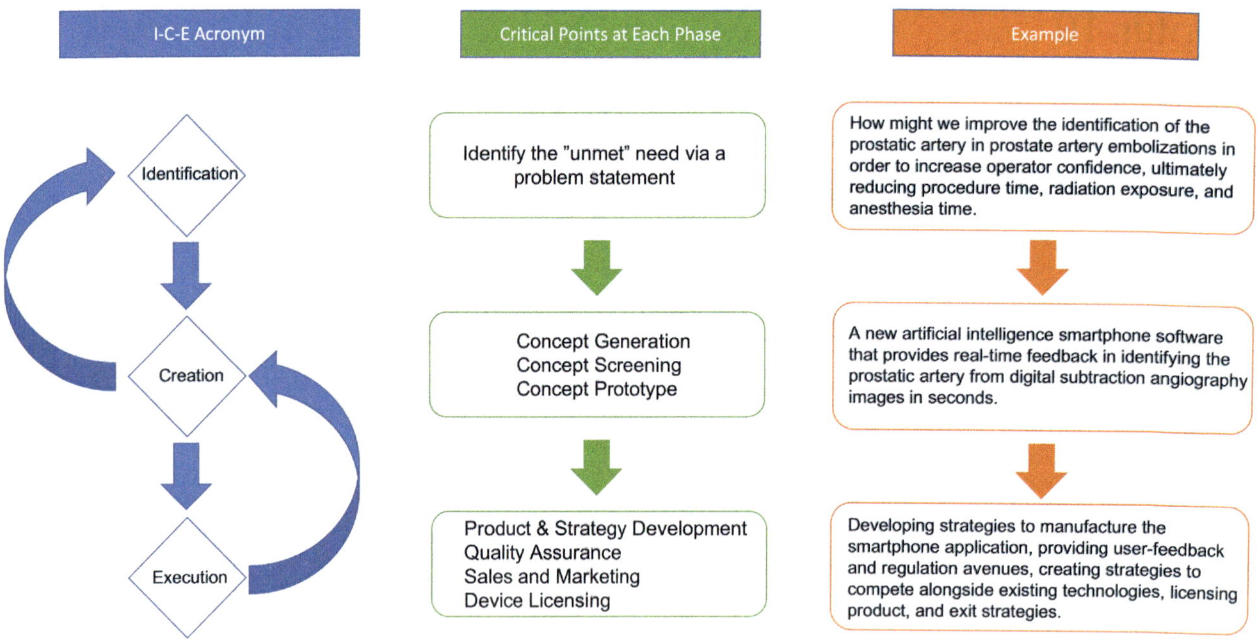

FIGURE 10.1 Depicts the dynamic nature I–C–E acronym while identifying critical points and examples at each phase.

Prototypes can fall into one of the following four major categories. A proof-of-principle prototype confirms important functional characteristics of the intended device without the actual functionality or visual appearance of the final product.[2,3] A working prototype is an early example of the design that consists of nearly all the utilities of the final product.[2,3] The emphasis of this prototype is to develop the closest "functional" rendition of your final product to test your concept. High importance is placed on this model as investors can gauge validity and market value from this design. Again, visual appearance is not as important in this schematic. A visual prototype serves to provide a visual design of the imagined model.[2,3] Functionality and utility are not assessed in this model. Finally, a functional prototype incorporates the visual and functioning components of the design, attempting to mimic the final design.[2,3] This is not to be confused with the final design as the techniques and scaling may be different than the product.

Execution. This phase is the final step in the device discovery journey. It incorporates developing strategies in tactfully introducing the product in the current market, quality management, sales and marketing, and ultimately licensing your device. This phase will be detailed extensively in future chapters.

References

1. Payne MM. Charles Theodore Dotter. The Father of Intervention. *Texas Heart Inst J*; 2001. U.S. National Library of Medicine www.ncbi.nlm.nih.gov/pmc/articles/PMC101126/.
2. Rapid prototyping of ceramics. In: *Handbook of Advanced Ceramics: Materials, Applications, Processing, and Properties, by Somiya Shigeyuki*. Academic Press; 2013, 491–491.
3. Yock P, Zenios S, Makower J, et al. *Biodesign: The Process of Innovating Medical Technologies*. Cambridge University Press; 2019.

Chapter 11

Device prototyping: Iterative refinement, optimization process

Austin-Marley Windham-Herman and Walter G. Bircher
Yale University, New Haven, CT, United States

Why it matters?

Design is an iterative process, represented by a repeating sequence of ideation, prototyping, and evaluation. You can imagine the process like a cone (Fig. 11.1), progressing from very broad to very narrow. It is helpful to follow discrete stages when designing a clinical solution; an example framework is detailed in Table 11.1.[1] The ideation phase begins with understanding not just the clinical problem you seek to solve, but the entire environment, including the competitive space, customer needs, and overall market structure. The prototyping and evaluation phases are the empirical methods then used to **Reduce to Practice** and narrow the scope of solutions, informing through successive improvements or failures. Having a complete understanding of the fundamentals is vital to informing the direction of the design process and can prevent costly mistakes and engineering misadventures down the road.

Customer requirements and technical benchmarking

Sourcing **Customer Requirements** is a key part of the ideation phase, and should be formalized through well-performed customer interviews and technical translation of customer needs into design parameters and constraints.[2] For example, if a customer desires a "lightweight" device, the resultant technical constraint might be "less than 100 g."

It is also important to establish key capabilities of competitor products early in the design process. Known as **Technical Benchmarking**, this process sets concrete design goals and creates useful constraints for engineers. Without such constraints, the design process can be unnecessarily prolonged, as it may appear that there are too many viable paths to a solution, and it can be unclear how to proceed. Understanding the nature of the problem prior to delving into detailed design is critical. Indeed, expert

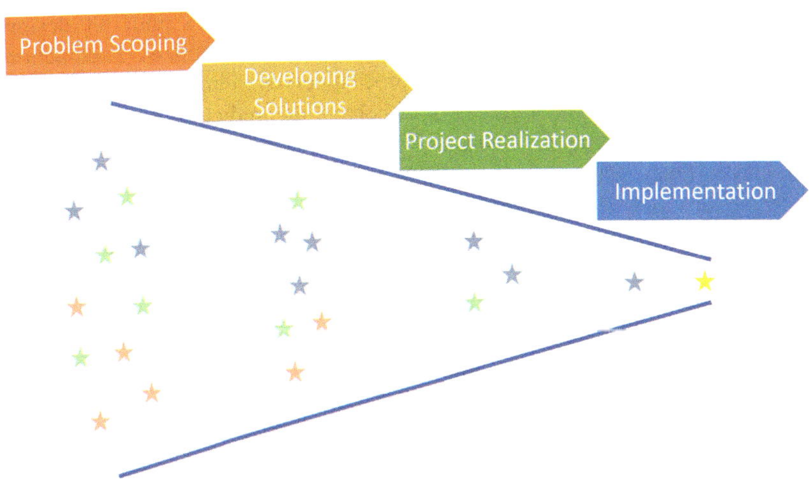

FIGURE 11.1 The design process refines a large space of initial possibilities, across successive iterations/phases, in order to achieve an optimized output.

Translational Interventional Radiology. https://doi.org/10.1016/B978-0-12-823026-8.00053-5

TABLE 11.1 Design process framework.[1]

Phases		Purpose/goal	Resources required
Problem scoping	Need identification	Identify opportunity areas that align with your personal interests and strengths	Experience and observation
	Problem definition	Set boundaries around the problem and its scope	Formalized requirements and constraints
	Gathering information	Develop a full understanding of the problem space and environment	Customer interviews, competitor technical benchmarking, market research
Developing alternative solutions	Generating ideas	Brainstorm a large number of potential approaches to solving the problem, including all methods, materials, and processes.	Sourcing diverse opinions from advisors and industry experts
	Modeling	Create physical version of design to be evaluated for function, ergonomics, durability, etc.	CAD programs, basic materials (wood, screws, springs, tubing), 3D printing
	Feasibility analysis	Determine if major functional elements of design are viable	CNC machining, preclinical testing, cost, and value modeling
	Evaluation	Prove function, ergonomic quality, durability, etc., of physical model	Customer beta testers, usability and efficacy studies
Project realization	Decision	Decide on final designs and major manufacturing processes	Expert manufacturer opinion, expert mold/tool makers, industrial designers
	Communication	Coordinate between stakeholders: Investors, manufacturers, distributors, customers, regulators	Expert legal counsel, skilled team to implement across each domain
	Implementation	Adjust design (DFA/DFM) to suit major processes, and initiation of at-scale manufacturing	Manufacturing facilities, distribution, and sales infrastructure

designers spend nearly twice as much time identifying requirements and constraints as novice designers.[1]

Types of prototypes

Prototypes serve many purposes and come in many forms. A prototype does not have to "look good" or be fully functioning to bring value to the design process. As an example, a prototype can be completely nonfunctioning, but its shape and texture can quickly provide deep insights into ergonomics that move the design process forward. Conversely, a prototype can bear little to no resemblance to the final device, but illustrate the function of a key mechanism, proving its viability. These bits of information can be collected in pieces, and later assembled into a final prototype that contains all design elements and functions as intended. It is important to keep an open mind during the prototyping process, remembering that each prototype brings information to the table—even if it does not work as expected. As detailed in *Biodesign: the Process of Innovating Medical Technologies*, prototypes can be thought of as *works-like*, *looks-like*, *feels-like*, *is-like*, or *looks-like/is-like* models.[3] The last category, *looks-like/is-like* is the best type of prototype to test user experience, as it is a robust, functional prototype that is close to the final look and design.

Iterative design

Each prototype is evaluated according to technical benchmarking specifications, changes to the design are ideated, and the cycle repeats itself as work begins on the next prototype. This process continues until all design goals are satisfied and the design is optimized. Each action taken in the design process serves to inform the next, even if the action fails to behave as predicted. A failed prototype can lead a designer to consider a different design strategy, which may lead to the discovery of new intellectual property. In this way, the design process can be viewed as a cascading series of small experiments, as the designer works to uncover the optimal solution. The cascading nature of the design process illustrates why it is important to properly define and constrain project requirements early on as an incomplete problem definition can lead to a series of design iterations in a direction that is not viable.

Prototype fidelity

Prototype fidelity should increase over time, with initial prototypes being detailed enough to demonstrate shape, feel, or functionality, but not exceeding this level of detail. It may be tempting to spend time in the beginning creating highly detailed prototypes, but inevitably it will slow the design process. This is likely to happen because large changes to the design typically occur early in the design process, causing premature design detail to be abandoned. Initial low-fidelity prototypes can be made from materials on-hand, such as wood, cardboard, or plastics, or they can be 3D printed on hobby grade machines. As the design process progresses, specific materials will be selected as candidates for the final design. At this point, parts can be machined from these materials to more closely approximate the feel and function of the final design. Finally, **Design for Manufacture/Design for Assembly (DMF/DFA)** considerations can be considered and the design transitioned from prototype to high volume production.

Real-world examples

Virtual Incision – Highly complex medical devices may take many years of prototyping and design iteration to come to fruition, some more than a decade. One example is the miniature surgical robot for abdominal procedures, pioneered by Virtual Incision Corporation. While the final product is a highly articulated robot capable of precisely resecting and cauterizing tissue, initial prototypes of this robot had no arms at all, and simply demonstrated that motors could operate *in vivo*.[4] A subsequent prototype added two simple hinged arms to the robot, demonstrating tissue resection and cautery within the limited reach of the arms.[5] Finally, highly articulated prototypes were built and tested, performing colectomies in porcine models, shortly followed by an autoclavable, highly miniaturized prototype capable of performing a human colon resection.[6–8] This process took over 10 years from first prototype to human trial, and tens of prototypes were iterated along the way, illustrating how prototypes increase in fidelity and complexity over time. Technical benchmarking was performed by comparing the robot to the da Vinci Surgical System, the most comparable commercially available device.

TVA Medical – Moderately complex devices may have a quicker path to market, but still rely on a process of iterative design. TVA Medical developed a system for creating percutaneous AV fistulae in the arm, starting in 2008 and completing initial clinical trials within 5 years.[9] An important piece of the solution was an armboard for the cath lab table that would immobilize the arm but not interfere with IV access or catheter access sites. Engineer Michael Hemati started by examining average sizes and shapes of human hands and arms, and cutting plywood models and velcro straps to size.[10] A few weeks were spent iterating sizes, strap positions, and left/right handed dual compatibility with wood models, and testing on office staff. Another 2–3 weeks were spent designing CAD models, and another 2 weeks to order and receive a machined Delrin armboard. The armboard worked very well, it was rigid enough to withstand repeated use, and could be easily cleaned. However, when this model was eventually used in clinical trials, it revealed a previously unknown use requirement which necessitated a major design change. The arm board was too large and heavy to be easily shipped to hospitals around the world. A lightweight, folding carbon–fiber version with the same fundamental design was then created and used for subsequent clinical efforts. The company received FDA approval for their percutaneous technology in 2018, and the armboard is now distributed to hospitals globally.[11] This example highlights the value of early stage, rapid prototyping (Fig. 11.2) and of conducting a broad **Failure Mode and Effects Analysis (FMEA)**.

Theranova – Engineers at Theranova were challenged to prepare an at-home neuromodulatory stimulator device for initial demos on a short timeline and strict budget.[10] They were tasked to create a handheld enclosure for the electrical components and interface but needed to decide what route they would take for manufacturing. Low-volume 3D printed parts could be made cheaply and be modified rapidly, compared with injection-molded parts which required extremely high upfront costs and were difficult to modify, but which had lower per unit costs at high volume. To make the decision, they needed to reach a

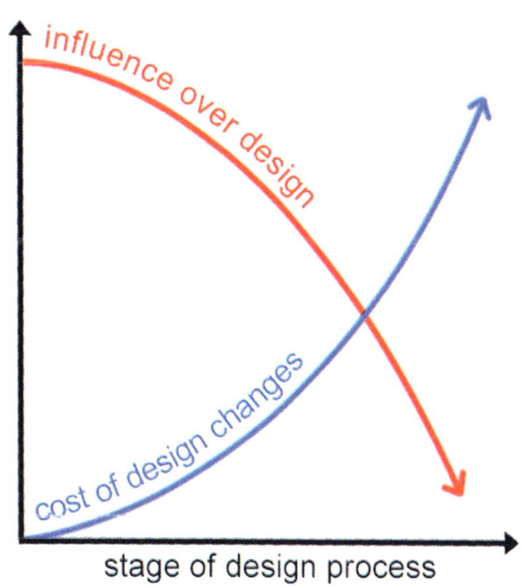

FIGURE 11.2 It is often much faster and less expensive to iterate earlier in the design process. Indeed, data show that expert designers spend almost twice as much time in early design phases compared to novice designers.

Value Inflection point at which the decision had been optimally risk reduced. With relatively little expense, they were able to quickly iterate on the design using in-house 3D printers. With postprocessing, sanding, and application of primer and paints to the printed housing, they had created a professional looking device. However, the use case of the device necessitated it be sturdy enough to withstand drops, and only molded parts could deliver the necessary strength. With the confidence of having a functional looks-like/is-like prototype, they could proceed with the more costly investment of creating injection molds.

Get started

- Develop a complete understanding of the problem space.
- Conduct customer interviews and technical benchmarking.
- Iteratively downselect solutions and prototypes.
- Develop and evaluate the most viable options until the appropriate value inflection point.

References

1. Atman CJ, Adams RS, Cardella ME, Turns J, Mosborg S, Saleem J. Engineering design processes: a comparison of students and expert practitioners. *J Eng Educ*. October 2007;96(4):359–379.
2. Doorley S, Holcomb S, Klebahn P, Segovia K, Utley J. *Design Thinking Bootleg*; 2018. Retrieved from: https://dschool.stanford.edu/resources/the-bootcamp-bootleg.
3. Yock PG, Zenios S, Makower J, et al. *Biodesign: The Process of Innovating Medical Technologies*. Cambridge University Press; February 2, 2015.
4. Rentschler ME, Dumpert J, Platt SR, Farritor SM, Oleynikov D. Natural orifice surgery with an endoluminal mobile robot. *Surg Endosc*. July 1, 2007;21(7):1212–1215.
5. Lehman AC, Dumpert J, Wood NA, et al. Natural orifice cholecystectomy using a miniature robot. *Surg Endosc*. February 1, 2009;23(2):260–266.
6. Dolghi O, Strabala KW, Wortman TD, Goede MR, Farritor SM, Oleynikov D. Miniature in vivo robot for laparoendoscopic single-site surgery. *Surg Endos*. October 1, 2011;25(10):3453–3458.
7. Wortman TD, Mondry JM, Farritor SM, Oleynikov D. Single-site colectomy with miniature in vivo robotic platform. *IEEE (Inst Electr Electron Eng) Trans Biomed Eng*. January 25, 2013;60(4):926–929.
8. Virtual Incision. *Virtual Incision Announces World's First Use of its Miniaturized Robot in Human Surgery*; 2016 [online] Available at: https://virtualincision.com/wp-content/uploads/2016/03/VIC-FIM-PR-03012016.pdf [Accessed 27 June 2020].
9. Rajan DK, Ebner A, Desai SB, Rios JM, Cohn WE. Percutaneous creation of an arteriovenous fistula for hemodialysis access. *J Vasc Intervent Radiol*. 2015;26(4):484–490. https://doi.org/10.1016/j.jvir.2014.12.018.
10. Hemati M. *Personal Interview*. June 28, 2020.
11. TVA Medical I. *TVA Medical Receives FDA De Novo Marketing Authorization for everlinQ® endoAVF System*; June 27, 2018. Retrieved from: https://www.prnewswire.com/news-releases/tva-medical-receives-fda-de-novo-marketing-authorization-for-everlinq-endoavf-system-300670950.html.

Chapter 12

Device classification

Gaurav Gadodia
Division of Vascular and Interventional Radiology, Department of Radiology, Medical College of Wisconsin, Milwaukee, WI, United States

Key points

- The Food and Drug Administration (FDA) classifies new devices into a three-tier system based mostly on the risk profile of the device, with Class I being the lowest risk devices and Class III being the highest.
- Device classification drives what regulatory pathway must be taken for the device to obtain premarket clearance or approval before marketing and human use, with more stringent controls placed on the highest risk devices.
- There are resources available to inventors allowing for evaluation of the appropriate classification and thus regulatory pathway for their device, which can help guide their design and testing process.

Why it matters?

Classification of novel medical devices by the FDA is based on risk profile, device complexity, and intentions and indications for use. It is this classification that guides the regulatory pathway that needs to be taken for the device to obtain premarket clearance for its use in humans, with significant differences in time course and resources needed between the pathways. This chapter will discuss the device classification process and system, with examples of each, along with its effect on the regulatory pathway needed for premarket clearance, and the processes by which inventors can determine the appropriate classification for their device and thus guide their design and development process toward their eventual regulatory requirements. Further details on regulatory pathways will be discussed in later chapters.

Device classification: basics[1–4]

In the United States, all medical devices are ultimately cleared or approved for marketing for in-human use by the Food and Drug Administration (FDA). The FDA utilizes a three-tier system to classify novel medical devices mostly based on their risk profile (both to patients and users), with some consideration given to device complexity, intended use, and indications for use. The classifications span from Class I, considered the lowest risk devices, to Class III, which are the highest risk devices. This classification system along with examples of each Class is further detailed in Table 12.1.

Depending on the source, roughly 35%–50% of devices fall into Class I, 40%–53% into Class II, and 5%–10% in Class III, with approximately 3% of devices receiving Unclassified status (discussed later in this chapter).[2,3]

Regulatory basics[1–5]

The classification of a novel medical device determines what type of regulatory controls and requirements must be applied to ensure safety and effectiveness to obtain premarket clearance or approval.

Unless exempt from specific portions, devices of any class submitted to the FDA must undergo general controls. These controls include device registration and listing, manufacturing requirements, prohibitions on adulteration or misbranding, and the recording and reporting of adverse events and other corrections. Low-risk, or Class I, devices are only subject to these general controls, as these controls are deemed sufficient to ensure safety and effectiveness of these low-risk devices. Once these general control requirements, the majority (74%–93%[1,2]) of Class I devices are exempt from further clearance before marketing.[6]

For moderate risk or Class II devices, general controls are deemed insufficient to appropriately ensure safety and efficacy, necessitating additional often device-specific Special Controls. These Special Controls can include the meeting of certain performance standards, postmarket surveillance, and special labeling and data requirements. While some are exempt from further clearance, most Class II devices, and a small minority of Class I devices, must submit a 510(k) premarket notification for clearance.[6,7]

Finally, the highest risk devices that fall into Class III are those for which general and any special controls are

Translational Interventional Radiology. https://doi.org/10.1016/B978-0-12-823026-8.00047-X

TABLE 12.1 FDA device classification classes, descriptions, and examples.

Class	FDA description[1–4]	Examples from the FDA database[14]
I	Low-to-moderate risk: Present minimal potential for harm, often external and noninvasive devices	Basic laboratory reagents, abdominal binders, bedpans, crutches, manual surgical instruments, suture removal kits
II	Moderate-to-high risk: Higher risk than Class I devices, often due to complexity	Complex laboratory reagents, central venous catheters, radiology PACS, endoscopes and colonoscopes, imaging machines (X-ray, CT, MR scanners), surgical sutures
III	High risk: Highest risk devices, including those that are life-sustaining, implanted, otherwise have potentially unreasonable risk, or those that are likely high-risk but without a predicate device.	Ablation systems, defibrillators and pacemakers, microspheres, surgical prostheses (e.g., for orthopedics and plastic surgery), stents

TABLE 12.2 Regulatory pathways by class[1–7,10].

Class	Regulatory controls	Example of requirements	Submission type
I	General Controls	• Device registration and listing • Manufacturing requirements • Prohibitions on adulteration or misbranding • Recording and reporting of adverse events	• Majority: Exempt from any premarket submission process • Minority: 510(k) premarket notification
II	General and Special Controls	General controls, and: • Specific performance standards • Postmarket surveillance • Special labeling and data requirements	• Majority: 510(k) premarket notification • Minority: Exempt from any premarket submission process
III	General Controls and PMA	General controls, and: • Rigorous requirements including clinical trials	• PMA

inadequate to ensure device safety and effectiveness, precluding these devices from obtaining FDA clearance (of note, historically some Class III devices could obtain 510(k) clearance if they were similar to a device type on the market before the Medical Device Amendments of 1976 which did not require a premarket approval (PMA), though this practice has been phased out as these "preamendment devices" have now been appropriately classified as of 2019[3,5,8,9]). Instead, devices in Class III require PMA, a rigorous process often involving clinical trials and other requirements befitting their high-risk designation.[10] This breakdown of regulatory controls by device class is summarized in Table 12.2.

While this basic discussion is crucial to understand the importance of device classification, there is much more complexity to device regulation. Further details on the regulatory pathways including exemptions, premarket notifications, PMAs, and other special circumstances can be found in Chapter 71 of this book.

Device classification: process[1–3]

It is imperative for inventors to determine the likely classification of their device early in the design and testing process so that these efforts can be tailored to the likely regulatory requirements that the device will ultimately undergo upon its submission for FDA clearance or approval. A few methods for evaluating which classification and submission status are appropriate for a novel device are available, often using FDA resources. For background, all medical devices regulated by the FDA fit into one of approximately 1700 generic device archetypes they have created, each of which has a unique product code.[11,12] These product codes are based on product classifications as designated in Parts 862–892 of Title 21 of the Code of Federal Regulations. These parts are broken up by medical specialty and are also known as device classification "panels," with mirroring offices within the FDA's Center for Devices and Radiological Health.[11–15] The structure of the FDA and its offices will be further detailed in Chapter 69.

Each product code/device archetype has associated regulatory requirements and device classifications. As such, the most common method to identify the likely classification of a novel device is to search the FDA's Product Classification Database[16] by device name/type, panel, submission type, product code, or other identifiers. Another method is to search the databases of submission/clearance type (e.g., databases of

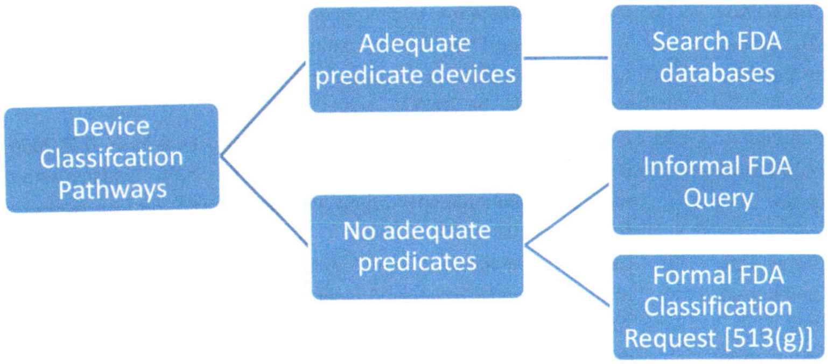

FIGURE 12.1 Available pathways to evaluate appropriate device classification.[1–3]

510(k)s or PMAs), which in turn can give the appropriate product code for classification.[17–19] Finally, lists of devices made or marketed by medical device manufacturers and companies can be searched—thanks to the fact that product registration and listing is a General Control applicable to all devices—and their product code and thus classifications reviewed.[20]

In unique cases, such as when a device does not fit appropriately into a single product code/archetype due to its intended use or novelty, further guidance can be obtained directly from the FDA, both informally, via contacting the FDA (specifically the Division of Industry and Consumer Education), or formally, via a 513(g) request.[21,22]

In general, if a device is so novel that it does not have an adequate predicate for comparison and thus does not fit into one of the generic device types (also termed "not substantially equivalent"), it will automatically be considered a Class III device as there are no existing General or Specific Controls that can be applied.[23] Inventors of such devices can either follow the PMA process as is normal for Class III devices or petition for reclassification if they feel there is an appropriate predicate device.[23,24] These various pathways are diagrammed in Fig. 12.1. Alternatively, innovators may submit a De Novo classification request initially, which is further discussed in Chapter 71.

Special device classifications[23,25]

Of note, certain devices and product codes will, when searched, return the device class of "unclassified." These are device types that did not get classified during the amendments process started in 1976 and are still unclassified to this time. This is opposed to "not classified" devices, which are ones that were submitted/created after 1976, or "postamendment" but have yet to be classified. Generally, these device types still have notes on the submission type required, based on previous devices in that category.

Conclusion

This chapter covered the FDA's risk-based classification system for medical devices in the United States. This classification guides which regulatory controls are applied to the device before it can be marketed. Understanding this system and its impact on regulation is crucial for inventors embarking on the design and production process. The next chapter will discuss FDA product categories other than devices.

Get started

- Learn more about medical device classification directly from the FDA on their "Classify Your Medical Device" page: https://www.fda.gov/medical-devices/overview-device-regulation/classify-your-medical-device
- Become familiar with the various FDA databases allowing for evaluating the appropriate product code and device classification
- Classify your medical device, and use the resources in this and later chapters to appropriately prepare for the appropriate submission and regulatory process

References

1. *Classify Your Medical Device*. U.S. Food and Drug Administration; February 7, 2020. Updated https://www.fda.gov/medical-devices/overview-device-regulation/classify-your-medical-device. Accessed July 2, 2020.
2. Piermatteo K. *How is My Medical Device Classified? Online presentation at: FDA Mediasite Catalog*; September 2019. http://fda.yorkcast.com/webcast/Play/17792840509f49f0875806b6e9a1be471d?catalog=6d6af3ca61754c3c869f7f556bbede9e21. Accessed July 2, 2020.
3. Yock PG, Zenios SA, Makower J, et al. 4.2 regulatory basics. In: *Biodesign: The Process of Innovating Medical Technologies*. 2nd ed. Cambridge, UK: Cambridge University Press; 2015:319–349.
4. *21CFR860.3*. 2019.
5. Center for Devices and Radiological Health. *Regulatory Controls*. U.S. Food and Drug Administration; March 27, 2018. Updated https://www.fda.gov/medical-devices/overview-device-regulation/regulatory-controls. Accessed July 28, 2020.
6. Center for Devices and Radiological Health. *Class I/II Exemptions*. U.S. Food and Drug Administration; July 1, 2019. Updated https://www.fda.gov/medical-devices/classify-your-medical-device/class-i-ii-exemptions. Accessed July 29, 2020.

7. Center for Devices and Radiological Health. *Premarket Notification 510(k)*. U.S. Food and Drug Administration; March 13, 2020. Updated https://www.fda.gov/medical-devices/premarket-submissions/premarket-notification-510k. Accessed July 29, 2020.
8. Center for Devices and Radiological Health. *515 Program Initiative*. U.S. Food and Drug Administration; December 19, 2017. Updated https://www.fda.gov/about-fda/cdrh-transparency/515-program-initiative. Accessed July 2, 2020.
9. Center for Devices and Radiological Health. *515 Project Status*. U.S. Food and Drug Administration; December 20, 2019. Updated https://www.fda.gov/about-fda/cdrh-transparency/515-project-status. Accessed July 2, 2020.
10. Center for Devices and Radiological Health. *PMA Approvals*. U.S. Food and Drug Administration; August 23, 2018. Updated https://www.fda.gov/medical-devices/device-approvals-denials-and-clearances/pma-approvals. Accessed July 2, 2020.
11. Center for Devices and Radiological Health. *Device Classification Panels*. U.S. Food and Drug Administration; August 31, 2018. Updated https://www.fda.gov/medical-devices/classify-your-medical-device/device-classification-panels. Accessed July 3, 2020.
12. Center for Devices and Radiological Health. *Product Code Classification Database*. U.S. Food and Drug Administration; March 22, 2018. Updated https://www.fda.gov/medical-devices/classify-your-medical-device/product-code-classification-database. Accessed July 3, 2020.
13. *21CFR862-892*. 2019.
14. Center for Devices and Radiological Health. *Reorganization of the Center for Devices and Radiological Health*. U.S. Food and Drug Administration; July 7, 2020. Updated https://www.fda.gov/about-fda/center-devices-and-radiological-health/reorganization-center-devices-and-radiological-health. Accessed July 13, 2020.
15. Center for Devices and Radiological Health. *Office of Product Evaluation and Quality*. U.S. Food and Drug Administration; May 1, 2019. Updated https://www.fda.gov/about-fda/cdrh-offices/office-product-evaluation-and-quality. Accessed July 13, 2020.
16. *Product Classification Database*. U.S. Food and Drug Administration; July 27, 2020. Updated http://www.accessdata.fda.gov/scripts/cdrh/cfdocs/cfpcd/classification.cfm. Accessed July 29, 2020.
17. *510(k) Premarket Notification Database*. U.S. Food and Drug Administration; July 27, 2020. Updated http://www.accessdata.fda.gov/scripts/cdrh/cfdocs/cfPMN/pmn.cfm. Accessed July 29, 2020.
18. *Premarket Approval (PMA) Database*. U.S. Food and Drug Administration; July 27, 2020. Updated http://www.accessdata.fda.gov/scripts/cdrh/cfdocs/cfPMA/pma.cfm. Accessed July 29, 2020.
19. *Device Classification under Section 513(f)(2)(De Novo) Database*. U.S. Food and Drug Administration; July 27, 2020. Updated http://www.accessdata.fda.gov/scripts/cdrh/cfdocs/cfPMN/denovo.cfm. Accessed July 29, 2020.
20. *Establishment Registration & Device Listing Database 7/27/2020*. U.S. Food and Drug Administration; July 27, 2020. Updated http://www.accessdata.fda.gov/scripts/cdrh/cfdocs/cfRL/rl.cfm. Accessed July 29, 2020.
21. Center for Devices and Radiological Health. *Contact Us – Division of Industry and Consumer Education (DICE)*. U.S. Food and Drug Administration; July 1, 2020. Updated https://www.fda.gov/medical-devices/device-advice-comprehensive-regulatory-assistance/contact-us-division-industry-and-consumer-education-dice. Accessed July 29, 2020.
22. Center for Devices and Radiological Health. *Procedures for Section 513(g) Requests for Information - Guidance*. U.S. Food and Drug Administration; December 13, 2019. Updated https://www.fda.gov/regulatory-information/search-fda-guidance-documents/fda-and-industry-procedures-section-513g-requests-information-under-federal-food-drug-and-cosmetic. Accessed July 13, 2020.
23. Center for Devices and Radiological Health. *Overview of Medical Device Classification and Reclassification*. U.S. Food and Drug Administration; December 19, 2017. Updated https://www.fda.gov/about-fda/cdrh-transparency/overview-medical-device-classification-and-reclassification. Accessed July 13, 2020.
24. Center for Devices and Radiological Health. *Reclassification*. U.S. Food and Drug Administration; December 20, 2019. Updated https://www.fda.gov/about-fda/cdrh-transparency/reclassification. Accessed July 13, 2020.
25. Center for Devices and Radiological Health. *Medical Device Classification Product Codes*. U.S. Food and Drug Administration; May 21, 2015. Updated https://www.fda.gov/regulatory-information/search-fda-guidance-documents/medical-device-classification-product-codes-guidance-industry-and-food-and-drug-administration-staff. Accessed July 3, 2020.

Further reading

1. Center for Devices and Radiological Health. *Evaluation of Automatic Class III Designation (De Novo) Summaries*. U.S. Food and Drug Administration; June 3, 2020. Updated https://www.fda.gov/about-fda/cdrh-transparency/evaluation-automatic-class-iii-designation-de-novo-summaries. Accessed July 13, 2020.

Chapter 13

Device testing: Safety and efficacy

Millie Liao and Kyle Cooper
Department of Interventional Radiology, Loma Linda University, Loma Linda, CA, United States

Before entering clinical testing to prove efficacy, let alone widespread commercial availability, medical devices in the United States must first undergo extensive preclinical testing and authorization by the Food and Drug Administration (FDA), including obtaining Investigational Device Exemption (IDE) approval. Through this rigorous regulatory process, the US FDA maintains its stated mission goals of guiding device innovation development and testing while simultaneously limiting unjustified risk and protecting patient safety. The 21 Code of Federal Regulations 860.7 considers a device as safe when "based on valid scientific evidence, [...] the probable benefits to health from use of the device for its intended uses and conditions of use, when accompanied by adequate directions and warnings against unsafe use, outweigh any probable risks." Many new devices developed for use in interventional radiology are categorized as Class 3 medical devices by the FDA. By definition, these "cutting edge" devices present a higher associated risk profile and require additional testing to assure safety and effectiveness. Until a medical device is deemed effective and safe for use in preclinical trials, it cannot advance to human clinical trials.[1]

Although validation of device safety and effectiveness is eventually based on data from clinical trials, device-related clinical testing first requires IDE approval. The preclinical testing and approval of the IDE application process can be long and time consuming, taking up to 5 years in some cases. Devices are divided into different categories, which have different application requirements; submission into the correct category will help save time and resources. The US FDA divides devices into significant and nonsignificant risk. A device that "presents a potential for serious risk to health, safety, or welfare of a subject" is considered significant risk, which requires a formal IDE application review. Meanwhile, nonsignificant risk devices only require Institutional Review Board (IRB) approval as sufficient for meeting IDE requisites.[2,3] Furthermore, the application needs to support and justify the need for an intended type of clinical studies, such as a feasibility study or pivotal trial. Pivotal studies are often performed by manufacturers to support a marketing application with the purpose to obtain data on safety and effectiveness for a specified intended use, usually with a statistically justified number of subjects. FDA review for pivotal studies is more complex compared to feasibility studies. Feasibility studies, also performed by manufacturers, are meant to obtain preliminary data on safety and effectiveness, typically with a small number of subjects, which subsequently guides the design for a pivotal study.[4] The application process can also be more complex for device modifications or novel devices similar to existing devices. Various paths are available, including submitting a new premarket approval (PMA) application with both preclinical and clinical data or panel-track PMA supplements.[1] Determination of which application pathway can be complex, however submission into the correct category will help expedite the process.

A successful IDE application entails thorough documentation of all prior clinical, laboratory, and animal testing data, with both safety and efficacy endpoints being met. After the establishment of the device concept and functional intent, preclinical testing can be conducted to meet specific performance result endpoints guided by published standards criteria for that category of medical devices. Upon satisfying these draft specification requirements, subsequent preclinical testing can be done to formally determine its safety and efficacy profile, typically via laboratory bench and animal testing. A variety of factors must be evaluated, including durability, functionality, biocompatibility, and protection against user error and device malfunction. With Class 3 medical devices, vital issues to address include determining the aspects of device malfunction, the potential patient risk with this mode of device failure, and the built-in strategy to minimize or address adverse events.[1]

Various laboratory bench tests are typically performed to evaluate the mechanical, physical, chemical, and

Translational Interventional Radiology. https://doi.org/10.1016/B978-0-12-823026-8.00034-1

toxicological profiles that need to be considered for a device that will be used in humans. Some FDA bench-testing requirements can be satisfied by a thorough review of current literature, taking into account previously approved devices and materials.[5] For durability testing, composite element analysis can help predict the device's mechanical performance over time, such as for intravascular stents. High-speed pulsatile fatigue models can simulate in vivo mechanical stresses and strains that would be acquired over decades, helping predict the long-term impact on a device. Meanwhile, functionality testing can assess the device mechanics during deployment and application in a simulated environment. Device testing should be performed using simulations that reflect both typical and suboptimal, challenging situations. Simulation models should strive to predict device performance in difficult, tortuous anatomy to ensure device functionality over a range of clinical settings that may be encountered. Other laboratory bench exams can evaluate for corrosion resistance and in vitro biocompatibility, including hemolysis, thrombogenicity, carcinogenicity, cytotoxicity, and inflammation.[5] Interventional radiology devices are typically utilized under image guidance, which necessitates an evaluation of device radiopacity, echogenicity, and magnetic resonance imaging compatibility. All medical devices must be tested to assess the device's ability to be sterilized, packaged, shipped, and stored while maintaining quality assurance and performance as intended by the time for patient use.[1,6]

Different factors must be taken into consideration when selecting the ideal animal model for device testing. Ideally, animal organ-specific anatomy should be near identical in size and structure to humans. For example, for intravascular stents, the ideal model would reflect similar angulation, radial stress, and compression forces as would be encountered in the human body. In addition, animal physiologic and biochemical factors also need to be considered. For example, for embolization devices that depend on the clotting cascade, selecting a model with a comparable clotting cascade and biochemical factors is crucial to help determine whether sufficient or even excessive clot is formed after deployment. Other commonly utilized interventional radiology device-related outcomes that can be evaluated through appropriate animal models include hemodynamic instability, hemorrhage, thrombosis, vasospasm, inflammatory response, and vascular remodeling/stenosis.[1]

IDE exempt situations also exist, which includes the practice of medicine. Section 1006 of the FD&C Act states that nothing "shall limit or interfere with authority of healthcare practitioner to prescribe or administer a legally marketed device to a patient for any condition or disease within a legitimate healthcare practitioner/patient relationship." If a physician determines that the use of a device is appropriate for their patient, the FDA will not interfere with the practice of medicine. However, in these situations, the patient must not be put at significant risk, and these circumstances should not be used to determine overall device safety or effectiveness. In the practice of medicine context, an IDE is not required; however, individual institutions may require IRB approval and informed consent before device application. These commercial devices, commonly diagnostic devices, are used in accordance with their labeling and utilized to evaluate consumer preference of either a modification or combination of devices. The physician should be well informed about the risks and benefits of the product and should use well-founded, scientific reasoning and medical judgment for the use of the device. A record of the devices use should be maintained, including any major or minor adverse events.[4]

Medical device advancement, which can be accomplished through cooperation between medical providers, industry, and the FDA, remains crucial in medicine for the benefit of patients. The US FDA maintains its aims of guiding device innovation development and testing while preventing any unnecessary risk and upholding patient safety. By offering well-defined guidelines and expectations for preclinical testing, the FDA remains involved in the device development process, relieving some of the uncertainty associated with new device applications for providers and patients.[1]

Key points

- The mission of the US FDA is to guide device innovation development and testing while preventing any unnecessary risk and upholding patient safety.
- Through preclinical laboratory and animal testing, a variety of factors must be evaluated, including durability, functionality, biocompatibility, and protection against user error and device malfunction.
- A successful IDE application entails thorough documentation of all prior clinical, laboratory, and animal testing data, with both safety and efficacy endpoints being met.

Why it matters?

Through cooperation between medical providers, industry, and the FDA, medical device advancement remains crucial in medicine for the benefit of patients. The mission of the US FDA is to guide device innovation development and testing while preventing any unnecessary risk and upholding patient safety. Through preclinical laboratory and animal testing, a variety of factors must be evaluated, including durability, functionality, biocompatibility, and protection against user error and device malfunction. If a medical device is deemed effective and safe for use in preclinical trials, it can advance to human clinical trials.

Get started

1. Conduct literature review including previously approved devices and materials

2. Determine appropriate bench laboratory testing profiles (i.e., durability, functionality, biocompatibility, and protection against user error, device malfunction)
3. Determine appropriate animal laboratory testing profiles (i.e., functionality, physiologic impact, user error, device malfunction)
4. Prepare and submit for IDE application

Potential pitfalls

Toxicity data: Be certain there is sufficient data for toxicity, mutagenicity, or genotoxicity for materials used. In the event of having insufficient data, toxicity testing will need to be conducted, which can be resource and time intensive.

Set clear metrics: Ensure to set defined standards regarding benefit and risk with data that is thorough enough to withstand unavoidable sample biases and variables (such as gender, body habitus, pathology).

Understanding of submission process: Working with a partner or agency that understands the regulatory submission and approval process can minimize delays and avoid unnecessary expenses.

Real-world examples

Intravascular filters investigations have been well conducted and standards are established by the FDA to include non-thrombogenicity of the device, filter efficiency without blood flow hinderance, and secure fixation.[7]

Submission for new vascular embolization devices, such as coils, require information including data on raw materials, components of any accessory devices (ie. detachment technique and deliver catheter), and coil tensile strength and fatigue.[8]

The use of ultra-high molecular weight polyethylene within orthopedics for total joint replacement has been thoroughly investigated for decades. The UHMWPE Biomaterials Handbook[9] provides a comprehensive review on various aspects of preclinical testing, including packaging, sterilization, fatigue/fracture, and microCT analysis.

Additional resources

- FDA.[2, 3, 4]
- International Organization for Standardization (ISO) 10993[10] provides standards and general principles for the biological compatibility and safety evaluation of medical devices.
- Nelson Labs[11], Toxikon[12], and Element[13] and NAMSA[14] are medical contract research organizations that assists with moving medical devices through the development process.
- Lubrizol[15] provides resources regarding biocompatibility and for biostability performance.

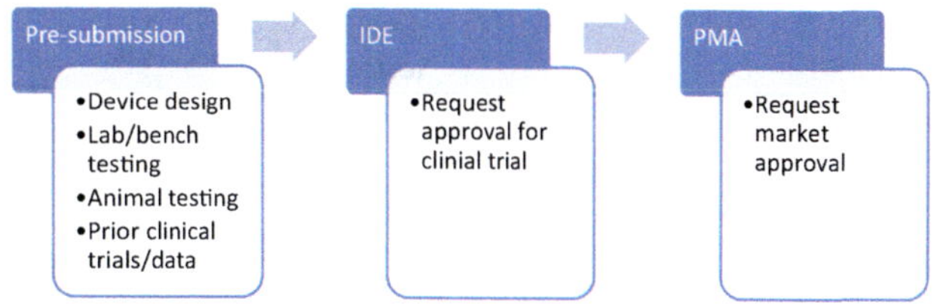

Adapted from Center for Devices and Radiological Health Investigational Device Exemption (IDE). U.S. Food and Drug Administration. FDA; 13 December 2019. www.fda.gov/medical-devices/how-study-and-market-your-device/investigational-device-exemption-ide.

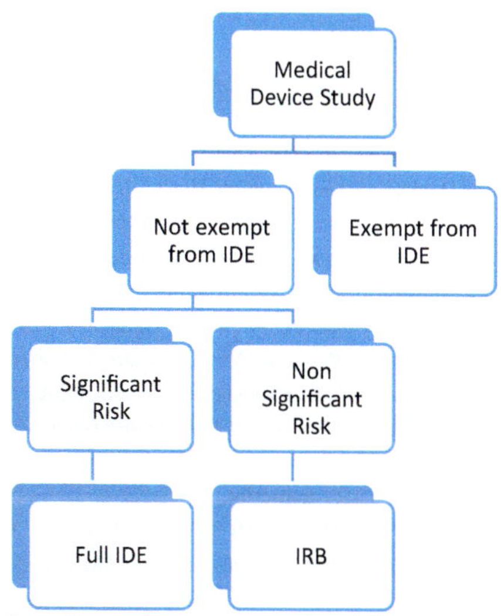

Adapted from Center for Devices and Radiological Health Investigational Device Exemption (IDE). U.S. Food and Drug Administration. FDA; 13 December 2019. www.fda.gov/medical-devices/how-study-and-market-your-device/investigational-device-exemption-ide.

Examples for **animal preclinical medical device testing**	Biocompatibility
	Physiologic compatibility
	Toxicology
	Tissue integration
	Functionality
	Drug release
	User error

Adapted from Center for Devices and Radiological Health Investigational Device Exemption (IDE). U.S. Food and Drug Administration. FDA; 13 December 2019. www.fda.gov/medical-devices/how-study-and-market-your-device/investigational-device-exemption-ide.

Examples for **laboratory bench preclinical medical device testing**	Durability
	Functionality
	Toxicology
	Mechanical strength
	Degradation
	Functionality
	Shelf life
	Sterilizeability
	Transportability
	Imaging compatibility
	Device failure
	User error

Adapted from Owen, F. Clinical Trials for Medical Devices: FDA and the IDE Process. Center for Devices and Radiologic Health.

References

1. Holzer R, Gobel J, Hijazi Z. The challenges surrounding preclinical testing in transcatheter device development and the implications on the clinic. *Expet Rev Med Dev*. 2017;14(2):87–92.
2. Center for Devices and Radiological Health. *IDE Institutional Review Boards (IRB). U.S. Food and Drug Administration*. FDA; 6 September 2018. www.fda.gov/medical-devices/investigational-device-exemption-ide/ide-institutional-review-boards-irb.
3. *Information Sheet Guidance for IRBs, Clinical Investigators, and Sponsors: Significant Risk and Nonsignificant Risk Medical Device Studies*. Center for Devices and Radiologic Health, FDA; January 2006.
4. Kalb S. *IDE Basics - Transcript. U.S. Food and Drug Administration*. FDA; 4 January 2018. www.fda.gov/training-and-continuing-education/cdrh-learn/ide-basics-transcript.
5. Shafer J, Hendershot M, Hubbell T. *Medical Device Development: Design Validation and Preclinical, Part 1*. NAMSA; 2015.
6. Shafer J, Hendershot M, Hubbell T. *Medical Device Development: Design Validation and Preclinical, Part 2*. NAMSA; 2015.
7. Center for Devices and Radiological Health. "Guidance for Cardiovascular Intravascular Filter 510(k) Submissions." *U.S. Food and Drug Administration*, FDA. https://www.fda.gov/regulatory-information/search-fda-guidance-documents/guidance-cardiovascular-intravascular-filter-510k-submissions-guidance-industry-and-fda-staff.
8. Center for Devices and Radiological Health. "Vascular and Neurovascular Embolization Devices Guidance." *U.S. Food and Drug Administration*, FDA. https://www.fda.gov/medical-devices/guidance-documents-medical-devices-and-radiation-emitting-products/vascular-and-neurovascular-embolization-devices-class-ii-special-controls-guidance-document-industry.
9. Kurtz Steven M. *UHMWPE Biomaterials Handbook: Ultra-High Molecular Weight Polyethylene in Total Joint Replacement and Medical Devices*. William Andrew; 2016.
10. ISO 10993-1:2018." *ISO*; 1 October 2018. https://www.iso.org/standard/68936.html.
11. *Microbiology Testing Lab."* Nelson Labs; 3 June 2021. https://www.nelsonlabs.com/.
12. Device." Medical Device and Diagnostic Solutions; 18 July 2022. https://medtech.labcorp.com/.
13. Materials and Product Testing, Inspection & Certification." Element. https://www.element.com/.
14. NAMSA Is the World's Leading Medtech CRO Offering Global End-to-End Development Services." NAMSA, https://namsa.com/.
15. Specialty Chemicals - the Lubrizol Corporation." Lubrizol, http://www.lubrizol.com/.

Chapter 14

Diagnostic testing

Shakthi Kumaran Ramasamy[1], R. Chitra[2] and Arun Kamireddy[3]
[1]*Department of Radiology, Stanford University School of Medicine, Stanford, CA, United States;* [2]*Department of Anatomy, Guntur Medical College, Guntur, Andhra Pradesh, India;* [3]*Department of Interventional Radiology, Medical College of Wisconsin, Milwaukee, WI, United States*

Why it matters?

The use of image guidance has proven critical in diagnostic procedures, allowing for the safe insertion of a needle into an organ or a lesion to acquire tissue for examination such as microbiology, cytology, histology, and/or molecular profiling.[1] Image-guided biopsies are increasingly performed when compared to open surgical biopsies and specimens acquired from biopsies are pivotal in the field of clinical oncology. Oncologic imaging has progressed from a simple assessment of tumor location and size to a sophisticated multimodality exploration of molecular, physiologic, genetic, immunologic, and biochemical events at microscopic to macroscopic scales, performed noninvasively and sometimes in real time.[2] There is a need for innovation to obtain increasingly more information about the tumor from the biopsy. Testing of an in vitro diagnostic (IVD) product is essential to obtain premarket approval and provide more information about the safety and efficacy of the product.

Overview

Biopsy is intended for use in the diagnosis of a disease. With the evolution of treatment toward precision medicine, various newer image guidance technologies such as optical imaging, steerable needle technologies, and robotic-assisted magnetic resonance-guided interventions might help with adequate sampling for next-generation genomic analysis.[3,4] Percutaneous needle biopsy (PNB) is usually performed under ultrasound or computerized tomography guidance. However, MRI guidance can be particularly helpful for breast or prostate tissue sampling. The diagnostic accuracy of newer interventions such as intravascular US-guided trans-venous biopsy done on difficult-to-reach perivascular sites in the abdomen and pelvis was comparable to recognized PNB thresholds.[5] Ultimately, this helps in improving the sensitivity and specificity of the biopsies performed by an interventional radiologist for the diagnosis of a disease. When a PNB supposedly gives false-positive results, it can result in unnecessary or even harmful treatment of the patient, whereas false-negative results cause a significant delay in determining the correct diagnosis and initiation of the relevant treatment.

In vitro diagnostic products are those reagents, instruments, and systems intended for use in the diagnosis of disease or other conditions, including a determination of the state of health, to cure, mitigate, treat, or prevent disease or its sequelae. Such products are intended for use in the collection, preparation, and examination of specimens taken from the human body [21 CFR 809.3]. A significant risk IVD device is generally one that is for use of substantial importance in diagnosing, curing, mitigating, or treating disease, or otherwise preventing impairment of human health and presents a potential for serious risk to the health, safety, or welfare of a subject or otherwise presents a potential for serious risk to health, safety, or welfare of a subject. 21 CFR 812.3(m). If the potential risk does not rise to the level described earlier, the study is not considered to pose a significant risk.

Sensitivity and specificity: measures of diagnostic accuracy

The proportion of true positive subjects in a whole set of disease patients is expressed as sensitivity (True Positive/True Positive + False Negative). Sensitivity is defined as the probability of getting a positive test result in subjects

Translational Interventional Radiology. https://doi.org/10.1016/B978-0-12-823026-8.00014-6

with the disease. In addition to sensitivity, specificity is a measure of diagnostic test accuracy. It is defined as the percentage of people without the disease who have a negative test result out of the total number of people without the disease (True Negative/True Negative + False Positive). Specificity represents the probability of a negative test result in a subject without the disease.[6] Diagnostic accuracy measurements are particularly sensitive to the study's design.

Regulatory classification

Regulatory authority: IVD products are devices as defined in section 201(h) of the Federal Food, Drug, and Cosmetic Act, and may also be biological products subject to section 351 of the Public Health Service Act. Like other medical devices, IVD products are subject to premarket and postmarket controls. IVD products are generally subject to categorization under the Clinical Laboratory Improvement Amendments (CLIA '88) of 1988.

How are IVD products classified?

The FDA classifies medical devices, including IVD products, into Class I, II, or III according to the level of regulatory control that is necessary to reasonably assure safety and effectiveness. The classification of IVD products (or other medical devices) determines the appropriate premarket process.

The Code of Federal Regulations lists the classification of existing IVDs in 21 CFR 862, 21 CFR 864, and 21 CFR 866.

Budgetary considerations and how to fund this stage

Funding Opportunity Announcement numbers RFA-HL-23-014, RFA-HL-23-010, RFA-HL-23-011, RFA-HL-23-012, RFA-HL-23-013 and RFA-CA-22-025 can be applied for and used to support the studies at this particular stage.

References

1. Sheth RA, Baerlocher MO, Baerlocher MO, et al. Society of interventional radiology quality improvement standards on percutaneous needle biopsy in adult and pediatric patients. *J Vasc Intervent Radiol.* 2020;31(11):1840–1848. https://doi.org/10.1016/j.jvir.2020.07.012.
2. Serkova NJ, Glunde K, Haney CR, et al. Preclinical applications of multi-platform imaging in animal models of cancer. *Cancer Res.* 2020;81(5):1189–1200. https://doi.org/10.1158/0008-5472.can-20-0373.
3. Tam AL, Lim HJ, Wistuba II, et al. Image-guided biopsy in the era of personalized cancer care: proceedings from the society of interventional radiology research consensus panel. *J Vasc Intervent Radiol.* 2016;27(1):8–19. https://doi.org/10.1016/j.jvir.2015.10.019.
4. Hiraki T, Kamegawa T, Matsuno T, et al. Robotic needle insertion during computed tomography fluoroscopy–guided biopsy: prospective first-in-human feasibility trial. *Eur Radiol.* 2020;30(2):927–933. https://doi.org/10.1007/s00330-019-06409-z.
5. Swenson C, Martin JG, Jaffe TA, et al. Intravascular ultrasound-guided transvenous biopsy of abdominal and pelvic targets difficult to access by percutaneous needle biopsy: technique and initial clinical experience. *J Vasc Intervent Radiol.* 2021;32(9). https://doi.org/10.1016/j.jvir.2021.04.029.
6. Šimundić AM. Measures of diagnostic accuracy: basic definitions. *EJIFCC.* 2009;19(4):203–211.

Chapter 15

Regulatory process

Discussion of FDA history, organization, and regulatory purview and application process

Delaney McGuirt[1], Sona Ghorashi[1] and Nariman Nezami[2,3]
[1]University of Maryland School of Pharmacy, Baltimore, MD, United States; [2]Division of Vascular and Interventional Radiology, Department of Diagnostic Radiology and Nuclear Medicine, University of Maryland School of Medicine, Baltimore, MD, United States; [3]Experimental Therapeutics Program, University of Maryland Marlene and Stewart Greenebaum Comprehensive Cancer Center, Baltimore, MD, United States

Background

The modern-day role of the Federal Food and Drug Administration (FDA) as a consumer protection agency began with the passage of the 1906 Pure Food and Drug Act, granting the bureau its regulatory functionality.[1] The Pure Food and Drug Act was formed both out of need for comprehensive legislative reform of the consumer marketplace and in response to Upton Sinclair's *The Jungle* exposing atrocities of Chicago's meatpacking industry.[1]

The FDA was known by several names and lived under various government departments over the years, official becoming the "Food and Drug Administration" in 1930 and moving to its current home, the Department of Health and Human Services, in 1979.[2] Much of FDA's powers we know today were granted by the 1938 Food, Drug, and Cosmetic Act, which added new consumer protections concerning "unlawful cosmetics and medical devices" in addition to stricter food and drugs regulations and increased powers of enforcement.[3]

What does the FDA do?

The overall mission of the FDA[4] is to ensure public health and safety through regulation of a broad scope of products and services including, but not limited to the following[5]:

- Human and veterinary drugs (prescription and over the counter)
- Biological products (blood, vaccines, tissues)
- Medical devices
- National food supply (except meat, poultry, and some egg products)
- Cosmetics
- Radiation-emitting products
- Tobacco product manufacturing, marketing, and distribution

Combination products

The FDA defines combination products as "therapeutic and diagnostic products that combine drugs, devices, and/or biological products."[6] 21 CFR 3.2(e) further defines a combination product as that comprised of two or more regulated components combined as one entity, or two or more separate regulated products packaged together.[7] Additionally, it also includes a regulated product packaged separately but intended for use exclusively with another approved regulated product to achieve its intended use.[7]

The FDA provides the following table (Table 15.1)[7] describing the nine distinct types of combination products:

With increasing technological advancements, innovations with less clear delineations of product type are becoming increasingly common. These pose regulatory challenges to the FDA, which is historically structured with separate departments for reviews of explicitly defined product types. The three centers that review combination products include the CBER, Center for Drug Evaluation and Research (CDER), and CDRH.[6] Additionally, the Office of Combination Products was created to oversee regulation and guidance of combination products.[6]

Translational Interventional Radiology. https://doi.org/10.1016/B978-0-12-823026-8.00102-4

TABLE 15.1 FDA categories of combination products.[7]

Type	Description	Common Example(s)
1	Convenience kit or copackage	Drug or biological product vials packaged with device(s) or accessory kits (empty syringes, auto-injectors, transfer sets), first aid or surgical kits containing devices and drugs
2	Prefilled drug delivery device/ system	Prefilled drug syringe, auto-injectors, metered-dose inhalers, dry powder inhalers, nasal spray, pumps, transdermal systems, prefilled iontophoresis system or microneedle "patch"
3	Prefilled biologic delivery device/System	Vaccine or other biological product in a prefilled syringe, autoinjector, nasal spray, transdermal systems, or microneedle patch preloaded with biological product
4	Device coated/impregnated/ otherwise combined with drug	Drug pills embedded with sensors, contact lens coated with a drug, drug-eluting stents, drug-eluting leads, condoms with spermicide, dental floss with fluoride, antimicrobial coated catheters/sutures, bone cements with antibiotics
5	Device coated or otherwise combined with biologic	Live cells seeded on or in a device scaffold, extracorporeal column with column-bound protein
6	Drug/biologic combination	Antibody-drug conjugates, progenitor cells combined with a drug to promote homing
7	Separate products requiring cross labeling	Light-activated drugs or biological products not copackaged but labeled for use with a specific light source device
8	Possible combination based on cross labeling of separate products	Drug/biological product under development utilizes a device, but unclear whether the final product will require that the two be cross-labeled
9	Other type of part 3 combination product (e.g., drug/device/ biological product)	All three articles are combined in a single product (e.g., a prefilled syringe containing an antibody-drug conjugate) device to manufacture a biologic, also includes a drug or biologic in the kit, or the product contains two different combination product types

FDA approved: what does it mean and what is covered?

There are common misconceptions around what the FDA reviews and approves. Approval refers to the process by which the FDA evaluates the safety and efficacy of a product *before* it is taken to market.[8] This is different from its regulation of the importation of products and postmarketing surveillance. It is important researchers and the general public understand what the FDA does and does not approve premarket (Table 15.2).

FDA structure and review

The FDA is located within the United States Department of Health and Human Services. In March 2019, the agency was restructured to better advance its mission and goals of public health and safety[9] (https://www.fda.gov/media/121839/download).

The FDA approval process varies with product type, and applications are submitted to the most relevant center. In general, the guiding principle of FDA review is a risk-benefit calculation to ensure new products are safe for public use and any proposed benefits outweigh associated risks.[10] To do this, the FDA conducts premarket reviews of human and animal data from company-led trials for safety and efficacy. This process is conducted for human drugs, biological products, complex medical devices, food and color additives, infant formulas, and animal drugs.[10] Of note, the FDA itself does not develop or test products.

Drug approval process

Approval of drug products, including prescription and over-the-counter drugs, is conducted by the FDA CDER, the largest of the FDA's six centers.[11] The FDA utilizes a two-tier system of review, consisting of both priority and standard reviews. Standard review typically takes 10 months whereas priority review is conducted in 6 months.[12] Priority review is granted to drugs that "would be significant improvements in the safety or effectiveness of the treatment, diagnosis, or prevention of serious conditions when compared to standard applications."[12] The FDA evaluates and decides on review designation on a per-application basis, with sponsor's having the option to directly request priority review. The FDA informs applicants of their review status designation within 60 days of receipt of the original application.[12]

There are five categories of drug product applications that the FDA reviews (Table 15.3).[13–17] In general, the FDA reviews applications and evaluates safety and efficacy, labeling, and manufacturing and conducts a benefit-risk calculation for the proposed drug product.

TABLE 15.2 Product categories covered by FDA.

FDA approves[8]:	FDA does not approve[8]:
• New drugs and biologics • Food additives in human food • Color additives used in FDA-regulated products	• Companies (health care facilities, laboratories, manufacturers, etc.) • Compounded drugs (drugs that are formulated by pharmacists who combine ingredients in a designated facility) • Tobacco products • Cosmetics • Medical foods (defined as food "used for the dietary management of a disease or health condition that requires special nutrient needs) • Infant formula • Dietary supplements • Food labels • Structure-function claims on dietary supplements and other foods • Animal drugs • Food additives in animal food

Orphan drug status

Additionally, a sponsor may request "orphan status" for their drug product when it will be utilized to treat a rare disease or condition.[18] The FDA Orphan Drug Act (ODA) provides sponsors specific criteria the product and disease state or condition must meet to obtain an orphan designation.[19] Requests are submitted to the Office of Orphan Products Development (OOPD) and do not exempt the sponsor from all standard regulatory requirements and applications.[18]

Device approval process

All marketed medical devices in the United States are subject to Title 21 of the Code of Federal Regulations (21 CFR) and the Federal Food, Drug, and Cosmetic Act. Specifically, the Medical Device Amendments to the Act outline the general controls under which the FDA may regulate all medical devices for safety and efficacy.[20] The level of regulatory control increases with risk classification (Table 15.4).[21] Class II devices are subject to additional "special controls" surrounding performance standards, postmarket surveillance, patient registries, special labeling requirements, premarket data requirements, and guidelines.[22]

Premarket notification 510(k)

Defined by 21 CFR Part 807 Subpart E, a Premarket Notification 510(k) must be submitted for most Class II devices before commercial distribution.[23] Sponsors must prove a device is "substantially equivalent to one legally in commercial distribution in the United States (1) before May 28, 1976; or (2) to a device that has been determined by FDA to be substantially equivalent".[24]

After submission of a premarket notification and acceptance for review, the application moves on to a comprehensive "substantive review". Ultimate decisions are supplied via email, typically within 90 calendar days of receipt of submission, and the product is deemed either substantially equivalent or not substantially equivalent.[23]

Premarket approval

Defined by 21 CFR Part 814, premarket approval is required of class III devices that are high risk and "pose a significant risk of illness or injury, or are found not substantially equivalent to Class I and II predicate through the 510(k) process."[24] More extensive than the premarket notification process, premarket approval includes submission of clinical evidence of safety and efficacy and also incurs review fees.

Premarket approval (PMA) review involves four steps[25]: (1) administrative review of completeness and limited scientific review; (2) substantive review—in-depth scientific, regulatory, and quality review; (3) panel review and recommendation from the advisory committee; (4) notification of FDA decision after final deliberations.

The entire process is completed within 180 days—see Table 15.5 for possible FDA responses to PMA.[25]

Investigational device exemption

Devices subject to investigational device exemption may be used in a clinical study for purposes of safety and efficacy data collection for a PMA application or Premarket Notification 510(k) submission.[26] Such clinical studies are subject to standard FDA, Institutional Review Board, and good clinical practices requirements. Particular modifications to, or new uses of, legally marketed devices may also be considered investigational use.[26]

TABLE 15.3 Categories of FDA drug applications.

Application type	Abbreviation	Definition and use	Subcategories
Investigational new drug	IND	Process for gaining exemption from federal law requiring approved marketing application before shipment of a drug across state lines. Allows sponsors to send drugs to investigators in other states. Applications can be submitted as "commercial" or "research," depending on the sponsor. Applications must include information for the following three categories: • Animal pharmacology and toxicology studies • Manufacturing information • Clinical protocols and investigator information After IND submission, the FDA reviews applications to verify research subject safety is ensured. Sponsors are required to wait 30 calendar days before clinical trial initiation. Center for Drug Evaluation and Research (CDER) provides a Pre-IND application consultation pprogram that guides sponsors and provides advice on the application process.	1. ***Investigator:*** submitted by a physician who initiates and conducts the investigation and who will immediately supervise use of the investigational drug. 2. ***Emergency use:*** allows expedited authorization of experimental drug in emergency situations during which there is not sufficient time for a full review. Also used for "patients who don't meet criteria of an existing study protocol or if approved study protocol does not exist." 3. ***Treatment:*** Allows use of an experimental drug before clinical trial and ongoing FDA review are complete when preliminary data has demonstrated promise for "serious or immediately life-threatening conditions".
New drug application	NDA	Submitted by a sponsor seeking to sell and market a new drug in the United States. Incorporates IND trial data.	N/A
Abbreviated new drug application	ANDA	Utilized for review of a generic drug product, or "one that is comparable to an innovator drug product in dosage form, strength, route of administration, quality, performance characteristics, and intended use." Typically not required to include preclinical and clinical data on safety and efficacy but instead must demonstrate product performs the same as the innovator drug, or "bioequivalence."	N/A
Therapeutics biologics application	BLA	Utilized for any biological products, defined by the FDA as: • Monoclonal antibodies for in vivo use • Cytokines, growth factors, enzymes, immunomodulators, thrombolytics • proteins intended for therapeutic use that are extracted from animals or microorganisms, including recombinant versions of these products (except clotting factors) • Other nonvaccine therapeutic immunotherapies	N/A
Over-the-counter (OTC) drug monograph	OTC	The FDA defines OTC drug as those available to consumers without a prescription. This process is overseen by CEDR'S office of nonprescription drugs which does not review individual drug products, but instead evaluates active ingredients and labeling. Proposed OTC products are reviewed based on 80 different drug classes and a monograph is developed for each category. The FDA defines a monograph as "a kind of 'recipe book' covering acceptable ingredients, doses, formulations, and labeling". If the product meets standards of the final monograph, it may be marketed without further FDA review, otherwise the sponsor must file an NDA.	N/A

TABLE 15.4 FDA medical device classification.[21]

	Risk	Regulatory control	FDA submission
Class I	Low	General	510(k) exempts
Class II	Moderate	General and special (if available)	510(k) required for most
Class III	High	General and PMA	PMA required

Abbreviates used: *PMA*, premarket approval.

TABLE 15.5 Expected responses to premarket approval applications.[25]

Approval order	FDA found no reasons for denial and approves contingent upon final labeling and notifies public of approval.
Approvable letter	FDA found approval may be granted if additional information or conditions are agreed upon by the applicant.
Not approvable letter	FDA found application met a reason for denial or cannot reach an approvable decision due to lack of information provided. FDA outlines conditions to make PMA approvable where possible. Applicant may amend and respond, request administrative review, or withdraw.
Order denying approval	FDA found application did not follow PMA requirements or meets any of the grounds for denial under the federal food, drug, and cosmetic Act.

TABLE 15.6 FDA regulated products used in IR.

FDA category	Example interventional radiology related products
Human drugs	Chemotherapy agents utilized in chemoembolization
	Novel radionucleotides utilized in radioembolization (i.e., Holmium-166)
	Iodinated contrast agents
Medical devices	XACT ACE (R) Robotic System (Hingham, MA, USA) assisted needle placement for biopsy and ablation
	WAVELINQ endovascular dialysis fistula creation device
Radiation-emitting electronic products	Fluoroscopy machines
	Computed tomography (CT) machines
	Nuclear medicine radionucleotide producing devices
	Ultrasound machines

Humanitarian device exemption

A Humanitarian Device Exemption (HDE) may be submitted for a Class III device that is developed to "benefit patients with rare disease or conditions."[27] The 1984 ODA defines a rare disease as "a disease or condition that affects fewer than 200,000 people in the United States."[27] Sponsors must first obtain humanitarian use device designation through application to the FDA's OOPD.[27] Devices granted HDE designation are exempt from certain requirements and incur profit and use restrictions.

De novo classification requests

De novo classification may be requested when a novel medical device is assured to be reasonably safe and effective for its intended use under general controls, or general and special controls, but there is no legally marketed predicate device.[28]

IR-related FDA regulated products

Many products relevant to Interventional Radiology fall under FDA regulation (Table 15.6). Nearly every tool used by the interventional radiologist is categorized as an FDA "medical device." Additionally, all relevant imaging equipment is regulated under the FDA's category of "radiation-emitting electronic products." See Table 15.6 for example interventional radiology products and their respective FDA regulatory category.

References

1. *When and Why Was FDA Formed?* FDA; November 3, 2018. Published https://www.fda.gov/about-fda/fda-basics/when-and-why-was-fda-formed. Accessed July 25, 2020.
2. *History of FDA's Internal Organization.* FDA; March 14, 2019. Published https://www.fda.gov/about-fda/history-fdas-fight-consumer-protection-and-public-health/history-fdas-internal-organization. Accessed July 25, 2020.
3. *How Did the Federal Food, Drug, and Cosmetic Act Come about?* FDA; November 3, 2018. Published https://www.fda.gov/about-fda/fda-basics/how-did-federal-food-drug-and-cosmetic-act-come-about. Accessed July 25, 2020.
4. *What We Do.* FDA; November 3, 2018. Published https://www.fda.gov/about-fda/what-we-do. Accessed July 25, 2020.
5. *What Does FDA Regulate?* FDA; October 16, 2019. Published https://www.fda.gov/about-fda/fda-basics/what-does-fda-regulate. Accessed July 25, 2020.
6. *About Combination Products.* FDA; June 18, 2019. Published https://www.fda.gov/combination-products/about-combination-products. Accessed July 25, 2020.
7. *Combination Product Definition Combination Product Types.* FDA; June 18, 2019. Published https://www.fda.gov/combination-products/about-combination-products/combination-product-definition-combination-product-types. Accessed July 25, 2020.
8. *Is it Really "FDA Approved?".* FDA; May 17, 2020. Published https://www.fda.gov/consumers/consumer-updates/it-really-fda-approved. Accessed July 25, 2020.
9. *FDA Organization.* FDA; January 17, 2020. Published https://www.fda.gov/about-fda/fda-organization. Accessed July 25, 2020.
10. *About FDA Product Approval.* FDA; November 3, 2018. Published https://www.fda.gov/news-events/approvals-fda-regulated-products/about-fda-product-approval. Accessed July 25, 2020.
11. *How Drugs Are Developed and Approved.* FDA; May 20, 2019. Published https://www.fda.gov/drugs/development-approval-process-drugs/how-drugs-are-developed-and-approved. Accessed July 25, 2020.
12. *Priority Review.* FDA; November 3, 2018. Published https://www.fda.gov/patients/fast-track-breakthrough-therapy-accelerated-approval-priority-review/priority-review. Accessed July 25, 2020.
13. *Investigational New Drug (IND) Application.* FDA; July 14, 2020. Published https://www.fda.gov/drugs/types-applications/investigational-new-drug-ind-application. Accessed July 25, 2020.
14. *New Drug Application (NDA).* FDA; June 27, 2019. Published https://www.fda.gov/drugs/types-applications/new-drug-application-nda. Accessed July 25, 2020.
15. *Abbreviated New Drug Application (ANDA).* FDA; December 20, 2019. Published https://www.fda.gov/drugs/types-applications/abbreviated-new-drug-application-anda. Accessed July 25, 2020.
16. *Therapeutic Biologics Applications (BLA).* FDA; February 24, 2020. Published https://www.fda.gov/drugs/types-applications/therapeutic-biologics-applications-bla. Accessed July 25, 2020.
17. *Drug Applications for Over-the-counter (OTC) Drugs.* FDA; May 7, 2020. Published https://www.fda.gov/drugs/types-applications/drug-applications-over-counter-otc-drugs. Accessed July 25, 2020.
18. *Commissioner O of the. Designating an Orphan Product: Drugs and Biological Products.* FDA; April 8, 2020. Published https://www.fda.gov/industry/developing-products-rare-diseases-conditions/designating-orphan-product-drugs-and-biological-products. Accessed July 25, 2020.
19. *Orphan Drug Act - Relevant Excerpts.* FDA; April 24, 2019. Published https://www.fda.gov/industry/designating-orphan-product-drugs-and-biological-products/orphan-drug-act-relevant-excerpts. Accessed July 25, 2020.
20. *A History of Medical Device Regulation & Oversight in the United States.* FDA; July 8, 2019. Published https://www.fda.gov/medical-devices/overview-device-regulation/history-medical-device-regulation-oversight-united-states. Accessed July 25, 2020.
21. *How to Study and Market Your Device.* FDA; January 28, 2020. Published https://www.fda.gov/medical-devices/device-advice-comprehensive-regulatory-assistance/how-study-and-market-your-device. Accessed July 25, 2020.
22. *Regulatory Controls.* FDA; February 8, 2019. Published https://www.fda.gov/medical-devices/overview-device-regulation/regulatory-controls. Accessed July 25, 2020.
23. *Premarket Notification 510(k).* FDA; March 13, 2020. Published https://www.fda.gov/medical-devices/premarket-submissions/premarket-notification-510k. Accessed July 25, 2020.
24. *Overview of Device Regulation.* FDA; February 8, 2019. Published https://www.fda.gov/medical-devices/device-advice-comprehensive-regulatory-assistance/overview-device-regulation. Accessed July 25, 2020.
25. *Premarket Approval (PMA).* FDA; July 9, 2019. Published https://www.fda.gov/medical-devices/premarket-submissions/premarket-approval-pma. Accessed July 25, 2020.
26. *Investigational Device Exemption (IDE).* FDA; December 13, 2019. Published https://www.fda.gov/medical-devices/how-study-and-market-your-device/investigational-device-exemption-ide. Accessed July 25, 2020.
27. *Humanitarian Device Exemption.* FDA; September 5, 2019. Published https://www.fda.gov/medical-devices/premarket-submissions/humanitarian-device-exemption. Accessed July 25, 2020.
28. *De Novo Classification Request.* FDA; November 20, 2019. Published https://www.fda.gov/medical-devices/premarket-submissions/de-novo-classification-request. Accessed July 25, 2020.

Chapter 16

Procedural technique development

Shakthi Kumaran Ramasamy[1], R. Chitra[3] and Kilari Sreenivasulu[2]
[1]*Department of Radiology, Stanford University School of Medicine, Stanford, CA, United States;* [2]*Departments of Radiology Mayo Clinic, Vascular and Interventional Translational Laboratory, Rochester, MN, United States;* [3]*Department of Anatomy, Guntur Medical College, Guntur, Andhra Pradesh, India*

Key points

1. Planning Research and Experimental Procedures on Animals: Recommendations for Excellence (PREPARE) guidelines are useful for planning animal research and testing.
2. The various methodologies are in vivo, ex vivo, simulation, in vitro, and virtual animal models.
3. The different stages of a procedure technique development are innovation, development, exploration, assessment, and long-term surveillance.

Why it matters?

The need to research, innovate, and develop new minimally invasive procedures for the treatment of different disease states is the foundation for clinical practice in Interventional Radiology (IR).

Overview

Procedure technique development is crucial because it helps in the successful translation of a concept to the bedside as a clinical procedure. The initial treatment using the new innovative procedure may be successful for a disease state for which the procedure might not be the treatment after all. After the first successful reports of the new clinical procedure for a primary clinical indication emerge, multiple different attempts are made to reproduce it. Once the primary clinical indication for a procedure has been established, the physicians and scientists assess the safety and feasibility of the procedure, generally using animal experiments. For a trial to be carried out in human patients, supporting data from preclinical studies are requisite. This chain of events in establishing a new procedure as the standard of care is demonstrated to be similar.

Preclinical considerations

Before a conceptual procedural technique can be explored further using preclinical testing, it is important to determine the novelty of the concept. Furthermore, it is essential to check if there are established prior animal experiments that have tested a similar concept. The PREPARE (Planning Research and Experimental Procedures on Animals: Recommendations for Excellence) guidelines are useful for planning animal research and testing. These guidelines have a 15-component checklist which includes recommendations for literature searches, legal issues, ethical issues, harm—benefit assessment and humane endpoints, experimental design and statistical analysis, objectives and timescale, funding and division of labor, facility evaluation, education and training, health risks, waste disposal and decontamination, test substances and procedures, experimental animals, quarantine and health monitoring, housing and husbandry, experimental procedures, humane killing, release, reuse or rehoming and necropsy.[1]

Methodologies

Live animal models (in vivo)

Animal models are essential for the establishment of new procedures and improving current methods and outcomes. Several animal experiments have been conducted and documented in the development of many IR procedures. It is beyond the scope of this chapter to cover the different animal models and various lab techniques employed.

Translational Interventional Radiology. https://doi.org/10.1016/B978-0-12-823026-8.00068-7

Choosing the right animal for the right reason[1]-selection of a suitable model for preclinical testing is dependent on the type of the IR procedure. Live animal studies utilization in preclinical research is necessary for the translation of a proof of concept to clinical care. However, challenges encountered are lab requirements and economic and technical feasibility. Interventional Oncology procedures are increasingly dependent on animal models for successful translation of a proof of concept to a clinical procedure. For example, radioembolization and chemoembolization of the prostate gland were first experimented in canine models that warrant a clinical trial in patients with prostate cancer.[2,3] The reproducibility of the tumor microenvironment and heterogeneity is more important than the size and type of the animal for tumor models. Hepatocellular carcinoma (HCC) animal models can be classified as follows—humanized models, xenograft models, syngeneic models, genetically engineered models, and chemically induced models. Rodents, swine, canines, rabbits, woodchucks, and primates have historically served as HCC animal models. However, transgenic pig models, VX2 rabbit models, mice, and rodents are currently used with the intent to maintain similarity to HCC in human patients.[2] Humanized patient-derived xenografts (PDX) mouse model hosts a human-specific tumor microenvironment.[3] Transgenic pig model (Oncopig) has been utilized for the induction, characterization of pancreatic cancer and to test new locoregional therapies.[4,5] Endovascular and neuro-endovascular research are preferred to be conducted in large animal models such as dogs, rabbits, and pigs when compared to rodents.[6,7] Animal models for hand-on training for IR procedures have been outdated and are replaced by simulation techniques and ex vivo models.[8,9]

Ex vivo models

Ex vivo models have been commonly used for various ablation technique development such as microwave,[10,11] cryo,[12] thermo-chemical,[13] and radiofrequency (RF).[14] They are either phantoms or specific explanted organs from an animal after their termination in the lab or at the slaughterhouse. Ethical approval is required for the termination of lab animals.

Simulation

Simulation in IR can be of many types such as phantoms, mannequins, computerized digital simulations, virtual reality (VR), augmented reality (AR). VIST-Lab (Mentice), and ANGIO Mentor™ (www.simbionix.com) Mentor (Simbionix 3D systems) are simulators that help in renal angioplasty and stenting, respectively. These simulations are aimed for training and technical skill improvement in general and not for patient-specific procedure technique simulation. However, for certain procedures such as endovascular aneurysm repair the 3D systems Simbionix (Tel Aviv, Israel) software helps in preclinical procedure planning using patience specific CT data to create a fully interactive fluoroscopic simulation of anatomy.[15] Planning and training of interventional procedures using VR, AR, and mixed reality (MR) is highly underutilized. The relative cost of a VR laboratory is less expensive than an animal laboratory.[16] For example, in a preclinical rat model of transarterial chemoembolization (TACE), 3D AR visualization of preprocedural imaging has been shown to aid in the reduction of procedural metrics.[17] AR platform on smartphones and smart glasses device demonstrated on a phantom has shown to assist in percutaneous needle-based interventions.[18] MR has been used to guide needle placement for radiofrequency ablation.[19] A preclinical evaluation MR navigation system for ultrasound-guided percutaneous punctures demonstrated using opaque ballistic gel phantom has been demonstrated to have improved parameters when compared to the standard approach.[20]

In vitro models

Bioengineering has pathed the way for the fabrication of biomimetic tissues and organs. Nozzle-based bioprinting and stereolithography have been achieved to fabricate biologically functional multi-component tissues due to their robustness and versatility.[21–31] Microfabrication using Polydimethylsiloxane (PDMS) is suitable for blood vessel modeling.[32–35] The sacrificial 3D bioprinting technology has helped to engineer blood vessels, vascular models, and vascular disorder models like thrombosis-on-a-chip. This chip model was created by injecting human whole blood into endothelialized bifurcation microchannels contained in a gelation methacryloyl hydrogel created utilizing a sacrificial bioprinting method, followed by in situ thrombosis development.[36] The biomimicry of the chip was accurate as of its counterparts in the human body and serves as a powerful tool for studying thrombosis, thrombolysis, and fibrosis. Khademhosseini and colleagues demonstrated a bioengineering approach to embed saccular structures on the surface of a PDMS device, which could be used as a model for vascular aneurysms with comparable morphologies in the future.[37]

The field of cancer engineering has grown into a potent tool for examining the biology, physiology, and angiogenesis of tumors by applying the knowledge gained from engineering biofunctional tissues to tumor models. Kamm and colleagues, for example, created linked blood vessel-like models with perfusable lumens to investigate cancer cell extravasation from the vasculature into the surrounding matrix.[38–40] Vascularized glioblastoma models were recently bioengineered using a combination of sacrificial bioprinting methods and tumor spheroids, which successfully replicated tumor proliferation and angiogenesis on a microfluidic chip.[41] These models can also predict personalized physiological and treatment responses if fabricated from patient-derived tumor cells or biopsies.

In vitro vascular models have not been reported to be used in interventional radiology, leave alone for the development of procedural techniques. It is anticipated that these models will have widespread adoption for assessing a wide range of endovascular interventions in the future. To accomplish accurate and personalized surgical planning, these models can be manufactured using patient data to replicate their structures, geometries, sizes, and tortuosity outside the human body in a way that is easily accessible and visible. Similar approaches could also be used to create biological tissues for the replacement of damaged equivalents and subsequent regeneration after postinterventional procedures.

Virtual animal models

With the development of virtual animal models like the virtual sea slug by the scientists from the University of Illinois at Urbana Champaign (IL, USA), the question arises whether animal models can be replaced by virtual models for preclinical research.[42] The Virtual Zoo (ViZoo) models can be paired with the already existent biological tissue parameter database which can provide electromagnetic, thermal, fluid, acoustic, and magnetic resonance values.[43,44]

Stages of procedure technique development

The stages for the development of an interventional radiology procedural technique is similar to the IDEAL recommendations.[45,46] The five stages are as described below:

Stage 0: innovation—proof of principle

The first use of a new procedure in a patient is spurred by the need for a solution to a clinical problem, which is referred to as the stage of innovation. For example, Ariel described a method in 1965 for administering Y90 through the celiac artery, which was then used in four patients.[47] However, Miller was the first to describe the technique of hepatic artery catheterization for liver perfusion.[48] Ethics approval is not required at this stage of the development; however, informed consent from a competent patient is required. Irrespective of the outcome, all new procedures must be reported for the prevention of similar adverse effects in the future.

Stage 1: development—refinement, definition, and early adoption

The planned use of a procedure in a limited group of animals to support experience with its first use to early adoption and, to refine and define the current approach is referred to as development. For example, in a study using rats as animal models, the preparation and administration of Adriamycin-loaded bovine albumin microspheres were first described by Willmott and colleagues.[49] Meade and colleagues determined that the microspheres with sizes of 15 and 32.5 μm were three times more likely to lodge in hepatic tumors than normal liver parenchyma in a group of 25 rats[50] whereas 50 μm microspheres, on the other hand, were found to be uniformly distributed in both normal and malignant tissue. These studies helped to determine the optimal microsphere size and tumor-to-liver dispersion ratios. At this stage, study design and protocols must be well established. American Association for Accreditation of Laboratory Animal Care (AAALAC) accreditation and the Institutional Animal Care and Use Committee (IACUC) approval for the animal studies.

Stage 2: exploration—dissemination

The stage of exploration begins after the key steps and technical aspects of the procedure have been established. However, the user experience with the procedure may be very limited, and more evident animal studies are required. At this stage, the procedure is translated into a clinical study involving human patients. The prospective database from the randomized control trials with documentation of the clinical outcomes plays a crucial role in the widespread clinical adoption of a procedural technique. For example, phase 1 trials for glass microspheres (TheraSphere, BTG International Group, London, UK) and resin microspheres (SIR-Spheres, Sirtex Medical Ltd., New South Wales, Australia).[51,52(p90)]

Stage 3: assessment—comparison with existing treatments

The development of a novel approach and the description of its consequences was the emphasis of the previous phases; this stage tries to evaluate effectiveness versus current standards of treatment. The evolution of clinical procedures can still happen at this stage. For example, this clinical study aims to evaluate and the compare outcome of stereotactic body radiation therapy, yttrium-90 radioembolization (TARE), RF ablation, or TACE as a bridge to liver transplant in patients with HCC.[53]

Stage 4: long term—surveillance, quality control, and improvement

In this stage, established procedures are evaluated for uncommon adverse events and long-term results, as well as variances in the outcome. The evolution and improvement of a procedure continue at this stage. For example, Simon and colleagues compared the long-term survival outcomes and tumor response to TACE, TARE, and transarterial ethanol ablation for HCC lesions $\geq$10 cm.[54] Segmental TARE,[55] combination with other locoregional therapies (LRTs) such as immunotherapies,[56,57] TARE for solitary, unresectable HCC lesions <8 cm,[58] newer animal models to evaluate LRTs[5] and better tumor response assessment guidelines[59–62] are all recent improvements in the evolution of radioembolization for HCC.

Get started

- Follow the PREPARE[1] guidelines to test the proof of concept
- Obtain AAALAC accreditation and IACUC approval for animal studies.
- Choosing the right animal for the right reason
- Consider using novel in vitro models and virtual animal models
- Report results using the ARRIVE guidelines 2.0[63]

Potential pitfalls

1. Before you start, it is essential to clearly define the goals of the new procedure
2. Build the right team
3. Funding is as essential as determining the unmet need

Real-world examples

1) Treatment modalities such as MR-guided Focused Ultrasound Ablation for the Treatment of Nonspinal Osteoid Osteoma are evaluated as a prospective multicenter cohort study following the IDEAL recommendations.[65,66]
2) The Role of the USA Food and Drug Administration (FDA) in the developmental pathways of new prescription drugs have contrasts to the invasive procedure process but both undergo the same stages of development.[65,67]
3) The first study to test the use of an articulated laparoscopic camera system for minimally invasive surgery in gynecology in humans, including both a trial and a series of cases, is classified as an IDEAL stage 1 and 2a study.[46,65,68]

Resources: additional reading

1. Guide for the Care and Use of Laboratory Animals[64]
2. The Puzzle People: Memoirs of A Transplant Surgeon by Thomas E. Starzl.

References

1. Smith AJ, Clutton RE, Lilley E, Hansen KEA, Brattelid T. PREPARE: guidelines for planning animal research and testing. *Lab Anim.* 2018;52(2):135–141. https://doi.org/10.1177/0023677217724823.
2. Obeid M, Khabbaz RC, Garcia KD, Schachtschneider KM, Gaba RC. Translational animal models for liver cancer. *AJIR.* 2018;2. https://doi.org/10.25259/AJIR-11-2017.
3. Macek Jilkova Z, Kurma K, Decaens T. Animal models of hepatocellular carcinoma: the role of immune system and tumor microenvironment. *Cancers.* 2019;11(10):1487. https://doi.org/10.3390/cancers11101487.
4. Boas FE, Nurili F, Bendet A, et al. Induction and characterization of pancreatic cancer in a transgenic pig model. *PLoS One.* 2020;15(9):e0239391. https://doi.org/10.1371/journal.pone.0239391.
5. Nurili F, Monette S, Michel AO, et al. Transarterial embolization of liver cancer in a transgenic pig model. *J Vasc Intervent Radiol.* 2021;32(4):510–517. https://doi.org/10.1016/j.jvir.2020.09.011. e3.
6. Herrmann AM, Meckel S, Gounis MJ, et al. Large animals in neurointerventional research: a systematic review on models, techniques and their application in endovascular procedures for stroke, aneurysms and vascular malformations. *J Cerebr Blood Flow Metabol.* 2019;39(3):375–394. https://doi.org/10.1177/0271678X19827446.
7. Namba K, Mashio K, Kawamura Y, Higaki A, Nemoto S. Swine hybrid aneurysm model for endovascular surgery training. *Intervent Neuroradiol.* 2013;19(2):153–158. https://doi.org/10.1177/159101991301900202.
8. Wright KC, Lunderquist A. Animal models for "hands-on" interventional radiology training. *Semin Intervent Radiol.* 2002;19(2):135–142. https://doi.org/10.1055/s-2002-32788.
9. Sandmann J, Müschenich FS, Riabikin A, Kramer M, Wiesmann M, Nikoubashman O. Can silicone models replace animal models in hands-on training for endovascular stroke therapy? *Intervent Neuroradiol.* 2019;25(4):397–402. https://doi.org/10.1177/1591019919833843.
10. Pfannenstiel A, Sebek J, Fallahi H, et al. Directional microwave ablation: experimental evaluation of a 2.45-GHz applicator in ex vivo and in vivo liver. *J Vasc Intervent Radiol.* 2020;31(7):1170–1177. https://doi.org/10.1016/j.jvir.2020.01.016. e2.
11. Siriwardana PN, Singh S, Johnston EW, et al. Effect of hepatic perfusion on microwave ablation zones in an ex vivo porcine liver model. *J Vasc Intervent Radiol.* 2017;28(5):732–739. https://doi.org/10.1016/j.jvir.2016.03.006.
12. Lu A, Favazza CP, Felmlee JP, et al. Experimental investigation of cryoneedle heating during MR-guided cryoablation in ex vivo tissue samples at 1.5-tesla. *J Vasc Intervent Radiol.* 2021;32(5):721–728. https://doi.org/10.1016/j.jvir.2021.01.272. e2.
13. Farnam JL, Smith BC, Johnson BR, et al. Thermochemical ablation in an ex-vivo porcine liver model using acetic acid and sodium

hydroxide: proof of concept. *J Vasc Intervent Radiol.* 2010;21(10):1573–1578. https://doi.org/10.1016/j.jvir.2010.06.012.

14. Wong SN, Santi GE, Nurjadin H, Aguilar R, Gosalvez-Pe S. Temperature-dependent electrode repositioning for multiple overlapping radiofrequency ablation in ex vivo porcine livers. *J Vasc Intervent Radiol.* 2010;21(11):1733–1738. https://doi.org/10.1016/j.jvir.2010.07.011.
15. Bartal G, Rundback JH. Simulation training in interventional radiology. In: Keefe NA, Haskal ZJ, Park AW, Angle JF, eds. *IR Playbook.* Springer International Publishing; 2018:7–15. https://doi.org/10.1007/978-3-319-71300-7_2.
16. Berry M, Hellström M, Göthlin J, Reznick R, Lönn L. Endovascular training with animals versus virtual reality systems: an economic analysis. *J Vasc Intervent Radiol.* 2008;19(2):233–238. https://doi.org/10.1016/j.jvir.2007.09.004.
17. Park BJ, Perkons NR, Profka E, et al. 3D augmented reality visualization informs locoregional therapy in a translational model of hepatocellular carcinoma. *J Vasc Intervent Radiol.* 2020;31(10):1612–1618. https://doi.org/10.1016/j.jvir.2020.01.028. e1.
18. Li M, Seifabadi R, Long D, et al. Smartphone- versus smartglasses-based augmented reality (AR) for percutaneous needle interventions: system accuracy and feasibility study. *Int J CARS.* 2020;15(11):1921–1930. https://doi.org/10.1007/s11548-020-02235-7.
19. Si W, Liao X, Qian Y, Wang Q. Mixed reality guided radiofrequency needle placement: a pilot study. *IEEE Access.* 2018;6:31493–31502. https://doi.org/10.1109/ACCESS.2018.2843378.
20. Federico Davrieux C, Giménez ME, Gonzalez C, et al. Mixed reality navigation system for ultrasound-guided percutaneous punctures: a pre-clinical evaluation. *Surgical Endoscopy and Other Interventional Techniques.* 2020. https://doi.org/10.1007/s00464-019-06755-5. Published online.
21. Li X, Moosavi-Basri SM, Sheth RA, et al. Bioengineered in vitro vascular models for applications in interventional radiology. *Curr Pharmaceut Des.* 2019. https://doi.org/10.2174/1381612824666180416114325. Published online.
22. Huang TQ, Qu X, Liu J, Chen S. 3D printing of biomimetic microstructures for cancer cell migration. *Biomed Microdevices.* 2014;16(1):127–132. https://doi.org/10.1007/s10544-013-9812-6.
23. Liu J, Hwang HH, Wang P, et al. Direct 3D-printing of cell-laden constructs in microfluidic architectures. *Lab Chip.* 2016;16(8):1430–1438. https://doi.org/10.1039/c6lc00144k.
24. Zhang YS, Oklu R, Dokmeci MR, Khademhosseini A. Three-Dimensional bioprinting strategies for tissue engineering. *Cold Spring Harbor PerspectMed.* 2018;8(2). https://doi.org/10.1101/cshperspect.a025718.
25. Liu W, Zhang YS, Heinrich MA, et al. Rapid continuous multimaterial extrusion bioprinting. *Adv Mater.* 2017;29(3):1604630. https://doi.org/10.1002/adma.201604630.
26. Miller JS, Stevens KR, Yang MT, et al. Rapid casting of patterned vascular networks for perfusable engineered three-dimensional tissues. *Nat Mater.* 2012;11(9):768–774. https://doi.org/10.1038/nmat3357.
27. Kolesky DB, Truby RL, Gladman AS, Busbee TA, Homan KA, Lewis JA. 3D bioprinting of vascularized, heterogeneous cell-laden tissue constructs. *Adv Mater.* 2014;26(19):3124–3130. https://doi.org/10.1002/adma.201305506.
28. Kolesky DB, Homan KA, Skylar-Scott MA, Lewis JA. Three-dimensional bioprinting of thick vascularized tissues. *Proc Natl Acad Sci USA.* 2016;113(12):3179–3184. https://doi.org/10.1073/pnas.1521342113.
29. Lee VK, Kim DY, Ngo H, et al. Creating perfused functional vascular channels using 3D bio-printing technology. *Biomaterials.* 2014;35(28):8092–8102. https://doi.org/10.1016/j.biomaterials.2014.05.083.
30. Bertassoni LE, Cecconi M, Manoharan V, et al. Hydrogel bioprinted microchannel networks for vascularization of tissue engineering constructs. *Lab Chip.* 2014;14(13):2202–2211. https://doi.org/10.1039/c4lc00030g.
31. Jia W, Gungor-Ozkerim PS, Zhang YS, et al. Direct 3D bioprinting of perfusable vascular constructs using a blend bioink. *Biomaterials.* 2016;106:58–68. https://doi.org/10.1016/j.biomaterials.2016.07.038.
32. Hou X, Zhang YS, Santiago GT de, et al. Interplay between materials and microfluidics. *Nat Rev Mater.* 2017;2(5):17016. https://doi.org/10.1038/natrevmats.2017.16.
33. Wang GJ, Ho KH, hui HS, Wang KP. Microvessel scaffold with circular microchannels by photoresist melting. *Biomed Microdevices.* 2007;9(5):657–663. https://doi.org/10.1007/s10544-007-9067-1.
34. Lima R, Oliveira M, Ishikawa T, et al. Axisymmetric polydimethysiloxane microchannels for in vitro hemodynamic studies. *Biofabrication.* 2009;1(3):035005. https://doi.org/10.1088/1758-5082/1/3/035005.
35. Avery RK, Albadawi H, Akbari M, et al. An injectable shear-thinning biomaterial for endovascular embolization. *Sci Transl Med.* 2016;8(365). https://doi.org/10.1126/scitranslmed.aah5533.
36. Zhang YS, Davoudi F, Walch P, et al. Bioprinted thrombosis-on-a-chip. *Lab Chip.* 2016;16(21):4097–4105. https://doi.org/10.1039/c6lc00380j.
37. Lee WG, Ortmann D, Hancock MJ, Bae H, Khademhosseini A. A hollow sphere soft lithography approach for long-term hanging drop methods. *Tissue Eng C Methods.* 2010;16(2):249–259. https://doi.org/10.1089/ten.tec.2009.0248.
38. Zervantonakis IK, Hughes-Alford SK, Charest JL, Condeelis JS, Gertler FB, Kamm RD. Three-dimensional microfluidic model for tumor cell intravasation and endothelial barrier function. *Proc Natl Acad Sci USA.* 2012;109(34):13515–13520. https://doi.org/10.1073/pnas.1210182109.
39. Jeon JS, Zervantonakis IK, Chung S, Kamm RD, Charest JL. In vitro model of tumor cell extravasation. *PLoS One.* 2013;8(2). https://doi.org/10.1371/journal.pone.0056910.
40. Jeon JS, Bersini S, Gilardi M, et al. Human 3D vascularized organotypic microfluidic assays to study breast cancer cell extravasation. *Proc Natl Acad Sci USA.* 2015;112(1):214–219. https://doi.org/10.1073/pnas.1417115112.
41. Lee V, Dai G, Zou H, Yoo S. Generation of 3-D glioblastoma-vascular niche using 3-D bioprinting. In: *2015 41st Annual Northeast Biomedical Engineering Conference.* NEBEC); 2015. https://doi.org/10.1109/NEBEC.2015.7117111. Published online.
42. Brown JW, Caetano-Anollés D, Catanho M, et al. Implementing goal-directed foraging decisions of a simpler nervous system in simulation. *eNeuro.* 2018;5(1). https://doi.org/10.1523/ENEURO.0400-17.2018. ENEURO.0400-17.2018.
43. Hasgall PA, Di Gennaro F, Baumgartner C, et al. *"IT'IS Database for Thermal and Electromagnetic Parameters of Biological Tissues," Version 4.0.* May 15, 2018. https://doi.org/10.13099/VIP21000-04-0 (itis.swiss/database).
44. Christ A, Kainz W, Hahn EG, et al. The Virtual Family—development of surface-based anatomical models of two adults and

two children for dosimetric simulations. *Phys Med Biol*. 2010. https://doi.org/10.1088/0031-9155/55/2/n01. Published online.

45. Hirst A, Philippou Y, Blazeby JM, et al. No surgical innovation without evaluation: evolution and further development of the IDEAL framework and recommendations. *Ann Surg*. 2019. https://doi.org/10.1097/sla.0000000000002794. Published online.
46. McCulloch P, Altman DG, Campbell WB, et al. No surgical innovation without evaluation: the IDEAL recommendations. *Lancet*. 2009;374(9695):1105–1112. https://doi.org/10.1016/S0140-6736(09)61116-8.
47. Irving M. Ariel treatment of inoperable primary pancreatic and liver cancer by the intra-arterial administration of radioactive isotopes (y90 radiating microspheres). *Ann Surg*. 1965. https://doi.org/10.1097/00000658-196508000-00018. Published online.
48. Miller TR, Griman 0 R. Hepatic artery catheterization for liver perfusion. *Arch Surg*. 1961;82:423.
49. Willmott N, Cummings J, Stuart JF, Florence AT. Adriamycin-loaded albumin microspheres: preparation, in vivo distribution and release in the rat. *Biopharm Drug Dispos*. 1985;6(1):91–104. https://doi.org/10.1002/bdd.2510060111.
50. Meade VM, Burton MA, Gray BN, Self GW. Distribution of different sized microspheres in experimental hepatic tumours. *Eur J Cancer Clin Oncol*. 1987. https://doi.org/10.1016/0277-5379(87)90416-0. Published online.
51. Wollner I, Knutsen C, Smith P, et al. Effects of hepatic arterial yttrium 90 glass microspheres in dogs. *Cancer*. 1988;61(7):1336–1344. https://doi.org/10.1002/1097-0142(19880401)61:7<1336::aid-cncr2820610711>3.0.co;2-k.
52. Herba MJ, Illescas FF, Thirlwell MP, et al. Hepatic malignancies: improved treatment with intraarterial Y-90. *Radiology*. 1988;169(2):311–314. https://doi.org/10.1148/radiology.169.2.3174978.
53. Mohamed MR, Katz AW, Tejani MA, et al. Comparison of outcomes between SBRT, yttrium-90 radioembolization, transarterial chemoembolization, and radiofrequency ablation as bridge to transplant for hepatocellular carcinoma. *Adv Radiat Oncol*. 2015. https://doi.org/10.1016/j.adro.2015.12.003. Published online.
54. Yu SCH, Hui JWY, Li L, et al. Comparison of chemoembolization, radioembolization, and transarterial ethanol ablation for huge hepatocellular carcinoma ($\geq$ 10 cm) in tumour response and long-term survival outcome. *Cardiovasc Intervent Radiol*. October 3, 2021. https://doi.org/10.1007/s00270-021-02777-6. Published online.
55. Gade TPF, Hunt SJ, Harrison N, et al. Segmental transarterial embolization in a translational rat model of hepatocellular carcinoma. *J Vasc Intervent Radiol*. 2015;26(8):1229–1237. https://doi.org/10.1016/j.jvir.2015.02.006.
56. Llovet JM, Kelley RK, Villanueva A, et al. Hepatocellular carcinoma. *Nat Rev Dis Prim*. 2021;7(1):6. https://doi.org/10.1038/s41572-020-00240-3.
57. Facciorusso A, Bargellini I, Cela M, Cincione I, Sacco R. Comparison between Y90 radioembolization plus sorafenib and Y90 radioembolization alone in the treatment of hepatocellular carcinoma: a propensity score analysis. *Cancers*. 2020;12(4):897. https://doi.org/10.3390/cancers12040897.
58. Salem R, Guy E, Johnson E Kim, et al. Yttrium 90 radioembolization for the treatment of solitary unresectable hepatocellular carcinoma the legacy study. *Hepatology*. 2021. https://doi.org/10.1002/hep.31819. Published online.
59. Josep M. Llovet, riccardo lencioni. mRECIST for HCC: performance and novel refinements. *J Hepatol*. 2020. https://doi.org/10.1016/j.jhep.2019.09.026. Published online.
60. Cucchetti A, Serenari M, Sposito C, et al. Including mRECIST in the Metroticket 2.0 criteria improves prediction of hepatocellular carcinoma-related death after liver transplant. *J Hepatol*. 2020;73(2):342–348. https://doi.org/10.1016/j.jhep.2020.03.018.
61. Seymour L, Bogaerts J, Perrone A, et al. iRECIST: guidelines for response criteria for use in trials testing immunotherapeutics. *Lancet Oncol*. 2017;18(3):e143–e152. https://doi.org/10.1016/S1470-2045(17)30074-8.
62. Vouche M, Kulik L, Atassi R, et al. Radiological-pathological analysis of WHO, RECIST, EASL, mRECIST and DWI: imaging analysis from a prospective randomized trial of Y90 $\pm$ sorafenib. *Hepatology*. 2013. https://doi.org/10.1002/hep.26487. Published online.
63. The ARRIVE guidelines 2.0. ARRIVE Guidelines. Accessed February 1, 2022. https://arriveguidelines.org/arrive-guidelines.
64. National Research Council (U.S. Institute for laboratory animal research (U.S.), national academies press (U.S.). In: *Guide for the Care and Use of Laboratory Animals*. 8th ed. National Academies Press; 2011.
65. McCulloch P. Developing appropriate methodology for the study of surgical techniques. *J R Soc Med*. 2009;102(2):51–55. https://doi.org/10.1258/jrsm.2008.080308. PMID: 19208868; PMCID: PMC2642873.
66. Geiger D, Napoli A, Conchiglia A, Gregori L, et al. MR-guided focused ultrasound (MRgFUS) ablation for the treatment of nonspinal osteoid osteoma. *J Bone Jt Surg*. 2014;96(9):743–751. https://doi.org/10.2106/JBJS.M.00903.
67. Glasser SP, Ashton CM, Wray NP. The role of the USA food and drug administration in clinical research. In: Glasser S, ed. *Essentials of Clinical Research*. Cham: Springer; 2014. https://doi.org/10.1007/978-3-319-05470-4_6.
68. Ng TY, Ngu SF, Kam TYD, et al. First in-human trial and prospective case series of an articulated laparoscopic camera system in minimally invasive surgery in gynecology: an IDEAL stage 1 and 2a study. *BMJ Surg Interv Health Technol*. 2022;4:e000117. https://doi.org/10.1136/bmjsit-2021-000117.

Chapter 17

Behavioral interventions

Conner D. Reynolds[1,2], Jonas Kruse[3] and Gregory J. Woodhead[2]
[1]Transitional Year Residency, John Peter Smith Health Network, Fort Worth, TX, United States; [2]Department of Medical Imaging, University of Arizona College of Medicine, Tucson, AZ, United States; [3]School of Medicine, Texas Christian University, Fort Worth, TX, United States

Summary

- Translating biomedical discoveries into clinical environments requires investigators to address and modify behavioral barriers to full implementation.
- This chapter outlines a four-step process for developing, designing, and testing behavioral interventions.
- This process involves identifying context-specific problem behavior, defining the mechanistic drivers of problem behavior using theoretical frameworks, collaborating with stakeholders to design interventions that work for them, and evaluating the effectiveness of these interventions in the clinical environments.

Introduction

In 1985, Drs. Michael S. Brown and Joseph L. Goldstein were jointly awarded the Nobel Prize in Physiology or Medicine for elucidating the role of low-density lipoproteins (LDL) in regulating cholesterol metabolism.[1] This triggered a flurry of clinical investigations demonstrating the safety and efficacy of β-Hydroxy β-methylglutaryl-CoA (HMG-CoA) reductase inhibitors ("statins") in cardiovascular disease prevention.[2–4] It appeared to be a true bench-to-bedside success story. The Nobel Laureates predicted the rise of statin medications would lead to the "*end of heart attacks*" and eliminate "*coronary artery disease as a major public health problem*".[5] The First International Heart Health Conference's Advisory Board declared statins' success a heralding sign for a future where heart disease and stroke are rarities.[6] However, over 30 years later, the prevalence of cardiovascular disease continues to rise. Heart disease remains the leading cause of death in the United States.[7] Ischemic heart disease and stokes remain the leading cause of death worldwide, accounting for approximately 27% of lives lost annually.[8]

Why were statins unable to deliver on the scientific community's confident predictions? According to Greenland and Lloyd—Jones, the root cause stems from a failure in fully translating biomedical science into clinical practice.[9] While clinical trials showed significant reductions in plasma LDL, vascular events, morbidity, and mortality,[2–4] such highly controlled processes for administrating medications do not reflect real-world conditions. A recent meta-analysis of 376,162 patients revealed poor adherence to any medical therapy for cardiovascular disease.[10] Among patients who have had at least one heart attack, nearly 33% did not adhere to effective, evidence-based treatments. This effect approaches 50% among patients without previous heart attacks.[10] Differences in patient behavior necessarily underscore the mismatch between outcomes of clinical trials and clinical practice.

Many health behaviors, such as smoking, activity level, diet, and adherence to medical therapy, have essential long-term consequences. The considerable variation in these behaviors is informed by patients' value systems, beliefs, and overall understanding of health consequences. Behavioral interventions in medicine methodically identify and affect patients' voluntary choices concerning their health.[11] This objective is distinct from medical interventions, where changes in physiologic markers define primary endpoints. It allows for the versatile implementation of behavioral interventions, ranging from the individual to the entire organization. Behavioral interventions targeting lifestyle modifications such as physical activity, smoking cessation, weight loss, and dietary choices improve clinical outcomes in clear, cost-effective ways.[12–19] Thus, designing research that assesses patient health behaviors will synergize with biomedical research efforts and enhance translational medicine.

Planning an effective behavioral intervention requires the same scientific rigor as biomedical research. Previous work from psychological and social sciences outlined the key elements of behavioral intervention design. Accomplishing each of these steps necessitates detailed planning,

Translational Interventional Radiology. https://doi.org/10.1016/B978-0-12-823026-8.00089-4

interdisciplinary collaboration, pilot testing, and gradual refinement throughout intervention development.

There are many issues in the field of interventional radiology that behavioral interventions could improve.[20–23] Previous work from psychology and sociology has outlined key elements in behavioral intervention design.[24,25] The following section will discuss critical steps for behavioral intervention design. These will be contextualized using a common behavioral problem among interventional radiologists—underreporting of needlestick injuries.[21]

Step 1: Define the problem and where it is happening

Solving any problem begins with identifying the components that define it. Exploratory and epidemiologic research are critical for determining these types of issues in the early stages. This can be accomplished through careful literature review or analysis of publicly available healthcare data. Another important consideration is deciding the most important level at which to intervene. In a recent study, Deipolyi et al. found that interventional radiologists tend to underreport needlestick injuries that occur during their cases. For this problem, an obvious interventional target would be the radiologists. Investigators could choose to intervene at the individual level by educating on the importance of reporting and receiving appropriate treatment. An alternate choice might involve intervening at the unit level by revising policies and promoting a culture of psychological safety that encourages reporting practices. For complex and pervasive problems, the solution may require multiple concurrent interventions to achieve resolution.[25] Defining problems and where they exist can be precarious and time-consuming for investigators but will yield dividends in future development stages.

Step 2: Understand the drivers of problem behavior

Mechanisms underlying problem behavior are foundational to the entire behavioral intervention design process. Many interventions remain unsuccessful because investigators do not fully understand the primary drivers of problem behavior.[26] Therefore, elucidating these mechanisms precisely and accurately is essential for engineering effective interventions. However, it is easier said than done, as individual variability makes exact behavioral drivers notoriously tricky to pin down. Extensive work in psychology and sociology has generated theoretical frameworks for understanding the variation in health behaviors. The most well-developed include the Health Belief Model, Theory of Planned Behavior, Social Cognitive Theory, and Transtheoretical Model.[27–32] Each of these frameworks has strengths, weaknesses, and limitations, which may not be suitable for all use cases. Investigators should balance these factors when selecting one or more appropriate frameworks to fit their problem. Additional considerations should include evaluating whether individuals possess sufficient knowledge, motivation, and skills to perform the new target behavior. Fisher's informational-motivation-behavioral skills model has been leveraged for many successful health behavior changes.[25,33–35]

Once these components have been meticulously appraised and unified into a cohesive mechanism, intervention design can begin. When surveyed about reporting needlestick injuries, investigators found that interventional radiologists perceived reporting systems as minimally useful (28%) and too lengthy (13%).[21] Defining the mechanism would require diving deeper into negative perceptions of the reporting system. Perhaps the systems utilize excessive screens, wording, checkboxes, and drop-down lists, leading to click fatigue.[36] Perhaps physicians receive minimal-to-no timely feedback from administration following submission, reducing their motivation to report future incidents.[37] Perhaps there is significant shame and a poor culture of psychological safety, which would solicit a punitive response for their mistake. Collaborative focus groups of unit, department, or hospital leadership representatives are a robust tool for generating highly accurate mechanisms.

Step 3: Design interventions with stakeholders in mind

The primary goal of behavioral interventions is to modify individuals' behavior in the natural environment. Their purpose should be simple and easily understood by the target population. If it is complicated or vague, individuals may not understand the problem behavior or how it should be changed. Interventions should provide solutions that are acceptable and achievable with high fidelity by the target population. Those that are significantly disruptive to workflow, time consuming, expensive, or incompatible with organizational goals are likely to be rejected.[25] Optimally addressing the intervention's comprehensibility, acceptability, and achievability requires collaborative input from stakeholders.

Interdisciplinary stakeholders are vital to the design process because they can reveal where things are not working. Tackling the problem of underreporting needlesticks should include interventional radiologists, other members of the procedure team (e.g., radiological technicians, residents, nurses, physician assistants, medical assistants), reporting system design team, and administrative personnel. Investigators should propose their interventions to these stakeholders and solicit feedback to refine interventions. Suppose radiologists find the reporting system to be minimally useful because it requires 30-min to complete. In that case, investigators may focus on simplifying incident reporting systems and streamlining administrative feedback rather than additional education or penalties.

Step 4: Assess feasibility and effectiveness in the real world

Once the intervention has been fine-tuned and potential failure modes assessed, it can be evaluated in the clinical setting. This process begins with a feasibility study to determine if an intervention can be pragmatically implemented with a small sample from the target population. Feasibility studies reveal design strengths and weaknesses that investigators must address before large-scale implementation. When combined with well-established improvement methodologies (e.g., Plan-Do-Study-Act (PDSA) Cycling, LEAN, Six Sigma, RE-AIM), these data generate invaluable insights and rapidly refine intervention design.[38–41]

The gold standard for evaluating efficacy and effectiveness involves designing a randomized clinical trial. Some individuals or organizational units should be randomly assigned to receive interventions, while others remain under baseline conditions. For early trials, primary endpoints are concerned with whether behaviors changed immediately following the intervention. In our example, if investigators created a new needlestick injury reporting and feedback system within a large hospital network, they could attempt implementation at several randomly selected radiology departments to evaluate whether reporting rates increased relative to others. However, with multifaceted or resource-intensive interventions, randomization is not always possible for clinical settings. In these cases, investigators could consider alternative evaluation methods such as matched sample trials, before and after studies, case series, and case studies. Previous work details the appropriate use cases and the effectiveness of these types of evaluation methods.[25,42]

After progressively upscaled iterations of the intervention, trial endpoints should become more comprehensive and longer range. These endpoints should harken back to Step 2 by determining the intervention's effect on each step of the mechanism driving problem behaviors. This recursive process will affirm whether the intervention works as initially intended, allowing further revision and refinement. Secondary clinical outcomes resulting from changes in behavior, such as satisfaction, infection rates, or mortality, should also be measured. Even if an intervention successfully changes problem behaviors, it may not be fully implemented if it fails to produce outcomes that are meaningful to the organization.

Investigators should also give special consideration to the implementation process, detailed in the Implementation Research section of this book. Understanding how an intervention works, which context it is most effective in, and whether it produces sustainable, meaningful change is critical for long-term success.

Conclusion

Developing high-fidelity behavioral interventions in healthcare can be notoriously tricky but vital for researchers aiming for full bench-to-bedside translatability. This chapter has provided a four-step process for developing, designing, and testing behavioral interventions. This process involves identifying context-specific problem behavior, defining the mechanistic drivers of problem behavior using theoretical frameworks, collaborating with stakeholders to design interventions that work for them, and evaluating the effectiveness of these interventions in the clinical environments. Investigators beginning this process should start by:

1. Defining problem behaviors and deciding whether an intervention is needed before investing time and other resources
2. Talking with key stakeholders across disciplines to understand their conception of the problem behaviors
3. Cross-referencing stakeholder feedback with one or more theoretical frameworks of health behavior to define the mechanisms driving problem behaviors
4. Designing behavioral interventions that address each of these mechanistic targets
5. Soliciting stakeholder feedback on intervention designs to ensure comprehensibility, acceptability, and achievability by the target population
6. Launching small scale pilot studies to evaluate whether interventions feasibly modify problem behaviors in clinical environments
7. Employing improvement methodologies (e.g., PDSA Cycling, LEAN, Six Sigma, RE-AIM) to rapidly refine and optimize the interventional design for implementation in clinical environments
8. Implementing the intervention with progressively larger groups to understand how it works, where it is most effective, and whether it produces sustainable, meaningful change is critical for long-term success

Potential pitfalls

1. If the behavioral intervention is not producing expected results, investigators are likely missing the principal drivers of problem behavior. Investigators should stop and reevaluate their proposed mechanisms and modify the intervention accordingly.

2. Behavioral interventions that are not understood, accepted, and reasonably achievable by the target population are destined to fail.
3. Even if a behavioral intervention successfully changes problem behaviors, it may not be fully implemented if it fails to produce outcomes that are meaningful to the organization.

References

1. *The Nobel Prize in Physiology or Medicine*. Nobel Media AB; 1985. https://www.nobelprize.org/prizes/medicine/1985/summary/.
2. Scandinavian Simvastatin Survival Study G. Randomised trial of cholesterol lowering in 4444 patients with coronary heart disease: the Scandinavian Simvastatin Survival Study (4S). *Lancet*. 1994;344(8934):1383–1389. https://doi.org/10.1016/S0140-6736(94)90566-5.
3. Sacks FM, Pfeffer MA, Moye LA, et al. The effect of pravastatin on coronary events after myocardial infarction in patients with average cholesterol levels. *N Engl J Med*. 1996;335(14):1001–1009. https://doi.org/10.1056/NEJM199610033351401.
4. Shepherd J, Cobbe SM, Ford I, et al. Prevention of coronary heart disease with pravastatin in men with hypercholesterolemia. *N Engl J Med*. 1995;333(20):1301–1308. https://doi.org/10.1056/NEJM199511163332001.
5. Brown MS, Goldstein JL. Heart attacks: gone with the century? *Science*. 1996;272(5262):629. https://doi.org/10.1126/science.272.5262.629.
6. Advisory Board of the International Heart Health C. Victoria declaration on heart health. Déclaration de Victoria sur la santé du coeur. *Can Med Assoc J*. 1992;147(12):1794–1797.
7. CDC. Leading Causes of Death. https://www.cdc.gov/nchs/fastats/leading-causes-of-death.htm.
8. WHO. The Top 10 Causes of Death. World Health Organization.who.int/news-room/fact-sheets/detail/the-top-10-causes-of-death.
9. Greenland P, Lloyd-Jones D. Time to end the mixed—and often incorrect—messages about prevention and treatment of atherosclerotic cardiovascular Disease Editorials published in the journal of the american college of cardiologyreflect the views of the authors and do not necessarily represent the views of JACCor the American college of cardiology. *J Am Coll Cardiol*. 2007;50(22):2133–2135. https://doi.org/10.1016/j.jacc.2007.05.055.
10. Naderi SH, Bestwick JP, Wald DS. Adherence to drugs that prevent cardiovascular disease: meta-analysis on 376,162 patients. *Am J Med*. September 2012;125(9):882–887. https://doi.org/10.1016/j.amjmed.2011.12.013. e1.
11. Cutler DM. Behavioral health interventions: what works and why? In: Anderson NB, Bulatao RA, Cohen B, eds. *Critical Perspectives on Racial and Ethnic Differences in Health in Late Life*. National Academies Press; 2004.
12. Knowler WC, Barrett-Connor E, Fowler SE, et al. Reduction in the incidence of type 2 diabetes with lifestyle intervention or metformin. *N Engl J Med*. 2002;346(6):393–403. https://doi.org/10.1056/NEJMoa012512.
13. Herman WH, Hoerger TJ, Brandle M, et al. The cost-effectiveness of lifestyle modification or metformin in preventing type 2 diabetes in adults with impaired glucose tolerance. *Ann Intern Med*. 2005;142(5):323–332. https://doi.org/10.7326/0003-4819-142-5-200503010-00007.
14. Writing Group of the PCRG. Effects of comprehensive lifestyle modification on blood pressure control main results of the premier clinical trial. *JAMA*. 2003;289(16):2083–2093. https://doi.org/10.1001/jama.289.16.2083.
15. McGuire HL, Svetkey LP, Harsha DW, Elmer PJ, Appel LJ, Ard JD. Comprehensive lifestyle modification and blood pressure control: a review of the premier trial. https://doi.org/10.1111/j.1524-6175.2004.03147.x. *J Clin Hypertens*. 2004;6(7):383–390. https://doi.org/10.1111/j.1524-6175.2004.03147.x.
16. Stevens VJ, Obarzanek E, Cook NR, et al. Long-term weight loss and changes in blood pressure: results of the trials of hypertension prevention, phase II. *Ann Intern Med*. 2001;134(1):1–11. https://doi.org/10.7326/0003-4819-134-1-200101020-00007.
17. Whelton PK, Appel L, Charleston J, et al. The effects of nonpharmacologic interventions on blood pressure of persons with high normal levels: results of the trials of hypertension prevention, phase I. *JAMA*. 1992;267(9):1213–1220. https://doi.org/10.1001/jama.1992.03480090061028.
18. Fu J, Liu Y, Zhang L, et al. Nonpharmacologic interventions for reducing blood pressure in adults with prehypertension to established hypertension. *J Am Heart Assoc*. 2020;9(19):e016804. https://doi.org/10.1161/JAHA.120.016804.
19. Wexler R, Aukerman G. Nonpharmacologic strategies for managing hypertension. *Am Fam Phys*. June 1, 2006;73(11):1953–1956.
20. Rosenkrantz AB, Prologo JD, Wang W, Hughes DR, Bercu ZL, Duszak R. Opioid prescribing behavior of interventional radiologists across the United States. *J Am Coll Radiol*. 2018;15(5):726–733. https://doi.org/10.1016/j.jacr.2018.01.003.
21. Deipolyi AR, Prabhakar AM, Naidu S, Oklu R. Needlestick injuries in interventional radiology are common and underreported. *Radiology*. 2017;285(3):870–875. https://doi.org/10.1148/radiol.2017170103.
22. Makary MS, da Silva A, Kingsbury J, Bozer J, Dowell JD, Nguyen XV. Noninvasive approaches for anxiety reduction during interventional radiology procedures. *Top Magn Reson Imag*. 2020;29(4).
23. DeJesus L, Echenique AM, Absich A. Importance of postoperative care and discharge instructions in patients undergoing a uterine fibroid embolization. *J Radiol Nurs*. 2016;35(4):323–325. https://doi.org/10.1016/j.jradnu.2016.10.003.
24. Eldredge LKB, Markham CM, Ruiter RAC, Fernandez ME, Kok G, Parcel GS. *Planning Health Promotion Programs: An Intervention Mapping Approach*. 4 ed. Jossey-Bass: A Wiley Brand; 2016.
25. Abraham C, Denford S. Design, implementation, and evaluation of behavior change interventions: a ten-task guide. In: Hagger MS, Cameron LD, Hamilton K, Hankonen N, Lintunen T, eds. *The Handbook of Behavior Change*. Cambridge University Press; 2020. chap. 19.
26. Kelly MP, Barker M. Why is changing health-related behaviour so difficult? *Public health*. 2016;136:109–116. https://doi.org/10.1016/j.puhe.2016.03.030.
27. Green EC, Murphy EM, Gryboski K. The health belief model. In: *The Wiley Encyclopedia of Health Psychology*. Wiley Online Books; 2020:211–214. https://doi.org/10.1002/9781119057840.ch68.
28. Godin G, Kok G. The theory of planned behavior: a review of its applications to health-related behaviors. *Am J Health Promot*. 1996;11(2):87–98. https://doi.org/10.4278/0890-1171-11.2.87.

29. Godin G, Bélanger-Gravel A, Eccles M, Grimshaw J. Healthcare professionals' intentions and behaviours: a systematic review of studies based on social cognitive theories. *Implement Sci.* 2008;3(1):36. https://doi.org/10.1186/1748-5908-3-36.
30. Bandura A. Social foundations of thought and action: a social cognitive theory. In: *Social Foundations of Thought and Action: A Social Cognitive Theory*. Prentice-Hall, Inc; 1986. xiii, 617-xiii, 617.
31. Bandura A. Health promotion from the perspective of social cognitive theory. *Psychol Health.* 1998;13(4):623−649. https://doi.org/10.1080/08870449808407422.
32. Cole TK. Smoking cessation in the hospitalized patient using the transtheoretical model of behavior change. *Heart Lung.* 2001;30(2):148−158. https://doi.org/10.1067/mhl.2001.111249.
33. Fisher JD, Fisher WA. Changing AIDS-risk behavior. *Psychol Bull.* 1992;111(3):455−474. https://doi.org/10.1037/0033-2909.111.3.455.
34. Rongkavilit C, Naar-King S, Kaljee LM, et al. Applying the information-motivation-behavioral skills model in medication adherence among Thai youth living with HIV: a qualitative study. *AIDS Patient Care STDS.* 2010;24(12):787−794. https://doi.org/10.1089/apc.2010.0069.
35. Chang SJ, Choi S, Kim S-A, Song M. Intervention strategies based on information-motivation-behavioral skills model for health behavior change: a systematic review. *Asian Nurs Res.* 2014;8(3):172−181. https://doi.org/10.1016/j.anr.2014.08.002.
36. Collier R. Rethinking EHR interfaces to reduce click fatigue and physician burnout. *Can Med Assoc J.* 2018;190(33):E994−E995. https://doi.org/10.1503/cmaj.109-5644.
37. Benn J, Koutantji M, Wallace L, et al. Feedback from incident reporting: information and action to improve patient safety. *Qual Health Care.* 2009;18(1):11. https://doi.org/10.1136/qshc.2007.024166.
38. Donnelly P, Kirk P. Use the PDSA model for effective change management. *Educ Prim Care.* 2015;26(4):279−281. https://doi.org/10.1080/14739879.2015.11494356.
39. Mazzocato P, Savage C, Brommels M, Aronsson H, Thor J. Lean thinking in healthcare: a realist review of the literature. *Qual Health Care.* 2010;19(5):376. https://doi.org/10.1136/qshc.2009.037986.
40. de Koning H, Verver JPS, van den Heuvel J, Bisgaard S, Does RJMM. Lean Six Sigma in healthcare. *J Healthc Qual.* 2006;28(2):4−11. https://doi.org/10.1111/j.1945-1474.2006.tb00596.x.
41. King DK, Shoup JA, Raebel MA, et al. Planning for implementation success using RE-AIM and CFIR frameworks: a qualitative study. Original research. *Front Public Health.* 2020;8(59). https://doi.org/10.3389/fpubh.2020.00059.
42. Evans D. Hierarchy of evidence: a framework for ranking evidence evaluating healthcare interventions. *J Clin Nurs.* 2003;12(1):77−84. https://doi.org/10.1046/j.1365-2702.2003.00662.x.

Part III

Clinical: fundamentals

Chapter 18

Introduction to clinical research: What is it? Why is it needed?

Shashidhara Murthy[1] and Sandeep Murthy[2]

[1]Department of Interventional Radiology, Icahn School of Medicine at Mount Sinai Hospital, New York, NY, United States;
[2]University of Virginia School of Medicine, Charlottesville, VA, United States

Key points

- Clinical research involves the understanding and testing of methods to improve human health and combat illness.
- The success of clinical research has directly improved every avenue of care including prevention, detection, diagnosis, and treatment.
- Clinical research typically involves many overseeing agencies such as the Food and Drug Administration (FDA) and the involvement of an Institutional Review Board (IRB).

What is it and why it matters?

The history of clinical research may date back to biblical times but commonly starts with Dr. James Lind. Dr. Lind was a surgeon who performed some of the first recognizable clinical trials in 1747 in efforts to combat scurvy among sailors. We have come a long way in the centuries since, with the introduction of placebos in the 1800s and the first randomized control trial studying streptomycin occurring in 1946.[2] The majority of medications and treatments we benefit from today are a result of clinical research.

In this chapter, we will define clinical research and explore the impact it has had on human health. Clinical research is a field of study that attempts to understand health and illness through a variety of different methods that consist of human participants. For example, clinical studies have shown the effectiveness of vascular closure devices in decreasing time to hemostasis almost 10-fold as well as highlighting the safety of using such devices.[1] Clinical research is the foundation for *evidence-based medicine* that continues to improve patient care. It is undisputed that clinical research has made a profound impact on clinical practice by advancing the prevention, detection, diagnosis, and treatment of human disease. Many clinical research studies are driven by discoveries in the basic sciences that are then translated to medical applications. Depending on the goal of the study, the research can either be observational or interventional.

Observational studies

Observational studies only involve monitoring a patient's health without any interventions. Data collection methods in observational studies can include bloodwork, questionnaires, and *health monitoring* (Fig. 18.1).[3]

Observational studies can be further classified based on the study characteristics including a timeline of data collection, distribution of participants, and method of data collection. A few examples of observational studies include the below designs, which we will go into more detail about in the following chapters.

- Case-control study: Case-control studies compare a group of patients with the disease to a group without the disease. These types of studies can be useful for discovering risk factors and the impact of social determinants of health.[4]
- Case series: Case series, or case reports, are descriptive studies that analyze a homogenous group of patients without a comparison group.[3]
- Cross-sectional study: Cross-sectional studies are focused on data that was collected at a single point in time.[4] A major limitation of this study is that many times, the single point is not representative of the full picture.
- Cohort study: Cohort studies follow a homogenous group of people through a certain period of time and perform cross-sectional data collection at specific intervals. These types of studies require a long-term commitment from participating subjects.[4]

Translational Interventional Radiology. https://doi.org/10.1016/B978-0-12-823026-8.00036-5

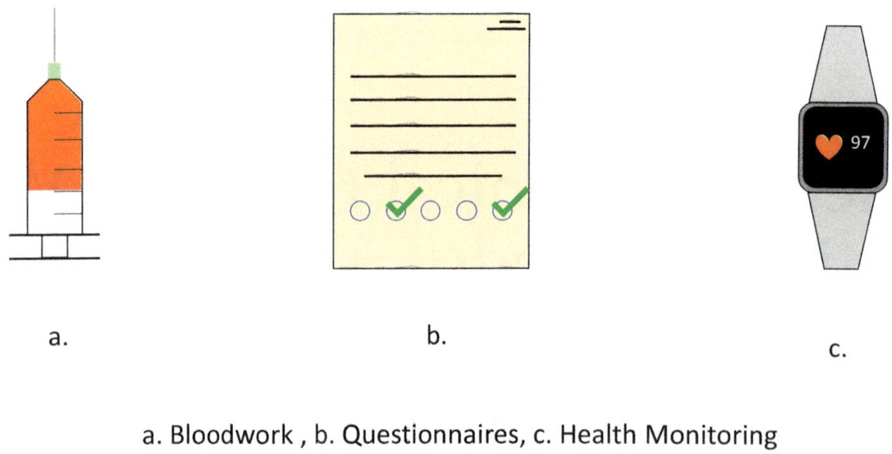

FIGURE 18.1 Examples of data collection in observational studies.

Interventional studies

Interventional studies are when a new variable, often a drug, is introduced and compared to a placebo or the current standard of care.[5] A real-life example of an interventional study includes "Prone Positioning in Severe Acute Respiratory Distress Syndrome," which was a randomized control trial published in the New England Journal of Medicine. This study yielded results highlighting an increased survival rate from prolonged prone positioning for patients experiencing severe acute respiratory distress.[6] Three notable *study designs* we will discuss in later chapters are introduced below.

- Randomized control trial: Randomized control trials take a group of participants and separate them into two groups. The goal of the random separation is to control for confounding variables and limit bias.[7]
- Nonrandomized trial: Nonrandomized trials, also known as single-arm trials, are study designs where a group that received an intervention is compared to a group where no intervention occurred. As the name implies, there is no randomization in the delivery of the intervention. The lack of randomization makes this study design open to bias and the results not as credible.[8]
- Crossover study: A crossover is a study design where participants "crossover" to the other treatment during the study. The benefit of such a design is to compare responses in individual subjects as each subject will receive all treatment options.[9]

Who is involved in clinical research?

There are two distinct groups involved in a clinical study, the research team conducting the research and the human subjects participating in the study. Human subjects are often compensated for their time and participation in a study. Subject qualifications for participation in the study will be clearly established in the study protocol by both inclusion and exclusion criteria. Clinical studies typically require both healthy and diseased patients to ensure credible results, for example, a study using Positron Emission Tomography imaging to look at histone deacetylase dysregulation in Schizophrenia also includes control patients that have not been diagnosed with Schizophrenia.[10]

Research teams consist of a principal investigator, often a clinician or clinical scientist, that leads the clinical study. The composition of the remaining team members typically includes clinical fellows, postdoctoral students, graduate students, clinical research coordinators, and/or research technicians. *We will discuss these roles more in-depth in* Chapter 48*: Research Team.*

Regulatory body

The high-risk potential of clinical research constitutes the involvement of governing boards to ensure ethical and accurate research is conducted, two of the most notable oversight organizations are the Food and Drug Administration (FDA) and an *Institutional Review Board.*[11] The study protocol needs to be approved by the relevant regulatory agencies, maintained, and updated if needed. The FDA will be further discussed in Chapter 69*: FDA overview.*

Funding

The cost of developing a new drug and bringing it to market can be in the hundreds of millions of dollars.[12] Funding for clinical research can typically be divided into public and private funds. Public funds include grants for research from government agencies such as the National Institute of Health and National Science Foundation.[13] Private funds include foundations as well as pharmaceutical and medical device companies such as Pfizer, Merck, and

Sanofi to name a few. Many academic principal investigators frequently partner with industry sponsors to oversee and execute their clinical trials.

What are clinical trials?

Before performing clinical trials, an Investigational New Drug (IND) application showing successful results in *preclinical research* must be completed. Once the IND is approved, clinical trials in humans can begin. Clinical trials are a type of interventional study that, over multiple phases or stages, attempts to prove that a drug, device, or biological is both safe to use, as well as effective in its desired outcome. The clinical trials we will discuss are an FDA regulated process to bring the above products to market. The term phases typically describe the process for biologicals and drugs, while the term stages are used for medical devices. Not all medical devices require FDA approval but all nonsupplement pharmaceuticals require FDA approval, especially prescription medication. There are typically four phases of clinical trials (Fig. 18.2). Phase I introduces the drug to a limited number of healthy patients to determine safety and dosage. Phase II continues to focus on safety and begins to look at efficacy by introducing the drug to diseased patients, typically this includes ~100 subjects. Phase III focuses on both the safety and efficacy of the drug by significantly increasing the sample size to the thousands. Phase IV is a postapproval phase that monitors the drug after entering the market for long-term adverse effects.[7] New Drug Applications (NDA) are submitted following Phase III if results prove the drug is both safe for patients and significantly effective in the desired outcome when compared to placebo or the current *standard of care*.

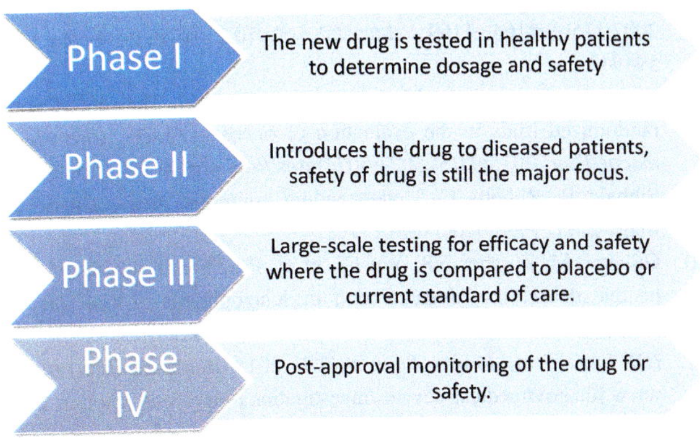

FIGURE 18.2 Clinical trial process for drugs and biologicals.

While there are four phases for drug clinical trials, there are only three stages for medical devices (Fig. 18.3). Not all medical devices require approval by the FDA. Medical devices are broken into three classes (Class I–III), with class I exempt from FDA approval and class II and III

Stage	Study type	Description
I	Exploratory	Commonly referred to as feasibility studies, these studies are done in early stages of development to showcase a basic level of safety and efficacy. This stage is important for continuation into the pivotal studies.
II	Pivotal	Pivotal studies are the most important part of this process, they demonstrate safety and effectiveness in desired outcome. The results of the pivotal study are used to achieve approval.
III	Post-market	When compared to phases of a drug trial, the post-market stage is similar to Phase IV. The device is followed closely after market approval for any sign of adverse effects and results of long term success

Pivotal studies are performed to demonstrate the device is safe and effective for a specific use within a defined patient population. The results of a pivotal study are used to gain regulatory approval to market the device.

FIGURE 18.3 Clinical trial process for medical devices.

requiring some sort of application to the FDA. Another main difference is that while drug clinical trials require participation from healthy volunteers, device clinical trials are only tested in the patient population in question.[14] Healthy subjects are not used as a control group in device clinical trials, rather, a device intervention is compared to standard of care in the control group. Where drugs require an NDA, devices require a Pre-Market Approval (PMA) application that is to be completed following the results of the pivotal study. An example of a pivotal study is the recent SCENT Trial, a nonrandomized medical device trial, conducted at Columbia University and published in Stroke. The study looked at the safety and efficacy of a Surpass Flow Diverter (Surpass; Stryker Neurovascular, Fremont, CA) in intracranial internal carotid arterial aneurysms. The trial results demonstrated the SCENT product was both safe and effective in its desired outcome when compared to historical endpoints.[15] Following the results of this pivotal study, the product can proceed to file a PMA.

Get started

- Visit www.ClinicalTrials.Gov, a website maintained by the US National Library of Medicine, and browse through open clinical trials. Look at the interventions used (if any) and familiarize yourself with the terms used in the study description section.
- There are extenuating circumstances regarding the use of unapproved drugs. Research what some of these reasons might be by visiting www.FDA.Gov and heading to the "Unapproved Drugs" section of the site.

References

1. Wood DA, Krajcer Z, Sathananthan J, et al. Pivotal clinical study to evaluate the safety and effectiveness of the MANTA percutaneous vascular closure device. *Circ Cardiovasc Interv*. 2019;12(7):e007258. https://doi.org/10.1161/circinterventions.119.007258.
2. Bhatt A. Evolution of clinical research: a history before and beyond james lind. *Perspect Clin Res*. 2010;1(1):6–10.
3. *Learn about Clinical Studies*; 2019. Retrieved from https://clinicaltrials.gov/ct2/about-studies/learn#ClinicalTrials.
4. Song JW, Chung KC. Observational studies: cohort and case-control studies. *Plast Reconstr Surg*. 2010;126(6):2234–2242. https://doi.org/10.1097/PRS.0b013e3181f44abc.
5. *Glossary of Common Terms*; 2020. Retrieved from https://clinicaltrials.gov/ct2/about-studies/glossary.
6. Guérin C, Reignier J, Richard JC, et al. Prone positioning in severe acute respiratory distress syndrome. *N Engl J Med*. 2013;368(23):2159–2168. https://doi.org/10.1056/NEJMoa1214103.
7. Stanley K. Design of randomized controlled trials. *Circulation*. 2007;115(9):1164–1169. https://doi.org/10.1161/circulationaha.105.594945.
8. Simon R, Blumenthal GM, Rothenberg ML, et al. The role of nonrandomized trials in the evaluation of oncology drugs. *Clin Pharmacol Ther*. 2015;97(5):502–507. https://doi.org/10.1002/cpt.86.
9. Sibbald B, Roberts C. Understanding controlled trials. Crossover trials. *BMJ*. 1998;316(7146):1719.
10. Gilbert TM, Zürcher NR, Wu CJ, et al. PET neuroimaging reveals histone deacetylase dysregulation in schizophrenia. *J Clin Invest*. 2019;129(1):364–372. https://doi.org/10.1172/jci123743.
11. *IDE Institutional Review Boards (IRB)*; 2018. Retrieved from https://www.fda.gov/medical-devices/investigational-device-exemption-ide/ide-institutional-review-boards-irb.
12. Morgan S, Grootendorst P, Lexchin J, Cunningham C, Greyson D. The cost of drug development: a systematic review. *Health Pol*. 2011;100(1):4–17. https://doi.org/10.1016/j.healthpol.2010.12.002.
13. *Grants & Funding*; 2019. Retrieved from https://grants.nih.gov/funding/index.htm.
14. *Overview of Medical Device Classification and Reclassification*; 2017. Retrieved from https://www.fda.gov/about-fda/cdrh-transparency/overview-medical-device-classification-and-reclassification.
15. Meyers PM, Coon AL, Kan PT, Wakhloo AK, Hanel RA. SCENT trial. *Stroke*. 2019;50(6):1473–1479. https://doi.org/10.1161/strokeaha.118.024135.

Chapter 19

The question: Types of research questions and how to develop them

Zain Badar[1] **and Joao Amaral**[2]

[1]*Interventional Radiology Department, University Health Network, University of Toronto, Toronto, ON, Canada;* [2]*Division of Interventional Radiology, Diagnostic Imaging Department, The Hospital for Sick Children, Toronto, ON, Canada*

- The type of research question to ask depends on whether quantitative or qualitative research is being conducted
- A research question should be *clear*, *focused*, *concise*, *complex*, and *arguable*
- Open-ended questions that consider the target audience is crucial in the development of a research question

Research questions play a crucial role in establishing a basic platform before developing a translational research study. Irrespective of whether it is a two-stage, four-stage, or five-stage scheme, the initial questions are utilized as a brainstorming technique to launch a platform.[1] The purpose behind the development of a research question is to provide a path for the writer to focus, enabling him or her to make a clear and concise point in their research.[2]

There are seven key types of research questions, which depend on whether the writer is conducting qualitative versus quantitative research. The type of question the writer asks depends on what they need to know to validate their research[2]

Real-world example

1. Descriptive Research—these questions are asked with the objective of attaining more quantitative information on a topic and are typically "how" or "why" questions. For example: what are the characteristics of a reverse curve catheter?
2. Comparative Research—these questions compare two or more variables or attributes, for example: what are the differences and similarities between a C1 and Simmons 1 catheter?
3. Correlational Research—these questions focus on the statistical relationship between two or more variables or attributes, for example: what is the relationship between microwave ablation and transarterial chemoembolization for the treatment of hepatocellular carcinoma?
4. Exploratory Research—while similar to descriptive questions, exploratory questions are designed to investigate and understand more qualitative information on a topic. For example: what are the main factors in the treatment of multifocal hepatocellular carcinoma? What is the role of transarterial radioembolization in multifocal hepatocellular carcinoma?
5. Explanatory Research—these questions are asked with the objective of interpreting relationships between variables, for example: does biliary stenting have an effect on long-term patency of malignant biliary strictures? What is the impact of malignant biliary strictures on morbidity? What are the indications of biliary stenting?
6. Evaluation Research—these questions aim to help the writer understand the audience's assessment of one or more variables, for example: what are the advantages and disadvantages of utilizing a balloon occlusion catheter for transarterial chemoembolization? How well do doxorubicin-eluting beads work with the use of a balloon occlusion catheter? How effective is the outcome in terms of recurrent or residual disease for doxorubicin-coated beads deployed via a balloon occlusion catheter in transarterial chemoembolization?
7. Action Research—these questions help the writer understand actions that need to be taken to achieve an outcome, for example: how can treatment of venolymphatic malformations be achieved? What are the most effective strategies to improve sclerotherapy?

Once the writer has selected the type of research question to ask, the next step is to develop a strong

Translational Interventional Radiology. https://doi.org/10.1016/B978-0-12-823026-8.00074-2

FIGURE 19.1 Fundamentals of developing a research question.

question. The process to develop and propose the most pertinent research questions is initiated with the fundamentals as described in Fig. 19.1.

Get started

Development of a research question not only depends on quantitative vs qualitative research but also depends on the target audience. Having a clear, focused, concise, complex and arguable research question can aid in establishing the basic platform prior to development of a research study.

Potential pitfalls

In step 1, preliminary research is conducted to identify any and all literature that has been published in this area. All literature is reviewed to provide the writer with a better understanding of the topic, and more importantly, to identify any gaps or issues that have yet to be addressed—this will form the crux of the research; the focus of the study.[2,3]

Step 2 is to consider the target audience in developing the research question. Specifically, the writer will need to consider what type of question would captivate and draw their target audience in and maintain their engagement throughout the paper.[2,3]

Step 3 revolves around drafting open-ended questions as part of one or more brainstorming exercises. In general, "how" and "why" questions should be favored over simple "yes" or "no" questions, as these would spark interest, thought, and conversation.

Once the initial open-ended questions have been drafted, the writer finalizes their research question, evaluating it against the criteria for a strong research question: Clear, Focused, Concise, Complex, and Arguable.[2,3]

Clarity aids the writer in directing his or her research, avoiding generalizations around the topic. A clear question enables the audience to easily understand the specifics of the research paper and its value proposition. A clear question is crucial especially in instances where other literature on the topic is widely available, setting the writers' research apart.[2,3]

Focus looks at getting to the heart of the topic behind the research paper. Like a clear question, a focused one reassures the audience that an answer is provided within the words of the paper.[2,3]

Concise questions are written in one short sentence and not long-drawn-out, causing the writer to lose the readers' interest.[2,3]

Complex questions are those that are answered and addressed by in-depth analysis and research, rather than "yes" or "no" statements or facts. A complex question, when written to its target audience, will attract them toward the research paper.[2,3]

Arguable questions leave the door open to debate; some of the best research papers include findings and analyses that are not factually conclusive and can be countered by others.[2,3]

Resources

Additional articles:

1. National Institutes of Health, Office of Portfolio Analysis and Strategic Initiatives. NIH Roadmap for Medical Research. Available at: http://nihroadmap.nih.gov/index.asp. Accessed October 15, 2008.
2. National Institutes of Health. RFA-RM-06-002: Institutional Clinical and Translational Science Award. Available at: http://grants.nih.gov/grants/rfa-files/RFA-RM-06-002.html. Accessed October 15, 2008.
3. National Institutes of Health. NIH Rapid Access to Interventional Development (NIH-RAID Pilot). Available at: http://nihroadmap.nih.gov/raid/. Accessed October 15, 2008.
4. National Institutes of Health. NIH Roadmap Centers for .Biomedical Computing. Available at: http://www.bisti.nih.gov/ncbc/index.cfm?. Accessed October 15, 2008.

References

1. Rubio DMG, Schoenbaum EE, Lee LS, et al. Defining translational research: implications for training. *Acad Med.* 2010;85(3):470–475. Published March 2010. https://journals.lww.com/academicmedicine/Fulltext/2010/03000/Defining_Translational_Research__Implications_for.21.aspx. Accessed July 29, 2020
2. Petrou M, Forester BR, Reich DS. Translational research in radiology: challenges and role in a patient-based practice. *Acad Radiol.* 2009;16(5):593–596. Published May 2009 https://www.academicradiology.org/article/S1076-6332(09)00051-8/fulltext. Accessed July 29, 2020.
3. How to Write a Research Question. The Writing Center. https://writingcenter.gmu.edu/guides/how-to-write-a-research-question. Updated August 8, 2018. Accessed July 29, 2020.

Chapter 20

Study population: Who and why them?

Zain Badar[1] and George Koshy Chiramel[2]
[1]Interventional Radiology Department, University Health Network, University of Toronto, Toronto, ON, Canada; [2]Department of Diagnostic Imaging, Hospital for Sick Children, Toronto, ON, Canada

Potential pitfall, real-world example: sample selection—the "how" and "why" of choosing the study population

The fundamentals of selecting participants for a research study include sample representativeness, sample frame, and choosing the appropriate type of sampling method.

- Statistical inference can be done meaningfully only after establishing adequate sample representativeness.
- Sampling frame defines a set of elements that help a researcher to select an adequate sample of the target population.
- Sampling methods can be categorized into either probabilistic sampling or nonprobabilistic sampling.

Sample representativeness is a concept that describes parameters utilized to estimate the sample size and composition. A sample is defined as a subset of the target population.[1] A retrospective approach is often considered in translational research. A target population can be identified with the help of clinicians selecting a sample of individuals for treatment. For example, urologists and interventional radiologists can jointly select patients with benign prostatic hypertrophy who could benefit from prostate artery embolization. This multidisciplinary approach is beneficial for the patient and is also one of the basic principles of translational research. Alternatively, researchers can utilize a registry system to identify a subset of individuals to represent the target population. Once a sample size has been established, *statistical inference* can be utilized to draw conclusions about the target population.[1] However, if the sample contains fewer number of participants than the defined threshold, statistical inference could be compromised in terms of precision and power, unless the representativeness of the sample is preserved. In cases where samples lack representativeness, statistical inference cannot be achieved and thus conclusions regarding the target population are invalid. Sampling bias and nonresponse bias can lead to a lack of representativeness and should be avoided during sample selection.[1,2]

Sample frame defines a list or set of elements utilized during a research study to aid in selecting a target population. A sample frame is specific, as compared to a target population that is more general.[1,2] For example, a researcher may utilize a registry to identify cases with peptic ulcer disease. However, as the sample should represent the target population, an in-depth analysis of the study objectives is required to avoid sample frame limitations.[1] An example of a sample frame could be either a hospital or a health record registry. Limitations on a hospital or health record sample frame would include incomplete or outdated data, type of health insurance provided, patients lost to follow-up, and ICD-10-specific diagnosis. Thus, obtaining a sample frame in interventional radiology can be a tedious process because a target population is difficult to define, as the sample usually includes patients with an established diagnosis.

Health research can either include the entire population (census) or a representation of the target population (sample). In interventional radiology, the latter is usually used. Sampling can be further subcategorized into either probabilistic sampling or nonprobabilistic sampling (Fig. 20.1). The former includes simple random sampling, systematic random sampling, stratified sampling, cluster sampling, and complex sampling. The latter includes convenience sampling, purposive sampling, quota sampling, and snowball sampling.[1–3]

Real-world example: probabilistic sampling

- *Simple random sampling:* From a complete list of participants, individuals are randomly selected utilizing a randomization table.[1] An example would be all individuals registered with an airline service, where there is a

Translational Interventional Radiology. https://doi.org/10.1016/B978-0-12-823026-8.00033-X

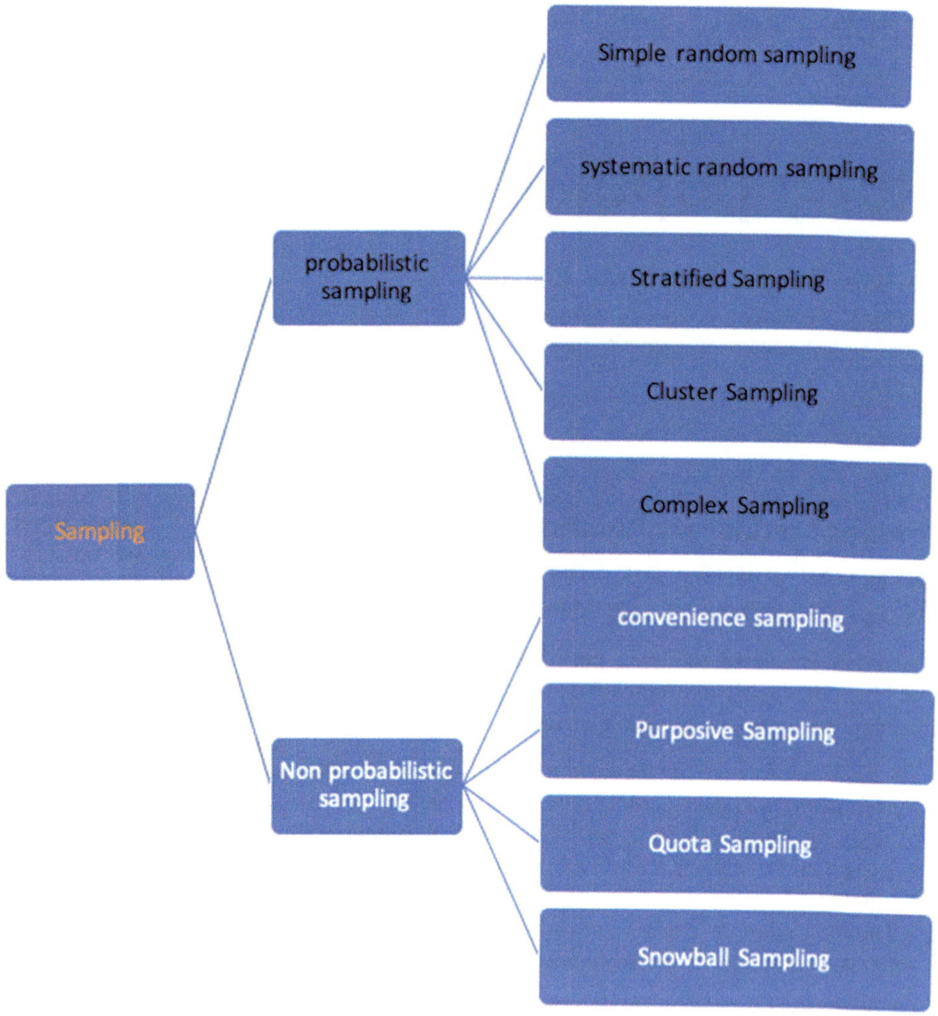

FIGURE 20.1 Categorizing probabilistic vs non-probabilistic sampling.

gift for the registered individuals in the form of a free ticket. The company would use random sampling to select 2 out of the 100 individuals to provide them with a free airline ticket.

- *Systemic random sampling:* Individuals are selected based on fixed intervals in a numbered population.[1] For example, to select 100 participants in a study at a clinic that sees 500 patients, every fifth patient who registers could be included.
- *Stratified sampling:* Sampling method in which the total target population is divided into smaller subgroups (strata). Then, the samples are chosen within each stratum (via simple or systemic sampling).[1] For example, interventional radiology procedure outcomes for a retrospective or prospective study can be categorized into subgroups by age: 10–19, 20–39, 40–59, 60, and above.
- *Cluster sampling:* Researchers divide the target population into multiple group (or clusters). This type of sampling is less precise than either simple or stratified random sampling. Cluster sampling is usually utilized in market research.[1] For example, a researcher could be interested in identifying patients suffering from diabetes in various geographical regions of a country.
- *Complex sampling:* Also referred to as multistage sampling, where smaller and smaller sampling units are obtained at each stage.[1] For example, the federal government of a country is interested in conducting a study regarding the average household expenses in a particular state or province. The study can initially take a census as a sampling frame, that can be systemically selected in the first-stage sampling unit. In the second stage, a certain number of households can be

selected. In the third stage, adult residents in these households are selected.

Potential pitfalls: nonprobabilistic sampling

- *Convenience sampling:* individuals are selected in order of appearance as they are convenient sources of data for researchers.[1] The sampling process is completed once sample saturation or time saturation is reached. Randomized clinical trials are mainly based on the concept of convenience sampling.
- *Purposive sampling:* also referred to "expert" sampling. The main objective is to produce a sample by referring to the opinions of experts established in a specific field.[1] For example, a group of interventional radiologists provides treatment options for multifocal hepatocellular carcinoma, including options such as transarterial chemoembolization, transarterial radioembolization, and microwave or radiofrequency ablation. These treatment options can then be utilized to retrospectively select a sample set.
- *Quota sampling:* is defined by obtaining a very tailored sample that reflects a characteristic or trait of a population.[1] The target population is first classified by characteristics such as gender and age. This is followed by sampling units to complete each quota. For example, a multicenter study observing transradial versus transfemoral approach uterine fibroid embolization might be required to have a specific quota of patients per center to qualify in the study.
- *Snowball sampling:* individuals are selected by the researcher at the initial stage. These individuals then aid in contacting other members with similar characteristics to take part in the research study.[1] For example, women with varicose veins might contact other women via an online forum, to take part in a research study where interventional radiologists perform endovenous laser therapy as a treatment regimen for varicosities.

The reason for selecting a subset of individuals can vary. It is usually based on various factors such as type of the study, hypothesis, research questions, and treatment outcomes. Various types of research studies, clinical trials for example, might select individuals that do not represent the entire population, as convenience sampling is usually utilized.[3–5] An observational study, on the other hand, provides more of a heterogeneous sample. As sample representation is critical for a cross-sectional study, the sampling process usually uses complex probabilistic sampling. A case-control study may suffer from selection bias due to the improper selection of control subjects. However, in a case-control study, the sample representation is based on the relationship between cases and controls.[4–6]

In conclusion, sample selection plays a critical role in health research. All three aspects of sample selection—sample representativeness, sample frame, and sample types are crucial in this process.

Get started

Sample Selection utilizes the concept of sample representativeness and sample frame. However, researchers must also understand the usage of probabilistic and non-probabilistic sampling to make the process of sample selection less cumbersome.

Resources

Additional articles:

1. Duquia RP, Bastos JL, Bonamigo RR, González-Chica DA, Martínez-Mesa J. Presenting data in tables and charts. *An Bras Dermatol.* 2014;89:280–285.
2. OpenEpi.com. Dean AG, Sullivan KM, Soe MM. OpenEpi: Open Source Epidemiologic Statistics for Public Health, Version. [updated 2013 Apr 6]. [2014 Mar 22].
3. Duquia RP, Baptista Menezes AM, Reichert FF, de Almeida Jr HL. Prevalence and associated factors with sunscreen use in Southern Brazil: a population-based study. *J Am Acad Dermatol.* 2007;57:73–80.
4. Porta M, ed. *A dictionary of epidemiology*. 5th. New York: Oxford University Press; 2008.
5. Jones SR, Carley S, Harrison M. An introduction to power and sample size estimation. *Emerg Med J.* 2003;20:453–458.

References

1. Martínez-Mesa J, González-Chica DA, Duquia RP, Bonamigo RR, Bastos JL. Sampling: how to select participants in my research study? *An Bras Dermatol.* 2016;91(3):326–330.
2. Martínez-Mesa J, González-Chica DA, Bastos JL, Bonamigo RR, Duquia RP. Sample size: how many participants do I need in my research? *An Bras Dermatol.* 2014;89:609–615.
3. Röhrig B, du Prel JB, Wachtlin D, Kwiecien R, Blettner M. Sample size calculation in clinical trials: part 13 of a series on evaluation of scientific publications. *Dtsch Arztebl Int.* 2010;107:552–556.
4. Suresh K, Thomas SV, Suresh G. Design, data analysis and sampling techniques for clinical research. *Ann Indian Acad Neurol.* 2011;14:287–290.
5. Rothman KJ, Gallacher JE, Hatch EE. Why representativeness should be avoided. *Int J Epidemiol.* 2013;42:1012–1014.
6. Krause M, Lutz W, Boehnke JR. The role of sampling in clinical trial design. *Psychother Res.* 2011;21:243–251.

Chapter 21

Outcome measurements: What data is being collected and why?

Zain Badar, Mark Nassar and Adam Christopher Krajewski
Interventional Radiology Department, University Health Network, University of Toronto, Toronto, ON, Canada

- Common outcome measurements for diagnostic studies include sensitivity, specificity, accuracy, NPV, and PPV.
- Other outcome measurements in therapeutic studies include overall survival, progression-free survival, disease control rates, objective response rates, time to progression, odds ratio, relative risk, and absolute risk.
- Intention to treat analysis: patients are analyzed according to their originally assigned groups, described as "once randomized, always analyzed."[1]

Potential pitfalls

Research variables play a crucial role in determining treatment outcomes, as they form the basis of clinical decision making. Statistical variables chosen must follow the initial study design. Peer-reviewed publications often have adequate study designs but can sometimes lack basic statistical principles or vice versa. This can result in misinterpretation of data and flawed treatment outcomes in translational research. Additionally, it is important to understand the difference between statistically and clinically significant data.[2] A data set might be statistically significant ($P < .05$); however, the transition to clinical practice might not be beneficial. For example, a researcher might obtain a data set of patients with Klippel Trenaunay syndrome, the majority of whom might have the AGGF1 gene. This data set could be statistically significant to the researcher conducting the study. Nonetheless, in terms of providing sclerotherapy to these individuals, the interventional radiologist might not find this data clinically relevant. Thus, in interventional radiology, even though statistical parameters play a crucial role in research, the presence of statistically significant differences might be of limited value in a clinical setting. A nonsignificant outcome should not lead one to assume that the treatment was ineffective clinically, as a low power study and measurement variables can influence statistical outcomes. Furthermore, confidence intervals have been utilized as pertinent measurement variables in translational research, often influencing treatment plans.[3]

Real-world examples

Sensitivity of a test can be defined as a screening test's probability of accurately identifying a group of people who are known to have a certain condition and who indeed have the condition (true positives) while labeling a group of people as not having this condition when they actually have it (avoiding false negatives). Thus, sensitivity is defined as the probability of a group of individuals with a certain condition who are accurately identified by a screening test as actually having the condition.[4] Similarly, in the same context, positive predictive value (PPV) is the probability that a group of individuals with a positive screening test will actually have the condition. Specificity on the other hand is defined as the probability of a group of individuals without a certain condition, accurately identified by a screening test as NOT having the condition. Negative predictive value (NPV) is the probability of a group of individuals with a negative screening test accurately identified as actually not having the condition. Sensitivity and specificity are related to the accuracy of a test. Thus, one can infer that the accuracy of a test is defined as the ability to differentiate between healthy versus nonhealthy cases while analyzing a population. To estimate the diagnostic accuracy of a test, the proportion of true positive and true negative cases is calculated.[4–6] Specific to radiology, in 2013, Chen et al. wrote a letter to the editor regarding an article published by Naraghi in 2012 with respect to the evaluation of MR imaging in the diagnosis of knee laxity. Chen et al. commented that the PPV and NPV of MR imaging were reported in the article;

Translational Interventional Radiology. https://doi.org/10.1016/B978-0-12-823026-8.00031-6

however, the PPV and NPV values were calculated without any disease prevalence.[7,8] Thus, the importance of disease prevalence in calculating PPV and NPV needs to be considered during statistical analysis.

Other outcome measurements in therapy studies include *overall survival (OS), progression free survival (PFS), disease control rates, objective response rate, time to progression, odds ratio, absolute risk, and relative risk.*

OS is defined as the time from randomization of a subject to time of death, as long as it takes into consideration the preservation of quality of life. OS can be precisely measured and evaluated on a continual basis. However, OS can be affected by subsequent therapies, including noncancer-related deaths, and can often involve large studies with significant time for follow-up.

On the other hand, PFS is defined as the time from randomization of a subject to tumor progression or death. PFS is becoming more prominent in clinical trial design as it can measure stable disease with a smaller sample size while taking into consideration short-term follow-up. PFS is based on objective and quantitative measurements and is not affected by subsequent therapies. Nonetheless, disadvantages of PFS can include evaluation bias and frequent diagnostic/radiologic or other assessments. For example, RECIST criteria (response evaluation in solid tumors) in diagnostic radiology formulate the basis of PFS.[9,10]

Additionally, relative risk or risk ratio is defined as the probability of an outcome in an exposed group compared to the probability of an outcome in an UN-exposed group. Thus, it is the risk of an event in the experimental group compared to the risk in a control group. On the other hand, the absolute risk is the probability of an event in a population of interest. It can be calculated by a number of events in a group divided by a number of people in that specific group. An odds ratio is a measure of association between an exposure and an outcome, in other words, it is the odds of an event in an experimental group relative to the odds of an event in a control group.[1,10]

Other outcome variables include time to treatment failure, defined as measuring time from randomization to the end-point of treatment (discontinuation or completion of treatment). Time to progression is defined as the time from randomization to tumor progression (excluding death). Overall response rate (ORR) is defined as the proportion of individuals who have either had a partial or complete response to treatment (excluding stable disease). On the other hand, disease control rate is a composite of ORR and stable disease, and thus is utilized to evaluate treatment efficacy relating to tumoristatic effects.[1,10]

Potential pitfalls

The intention-to-treat analysis is a method for analysis in a prospective randomized study. Randomly selected participants assigned to various groups are included in the statistical analysis, irrespective of any treatment they have received. Thus, the principal investigator is able to evaluate the efficacy of a specific intervention with nonbiased accurate results, as randomization is completely preserved. However, bias can become a part of the analysis if treatment groups are not analyzed according to the initially assigned/designated group.[1]

The presentation of data in a clear, efficient, and effective manner aids in establishing a well-defined approach for statistical parameters. The primary basic statistical principles include *P*-values, sensitivity, specificity, positive and negative predictive values, incidence, and prevalence. Furthermore, an understanding of power, regression, *t*-test, hypotheses, error types, chi-square, analysis of variance, correlation, biases, and distributions can aid in clinical decision making.[11,12]

Real-world examples and potential pitfalls

A study by Huang et al. concluded that 20 statistical tests were listed as basic tests in interventional radiology research. The authors reviewed major clinical interventional radiology articles from JVIR and *Radiology*. The outcomes of the studies were either cross-sectional (at a single time point) or longitudinal (measured over time) in nature. 56% of articles utilized basic tests of association such as *t*-test, χ^2 test, Fisher's exact test, Mann–Whitney U test, Wilcoxon signed-rank test, one-way ANOVA test, Kruskal–Wallis test, Mantel–Haenszel test, and Bonferroni correction. On the other hand, only 20% of articles utilized regression analysis as an advanced test of association. Advanced tests of association utilized in the interventional radiology literature included logistic regression, Pearson correlation coefficient, simple linear regression, spearman rank correlation, and multiple linear regression. Survival analysis was applied in 27% of cases. Examples of survival analysis included Kaplan–Meier test, log-rank test, and cox-proportional hazards test. Decision statistics such as sensitivity and/or specificity were utilized in a minority of cases in the interventional radiology literature. Confidence intervals and power calculations, however, were reported frequently. Data collection was subcategorized into prospective and retrospective studies. Thus, in the setting of translational research in interventional radiology, it is highly recommended to review the article and study it in detail to obtain a thorough understanding of the basic statistical parameters required for translational research.[13]

Even though it is important to understand all statistical parameters, it is more crucial to understand their clinical relevance to aid physicians in improving clinical decision making.

Get started

Statistical analysis in the context of clinical parameters provides more essential information in outcome analysis/measurements rather than statistical analysis alone. Understanding clinical relevance can aid in implementing statistical parameters as a means of improving clinical outcomes.

Resources

Additional articles:

1. Casarella WJ. Vascular and interventional radiology from Oregon to Zurich to Louisville: the making of a specialty. *AJR Am J Roentgenol.* 2000;174:16–17.
2. Malone DE, MacEneaney PM. Applying 'technology assessment' and 'evidence based medicine' theory to interventional radiology. Part 1: suggestions for the phased evaluation of new procedures. *Clin Radiol.* 2000;55:929–937.
3. Goldin J, Sayre JW. A guide to clinical epidemiology for radiologists: part I—study design and research methods. *Clin Radiol.* 1996;51:313–316.

References

1. McCoy CE. Understanding the intention-to-treat principle in randomized controlled trials. *West J Emerg Med.* 2017;18(6):1075–1078. https://doi.org/10.5811/westjem.2017.8.35985.
2. Casarella WJ. Vascular and interventional radiology from Oregon to Zurich to Louisville: the making of a specialty. *Am J Roentgenol.* 2000;174:16–17. https://doi.org/10.2214/ajr.174.1.1740016.
3. Malone DE, MacEneaney PM. Applying 'technology assessment' and 'evidence based medicine' theory to interventional radiology. Part 1: Suggestions for the Phased Evaluation of New Procedures. *Clin Radiol.* 2000;55(12):929–937. https://doi.org/10.1053/crad.2000.0598.
4. Trevethan R. Sensitivity, specificity, and predictive values: foundations, pliabilities, and pitfalls in research and practice. *Front Public Health.* 2017;5:307. https://doi.org/10.3389/fpubh.2017.00307. Published 2017 Nov 20.
5. Mallett S, Halligan S, Thompson M, Collins GS, Altman DG. Interpreting diagnostic accuracy studies for patient care. *BMJ.* 2012;345:e3999. https://doi.org/10.1136/bmj.e3999.
6. Pewsner D, Battaglia M, Minder C, Marx A, Bucher HC, Egger M. Ruling a diagnosis in or out with "SpPIn" and "SnNOut": a note of caution. *BMJ.* 2004;329:209–213. https://doi.org/10.1136/bmj.329.7459.209.
7. Naraghi AM, Gupta S, Jacks LM, Essue J, Marks P, White LM. Anterior cruciate ligament reconstruction: MR imaging signs of anterior knee laxity in the presence of an intact graft. *Radiology.* 2012;263(3):802–810.
8. Chen F-yao, Yang CP, Chen PY. Comment on the correct use of predictive values for evaluating diagnostic tests. *Radiology.* 2013;266(1):364–365.
9. Banoo S, Bell D, Bossuyt P, et al. Evaluation of diagnostic tests for infectious diseases: general principles. *Nat Rev Microbiol.* 2007;5:S21–S31. https://doi.org/10.1038/nrmicro1523.
10. Villaruz LC, Socinski MA. The clinical viewpoint: definitions, limitations of RECIST, practical considerations of measurement. *Clin Cancer Res.* 2013;19(10):2629–2636. https://doi.org/10.1158/1078-0432.CCR-12-2935.
11. Goldin J, Sayre JW. A guide to clinical epidemiology for radiologists: Part I – study design and research methods. *J Radiol Oncol.* 1996;51(5):313–316. https://doi.org/10.1016/S0009-9260(96)80106-1.
12. Elster AD. Use of statistical analysis in the AJR and radiology: frequency, methods and subspecialty differences. *Am J Roentgenol.* 1994;163(3):711–715. https://doi.org/10.2214/ajr.163.3.8079874.
13. Huang W, La Berge JM, Lu Y, Glidden DV. Research publications in vascular and interventional radiology: research topics, study designs, and statistical methods. *Clin Stud.* 2002;12(3):247–255. https://doi.org/10.1016/S1051-0443(07)61717-5.

Chapter 22

Optimizing the question: Balancing significance and feasibility

Travis William McCain Pebror
Interventional and Diagnostic Radiology, Department of Radiology and Imaging Sciences, Indiana University School of Medicine Indianapolis, IN, United States

Key points

- Clinical studies are an opportunity to understand and refine the potential benefits of a novel intervention; however, the costs, time, and risks of clinical studies impose important limitations on the ability of a novel intervention to demonstrate and affect significant change in clinical outcomes.
- An ideal clinical trial will demonstrate significance via select outcomes that thoughtfully align with the overall clinical strategy and clearly demonstrate the value proposition of the novel intervention to patients, providers, and other stakeholders.
- Given the physiological, practical, financial, regulatory, and ethical constraints of clinical trials, it is necessary to make tradeoffs when deciding which outcomes to pursue while planning a clinical trial protocol.

Why it matters?

Not all novel interventions will offer appreciable benefits to patients, providers, or other stakeholders over existing treatment options. This can be because the new intervention truly is no different than existing options or because that difference is insignificant enough to go unnoticed by patients and physicians. Clinical trial planning is an opportunity to understand the significance of a novel intervention and mitigate the clinical risk involved in its development. Practical considerations regarding measurement, logistics, statistics, budgets, regulations, and human subjects' protections must be taken into account when planning clinical research. These considerations require careful iteration of the research question and selection of the most appropriate clinical trial design and outcome measurements to align the limited results of the clinical trial with the overall development strategy to ensure the innovation will effect meaningful change.

Real-world examples: defining a value proposition to guide clinical strategy

What is unique about a novel intervention that would make patients or clinicians take notice? Is this the first treatment for a condition? Or perhaps a more effective, tolerable, efficient, or affordable option than what is currently available? Before investing resources in developing an idea, it is important to clearly articulate a value proposition. A **value proposition** conveys why a novel intervention is unique and useful from the perspective of *specific* group of patients or other stakeholders though it is often costly, risky, and difficult to prove the value proposition of a novel intervention in the initial stages of a clinical strategy. For example, some patients with pulmonary embolism carry a high risk of provoked hemorrhage with systemic lytic therapies. The Indigo Aspiration System (Penumbra, Inc) offers a transcatheter means of mechanically extracting a pulmonary embolism while avoiding the risk of systemic lytic therapy. The potential value proposition for mechanical embolectomy in this specific group of patients is risk reduction versus lytic therapy; however, initial efficacy and safety of the device were established in a single-arm study (EXTRACT-PE)[1] that excluded patients who had a contraindication to systemic anticoagulation.[2] Given the substantial risk and resources involved in procuring clinical data, oftentimes the unique value proposition is not directly pursued in the initial stages of the clinical strategy until the preliminary safety and efficacy of the device are first established.

Real-world examples: constructing a clinical strategy

There are a variety of compelling uses of clinical studies in translational research besides characterizing the safety and

Translational Interventional Radiology. https://doi.org/10.1016/B978-0-12-823026-8.00099-7

effectiveness of an innovation. Early clinical data can better characterize technical risk; inform the design of prototypes; demonstrate cost-effectiveness for payor approval; etc. With the myriad of possible uses of clinical studies, it is important to align the pursuit of clinical outcomes or endpoints toward an overarching strategy. A **clinical strategy** defines the overall clinical pathway for the development of an innovation and helps define and prioritize each study's endpoints.[3] Collectively, these endpoints characterize the safety and effectiveness of the new innovation in the near term while demonstrating the unique value proposition of innovation in the long term (see Fig. 22.1). Early studies often focus on characterizing the safety and effectiveness of a device in one group of patients before latter studies demonstrate its value proposition in another. For example, transcatheter aortic valve implantation offers a far less invasive means of addressing aortic stenosis in native valves versus open surgery. An early clinical trial, PARTNER demonstrated the safety and effectiveness of TAVI in patients with severe aortic stenosis who were deemed moderate-to-high risk for surgery.[4] Ongoing clinical studies aim to demonstrate the effectiveness of the device in younger, healthier patients who might also value the reduced morbidity of a minimally invasive approach.[5] Overtime, carefully selected clinical outcomes may demonstrate the value proposition of the intervention that might differ significantly from the clinical outcomes pursued initially.

Real-world examples: understanding the significance of clinical outcomes

Translational research determines which theoretical benefits of an innovation are actually meaningful. Significance is heavily predetermined by a selection of the appropriate clinical study design to ensure observation of both meaningful outcomes for the question at hand and significant change in those outcomes. Selection of meaningful outcomes is detailed in Chapter 42, but in brief the most compelling clinical outcomes directly measure how a patient feels, functions, or survives and are already clearly understood by patients and/or clinicians to direct treatment decision making.

Given an outcome, there are various ways to measure the significance of the change in that outcome following intervention; namely statistical significance and clinical significance. **Statistical significance** describes the probability that an observed difference is attributable to the intervention versus chance through measures such as the *P*-value and confidence interval (see Chapter 58). Unfortunately, many clinicians equate statistical significance with clinical importance; however, the importance of an intervention is better conveyed by clinical significance. **Clinical significance** describes the magnitude of a treatment's effect size in clinical practice through the measurement of the absolute or relative change in clinical outcomes. Many additional measures of clinical significance have been constructed, including absolute risk reduction, relative risk reduction, number needed to treat, and odds ratio.[6] Statistical significance is required to demonstrate clinical significance, but not vice versa, that is, there are many statistically significant changes that are clinically irrelevant. See Fig. 22.2 for a methodology to approach significance by analyzing the results of a randomized control trial comparing yttrium 90 (Y90) radioembolization to transarterial chemoembolization for treatment of hepatocellular carcinoma.[7]

Taken together, an ideal strategy for characterizing an innovation is to measure the clinically significant change in meaningful outcomes. Unfortunately, however, there are

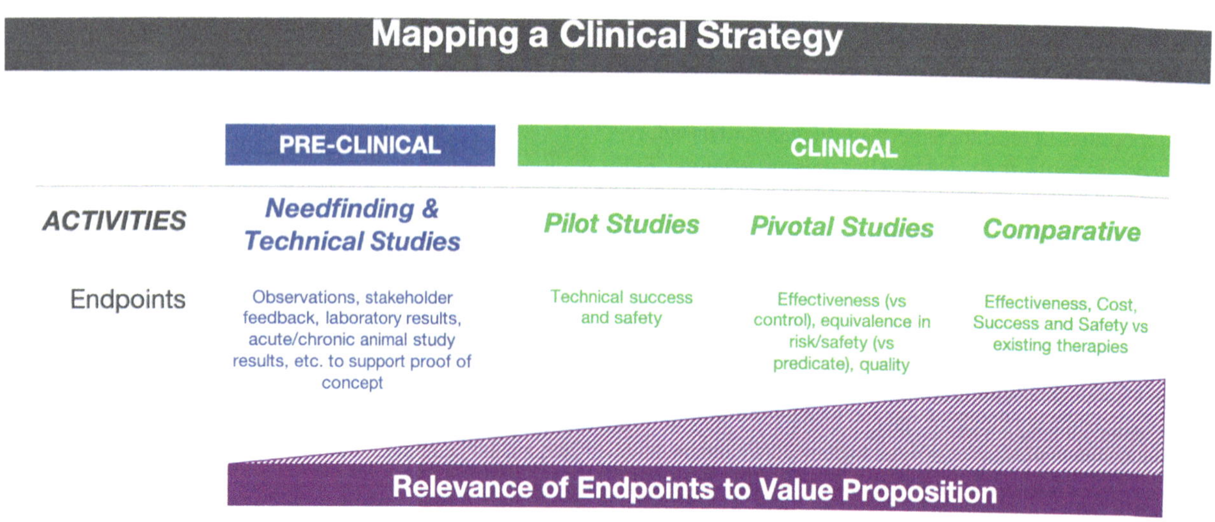

FIGURE 22.1 A clinical strategy outlines and coordinates the endpoints necessary to eventually demonstrate the value proposition of an innovation.

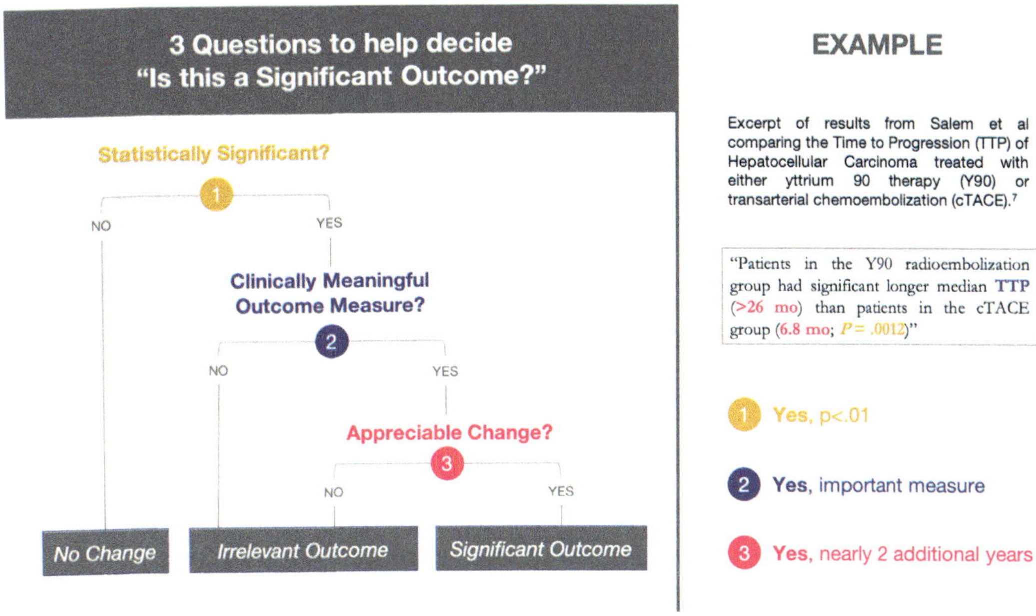

FIGURE 22.2 Three questions help answer whether an outcome is clinically significant. First, it is important to ask whether the observed difference is statistically significant by ensuring a low likelihood of a type 1 statistical error (see Chapter 60). Next, it is important to ask whether the outcome observed is clinically meaningful (see Chapter 42). Finally, it is important to consider the magnitude of the change observed. If all three conditions are met, it is a significant outcome as illustrated in the example from Salem et al.[7]

important considerations that often limit the feasibility of pursuing the most compelling outcomes.

Feasibility considerations

Aside from whether the problem being pursued has large enough market potential to garner sufficient funding to execute the clinical strategy, there are several factors that affect feasibility and must be taken into consideration when planning. The most obvious consideration is which study design is selected. Thereafter, the feasibility of the proposed study becomes substantially more evident when drafting a research protocol. A **research protocol** is a document that details the background, rationale, objectives, methods, and organization of a clinical study along with statistical and ethical considerations. Several factors pertinent to the research protocol that determine the feasibility of a study are detailed in subsequent chapters of the Clinical: Preparation section of this book, including the selection of measurable outcomes, powering study sufficiently, regulatory requirements, budgeting, staffing, data management, harm reporting, adherence, and human subjects considerations.

A two by two of significance versus feasibility conceptualizes four categories of outcomes that a clinical study can pursue (see Fig. 22.3). An often-critical link between the feasibility of a study and the significance of outcomes is a selection of endpoints that determine the methods of observation and how the study is powered. A **primary endpoint** is a precisely defined variable that reflects the clinical outcome most important for a study to observe and thus dictates how the study is powered. Endpoints that are rare, difficult to measure and large all make it more difficult to sufficiently power a study (see Chapter 43). **Secondary endpoints** are additional clinical outcomes of interest to investigators that exercise less bearing on the research protocol than primary endpoints.

Potential pitfalls: less is more

Given the tradeoffs between significance and feasibility of clinical outcomes, it is important to consider the study being planned within the context of the broader clinical strategy. Rather than pursue several of the most compelling outcomes now, ask what is the least significant outcome that is sufficient for this stage of the clinical strategy? Although important, demonstrating the value proposition of an innovation might be one of the latter aims of a clinical strategy. Finally, while clinical planning can seem tedious, it is important to remember that if an innovation cannot demonstrate compelling results in a well-controlled clinical trial, it is highly unlikely the innovation would do so in the real world.

Get started and resources

- **Refine the key value proposition** for your novel intervention related to one or more clinical outcomes. Try to

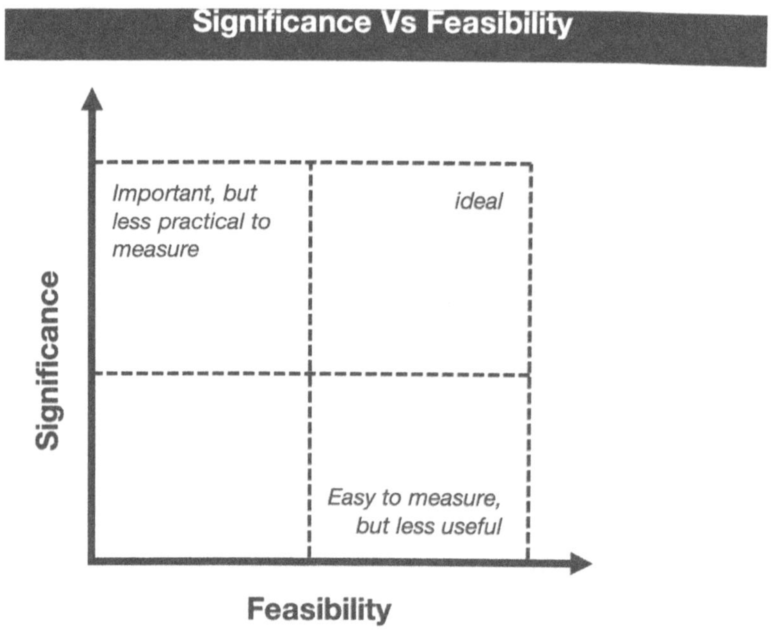

FIGURE 22.3 Ideally, clinical strategy would pursue endpoints that are both feasible and significant. However, there are often tradeoffs that must be made at various points in developing an innovation. Sometimes, it may be expedient to sacrifice some significance to make a pursue a measure that is appreciably more practical to measure.

articulate the value proposition of your novel intervention from the patient's perspective rather than that of the provider, payor, or other stakeholder. For more information on crafting a value proposition, visit:
- National Science Foundation Innovation Corps Handbook.
- NIH Center for Accelerated Innovations at Cleveland Clinic (http://www.ncai-cc.ccf.org/skills/NSF.php).
- *Value proposition design: How to create products and services customers want* by Alexander Osterwalder et al. (John Wiley & Sons, 2014).

- **Create a list of additional uncertainties** by reflecting on feedback and data collected thus far. What areas of uncertainty must be resolved to demonstrate the value proposition of your innovation? What other uncertainties can you anticipate relating to effectiveness, regulatory approval, payor reimbursement, etc. that could be addressed with clinical data? The FDA 510(k) and De Novo databases are a great place to see what types of data manufacturers have submitted to resolve questions regarding the safety and effectiveness of previously approved devices.
- **Draft a clinical development strategy.** Reframe these uncertainties into questions (see Chapter 11). Prioritize these questions such that the clinical uncertainties can be addressed subsequently in a series of clinical studies. The clinical strategy should be updated regularly as the myriad of risks addressable by clinical data is identified, addressed, or resolved throughout the development process. A great resource for exploring clinical strategy more:
  - *Biodesign: the process of innovating medical technologies* by Paul Yock et al. (Cambridge University Press, 2015).
  - A companion website for the clinical strategy chapter is available online at http://ebiodesign.org/chapter/5-3-clinical-strategy/
- **Identify the highest priority question in the clinical development strategy.** After reading Chapter 42, determine the most and the least compelling outcomes that could answer this clinical question. Review Clinical Study Types and Clinical Trials designs in prior chapters comparing the possible outcomes and limitations of each research design to address the question. Repeat this process for several questions to map out an expected series of clinical studies and/or trials to implement the clinical strategy.
- **Begin a research protocol sketch of one or more clinical study.** Subsequent chapters will introduce methods of clinical planning that have bearing on the feasibility of a proposed clinical study. Update each draft with details regarding the strength of expected outcomes for the selected study design and feasibility of implementing the draft protocol. Drafting more than one protocol will allow you to compare the tradeoffs between expected outcomes and feasibility of different clinical study approaches to resolve the same clinical question. A good template for starting a clinical protocol can be found here.

- Standard Protocol Items: Recommendations for Interventional Trials (https://www.spirit-statement.org/title/)
- National Institutes of Health Clinical E-Protocol Writing Tool (https://e-protocol.od.nih.gov/#/home)

References

1. Sista A. *Late-Breaking Clinical Trials. Presented at: VIVA 19*. Las Vega; Nov. 4–7, 2019.
2. Penumbra Inc. *Evaluating the Safety and Efficacy of the Indigo Aspiration System in Acute Pulmonary Embolism (EXTRACT-PE). Clinicaltrials.gov Identifier: NCT03218566*. 2017.
3. Yock P, Zenios S, Makower J, et al. *Biodesign: The Process of Innovating Medical Technologies*. Cambridge University Press; 2015.
4. Leon M, Smith C, Mack M, et al. Transcatheter aortic-valve Implantation for aortic stenosis in patients who cannot undergo surgery. *N Engl J Med*. 2010;363:1597–1607.
5. Harold JG. Harold on history: the evolution of transcatheter aortic valve replacement. *Cardiol Mag*. 2017.
6. Laupacis A, Sackett DL, Roberts RS. An assessment of clinically useful measures of the consequences of treatment. *N Engl J Med*. 1988;318.26:1728–1733.
7. Salem R, Gordon A, Mouli S, et al. Y90 radioembolization significantly prolongs time to progression compared with chemoembolization in patients with hepatocellular carcinoma. *Gastroenterology*. 2016;151.6:1155–1163.

Part IV

Statistical principles

Chapter 23

Common issues in analysis

Christopher Ovanez
Russell H. Morgan Department of Radiology and Radiological Science, Johns Hopkins Hospital, Baltimore, MD, United States

Within this chapter, I would like to focus on the more common issues that could be encountered during the process of data analysis. Please note that details regarding the process of data gathering, as well as the statistical side of data analysis, are covered in depth in other portions of this manuscript.

When it comes to data analysis, there are a number of things that a researcher needs to be cognizant of, and these will differ considerably depending upon a number of potential scenarios and variables—for instance, the type of research study design you are pursuing (i.e., prospective, retrospective, observational, case-control, etc.), and what types of variables you are looking at (i.e., patient clinical data, animal data, product data, safety/quality improvement data, retrospective or prospective data, etc.).

Taking a step back, before even beginning to analyze your data, you want to absolutely ensure that the data that have been gathered are of the utmost quality—being free of error, bias, or manipulation, and is as complete as possible for the needs of your project, to ensure that analysis is performed on a set of data that is the most ideal, so that the results that ensue are of the highest quality, and are truthful and accurate. As the common saying goes—"garbage in, garbage out".

Once you have a proper set of data gathered, starting to analyze the data might not be as easy as one might think, and could potentially be quite a complex task.

So, what would be the best way to start?

It would certainly be a good idea to work with an independent biostatistician in analyzing your data, even if you are fully capable of accomplishing the tasks of a biostatistician in analyzing your own data. Analyzing your own data could potentially lead to unwanted issues and consequences, namely in terms of introducing unconscious bias or manipulation of the data, perhaps without even realizing it, also you might utilize an inappropriate or nonideal type of method or test for data analysis. This is dangerous and could lead to misinterpretation of data during the analysis and interpretation phases of the study and lead to conclusions that are not truthful or accurate, and potentially lead to devastating issues if applied to patient care.

A word of advice—people usually seek the assistance of a biostatistician after the data are already gathered and they are in need of analyzing their data; however, seeking the assistance of a biostatistician early on, even before starting data gathering, could prove quite beneficial. Biostatisticians possess a wealth of knowledge and usually have an alternative set of perspectives from your own, and this could be very helpful during the design phase of a project. For instance, let us say you have come up with a grand idea for a project and absolutely cannot wait to get started. Working with a biostatistician early on could prove quite beneficial in thinking about what type of data variables you might want to collect, how to organize these data, and how many subjects to include in your study to ensure an appropriately powered study, etc.

Before beginning the process of data analysis, an investigator must make sure that their data set does not possess any obvious gaps, missing data, or errors, and the data have been accurately allocated into the appropriate groups and categories. Not taking the time to double-check and ensure data completeness and accuracy before beginning analysis of the data could lead to quite disastrous consequences, for instance, inaccurate and confusing results, as well as a waste of time and resources.

What are common types of issues one could run into during data analysis?

Misclassification is an issue one could run into during data analysis, and in essence refers to the false characterization of a particular study characteristic secondary to an error in measurement.[1] To break it down further, you could have differential misclassification or non-differential misclassification.[1] Differential misclassification refers to misclassification that results from a non-systematic error in data measurement, and non-differential misclassification

Translational Interventional Radiology. https://doi.org/10.1016/B978-0-12-823026-8.00059-6

refers to a non-systematic error in data measurement, with non-differential misclassification tending to lead to a relative risk that nears one more so than if data were obtained under more ideal circumstances of measurement.[1]

Confounding is another type of issue that one could run into during data analysis. Confounding is essentially a form of bias that leads to misinterpretation of the observed association between a particular exposure of interest and a disease process, when you have a variable (or variables) that leads to an apparent association.[2] By definition, a confounding factor is something that is associated with a particular exposure as well as the outcome of interest, but is not directly part of the causal pathway between the particular exposure or outcome of interest.[2] A common example in epidemiology of a confounder is smoking, which for instance can lead to an apparent association between alcohol usage and lung cancer.[3] Taking a step back, you might wonder how alcohol itself could lead to lung cancer, but with smoking as a confounder, the reason seems to be more obvious, in that an individual who drinks alcohol is probably more likely to smoke, and individuals who smoke and more likely to develop lung cancer. This example is a relatively simple one, but during the course of research design involving more complex issues and variables, such associations might not as readily pop out at you when you take a step back as in this particular example. A confounder could be potentially discovered when you remove a particular variable from the analysis, and the size of a particular association between a particular variable and outcome significantly changes due to this particular change.

Effect modification is another type of issue one could run into during data analysis and occurs when the association between a particular variable and a particular outcome changes due to the manipulation of another variable. You might be thinking to yourself that effect modification kind of sounds similar to a confounding variable, but it is in fact different. With effect modification, there is actually an association between a particular exposure and outcome, and that association is altered by another variable, whereas in the process of confounding, there is no real association between a particular exposure and outcome (the example we talked about between the exposure being alcohol use, and outcome of lung cancer), with no true association between these two things, and the association being thought to be real due to a confounding variable such as smoking. If you remove the confounding variable of smoking from the aforementioned example, there is no true association between them, whereas in effect modification, if you remove a particular variable, there is still a real association between a particular exposure and outcome.

Lead time bias is another common issue when it comes to data analysis and interpretation and essentially means that the early detection of a particular disease process is misconstrued for improved survival.

There are several other types of bias that could misconstrue the data and process of analysis, for instance, there is *recall bias*, which occurs when a study participant does not accurately or completely recall a certain event or piece of data, leading to issues in data gathering and ultimately analysis and interpretation. *Selection bias* is another common type of bias whereby the composition of participants of a study is very different from the larger population to which it attempts to generalize, which leads to errors and many issues with regard to the accuracy of a particular association, as well as generalizability of data and its interpretation to a wider more realistic patient population.

In conclusion, the analysis phase of a research study can be complex, with several variables that one needs to be cognizant of depending upon the type of study one is pursuing, as discussed earlier. During data analysis, recruiting the assistance of an independent biostatistician is ideal for many reasons, especially early on in the research design process, namely to steer clear of unconscious bias and manipulation of data, as well as for another perspective, which is always helpful.

There are many other types of bias and error that may occur, but I wanted to touch upon the more common ones for this chapter. I encourage you to explore the resources listed within the reference section for additional information.

References

1. *Epidemiology and Biostatistics: An Introduction to Clinical Research by Byran Kestenbaum.* 2nd ed. Springer; 2019.
2. Yan F, Robert M, Li Y. Statistical methods and common problems in medical or biomedical science research. *Int J Physiol Pathophysiol Pharmacol.* November 1, 2017;9(5):157–163. PMID: 29209453; PMCID: PMC5698693.
3. Steinke DT. *In Clinical Pharmacy Education, Practice and Research.* 2019.

Further reading

1. *Epidemiology: Principles and Practical Concerns by Jan Van Den Broek, Jonathan Brestoff.* Springer; 2013.

Chapter 24

Everyday statistics for the clinician

Linzi Arndt Webster
Emory University School of Medicine Department of Radiology, Division of Interventional Radiology and Image Guided Medicine, Atlanta, GA, United States

Statistics tell your research story. Understanding data analysis and its nuances guides informed consumption of data. In this chapter, we will discuss basic statistical principles consisting of significance, precision, and accuracy. Then, we will discuss test attributes, such as sensitivity and specificity, positive and negative predictive values, incidence, and prevalence. Lastly, we will delve into ways to quantify risks, including odds ratios, relative risk, attributable risk, absolute risk reduction, number needed to treat, and number needed to harm.

What is significance?

Significance, specifically statistical significance, describes the possibility that the relationship between two or more variables is caused chance. All tests of significance start with a null hypothesis, which states that there is no difference between the variables of interest. This null hypothesis is determined before any tests of significance. For example, if I wanted to explore the average temperatures between February and May, then I would measure the outside temperature each day of the respective months. When thinking about how to assess if there was a significant difference between the two, I would state my null hypothesis as *"there is no difference between the temperatures observed in March and those observed in May."* When assessing the relationships between different variables, the choice will be made for the most appropriate tests. In my example above, where we are observing quantitative, nominal values, we should use student's *t*-test to assess for significance. In another, clinically applicable example, we could ask if the numbers of days between a patient diagnosis of cancer and first treatment were different depending on minority status. The null hypothesis would be "*there is no difference in the number of days between patient diagnosis and first treatment for majority race and minority race patients.*" This example could also employ a t-test because we are comparing two nominal variables.

The universally accepted way to report the statistical relationship between two groups is through a *P*-value. "*P*-value" is an abbreviation for probability value, which is the probability that the relationship observed is due to chance alone. Cutoffs for *P*-value can vary depending on the number of hypotheses tested but, the generally accepted *P*-value is expressed as a decimal with a threshold value of <0.05 accepted as "significant." This "0.05" value does not have any special property except to state that "the probability that this result is not due to the relationship between the two variables is 5%." Another way to state this result is "there is a less than 5% chance the results observed is due to chance." For those values where $P < .01$, we can state these results are "highly significant" because the chance that the results are due to chance is only 1%.

What is precision and what is accuracy? Why does this matter for my research?

Precision is how closely two measurements are to each other. Accuracy is how closely two measurements are to the "truth." An example of something precise but not accurate may be a medical student attempting radial access for an arterial line. They may poke and prod the same general area, two-finger breaths above the styloid process of the radius. They may attempt in the same area multiple times, but not appropriately access. Accuracy would reflect the medical student that attempted access under ultrasound guidance, who was able to visualize access and ensure a proper or "true" stick in the radial artery. It is possible to have precision without accuracy, accuracy without high precision, and precision with accuracy. To continue our example, if you were researching best radial access practices and measuring from the anatomic snuffbox to the stick, an example of precision without accuracy would be a medical student attempting a radial access in the same

Translational Interventional Radiology. https://doi.org/10.1016/B978-0-12-823026-8.00098-5

location each time but never correctly accessing the radial artery. Accuracy without precision would similar to someone attempting access in someone who is a "tough stick" where ultrasound is used to visualize the best place along the forearm instead of the distal radius. Lastly, precision with accuracy would be an attending performing the access in the same place with a viable access created.

Precision is important in research because high precision minimizes data variability, which enables you to detect differences between data sets with a smaller sample size. Random errors in data collection reduce precision and increase the standard deviation of the data. Accuracy in measurement is how research uncovers new biological truths and delivers high-quality patient care. Accuracy decreases when there are methodological or systematic errors present.

What are sensitivity and specificity? How can i use these properties to explain what a positive and negative test mean?

Sensitivity and specificity are intrinsic properties of a test. In a clinical test for a disease, sensitivity is how well the test identifies those with a condition, while specificity is how well a test identifies those that are healthy. A 100% sensitive test is one that detects all those with a condition, meaning there are no "false negatives." A 100% specific test identifies all healthy patients and has no "false positives." Of course, there are few such perfect tests in real life, which is why it's important to understand how to calculate what the odds of someone actually having a condition if the test is positive, the odds of someone not having the disease if they test positive, the odds of someone not having a disease if they test negative, and the odds of someone having a disease if they test negative. In research, comparing a new test to an FDA-approved "gold standard" is how sensitivity and specificity are measured.

	"Gold standard" test positive	Negative
New test positive	50	2
Negative	4	60

Sensitivity is calculated by how well a test identifies the sick

$$\text{sensitivity} = \frac{\text{true positive}}{\text{true positive} + \text{false negatives}} = \frac{50}{50+4} = 92\%$$

Specificity is how well a test identifies the healthy

$$\text{specificity} = \frac{\text{true negative}}{\text{true negative} + \text{false positives}} = \frac{60}{60+2} = 97\%$$

In this example, we can calculate the positive predictive value, that is, how likely it is that someone has a disease if they test positive.

$$\text{Postive predictive value} = \frac{\text{all who test positive with the disease}}{\text{all with positive screening test}} = \frac{\text{TP}}{\text{TP}+\text{FP}} = \frac{50}{50+2} = 96\%$$

Therefore, if a patient gets a positive test, there is a 96% chance they have the disease and a 4% they do not.

Negative predictive value tells us the odds that those who test negative do not actually have the disease

$$\text{negative predictive value} = \frac{\text{all with negative test and no disease}}{\text{all with negative test}} = \frac{\text{TN}}{\text{TN}+\text{FN}} = \frac{60}{64} = 93\%$$

If a patient has a negative test, we know there is a 7% chance they actually have the disease.

How do incidence and prevalence change my positive predictive values and negative predictive values?

Incidence is the number of new people with a disease divided by the number of people at risk for the disease at baseline, while prevalence is the number of all cases of people with a disease divided by the disease at baseline. Incidence and prevalence change our positive predictive values and negative predictive values. When a disease is prevalent, the positive predictive value for a test increases, while the negative predictive value decreases. The opposite is true for a rare disease. With a rare disease, positive predictive value is low (you expect a higher ratio of false positives to true positives) and negative predictive value is higher (you expect few false negatives compared to true negatives).

How do I quantify risk? What formulas should I use? What pitfalls should I be aware of?

Quantifying risk is how we communicate what new interventions can do. Odds ratios, relative risk, number needed to treat (NNT), and number needed to harm (NNH) are calculations that aid in articulation of an intervention's impact.

The odds ratio explains the ratio of the odds of an event occurring in an exposed group compared to an unexposed group. This ratio helps explain how strongly a variable is associated with an outcome of interest.

	Cases	Controls
Exposed	45	30
Not exposed	10	22

$$\text{odds ratio} = \frac{\frac{\text{cases exposed}}{\text{cases not exposed}}}{\frac{\text{controls exposed}}{\text{controls not exposed}}} = (45 / 10)/(30 / 22) = 3.3$$

An odds ratio of 1 means the exposure does not affect the odds of cases. An odds ratio of greater than 1 means there are higher odds of a cases in the exposure group than the unexposed group. An odds ratio of less than 1 means the exposure decreases the odds of a case (is protective). When looking at odds ratios, there should be a 95% confidence interval reported. If that value does not cross 1, then it is considered a proxy of significance. When analyzing odds ratios, noncausal relationships may be observed if there is confounding in the data, which occurs when another factor (that is not measured in the exposed group) influences the development of a case. This is important to think about when drawing a conclusion based on odds ratios. For more information on odds ratios, see this paper.[1] Odds ratios can overestimate risks in common diseases, and therefore is more appropriate when studying rare outcomes.

Relative risk is the ratio of a probability of an event occurring in the exposed group versus the probability of an event in the unexposed group. However it specifies the likelihood in a way that does not speak to an absolute risk.

$$\text{Relative risk} = \frac{\text{risk of group A}}{\text{risk of group B}}$$

As a hypothetical example, the risk of developing complications with femoral access for patients with blood pressure measurements >180/100 mmHg is 20%, whereas for those with blood pressure measurements of >180/100 mmHg is 2%.

$$\text{Relative risk} = \frac{20\%}{2\%} = 10$$

Here we see that patients with blood pressure measurements of >180/100 mmHg are 10 times more likely to have a femoral access complication than those with blood pressure measurements of <180/100 mmHg. In applying relative risk, it is important to know the exposure status for the variable of interest, so this is an appropriate computation in studies where the exposure is known with certainty, like prospective cohort studies.[2]

Lastly, attributable risk (synonym: absolute risk reduction) is the difference in the probability of an event happening in the exposed group compared to the unexposed group. Using the example above,

$$\text{attributable risk} = 20\% - 2\% = 18\%$$

Attributable risk helps us determine the number of cases that could be prevented with the elimination of the risk factor. Attributable risk is commonly used in public health research. For further reading, this paper by Feng et al. includes graphical depictions comparing and contrasting the three concepts, odds ratio, relative risk, and attributable risk.[3]

NNT and NNH are two other ways to quantify the impact of a therapy or intervention.

Number needed to treat is the reciprocal of absolute risk reduction.

$$\text{Number needed to treat} = \frac{1}{\text{Absolute Risk Reduction}}$$

For a new blockbuster drug that reduces the rate of heart attacks by 10%, we can calculate

$$\frac{1}{10\%} = 10$$

The NNT is 10. Ten patients would need to be treated to prevent one heart attack with the new drug.

Similarly, the number needed to harm is calculated using the reciprocal of the attributed risk. Returning to our femoral access example where the attributable risk is 18%:

$$\text{Number needed to harm} = \frac{1}{\text{Attributable Risk}} = \frac{1}{18\%} = 5.5$$

Here we see that for every six patients who have a blood pressure greater than 180/100 mmHg before femoral access, one will be harmed.

Leveraging these statistics like absolute risk, relative risk, NNT, and odds ratios to communicate risks to the medical community and to patients can be particularly impactful because these values may change how information is interpreted. Explaining that a patient needs blood pressure management before a procedure can be more impactful if a provider can relate that high blood pressure increases the risk of complications by 10-fold. This same statistic may be perceived differently if the interventionalists explain that high blood pressure increases a patient's risk for femoral access complications by 18%, even though the same data were used to arrive to those numbers. Combinations of explaining these basics statistics can aid in both scientific reporting and, even more importantly, patient communication.

References

1. Szumilas M. Explaining odds ratios. *J Can Acad Child Adolesc Psychiatry*. 2010;19(3):227–229. https://pubmed.ncbi.nlm.nih.gov/20842279.
2. Tenny S, Hoffman MR. *Relative Risk*. Stat Pearls; 2020. https://www.ncbi.nlm.nih.gov/books/NBK430824/.
3. Feng C, Wang B, Wang H. The relations among three popular indices of risks. *Stat Med*. 2019;38(23):4772–4787. https://doi.org/10.1002/sim.8330.

Chapter 25

Distributions

Jason A. Fisher
Hartford Hospital Radiology, Hartford, CT, United States

Key points

Distributions describe the key features of a tested sample or population. In doing so, distributions put calculated values into context. They allow for further evaluation of the variation within a sample set and rapid comparison between groups.

Distributions describe the spread of values for a given variable across a measured group. Typically, the group at hand is the sample being measured. However, the population as a whole can be represented in a distribution by inference from the sample. To revisit these definitions, remember that the population encompasses the total number of individuals meeting the criteria of interest. Often there is no way to gather data on these individuals in total and so we seek to define a representative *sample* of the subjects. It is of particular importance to translation research to point out the *population* can be a hypothetical group. For example, if we are planning on studying the implementation of a new embolic being brought to market, then our population of interest may be the patients who demonstrate persistent occlusion of the target vessel at 1 year. This population does not exist in reality, but we can generate a reflective sample of participants based on existing data. Generating such a sample is incredibly important to the validity of a study. The simplest form of choosing a representative sample is termed random sampling. By simply selecting random subjects from a population, we assume that they will stand an equal chance of representing the diverse characteristics of the population. The larger number of subjects in the sample, the truer and estimation this becomes. However, it is obvious that the more detailed we become in our study population the less likely a random selection of subjects is to be representative. Returning to the previous example, an individual walking by your lab is not likely to match the population of patients who will wind up receiving this new embolic during a procedure. This introduces the need for the stratification of sample subjects. Starting with the population of interest, we may outline the characteristics they share, such as level of vascular disease, certain comorbidities, and hospital admissions. By then choosing a "random" sample of subjects from only those which these additional criteria we can generate a sample that while not representative of the local population at large, *is more* representative of the population to whom our new device will apply.

To transition into building a distribution of the values of interest, we need to define central tendency and probability density. While entire books focus on probability density functions, it is sufficient to say that the probability density function represents the curve of our distribution when the measured variable is continuous. As there are unending positive and negative possibilities, in theory, for a continuous variable in a population, the probability of any one value being selected in a sample approaches zero. However, when choosing to compare two groups of a population, we can discern the probability of one group's value relative to the other. Another key feature of the probability density function is the area under a range of its curve. Calculating the area under the curve for a region of the distribution yields the likelihood that a sample value will be in the same range of magnitude. This is the connection between significance values and the distribution of a sample (see chapter 60.) where setting a threshold for significance in the population can increase our confidence in the result due to nonrandom effects.

For any given variable of interest within a sample, we would expect multiple results of matching values by nature of having chosen an appropriately representative portion of the population. The most commonly occurring value within the sample is the central tendency and is represented by the peak of a distribution graph. Many different measures can be chosen as the central tendency. Most commonly, the average, or mean, becomes the measure of central tendency evaluated in a study as the "average" result should apply to the largest segment of the population. However, any measured value, can be selected and applied including simple values such as mode, median, or more complex

Translational Interventional Radiology. https://doi.org/10.1016/B978-0-12-823026-8.00069-9

features, such as the geometric and harmonic means of a series. While we will avoid a discussion of the calculation of these values, we will address the cases where choosing a different central tendency variable will adjust our study distribution. For example, utilizing the median over the mean as a measure of central tendency may help maintain continuity in skewed distributions containing few high magnitude outliers. These aberrant samples may cause undue influence on the mean however, the median will be less affected.

While a given central tendency variable yields the peak of our distribution, the variability around that measure will define the width and slope of our sample. This is the dispersion of the sample or variance. The standard deviation is the most commonly used example of a measure of dispersion, particularly in the normal distribution that we have been describing. Standard deviation is characterized as the square of the sample variance, where variance is the spread of the values from the measure of central tendency. Simplified equations are provided in the attached appendix. As mentioned, for a normal (or Gaussian) distribution, we can draw significant information from the standard deviation. Colloquially, the normal distribution is often described as a "bell curve." This comes from its symmetric nature about its mean. This relates to the standard deviation of the sample in that 68% of the sampled values will fall within the 1-standard deviation of the mean value, 34% above and 34% below. This pattern extends to include 95% of values within 2 standard deviations and 99% of values within three (Fig. 25.1).

To summarize, we have defined the measures for the center of a sample set as well as the spread of values we might expect to define surrounding the central measure. We

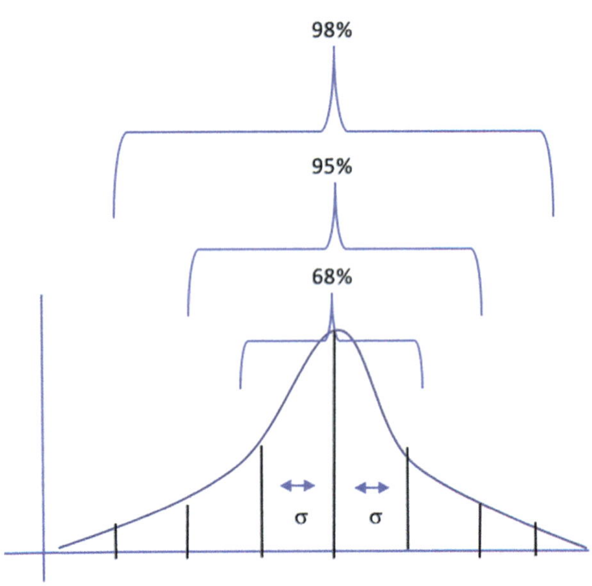

FIGURE 25.1 Standard normal distribution.

mentioned the probability function as describing our distribution curve but have not adequately described the features of the curve itself. For graphical comparison of the data, we measure the degree to which the peak value of the curve relative to the spread of values around the peak, a feature termed Kurtosis. The higher a distribution's Kurtosis, the sharper the peak of the curve. Be careful not to equate this with a higher, or larger magnitude, peak value. Rather it is the values clustered around the peak that demonstrate the lower frequency and result in a lower area under the curve. Put another way, a higher kurtosis represents a higher ratio of the frequency of the central value relative to the values surrounding the peak. A second characteristic of the distribution curve is the degree to which the graph is skewed. Skew measures differences in the mean and median values. A distribution can be said to be right skewed or positively skewed when its median value, or peak, is shifted to the left, and there are more outlying data points on the right side of the curve. This results in the mean value being higher than the median. The converse is true for a negatively skewed distribution, the peak will lay to the right and there will be weighting of the average to the left of the curve (Fig. 25.2).

Keeping in mind the definition of skew, we can formalize our description of the standard normal distribution above. The normal distribution is commonly found in nature and health sciences. An average value exists in the data set and the median value is equal to the average. A single peak is formed with a standard deviation as described above. The range of values is often represented as extending to approach, but not reach, zero. While this defines a normal distribution, statistical methods will very frequently utilize a construct known as *the standard normal* distribution. The standard normal utilizes a standard deviation of one about a mean value of zero. From this distribution, a table of *Z-scores* is derived and used to describe a data point's distance from the mean. By describing the data in this way, we can establish a universal language for where a value falls relative to its population. When performance on an exam or even response to a therapy is compared to its peers, we often describe its percentile or relative position in a similar population. This utilizes the standard normal distribution to calculate a Z-score for that value. By way of example, a z-score of +2 indicates the value to be 2 standard deviations from the mean in the positive direction. By using the features of the normal distribution regarding deviation, there are only ~2% of values in the distribution greater in magnitude than the tested value. Put another way that value is in the 98th percentile. A z-score of −1.0 would equate to roughly the 16th percentile. It is important to reiterate that, particularly in health translational research, the population distribution and characteristic values are not definitely known. A thorough review of available literature allows

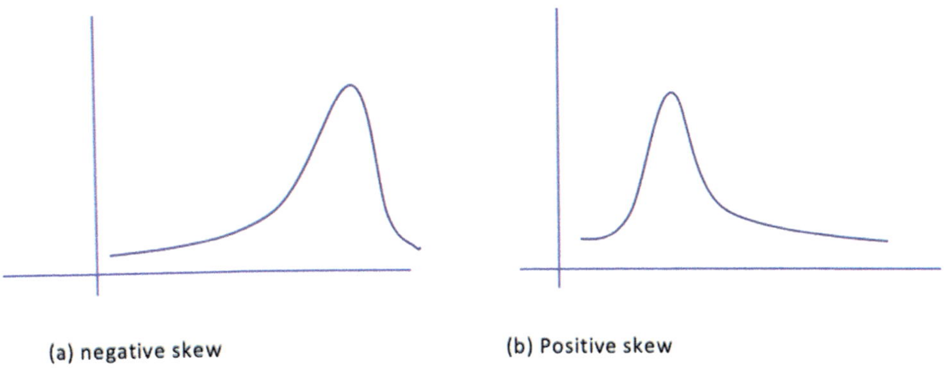

FIGURE 25.2 (A) Negative skew. (B) Positive skew.

for appropriate assumptions to be outlined, and a stratified sample to be obtained. The larger the tested sample, the more confident we can be with translating the results to the population of interest. For numerous reasons, we are constrained from reaching the sample size that would approach the population and so other distributions are employed to address these concerns.

In the setting of a small sample size relative to the population of interest, a *t-distribution* is often utilized. The t-distribution seeks to mirror the normal distribution criteria and therefore holds for the same statistical tests. So how is the t-distribution different from the standard normal distribution? Well, they are the same in one specific case; where the *degrees of freedom* of the t-distribution extend to infinity. While, unlikely to be helpful, this scenario is useful in working backward to describe the t-distribution. In the opposite scenario where the degrees of freedom approach zero, the ratio of the area under the curve for the tails of the distribution approaches the area under the curve for the central distribution resulting in the flattening of the distribution to the point of a uniform line. As we add degrees of freedom, we increase the kurtosis of the t-distribution. Degrees of freedom are generated from the number of samples we observe in a trial minus one. This is derived from a principle of predetermined variables. By knowing the overall average value of the sample as well as the actual values of all data points *less one* we can figure out the remaining variable and therefore its value is *predetermined*. Put another way, it has no degrees of freedom to assume any other value on the continuum. Instead of using a Z-score, in the t-distribution, a t-statistic is calculated and will be discussed in subsequent chapters (see Chapter 63.).

Thus far we have built a description of statistical distributions using the most common forms for variables that may take on any value (though may be bounded by upper and lower limits in a clinical sense). However, this approach neglects a large portion of datasets containing discrete variables. Discrete variables are values that fit a well-defined category. The most basic examples include "yes versus no" data. A collected data point either fits the described category or does not. Transferred to the clinical realm, a new drug-coated embolic will either meet a primary efficacy end point or not. We may choose to analyze the scenarios in-between using continuous statistics or analyze the discrete groups. References in literature to "*technical success*" of a procedure form another concrete example. However, discrete variables need not be contained in only two categories. There may be several depending on how we partition the outcome variable. An evaluation of the preprocedural risk of a technique may be discrete if we choose to evaluate clinical cutoff values based on outcomes. Let us take the evaluation of functional status before trans-arterial therapies for hepatic malignancy. Often patients will be grouped into the appropriate Eastern Cooperative Oncology Group Performance Status Scale (ECOG) level and outcomes will be evaluated between groups despite individual patients having met their ECOG criteria in different ways.

The most common discrete variable distributions include the binomial and Poisson distributions. The binomial distribution seeks to describe the scenario where several observations are performed and the probability of success versus failure is analyzed. Using our example above, if success is categorized as a threshold of partial tumor response we can evaluate the number of patients meeting the defined success category versus those that do not. We can use an assumed success rate in the population of 50% as the standard. If an improvement in device or technique is then being studied that results in 70% success, and the same number of observations are performed, the central tendency of our distribution will shift right (increase) and generate a small positive skew. As there are more successes, there are an equal number of losses from the outlying variables. Our kurtosis will increase, resulting in a sharper peak.

The Poisson distribution is used to characterize a scenario where a probability of an individual event occurring is evaluated over a set time. In such a setting, the outcome variable must be assessed as a discrete binomial variable, success versus failure, the definitions of which can be set by the investigator. Using the average number of successful results in a given time period, we can compare the Poisson distribution to the binomial. The success rate observed is defined as the variable lambda, λ, and is equal to the number of trials multiplied by the probability of success of each trial. Note the probability of success was the outcome of interest in the binomial distribution. Restated, the success rate observed over time is our variable of interest in the Poisson distribution. It is equal to the number of trials times the probability of success (the variable of interest from the binomial distribution). As the number of trials increases and the chance of an individual success is small, the Poisson distribution will approximate a binomial distribution.

Appendix of equations

Test	Description/Equation
Mean	Average value of a set $\frac{\sum x}{n} = \mu$
Mode	Most frequent occurring value in a set
Median	Central value in a set
Variance	The sum of each value less the mean value divided by the number of values in the set $\sigma^2 = \frac{\sum (x-\mu)^2}{n}$
Standard deviation	Square root of the variance $\sigma = \sqrt{\sigma^2}$
Standard normal transform	Individual value less the mean value divided by the standard deviation $Z = \frac{x-\mu}{\sigma}$

Chapter 26

Hypotheses and types of error

Jason A. Fisher
Hartford Hospital Radiology, Hartford, CT, United States

Key points

Formation of an acceptable hypothesis sets the stage for the remainder of the statistical evaluation of your data. When testing a hypothesis, we relate our results back to the null hypothesis and seek to accept or overturn this lemma. An understanding of inherent error types is crucial to result interpretation as well as study design via their connections to significance and study power.

By this point, you have seen that clinical statistics seek to separate effects generated by chance from those which directly result from the proposed trial. Hypothesis generation and hypothesis testing are frameworks for formalizing this difference into a statement that we can accept or reject with the evidence from a study. In the simplest case, two statements can be generated to this end. We term these statements the *null hypothesis* and the *alternative hypothesis*. Typically, the null hypothesis defines the scenario where the tested intervention yields no change in the outcome variable. In clinical and translational research, the alternative hypothesis typically defines the improvement in a clinical scenario. By convention, statistical methods seek to *reject the null hypothesis* to favor the alternative hypothesis. That is, we can show that a statistical difference was found and postulate it was our intervention that directed this result, not simple chance. Furthermore, failing to reject the null hypothesis is not equivalent to the null hypothesis statement being fact. Rather, our design failed to prove it otherwise. Additional methods for increasing the certainty in a given trial are addressed in upcoming chapters. Several parameters will be introduced, which create threshold values for ruling out the effects of chance and false results, both negative and positive.

To start defining a hypothesis test, it is best to work in reverse from what the trial seeks to prove. Our null hypothesis will be the opposite (or the case where no change is demonstrated between tested groups). Therefore, the null hypothesis (H_0) is most often defined as the measure of the tendency for group 1 being equal to that of group 2. A simple example in which the tested value is the mean of the groups follows. The statement holds if the tested value were to be the group mode, median, etc.

$$H_0 \mu_1 = \mu_2$$

or

$$H_0 \ \mu_1 \geq \mu_2$$

While common, it is not required that the baseline hypothesis be that no difference exists. It may often be the case, particularly in translational research, that the improvement is iterative and we want to show a difference compared to a standard threshold value. We can adjust our equation to set H_0 greater than, equal to, or lesser than the accepted value. Consider evaluating the intraprocedural pulmonary pressure for patients undergoing lytic infusion for pulmonary embolism. We may hypothesize that the test group pulmonary pressure decreases 1.5-fold compared to that of the control. Therefore, our null hypothesis may state group 1 pressures are greater than or equal to 1.5 times that of group 2.

Defining an alternate hypothesis is then made simple. One needs to define the case where the null hypothesis is not true. For example,

$$H_1 \ \mu_1 \neq \mu_2$$

Once we have defined the hypothesis statements, it follows that we need a mechanism to test these statements against one another. Many tests are described in the following chapters. However, before we can go about seeking an answer, we need to understand the possible outcomes of our statistical test. In total, there are four ways to reach three possible conclusions in hypothesis testing. The following table is an often-used graphical representation. In the case that the null hypothesis is indeed true, we may correctly not reject it and move on; however, we may incorrectly reject the null hypothesis anyways, resulting in a false-positive result. This scenario defines a Type I error. Similarly, when the null hypothesis is indeed false, our

Translational Interventional Radiology. https://doi.org/10.1016/B978-0-12-823026-8.00056-0

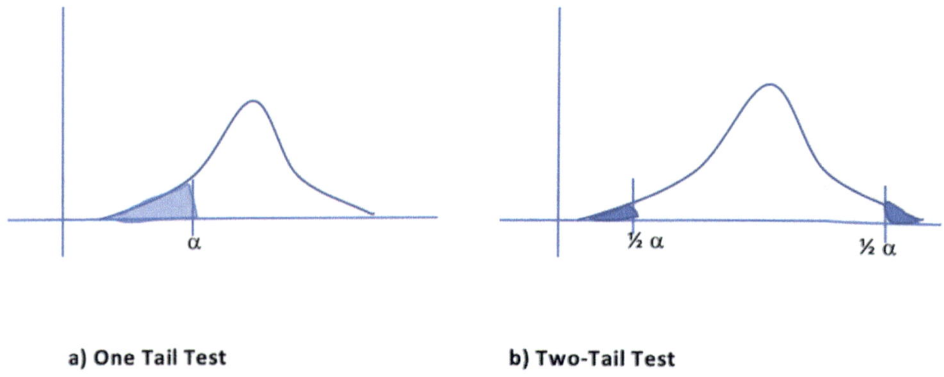

FIGURE 26.1 Significance thresholds.

study may lead us to correctly reject the null hypothesis, or we are doomed to fail to reject the null hypothesis, generating a false negative result. This is a Type II error.

	H_0 true	H_1 false
Reject null hypothesis	Type I (false positive)	Correct result
Fail to reject null hypothesis	Correct result	Type II (false negative)

Both Type I and Type II errors are inherent to the practice of statistics. However, appropriate study design does allow the investigator to minimize the risk of either error type at the expense of the other. The relationship therein is governed by the significance value of the study. When our test statistic exceeds our significance threshold in absolute magnitude, we have proven that the detected effect is unlikely due to chance. Commonly, significance is set to a standard threshold of 0.05 or 0.01 and is represented by the character α. The significance represents the risk of committing a Type I error, rejecting the null hypothesis when it is true, that is, detecting an effect where none exists. In other words, the α level we set helps generate confidence in the decision to overturn the null hypothesis. By choosing a lower α value (i.e., 0.01), we make it harder to detect an effect, but we can be 99% certain that a detected effect is nonrandom and that we have successfully rejected our null hypothesis. The variable, β, similarly represents the probability of reaching a Type II error (failing to reject the null hypothesis when it is indeed correct). Put another way, failing to detect a difference in tested values. Failing to detect this difference is mitigated by the power of our study discussed in subsequent chapters (Fig. 26.1).

Lastly, we must consider the directionality of our studied variable. Let us revisit the initial example of our null hypothesis. We stated the measured means of the two groups to be equal, therefore our alternate hypothesis would be that they are not equal. If our outcome measure is anything other than the population measure (higher or lower), we will overturn our null hypothesis provided we meet our chosen significance level. Such a test would be a bidirectional, or two-tailed, test. Critically, in bidirectional tests, our significance threshold is split over two extremes of the distribution. A 5% α threshold would therefore encompass 2.5% of the area under the curve of lower extreme and 2.5% area under the curve of the upper extreme. We would need to generate a larger magnitude of effect to meet the significance level. Conversely, when we utilize a one-tailed test, we can demonstrate a smaller magnitude of effect. Therefore, a one-tailed test is "easier" to perform and successfully overturn the null hypothesis. When we discuss the area under the curve in hypothesis testing, there is an underlying assumption that the sample and population both adhere to the Gaussian distribution.

Further reading

1. Chin R, Lee B. *Principles and Practice of Clinical Trial Medicine*. 1est ed. London, UK: Elsevier; 2008.
2. Urdan T. *Statistics in Plain English*. 4th ed. New York, NY: Taylor and Francis; 2017.

Chapter 27

Power

Jason A. Fisher
Hartford Hospital Radiology, Hartford, CT, United States

Key points

Statistical power is a key determinant of a study's practical viability and statistical validity. This is especially important in translational research in healthcare as a result of the common necessity of smaller samples sizes.

Like most of the measures we have discussed, statistical *power* is underpinned by probability: in this case, the probability of correctly rejecting the null hypothesis. As we have just defined in the preceding chapter, being statistically correct refers to successfully rejecting our null hypothesis when it is indeed false. Put another way, the power of a study refers to the study parameters (e.g., sample size, effect size) being robust enough to notice a meaningful difference in the set of data provided, when in fact, there is a difference. It is important to note at the outset that power only plays a role in correctly rejecting the null hypothesis while favoring the alternative hypothesis, that is, when we are confident there is a difference to detect.

Before addressing how to ensure a given study is prepared to achieve this task, we need to frame power into the discussion of the error. We just defined power to be the ability to correctly reject the null hypothesis. It is a small leap, then, to our formal definition and mathematical expression of a study's power. Failing to reject the null hypothesis when in fact it is false is known as Type II error, or a false negative. From the complete set of possibilities for the null hypothesis being false, we can subtract our Type II error to yield the cases where the null hypothesis is rejected appropriately. Therefore, the power of a study is equivalent to (1 − Type II error), which is (1 − probability of a false negative) or $(1-\beta)$. You can see from Fig. 27.1 and from the previous equation that the smaller β is, the larger will be our power or our ability to correctly pick up on a statistical difference between groups in our data.

So how can we adjust power during study design? We need to understand how power is related to the size of our sample (n) and to the effect size of our intervention. Several variables relate to a study's power, and it is important to evaluate these relationships to formalize our definition of power thus far. However, you will see that often these values are preset, leaving sample size as the simplest way to achieve a desired study's power.

When designing a study, we would like to set an adequate power. For illustration, in a simple test of two new catheter designs with respect to their flow rate with a sample size of one trial in each group to reject our null hypothesis, despite it being false, is little more than chance. As we add five trials to our sample, we may see a trend begin to emerge but upon statistical testing, there is no significance. Next, extend our sample size to 1000 trials, and now the average values for our measured flow rates between catheters may have sufficiently separated that a difference will hold up to testing and prove significant. Increasing the number of subjects in a trial often reaches a practical ceiling or plateau based on logistics, overhead costs, and other factors. Combining data sets from multiple study sites can increase the power of a statistical test if the study designs are similar enough to each other.

The effect size is a descriptive value that addresses the magnitude of our results relative to each other (as opposed to the significance with which we describe the relationship of the results to a hypothesized null or zero magnitude). Effect size exhibits a direct relationship with power: as effect size increases, power will increase. Effect size can increase in two ways: when the difference between the average values of the null group and the alternative group increases (with a set standard deviation) or when the standard deviation of the sample observations is decreased (see Fig. 27.2). Similarly, the significance threshold for a study (Type I error or α) will result in a larger power as it increases, from example, from $\alpha = 0.05$ to $\alpha = 0.10$. This is a bit more intuitive considered in the reverse. If we are willing to accept a higher probability of incorrectly rejecting our null hypothesis despite it being true (a Type I error or false positive), our study will have a higher power or ability to avoid false negatives (Type II errors).

Translational Interventional Radiology. https://doi.org/10.1016/B978-0-12-823026-8.00067-5

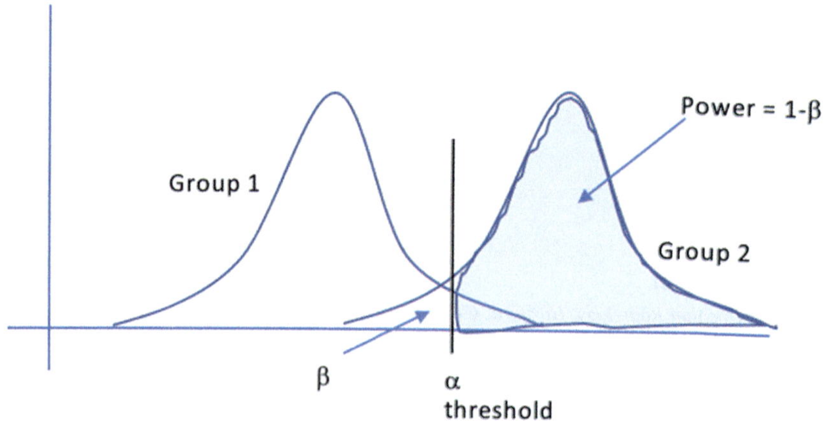

FIGURE 27.1 Graphical representation of power.

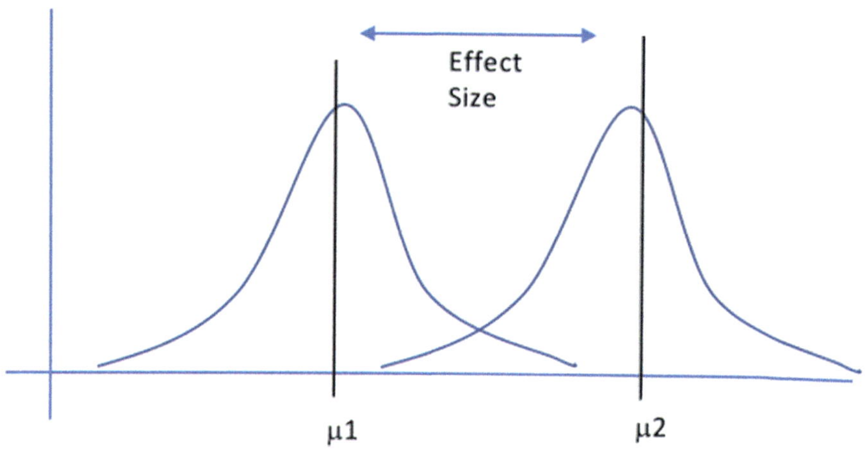

FIGURE 27.2 Graphical representation of effect.

With the knowledge of the variables that exert an effect on statistical power, we can turn to a discussion of using power as a functional measure, termed *Power Analysis*. While power analysis can involve calculating any of the variables above, two main examples are crucial for a foundation in translational research. The first is the simple question of "*how much power does a study have?*" which we addressed in the definition. However, far more important to the topic of study design is "*what sample size do I need to obtain appropriate power in my study?*"' Given that this is central to study design, we need to determine power at the study design phase (i.e., before the study is carried out). As such, typical standards exist for the defining variables above. While this can be adjusted as necessary, an α value of 0.05 to define a significant result and a power of 0.80 are commonplace. Phase 3 trials and multicenter studies usually require a higher power, while pilot studies are allowed to have a lower one. Using these standards, we will need to determine the effect size we wish to detect and from there calculate a sample size we are required to enroll. Common systems for determining effect size include the Cohen's *D* test and the Pearson *r* correlation. While the mechanisms specific to each statistical test will be described in detail in the future chapters, we will address one of the most common examples for evaluating averages between two groups and determining a study sample size.

For evaluating significant differences between two mean values, the student's *t*-test is a common framework. This would be a good choice for a study that compares two devices, protocols, medications, and so on. The average value for the variable of interest within each group can be compared for statistical significance and translated to clinical significance thereafter. We can accept a standard α value of 0.05. The Cohen's *D* test describes the magnitude between the average values after adjustment for their standard deviations. Similar to the significance level, a standard threshold can be applied. In the case of Cohen's *D*, the threshold for a large effect is set to 0.80. Lower values may be used to describe mild or moderate effects. As we are seeking to determine our study sample size, we need to set a similar threshold for the desired power of our study. Remembering that power is directly resolved from our β value we can solve for the threshold power using a standard β value of 0.2. By doing so, we are accepting a 20% chance of committing a Type II error (false negative). This can be adjusted as necessary in the study design. Our formula, $1-\beta$, yields a power of 0.8, or, an 80% chance of detecting an effect where one exists (a true positive).

With these parameters in place, we are ready to determine our sample size. The following formula for sample size includes constant values from the Z tables of standard normal distribution. The first Z_α is a constant derived from our chosen significance level and the type of outcome variable being studied. The second, $Z_{1-\beta}$, is directly related to the chosen power of the study. For our example, we will use a comparison of two study groups, evaluating a continuous outcome variable. While the derivation of the following equations and Z value tables are outside the scope of our text, standard values are readily available. In this case, our Z table values will yield an Z_α of 1.96 for two-sided $\alpha = 0.05$ and $Z_{1-\beta}$ of 0.84 for $\beta = 0.2$. We will be enrolling an equal number of patients into each study group and therefore define one value, *n*, as the number of participants needed per group (a variable K can be used as the ratio of the participants in each group when they are unequal).

$$n = \frac{\left(\sigma_1{}^2 + \frac{\sigma_2{}^2}{K}\right)\left(Z_{1-\alpha/2} + Z_{1-\beta}\right)^2}{\Delta^2}$$

We need to make a few more assumptions about our sample to finish defining these variables. σ, which is the standard deviation of a sample, here represents our best knowledge about the distributions within populations being studied based on available literature. Δ will represent the absolute magnitude of the effect size between the estimated mean values we study. Now expand our example to the clinical realm. Using the above standard threshold values and assumptions let us evaluate two groups who are reporting an average pain score using a standard versus an experimental pain regimen during uterine fibroid embolization. Our study will evaluate the Visual Analogue Pain Score responses in each group. Based on existing data we may expect that the standard treatment group reports a pain score of 40 mm. We expect our treatment to demonstrate a 20% (8 mm) improvement or a pain score of 32 mm. The standard deviation in reported pain scores was determined to be ±10 mm. Substituting for our equation:

$$n = \frac{\left(10^2 + \frac{10^2}{1}\right)(1.96 + 0.84)^2}{|40 - 32|^2}$$

$$n = \frac{(200)(7.84)}{64} = 24.5 \text{ or } \sim 25 \text{ subjects}$$

As we cannot sample a fraction of a subject, we must round up to the nearest whole to ensure an appropriate sample group. Here we calculate a sample group of 25 subjects per study arm (total 50 subjects) to yield an 80% chance of detecting our true positive if a difference exists.

Take some examples from the literature as well, where sample sizes are set at practicable sizes for a clinical setting. For example,[1] compared fewer than 100 patients in each group, observing first attempt success rates for central venous access in pediatric patients to be 20 percentage points higher in the group that used ultrasound guidance as compared to landmark guidance (65% vs. 45%). The study's design used an α of 0.05 and a β of 0.20, or a power of 0.80. A similar study by[2] enrolled fewer than 75 Emergency Department patients in each group and observed, among other outcomes, a statistically significant 1.9 fewer attempts required to gain central venous access when ultrasound guidance was used as compared to landmark guidance. An unrelated study by[3] retrospectively investigated 36 total pediatric patients'' status postpartial splenic embolization procedure. This study, investigating a less frequently encountered clinical presentation and procedure, will understandably have smaller sample sizes. Nio et al. were still able to observe statistically significant improvements in hematologic indices likely related to the larger effect size of partial splenic embolization in patients with hypersplenism from portal hypertension. In these ways, statistical power can be achieved in different circumstances with different sample sizes and different effect sizes.

References

1. Bruzoni M, Slater BJ, Wall J, et al. A prospective randomized trial of ultrasound- vs landmark-guided central venous access in the pediatric population. *J Am Coll Surg*. May 2013;216(5):939–943. https://www.journalacs.org/article/S1072-7515(13)00112-9/fulltext.
2. Miller AH, Roth BA, Mills TJ, et al. Ultrasound guidance versus the landmark technique for the placement of central venous catheters in the emergency department. *Acad Emerg Med*. August 2002;9(8):800–805. https://pubmed.ncbi.nlm.nih.gov/12153885/.
3. Nio M, Hayashi Y, Sano N, et al. Long-term efficacy of partial splenic embolization in children. *J Pediatr Surg*. December 2003;38(12):1760–1762. https://pubmed.ncbi.nlm.nih.gov/14666461/.

Chapter 28

Regression analysis: Applications to radiological research and recent examples

Tao Liu[1], Abraham Liu[1] and Christopher Ovanez[2]

[1]Department of Biostatistics, Brown University, Providence, RI, United States; [2]Russell H. Morgan Department of Radiology and Radiological Science, Johns Hopkins Hospital, Baltimore, MD, United States

Key points

- Linear models and generalized linear models form a big class of regression methods that are widely used in clinical research.
- Poisson regression, logistic regression, and Cox proportional hazard models can be used to analyze nonnormally distributed data, such as count, binary, categorical, ordinal, or time-to-event data.
- Semiparametric methods (such as Cox PH model) and nonparametric methods (such as tree-based methods) require weaker assumptions and can be used when parametric distribution assumptions do not hold.
- Regression analyses reveal association. Under certain conditions, regression can be used to account for confounding effects to infer causation.
- Regression models can be used for both estimation and prediction.

Introduction

Regression analysis is a commonly used quantitative analysis technique in clinical and various scientific research. A general form of the regression model can be expressed as follows.

$$h\{E(Y)\} = f(X;\beta)$$

where Y denotes the outcome variable of interest, $X = (X_1, X_2, \cdots, X_p)^\top$ is a vector of predictor variables (Y and X are also referred to as dependent and independent variables, respectively), $f(\cdot;\cdot)$ is a known function defined by the unknown parameter(s) β, and $h(\cdot)$ (often referred to as link function) is a known transformation function on the expected value of Y.

A widely used regression model is a simple linear model that describes the relationship between the outcome Y and a single predictor X,

$$E(Y) = \beta_0 + \beta_1 X$$

or

$$Y = \beta_0 + \beta_1 X + \varepsilon,$$

where ε is an error term presumably independent of X's. Using the least-squares estimation of the model, we require $E(\varepsilon) = 0$. For maximum likelihood estimation, we typically make a distributional assumption of $\varepsilon \sim N(0, \sigma^2)$. For this linear model, $h(u) = u$ is the identity function; $f(X;\beta)$ is a linear function of X, and $\beta = (\beta_0, \beta_1)^\top$ is the model coefficients. The parameter β_0 is called the intercept, denoting the expected value of Y when $X = 0$; and β_1 is the regression coefficient (slope) which describes the expected change of Y as X increases by one unit; see Fig. 28.1.

The association between Y and X is described by $\beta_1 = E(Y|X+1) - E(Y|X)$. Having $\beta_1 = 0$ means that Y and X are linearly independent or mean independent. The coefficient β_1 is closely related to the Pearson correlation coefficient r between Y and X,

$$\beta_1 = r\frac{\sigma_Y}{\sigma_X},$$

where σ_X and σ_Y are the standard deviations of X and Y, respectively.

Linear and generalized linear models

Linear models can include multiple predictor variables and belong to a large class of models, called generalized linear

Translational Interventional Radiology. https://doi.org/10.1016/B978-0-12-823026-8.00026-2

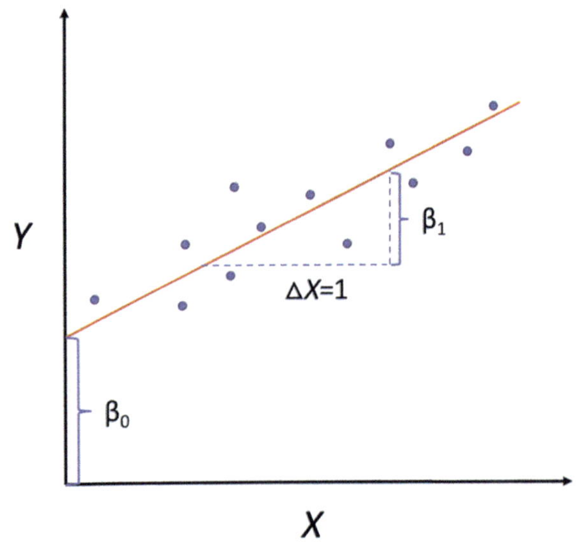

FIGURE 28.1 A hypothetical example of linear model fitted to simulated dataset.

models (GLMs).[1] Multiple/multivariable linear regressions are specified as

$$E(Y) = \beta_0 + \beta_1 X_1 + \cdots + \beta_p X_p,$$

or

$$Y = \beta_0 + \beta_1 X_1 + \cdots + \beta_p X_p + \varepsilon,$$

where ε is an error term independent of X's. We often write the right-hand side of the equations as

$$\beta^\top X = \beta_0 + \beta_1 X_1 + \cdots + \beta_p X_p,$$

where $\beta = (\beta_0, \beta_1, \cdots, \beta_p)^\top$ is a vector of model parameters and $(1, X_1, \cdots, X_p)^\top$ is a vector of model predictors. The first element of β is the intercept, and the rest elements represent the expected change in Y for a unit increase in the corresponding predictor of X while fixing the other predictor variables unchanged. We often refer to $\beta^\top X$ as the linear predictor. Linear models are widely used in scientific research when the outcome of interest is continuously and normally distributed.

Linear models are a special case of generalized linear models. Beside normally distributed outcome data, GLMs can be used to analyze data that are counts categorical, or positively distributed data.[1] Typically, GLMs are used when the outcome data distribution belongs to the exponential distribution family (e.g., normal, exponential, and Poisson distributions).

GLMs take a general form as

$$h\{E(Y)\} = \beta^\top X,$$

where the link function $h(\cdot)$ is chosen based on the distributional properties of Y. For example, for count or nonnegative data, we use a log link function $h(w) = \log(w)$. The resulting model is called Poisson regression or log-linear model. For binary outcome data, we use a logit link function $h(w) = \log\{w/(1-w)\}$. The resulting model is called logistic regression model. In the following section, we will go over these two commonly used GLMs.

Poisson regression: The Poisson distribution is primarily used when the outcome Y is count data or frequency data. The Poisson regression model assumes that

$$\log\{E(Y)\} = \mu(X) = \beta^\top X,$$

where μ represents the log(population mean) for a given vector of predictor variables X. Due to the construction of the model, Poisson regression is also referred to as log-linear regression. The model coefficients β are interpreted as the log incidence rate ratio (relative risk). Specifically, note that

$$\log\left(E\left(Y \middle| X_1, \cdots, X_j + 1, \cdots, X_p\right)\right. - \log\left(E\left(Y \middle| X_1, \cdots, X_j, \cdots, X_p\right)\right) = \beta_j$$

which implies

$$\frac{E\left(Y \middle| X_1, \cdots, X_j + 1, \cdots, X_p\right)}{E\left(Y \middle| X_1, \cdots, X_j, \cdots, X_p\right)} = \exp(\beta_j).$$

While fixing the values of other predictor variables, $\exp(\beta_j)$ represents how much the rate (frequency) for an event will increase for each unit increase of X_j while keeping other predictor variables fixed.

An analyst should keep in mind that due to the properties of the Poisson distribution, the Poisson regression model assumes that the conditional variance is the same across X. When this assumption does not hold, it is called "overdispersion". In this case, a negative binomial regression model, which is a generalization of the Poisson regression model, should be used. Another useful extension is zero-inflated Poisson regression which is used in situations where there are unproportionally high numbers of zero counts in the outcome data.

Logistic Regression: Logistic regression is a simple but powerful model for discrete or categorical outcome data. It is based on the logit (**logi**stic **unit**) function defined as $logit(w) = \log\{w/(1-w)\}$. There are three types of logistic regression models, respectively for binary, multinomial, and ordinal outcomes. For the more common binary outcomes, the logistic model has the form:

$$logit(p) = \beta^\top X,$$

where $p = E(Y) = Pr(Y = 1)$ represents the probability that the event of interest occurs and β represents the model coefficients of these predictor variables. The coefficients β

can be interpreted in a similar manner to the Poisson regression model coefficients, but as log odds ratios. (Not that $logit(p) = \log\{p/(1-p)\} = \log\{\text{Odds}\}$).

The multinomial and ordinal logistic regression models are a generalization of the binary logistic regression model. Multinomial logistic regression is for data with three or more categories, and ordinal logistic regression is for data with three or more outcomes that have order. These models will not be further covered in this section.

Regression analysis for time-to-event data

Time-to-event data as known as survival data records the time until an event of interest (e.g., death) occurs. There are two important features in time-to-event data (1) the outcome of time to event is positively distributed and usually not following a normal distribution; and (2) censoring is commonly observed. Censoring means that the outcome is only known to fall into an interval. For example, right censoring occurs when a study subject does not experience any event at a study terminal, say time C. In this case, the event time for the subject is not completely missing but known only to be a value in (C, ∞). There are other types of censoring such as left censoring and interval censoring. In this section, we focus on two regression models for analyzing such data: Cox proportional hazard models and accelerated failure time models.

Cox proportional hazard (PH) model: The Cox PH model is expressed by a hazard function $\lambda(t)$, which is defined as the immediate risk of a subject experiencing an event at time t given that the subject has "survived" to the time t, $\lambda(t) = f(t)/S(t)$ where $f(t)$ and $S(t) = \int_{u>t} f(u)du$ are density and survivor functions, respectively. The Cox PH model assumes that

$$\lambda(t) = \lambda_0(t)\exp(\beta^\top X)$$

where $\lambda_0(\cdot)$ is the baseline hazard function which is left to be nonparametric with only the constraint that it is a nonnegative function. Note that for the Cox model, the linear predictor $\beta^\top X$ does not include an intercept (the baseline hazard function can be regarded as the "intercept"). It is easy to verify that the vector $exp(\beta)$ has the interpretation of hazard ratios (HRs). A hazard ratio greater than 1 (or equivalently, any β_i greater than 0) indicates that as corresponding predictor variables increases, the risk of having an event (e.g., death) increases, and the expected survival time decreases. On the other hand, an HR of less than one is considered a "protective" prognostic effect. It should be noted that the Cox PH model relies on two assumptions to be valid. First, the hazard functions are proportional over time for different levels of X. Second, censoring is noninformative, in the sense that the censoring time is independent of event time among those at risk over time.

Accelerated failure time (AFT) model is an alternative parametric model to the Cox PH model. Instead of making the proportional hazards assumption, the AFT model instead assumes that predictor variables affect survival time T multiplicatively by a factor, expressed in the following form:

$$log(T) = \beta^\top X + w,$$

where w represents an "error" following a parametric distribution. Common distributions used include exponential, Weibull (which is a generalization of the exponential distribution), and normal distributions.

Although the AFT model does not make a proportional hazards assumption, it does assume that the data has a parametric distribution. That is a strong assumption and may be violated in many cases, which is why the AFT model is not used as widely as the more flexible semiparametric Cox PH model. Readers can find more details about survival data analysis techniques in Klein.[2]

Tree-based regression methods

Classification and regression trees form the basis of a big class of tree-based methods. The tree-based methods are generally regarded as nonparametric methods. A simple tree involves a sequence of binary splits of the sample space. See Fig. 28.2 for example: At each split, a local optimal cutoff is chosen on a variable such that the resulting two leaves differ most from each other while within each leaf, the data are relatively homogeneous. The splits are made iteratively to minimize a prespecified loss (impurity) function such as misclassification error, Gini index, or cross-entropy (deviance). At the terminal nodes (no further splits), a simple model (e.g., a constant) is fit to the data in each of the nodes.

Tree-based methods are simple yet powerful. They are particularly useful when the relationship between the outcome variable and predictor variables cannot be described adequately using some simple function form (e.g., by a linear function) and when there are complicated interaction effects among predictor variables. A simple (and sparse) tree can be easily implemented for clinical use for quick risk assessment, triage, and classification. However, simple trees also have drawbacks. First, a simple tree can overfit the data if no pruning is properly performed. Second, simple trees do not yield a smooth prediction. Third, their prediction/classification can be unstable if the trees are developed by a small dataset.

Random forest (RF) models can address some of these drawbacks. RF uses the bagging/bootstrap method to generate multiple datasets (sample with replacement). From

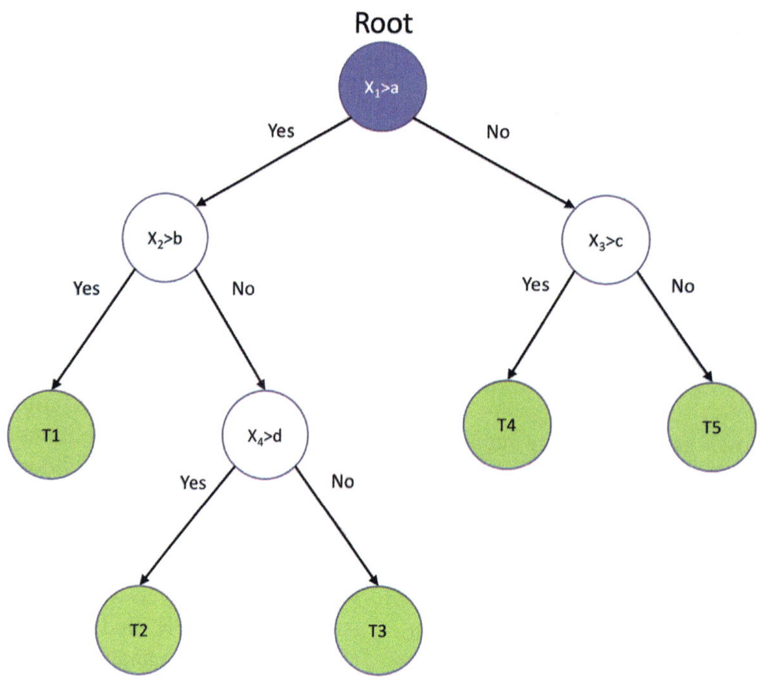

FIGURE 28.2 A binary classification tree. Green: terminal nodes/leaves.

each dataset, a tree is generated and the resulting trees are averaged at the end for prediction or estimation. Through bagging average, the prediction is smoothed and the variance is reduced. RF is generally less subject to overfitting, i.e., random errors in individual trees tend to be canceled out through averaging.

Boosted trees are another method that involves fitting multiple trees to a dataset. Unlike RF, boosted trees are fitted sequentially where each tree is fit to the "residuals" of data that cannot be explained from former trees. XGBoost (eXtreme Gradient boosting) method uses Taylor polynomials and model regularization to accelerate model fitting and stability. XGBoost is particularly useful for analyzing large datasets and is often one of the best choices for solving prediction problems.

Unstructured data and analysis methods

Unstructured data are data that do not have a defined structure to it, like well-defined columns, attributes, or features (e.g., a random assortment of images, video, audio, or text). Unstructured data are often encountered in radiological research. Analysis of such data can be difficult for parametric models, but analysis can be done through nonparametric methods like neural networks.

The idea of neural networks is inspired by the human nervous system. A neural network is composed of three types of layers: input layers, hidden layers, and output layers. All layers are composed of neurons. From the neurons in the input layer, calculations are made through the given information and are passed to the neurons in the next layer, forming many types of interactions that exponentially increase with the number of total neurons. See Fig. 28.3 for an illustration.

The benefit to neural networks is their ability to identify inputs on their own. Common types of neural networks that can deal with unstructured data are convolutional neural networks commonly used for analyzing image data and recurrent neural networks for analyzing unstructured time-series data in natural language processing. These methods will not be further covered in this section.

Association, causation, and prediction

Without additional assumptions or conditions, regression analysis can be used to reveal associations between predictor variables and outcomes. For example, for a simple linear model $E(Y) = \beta_0 + \beta_1 X$, the regression coefficient β_1 is associated with the Pearson correlation between the Pearson correlation between X and Y. However, the existence of association between X and Y does not imply that X has a causal impact on Y. Fig. 28.4 is a directed acyclic graph showing a simple case that another factor W that affects both X and Y. Without accounting for the effect of W, we may find that marginally (without adjusting for W) X and Y are associated, even though there is no causation relationship between them (in the sense that if there is no arrow from X to Y, interventions on X will not lead to changes in Y). Here, W is referred to as a confounder.

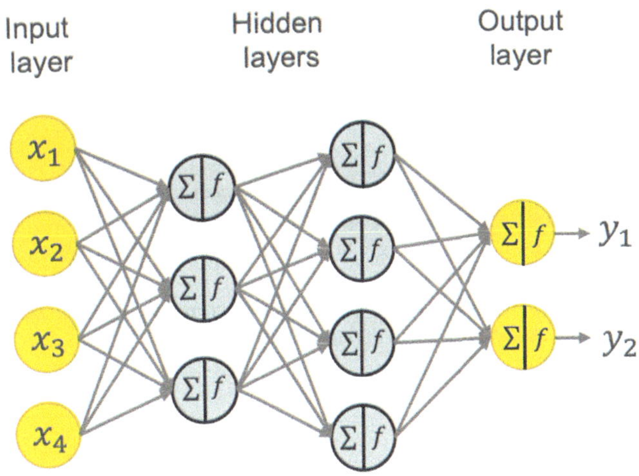

FIGURE 28.3 A simple example of the layers in a neural network (https://www.knime.com/blog/a-friendly-introduction-to-deep-neural-networks).

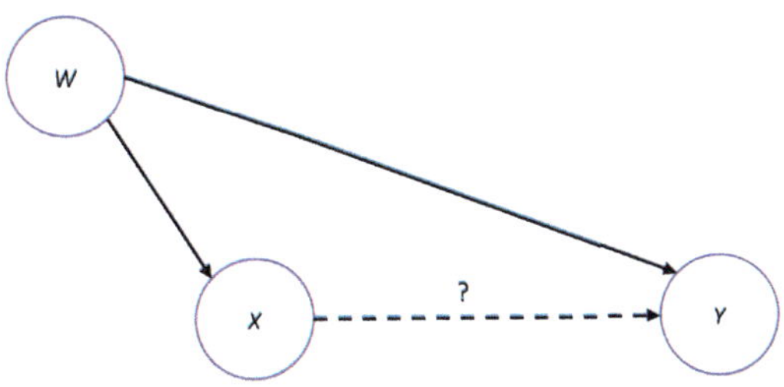

FIGURE 28.4 A directed acyclic graph (DAG) showing a potential causal relationship between X and Y with a confounder W.

There are certain conditions under which association from regression analysis can be inferred as causation. One such condition is that the level of X is totally random among the study sample. For example, X can be an indicator of study treatment randomization in a randomized clinical trial. Under perfect compliance, the regression coefficient β_1 can be interpreted as the causal effect of comparative treatments.

Another condition is when we can assume that there is no unmeasured confounding. More formally, let $Y(x)$ denote the outcome that would potentially (counterfactually) be observed if we can set $X = x$. We assume that

$$\{All\ Y(x)\} \perp X \mid W,$$

which is also called the strong ignorability assumption. That means that conditional on W, X is independent of the potential values of Y, that is, can be regarded as being randomization at each level of W. Under this assumption, an adjusted regression analysis can be used to account for the confounding effect

$$E(Y) = \beta_0 + \beta_1 X + \beta_W^T W$$

and if the linear assumption still holds, the regression coefficient β_1 has a causal interpretation.

Regression models are also widely used for prediction. Instead of focusing on estimation and inference of model parameters, prediction models are typically trained and selected using loss functions that focus on minimizing out-of-sample (out-of-bag) prediction "error." For this reason, development of prediction models often involves splitting data into three parts as training, validation, and test datasets, or involves cross-validation. To make a more flexible and accurate prediction, nonparametric regression models are typically used. At the same time, to avoid overfitting, some model regularization techniques are implemented to increase prediction stability and to improve out-of-sample prediction performance.

Examples of recent studies

Most quantitative research in radiology involves regression analyses. The following are examples of recent studies that were devoted to COVID-19 diagnosis and prognosis using artificial intelligence and survival analysis models. Study by Wang et al.[3] developed a deep learning algorithm to differentiate between benign and malignant ovarian lesions through routine MR imaging via convolutional neural networks. Study by Tang et al.[4] used univariate and multivariate Cox regression survival analysis to determine that encephalopathy at admission predicts later progression to death in SARS-CoV-2 infection, which may have important implications for risk stratification in clinical practice. More recently, Kim et al.[5] designed and analyzed a diagnostic AI model that addresses selection bias, increased accuracy in diagnosis, and performed a comparative assessment against radiologists for early disease detection. This work can significantly expedite and improve COVID-19 patient care. These recent examples demonstrate the importance of statistical regression analysis in radiological research for advancing clinical care and public health.

In summary, regression is a broad topic. Details about different regression methods, their nuances, and applications are better covered in separate chapters and in dedicated statistics textbooks. Readers can refer to these as a reference to garner a more robust understanding of the topic. As a researcher, it is a great idea to familiarize with the nomenclature and at least the basics when it comes to regression, as well as to quantitative analyses in general. While it is equally a good idea to work with a dedicated biostatistician during the course of research endeavors, it is wise to be well versed in statistical methods and applications to have meaningful discussions with colleagues across various medical fields and to conduct quality research.

References

1. McCullagh P, Nelder JA. *Generalized Linear Models*. Routledge; January 22, 2019.
2. Klein JP. *Survival Analysis: Techniques for Censored and Truncated Data*. Moeschberger ML. New York: Springer; February 2003.
3. Wang R, Cai Y, Lee IK, et al. *Evaluation of a Convolutional Neural Network for Ovarian Tumor Differentiation Based on Magnetic Resonance Imaging*. Springer; 2021.
4. Tang by L, Liu S, Xiao Y, et al. *Encephalopathy at Admission Predicts Adverse Outcomes in Patients with SARS-CoV-2 Infection*. Wiley; 2021.
5. Kim CK, Choi JW, Jiao Z, et al. An automated COVID-19 triage pipeline using artificial intelligence based on chest radiographs and clinical data. *npj Digit. Med.* 2022;5:5. https://doi.org/10.1038/s41746-021-00546-w.

Further reading

1. *Epidemiology and Biostatistics: An Introduction to Clinical Research by Byran Kestenbaum*. 2nd ed. Springer; 2019.
2. Dictionary of epidemiology. In: *Miguel Porta*. 6th ed. Oxford University Press; 2014.
3. *Epidemiology: Principles and Practical Concerns by Jan Van Den Broek, Jonathan Brestoff*. Springer; 2013.

Chapter 29

t-test

Siddhant Thukral[1], Stefan Kovac[2] and Mounica Paturu[3]

[1]*Division of Vascular and Interventional Radiology, Department of Radiology, Duke University Medical Center, Durham, NC, United States;* [2]*Heersink School of Medicine, The University of Alabama at Birmingham, Birmingham, AL, United States;* [3]*Department of Neurosurgery, Duke University Medical Center, Durham, NC, United States*

Key points

- The t-test is a parametric test designed to compare means between two sets of data.
- The t-test relies on a series of assumptions that must be verified before being utilized.
- There are three versions of the t-test that clinicians should be familiar with: (1) one-sample t-test, (2) two-sample t-test, and (3) paired samples t-test.

Why it matters?

The t-test or student's t-test is one of the most widely used statistical hypothesis tests researchers use. The t-test compares the means of two groups while accounting for the potential for variability within them to determine if two sets of data are significantly different from one another. Like most statistical tests, the t test relies on a series of assumptions that must be verified for the results to be valid. A greater foundational understanding of each of the variations of the t-test can aid clinicians and researchers in appropriately designing experiments and interpreting data results with greater statistical accuracy.

Introduction

The t-test is a method to assess the center and variations of two individual sample groups and test a hypothesis based on the difference between their sample means to determine statistical significance.[1] It is one of the most widely used statistical hypothesis tests in clinical and basic science research encompassing all fields of medicine. Given its breadth of utilization within scientific literature, a foundational understanding of the t-test can aid clinicians and researchers in appropriate experimental design and accurate data interpretation. There are three forms of the t-test that clinicians and researchers should be familiar with: (1) one-sample t-test; (2) two-sample t-test; and (3) two-sample paired t-test.[1] This chapter will elaborate upon these t-tests, providing examples on appropriate usage for each, and offer information on how to interpret t-test results successfully.

Inferential statistics and T-distribution

In statistics, "inference" is the process of making judgments about the unknown based on assumptions of what is already known.[2] Inferential statistics utilize sample data to define a more generalized conclusion about a larger yet similarly representable dataset.[2] Statistical inferences can be divided into parametric and nonparametric methods.[3] Parametric methods refer to a statistical technique whereby assumptions are made about the distribution parameters and the distribution of probability variables is defined.[3] The normal distribution, or Gaussian distribution, is a well-known component of parametric inferential statistics, with 68% of the data falling within one standard deviation of the mean and 95% of the data falling within two standard deviations of the mean.[2] T-tests are a common parametric test where the t-distribution is continuous, symmetrical, and features heavier tails with a shorter and flatter center than a normal distribution.[2] The shape of the data distribution is dependent on the degrees of freedom (DF), and as sample size increases, the T-distribution curve approaches normal distribution curve parameters.[2] In situations where these assumptions cannot be made, nonparametric methods are used and are considered as more conservative approaches resulting in a lower probability that a test will correctly reject a false null hypothesis (i.e., type II/beta error).[3]

Assumptions for a t-test

To prevent critical errors due to improper use of statistical tests and methodology[4] in medical research, especially when employing the ubiquitous t-test analysis, the

Translational Interventional Radiology. https://doi.org/10.1016/B978-0-12-823026-8.00104-8

assumptions and limitations for each variant t-test should be understood. The t-test relies on a series of assumptions about the sample being analyzed when producing meaningful output. As t-tests are a type of parametric method, they can typically be used when the sample satisfies the assumptions of (1) independence of data, (2) normality, and (3) equal variances.[3] For t-tests, the population standard deviation is unknown, so the sample standard deviation is applied as a substitute when comparing the difference in means.[2,3] Furthermore, the t-test is used to compare means among only two groups; if the means of more than two groups need to be analyzed, then the F-test for analysis of variance (ANOVA) may be appropriate.[5] Before beginning data analysis with any t-test, data assumptions must be verified as summarized in Table 29.1.

Key differences between a t-test and z-test

When working with means from two sample populations with parametric methods, it is important to delineate the appropriate usage of the t-test compared to the z-test. The t-test refers to a univariate hypothesis test based on a t-distribution, as discussed above, in which the mean is known, sample size is small (i.e., $n < 30$), and population variance is unknown.[3] The z-test is a univariate test differing from the t-test as it's based on population standard deviation rather than sample deviation.[2] In situations where population variance is unknown or not given, a t-test can be used to calculate whether the means of two different data sets differ by substituting sample variance for population variance.[3] When using sample variance, the sample distribution follows a t-distribution, dependent on degrees of freedom, rather than a normal distribution like the z-test.[3,7]

If the population variance is known for a data set with a large sample size (i.e., greater than 30), then the z-test can be used.[7] An example of appropriate use for a z-test can be highlighted in a study by Hartman et al.[8] which sought to determine if the incidence of thrombotic events increases in patients who undergo CT-guided percutaneous renal tumor ablations. Comparisons were made with a z-test using person-time incidence rates of a 2445 patient control population.

TABLE 29.1 Test assumptions and requirements.[6]

One-sample t-test
1. Test variable is measured as interval or ratio (continuous)
2. Independence of observation (values analyzed are independent of each other)
3. Random sample of data from the population
4. Normally distributed population and sample (exception: Large sample size)
5. Homogeneity of variances (variances are about equal in both sample and population)
6. No outliers
Independent Samples t-test
1. Test variable is measured as interval or ratio (continuous)
2. Independent variable is categorical
3. Cases have values on both the independent and dependent variables
4. Independence of observation (values analyzed are independent from each other)
5. Random sample of data from the population
6. Normally distributed population and sample (exception: Large sample size)
7. Homogeneity of variances (variances are about equal in both sample and population)
8. No outliers
Paired Samples t-test
1. Test variable is measured as interval or ratio (continuous)
2. Samples or groups are related
3. Random sample of data from population
4. Normal distribution of the difference between paired values
5. No outliers observed in the differences of related values

One-sample t-test

One sample t-test is used to test if a sample mean differs significantly from the population mean or a given constant.[9] As summarized in Table 29.2, the one-sample t-test relies on a reference mean in its calculation compared to the independent samples t-test which compares the mean of two defined samples. One sample t-test is commonly used when the effect of a new treatment is being assessed on a single group of patients.[10] Steps to run a one-sample t-test and how to interpret the results are summarized in Fig. 29.1.

Independent sample t-test

The independent sample t-test, also known as an unpaired two-sample t-test, is commonly employed when the two independent variables are discrete levels or groups, while the dependent variable is continuous and normally distributed.[11,12] This test aims to determine if the means of two different populations being compared are the same.[2] This t-test assumes that samples have been independently selected from their population and that the populations from which the samples were selected have a normal distribution (unless samples are large enough to invoke the central limit theorem), and the population is not assumed to have equal variance.[2,3] For example, in a study by DiPietro et al.,[13] which investigated the experiences of 61 matched IR applicants and 34 program directors, a comparison of Likert scale responses was analyzed by a two-tail t-test. Steps to run a two-sample t-test and how to interpret the results are summarized in Fig. 29.1.

Paired t-test

Unlike the independent sample t-test, which analyzes means from two statistically independent samples, the paired t-test is used when data in a sample comes in the form of matched pairs.[11] For example, a study by Al-Hakim et al.[14] investigated differences in acute venographic changes of the external iliac vein before and after ipsilateral common iliac vein stent placement. The matched diameters before and after the intervention were compared using a paired sample t-test.[14] Compared to a two-sample t-test that assumes that the data from both samples are normally distributed, independent, and does not assume equal variances, the paired t-test only assumes that the data is normally distributed.[11] Additionally, it is important to note that the paired t-test calculates DF as a sample size of matched pairs (n) minus 1 (n−1) as compared to the independent sample t-test which calculates DF as 2(n−1).[2,11] Steps to run a paired sample T-test and interpret the results are summarized in Fig. 29.1.

TABLE 29.2 Equations and formulas.[2]

One Sample T-test	$t = \frac{\overline{x} - \mu_0}{\frac{s}{\sqrt{n}}}$
Standard deviation calculation	$s = \sqrt{\frac{\sum_{i=1}^{n}(x_i - \overline{x})^2}{n-1}}$
Degrees of freedom for one sample T-test	DF = n 1
Independent samples T-test	$t = (\frac{\overline{x_1} - \overline{x_2}) - (\mu_1 - \mu_2)}{\sqrt{S_p^2\left(\frac{1}{n_1} + \frac{1}{n_2}\right)}}$
Pooled variance	$S_p^2 = \frac{(n_1 - 1)S_1^2 + (n_2 - 1)S_2^2}{n_1 + n_2 - 2}$
Paired samples T-test	$t = \frac{\overline{d} - (\mu_1 - \mu_2)}{\frac{s_d}{\sqrt{n}}}$
Degrees of freedom for two-sample T-test	$(n_1 + n_2 - 2)$

$\overline{x}$ = Sample mean.
x_i = Single x value.
μ_0 = Reference mean.
s = Standard deviation of sample.
n = Sample size.
$\overline{x_1}$ = Mean of sample 1.
$\overline{x_2}$ = Mean of sample 2.
μ_1 = Mean of population 1.
μ_2 = Mean of population 2.
S_p^2 = Pooled variance.
S_1^2 = variance of sample 1.
S_2^2 = variance of sample 2.
$\overline{d}$ = mean of difference in scores.
s_d = standard deviation of the difference scores.

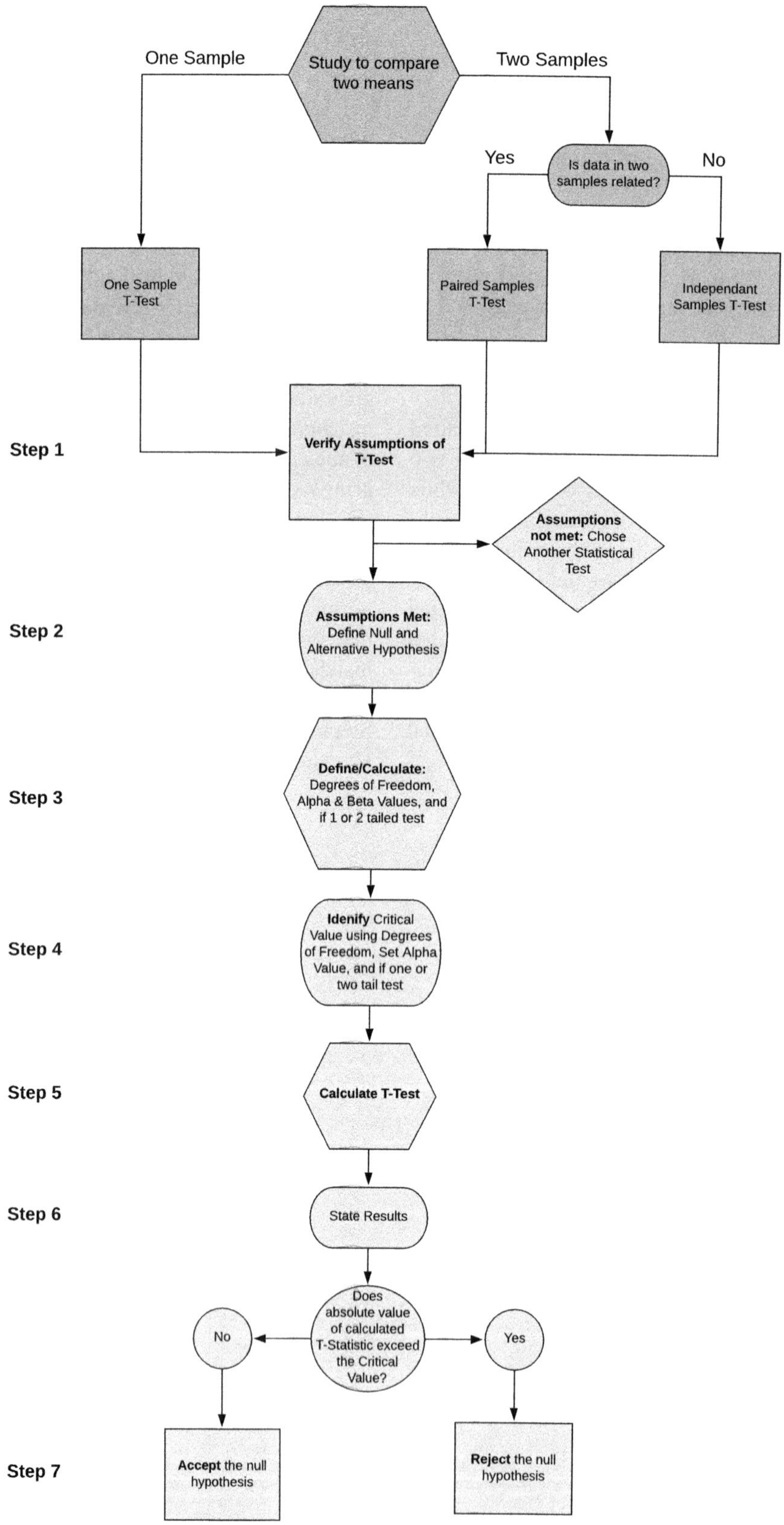

FIGURE 29.1 Steps to run and interpret a T-test.

Conclusion

The t-test is a parametric test that is used to compare means from two different categories. Like most statistical tests, the t-test relies on a series of assumptions about the population and sample it is analyzing. It is imperative to verify these assumptions prior to conducting the t-test to prevent false interpretation of the results. Depending on the dataset being analyzed, there are three well-known versions of the t-test that can be used: (1) one-sample t-test, (2) independent samples t-test, and (3) paired t-test. Understanding the utility and limitations of these tests will allow for improved study designs and increased transparency in research studies.

Get started

When comparing the means of two independent variables, a t-test may be considered. Before conducting the t-test, it is crucial to verify that all assumptions of the t-test have been met; if not, another test should be considered. The critical value can be located on most statistical software and textbooks after deciding on an alpha value and calculating the degrees of freedom. If the calculated t-statistic is greater than the critical value, then the null hypothesis can be rejected.

References

1. Wadhwa RR, Marappa-Ganeshan R. t test. In: *StatPearls*. Treasure Island (FL): StatPearls Publishing. Copyright © 2020, StatPearls Publishing LLC; 2020.
2. Boslaugh S. Statistics in a Nutshell. In: *A Nutshell*. 2nd ed. Farnham, Surrey, England: O'Reilly; 2012:569; xix.
3. Kim TK. T test as a parametric statistic. *Kor J Anesthesiol*. 2015;68(6):540–546.
4. Rochon J, Gondan M, Kieser M. To test or not to test: preliminary assessment of normality when comparing two independent samples. *BMC Medical Research Methodology*. 2012;12. pp. 81–81.
5. Mishra P, Uttam S, Pandey CM, Mishra P, Gaurav P, et al. Application of student's t-test, analysis of variance, and covariance. *Ann Card Anaesth*. 2019;22(4):407–411.
6. SPSS Tutorials. *Analyzing Data*; June 24, 2020 [cited 2020 June 30, 2020]; Available from: https://libguides.library.kent.edu/SPSS.
7. Meier KJ, Brudney JL, Bohte J. *Applied Statistics for Public and Nonprofit Administration*. Nelson Education; 2011.
8. Hartman JB, Bhojwani N, Corn DJ, et al. Increased risk of thrombosis following image-guided percutaneous cryoablation of renal tumors. *J Vasc Intervent Radiol*. 2013;24(4):S19–S20.
9. Skaik Y. The bread and butter of statistical analysis "t-test": uses and misuses. *Pak Med Sci*. 2015;31(6):1558–1559.
10. Cressie NAC, Sheffield LJ, Whitford HJ. Use of the one sample t-test in the real world. *J Chron Dis*. 1984;37(2):107–114.
11. Xu M, Fralick D, Zheng JZ, Bokai W, Tu XM, Feng C, et al. The differences and similarities between two-sample T-test and paired T-test. *Shanghai Arch Psychiatr*. 2017;29(3):184–188.
12. De Muth JE. Overview of biostatistics used in clinical research. *Am J Health Syst Pharm*. 2009;66(1):70–81.
13. DePietro DM, Kiefer RM, Redmond JW, Hoffmann JC, Trerotola SO, Nadolski GJ, et al. The 2017 integrated IR residency match: results of a national survey of applicants and program directors. *J Vasc Intervent Radiol*. 2018;29(1):114–124.
14. Al-Hakim RA, Kaufman JA, Farsad K. Iliac vein stent placement: acute venographic changes and relevance to venous biomechanics. *J Vasc Intervent Radiol*. 2018;29(7):1023–1027.

Chapter 30

Chi square

Siddhant Thukral[1], Stefan Kovac[2] and Mounica Paturu[3]

[1]*Division of Vascular and Interventional Radiology, Department of Radiology, Duke University Medical Center, Durham, NC, United States;* [2]*Heersink School of Medicine, The University of Alabama at Birmingham, Birmingham, AL, United States;* [3]*Department of Neurosurgery, Duke University Medical Center, Durham, NC, United States*

Key points

- The chi-square test is a nonparametric test designed to analyze categorical data.
- Like most statistical tests, the chi-square test has a series of assumptions that must be verified before being utilized.
- There are three versions of the chi-square test that clinicians and researchers should be familiar with: (1) chi-square test of goodness of fit, (2) chi-square test of independence, and (3) chi-square test of proportion.

Why it matters?

The chi-square test is a nonparametric test frequently utilized in medical research due to its ability to analyze categorical data and test multiple hypotheses: independence of categories, the goodness of fit, and equality of proportions. Therefore, it is crucial to understand the statistical principles, deviations, and mathematics behind the different variations of the chi-square to increase the transparency and accuracy of study design and result interpretations.

Introduction

The chi-square (χ^2) test is a nonparametric statistical test utilized in analyzing data comprised of frequencies or counts.[1] There are three versions of the chi-square test that clinicians and researchers should be familiar with: (1) chi-square test of goodness of fit, (2) chi-square test of independence, and (3) chi-square test of proportions. When conducting research of any magnitude, understanding the statistical principles, derivations, and mathematics behind each of these unique chi-square tests can increase the transparency and accuracy of study design and results from interpretation.[2] Therefore, the purposes of this chapter are to detail the foundation of the various types of chi-square tests, provide examples of appropriate usage of each test, and share guidance on how to interpret results properly.

Nonparametric methods, chi-square distribution, and R×C tables

The chi-square test is designed to analyze categorical variables and not continuous or parametric data—therefore, it is considered a nonparametric test.[3,4] In contrast to parametric statistical tests, nonparametric or "distribution-free" statistical tests are devoid of stringent assumptions about the population from which a study sample is drawn, thereby enabling them to analyze nonnumerical data.[5] To best conduct nonparametric analysis, the categorical values for a chi-square test can be organized via R×C tables.

The R×C table, or contingency table, is a tool used to analyze the relationship between two categorical variables, where the "R" and "C" refer to the rows and columns, respectively, that the specific table contains.[2] When using the R×C table, investigators must first decide how their data can be grouped into distinct categories for analysis.[2] Incorporating these concepts, an example of a contingency table is provided in Table 30.1 for an investigator seeking to correlate the relationship between the number of previous deep vein thrombosis events in the ipsilateral limb and the development of postthrombotic syndrome.

Subsequently, researchers can utilize the known properties of the chi-square distribution for each R×C table to identify whether their results are significant.[2] A special case of the gamma distribution, the chi-square distribution, is a continuous theoretical probability distribution used to describe the sum of squared random variables. Beyond significance testing, it can also be used to test the goodness of fit of a data distribution, independence of data series, and determine confidence intervals for variance and standard deviation of a random variable from a normal distribution.

Translational Interventional Radiology. https://doi.org/10.1016/B978-0-12-823026-8.00028-6

TABLE 30.1 Contingency table example.

		Development of post thrombotic syndrome (PTS)			
Previous episodes of DVT in ipsilateral limb		No PTS (Villalta Score of <5)	PTS (Villalta Score of ≥ 5 and < 10)	Moderate PTS (Villalta Score of ≥ 10 and < 15)	Severe PTS (Villalta Score of ≥ 15)
	0	$Cell_{11}$	$Cell_{12}$	$Cell_{13}$	$Cell_{14}$
	1–2	$Cell_{21}$	$Cell_{22}$	$Cell_{23}$	$Cell_{24}$
	>3	$Cell_{31}$	$Cell_{32}$	$Cell_{33}$	$Cell_{34}$

This R×C table is considered a 3 × 4 table due to its three rows and four columns. Cells in R×C tables are notated using *ij* subscripts that are commonly used throughout statistics including statistical formulas. The *i* and *j* designate the rows and columns respectively.

Chi-square distribution is right-skewed as it is based on a sum of squared variables and has only one parameter, *k*, degrees of freedom, which must be positive.[2] The shape of the chi-square distribution is strongly influenced by the degrees of freedom, but as degrees of freedom approach infinity, it begins to resemble Gaussian distribution.[2]

Finally, as the chi-square test is discrete and the chi-square distribution is continuous, investigators may elect to apply Yate's correction to adjust for this discrepancy.[2] The Yate's correction is performed by subtracting 0.5 from the absolute value of (observed-expected) before squaring in the formula for the chi-squared statistic as outlined in Table 30.2. Yate's correction reduces the probability of Type 1 error. However, it also increases the probability of Type 2 error and reduces study power; consequently, it is not universally endorsed.[2,6]

Assumptions and limitations of a chi-square test

Unlike most statistical tests, the chi-square test provides information on the significance of observations and detailed information on which categories account for any of the identified differences.[7] However, similar to other statistical tests, a series of requirements or assumptions need to be satisfied before utilizing the chi-square test.[7] In addition to verifying appropriate test assumptions, the chi-square test should always be coupled with an appropriate strength test as the chi-square test is also a significance test.[7]

Two critical limitations of the chi-square test include: (1) no analyzed categories have an expected count less than one, (2) and no more than 20% of the expected counts are less than 5.[4,7] To correct the latter limitation, investigators may choose to increase their sample size or merge discrete categories until their combined value is greater than or equal to 5.[7] Assumptions of the chi-square test of independence are summarized in Table 30.3, and the stepwise process for calculating and interpreting the results of this test is summarized in Fig. 30.1.

Chi-square test for goodness of fit

For studies seeking to determine if the distribution of a categorical variable within a population follows a specific

TABLE 30.2 Equations and formulas.

Expected Value for Cell	$Eij = \frac{\text{ith row total} \times \text{jth column total}}{\text{Sum total of all cells}}$
Chi-square test for goodness of fit	$X^2 = \sum_{i=1}^{k} \frac{(\text{Observed}_i - \text{Expected}_i)^2}{\text{Expected}_i}$
Degrees of freedom for the chi-square goodness of fit	df = k − 1
Chi-square test for independence Chi-square test for equality of proportions	$X^2 = \sum_{ij} \frac{(\text{Observed}_{ij} - \text{Expected}_{ij})^2}{\text{Expected}_{ij}}$
Degrees of freedom for chi-square test for independence Degrees of freedom for chi-square test for equality of proportions	(r-1) (c-1)
Yates correction for continuity	$X^2 = \sum_{ij} \frac{(\lvert \text{Observed}_{ij} - \text{Expected}_{ij} \rvert - 0.5)^2}{\text{Expected}_{ij}}$

E_{ij} = Specific Cell, k = number of groups, levels, or categories of the categorical variable, *i* = Row, *j* = Column, r = number of rows, c = number of columns.

TABLE 30.3 Assumptions of a chi-square test of independence.[4]

1. Data included in the cells represent frequencies or counts and not percentages.
2. Each category is independent of other categories.
3. A subject counted in a category is not counted in another category.
4. Study groups analyzed are independent and not related
5. At least two categories being analyzed.
6. No cell has a value less than one, and the expected value in the cells is 5 or more in at least 80% of the cells analyzed.

pattern of proportions, the chi-square goodness of fit test may be appropriate.[2] In a study by Johnston et al.,[8] which sought to investigate the use of discrete event simulation coupled with a visual display of results to reduce patient flow in hospitals' emergency and radiology departments, chi-squared goodness of fit test was performed to test the hypothesis that procedure times have a γ distribution function. This test is calculated using expected values derived from hypothesized proportions. Because the data analyzed can usually be arranged into a single row, only single subscripts are in the chi-square equation (Table 30.2).[2] Degrees of freedom for the chi-square goodness of fit test is calculated using the equation (g-1).[2] The primary assumption of this test is the independence of sample selection, where choosing one participant should not directly or indirectly influence the selection of another participant.[9] The null hypothesis for this test is that the sample being analyzed is from the population to which it is being compared, whereas the alternative hypothesis states that the sample is not from the population to which it is being compared.[10]

Chi-square test of independence

The chi-square test of independence, or Pearson's chi-square, is a test that utilizes contingency tables to determine whether the observed pattern between two categorical variables in the contingency table is strong enough to deem those chosen variables as dependent.[11] For example, Gottumukkala et al.[12] used Pearson's chi-square test to compare disparities such as time, categorical outcomes, and proportions between paracentesis and thoracentesis performed by radiologists and nonradiologists in a national sample of Medicare beneficiaries. Mathematically, the Pearson's chi-square test calculates the differences between observed and expected cell values within a contingency table as follows: (1) observed values are used to calculate expected values for each cell of the table with the

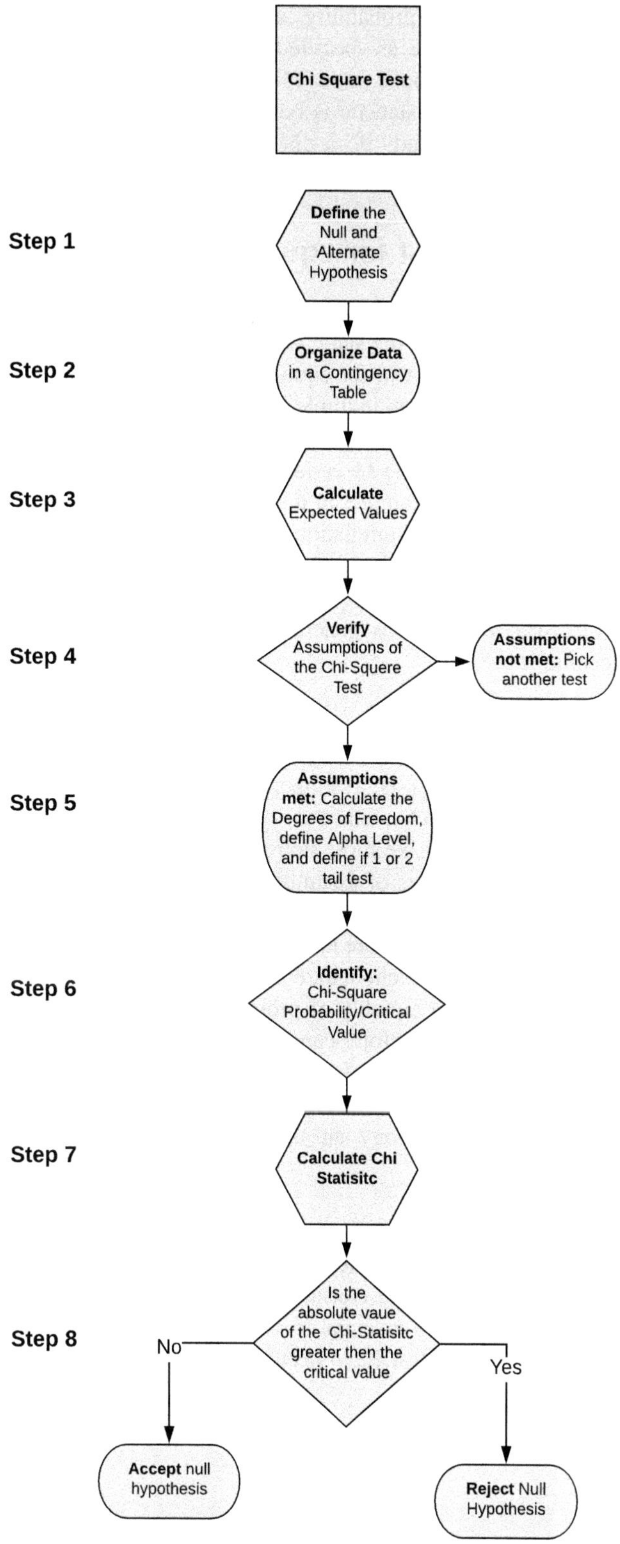

FIGURE 30.1 Stepwise approach to calculate and interpret chi-square.

assumption that the two tested variables are independent, (2) the Pearson chi statistic is calculated as highlighted in Tables 30.2 and 30.3, (3) Pearson's chi statistic is compared to the chi-squared probability, also known as the critical value, for the value as specified by established statistical software or textbooks, and (4) the null hypothesis is accepted if the chi statistic is below the chi-squared probability and rejected if it is above the chi-squared probability.[6]

Chi-square test for equality of proportions

Although computed the same way as Pearson's chi-square test, the chi-square test for equality of proportions is used for data collected from multiple independent populations. It tests the validity of the null hypothesis that the distribution of the chosen variable or variables is the same among all populations,[2] for example, in a study by Hasiloglu et al.[13] assessing the iron accumulation in the choroid plexus of 18 patients with Beta Thalassemia versus Control via MRI, the chi-square test for equality of proportions compared data from over 90 locations generated from multiple imaging techniques.

Conclusion

The chi-square test is a nonparametric test that can be valuable in research methodology when comparing categorical variables. Like most statistical tests, the chi-square test relies on a series of assumptions about the population from which the sample was drawn. Once these assumptions are verified, the chi-square test can be used. Chi-square test for goodness of fit, chi-square test of independence, and chi-square test for equality of proportions all test different hypotheses and, therefore, can be used in various studies. A good understanding of these chi-square tests will increase the accuracy and transparency of research studies in Interventional Radiology and the entire field of medicine.

Get started

When comparing two independent categorical variables, a chi-square test may be considered. Before conducting the chi-square test, it is essential to verify that all assumptions of the chi-square test have been met; if not, another test should be considered. After deciding on an alpha value and calculating the degrees of freedom, the chi probability, also known as the critical value, can be located on most statistical software and textbooks. If the calculated chi-statistic is greater than the chi probability, then the null hypothesis can be rejected.

References

1. Zibran MF. *Chi-squared Test of Independence*. Alberta, Canada: Department of Computer Science, University of Calgary; 2007.
2. Boslaugh S. Statistics in a nutshell. In: *A Nutshell*. 2nd ed. Farnham, Surrey, England: O'Reilly. xix; 2012:569.
3. Patra NK. Chapter 7 - implementation of electronic resource management in libraries: a case study. In: Patra NK, ed. *Digital Disruption and Electronic Resource Management in Libraries*. Chandos Publishing; 2017:83–133.
4. McHugh ML. The chi-square test of independence. *Biochem Med*. 2013;23(2):143–149.
5. Siegel S. Nonparametric statistics. *Am Statistician*. 1957;11(3): 13–19.
6. Hess AS, Hess JR. Understanding tests of the association of categorical variables: the Pearson chi-square test and Fisher's exact test. *Transfusion*. 2017;57(4):877–879.
7. Musa AZ. Statistical issues II: chi-square test statistics. *NJCBR Online*. 2017;7(9):29–35.
8. Johnston M, Samaranayake P, Dadich AM, Fitzgerald A. Modelling radiology department operation using discrete event simulation. In: *MODSIM, International Congress on Modelling and Simulation*. Modelling and Simulation Society of Australia and New Zealand; 2009.
9. Dasgupta N, Alldredge JR. A chi-square goodness-of-fit analysis of dependent resource election data. *Biometrics*. 2000;56(2):402–408.
10. Cochran WG. *The χ2 Test of Goodness of Fit*. The Annals of mathematical statistics; 1952:315–345.
11. Boduszek D. *Chi-square Test of Independence & McNemar Test*. Verkregen op; 2016. pp. 18–06.
12. Gottumukkala RV, Prabhakar AM, Hemingway J, Hughes DR, Duszak R. Disparities over time in volume, day of the week, and patient complexity between paracentesis and thoracentesis procedures performed by radiologists versus those performed by nonradiologists. *J Vasc Intervent Radiol*. 2019;30(11):1769–1778. e1.
13. Hasiloglu ZI, Asik M, Ure E, Ertem F, Apak H, Albayram S. The utility of susceptibility-weighted imaging to evaluate the extent of iron accumulation in the choroid plexus of patients with β-thalassaemia major. *Clin Radiol*. 2017;72(10):903.e1–903.e7.

Chapter 31

Analysis of variance

Tony H. Rizk
Medical University of South Carolina, Charleston, SC, United States

Key points

- Analysis of Variance (ANOVA) is the statistical test of choice to determine if there are differences in the means of three or more randomly selected, normally distributed, populations groups with homogenous variances.
- One-way ANOVA is most used and is characterized by a single factor to categorize populations.
- Post hoc comparisons must be performed following a statistically significant ANOVA test to determine which groups are different.

Why it matters?

Interventional radiology has experienced a significant increase in clinical trials and observational studies in recent years as a response to the emergence of novel minimally invasive therapies, highlighting the importance of describing and utilizing effective statistical analysis methods to promote informed patient care and clinical decision making. ANOVA is a commonly used method, and it is imperative to understand when this test is indicated, when it should not be used, how to interpret the results, and the potential pitfalls. Real-world examples of the use of ANOVA in interventional radiology will be described in this chapter. Furthermore, general guidelines on how this test is performed will increase the quality of analysis by helping researchers utilize and interpret this method in future

Introduction

One of the most utilized methods of statistical analysis in the medical literature is Analysis of Variance (ANOVA). ANOVA is a single statistical test to detect differences in the means of three or more mutually independent groups or treatments.[1] Using conditions defined by discrete classification variables, ANOVA can analyze variation in a continuous random variable and determine whether the groups have equivalent mean values (null hypothesis).[2] Furthermore, to use the ANOVA test, the following assumptions must be made[3]:

- Normal distribution of each population group
- Homogenous population variances for each treatment group
- Treatment group observations represent a random sample from that population

Although it may be difficult to verify that all these assumptions are met, the one-way ANOVA is generally acceptable if the standard deviation of the largest sample is no more than twice the standard deviation of the smallest sample.[3] The most basic form of ANOVA is the one-way analysis, which is an extension of the Student 2-independent-samples *t*-test and a single property (the variance) is used to categorize populations.[4] The Student 2-independent-samples *t*-test is calculated by taking the difference between sample means and dividing by the standard error of the difference, which is the estimated variability of both groups. As a result, you can compare the variability between the groups (difference between samples means) and the variability within groups (random error). One-way analysis of variance follows this principle but uses variances to measure variability, rather than means and standard deviations, to test equality among several means.[4] ANOVA utilizes the F-test to study the hypothesis rather than the *t*-test. While the *t*-test is calculated from the difference between two means, the F-test uses the squared sums of the differences between observations and means (within and between groups).[5] In other words, the F-test is used to evaluate the sources of variability of measurements within groups and between groups.[6]

Calculation and interpretation

Table 31.1 is an example used to show the summary of the statistics that are used as the input to the ANOVA test. This example uses three groups that differ by one factor to evaluate the null hypothesis, H0: U1 = U2 = U3 (Table 31.1).

Translational Interventional Radiology. https://doi.org/10.1016/B978-0-12-823026-8.00024-9

Table 31.2 is an example of how to display the results of the ANOVA from the theoretical sample data derived from Table 31.1.

SS(B) represents the sum of squares between the groups, which then can be divided by degrees of freedom ($k - 1$) to produce MS(B), or mean square between the groups. On the other hand, SS(W) represents the sum of squares within the groups, which then can be divided by degrees of freedom ($N - k$) to produce MS(W), or mean square within the groups. Finally, MS(B) can be divided by MS(W) to produce a ratio of between-group variance and within-group variance. This ratio is also known as F-test. The F-test is then compared to the value at an α error level of 0.05, also known as the critical value. If the F-test is greater than the critical value, then the null hypothesis is rejected, and the test is deemed statistically significant at this error level.[7] Thus, if the P-value is < 0.05, then on average at least two of the groups are different.

Examples in interventional radiology

Table 31.3 summarizes three real-world examples to illustrate how ANOVA is used in interventional radiology research.

Post hoc analysis

The problem with one-way ANOVA is that it does not provide detailed information about which groups are different. Several types of post hoc comparisons described below can be used to clarify the differences between groups.[11].

- Tukey's method tests every possible pair of the groups to determine if there is a difference between the means. For example, the test can show that U1 and U3 are different, while U1 and U2 are not. If the interpretations are different, then the results could have resulted from insufficient evidence. This method is more tolerant of type 1 error compared to the other types of post hoc analysis.[11]
- Bonferroni method is a rigorous test that compares the groups using thresholds based on the t-distribution. This method is less tolerant of type 1 error, but this is at the expense of type 2 errors.[12]
- Dunnett method is an effective test that can uncover subtle but significant differences between groups in studies with control groups based on the t-distribution. The drawback to this method is that it can only test the experimental groups against the control groups but cannot compare across experimental groups.[11]
- Scheffe's method is an exploratory post hoc method that can be utilized when the theoretical differences between groups cannot be established due to a lack of prior literature. This test is based on the F-distribution and compares all pairwise combinations to generate a hypothesis to confirm significance.[11]

Alternatives

If populations do not meet the normality and equal variance assumptions, the one-way ANOVA cannot be utilized. In this scenario, the Kruskal–Wallis is a viable analog to the one-way ANOVA. It is a nonparametric test to compare ordinal variables between three or more populations from independent samples.[13,14] Kruskal–Wallis compares population medians rather than population means and can test the null hypothesis of equal medians among the populations. Furthermore, the Kruskal–Wallis test uses ranks

TABLE 31.1 Summary of statistics for the one factor generated by three groups (U1, U2, U3).

Groups	n	Mean	SD	Minimum	Maximum
U1					
U2					
U3					

TABLE 31.2 ANOVA output and table explanation.

	ANOVA				
Source	Sum of squares (SS)	Degrees of freedom (df)	Mean square (MS)	F-test	*P*-value
Between groups (B)	SS(B)	$k - 1$	SS(B)/df	MS(B)/ MS(W)	P(F > F*| H0)
Within groups (W)	SS(W)	$N - k$	SS(W)/df		
Total	SS(B) + SS(W)	$N - 1$			

k = total number of groups; N = total number of specimens from all groups;
SS(B)/k −1 = MS(B); SS(W)/N − k = MS(W); F-test = MS(B)/MS(W)
P-value calculated from the F-distribution for the df between and within groups ($k - 1$, $N - k$)

TABLE 31.3 ANOVA used in three interventional radiology research publications.

Article title	ANOVA study design	ANOVA results
Importance of variance in operator technique toward reducing the radiation dose during fluoroscopic procedures[8]	Factor: Radiation dose from permanent and temporary dialysis placement Groups: Three operator-dependent fluoroscopy techniques	Differences in fluoroscopy technique resulted in statistically significant variance in radiation dose. Permanent catheter—$F(2,111) = 4.64$, $P < 0.05$ Temporary catheter—$F(2,112) = 3.58$, $P < 0.05$
Intraarterial lidocaine for pain control in uterine artery embolization: a prospective, randomized study[9]	Factor: Reported pain at 4, 7, and 24 h Groups: Lidocaine during UAE (group A), lidocaine after UAE (group B), control (group C)	Mean pain at 4 h was 28.6 (SD, 24.5) for group A, 35.8 (SD, 22.6) for group B, and 59.4 (SD, 30.3) for group C, the difference was statistically significant ($P = 0.001$) Differences in pain between the groups at 7 and 24 h were not statistically significant.
Multiple-electrode radiofrequency ablation: simultaneous production of separate zones of coagulation in an in vivo porcine liver model[10]	Factor: Isoperimetric ratio, mean minimum and maximum diameter, length, and volume of coagulation Groups: Conventional technique using 9 single control zones of ablation (group 1), 12 zones of ablation using two ablations at a time (group 2), 18 zones of ablation using three ablations at a time	None of the ANOVA *P*-values were statistically significant for any of the ablation zone size comparisons. Multiple-electrode radiofrequency ablation can be performed without sacrificing individual electrode efficacy.

of the data, like other nonparametric tests, to analyze data in the order of size rather than value.[15]

Get started

If the study of interest consists of three or more population groups, you must first determine if the requirements of ANOVA are met. The population groups must be normally distributed, have homogenous variances, and represent random samples from the population. Then you must determine the factor to compare the groups against. If the ANOVA test is found to be statistically significant, meaning the F-test statistic was greater than the critical value at an α error level of 0.05, there is a difference in at least two of the population groups. You must then use post hoc analysis to demonstrate which of the groups are different. There are many different types of post hoc analysis that should be selected based on the characteristics of the study.

ANOVA pitfalls

There are a few pitfalls to be aware of when utilizing ANOVA for statistical analysis. ANOVA is subject to bias, particularly in relation to pretest-posttest correlation, and the direction of baseline imbalance.

Egbewale BE, Lewis M, Sim J. Bias, precision and statistical power of analysis of covariance in the analysis of randomized trials with baseline imbalance: a simulation study. *BMC Med Res Methodol.* 2014;14:49. https://doi.org/10.1186/1471-2288-14-49.

In addition, ANOVA is intended to test against all alternatives to the null hypothesis, which may make it suboptimal for testing against a specific hypothesis. It is also suboptimal when the losses are not proportional to the square of the differences among the unknown population means. ANOVA is also intended to be employed when the observations are normally distributed but may not yield exact p-values when observation distributions have heavier tails than the normal. Although ANOVA may yield almost exact p-values, it is less powerful than the corresponding permutation test if observations derive from nonnormal distributions under the alternative.

Good, Phillip I, Lunneborg, CE. Limitations of the analysis of variance. *J Modern Appl Stat Methods.* 2006;5(1): Article 5. https://doi.org/10.22237/jmasm/1146456240 Available at: http://digitalcommons.wayne.edu/jmasm/vol5/iss1/5.

References

1. Kim TK. Understanding one-way ANOVA using conceptual figures. *Korean J Anesthesiol.* 2017;70(1):22–26.
2. Larson MG. Analysis of variance. *Circulation.* 2008;117(1): 115–121.

3. Kiernan D. *Natural Resources Biometrics*. NY: Open SUNY Textbooks: Geneseo; 2014.
4. Bewick V, Cheek L, Ball J. Statistics review 9: one-way analysis of variance. *Crit Care*. 2004;8(2):130–136.
5. Kao LS, Green CE. Analysis of variance: is there a difference in means and what does it mean? *J Surg Res*. 2008;144(1):158–170.
6. Pandis N. Analysis of variance. *Am J Orthod Dentofacial Orthop*. 2015;148(5):868–869.
7. Kim H-Y. Analysis of variance (ANOVA) comparing means of more than two groups. *Restorat Dentist Endodont*. 2014;39(1):74–77.
8. Miller NG, Leite LV, Laroia S. Abstract No. 334 - importance of variance in operator technique towards reducing the radiation dose during fluoroscopic procedures. *J Vasc Intervent Radiol*. 2014;25(3, Supplement):S151.
9. Noel-Lamy M, Tan KT, Simons ME, Sniderman KW, Mironov O, Rajan DK. Intraarterial lidocaine for pain control in uterine artery embolization: a prospective, randomized study. *J Vasc Intervent Radiol*. 2017;28(1):16–22.
10. Laeseke PF, Sampson LA, Haemmerich D, et al. Multiple-electrode radiofrequency ablation: simultaneous production of separate zones of coagulation in an in vivo porcine liver model. *J Vasc Intervent Radiol*. 2005;16(12):1727–1735.
11. Lee S, Lee DK. What is the proper way to apply the multiple comparison test? *Korean J Anesthesiol*. 2018;71(5):353–360.
12. Armstrong RA. When to use the Bonferroni correction. *Ophthalmic Physiol Opt*. 2014;34(5):502–508.
13. Bewick V, Cheek L, Ball J. Statistics review 10: further nonparametric methods. *Crit Care*. 2004;8(3):196–199.
14. Huang W, LaBerge JM, Lu Y, Glidden DV. Research publications in vascular and interventional radiology: research topics, study designs, and statistical methods. *J Vasc Intervent Radiol*. 2002;13(3): 247–255.
15. Nahm FS. Nonparametric statistical tests for the continuous data: the basic concept and the practical use. *Korean J Anesthesiol*. 2016;69(1):8–14.

Chapter 32

Correlation

Tony H. Rizk
Medical University of South Carolina, Charleston, SC, United States

Key points

- Correlation is a statistical method used to investigate the strength of the relationship between two continuous or categorical variables.
- The correlation coefficient is used to demonstrate the direction and strength of the relationship between bivariate random variables. Pearson's coefficient is designated for normally distributed, continuous variables. Spearman's coefficient is derived from nonnormal continuous data ranks or ordinal data (e.g., data collected in the Linkert scale).
- The coefficient of determination quantifies how much of the observed correlation is due to the independent variable.

Introduction

Correlation analysis is a very commonly used method in interventional radiology research. It is imperative to understand when different correlation coefficients are indicated, when they are not, and how to interpret the results. In addition, information derived from a correlation relationship can be translated into a regression analysis to model the relationship. It is important to note that correlation analysis cannot be used to draw conclusions but can serve as a helpful way to establish potential hypotheses. Furthermore, this review will outline how these tests are performed, hoping to increase the quality of analyses by helping researchers utilize and interpret this method in future publications.

Correlation

Correlation is used to investigate if there exists a relationship between two phenomena that is not due to chance alone. Correlation analysis is frequently used in interventional radiology to examine the strength and direction of the relationship between two quantitative continuous variables using a set of data.[1] Typically, correlation analysis is used in the context of retrospective or observational studies and helps generate hypotheses.[2] However, even when there is a very strong correlation between two variables, correlation analysis cannot be used to establish causation.[2] For example, if the change in the magnitude of one variable influences the other variable to increase or decrease, there is a relationship between the two variables, but a third variable could be influencing this observation. In most cases, research studies utilize correlation in establishing linear relationships.[3]

Correlation coefficient

A correlation coefficient is a measurement without units used to help interpret the strength of the relationship between bivariate random variables. Different tests have been developed to analyze linear and nonlinear relationships to produce bivariate correlation coefficients.

- Linear correlation: Pearson's r is a parametric test requiring normal distributed, continuous, random variables.[3,4] This is the most used correlation coefficient.
- Rank correlation: Spearman's ρ (rs) is a nonparametric test that can be utilized in nonnormal distributions, which calculates the correlation coefficient from continuous data ranks (the relative order) rather than values.[4] This type of correlation is less impacted by skewed or highly variable data.
- Kendall's tau (T) represents an extension of spearman's ρ, employed in small datasets when there is an excessive repetition of the same rank.[4]

If X represents the independent variable, and Y represents the dependent variable, to determine the Pearson's

Translational Interventional Radiology. https://doi.org/10.1016/B978-0-12-823026-8.00035-3

correlation coefficient you must first multiply the deviation from mean for X and Y for each pair of samples, sum over the products, and divide it by the product of the squared-root sum-of-squares of X and Y. This ratio produces the sample value r.[5] The letter r is used to represent the direction of correlation, with positive r indicating a direct relationship, and a negative r demonstrating an inverse relationship.[4] The value assigned to r can range from -1 to 1, and this value indicates the strength of the relationship (Table 32.1). The closer a value is to one or -1, the stronger the two variables are correlated, and the more closely the data points will plot on a straight line. Conversely, the closer the values are to 0, the weaker the correlation, and the more scattered the data points, with a value of 0 occurring when X and Y are independent random variables.[4] Fig. 32.1 demonstrates a graphical representation of linear relationships using scatterplots with regression lines. If the P-value is < 0.05, then the association between the two variables is probably not due to chance, and this value is affected by sample size.[6]

Coefficient of determination

The coefficient of determination is a percentage that quantifies how much the variance of Y is explained by X.[7] It is often used in linear regression analysis to determine how well the slope fits the data. This is calculated by squaring the r value (r^2).[7] The closer the r^2 value is to 1.0, the better the line fits the data. The difference between r^2 and 1.0 represents the variability of outcome data that is not explained by the linear model.[7] For example, if $r^2 = 0.6$, then at least 40% of the variability of outcome data is most likely explained by factors outside the model. A partial correlation can be performed to control for a factor that could be influencing the correlation between X and Y. If controlling for this factor does not result in a change in the correlation, then the factor was not influencing the correlation.[6]

TABLE 32.1 Strength of linear relationship.

Correlation coefficient value	Strength of linear relationship
<0.3	Poor
0.3–0.5	Fair
0.6–0.8	Moderately strong
0.8–1.0	Very strong

Regression model analysis

Regression analysis is used to model the relationship between X and Y. The type of regression analysis is chosen based on whether Y is normally distributed.[8] If Y is continuous and normally distributed, simple linear regression can be used to relate X and Y to the equation $Y = a + b \times X$. In this equation, "a" represents where the line meets the y-axis and "b" represents the slope of the line, also known as the regression coefficient.[9] Using this equation, you can estimate the value of a continuous dependent variable (Y) by inputting a continuous or categorical independent variable (X). Regression analysis comes into play in the clinical setting when determining individual prognostic information from risk factors. Fig. 32.2 takes the example of a positive correlation from Fig. 32.1 and generates an equation for simple linear regression. If Y is dichotomous, logistic regression is the preferred regression model. Based on estimated logistic regression coefficients, odds ratios can be estimated for every X variable.[8] Regression analysis can be taken a step further to quantitatively estimate the relationship between multiple independent variables ($X1$, $X2$, $X3$) and a dependent variable. This type of analysis is called multivariate regression analysis.[10] In multivariate regression, Y is a continuous variable, while the independent variables can be continuous, categorical, and/or grade variables.[10]

Interventional radiology examples

Table 32.2 demonstrates examples of correlation analysis and regression analysis used in interventional radiology research. These examples serve to better illustrate the concepts outlined in this chapter (Fig. 32.3).

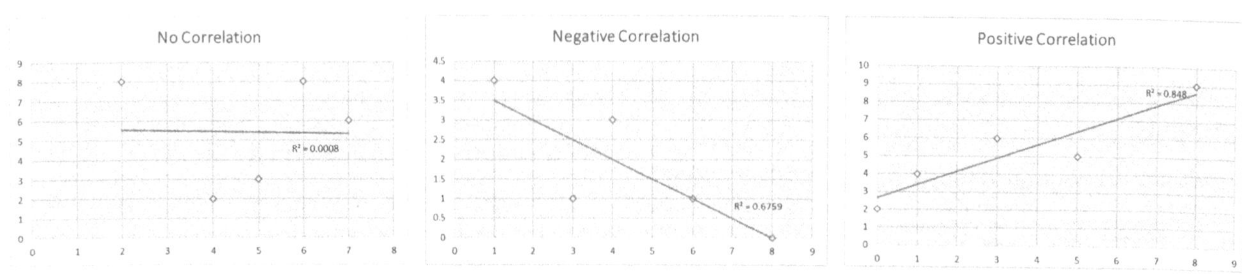

FIGURE 32.1 Examples of scatterplots used to show linear relationships: (A) poor correlation close to zero; (B) a moderate negative correlation; (C) a strong positive correlation. r^2 is the coefficient of determination.

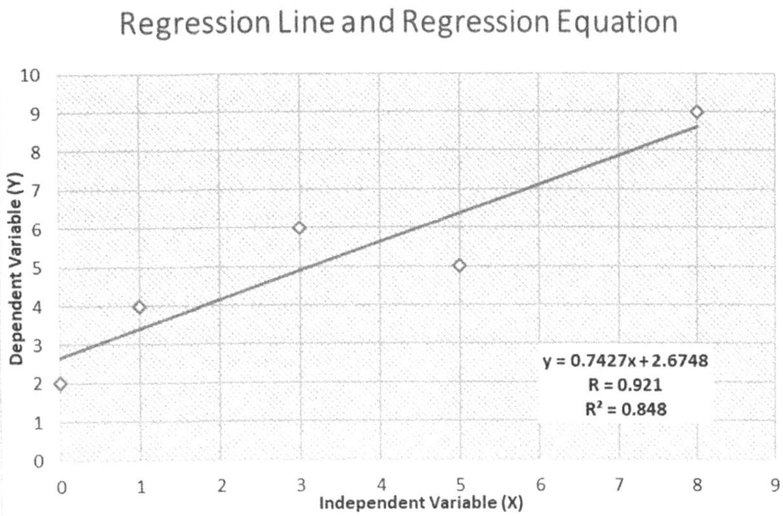

FIGURE 32.2 A sample scatterplot showing a positive linear correlation and the associated simple linear regression equation for the relationship between X and Y; r = correlation coefficient; r^2 = coefficient of determination.

TABLE 32.2 Examples of correlation and regression analysis in published literature.

Title	Method	Variables	Results
Bilateral versus unilateral femoral access for uterine artery embolization: results of a randomized comparative trial[11]	Spearman correlation used to investigate relationships between patient or procedure factors and radiation exposure, regardless of which type of femoral approach was used for uterine artery embolization	Independent: BMI, uterine size, number of angiographic images, fluoroscopy time Dependent: fluoroscopy time, dose-area product (DAP), procedure time, puncture site pain after treatment	BMI and DAP: positive correlation (Spearman $\rho = 0.67$; $P < 0.0001$) Uterine size and DAP: weak positive (Spearman $\rho = 0.32$; $P = 0.016$) Fluoroscopy time and DAP: weak positive (Spearman $\rho = 0.37$; $P = 0.005$) The rest of the variables did not correlate
Validation of an angiographic method for estimating resting blood flow to distal tissue beds in the lower extremities[12]	**Pearson's correlation** used to assess intra and interobserver consistency when scoring for estimated flow in lower extremity vessel segments	Independent: interobserver variability, interobserver variability Dependent: consistency of scoring for each vessel segment	For most cases, there was a strong positive correlation of $r > 0.9$. The correlation was stronger for proximal vessels than distal vessels.
Influence of statin therapy on aneurysm sac regression after endovascular aortic repair[13]	**Univariate logistic regression** analysis to determine independent predictors of aneurysm sac regression at 2 years using odds ratio	Independent: statin use, age, male sex, hypertension, diabetes mellitus, smoking, coronary artery disease, peripheral artery disease, chronic kidney injury, baseline maximum AAA diameter Dependent variable: AAA sac regression	Chronic kidney injury was associated with decreased odds (OR, 0.315; 95% CI, 0.105–0.944; $P = 0.039$) Baseline maximum diameter was associated with increased odds (OR, 1.070; 95% CI, 1.008–1.135; $P = 0.026$). The rest of the independent predictors were not statistically significant

Get started

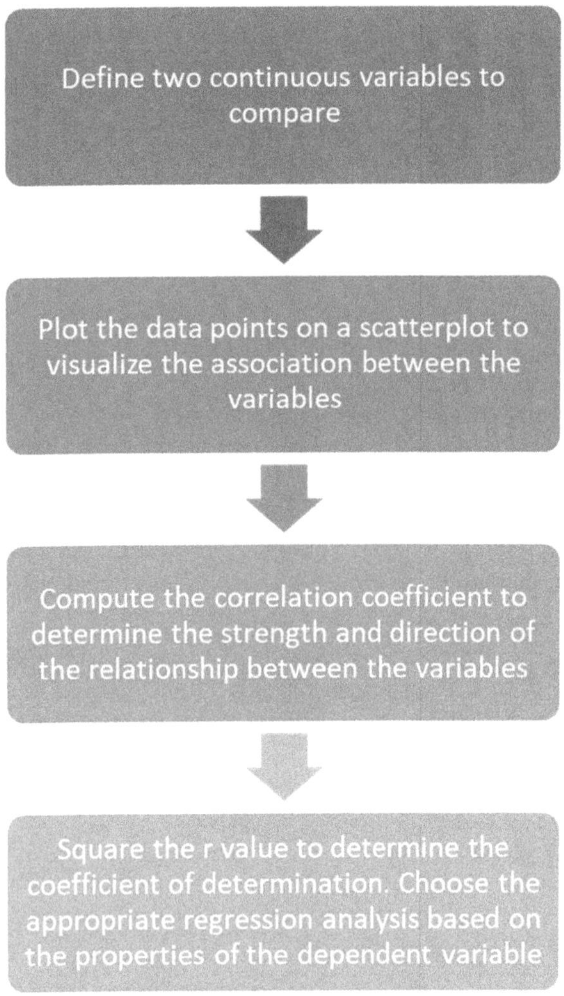

FIGURE 32.3 Correlation analysis step-by-step overview.

References

1. Franzese M, Iuliano A. Correlation analysis. In: Ranganathan S, Gribskov M, Nakai K, Schonbach C, eds. *Encyclopedia of Bioinformatics and Computational Biology*. Amsterdam; Oxford; Cambridge: Elsevier; 2019:706–721.
2. Zou KH, Tuncali K, Silverman SG. Correlation and simple linear regression. *Radiology*. 2003;227(3):617–628.
3. Schober P, Boer C, Schwarte LA. Correlation coefficients: appropriate use and interpretation. *Anesth Analg*. 2018;126(5).
4. Akoglu H. User's guide to correlation coefficients. *Turkish J Emerg Med*. 2018;18(3):91–93.
5. Armstrong RA, Eperjesi F, Gilmartin B. The use of correlation and regression methods in optometry. *Clin Exp Optom*. 2005;88(2):81–88.
6. Chan YH. Biostatistics 104: correlational analysis. *Singap Med J*. 2003;44(12):614–619.
7. Hamilton DF, Ghert M, Simpson AHRW. Interpreting regression models in clinical outcome studies. *Bone Joint Res*. 2015;4(9):152–153.
8. Alexopoulos EC. Introduction to multivariate regression analysis. *Hippokratia*. 2010;14(Suppl 1):23–28.
9. Schneider A, Hommel G, Blettner M. Linear regression analysis: part 14 of a series on evaluation of scientific publications. *Deutsches Arzteblatt Inter*. 2010;107(44):776–782.
10. Yu SC, Qi X, Hu YH, Zheng WJ, Wang QQ, Yao HY. Overview of multivariate regression model analysis and application. *Zhonghua Yufang Yixue Zazhi*. 2019;53(3):334–336.
11. Costantino M, Lee J, McCullough M, Nsouli-Maktabi H, Spies JB. Bilateral versus unilateral femoral access for uterine artery embolization: results of a randomized comparative trial. *J Vasc Intervent Radiol*. 2010;21(6):829–835.
12. Crawford ST, Olsen RV, Pilgram TK, Duncan JR. Validation of an angiographic method for estimating resting blood flow to distal tissue beds in the lower extremities. *J Vasc Intervent Radiol*. 2003;14(5):555–565.
13. Kim W, Gandhi RT, Pena CS, et al. Influence of statin therapy on aneurysm sac regression after endovascular aortic repair. *J Vasc Intervent Radiol*. 2017;28(1):35–43.

Chapter 33

Biases

Siddhant Thukral[1], Stefan Kovac[2] and Mounica Paturu[3]

[1]*Division of Vascular and Interventional Radiology, Department of Radiology, Duke University Medical Center, Durham, NC, United States;* [2]*Heersink School of Medicine, The University of Alabama at Birmingham, Birmingham, AL, United States;* [3]*Department of Neurosurgery, Duke University Medical Center, Durham, NC, United States*

Key points

- Bias is any trend or systemic error that decreases objectivity and directs a study knowingly or unknowingly toward false conclusions.
- Bias can be present at any stage of study, from the initial study design to the final publication.
- Bias before the study investigation can include selection bias.
- Bias during the study investigation can include crossover, investigator, or operator biases.
- Bias after the study investigation can include loss of follow-up, recall, or reporting biases.

Why it matters?

With the increasing dependence on evidence-based medicine to inform good clinical decision making for patients in the healthcare setting, research needs to be devoid of any systemic error that would inadvertently lead to false conclusions. Therefore, awareness of different forms of bias and efforts toward conscious mitigation via appropriate study design, investigation, and analysis is essential for achieving greater accuracy and validity to best guide clinical decision making.

Introduction

The core of scientific advancement is the creation and dissemination of accurate and reproducible research. Thus, the goal of every scientist is to design, conduct, and report research findings honestly and transparently, devoid of partiality altering the truth.[1] However, often unknowingly, many studies fail to meet these goals and inadvertently lead to the generation of distorted assumptions and false conclusions leading to ill-informed clinical decision making and detrimental patient care.[1,2] Any trend or systemic error that decreases objectivity and directs a study knowingly or unknowingly toward false conclusions is called bias (Fig. 33.1).[1]

As demonstrated by Table 33.1, bias can be present at any stage of the study, from the initial study design to the final publication.[2] Noting its prevalence, scientists, researchers, and clinicians need to develop a strong understanding of the various types of biases, hone the tools to identify existing bias in the literature, understand how that particular bias might influence the study's conclusion, and consciously focus on preventing bias when designing and conducting their investigations.[2] This chapter seeks to stratify common types of research biases based on when they are likely to be encountered, distinguish the causes of common biases, and highlight methods to minimize bias in a scientific investigation.

Bias before the study investigation

Bias before the study investigation can often be traced to errors in study design or patient recruitment strategies leading to inaccuracies in assessing the validity of the null hypothesis. For example, a poor study design can potentially introduce confounders into a study related to the exposure and outcome under investigation, as visualized in Fig. 33.2.[2] This type of bias can create generalizability issues that cannot be overcome with subsequent statistical analysis, so researchers must pay attention to pre-investigation bias when designing their study to prevent the loss of valuable time and resources.

Selection bias

In a well-designed study, the obtained sample population should be closely representative of the target population; therefore, sampling should be random and the sample size should be adequate for the study's design.[1] When there is

Translational Interventional Radiology. https://doi.org/10.1016/B978-0-12-823026-8.00055-9

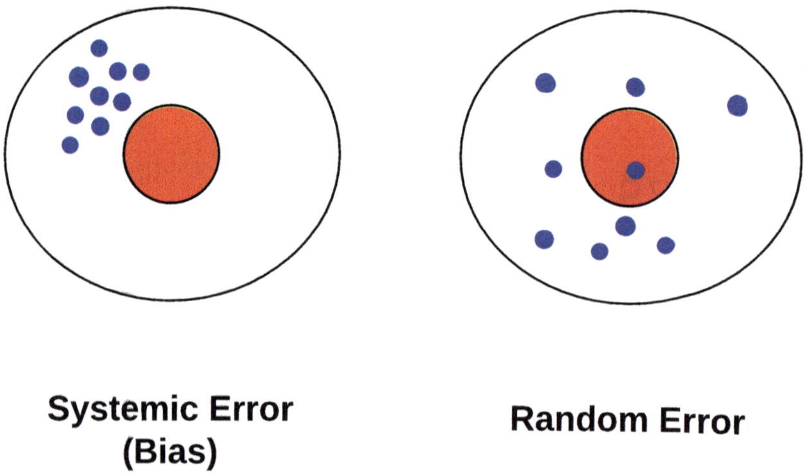

FIGURE 33.1 Systemic error, represented on the left, is one which results (blue dots) deviate from the truth (red target) secondary to influencers not be attributed to chance alone. Random error, represented on the right, is one in which the data deviates from the truth, but the spread of data is secondary to chance.

TABLE 33.1 Bias in research.[2]

Type of bias	When likely to encounter	Ways to reduce
Selection bias	Preinvestigation	Prospective study design in which outcome is unknown during enrollment. Rigorous participant selections with avoidance of known confounders.
Flawed study design	Preinvestigation	Clearly define study goals and outcomes.
Interviewer bias	During an investigation	Blind interviewer to exposure and standardize interactions.
Recall bias	During and following an investigation	Prospective study design in which outcome is unknown during enrollment. Objective measures instead of subjective measures.
Reporting bias	After study investigation	Register trial with accepted clinical trials registry.

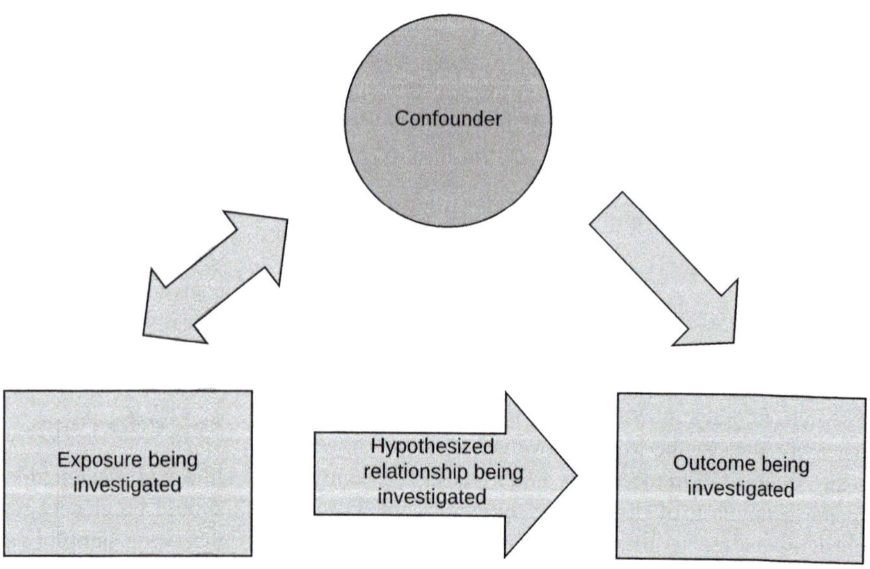

FIGURE 33.2 Counfounders are external influencers to the study aim(s) and may potentially skew the outcome being investigated by affecting both the independent and dependent variables.

a failure in this process, selection bias can be introduced to the study. Selection bias results from any error affecting the selection of study participants and/or factors that would affect study participation by altering the likelihood of participation for one particular demographic or sample group.[3] Consequently, it can considerably impact the external validity of a study, which is the generalization of study results to different measures, persons, settings, and times.[4] Although present in many various studies, selection bias is most likely to be a problem with case-control and retrospective cohort studies where exposure and outcomes have already occurred at the time of participant inclusion.[2]

1. Healthcare access bias

As discussed in a study by Gabr et al.,[5] healthcare bias results when patients analyzed at a certain hospital(s) are not representative of cases in the overall community population. Several factors may contribute to healthcare access bias, such as (1) hospital admissions being dependent on the professional or personal interests of the medical staff (popularity bias), (2) patient attraction to the prestige of particular clinicians (centripetal bias), (3) complexity and available resources of the healthcare organization itself—for example, "complicated cases" are referred to tertiary care, (4) patient cultural, geographic, and economic preferences (diagnostic/treatment access bias).[6] Economic disparities such as varying insurance plan coverage options are of particular concern in retrospective or observationally acquired data where bias would favor patients with treatment coverage compared to those without treatment coverage.[5]

2. Volunteer bias (self selection bias)

Volunteer bias occurs when individuals who volunteer to participate in a study have different relevant clinical characteristics than those who elect not to join the study.[3] One way of minimizing volunteer bias is by randomizing studies so that patients are assigned to treatment or placebo groups only after agreeing to participate in the study.[3] In this manner, the NIH-funded, phase III, multicenter ATTRACT trial (Active Venous Thrombosis: Thrombus Removal with Adjunctive Catheter-Directed Thrombolysis) trial was able to diminish selection bias by randomizing patients in a 1:1 ratio to pharmacomechanical catheter-directed thrombolysis group or control group only after determining eligibility and receiving consent.[7]

3. Sample Bias

When a sample in a clinical study is not adequately representative of the target population, this is known as sample bias and can skew the efficacy data of a procedure or treatment.[5] In prospective studies, sample bias can be mitigated by creating well-balanced arms using randomization. For example, in the SOX (Compression Stockings to Prevent the Post-Thrombotic) trial by Kahn et al.,[8] which sought to investigate the efficacy of elastic compression stockings for the prevention of post-thrombotic syndrome, sample bias was mitigated by presenting a multicenter randomised placebo-controlled trial with a wide variety of baseline factors, including the most proximal extent of presenting deep vein thrombosis.

Bias during investigation

Despite having a highly scrutinized study design focused on mitigating bias before recruitment or intervention, studies are still prone to bias during the actual treatment, intervention, or procedure process.

1. Crossover bias (contamination)

Crossover bias in randomized control trials (RCT) occurs when the control group is mistakenly exposed to the proposed intervention or procedure. This typically reduces the point estimate of an intervention's effectiveness and may lead to greater type 2 error (acceptance of null hypothesis when a difference actually exists).[5,9,10] In a study by Lee et al.,[11] which sought to compare chemoembolization versus hepatic resection for hepatocellular carcinoma, 31% of patients underwent crossover treatment despite data censoring occurring for those patients at the time of resection or transplant. This crossover was noted to positively influence survival outcomes in the chemoembolization group inaccurately.[5]

2. Investigator bias

When an investigator has strong feelings toward a particular treatment or intervention, this unconscious conflict of interest may influence the general attitude toward a research project and affect results. This is known as investigator bias.[5,12] A biased researcher will be more driven to find evidence in agreement with his or her hypothesis, thereby reducing objectivity in overall study results. Investigator bias can be managed by separating investigator roles and incorporating blinding into the study design.[12] For example, investigator bias was mitigated by the PREPIC trial by having all documented events reviewed by an independent committee separate from the medical professionals conducting the intervention.[13]

3. Operator bias

Operator bias can be introduced into the study when multiple operators performing the same interventional procedure are utilized.[5] Operators with differences in expertise, familiarity with the procedure in question, and institutional volume can decrease both the accuracy and precision of data collection and acquisition in a study. For example, an observational study by Jarrett et al.[14] analyzed the relationship between institutional volumes and catheter-directed thrombolysis outcomes for proximal lower-extremity deep vein thrombosis, reported that increased institutional volume was associated with lower in-hospital mortality and intracranial hemorrhage.

Bias after study investigation

Known as postinterventional bias, bias occurring after the treatment, intervention, or procedure being investigated can be evident in subsequent data-analysis or after-care, and follow-up settings.[5]

1. Loss of follow-up bias

Loss of patients to follow-up can present a challenge for most research studies, both prospective and retrospective. Ensuring patient participation for the entirety of the trial is important for a study's validity as patients lost to follow-up can have a different prognosis than those who complete the study.[15] When calculating the loss of patient follow-up, it is essential to include everyone randomly assigned in an RCT and those who have had the intervention during the prespecified time in a retrospective cohort study in the denominator.[15]

2. Recall bias

Recall bias is introduced when there is an error in recalling information, such as forgetting an event, claiming the occurrence of a false event, or recalling an event before or after the specified period of time in question.[16] Recall bias is inherent in retrospective analyses, and studies may be more susceptible when there are long intervals between study visits.[5] For example, in a retrospective case series, Kurklinsky et al.[17] reported recall bias in patients with chronically occluded iliofemoral venous thrombotic lesions treated with stent placement. The retrospective study design partly contributed to this. Recall bias is best avoided by having well and intentionally formatted recall questions, obtaining larger sample sizes to increase the signal-to-noise ratio, and verifying information with external sources when possible.

3. Reporting Bias

Reporting bias is a combination of publication bias and citation bias.[5] Publication bias can occur when the media publishes studies showing positive or significant findings with exaggerated headlines or far-reaching implications of study implications, which may falsely lead to increased perceived efficacy of an intervention.[5] Relatedly, citation bias occurs when particular articles are cited more frequently than others, which can lead to an over- or under-estimate of the clinical efficacy of an intervention's effect dependent on popularity.[5]

Conclusion

Bias is any trend or systemic error that decreases objectivity and directs a study toward false conclusions. Despite an investigator's best intentions, bias can permeate a study before its initiation, throughout the intervention, or persist beyond the follow-up period. Therefore, awareness of different forms of bias and efforts toward conscious mitigation via appropriate study design, investigation, and analysis is essential for achieving greater accuracy and validity to best guide clinical decision making.

Get started

It is crucial to understand the bias that may be present in the published literature. Therefore, continuous reading of published articles and study of research methodologies can aid investigators in distinguishing various forms of bias and how they may be mitigated.

References

1. Simundić A-M. Bias in research. *Biochem Med*. 2013;23(1):12–15.
2. Pannucci CJ, Wilkins EG. Identifying and avoiding bias in research. *Plast Reconstr Surg*. 2010;126(2):619–625.
3. Tripepi G, et al. Selection bias and information bias in clinical research. *Nephron Clin Pract*. 2010;115(2):c94–c99.
4. Steckler A, McLeroy KR. The importance of external validity. *Am J Publ Health*. 2008;98(1):9–10.
5. Gabr A, et al. Types of research bias encountered in IR. *J Vasc Intervent Radiol*. 2016;27(4):546–550.
6. Delgado-Rodríguez M, Llorca J. Bias. *J Epidemiol Communi Health*. 2004;58(8):635–641.
7. Vedantham S, et al. Pharmacomechanical catheter-directed thrombolysis for deep-vein thrombosis. *N Engl J Med*. 2017;377(23):2240–2252.
8. Kahn SR, et al. Compression stockings to prevent post-thrombotic syndrome: a randomised placebo-controlled trial. *Lancet*. 2014;383(9920):880–888.
9. Simmons N, et al. Assessment of contamination and misclassification biases in a randomized controlled trial of a social network peer education intervention to reduce HIV risk behaviors among drug users and risk partners in Philadelphia, PA and Chiang Mai, Thailand. *AIDS Behav*. 2015;19(10):1818–1827.
10. Torgerson DJ. Contamination in trials: is cluster randomisation the answer? *BMJ*. 2001;322(7282):355–357.
11. Lee YB, et al. Comparison of transarterial chemoembolization and hepatic resection for large solitary hepatocellular carcinoma: a propensity score analysis. *J Vasc Intervent Radiol*. 2015;26(5):651–659.
12. Ernst E, Canter PH. Investigator bias and false positive findings in medical research. *Trends Pharmacol Sci*. 2003;24(5):219–221.
13. Eight-year follow-up of patients with permanent vena cava filters in the prevention of pulmonary embolism. *Circulation*. 2005;112(3):416–422.

14. Jarrett H, et al. Impact of institutional volume on outcomes of catheter directed thrombolysis in the treatment of acute proximal deep vein thrombosis: a 6-year US experience (2005-2010). *Circulation.* 2015;132(12):1127–1135.
15. Dettori JR. Loss to follow-up. *Evid Base Spine Care J.* 2011;2(1):7–10.
16. Brusco NK, Watts JJ. Empirical evidence of recall bias for primary health care visits. *BMC Health Serv Res.* 2015;15(1):381.
17. Kurklinsky AK, et al. Outcomes of venoplasty with stent placement for chronic thrombosis of the iliac and femoral veins: single-center experience. *J Vasc Intervent Radiol*, 2012;23(8):1009–1015.

Chapter 34

Basic science statistics

Nicholas Vollano
College of Osteopathic Medicine, Kansas City University, Joplin, MO, United States

Understanding the fundamental concepts of statistical methodology is crucial when designing a research study and in interpreting reported results. An appropriately selected statistical analysis allows for quick and meaningful interpretation of a large amount of data. However, the inferences derived from a data set are only as precise as the analysis selected and incorrect selection of the statistical method poses a challenge in interpreting the findings and constructing a conclusion. This chapter introduces commonly used statistical analyses for basic science research and incorporates information from preceding chapters to determine the most appropriate tests for a set of data.

The first step in making this determination involves understanding descriptive and inferential statistics. Descriptive statistics condense large amounts of data into a simplified form for the reader to easily understand the relationships between variables within a sample.[1,2] These are frequently reported in studies to characterize the demographical distribution of the participants (e.g., gender, age, ethnicity). Descriptive statistics fall into three major categories: measures of variation (e.g., standard deviation, dispersion, range), frequency of categorical variables (e.g., percent, frequency), and measures of central tendency (e.g., mean, mode and median).[3,4] These can be displayed in a graphical or table format for a reader to quickly determine the application of a study's findings.[5]

However, despite its usefulness, there are limitations in the usage of descriptive statistics, especially when formulating or testing a hypothesis. Frequently, the objective of a study is to apply conclusions from the results of a sample to a larger population.[1] For example, if a researcher is inquiring whether medical students display a higher likelihood of developing neck pain than college students from other disciplines in the United States (US). To do this, the researcher could survey every US college student individually or observe an adequately representative group of participants and generalize the findings to the target population. This is the premise of inferential statistics. Inferential statistics analyze data derived from a sample group to draw inferences about a target population.[1,2,5] Typically, a research question can be answered by using several statistical tests that combine the usage of descriptive and inferential statistics. The choice of analysis is largely dependent on a study's outcome variable and the purpose of the analysis.[5]

Parametric and nonparametric tests

Statistical tests are grouped into two general categories: parametric and nonparametric.

Parametric tests or methods are used to analyze data sets of quantitative variables that are normally distributed.[2,6] They also assume that the samples have variance that is equivalent to the target population. Larger sample sizes are more likely to display a normal distribution and commonly utilize these types of statistical analyses.

Nonparametric methods are utilized when a sample has an unknown or skewed distribution, typically demonstrated with smaller sample sizes.[2,6] These types of tests are also useful for analyzing nominal and ordinal data.[2] Fortunately, most parametric tests have a nonparametric counterpart (see Table 34.1 for examples). This can be helpful as a secondary analysis when the requisites of a parametric test are broken or cannot be achieved.

The most frequently used parametric analyses are the *t*-test and analysis of variance (ANOVA). The *t*-test and ANOVA are used when evaluating the significance of continuous outcomes or outcomes that have a numerical value (e.g., blood pressure, body mass index). The *t*-test is arguably the most common statistical technique used and determines if there is a statistically significant difference between the means of two data sets.[2] The *t*-test exists in three forms as one sample, unpaired, and paired *t*-tests.

The one-sample *t*-test is used to determine if the mean of a sample is significantly different from the mean of a

Translational Interventional Radiology. https://doi.org/10.1016/B978-0-12-823026-8.00004-3

TABLE 34.1 Parametric tests with their nonparametric test analogs.

Parametric	Nonparametric
Standard deviation, mean	Interquartile range, median
One sample *t*-test	Wilcoxon signed rank test
Paired *t*-test	Wilcoxon signed rank test
Unpaired *t*-test	Mann–Whitney U-test
One-way analysis of variance (ANOVA)	Kruskal–Wallis test
Repeated measures ANOVA	Friedman test

chosen value (e.g., mean of the population the sample represents, zero, a reference value). For example, this analysis could be used to determine if a sample's mean age is truly representative of the population's mean age or if it is significantly different. This information could factor into how the results of a study can be applied to the general population. The unpaired *t*-test demonstrates if there is a significant difference in the means of a continuous variable between two independent sample groups.[7] The most used form of the three is the paired *t*-test. The paired *t*-test determines if there is a statistically significant change in the means between two dependent samples.[2] This translates to something being measured on every participant at two points in time or conditions. A typical setting for this can be seen when determining if there is a significant difference in a continuous variable before and after a treatment (e.g., measuring heart rate before and after consuming four cups of espresso).[2]

ANOVA is an extension of the *t*-test which can determine if there is a significant difference in the means of two or more groups.[5] When the ANOVA test is used to analyze the means of only two groups it yields similar results to the *t*-test. A significant ANOVA test indicates that the mean difference of at least one pair within the collection of groups is significantly different.[7] Additional pairwise comparison testing of means is needed to determine which pair differs. Multiple types of ANOVA exist but the two main categories are the one-way ANOVA and the one-way repeated measures ANOVA. The one-way ANOVA is an extension of the unpaired *t*-test, capable of comparing the means of two or more independent groups.[8] Similarly, the one-way repeated measures ANOVA builds upon the paired *t*-test but allows for comparison of means between measurements of two or more groups that were repeated at different times or conditions.[2] Fig. 34.1 summarizes the above and provides a quick method for choosing the most appropriate parametric statistical analysis.

To demonstrate how parametric statistical tests can be used in practice let us work through an example scenario:

A researcher organizes a study to evaluate if a person's baseline heart rate is significantly impacted after walking for 60 min. 20 participants are recruited. Their heart rates are measured upon arrival and again after 60 min of walking on a treadmill at a pace of three miles per hour. Their heart rates are detected using a heart rate monitor worn on the left wrist. In addition, age category, sex, and height are recorded. The data is collected and outlined in Table 34.2.

Age categories: 1 (21–30 years old), 2 (31–40 years old), 3 (41–50 years old). HR = heart rate. [a]Sex in this chart refers to the sex assigned at birth based on phenotypic features.

1. What test can we use to determine if walking for 60 min causes a significant change in heart rate for all 20 participants? For this example, we want to know if there is a statistically significant change in the means of paired heart rate measurements of one sample. Since no control group was included, we are only comparing the means of one group measured at different points in time. Therefore, the best choice is the paired samples *t*-test. Although a one-way repeated measures ANOVA may yield similar results, this should be reserved for comparing the means of three or more groups.
2. What if the researcher wants to determine whether there is a significant difference in the heart rate of the age categories after 60 min of walking? In this example, the participants were separated into three age groups, so any of the *t*-tests would not be appropriate. To decide between the ANOVA options discussed we must think about the question being asked. Although heart rate was measured twice in these participants, the question is only referring to the heart rate measured after 60 min of walking. Therefore, the one-way ANOVA is the best way to analyze if a statistically significant difference between the three age groups exists. To further determine which pair of means has a significant difference, we would perform additional pairwise comparison testing.

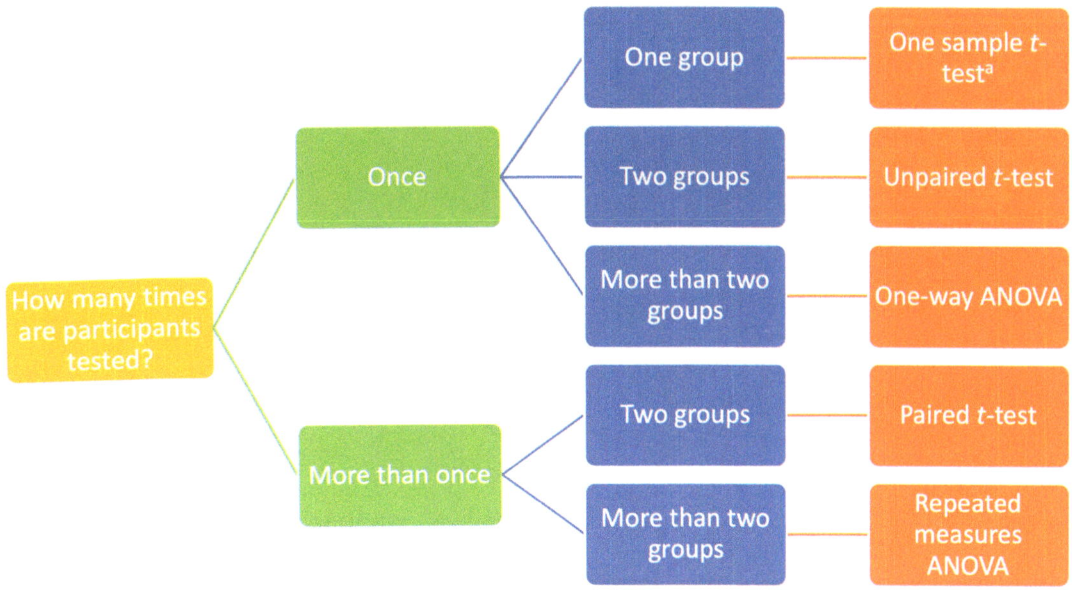

FIGURE 34.1 Common parametric statistical analyses. This flow chart demonstrates a simple method to decide which parametric test to utilize when comparing data. ANOVA: analysis of variance. [a]One-sample *t*-tests compare the mean of one sample group to a chosen value (e.g., theoretical value, reference value, zero, etc).

3. While enrolling participants in the study, the researcher is considering other variables (e.g., height) that may play a role in how walking impacts a person's heart rate and wants to make sure that there is not a significant difference between the male/female participants, so it does not impact the study's results. Which test would be best to evaluate this? In this question, the researcher is inquiring about height differences between two groups (the male and female participants). As height is only measured once, we can rule out the paired samples *t*-test. We can also rule out ANOVA tests because the comparison is between two groups (males and females). This leaves us with the best option being the unpaired samples *t*-test. Remember that although the one-way ANOVA is an extension of the unpaired samples *t*-test and may yield similar results, it is not the best analysis for this question.

It is important to note that parametric methods are more powerful in finding true differences between groups and are less likely to make type II errors in comparison to their nonparametric counterparts. However, due to the required assumptions, it is not always possible to use, and the researcher may elect for nonparametric analyses to be performed. Unlike their parametric counterparts, these methods compare the medians or rank of data to draw inferences. The median is not impacted as much as mean by outliers in a data set and is, therefore, less representative of the sample data.

The most common nonparametric methods used are the Wilcoxon signed rank test, Mann–Whitney U-test, and the Kruskal–Wallis test. They are analogs of the paired *t*-test, unpaired *t*-test, and ANOVA, respectively. The Wilcoxon signed rank test converts the difference of measurements taken at two time points or conditions into ordered ranks (e.g., 1, 2, 3, 4, etc.), ignoring any calculated zeros.[9] The ordered ranks of the two samples are then compared to determine if they are equal (accepting the null hypothesis) or statistically different (rejecting the null hypothesis). The Mann–Whitney U test determines whether the medians of the two independent samples are the same or if one is larger than the other. The Kruskal–Wallis test expands on the Mann–Whitney U-test and determines if there is a significant difference in the median of two or more independent samples.[2]

There are many other statistical methods utilized to analyze data including the Pearson Chi-Square test and regression analyses. The Pearson Chi-Square test determines how related two or more categorical variables are with each other.[10] Regression analyses create an equation that quantifies the relationship between two or more variables and allows one to predict the value of one variable with knowledge of another variable.[11]

Conclusion

Selecting the most appropriate statistical analysis is crucial in reporting quality study results that contribute to the science community. There are many statistical methods to choose from and each option assumes specific characteristics about the data. When determining the best test for any data set it is imperative to keep these assumptions in mind. Improper selection can mislead someone to make decisions based on inaccurate conclusions. Although the astute reader

TABLE 34.2 Theoretical data of 20 study participants.

Participant number	1	2	3	4	5	6	7	8	9	10	11	12	13	14	15	16	17	18	19	20
Age category	3	1	2	3	2	2	3	1	3	1	3	3	1	1	1	2	2	2	3	2
sex[2]	F	F	F	F	F	F	F	F	F	F	M	M	M	M	M	M	M	M	M	M
Height (inches)	60	63	62	61	63	61	59	64	64	62	65	67	68	72	71	69	70	68	70	69
HR at baseline	74	72	71	85	71	81	73	69	83	67	83	83	74	68	70	71	81	70	81	74
HR after 60 min walk	129	110	118	130	122	126	128	112	131	107	131	130	120	105	109	118	125	131	128	123

is careful to cross-reference information from one study to others before making important decisions (e.g., drafting new or editing old policies, creating treatment plans), this is an avoidable outcome by choosing the correct analysis. As mentioned in the introduction, a study's results are only as strong as the statistical method used. Having a basic understanding of the most frequently used analyses is an important step in improving and maintaining quality research but it is also crucial to recognize one's own limits. Consultation from expert statisticians can ensure appropriate statistical analyses are utilized and should be considered in the face of uncertainty.

References

1. Satake E. *Statistical Methods and Reasoning for the Clinical Sciences: Evidence-Based Practice*. 1st ed. San Diego, CA: Plural Publishing Inc.; 2015.
2. Ali Z, Bhaskar SB. Basic statistical tools in research and data analysis [published correction appears in Indian J Anaesth. 2016 Oct;60(10):790. *Indian J Anaesth*. 2016;60(9):662–669. https://doi.org/10.4103/0019-5049.190623.
3. Mishra P, Pandey CM, Singh U, Gupta A, Sahu C, Keshri A. Descriptive statistics and normality tests for statistical data. *Ann Card Anaesth*. 2019;22(1):67–72. https://doi.org/10.4103/aca.ACA_157_18.
4. Vetter TR. Descriptive statistics: reporting the answers to the 5 basic questions of who, what, why, when, where and a sixth, so what? *Anesth Analg*. 2017;125(5):1797–1802. https://doi.org/10.1213/ANE.0000000000002471.
5. Sebastião YV, St Peter SD. An overview of commonly used statistical methods in clinical research. *Semin Pediatr Surg*. 2018;27(6):367–374.
6. Mishra P, Pandey CM, Singh U, Keshri A, Sabaretnam M. Selection of appropriate statistical methods for data analysis. *Ann Card Anaesth*. 2019;22(3):297–301. https://doi.org/10.4103/aca.ACA_248_18.
7. Mishra P, Singh U, Pandey CM, Mishra P, Pandey G. Application of student's t-test, analysis of variance, and covariance. *Ann Card Anaesth*. 2019;22(4):407–411. https://doi.org/10.4103/aca.ACA_94_19.
8. Ledolter J, Gramlich O, Kardon R. Parametric statistical inference for comparing means and variances. *Invest Ophthalmol Vis Sci*. 2020;61(8):25. https://doi.org/10.1167/iovs.61.8.25.
9. Skene SS, Bunce C, Freemantle N, Doré CJ. Ophthalmic statistics note 9: parametric versus non-parametric methods for data analysis. *Br J Ophthalmol*. 2016;100(7):877–878.
10. Schober P, Vetter T. Chi-square tests in medical research. *Anesth Analg*. 2019;129(5):1193. https://doi.org/10.1213/ANE.0000000000004410.
11. Vetter T, Schober P. Regression: the apple does not fall far from the tree. *Anesth Analg*. 2018;127(1):277–283. https://doi.org/10.1213/ANE.0000000000003424.

Part V

Clinical: study types

Chapter 35

Design principles: Hierarchy of study types

Alex J. Solomon[1] and Nariman Nezami[2]

[1]The Johns Hopkins University School of Medicine, Baltimore, MD, United States; [2]Division of Vascular and Interventional Radiology, Department of Diagnostic Radiology and Nuclear Medicine, University of Maryland School of Medicine, Baltimore, MD, United States

Key points

- Quality of evidence relies upon both study design and an analysis of a study's strengths and limitations.
- Developing clinical guidelines requires a systematic and well-defined approach to generate clinically meaningful, evidence-based recommendations.
- Appropriate implementation of evidence-based medicine can improve patient outcomes by integrating the best available evidence with the art of medicine and patient values.

Why knowing hierarchy of study types matters

Evidence-based medicine (EBM) represents the integration of research and clinician expertise in providing optimal patient-centered care. EBM is founded on the principle that medical practice should be guided using the best available objective evidence. This necessitates a hierarchical classification of scientific studies, as poor research practices and false assumptions can lead to negative or even fatal patient outcomes.[1] As the body of literature across all medical specialties rapidly expands, it is critical for providers to have a basic understanding of study design, the hierarchy of studies, and the process by which recommendations are generated in order to appropriately apply the literature in their daily practice.

The hierarchy of studies: introduction to study design

The quality of medical literature can be defined as the ability of a study or studies to credibly estimate a treatment or exposure effect. Overall study design was used as the original system to grade the quality of evidence and led to a hierarchical classification system, often referred to as the levels of evidence, with higher levels representing higher quality data (Fig. 35.1). Medical literature can be grouped into randomized or observational studies, with the former traditionally viewed as superior. Randomized controlled trials (RCTs) are prospective and hypothesis-driven studies which attempt to minimize bias by distributing known and unknown patient-related factors evenly between treatment and control groups. Results from well-designed RCTs are considered the "gold-standard" of EBM as they most accurately estimate the true effect of the tested intervention. In 2002, Llovet et al. published a landmark RCT demonstrating improved survival for patients with unresectable hepatocellular carcinoma (HCC) following transarterial chemoembolization (TACE), spurring a paradigm shift in the treatment of HCC.[2]

Observational studies are perceived as inferior to RCTs due to their inherently increased potential to introduce bias and confounding variables.[3] A cohort study, which can be performed either prospectively or retrospectively, compares an outcome of interest between two separate groups (cohorts) based on the presence or absence of an exposure. Case-control studies group subjects into two cohorts based on the presence or absence of a known outcome of interest and retrospectively compare the potential exposures between the groups. Case-control studies are subject to a greater degree of bias as compared to cohort studies. Case series', case reports, and expert opinions are descriptive in nature and do not involve hypothesis testing.

Systematic reviews and meta-analysis, while sometimes excluded from the hierarchy of evidence pyramid as they represent data summaries rather than a study design, are critical for developing clinical practice guidelines. A systematic review appraises and summarizes all the literature

Translational Interventional Radiology. https://doi.org/10.1016/B978-0-12-823026-8.00008-0

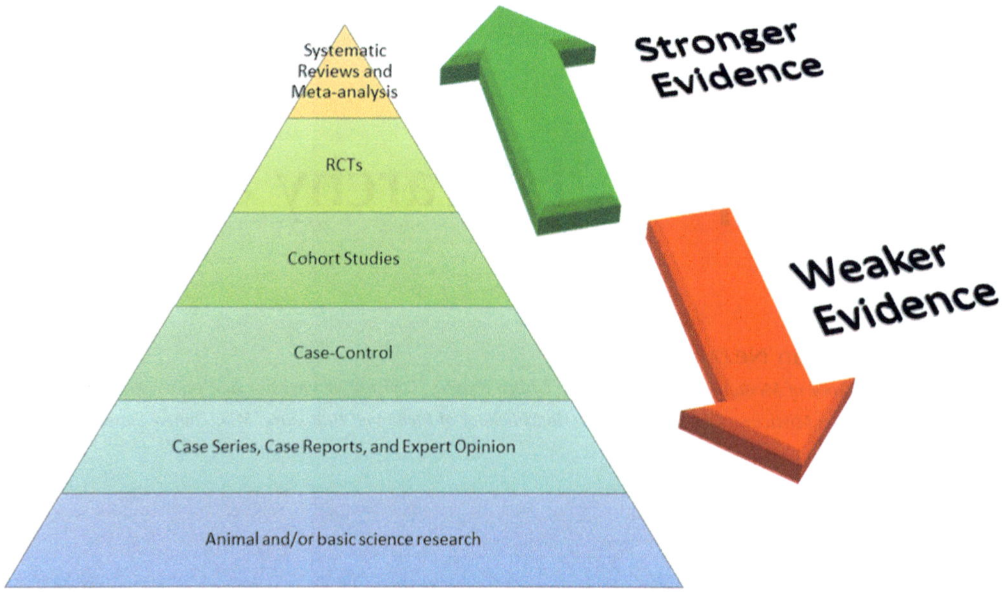

FIGURE 35.1 The traditional hierarchy of levels of evidence.[1]

available on a specific topic of interest. A meta-analysis similarly reviews the literature and then uses quantitative methods to demonstrate both a direction and magnitude of effect. Both approaches can evaluate randomized and observational studies to summarize the best available evidence, but are inherently limited by the quality of evidence presented in the individual studies and are subject to potential reporting/publication bias. Future chapters provide more detailed descriptions of the various study designs and the effect of biases on outcomes and generalizability.

Going beyond the pyramid: modified hierarchy of studies

From its inception, the traditional hierarchy was extensively modified as it became apparent that the level of evidence assigned to a study depends not only on the study design, but also on the clinical/research question and the quality of the study itself. Most clinical questions can be broadly categorized into diagnostic, therapeutic, prognostic, etiological, or economic. The level of evidence for a given study is dependent on the clinical question. For example, RCTs do not provide the highest level of evidence for questions related to disease prognosis as there are no comparison groups. Cohort studies and systematic reviews/meta-analyses of cohort studies therefore represent the highest level of evidence in this context, illustrated by Eltawil et al. in their prospective cohort study evaluating quality of life and survival following TACE.[4] In circumstances when RCTs would provide the highest quality evidence, nonrandomized controlled trials or observational studies may offer the best available evidence if an RCT would be unethical or impractical to conduct. Low-level evidence should not be immediately discredited as inferior, but rather critically appraised and potentially used to generate a higher level research and evidence.

Similarly, level of evidence is not synonymous with the quality of research. Inappropriately designed studies do not confidently estimate the effect of interventions or the magnitude of associations and thus should be viewed as lower quality. While a detailed analysis of how to critically evaluate each type of study is beyond the scope of this chapter, the Grading of Recommendation Assessment, Development, and Evaluation (GRADE) system, introduced in 2004, provides a systematic approach to assess study quality.[5–7] The GRADE system expands beyond the traditional pyramidal model to create a more complex and fluid classification of the quality of evidence (Fig. 35.2).[1,5,6,8]

GRADE stratifies and codes the levels of evidence into high, moderate, low, or very low quality based on the confidence a reader can have in the estimates of effect presented. Studies are initially classified according to overall study design, analogous to the traditional pyramidal model, but are then appraised for methodology, execution, precision, directness, publication bias, and outcomes including magnitude of treatment effect and dose–response gradients.[5,6,8] Study quality is cumulatively upgraded or downgraded accordingly to arrive at a final determination of the quality of evidence (Fig. 35.2).[1,5–8] When evaluating a systematic review or meta-analysis, consistency across studies is also incorporated. This dynamic process assigns a more appropriate level of evidence based on a study's strengths and limitations.

Quality (Level) of Evidence	Grade	Definition	Study Design
High	A	Further research is unlikely to change estimate of effect	RCT
Moderate	B	Further research may change estimate of effect	
Low	C	Further research is likely to change estimate of effect	Cohort or Case-control studies
Very Low	D	Estimate of effect is uncertain	Case Series, Case Reports, Expert Opinion

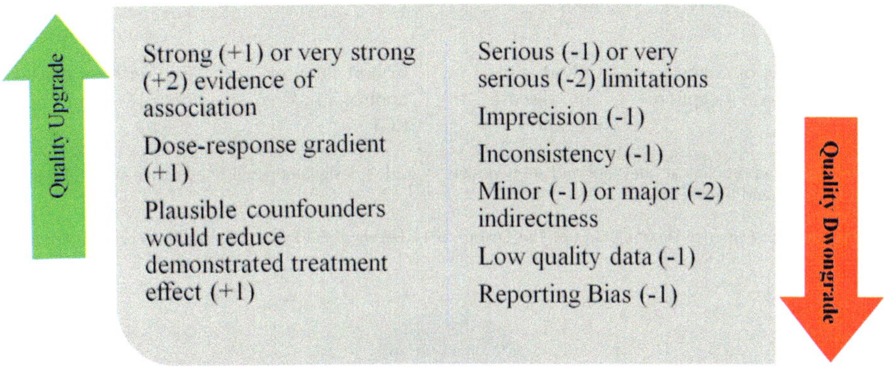

FIGURE 35.2 Modified levels of evidence. A dynamic approach to determining the quality of evidence according to the GRADE working group.[1,5,6,8] *GRADE*, Grading of Recommendation Assessment, Development, and Evaluation.

The Oxford Centre for Evidence-Based Medicine developed quick-reference levels of evidence table denoting which study designs are likely to represent the best quality evidence based on the specific clinical question (Table 35.1).[9] While this system acknowledges the relative merits of different study designs in various clinical situations, it does not represent clinical recommendations or guidelines and requires the clinician to judge study quality and generalizability.

From evidence to recommendations: clinical guidelines and strengths of recommendations

Clinical guidelines, if appropriately developed and followed, can help direct medical care while reducing reliance on individual physician experience, improve patient outcomes, and decrease healthcare expenditures.[1,10] Many professional organizations offer guidelines and recommendations for their specific areas of expertise based on specific clinical questions within a defined patient group and care setting.[5,6] The GRADE system again provides a systematic approach to generating clinical guidelines which is widely adopted and endorsed within the medical community by many prominent organizations including the World Health Organization, the Cochrane Collaboration, and UpToDate.[1,5,7]

After identifying all relevant outcomes, both positive and negative, a systematic review is conducted to synthesize and evaluate the quality of evidence. The magnitude of potential benefits, harms, and uncertainties are assessed alongside expected patient preferences, resource utilization, and feasibility to generate a clinical recommendation for treatment (Fig. 35.3).[11,12] A strength of recommendation (SOR), representing the degree to which recommendation adherence will confidently result in the desired clinical outcome, should accompany all clinical guidelines and is typically presented with the overall quality of evidence. High, moderate, and low-quality evidence can result in strong and weak recommendations as the evidence is balanced with the clinical outcomes, treatment alternatives, uncertainties, patient values and preferences, and resource utilization.[12] This is perhaps most evident when negative clinical outcomes can be catastrophic, as is the case with periprocedural thromboembolic or hemorrhagic complications in patients undergoing percutaneous interventions, where a multidisciplinary management approach is strongly recommended although supported by weak evidence.[13]

TABLE 35.1 Modified Oxford Centre for Evidence-Based Medicine 2011 Levels of Evidence.[a]

Clinical question	Levels of evidence				
	Level 1	Level 2	Level 3	Level 4	Level 5
Diagnosis	Systematic review of cross-sectional studies	Cross-sectional study	Nonconsecutive or inconsistent study	Case-control study	Physiologic-based reasoning
Treatment benefits	Systematic review of RCTs or *n*-of-1 trials	RCT or observational study with dramatic effect	Nonrandomized controlled cohort/follow-up study	Case series or case-control studies	Physiologic-based reasoning
Treatment harms	Systematic review of RCT, systematic review of nested case-control studies, *n*-of-1 trials, or observational study with dramatic effect	Individual RCT or impactful observational study	Nonrandomized controlled cohort/follow-up study	Case series, case-control, or historically controlled studies	Physiologic-based reasoning
Prognosis	Systematic review of cohort studies from disease inception	Cohort study from disease inception	Cohort study or control arm from RCT	Case series or case-control	–

Expected level of of evidence based upon clinical questions and study design. A study's designated level of evidence should be upgraded or downgraded according to its individual strengths and limitations.
RCT, Randomized controlled trial.
[a]*Modified from the OCEBM Levels of Evidence Working Group.* The Oxford 2011 Levels of Evidence. *https://www.cebm.net/index.aspx?o=5653.*

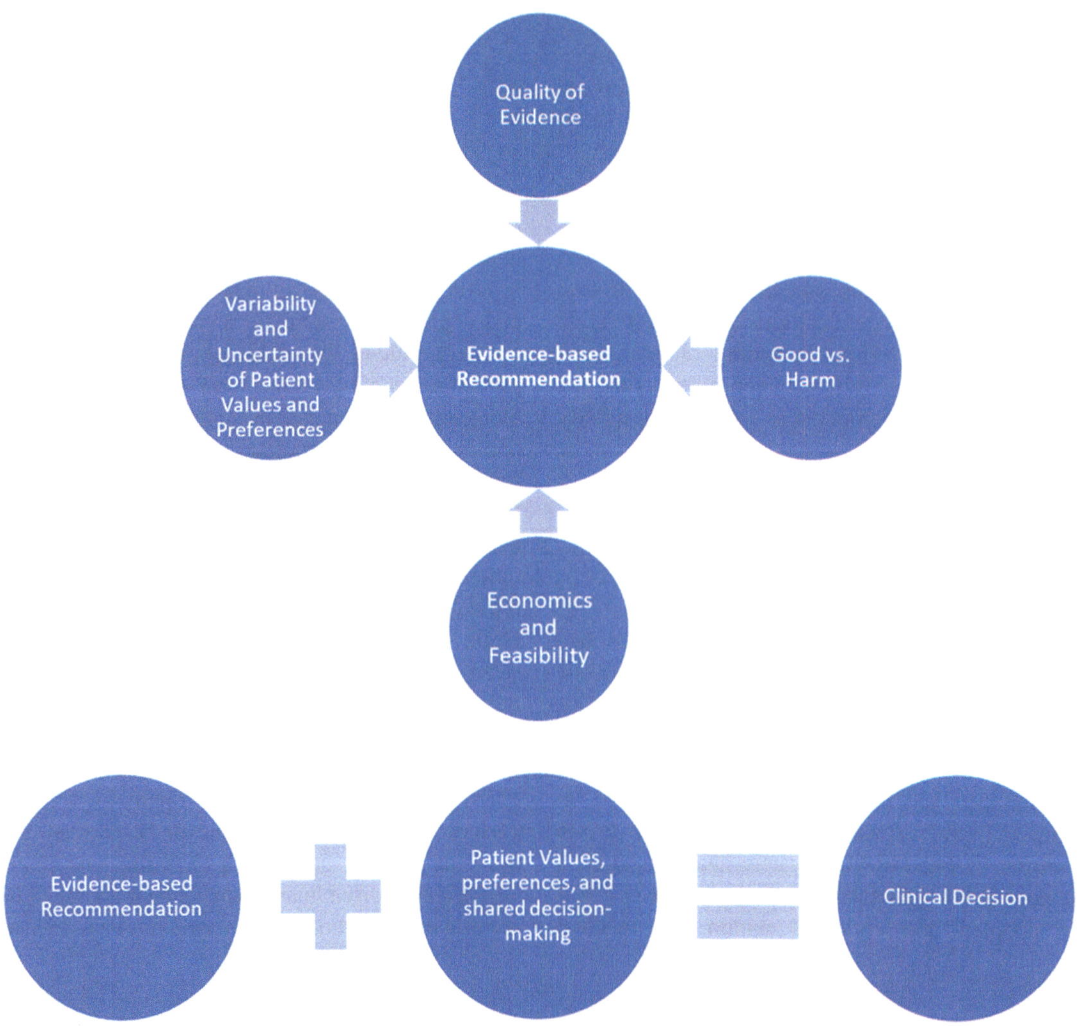

FIGURE 35.3 Integrating EBM, the art of medicine, and patient preferences.[1,5,6,11,12] *EBM*, Evidence-based medicine.

Different societies and organizations may offer their own SOR categorization. The GRADE working group proposes a binary system representing strong (Grade 1) or weak (Grade 2) recommendations which simplifies the myriad of existing grading schemes.[6,7,12] Grade 1 recommendations represent confidence that most or all patients will benefit from adherence to the recommendation, the outcome is of high patient importance, and the intervention is feasible.[6,12] Conversely, weak recommendations suggest adherence to recommendations and may provide benefits although there is less confidence in the risk/benefit ratio, uncertainty regarding patient values and preferences, and/or resource limitations.[6,12]

Recommendations are not absolute but rather represent an intervention that should be offered by well-informed clinicians. Organizations publishing clinical recommendations make inherent assumptions of patient preferences which may or may not apply to each specific clinical scenario, underscoring the importance of a patient-centered, shared decision-making process. Interpreting clinical recommendations requires knowledge of the methodology and definitions used to generate the guidelines and an understanding a patient's values, goals, preferences, and unique clinical circumstances.

Conclusion

Evidence-based medicine is an evolving field integrating clinical medicine with scientific discovery. Over the last 2 decades, EBM has improved upon the traditional pyramidal hierarchy of studies to create a dynamic, more accurate evaluation of study quality. Clinical practice guidelines, when generated using appropriate methodology, can help guide clinicians and result in improved patient outcomes. The integration of EBM into daily practice requires an understanding of the levels of evidence and the process by which clinical recommendations are generated. It is incumbent upon the provider to appropriately apply evidence and guidelines to their individual patients and participate in the shared decision-making process, incorporating patient values and preferences, before jointly arriving at a final treatment decision.

Get started, step by step evaluation

Recall the last individual study you used to make a clinical decision. How would you classify the level of evidence based on the traditional pyramidal hierarchy? Now apply the dynamic GRADE process—does the quality of evidence change?

What are your most common sources of clinical guidelines and recommendations? Evaluate the methodology by which those recommendations are made and how the strengths of recommendations are defined.

Useful resources

1. Grade of Recommendation Assessment, Development, and Evaluation (GRADE): https://www.gradeworkinggroup.org/
2. Oxford Centre of Evidence-Based Medicine: https://www.cebm.net/
3. UpToDate Grading Guide: https://www.uptodate.com/home/grading-guide

References

1. Djulbegovic B, Guyatt GH. Progress in evidence-based medicine: a quarter century on. *Lancet*. 2017;390(10092):415–423. https://doi.org/10.1016/S0140-6736(16)31592-6.
2. Llovet JM, Real MI, Montaña X, et al. Arterial embolisation or chemoembolisation versus symptomatic treatment in patients with unresectable hepatocellular carcinoma: a randomised controlled trial. *Lancet*. 2002;359(9319):1734–1739. https://doi.org/10.1016/S0140-6736(02)08649-X.
3. Guyatt GH. Users' guides to the medical literature. IX. A method for grading health care recommendations. Evidence-Based Medicine Working Group. *J Am Med Assoc*. 1995;274(22):1800–1804. https://doi.org/10.1001/jama.274.22.1800.
4. Eltawil KM, Berry R, Abdolell M, Molinari M. Quality of life and survival analysis of patients undergoing transarterial chemoembolization for primary hepatic malignancies: a prospective cohort study. *HPB*. 2012;14(5):341–350. https://doi.org/10.1111/j.1477-2574.2012.00455.x.
5. Atkins D, Best D, Briss PA, et al. Grading quality of evidence and strength of recommendations. *BMJ*. 2004;328(7454):1490. https://doi.org/10.1136/bmj.328.7454.1490.
6. Guyatt GH, Oxman AD, Vist GE, et al. GRADE: an emerging consensus on rating quality of evidence and strength of recommendations. *BMJ*. 2008;336(7650):924–926. https://doi.org/10.1136/bmj.39489.470347.AD.
7. GRADE. The GRADE Working Group. https://www.gradeworkinggroup.org/.
8. Guyatt GH, Oxman AD, Kunz R, Vist GE, Falck-Ytter Y, Schünemann HJ. What is "quality of evidence" and why is it important to clinicians? *BMJ*. 2008;336(7651):995–998. https://doi.org/10.1136/bmj.39490.551019.BE.
9. OCEBM Levels of Evidence Working Group. *The Oxford 2011 Levels of Evidence*. Oxford Centre for Evidence-Based Medicine. https://www.cebm.net/index.aspx?o=5653.
10. Pronovost PJ. Enhancing physicians' use of clinical guidelines. *JAMA*. 2013;310(23):2501. https://doi.org/10.1001/jama.2013.281334.

11. Guyatt GH, Oxman AD, Kunz R, et al. Incorporating considerations of resources use into grading recommendations. *BMJ*. 2008;336(7654): 1170–1173. https://doi.org/10.1136/bmj.39504.506319.80.
12. Guyatt GH, Oxman AD, Kunz R, et al. Going from evidence to recommendations. *BMJ*. 2008;336(7652):1049–1051. https://doi.org/10.1136/bmj.39493.646875.AE.
13. Patel IJ, Rahim S, Davidson JC, et al. Society of interventional radiology consensus guidelines for the periprocedural management of thrombotic and bleeding risk in patients undergoing percutaneous image-guided interventions—Part II: recommendations. *J Vasc Interv Radiol*. 2019;30(8). https://doi.org/10.1016/j.jvir.2019.04.017, 1168.e1–1184.e1.

Chapter 36

Case series: Design, measures, and classic example

Wylie T. Foss
University of Texas Houston MD Anderson Cancer Center, Houston, TX, United States

Key points

- A case series is a retrospective, noncomparative investigation that evaluates a group of patients with a known medical condition, disease state, exposure, or who have undergone a similar procedure.[1]
- The primary outcome of a case series is to identify a testable hypothesis that may be further evaluated with a more rigorous study design; secondary outcomes include commentary on treatment safety, diagnostic accuracy, or technical success.
- Case series are inexpensive, relatively easy to perform when compared to other types of studies, and are well suited to rare disease states or procedures.

Why it matters?

A case series is one of the most common studies performed in medical research. The argument could be made that the case series provides the foundation for nearly all of the more rigorous or lauded medical research studies. The case series is often the first form of medical research that trainees become involved in as they progress through their medical education. The relative ease with which a case series may be performed ensures that they will continue to be ubiquitous and impactful in the future.

A case series investigates a group of patients with known medical condition(s), exposure(s), disease state, or who have undergone a similar procedure over a specified time period. A case series offers a larger group of patients than a case report, which follows a cohort of up to three patients. Case series are typically retrospective in nature with the investigator performing chart review on a specific set of patients with a common interventional procedure, medical condition, etc. A case series may be consecutive, where all eligible patients that presented during the specified time period are included, or they may be nonconsecutive where only some eligible patients meeting certain criteria are included in series. A case series only has a study or investigation group. It does not contain a comparison group of control patients to perform comparisons against. Therefore it can be described as a noncomparative study. Only descriptive statistics such as measures of central tendency, dispersion or variation, and percentiles/ranges may be obtained from a case series.

The primary outcome of a case series is the creation of new hypotheses that can be meticulously tested in studies designed for this purpose such as a cohort study or a clinical trial. Case series are useful for learning about the progression of a disease state and postprocedural course. Investigators cannot draw definitive conclusions about treatment safety or diagnostic accuracy based only upon a case series. They can merely comment on potential relationships that may be the topic of further hypothesis-based research projects.

There are multiple benefits of a case series; namely, they are inexpensive, take significantly less time to perform than other investigations, are well suited to rare disease states, and may include a diverse patient cohort.[2] Given their retrospective nature, the patient data already exists and most hospitals have robust systems in place that enable clinicians to gather data based on billing codes or other search queries. When greater than three patients are involved, Institutional Review Board (IRB) approval is required from any involved institutions. Case series are inexpensive to perform because they usually only require work on the part of the researcher(s) who may or may not be paid for their work depending on the situation. Rare disease states or procedures are well suited to case series because small patient or case numbers prevent a comparative study from achieving enough statistical power to achieve statistical significance. Patient selection in a case

Translational Interventional Radiology. https://doi.org/10.1016/B978-0-12-823026-8.00020-1

series has significantly less inclusion and exclusion criterion than a typical randomized controlled research trial. Patients simply need to fit within the investigation group's characteristics (same procedure, exposure, or medical condition) to be included in a case series. This can result in a more diverse cohort, more closely representing what clinicians see in their clinical practice.

The following example has been fabricated to review the design of a case series study and is illustrated in Fig. 36.2. This single institution's consecutive retrospective case series includes all patients who underwent a kyphoplasty for vertebral compression fractures or bone metastasis at Hospital X from June 2009 to June 2019. IRB approval needs to be obtained before data collection. A total of 250 patients undergo kyphoplasty for these indications at Hospital X during this time period. Data collection includes patient demographics, indication for procedure, technical success of the procedure, and clinical success of the procedure specifically the improvement of pain immediately as well as 2 months postprocedure. Descriptive statistical analysis is performed on the collected data. A manuscript is prepared with a target journal or conference in mind. The goal of this investigation is to identify a testable hypothesis for further studies.

Get started

Action Items for Case Series Preparation:

1. Identify a clinical focus
   - This may be a certain medical condition, disease state, procedure, or exposure
2. Select a time period to focus your inquiry
3. Obtain IRB approval
4. Extract data to evaluate your clinical focus
5. Perform descriptive statistical analysis
6. Prepare a manuscript for submission
7. Pursue further studies, if applicable, to test any hypotheses generated

Potential pitfalls

Case series investigations are hindered by several limitations including bias susceptibility, the potential for incomplete data, and the absence of a comparison group.[2] Case series may be vulnerable to selection bias, a broad form of bias caused by the manner in which patients are selected to form a study group.[3] Attrition bias, a form of selection bias where patients are lost to follow-up, occurs when patients die, move away, or otherwise cannot be reached in a case series.[4] Data collection may not be complete in a case series either from patient attrition or variability in the clinical documentation for each patient. The retrospective nature of most case series means that if a variable of interest was not documented at a follow-up visit, it cannot be included. The absence of a comparison group is the most significant and inherent limitation of a case series. The lack of a comparison group precludes any sort of comparative statistical analysis. The benefits and limitations of a case series are reviewed in Fig. 36.1.

Examples of case series

Three example case series that have been published in the Journal of Vascular and Interventional Radiology within the last 8 years are discussed next and summarized in Fig. 36.3. Dr. Aparna Baheti and colleagues performed a case series investigation of all retrievable suprarenal Inferior Vena Cava filters placed from January 2008 to December 2017 at their institution with a focus on safety analysis in patients who had their inferior vena cava filters removed.[5] Dr. Bo Zhang and colleagues performed a case series investigation of 15 consecutive patients with cesarean scar pregnancy (a form of ectopic pregnancy) treated with uterine artery embolization and methotrexate.[6] Dr. Andrés Camacho and colleagues published a case series investigation of uterine artery embolization with Gelfoam for acquired symptomatic uterine arteriovenous shunting with a focus on technical and clinical success.[7]

Table 1. Benefits and Limitations of Case Series

Benefits	Limitations
Inexpensive to complete	Susceptibility to selection bias
Less time intensive than other designs	Potential for attrition bias
Suitable for the study of rare diseases	No comparison group
Diverse patient cohort may more closely represent clinical practice	Vulnerable to incomplete data collection or missing data

FIGURE 36.1 Benefits and limitations of a case series.

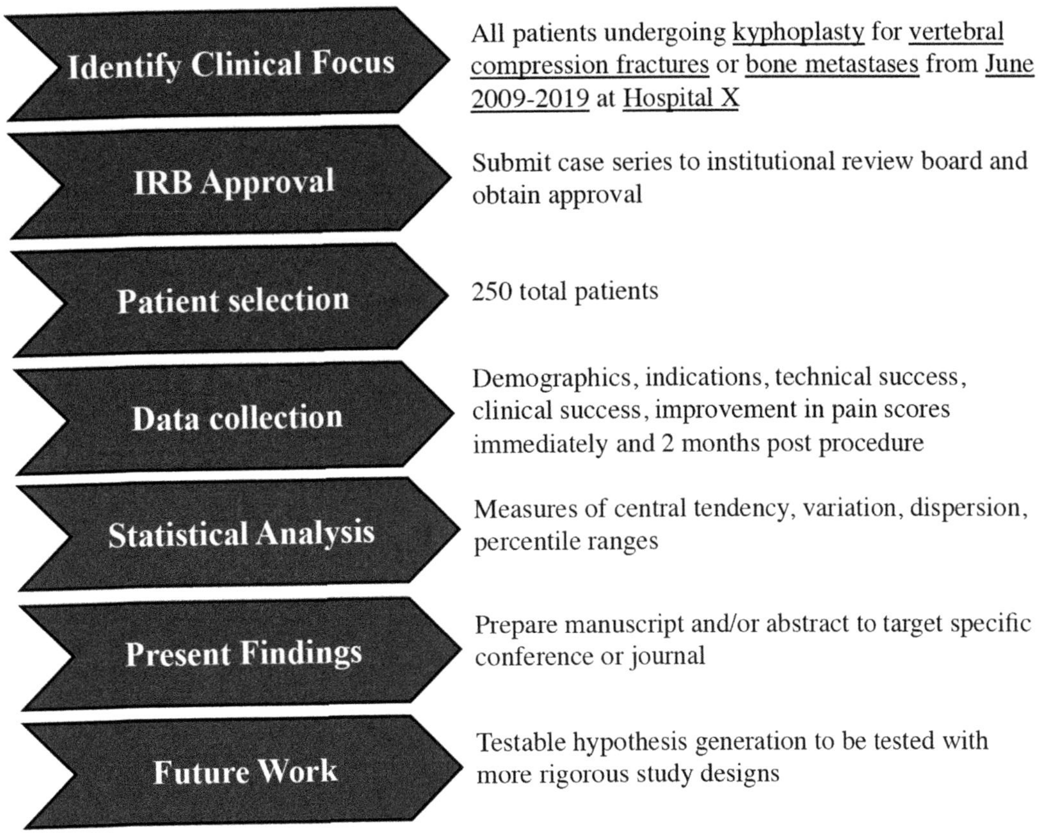

FIGURE 36.2 Case series study design steps illustrated with a fabricated example of all patients undergoing kyphoplasty at Hospital X from 2009–2019.

Table 2. Selected Examples of Case Series from IR Literature

Authors	**Topic**	**Focus**	**Conclusions**
Dr. Aparna Baheti et al[5].	Retrievable Suprarenal IVC filter placement and retrieval	Technical success and complication rate	Indicated suprarenal IVC filters may be placed and retrieved with a low complication rate
Dr. Bo Zhang et al[6].	Uterine artery embolization and Methotrexate in the treatment of Cesarean scar pregnancy	Technical success, clinical success, and resolution without need for hysterectomy	Uterine artery embolization might be an effective treatment for treating Cesarean scar pregnancy, further study needed
Dr. Andrés Camacho et al[7].	Gelfoam uterine artery embolization for acquired and symptomatic uterine arteriovenous shunting	Technical success, clinical success, and complication rates	Single embolic agent uterine artery embolization with Gelfoam has a low complication rate with high technical and clinical success rates

FIGURE 36.3 Selected examples of recent case series studies from IR literature.

Resources

National Cancer Institute Dictionary of Cancer Terms, Journal of Vascular and Interventional Radiology, and Pubmed query with case series study for more real world examples.

References

1. National Cancer Institute Dictionary of Cancer Terms. *Case Series*; 2019. https://www.cancer.gov/publications/dictionaries/cancer-terms/def/case-series. Accessed June 1, 2020.
2. Kooistra B, Dijkman B, Einhorn T, et al. How to design a good case series. *J Bone Joint Surg*. 2009;91:21–26.
3. National Cancer Institute Dictionary of Cancer Terms. *Selection Bias*; 2019. https://www.cancer.gov/publications/dictionaries/cancer-terms/def/selection-bias. Accessed June 1, 2020.
4. Mansournia M, Higgins J, Sterne J, et al. Biases in randomized trials: a conversation between trialists and epidemiologists. *Epidemiology*. 2017;28(1):54–59.
5. Baheti A, Sheeran D, Patrie J, et al. Suprarenal inferior vena cava filter placement and retrieval: safety analysis. *J Vasc Intervent Radiol*. 2020;31:231–235.
6. Zhang B, Jiang Z-B, Huang M-S, et al. Uterine artery embolization combined with Methotrexate in the treatment of cesarean scar pregnancy: results of a case series and review of the literature. *J Vasc Intervent Radiol*. 2012;23:1582–1588.
7. Camacho A, Ahn E, Appel E, et al. Uterine artery embolization with Gelfoam for acquired symptomatic uterine arteriovenous shunting. *J Vasc Intervent Radiol*. 2019;30:1750–1758.

Chapter 37

Case-control study: Design, measures, classic example

Wylie T. Foss
University of Texas Houston MD Anderson Cancer Center, Houston, TX, United States

Key points

- A case-control study is an investigation involving two groups of patients: a study ("case") group who have known medical condition(s), disease state, and a control group of similar patients without the focus of inquiry.[1]
- Exposure data are retrospectively collected for both groups.
- The primary outcome of a case-control study is the calculation of an odds ratio, a comparative statistical measure of the association between an exposure and an outcome.
- Case-control studies are relatively inexpensive compared to other forms of medical research, are suitable for the study of rare diseases, and allow for comparative statistical analysis of many risk factors or exposures.

Why it matters?

The case-control study is an important modality for medical research and is frequently performed. A case-control study requires a similar amount of work as a case series but allows the investigator to perform comparative statistics instead of only descriptive statistics as in a case series. Case-control studies also allow for efficient study of rare diseases, which may not be feasible with other forms of medical research. A well-executed case-control study that yields clinically relevant and statistically significant results is the basis for performing more rigorous investigations like randomized controlled trials.

A case-control study investigates two groups of patients: a study ("case") group who have known medical condition(s), disease state, and a control group who are a similar group of patients without the medical condition(s), disease state, to be investigated in the study group. Case-control studies have a control group whereas case series do not. Case-control studies are performed in a retrospective nature. A retrospective case-control study is performed in a similar manner to a case series; chart review is performed on both the study group and the control group.

The primary outcome of a case-control study is an odds ratio, a statistical measure of the association between an exposure and an outcome.[2] Put another way, an odds ratio functions as a risk estimate of an exposure leading to an outcome. Odds ratios may be calculated for an outcome of interest to the investigator (disease state, medical condition(s), procedure(s)) and consider the patient's exposure to a variable of interest. Variables of interest may include any aspect of a patient's medical history. An odds ratio of 1.0 indicates that the prevalence of a risk factor within the case group is the same as the prevalence in the control group. An odds ratio of 2.0 would indicate that the prevalence of a risk factor within the study group is two times greater than in the control group, which indicates a positive association between the risk factor and the outcome.

There are a number of benefits to a case-control study including their low cost compared to other more rigorous study designs, the suitability for the study of rare diseases, the ability to evaluate multiple risk factors in a single study, and the ability to increase the statistical power of the study by increasing the number of control patients.[3] Case-control studies are retrospective in nature; patient data already exist and typically only require a researcher to sift through it, limiting costs. The suitability for the study of rare diseases cannot be understated, for a disease with prevalence, such as congenital heart disease,[3] investigators would need to enroll 10,000 patients from a population to have approximately 100 cases of this disease. Utilizing a case control, investigators can collect 100 cases of the rare disease and 100 controls. Both studies would have similar analysis power; however, the case control would come at a lower

Translational Interventional Radiology. https://doi.org/10.1016/B978-0-12-823026-8.00058-4

cost and offer investigators significant time savings. Cases of rare diseases may be identified retrospectively and evaluated with a case-control study to evaluate risk or protective factors. Frequently, medical conditions or disease states are associated with a number of risk factors or exposures. A case-control study allows researchers to evaluate as many risk factors or exposures as they want and compare them with the control group. Investigators may increase the number of control patients to increase the statistical power of the case-control study to obtain statistically significant results.

The following example has been fabricated to review the design of case-control study and is summarized in Fig. 37.2. The interventional radiology department at Hospital Y performed a single institution retrospective case-control study in an effort to better understand the impact of uterine leiomyomas on patients' reproductive results from June 2009 to June 2019. The study group involved patients with known uterine leiomyomas as of June 2009 with reproductive plans while the control group included patients without uterine leiomyomas as of June 2009 with reproductive plans. Institutional Review Board (IRB) approval was obtained before data collection. A total of 180 patients were retrospectively selected with 90 in the case group and 90 in the control group. Data collection for this case-control study includes patient demographics, comorbidities, obstetric/gynecologic history, successful deliveries, pregnancy losses, and other pertinent obstetric/gynecologic complications. Descriptive and comparative statistical analyses are performed on the collected data. A manuscript is prepared for a specific target journal or conference. The goal of this investigation was to evaluate reproductive results in patients with uterine leiomyomas and patients without uterine leiomyomas.

There are two important variants of a case-control study, a matched case-control and a nested case-control, which are important to discuss as they offer certain advantages to investigators. A matched case-control study indicates that the cases and controls are "matched" based upon any desired characteristic, for example, body mass index. This is done to avoid confounding bias. The nested case-control study is generated from within a larger cohort study, hence the name "nested." The cohort study examines a large number of individuals across a population for a risk factor or disease etc. at the onset of the study. Once individuals with a risk factor or disease are identified, a nested case-control study may be created where cases are selected from this pool and controls are selected from individuals in the cohort study without the risk factor or disease of interest. The benefits of a nested case-control include offering investigators the ability to test a hypothesis that may not have been created at the onset of the cohort study as well as bias reduction, particularly information and selection biases. A potential disadvantage is the reduced statistical power due to the loss of sample size compared to the larger cohort study.

Get started

Action Items for Case-control Study Preparation:

1. Identify a clinical focus
   - This may be a certain medical condition, disease state, procedure or exposure.
2. Select a time period to focus your inquiry.
3. Obtain IRB approval.
4. Carefully select control patients to try to limit selection and confounding bias.
5. Collect data on the study and control groups to evaluate your clinical focus.
6. Perform descriptive and comparative statistical analysis.
7. Prepare a manuscript for submission and pursue further studies, if applicable.

Potential pitfalls

Case-control studies have several limitations that researchers must keep in mind, including but not limited to: the inability to estimate disease incidence, susceptibility to

Table 1. Benefits and Limitations of Case-control Studies

Benefits	Limitations
Less time intensive and expensive than other more rigorous study designs	Inability to calculate disease incidence
Suitable for the study of rare diseases	Susceptibility to recall bias
Ability to evaluate multiple risk factors or exposures	Prone to selection bias
May increase control patients to increase a study's statistical power	Confounding can skew investigators' conclusions

FIGURE 37.1 Benefits and limitations of case-control study design.

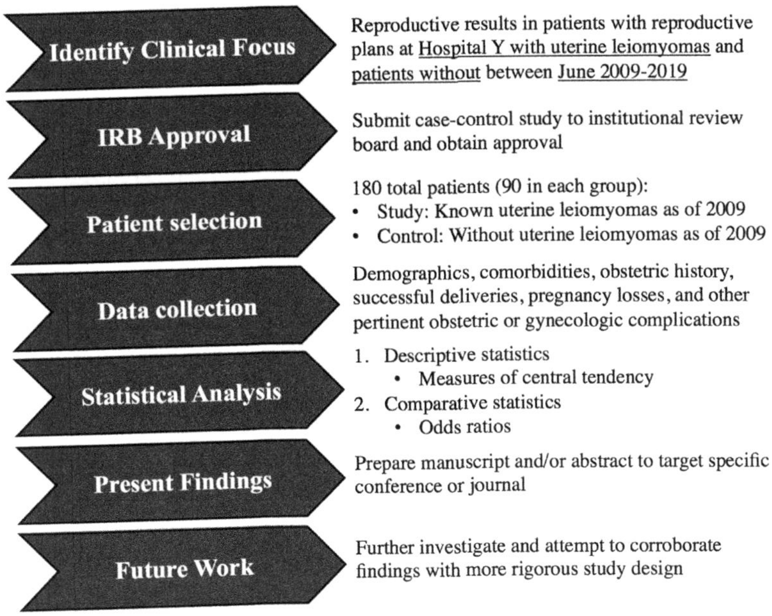

FIGURE 37.2 Fabricated example of case-control study design at Hospital Y including patients with uterine leiomyomas and a control group without.

multiple forms of bias, and difficulties associated with selecting control patients.[4] A cohort study should be used to evaluate disease incidence, it follows two groups of patients over a period of time: one with exposure to a risk factor and another without this exposure. A cohort study allows one exposure can be investigated with multiple outcomes while a case-control study looks at one outcome but multiple exposures. A cohort study allows for the calculation of risk difference, risk ratio, or odds ratio where a case-control study allows only for the calculation of an odds ratio [other statistics are inconsistent (asymptotically biased)]. Case-control studies are particularly susceptible to a form of bias known as recall bias. Recall bias is caused by differences in study patients' ability to recall information about their medical history such as exposures, treatments, and disease states. Patients who have developed a medical condition such as hepatocellular carcinoma may be able to recall more possible risk factors than patients in the control group (without hepatocellular carcinoma in this example). The selection of control patients can introduce confounding, a phenomenon occurring when two or more exposures have not been fully separated, causing investigators to conclude that the clinical outcome is attributable to one rather than a synergistic effect of both exposures. The benefits and limitations of case-control studies are reviewed in Fig. 37.1.

Real-world examples

Three example case-control studies that have been published in the Journal of Vascular and Interventional Radiology within the last 3 years are discussed next and summarized in Fig. 37.3. Dr. Raj Ayygari and colleagues performed a case-control study of patients undergoing prostate artery embolization from April 2013 to August 2018 to evaluate clinical outcomes and procedural measures with balloon occlusion microcatheters versus traditional, end-hole microcatheters.[5] Dr. Simon Chun Ho Yu and colleagues completed a prospective case-control study to evaluate the use of intra-arterial administration of verapamil for patients with intraprostatic anastomoses to extraprostatic arteries from June 2015 to March 2, 016.[6] Dr. Susan Shamimi-Noori and colleagues used a case-control study to evaluate clinical outcomes for patients undergoing treatment for nonmaturing hemodialysis fistulae via transradial access from May 2015 to November 2018.[7]

Resources

National Cancer Institute Dictionary of Cancer Terms, Journal of Vascular and Interventional Radiology, and Pubmed query with Case-control study for more real world examples.

FIGURE 37.3 Selected examples of case-control studies recently published in the Journal of Vascular and Interventional radiology (JVIR).

Table 2. Selected Examples of Case-control Studies from IR Literature

Authors	Topic	Focus	Conclusions
Dr. Raj Ayyagari et al[4].	Conventional end hole versus balloon occlusion microcatheter prostate artery embolization for BPH	Procedural metrics, improvement in symptoms, and adverse events	Balloon occlusion microcatheter use did not result in any technical benefits, reduced adverse events or improved clinical outcomes
Dr. Simon Chun Ho Yu et al[5].	Intra-arterial Verapamil for the treatment of intraprostatic anastomoses to extraprostatic arteries during prostate artery embolization for BPH	Adverse events and ischemic complications from Verapamil Technical and clinical success	Intra-arterial Verapamil administration is likely safe and effective in causing flow reversal in some types of anastomoses as well as preventing ischemic complications
Dr. Susan Shamimi-Noori et al[6].	Nonmaturing hemodialysis fistulae diagnosis and treatment via a transradial approach	Safety, rates of fistula abandonment, patency rates, technical and clinical success	Treating nonmaturing hemodialysis fistulae with a transradial approach is safe and had lower rates of fistula abandonment

References

1. National Cancer Institute Dictionary of Cancer Terms. *Case-control Study*; 2019. https://www.cancer.gov/publications/dictionaries/cancer-terms/def/case-control-study. Accessed June 1, 2020.
2. Szumilas M. Explaining odds ratios. *J Can Acad Child Adolesc Psychiatry*. 2010;19(3):227−229.
3. Reller M, Strickland M, Riehle-Colarusso T, et al. Prevalence of congenital heart defects in Atlanta, 1998-2005. *J Pediatr*. 2008;153: 807−813.
4. Tenny S, Kerndt C, Hoffman M. Case control studies. In: *StatPearls. Treasure Island (FL)*. StatPearls Publishing; 2020.
5. Ayyagari R, Powell T, Staib L, et al. Case-control comparison of conventional end-hole versus balloon-occlusion microcatheter prostatic artery embolization for treatment of symptomatic benign prostatic hyperplasia. *J Vasc Intervent Radiol*. 2019;30: 1459−1470.
6. Chun Ho Yu S, Cho C, Hung E, et al. Case-control study of intra-arterial Verapamil for intraprostatic anastomoses to extraprostatic arteries in prostatic artery embolization for benign prostatic hypertrophy. *J Vasc Intervent Radiol*. 2017;28:1167−1176.
7. Shamimi-Noori S, Sheng M, Mantell M, et al. Diagnosis and treatment of nonmaturing fistulae for hemodialysis access via transradial approach: a case-control study. *J Vasc Intervent Radiol*. 2020;31(6): 993−999.
8. National Cancer Institute Dictionary of Cancer Terms. *Prospective*; 2019. https://www.cancer.gov/publications/dictionaries/cancer-terms/def/prospective. Accessed June 1, 2020.
9. Dorey F. Statistical power: what is it and when should it be used. *Clin Orthop Relat Res*. 2011;469:619−620.
10. McNamee R. Confounding and confounders. *Occup Environ Med*. 2003;60:227−234.

Chapter 38

Cohort study: Design, measures, classic example

Rayan Abboud
Wake Forest School of Medicine, Winston-Salem, NC, United States

Design

Cohort studies are a fundamental type of observational study, where subjects are first chosen based upon a cohort-defining trait, rather than disease status. Upon being designated to their respective cohort, subjects are observed across time to determine if a study outcome was met. Common types of traits that define cohorts include exposure status or an event experienced by subjects. The study outcome is commonly defined as the development of disease during follow-up. The primary objective of a cohort study is to determine the association between the cohort-defining trait and the incidence of the outcome. In a practical sense, these types of studies can be used before clinical trials to identify traits as potential predictive factors (risk factors or protective factors) for a given disease process. An assumption of these studies is that the subjects are representative of the population of interest. The outcome measures obtained from such studies include relative risk and odds ratio.

When determining their cohort, investigators first choose a trait they are interested in studying. Some clinical examples of such traits include radiation exposure, treatment with a novel medication, and obesity. Subjects that do not exhibit this trait are termed *unexposed* and serve as the control of the study. The unexposed subjects are compared to the *exposed* group that is composed of subjects who share the cohort-defining trait. After some time, disease status is determined for all study subjects and the rate of disease is compared between the unexposed and exposed groups. In some cohort studies, rather than only studying a single exposure, some researchers may opt to investigate the association of multiple exposures on the outcome or multiple outcomes. In such cases, multiple exposures are defined at the onset of the study, all of which are compared to the study outcome. Cohort studies can also be conducted in a retrospective manner, where the exposed and unexposed groups are defined by existing data from medical records and outcomes can be defined as clinical event(s) during a certain period based on available data.

Measures

Outcome measures or endpoints for a cohort study are similar to those of other observational studies (case-control, cross-section, etc.). For the main question being investigated, one primary outcome is chosen per study. Any auxiliary hypotheses or questions of interest are conducted through secondary outcomes. While disease status is typically used when describing the primary outcomes of cohort studies, there are a number of other outcome measures that can be examined.[1] Some other examples include immunogenicity of a vaccine, quality-adjusted life years, and survival. Regardless of the outcome measure, the analysis remains mostly the same—the null hypothesis states that there is no statistically significant difference between the incidence of the outcome in the exposed versus the unexposed group, that is, there is no *association* between study outcome and exposure.

Types of cohort studies

Cohort studies can either be prospective or retrospective in nature. Prospective cohort studies are also known as concurrent or longitudinal cohort studies. Retrospective cohort studies are also known as nonconcurrent or historical cohort studies.[2] The difference between these two types lies in when the exposure data and outcome data are collected. In retrospective studies, the exposure status and outcome data are ascertained simultaneously at the start of the study, as both events have already occurred before this time. In other words, both data points are collected retrospectively.

Translational Interventional Radiology. https://doi.org/10.1016/B978-0-12-823026-8.00045-6

Whereas in prospective studies, the exposure status can be captured retrospectively, but the outcome measure has yet to occur and must be collected prospectively. Thus, most of the cohort data including demographics, baseline traits, and exposures can be obtained before the onset of the study outcome, but the outcome data are obtained after following patients over time.

Cohort study advantages and disadvantages

Retrospective cohorts are advantageous in their low cost and the time required for gathering data. Thus, a long follow-up period is a potential advantage of these studies as most data can be acquired at the advent of the study. Large datasets that are readily available and accessible to investigators are especially conducive to retrospective cohort studies. Such studies are also useful when examining unusual exposures or workplace exposures, where the population in question shares a specific trait, for example, a group of agricultural workers utilizing a novel off-label pesticide.

Prospective cohort studies have the added benefit of regularly collecting new data while subjects are observed over time. These studies generally aim to answer multiple questions about the representative population that is being observed. Thus, multiple potential predictive factors are monitored over time. As such, these studies are considered to be superior in quality compared to retrospective studies. However, this comes at the expense of being more time-consuming and expensive. Due to the long follow-up period, these studies are more likely to face an attrition bias, where there is a failure to follow-up commonly due to death, lack of interest, and loss of contact. Attrition bias would be problematic if the failure to follow up is associated with the exposure or the outcome. There are several approaches to mitigate attrition in prospective studies—those include multiple means of contacting subjects, recruiting altruistic subjects who are more likely to follow up, and other techniques that optimize communication.

Notable disadvantages of cohort studies include selection bias which alludes to the inherent confounding effects that exist with such approaches. If the two groups are similar in all respects besides their exposure status, then confounding is less likely. However, in many real-world scenarios, this is not the case as there can be several other traits that differ between the unexposed and exposed groups. If those traits are not considered in the statistical analysis, the apparent association in the exposure with the outcome may be biased by a confounder. In our previous example, suppose agricultural workers in the exposed group were found to have increased rates of squamous cell carcinoma as compared to workers at a different farm without that exposure. Furthermore, consider if the exposed group was composed of elderly individuals, whereas the unexposed group was primarily young individuals. If the age of the subjects was not available for the statistical analysis, then one may have falsely arrived at the association between the novel pesticide and the development of squamous cell carcinoma. Due to the retrospective approach of collecting baseline traits, investigators may not be able to capture those confounders of interest, such as patient demographics, comorbidities, and risk factors. Information on exposures and outcomes as well as clinical questions that can be answered using retrospective cohorts are limited by the available data. Investigators may not be able to control the quality of their data (e.g., missing values, attrition), making suboptimal data quality another limitation of these studies.

Methods

As with most observational studies, it is important to define baseline variables, exposures, and outcomes at the onset of the study before data collection or analysis. Some exceptions to this include criteria for a disease that are periodically updated in clinical practice guidelines, such as the coronary calcium score.[3] Other exceptions to defining exposures or outcomes at the start of the study include those defined across time in prospective studies with long follow-up periods, some of which span over 70 years.[4] General guidelines for outcomes are to utilize classification criteria from existing peer-reviewed studies, or in the event of a novel outcome, providing a highly specific definition of a positive outcome along with a rationale for choosing the definition.

Several statistical approaches can be used to measure the magnitude of association between exposure and outcome in cohort studies. Commonly reported measures of outcome incidence include absolute risk, relative risk, odds ratio, hazard ratio, and attributable risk. These topics are covered in the *Statistical Principles* section.

Classic example

Several cohort studies have studied the impact of Agent Orange exposure during the Vietnam War. A prospective cohort study published in Cancer followed 180,251 Vietnam veterans to examine the association between Agent Orange exposure and incidence of cancer.[5] Investigators utilized The Korea National Cancer Incidence Database to confirm the veteran's cancer diagnosis. The outcome of cancer was defined using the *International Classification of Diseases, 10th Revision* (ICD-10). Subjects were stratified into four groups, no exposure, low exposure, moderate exposure, and high exposure. Potentially confounding factors were taken into consideration and included age, deployment unit, and military rank. Analysis was carried out using Cox regression to determine the hazard ratio, similar to the risk ratio, of each exposure. After controlling for potential confounders, it was determined that veterans exposed to Agent Orange experienced an increased risk of cancers of all types combined with an adjusted hazards ratio of 1.08.

References

1. Smith PG, Morrow RH, Ross DA. *Outcome Measures and Case Definition*. Oxford University Press.
2. Gallin JI, Ognibene FP, Johnson LL. *Principles and Practice of Clinical Research*. Academic Press; 2017.
3. Greenland P, Blaha MJ, Budoff MJ, Erbel R, Watson KE. Coronary calcium score and cardiovascular risk. *J Am Coll Cardiol*. 2018;72(4):434–447. https://doi.org/10.1016/j.jacc.2018.05.027.
4. Medical Research Council MRC. *MRC National Survey of Health and Development Cohort/1946 Birth Cohort (NSHD/1946BC)*; May 21, 2015. Published https://mrc.ukri.org/.
5. Yi S-W, Ohrr H. Agent Orange exposure and cancer incidence in Korean Vietnam veterans: a prospective cohort study. *Cancer*. 2014;120(23):3699–3706. https://doi.org/10.1002/cncr.28961.

Chapter 39

Cross-sectional study: Design, measures, classic examples

Jalil Kalantari[1] and Kyle Cooper[2]

[1]Department of Radiology, Section of Interventional Radiology, Loma Linda School of Medicine, Loma Linda, CA, United States; [2]Department of Radiology, University of Michigan Health Systems, Ann Arbor, MI, United States

Key points

- Cross-sectional studies are characterized by the collection of all study data from a single point in time. These studies may have multiple outcomes and are relatively inexpensive and quick to conduct.
- These studies are the best studies to determine prevalence; however, they cannot evaluate the cause—effect relationship or the sequence of events.
- They can be conducted before or as a baseline for a cohort or a randomized control study.

Why it matters?

Interventional radiology is one of the most technologically advanced and fastest-growing specialties in modern medicine. With the rising innovative interventions and emphasis on evidence-based medicine, there is more need than ever for robust clinical trials. However, before formulating an expensive and time-consuming prospective clinical trial, an observational study, where no intervention is made, is needed to better understand a disease and exposure process. Observational studies include cross-sectional, case-control, and cohort studies[1] (Fig. 39.1). In a cross-sectional study design, an investigator evaluates outcomes and exposures in the participants at a given point in time (not necessarily in calendar time), for example, in the year 2020, at study enrollment, at age of 60, etc. As a result, they are primarily used to determine the prevalence and to conduct population-based surveys. The single acquisition of the data in a cross-sectional study makes this a relatively inexpensive and quick study to conduct. However, determining a trend and cause—effect relationship is difficult, as these studies lack chronological analysis. Finally, an investigator can estimate the odds ratio to evaluate potential associations between the exposure and outcome, which can be further evaluated by a cohort study or a randomized trial.[2]

Benefits and limitations

A cross-sectional study design is an evaluation of one group after a single data collection. This makes the study relatively quick and inexpensive. The study is predominantly used to determine prevalence,[1] which is important in clinical practice as it defines the likelihood of a diagnosis and the predictive value of any investigation. For example, armed with the knowledge of the low prevalence of glioblastoma in children, a radiologist more strongly considers other diagnoses of a brain mass in this patient population.

A cross-sectional study provides a “snap-shot” of the disease of interest at a specific time. This data can provide valuable information to healthcare administrators in assessing the health care needs of a population, and subsequent reallocation of resources. For instance, in the early 1960s, the Centers for Disease Control and Prevention designed the National Health and Nutrition Examination Survey, a cross-sectional study designed to assess the health and nutritional status of adults and children in the United States.[3] This interview-based survey of about 5000 persons each year across the country includes demographic, socioeconomic, dietary, and health-related questions, and determines the prevalence of major diseases and risk factors for diseases.

Translational Interventional Radiology. https://doi.org/10.1016/B978-0-12-823026-8.00040-7

FIGURE 39.1 Overview of the observational study.

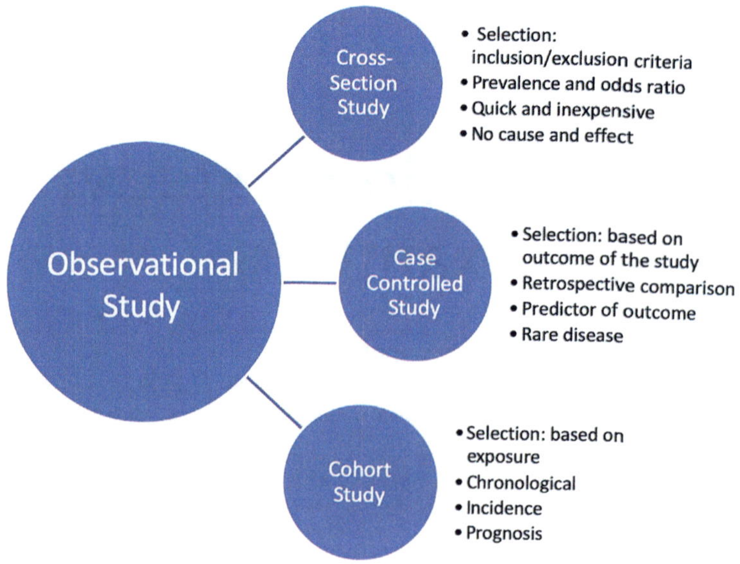

A common type of cross-sectional design is a diagnostic accuracy study. This can be helpful in clinical decision-making process as it compares a new diagnostic method with the current gold standard. The result of this study can provide accuracy, sensitivity, specificity, positive predictive value, negative predictive value, positive likelihood ratio, negative likelihood ratio, and diagnostic odds ratio.[4] For instance, in a study by Granados, the diagnostic accuracy of ^{18}F-FDG PET/CT in infective endocarditis and implantable cardiac electronic device (ICED) infection was evaluated and compared to the gold standard, a combination of clinical, echocardiographic, and microbiologic findings.[5] In addition to acquiring the mentioned parameters of the diagnostic testing, the study demonstrated the nuclear study can be useful in early detection of patients with prosthetic valves and ICEDs. However, it does not provide useful diagnostic benefits among patients with native valves.

Cross-sectional studies can provide associations in the interest group, which can be further studied in a cohort or a randomized trial study. For example, a detailed questionnaire was sent to all the workers at a rubber tire manufacturing company asking for symptoms related to obstructive lung disease, as well as detailed work history including areas of high-risk exposure to potentially hazardous conditions.[6] The result showed an increase in the prevalence of respiratory symptoms in those areas, leading to a cohort study that concluded smoking and occupational exposures in the rubber workers led to a higher incidence of pulmonary diseases.[7] It is noteworthy that the questionnaire was sent to all the current employees at a specific time and categorized based on exposure. On the other hand, if the study evaluated former and current employees and evaluated their history of symptoms and their exposure, in other words, based on the outcome, it would a retrospective cohort design.

The most important limitation of these study designs is the lack of chronological evaluation, which makes it difficult to determine cause and effect from the collected data. For instance, the negative correlation of height and age can be misinterpreted as people losing height as part of aging, higher life expectancy in shorter individuals, or younger generations getting taller.[1] Similarly, this design cannot evaluate disease trends, as the prevalence depends on the incidence (new cases) and length of survival or recovery (previously diagnosed cases). In other words, if the incidence of disease decreases, the prevalence may increase due to prolonged survival or delayed recovery. A solution to this problem is the implantation of a serial cross-sectional study or serial survey.[8] This can be an inexpensive alternative method to a cohort study to access trends in a population (see the example at the end of the chapter).

Another limitation of these studies is a low response rate, particularly in questionnaire-based studies; this results in lower study power, and subsequently may miss significant differences. Furthermore, the use of volunteers can add sample bias, as the individuals may not represent the general population. A solution to create a valid representative sample is to randomly select subjects to complete questionnaires or for interviews.

Lastly, if the disease of interest is rare, despite a large sample size, there may be no one with the disease. In such a circumstance, a sample of individuals with the disease can be studied in a case series study or a case-control study design.

Design

It is important to have a clear aim of the study. If the study aims to evaluate possible associations, the first step is to develop a testable hypothesis. There are three methods, described by Stuart Mill,[9] to create a hypothesis:

1. Method of difference: if the frequency is much higher in sets of circumstances. For instance, higher gastric cancer in Japan versus United Kingdom can lead to potential lifestyle factors increasing risk of gastric cancer.
2. Method of agreement: single factor common between number of circumstances. For instance, higher HIV rates in IV drug users, hemophiliacs, and patients with regular transfusions can lead to likely transmission of HIV through the bloodstream.
3. Method of concomitant variation: frequency of a factor with the frequency of the disease. For instance, an increasing rate of cigarette use and cardiovascular diseases can lead to a correlation between the factor and the disease.

The participants' selection in this study design is based on inclusion and exclusion criteria set by the study's objective. This is in contrast to the case-control study, where the selection is based on the outcome of the study, or the cohort study, where the selection is based on the exposure.

The selection can be based on person (who has the disease?) or place (where is the highest or lowest prevalence of the disease).[2] Lastly, it is important to randomly select individuals from the population to have a true representative sample from a population. There are two methods in the evaluation of the exposure and outcomes: questionnaire, or interview. A questionnaire-based study requires less expense and can lead to a larger sample size; however, the disadvantage is a lower response rate. On the other hand, an interview-based study provides a higher response rate, however requires higher expenses, and subsequently a smaller sample size.

Measurement

Once the selected sample is entered into the study, the outcome and exposure are measured at the same given time. For example, an investigator is interested to know the prevalence of nephrolithiasis in a small neighborhood in Irvine, California. Every single house included in this neighborhood was examined and a total of 2352 individuals were surveyed. Of these, 209 individuals have nephrolithiasis. The prevalence of nephrolithiasis in this small neighborhood is 209/2352 or 88.9/1000 population.

Depending on the results, the investigator can evaluate the association among the variables and the odds ratio. This

TABLE 39.1 Calculating odds ratio: odds of a condition occurring in the exposed group divided by the odds of it occurring in the nonexposed group.

	Disease	No disease
Exposed	A	B
Unexposed	C	D
Odds ratio = $\frac{A}{C} \div \frac{B}{C}$ = AD/BC		

TABLE 39.2 Prevalence of nephrolithiasis in a small neighborhood in Irvine, California.

	Nephrolithiasis	No nephrolithiasis	Total
Men	**129**	**1089**	1218
Women	**80**	**1054**	1134
Total	209	2143	2352
The odds ratio = $\frac{129/1089}{80/1054}$ or (129 × 1054)/(1089 × 80) = 1.6			

is calculated as odds of the exposed group divided by the odds of the unexposed group (see Table 39.1). As part of the earlier mentioned survey, 1218 were men and 1134 were female. Of the individuals with nephrolithiasis, 129 are males, and 80 are females (Table 39.2).

The interpretation of these odds ratios indicates males have higher odds of nephrolithiasis compared to females.

Examples

A large cross-sectional study by Lesmana et al.[10] was designed to evaluate the prevalence and risk factors of nonalcoholic fatty pancreas disease, in a private hospital. The inclusion criteria were adults aged greater than 18 years, no serious illness at the time of the study, and underwent an abdominal ultrasound assessment. The exclusion criteria were patients with history of significant alcohol drinking, with nonvisualization of the pancreas on ultrasound (153 patients). Fatty pancreas was found in 315 patients (35%). Additionally, the study further evaluated associated risk factors with the highest odds ratio, 4.129, for body mass index greater than 25 kg/m^2 (Table 39.3).

In another example of a cross-sectional study, Kuy S et al. developed a study to characterize utilization of inferior vena cava (IVC) filters in United States.[11] The study utilized a serial survey to evaluate the national trend of IVC filter placement for a decade, between 2000 and 2009. The study showed a 234% increase in IVC filter placements in the United States from 2000 to 2009.

TABLE 39.3 Cross-sectional study data from BMC Gastroenterology by Lesmana et al.[10]

Associations among risk factors and the presence of fatty pancreas ($n = 901$)			
	Fatty pancreas		Odds ratio
Risk factor	Yes ($n = 315$)	No ($n = 586$)	(95% CI)
Male sex ($n = 574$)	228 (72.4%)	346 (59.0%)	1.818 (1.351–2.446)
Age >35 years ($n = 612$)	268 (85.1%)	344 (58.7%)	4.011 (2.824–5 0.697)
Systolic BP ≥ 130 mmHg ($n = 201$)	99 (31.4%)	102 (17.4%)	2.175 (1.580–2.994)
Diastolic BP ≥ 85 mmHg ($n = 111$)	54 (17.1%)	57 (9.7%)	1.920 (1.286–2.866)
Body mass index ≥ 25 kg/m^2 ($n = 390$)	206 (65.4%)	184 (31.4%)	4.129 (3.088–5.520)
Type 2 diabetes mellitus ($n = 62$)	31 (9.8%)	31 (5.3%)	1.954 (1.164–3.280)
FPG ≥ 100 mg/dL ($n = 187$)	99 (31.4%)	88 (15.0%)	2.594 (1.867–3.603)
Triglycerides ≥ 150 mg/dL ($n = 226$)	105 (33.3%)	121 (20.6%)	1.921 (1.412–2.615)
Total cholesterol ≥ 200 mg/dL ($n = 491$)	203 (64.4%)	288 (49.1%)	1.875 (1.415–2.486)
LDL-C ≥ 100 mg/dL ($n = 742$)	275 (87.3%)	467 (79.7%)	1.752 (1.189–2.582)
HDL-C < 40 (M) or < 50 (F) mg/dL ($n = 194$)	86 (27.3%)	108 (18.4%)	1.662 (1.202–2.298)

Get started

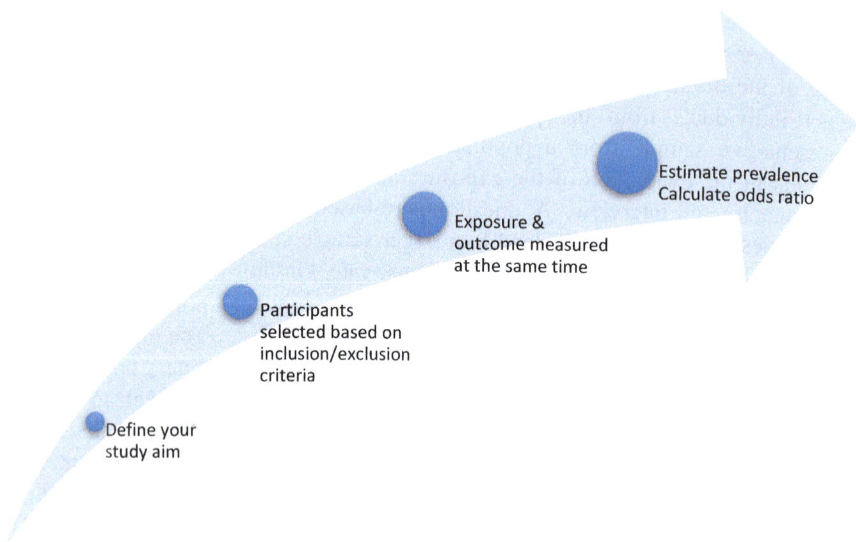

References

1. Mann CJ. Observational research methods. Research design II: cohort, cross sectional, and case-control studies. *Emerg Med J.* 2003;20(1):54–60. https://doi.org/10.1136/emj.20.1.54.
2. Hennekens C, Buring J, Mayrent S. *Epidemiology in Medicine.* 1st ed. Boston: Little, Brown; 1987:101–109.
3. NHANES - About the National Health and Nutrition Examination Survey. Centers for Disease Control and Prevention. https://www.cdc.gov/nchs/nhanes/about_nhanes.htm. Published September 15, 2017. Accessed June 10, 2020.
4. McMichael AJ, Gerber WS, Gamble JF, Lednar WM. Chronic respiratory symptoms and job type within the rubber industry. *J Occup Med.* 1976;18(9):611–617.
5. Thiese MS. Observational and interventional study design types; an overview. *Biochem Med.* 2014;24(2):199–210. https://doi.org/10.11613/bm.2014.022.
6. Granados U, Fuster D, Pericas JM, et al. Diagnostic accuracy of 18F-FDG PET/CT in infective endocarditis and implantable cardiac electronic device infection: a cross-sectional study. *J Nucl Med.* 2016;57(11):1726–1732. https://doi.org/10.2967/jnumed.116.173690.
7. Lednar WM, Tyroler HA, McMichael AJ, Shy CM. The occupational determinants of chronic disabling pulmonary disease in rubber workers. *J Occup Med.* April 1977;19(4):263–268. https://doi.org/10.1097/00043764-197704000-00007. PMID: 758090.
8. Setia MS. Methodology series Module 3: cross-sectional studies. *Indian J Dermatol.* 2016;61(3):261–264. https://doi.org/10.4103/0019-5154.182410.
9. Mill J, Parker J, Strand W. *A System of Logic, Ratiocinative and Inductive: Being A Connected View of the Principles of Evidence, and the Methods of Scientific Investigation.* 3rd ed. London: Cambridge University Press4; 2008:410–412.
10. Lesmana CR, Pakasi LS, Inggriani S, Aidawati ML, Lesmana LA. Prevalence of Non-Alcoholic Fatty Pancreas Disease (NAFPD) and its risk factors among adult medical check-up patients in a private hospital: a large cross sectional study. *BMC Gastroenterol.* 2015;15:174. https://doi.org/10.1186/s12876-015-0404-1. Published 2015 Dec 12.
11. Kuy S, Dua A, Lee CJ, et al. National trends in utilization of inferior vena cava filters in the United States, 2000-2009. *J Vasc Surg Venous Lymphat Disord.* 2014;2(1):15–20. https://doi.org/10.1016/j.jvsv.2013.08.007.

Chapter 40

Longitudinal study: Design, measures, and classic example

Karishma Shah[1], Zahi Qamhawi[2] and Gregory C. Makris[3]

[1]*Botnar Research Centre, Nuffield Department of Orthopaedics, Rheumatology and Musculoskeletal Sciences, University of Oxford, Oxford, United Kingdom;* [2]*Department of Radiology, Oxford University Hospitals, Oxford, United Kingdom;* [3]*Department of Endovascular and Interventional Radiology, Guys and St. Thomas' Hospitals NHS Foundation Trust, London, United Kingdom*

Study design

A longitudinal study is a study that repeatedly measures observations (collects data) over time.[1] It often involves following up with patients for a prolonged period, such as years, and measuring both explanatory and outcome variables at multiple points, usually more than two, of follow-up. [2] Longitudinal studies are most commonly observational, meaning that there is no external influence on study participants. However, they can also be experimental or randomized. A cohort study is the most common type of longitudinal study.[3]

The aim of a longitudinal study is to assess the relationship between explanatory or exposure variables (for example, risk factors) and outcome variables. Data for explanatory variables are collected at a time period before data for outcome variables. As a longitudinal study has a time interval between the explanatory and outcome variables, it is useful in assessing whether there is causation between the two variables. A causal relationship is demonstrated using the Bradford Hill criteria, which are nine principles used to provide evidence of a causal relationship in epidemiological studies, of which temporality is one.[4]

There are many different designs of longitudinal studies. The most common is a prospective study, where participants are followed up prospectively over a period of time (Fig. 40.1A). The participants can either be selected from a cohort (a group of participants with similar characteristics, for example, all from the same population or all with the same year of birth), or from a panel (a sample of participants which is randomly selected from the same population). Alternatively, a longitudinal study can be retrospective, in which the participants have already experienced the outcome and historic data are collected on potential explanatory variables (Fig. 40.1B). Rarely, longitudinal studies can be formed from repeated cross-sectional studies, in which random samples of patients are repeatedly assessed over a period of time.

Measures

Data are collected on explanatory variables and outcome variables. The choice of these variables should be driven by the aim of the study and possible confounders which will need to be accounted for in the analysis. Explanatory variables can include a large range of data, for example, anthropometric measurements (such as age, gender, socioeconomic background), and clinical or medical data (such as medical history, medication use, disease state). Explanatory variables can be either categorical data (binary/ordinal/nominal) or continuous. Outcome variables should focus on the presence or absence of the event being studied (such as a particular disease). Outcome variables are most often categorical data types. A complete list of which data are collected and how is it collected should be stored within a data dictionary. The data dictionary should include the exact questions the participants in the study are asked, the response options, and how these data are coded in the dataset (Fig. 40.2).

Describing data

When the study is analyzed, some participants will and some participants will not have the explanatory variable/s; and similarly, some participants will, while others will not, have the outcome variable/s. Often, the association between an outcome variable and multiple explanatory variable will

Translational Interventional Radiology. https://doi.org/10.1016/B978-0-12-823026-8.00057-2

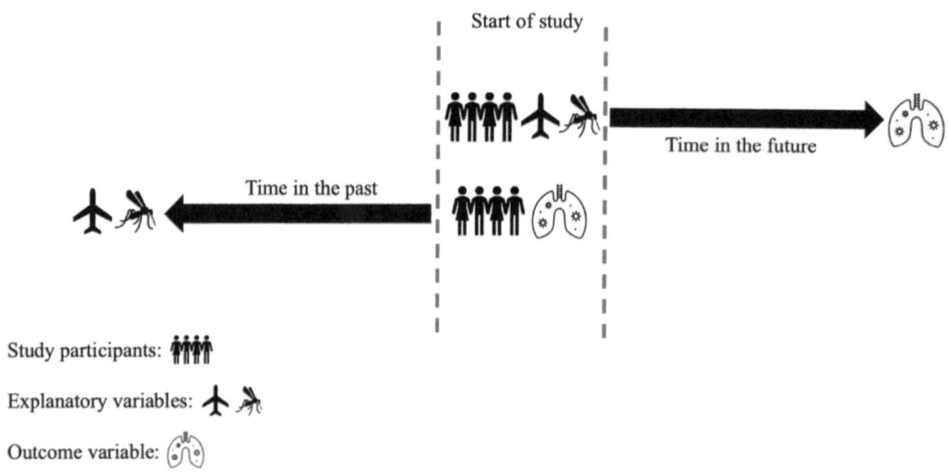

FIGURE 40.1 Hypothetical examples of prospective (1A) and retrospective (1B) longitudinal studies.

Participant_ID	String	NNNN	Participant ID	
Q_General_DOB	String	d/mm/yyyy	Date of birth	
Q_General_Gender	Text	Text	Gender	0=Male 1=Female 3=Other 4=Did not disclose
Q_Medical_Diabetes	Text	Text	Do you have diabetes?	0=No 1=Yes 2=Missing
Q_Pain_R_Knee	String	Real number	How severe is the pain in your right knee?	0 to 5 (5 point Likert Scale)
Exam_Swelling_R_Elbow	String	Real number	Swelling in right elbow on examination	0=No swelling 1=Minimal 2=Moderate 3=Severe 4=Not examined

FIGURE 40.2 Hypothetical example of data dictionary.

be assessed. It is important to first describe the distribution of explanatory variables across the participants, often stratified by the presence and absence of the outcome. For categorical variables, this can be done by a two-by-two table (2 × 2) table (also known as contingency table) (Fig. 40.3).[5] If a variable has multiple response options (referred to as levels), the number of rows or columns in the table can be adjusted accordingly.

To describe the distribution of a continuous explanatory variable, the spread should first be visualized. If the data from the explanatory variable are normally distributed (Gaussian distribution), the mean and standard deviation of the explanatory variable can be reported for each level of the outcome variable. If the data for the explanatory variable are not normally distributed, the median and interquartile range for each outcome level should be given. These descriptions of the data are often reported in tables of published manuscripts (Table 40.1). These tables also commonly include routine baseline data, such as age and gender, in addition to the explanatory variables. Historically, *P* values are provided in these tables, to compare the explanatory variables between the participants who do and do not have the outcome variable. However, more recently *P* values are not provided, as they can be misleading,[6] and the aim of these tables is not to test for differences between participants, but merely to describe the distribution of data.

		Outcome variable (Number of participants)	
		Yes	No
Explanatory variable (Number of participants)	Yes		
	No		

FIGURE 40.3 Examples of how to describe categorical data.

TABLE 40.1 Hypothetical example of how data can be described in a typical manuscript.

	Outcome variable (N = 100)	
Explanatory variables	**Yes (n = 40)**	**No (n = 60)**
Age (years) (Mean, standard deviation)	54.5 (3.2)	62.9 (3.1)
Gender Male (n, %) Female (n,%)	30 (75) 10 (25)	55 (91.7) 5 (8.3)
Body mass index (kg/m^2) (Median, interquartile range)	27.5 (18.3–33.6)	32.5 (22.3–37.7)
Alcohol intake (units/week) (n, %) <5 5–14 >14 Missing	10 (25) 27 (68) 3 (8) 0	0 (0) 55 (92) 1 (2) 4 (7)

Analyzing data and missing data

When analyzing data, the aim is to test for an association between the explanatory variable/s and the outcome variable. The measurements of multiple variables, which could be confounders,[7,8] allow for them to be adjusted in the analysis. If the explanatory variables and the outcome variable are measured at one time point each, by the end of the study, the participants either will or will not have the outcome variable. The statistical analysis will be relatively straight forward and could include tests (effect measures) such as odds ratios and relative risks, though the statistical test used should be guided by the aim and hypothesis of the study. However, if the explanatory or outcome variables are measured repeatedly at different time intervals, analysis becomes much more complicated. In such situations, generalized estimating equation (GEE) models or mixed effects regression (MER) models can be used.[9] GEE models assume that variables are not related to each other.[10] MER models take into account "fixed effects" (explanatory variables) and "random effects" (other factors that could affect the outcome and that we are trying to control).[11]

Longitudinal studies might have missing data. Missing data can reduce sample size and power and can lead to bias of the results. It is important to understand why data are missing, or what has caused it.[12] There are three patterns of missing data: (1) Missing not at random (the missing data are related to specific factors); (2) Missing at random (the missing data are related to other data that are collected in the study.[12] This can be tested by running t-tests or chi-squared tests between the variable that is missing data and another variable that might explain why this is). (3) Missing complete at random (MCAR) (the missing data are random and not related to any data collected or not collected in the study, the data can be assumed to be MCAR or Little's test can be used[13]). Patterns of missing data are often assessed through visualizing graphically.[14] If there is less than 5% of missing data overall and it is MCAR, a "complete case analysis" is used.[15,16] If participants with missing data are kept in the analysis, the missing data can be substituted through single imputation,[17] or multiple imputation[18–20] with analysis pooled across the datasets.[21]

Benefits

The primary advantage of a longitudinal study is the time period over which data are collected. This provides a temporal relationship between the explanatory and outcome variables, providing data to analyze whether events are related to exposures, and in particular where the time interval between the exposures and outcome are important. Longitudinal studies are also able to describe sequences of events a participant might experience, and how variables might change over time. For example, a participant might begin the study as a smoker, but after a few years become a nonsmoker. This could affect the outcome being studied, and the time between the change in smoking and the outcome might influence the strength of the effect measure.

Cohort studies are particularly useful in accounting for the "cohort effect" (also known as generation effects).[22] The cohort effect describes an outcome that is affected by the exposures at that time, most commonly an age-specific exposure.[23] For example, when assessing whether study participants use computerized technology, it is more likely that participants in their 20s and 30s use computerized technology compared to participants in their 60s and 70s, by nature of their age and therefore exposure to technology. In longitudinal studies, the prolonged time period of follow-up accounts for the changes in exposures over time. This ensures that the outcome seen in a longitudinal study is not due to the characteristic of the study participants at the beginning of the study. In the above example, if the participants in their 20s and 30s are followed up longitudinally for 50 years, the use of computerized technology could be more appropriately assessed.

Potential pifalls

Longitudinal studies are much more expensive to perform than cross-sectional studies. They also take a longer period of time before results are available. There is a risk of participant attrition and loss to follow-up, before the end of the study or before the study outcome is reached. As participants are followed up for a prolonged period of time and repeated measurements might be taken, there is a risk of intraparticipant correlation and clustering.

Often, multiple analyses can be performed from one longitudinal study, due to the large amount of data collected. However, if an analysis is performed for data that have not been collected for that specific purpose, the results can be biased and provide limited insight. For example, if a longitudinal study was designed to look at the effect of smoking on cancer, and data were also collected on alcohol intake, a secondary analysis could be performed analyzing the relationship of alcohol intake with cancer. However, the way alcohol intake was assessed in the study might not have been validated and might not provide the most appropriate data to answer this question. Finally, longitudinal studies often test association, not causation.

Real-world example

The Chingford 1000 Women Study is a longitudinal study set up in 1989, in Chingford, England.[24] It was initially designed to assess osteoporosis and osteoarthritis, and participants have been followed up annually (currently in the 31st year of follow-up). Participants were recruited from a general practice in Chingford, east London, and must have been female and between the ages of 45 and 65 at recruitment.[25] At baseline (the start of the study), each participant was interviewed using a questionnaire to assess medical history, a physical examination was performed, and radiographs of their joints were taken. At follow-up, repeated questionnaires were used, examinations were performed, and radiographs were taken.

Definition of the study type

The Chingford 1000 Women Study is prospective population-based cohort study. It is a prospective study as the participants were followed up over time. The participants were chosen from the general population of a particular area, and the study aimed to test hypotheses on the population, not the individual, level. The study participants have been shown to be representative of the middle-aged Caucasian female United Kingdom (UK) population.[26] This study is a type of cohort study, itself a type of longitudinal study, as the participants are being followed up over a period of time and all share the same characteristic of belonging to one general practice.

Benefits

The Chingford 1000 Women Study is one of the largest and longest cohort studies of osteoarthritis worldwide. Over 1000 women were recruited, and it has been listed as the National Institute of Health as an important resource in epidemiology.[24] In particular, it includes repeated risk factor measurements and radiographs of joints, which can be assessed for the incidence and prevalence of osteoarthritis. This allows investigators to assess risk factors for osteoarthritis and changes in risk factors over time. At the 15th year follow-up, 77% of the participants were still included in the study.[24] This high rate of retention and phenotypically rich cohort allow for a number of studies to be conducted, and for multiple confounders to be adjusted for.

Limitations

The Chingford 1000 Women Study includes a very specific population, in particular middle-aged women. This means that results from the study might not be transferable to men, to other age groups, and to populations whom might have different baseline demographics. It is also an uni-centre study, while more commonly multicentre studies are now being used in research. Radiographical data have been collected and can be assessed for osteoarthritis. However, there is no linkage to hospital data, such as the Clinical Practice Research Datalink[27] or the Hospital Episodes Statistics database,[28] which would allow validation of the results and the use of linked medical official medical records. The Chingford 1000 Women Study is also expensive and time consuming to run. The participants were recruited in middle age, and it is likely that at the time of writing, a proportion of those patients would have become decreased.

Get started

Designing a longitudinal study takes a lot of forethought, and similarly running the study is, by the nature of the design, time consuming. Below, we provide two recent examples of longitudinal radiological studies from the literature (Table 40.2). The first is a prospective cohort study investigating changes over time on chest computerized tomography scans of 90 patients hospitalized with novel coronavirus-19 (COVID-19).[29] The second is a

TABLE 40.2 Step-by-step approach with examples of longitudinal studies from the academic medical literature.

Steps:	Example 1[29]:	Example 2[30]:
1) Build study team and identify aims of the study	Analyze serial chest CT changes in patients with COVID-19	Describe infectious complications post-EAIRP
2) Describe the variables to be collected, and consider measures and effects. Ensure data dictionary is developed.	Ground glass opacity, Consolidation, Reticular pattern, Mixed pattern, Honeycomb pattern, Effusion, Pneumothorax, Mediastinal emphysema, Mediastinal lymphadenopathy (all quantified)[31]	Incidence of IC, Mortality at 30 days (if no so implant inserted) or 1 year (if implant inserted), Surgical repair, SAPS II and ICU stay
3) Gain ethical approval	IRB approval from Union Hospital, Tongji Medical College, Huazhong University of Science and Technology	Informed consent not required
4) Gain funding	Nil	Nil
5) Identify study population and recruitment plan	Hospitalized patients with suspected COVID-19 pneumonia	Hospitalized patients after first EAIRP,[32] taken from the French PMSI database
6) Define start of study	First day of symptoms	Admission to hospital between 2009 and 2014 after first EAIRP
7) Define end of study	March 8, 2020 (or patient discharged or deceased)	One year from admission
8) Recruit study participants	Patients admitted to Union Hospital between January 16, 2020 and February 17, 2020, with 1 rRT-PCR positive result, 1 abnormal CT scan	Identified by searching for a PMSI for EAIRP
9) Train study team, and begin collection of data (repeat this at multiple time intervals)	CT scans before admission, and during admission were collected and read by five experienced radiologists.	Extract data from PMSI, between hospital admission after first EAIRP to at least 1 year later
10) Manage data, and perform statistical analyses based on measures and effects	a) Kruskal–Wallis rank sum test to assess median values of CT quantification over time[31] b) Chi-square test to assess the frequency of CT patterns over time	Multivariable logistic regression

protocol that has been developed to assess for infectious complications following endoarterial interventional radiology procedures, from a longitudinal cohort database.[30]

COVID-19: Novel coronavirus 19; CT: Computerized tomography; EAIRP: Endoarterial interventional radiology procedures; ICU: Intensive care unit; IRB: Institutional review board; PMSI: Programme de Médicalization des Systèmes d'Information, r-RT-PCR: real-time reverse transcriptase-polymerase chain reaction for severe acute respiratory syndrome coronavirus 2 (SARS-CoV-2); SAPS: Simplified Acute Physiology Score.

Standard research team personnel and roles

Longitudinal studies often require a research team to maintain the study. However, in comparison to trials, observational studies often require much less involvement from the research team.

1) **Principle investigator:** The individual with overall responsibility for the study. They often hold the grant to fund the study. They are responsible for the study design and conduct, and oversee the recruitment, and data collection with analysis and interpretation.
2) **Regulatory coordinator:** In charge of the study protocol, including drafting, editing, amending, and updating. They also ensure reports are appropriate for ethical review.
3) **Data coordinator:** Ensure data collected is accurate and done so appropriately. They often store the data dictionary and are able to answer any data-related queries.
4) **Research coordinator:** Oversees the daily study requirements and data collection at appropriate follow-up intervals. They are responsible for ensuring participants are recruited, consent is obtained, and participants are adhering to the protocol.
5) **Statistician:** Responsible for data analysis and using appropriate statistical tools. Published results from studies should adhere to reporting guidelines, such as the Strengthening the Reporting of Observational Studies in Epidemiology guideline.[33]

Budgetary considerations and funding options

Observational studies are considerably cheaper to set up and run, in comparison to trials. Initial costs include renumeration of the research team, use of data collection tools (such as those which require a fee to use), and data storage facilities (including electronic methods). Ongoing field costs relate to the repeated follow-up visits of study participants. Large longitudinal studies have been known to decrease their field costs by reducing periodicity, such as by prolonging the time intervals between questionnaires.[34] Costs can also be managed by controlling the number of participants in the study and controlling the number of data variables collected.

Funding options are dependent on the geographical location of the study. The National Institute for Health Research (NIHR) in the UK, and the National Institutes of Health in the United States, are large funders of academic clinical research. Often large studies are formulated as collaborations between multiple institutes, including governmental agencies, research councils (such as the Medical Research Council), and charitable organizations (such as the Wellcome Trust).[35] For individuals who are interested in performing secondary analyses of the data collected in longitudinal studies, smaller grants are available. These include those from the NIHR, from local hospitals or institutes, or professional bodies (such as the British Society of Interventional Radiology Industry Funded Research Scholarship).[36]

Resources

1. UK Longitudinal Studies Centre
2. Menard D. Longitudinal Research. SAGE; 2002.
3. Coggon D, Rose G, Barker DJP. Longitudinal studies in Epidemiology for the uninitiated. BMJ; 2003.

References

1. Wang M, Beal DJ, Chan D, Newman DA, Vancouver JB, Vandenberg RJ. Longitudinal research: a panel discussion on conceptual issues, research design, and statistical techniques. *Work, Aging Retirement*. 2017;3(1):1–24.
2. Chan D. The conceptualization and analysis of change over time: an integrative approach incorporating longitudinal mean and covariance structures analysis (LMACS) and multiple indicator latent growth modeling (MLGM). *Organ Res Methods*. 1998;1(4):421–483.
3. Barrett D, Noble H. What are cohort studies? *Evid Base Nurs*. 2019;22(4):95–96.
4. Hill AB. The environment and disease: association or causation? *Proc Roy Soc Med*. 1965;58(5):295–300.
5. Sauerbrei W, Blettner M. Interpreting results in 2 × 2 tables: part 9 of a series on evaluation of scientific publications. *Dtsch Arztebl Int*. 2009;106(48):795–800.
6. Price R, Bethune R, Massey L. Problem with p values: why p values do not tell you if your treatment is likely to work. *Postgrad Med*. 2020;96(1131):1–3.
7. Skelly AC, Dettori JR, Brodt ED. Assessing bias: the importance of considering confounding. *Evid Base Spine Care J*. 2012;3(1):9–12.
8. McNamee R. Confounding and confounders. *Occup Environ Med*. 2003;60(3):227–234. quiz 164, 234.
9. Garcia TP, Marder K. Statistical approaches to longitudinal data analysis in neurodegenerative diseases: huntington's disease as a model. *Curr Neurol Neurosci Rep*. 2017;17(2):14.
10. Wilson JR, Lorenz KA. Generalized estimating equations logistic regression. Modeling binary correlated responses using SAS. In: *SPSS and R. ICSA Book Series in Statistics*. Vol. 9. Cham: Springer; 2015.

11. Van Dongen HP, Olofsen E, Dinges DF, Maislin G. Mixed-model regression analysis and dealing with interindividual differences. *Methods Enzymol.* 2004;384:139–171.
12. Rubin DB, Rubin D. *Multiple Imputation for Nonresponse in Surveys*. New York: John Wiley & Sons, LTD; 1987. Canada: John Wiley & sons 1987.
13. Little RJA. A test of missing completely at random for multivariate data with missing values. *J Am Stat Assoc.* 1988;83(404):1198–1202.
14. Zhang Z. Missing data exploration: highlighting graphical presentation of missing pattern. *Ann Transl Med.* 2015;3(22):356.
15. Jakobsen JC, Gluud C, Winkel P, Lange T, Wetterslev J. The thresholds for statistical and clinical significance - a five-step procedure for evaluation of intervention effects in randomised clinical trials. *BMC Med Res Methodol.* 2014;14:34.
16. Jakobsen JC, Wetterslev J, Winkel P, Lange T, Gluud C. Thresholds for statistical and clinical significance in systematic reviews with meta-analytic methods. *BMC Med Res Methodol.* 2014;14:120.
17. Dziura JD, Post LA, Zhao Q, Fu Z, Peduzzi P. Strategies for dealing with missing data in clinical trials: from design to analysis. *Yale J Biol Med.* 2013;86(3):343–358.
18. Sinharay S, Stern HS, Russell D. The use of multiple imputation for the analysis of missing data. *Psychol Methods.* 2001;6(4):317–329.
19. Huque MH, Carlin JB, Simpson JA, Lee KJ. A comparison of multiple imputation methods for missing data in longitudinal studies. *BMC Med Res Methodol.* 2018;18(1):168.
20. Li P, Stuart EA, Allison DB. Multiple imputation: a flexible tool for handling missing data. *JAMA.* 2015;314(18):1966–1967.
21. Marshall A, Altman DG, Holder RL, Royston P. Combining estimates of interest in prognostic modelling studies after multiple imputation: current practice and guidelines. *BMC Med Res Methodol.* 2009;9:57.
22. Last JM. *A Dictionary of Epidemiology*. Oxford University Press; 2001.
23. Keyes KM, Utz RL, Robinson W, Li G. What is a cohort effect? Comparison of three statistical methods for modeling cohort effects in obesity prevalence in the United States, 1971-2006. *Soc Sci Med.* 2010;70(7):1100–1108.
24. Chingford 1000 Women Study. Study Background [Available from: http://www.chingfordstudy.org.uk/collab/background.htm.
25. Hart D, Spector T, Egger P, Coggon D, Cooper C. Defining osteoarthritis of the hand for epidemiological studies: the Chingford Study. *Ann Rheum Dis.* 1994;53(4):220–223.
26. Hart DJ, Spector TD. The relationship of obesity, fat distribution and osteoarthritis in women in the general population: the Chingford Study. *J Rheumatol.* 1993;20(2):331–335.
27. Crown Copyright. *Clinical Practice Research Datalink.* 2021.
28. NHS Digital. *About the HES Database*; 2020 [Available from: https://digital.nhs.uk/data-and-information/data-tools-and-services/data-services/hospital-episode-statistics.
29. Wang Y, Dong C, Hu Y, et al. Temporal changes of CT findings in 90 patients with COVID-19 pneumonia: a longitudinal study. *Radiology.* 2020;296(2):E55–e64.
30. Aid Mellouk K, Soulaymani A, Gao F, Astagneau P, Misset B. Infectious complications of endoarterial interventional radiology: protocol for an observational study of a longitudinal national cohort of patients assessed in the French hospital discharge database (MOEVA study). *BMJ Open.* 2019;9(6):e024181.
31. Ooi GC, Khong PL, Müller NL, et al. Severe acute respiratory syndrome: temporal lung changes at thin-section CT in 30 patients. *Radiology.* 2004;230(3):836–844.
32. Vergnenègre A, Druet-Cabanac M, Rymer R. Interventional radiology and the medical applications of information systems: what consequences for the radiology department? *J Radiol.* 1999;80(6):565–568.
33. von Elm E, Altman DG, Egger M, Pocock SJ, Gøtzsche PC, Vandenbroucke JP. The strengthening the reporting of observational studies in Epidemiology (STROBE) statement: guidelines for reporting observational studies. *Int J Surg.* 2014;12(12):1495–1499.
34. Weir DR, Ofstedal MB. *Evaluating Strategies for Reducing Field Costs in a Longitudinal Study*; 2016. Available from: https://hrs.isr.umich.edu/sites/default/files/biblio/Weir_Ofstedal_costs_0.pdf.
35. Doyle A, Golding J. The costing and funding of longitudinal birth cohort studies. *Paediatr Perinat Epidemiol.* 2009;23(Suppl 1):86–92.
36. British Society of Interventional Radiology (BSIR). *BSIR/Industry Funded Research Scholarship*; 2018 [Available from: https://www.bsir.org/mediacentre/news/bsirindustry-funded-research-scholarship/.

Chapter 41

Clinical trials: Design, measures, classic example

Babak Mohit[1] **and Nariman Nezami**[2,3]

[1]*Sleep Disorders Center, Division of Pulmonary and Critical Care Medicine, Department of Medicine, University of Maryland School of Medicine, Baltimore, MD, United States;* [2]*Division of Vascular and Interventional Radiology, Department of Diagnostic Radiology and Nuclear Medicine, University of Maryland School of Medicine, Baltimore, MD, United States;* [3]*Experimental Therapeutics Program, University of Maryland Marlene and Stewart Greenebaum Comprehensive Cancer Center, Baltimore, MD, United States*

Introduction and history

Even though clinical trials are the most sophisticated studies in medicine, in essence, they are also the most fundamental and rudimentary method of experimental studies. The history of clinical trials in the development of humankind can be traced as far back to the time before our species was differentiated from our primate cousins[1,2]. The unique advancement that modern medicine has added to this history is its distinctive ability to better detect and discern its own biases and devise methods to overcome them. Therefore, it is of no surprise that the medical subject heading (MeSH) term "clinical trial" has been part of the MeSH vocabulary since its initiation in 1965.[3] Naturally, as methods of detecting bias have evolved over time, the concept of clinical trials has also become increasingly specialized. What is famously today the gold standard of clinical trials, the randomized controlled trial (RCT), was only introduced to the MeSH vocabulary as recent as 1990.[4] The objective of this chapter is to review common biases and how the clinical trial design can improve the possibility to overcome these biases.

Clinical trial

A clinical trial is an experiment, in which patients are the subjects.[5] By this definition, the objective of a clinical trial, as an experiment, is to test a hypothesis on the safety, efficacy, or efficiency of an intervention,[6] which can either be diagnostic or treatment interventions. The purpose of an experiment is to support a unique causal relationship between an exposure variable and an outcome variable while eliminating the role of any extraneous variables by controlling for confounding. Furthermore, given that all subjects in a clinical trial are patients, the ultimate goal of the experimental intervention is to find the intervention(s) that best reduce(s) morbidity or mortality.

The logical ability of an experimental design to eliminate the role of extraneous variables and establish a valid relationship between the exposure and outcome variable depends on the experimental design. Common threats to the validity of an experimental design are outlined in Table 41.1.

In the common nomenclature for communicating experimental design, observations are noted by O, intervention is scripted by X, randomization is noted by R, where each experimental group is written on a separate line and is noted from left to right where the left designates an earlier time.[8] According to this nomenclature the simplest experiment is noted by XO, which represents the case that a patient with no prior verifiable history is given an intervention (e.g., : a drug), and after a predetermined time goes back to report symptom change, and the experimenter observes and concludes that the impact of the intervention. Most case report studies and Phase I trials are based on this logic. However, as simple as this design is, there are numerous threats to the validity of its conclusions. Logical questions doubting the validity of this design can include what was the patients' history (History)? Was the change the impact of the drug or did the disease get better on its own (Maturation; Regression)? How do we know there was a change in symptoms (Testing; Instrumentation)? Did the patient come into the office randomly or did they select the doctor because they knew the doctor would provide that treatment (Selection)? Did the symptoms of the patient change because the patient believed in advance in the

Translational Interventional Radiology. https://doi.org/10.1016/B978-0-12-823026-8.00001-8

TABLE 41.1 Threats to design validity.

Statistical conclusion validity:
Low statistical power; violated assumptions of statistical tests; reliability
Internal validity:
History; maturation; testing; attrition; instrumentation; regression; selection; interaction of maturation, history, or instrumentation with selection; ambiguity about the direction of causal influence; diffusion or imitation of exposure; compensatory equalization of exposure; compensatory rivalry or resentful demoralization of patients.
Construct validity of exposures and outcomes:
Operational definitions of the exposure and outcome variables; timeframe within operational definitions; multiple treatment interactions; experimental bias; Hawthorne effect
External validity:
Interaction of treatment with selection, setting, or history.

Modified from Gross Portney, L., Watkins, M.P. *Foundations of Clinical Research: Applications to Practice.* Upper Saddle River, NJ.: Prentice-Hall Inc.; 2000.

impact of the treatment (Hawthorne effect)? The answers to each of these questions, and other points highlighted in Table 41.1, give an insight into the numerous confounders that could impact understanding of the true effect of the intervention. More advanced study designs can help to avoid the deflation or inflation of the true impact of the intervention on patient outcomes.

Subtypes

Study design can help overcome the validity doubts outlined in Table 41.1. The most simple addition to the basic XO study design is to have a comparison observation point. Comparison observation points can be across time such that there is at least one observation before and after the intervention O_1XO_2. The addition can also be across groups, where one group with similar symptoms receives the intervention while the comparison group does not:

$$O_1XO_2$$

$$O_1\ O_2$$

While having comparison (control) groups reduces some of the biases, the most powerful element of study design is randomization. Randomization is so important that all study designs that do not include randomization are quasiexperimental trials, while at least one phase (usually Phase II or Phase III) of clinical trials always includes randomization. The scientific standard in clinical research for establishing a causal relationship is design commonly known as the RCT and its design nomenclature is:[7]

$$RO_1XO_2$$

$$O_1\ O_2$$

Since the medical ethics and regulation prohibit withholding treatment, the control group of a randomized clinical often receives "standard of care" treatment instead of no treatment. This can also include placebo treatment. The design nomenclature therefore is:[7]

$$RO_1X_1O_2$$

$$O_1X_2O_2$$

In this case X_1 is the intervention of interest while X_2 is the comparator, which could be standard of care or placebo treatment.

Closely related to the intervention and placebo treatment, are the issues of adherence to treatment and blinding. Adherence to treatment is to ensure that each member of the intervention group has certainly received the treatment in the dose and quality described in the study protocol. Deviations from adherence to treatment may lead to deep miscalculations and bias in the results of the trial.[9]

Another source of observation bias is when participants know of their treatment status or the investigator's expectations. This may consciously (or unconsciously) influence the participants performance or their reporting of outcomes. To protect from this form of biased reporting, it is best that participants and treatment givers are unaware (or blind) of whether the treatment is the intervention or the standard of care/placebo. This strategy is known as blinding. If both the patient subject and the treatment giver are unaware of their treatment status, the study is known to be double blinded. If only the patient is unaware of whether they are receiving the treatment or placebo, this is known as a single-blinded study. However, the more objective the outcome is, for example, mortality in survival times studies, the need for blinding becomes less urgent. For more subjective outcomes, the need to adherence becomes a priority and may also involve the necessity that subjects, whether from the intervention group or the control group, not have access to each other so they are unable to communicate and share information with one another.[7]

The multiphases of clinical trials

Although the basic design of an RCT seems straightforward, modern-day RCT requires a large amount of preparation and investment. Furthermore, many interventions fail to demonstrate safety and/or efficacy during an experimental trial. To minimize resource loss, most scientific organizations do not require the entire RCT to be done in one study. Instead, the most common method of performing an RCT is a cumulative multi-phase trial. The phases of a clinical trial as well as the main objective of each phase are highlighted in Table 41.2.

It should be noted that given the multiphase nature of an RCT, some variations of the original RCT design can be accepted. For example, since from Phase 2 onward, all participants in the trial are patients with the disease/condition, the original design can be modified from:

$$RO_1X_1O_2$$

$$O_1X_2O_2$$

to:

$$RX_1O_2$$

$$X_2O_2$$

Note that omission of O_1 observation is only acceptable because O_1 is already established as part of the inclusion criteria in the Phase 2 trial.

Design

To better understand the elements of the design of an RCT, in the context of interventional radiology, we use the 2016 report of Salem, Gordon, Mouli et al.[11] on the comparison of yttrium-90 (Y90) radioembolization versus conventional transarterial chemoembolization (cTACE) in patients with Hepatocellular Carcinoma (HCC) as an example.

In this study, the population of the study is all patients with HCC. The two interventions that are being compared are (1) Y90 radio embolization and (2) cTACE. The main outcome is time to progression. In terms of the experimental nomenclature, the main outcome is the length of time of O_2. Other outcomes are safety, rate of response (based on biomarkers of tumor size and necrosis criteria), and survival time. The study is a Phase 2 study, so there is no longer a need to establish that the patients had HCC before randomization. To ensure that randomization was successful, all relevant clinical confounders have been noted in Table 1 of the reference. It is important to notice that all confounders were the same in the two treatment groups ($X_1 = $ Y90 radio embolization $X_2 = $ cTACE), which is highlighted by the lack of a significant P value of the difference (all $P > .05$). The results in Figure indicate that 25 months after randomization, nearly 75% of patients who received Y90 were progression free, while only 25% of patients received cTACE were progression free and the difference between the two groups is highly significant ($P = .001$). The authors rightfully claim that Y90 is better in preventing the progression of HCC compared to cTACE. The fundamental reason behind the validity of their claim is that through randomization they demonstrated that all other confounders had been successfully controlled for, as presented in Table 1 of the reference.

Benefits (advantages)

Clinical trials, and especially the RCT, are powerful design tools that can eliminate various biases, and threats to validity (see Table 41.1), through the use of the sequencing of

TABLE 41.2 Phases of clinical trial.

Phase name	Participants	Length of study	Purpose	Success rate
Phase 1	20–100 healthy volunteers or people with the disease/condition.	Several months	Is open label and single arm to study safety and dosage	Approximately 70% of drugs move to the next phase
Phase 2	Several hundred people with the disease/condition.	Several months to 2 years	Efficacy and side effects	33% of drugs move to the next phase
Phase 3	300–3000 volunteers with the disease/condition.	1–4 years	Efficiency and monitoring of adverse reactions	25%–30% of drugs move to the next phase
Phase 4	Several thousand volunteers with the disease/condition.	As long as the drug is on the market	Safety and efficacy	Ongoing as long as drug remains in market

Modified based on U.S. Food & Drug Administration (FDA). The Drug Development Process Step 3: Clinical Research. 2018; https://www.fda.gov/patients/drug-development-process/step-3-clinical-research. Accessed August 1, 2020 observation, randomization, intervention, blinding, and controlling the adherence to treatment. Nearly all modern medical products and interventions have come to us due to the powerful design of the clinical trial offers.

Limitation (downside)

As powerful as clinical trials are, they are unable to eliminate all biases, and threats to validity. For example, a clinical trial cannot overcome invalid operational definitions of the exposure and outcome variables. A clinical trial may improve how we measure phenomenon but is unable to answer why we measure phenomenon. After all, we cannot measure correctly, if what we are measuring is the wrong thing to measure. Furthermore, clinical trials are exceptionally expensive and risky. While the phasing of clinical trials has helped to manage some of the risks and to avoid some of the costs, it is alarmingly eye opening to consider the time and resources that go into the 75% of Phase 3 who do not yield a drug approval after recruiting up 3000 patients for a period of up to 4 years.

How to improve future clinical trials

Given that the data that emerge from clinical trials are considered the gold standard of science, there are several ways to improve the clinical trial.

1) Further reducing the probability of bias by improving methods of single blinding (where the patient does not know if s/he is in experimental or control group) or double blinding (where the patient and data recorder do know whether the patient is in experimental or control group). This may require innovative thinking to ensure that', for example, the experimental drug and the placebo pill both have the same shape, size, taste, and other features so they are indistinguishable. In the case of more complex trials, this may require the gathering of the experimental and control data in separate treatment facilities, so the experimental and control arms patients and personnel are less likely to communicate with each other.
2) Traditionally, the protocol paradigm for experimental studies has been to collect one patient group for one study. Furthermore, it has been so that successive trials are scheduled sequentially. Basket trials[12] are innovative new methods that increase efficiency by avoiding the duplication of regulatory and infrastructure efforts. When running a basket trial instead of multiple clinical trials in different diseases, one trial (the basket) can be designed with multiple targets, which allows patients with multiple diseases to enroll in cohorts or groups. If one group is showing a good response, this group is expanded to evaluate whether others could benefit from the new therapy. If another group is not revealing evidence of effectiveness, this group is closed and the patients are moved to consider another therapy. Basket trials have accelerated the development of vaccines during the recent Coronavirus Disease 2019 (COVID-19) pandemic.[13]
3) Bundling clinical trials with health economic studies. Given that clinical trials are the richest data source for comparative effectiveness studies, the synergies and economies of scale of data collection make these studies an invaluable resource for conducting cost-effectiveness studies. Economic data on direct and indirect costs of the intervention can be collected in parallel with the trial and used alongside effectiveness data to build cost-effectiveness models, which are important decision-making tools for healthcare policymakers (see chapter on Cost-effectiveness).

Conclusion

The unique study design features discussed in this chapter reveal the reasons of why clinical trials, and RCTs, are the gold standard in scientific inquiry and discovery. While the contents of the chapter provide an introduction to clinical trials, similar to other fields of specialized practice, practitioners can further their in-depth knowledge only through the study of specialized texts, and mentored practice in the field and clinic. The science of conducting clinical trials is continuously expanding the quantity and quality of the field, in which practitioners of the field should keep themselves updated with.

References

1. Sagan C. *The Demon-Haunted World*. New York, NY: Ballantine Books/Random House Inc.; 1996.
2. Sagan C. *The Dragons of Eden*. New York, NY: Ballantine Books/Random House Inc.; 1997.
3. National Library of Medicine (NLM). *Clinical Trials as Topic*; 1965. https://www.ncbi.nlm.nih.gov/mesh/68002986.
4. National Library of Medicine (NLM). *Randomized Controlled Trials as Topic*; 1990. https://www.ncbi.nlm.nih.gov/mesh/68016032.
5. Rothman KJ, Greenland S, Lash TL. *Modern Epidemiology*. 3rd ed. Philadelphia, PA: Lippincot, Williams, & Wilkins; 2008.
6. Evans SR. Fundamentals of clinical trial design. *J Exper Stroke Transl Med*. 2010;3(1):19–27.
7. Gross Portney L, Watkins MP. *Foundations of Clinical Research: Applications to Practice*. Upper Saddle River, NJ: Prentice-Hall Inc.; 2000.
8. Cambell DT, Stanley JC. *Experimental and Quasi-Experimental Designs for Research*. Chicago, IL: Rand McNally; 1963.
9. Zhang Z, Peluso MJ, Gross CP, Viscoli CM, Kernan WN. Adherence reporting in randomized controlled trials. *Clin Trials*. 2014;11(2):195–204.
10. U.S. Food & Drug Administration (FDA). *The Drug Development Process Step 3*. Clinical Research; 2018. https://www.fda.gov/patients/drug-development-process/step-3-clinical-research. Accessed August 1, 2020.

11. Salem R, Gordon AC, Mouli S, et al. Y90 radioembolization significantly prolongs time to progression compared with chemoembolization in patients with hepatocellular carcinoma. *Gastroenterology*. 2016;151(6):1155–1163. e1152.
12. Sabbatini, P. Making Clinical Trials More Efficient, Informative, and Effective. Retrieved from https://www.mskcc.org/news/making-clinical-trials-more-efficient-informative-and-effective [press release]. July 26, 2013.
13. Callaway E. The race for coronavirus vaccines: a graphical guide. *Nature*. 2020;580(7805):576–577.

Chapter 42

Meta-analysis: Design, measures, classic example

John T. Moon[1], Menelaos Konstantinidis[2] and Nariman Nezami[3]

[1]*Division of Interventional Radiology and Image-Guided Medicine, Department of Radiology and Imaging Sciences, Emory University School of Medicine, Atlanta, GA, United States;* [2]*Division of Biostatistics, Dalla Lana School of Public Health, University of Toronto, Toronto, ON, Canada;* [3]*Division of Vascular and Interventional Radiology, Department of Diagnostic Radiology and Nuclear Medicine, University of Maryland School of Medicine, Baltimore, MD, United States*

Key points

- Systematic reviews and meta-analyses require rigorous methodological planning.
- The assembly of a diverse team of relevant experts, detailed protocol development, and exacting execution of the proposed protocol are important to the study's success.
- Utilize and modify existing tools to streamline as well as minimize the introduction of bias to the study.

Background

Up until the last few decades, authoritative evidence synthesis was typically conducted through a narrative review in which an expert would provide an account of what they believed to be the state of the art and consequently, conclusions against a set of evaluation criteria were made. This approach introduces a variety of biases, for example, the reviewer's personal opinions, inconsistent evaluation criteria for evaluating the certainty of the evidence, and the lack of a transparent process. In this regard, the joint framework of systematic reviews and meta-analyses presents a methodologically rigorous solution to combat biases intrinsic to narrative reviews and provide authoritative evaluations of a clinical question.

Systematic reviews, often thought of as precursors to meta-analyses, are a collection of methodologies to identify, select, evaluate, and synthesize results from a pool of studies aimed at answering a scientific question. Meta-analyses take an additional step to statistically quantify the available evidence pertaining to a given question. More specifically, the key objective of a meta-analysis is the estimation of an effect (e.g., of an intervention) across separate, but similar studies aimed at answering a clinical question.

Designing and conducting a meta-analysis

We describe a basic framework for conducting meta-analyses in a stepwise fashion in the context of an Interventional Radiology-specific example.

Framing the clinical question

In designing a systematic review and meta-analysis, perhaps the most important piece is to determine the clinical question and only then, identify whether meta-analysis would constitute an appropriate evidence synthesis product. Take, for example, the following:

Example clinical question

What is the rate of *post-thrombotic syndrome* (outcome) following *catheter-directed thrombolysis* (Intervention A) as compared to *pharmacomechanical thrombectomy* (Intervention B) in the treatment of *acute proximal deep vein thrombosis (DVT)* (condition) in *patients aged 16–75* (setting)?

Often, meta-analyses are conducted to strengthen the evidence for or against an intervention (as opposed to basing decisions on a single study). It can take disparate data and stratify outcomes to specific clinical contexts or patient groups. With an increasing body of case reports, retrospective studies, and prospective clinical trials, it is important to critically evaluate and synthesize these data through the quantitative analyses within a meta-analysis.

Translational Interventional Radiology. https://doi.org/10.1016/B978-0-12-823026-8.00037-7

Once the need for a meta-analysis is established, the clinical question should be refined within relevant clinical parameters. "PICO" (standing for Population, Intervention, Comparison, and Outcome) is a commonly utilized mnemonic tool to limit the scope of the clinical question, and in turn, the search parameters. Additionally, having a well-defined set of PICOs will allow a better selection of appropriate method for meta-analysis, thereby strengthening the rigor and uptake of subsequent findings.

Example PICO

Population: Males and females between the ages of 16 and 75 with acute proximal deep vein thrombosis involving the iliac, common femoral, and/or femoral vein.
Intervention: DVT thrombolysis.
Comparators: Catheter-directed thrombolysis; pharmacomechanical thrombectomy; control.
Outcomes: Villalta score, clot reduction grade, bleeding, length of hospital stay.

Note that in defining the outcomes, it is necessary to identify how they will be reported in the review. For example, categorical outcomes should be reported as Odds Ratios or Relative Risk (Relative Risk is often preferred due to the ease of interpretation). Continuous variables should be reported as Mean Differences or Standardized Mean Differences, the latter more suitable for continuous outcomes in which several studies have differing scales.

Assembling a team

The priorities of a systematic review and meta-analysis are determined by its members. In assembling a team, mapping out all potential stakeholders specific to the review question can be informative. Stakeholders may include healthcare practitioners, referring physicians, patients themselves, policy advocates, guideline developers, professional organizations, and funding organizations. A successful team will typically include domain experts, statistical expertise, review expertise, and an information specialist. At a minimum, a review requires two authors, and ideally three or more.

Protocol development

A priori protocol development and dissemination are fundamental to a successful and reproducible systematic review and meta-analysis. The protocol requires determining eligibility criteria for study inclusion (e.g., PICOs and types of studies), a search strategy, a plan for data collection and management, methods to be used for the synthesis (i.e., meta-analysis for a quantitative synthesis), and a strategy to minimize biases throughout the systematic review and meta-analysis.

Bias may develop during the review process, thereby compromising the validity of the review and the subsequent results. For this reason, it is important that a predetermined methodology is defined (i.e., the protocol) and is followed throughout. However, on occasion, previously unforeseen challenges will present themselves. When this happens, it is important that deviations from the protocol are recorded and disclosed at the time of publication of the final review.

The Preferred Reporting Items for Systematic Reviews and Meta-Analysis (PRISMA) is a widely utilized reporting guideline for systematic reviews and meta-analyses. In developing a protocol, the PRISMA for systematic review protocols (PRISMA-P) guidelines should be used.[1] These guidelines provide a minimum set of items that should be included in a protocol. There are also a variety of quality assurance software programs that can streamline the protocol across team members. Some example programs include Covidence, Revman, and the Evidence for Policy and Practice Information and Co-ordinating Centre (EPPI)-Reviewer.[2] Once the protocol is ready, it should be published, either in a peer-reviewed journal, in an archive, or at a minimum, have it registered as is common in PROSPERO.[3]

Database search

Pubmed, EMBASE, the US National Library of Medicine, and the Cochrane Library are some of the databases that may be utilized during the literature search. Predefine search parameters to include results by year of publication, text availability, article type, and language. An understanding of Boolean search operators (Fig. 42.1) and Medical Subject Headings will also be useful in developing the Pubmed search. While specificity is necessary, it is to the reviewers' advantage to cast a wide net to ensure the capture of all potentially relevant studies.

Example Boolean search

((("catheter directed thrombolysis") AND ("pharmacomechanical thrombectomy")) AND ("deep vein thrombosis"))) OR (("catheter-directed thrombolysis") AND ("deep vein thrombosis")) OR (("pharmacomechanical thrombectomy") AND ("deep vein thrombosis"))

Screening and collection

The first pass of screening is to remove duplicates that may be found across different databases (software such as Covidence and Endnote generally do a good job of automatically doing this). Thereafter, studies are screened on relevance to the review using only the title and abstract. Subsequently, a full-text review is conducted, including (or excluding) studies after a detailed assessment of each study against the predefined eligibility criteria set out in the protocol. When

Example Boolean Search:
((("catheter directed thrombolysis") AND ("pharmacomechanical thrombectomy")) AND ("deep vein thrombosis"))) OR (("catheter-directed thrombolysis") AND ("deep vein thrombosis")) OR (("pharmacomechanical thrombectomy") AND ("deep vein thrombosis"))

FIGURE 42.1 The Boolean search operators for the example study.

excluding studies, note the reason for exclusion (e.g., wrong comparison, wrong patient population, wrong intervention, or wrong study design). These reasons will be used to populate the PRISMA diagram (Fig. 42.2).

Once the full-text review is completed, relevant information will be extracted from the included studies. Finally, each included study will undergo a quality assessment to evaluate the risk of possible biases. The Cochrane ROBINS-I tool is the most used for quality assessment of nonrandomized studies, and the Cochrane Risk of Bias 2 tool is mostly used to assess the quality of randomized studies.[4,5] At each stage, every item should be independently performed by two authors. Should conflicts between decisions arise, a resolution must be arrived at through discussion, or when this is not possible, an adjudicating decision should be made by a third author.

Data synthesis

Meta-analysis is ultimately a statistical method that synthesizes multiple effect estimates into a single summary estimate. This is usually done through fixed-effects or random-effects models. With such methods, the summary effect is estimated as a weighted average of the individual study effect estimates, where the weights are inversely proportional to a given study's precision (i.e., larger sample size or smaller confidence interval). In conducting a meta-analysis, it is important to consider possible sources of clinical and statistical heterogeneity, determine an appropriate method for quantifying heterogeneity, and methods to deal with meta-analyses presenting significant heterogeneity. To do so, a random-effects model is often used. Subgroup analyses with respect to one or more effect modifiers (variables inducing heterogeneity) may be

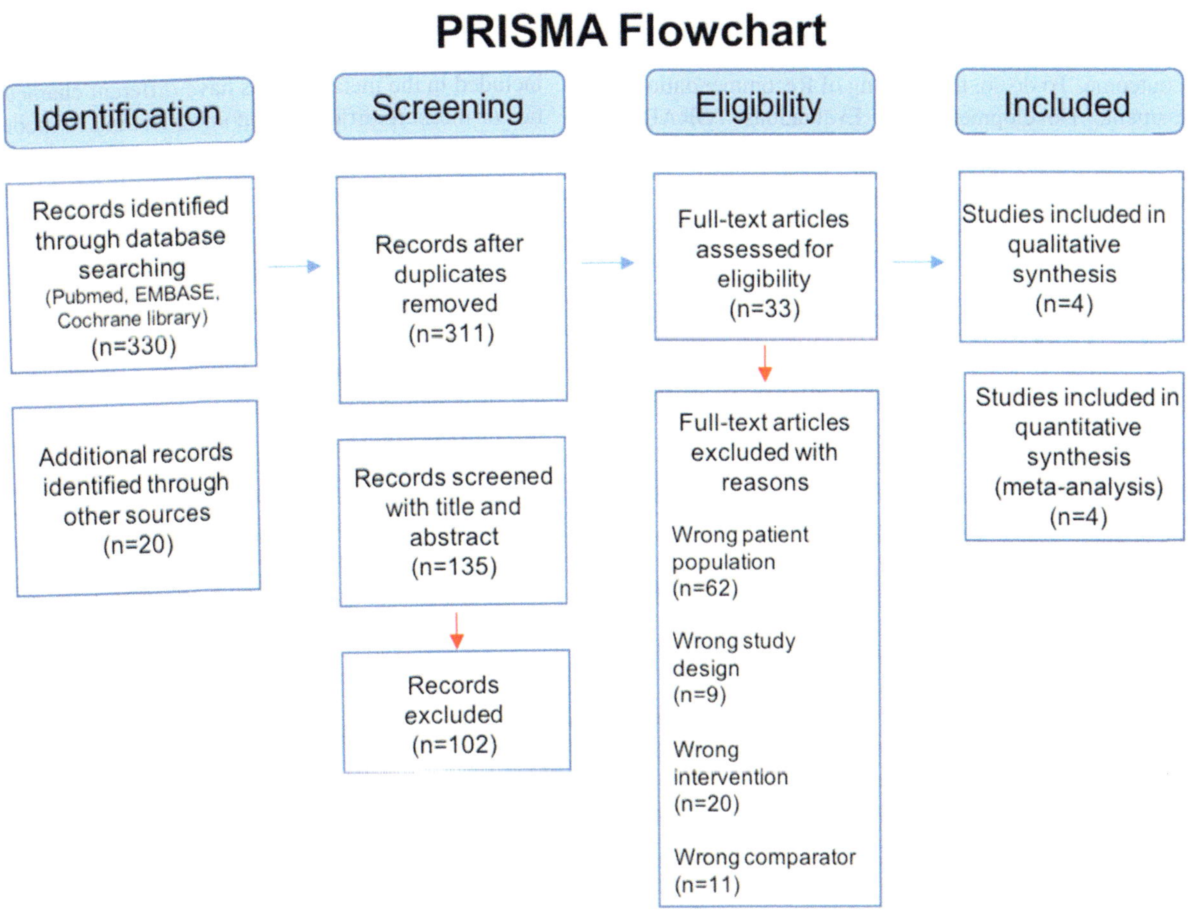

FIGURE 42.2 Preferred Reporting Items for Systematic Reviews and Meta-Analysis (PRISMA) flowchart.

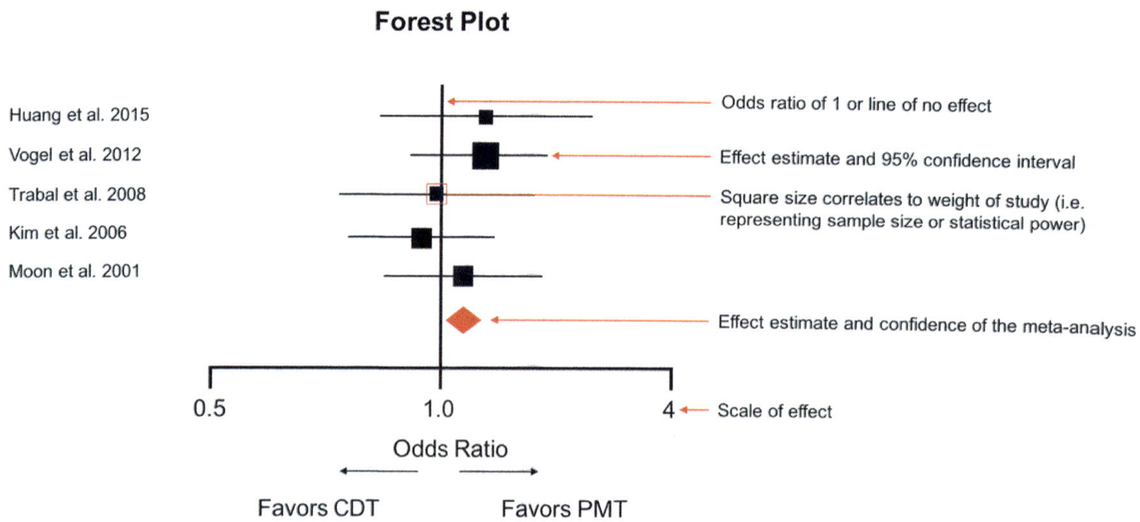

FIGURE 42.3 Description of a forest plot.

conducted to account for the heterogeneity. The final meta-analysis data are often presented in a forest plot, the anatomy of which is illustrated in Fig. 42.3:

Certainty of the evidence

The last step in conducting a systematic review and meta-analysis is evaluating the certainty of the evidence for each outcome. To do so, the Grading of Recommendations, Assessment, Development and Evaluations (GRADE) framework should be applied to each outcome and used to populate a Summary of Findings figure.[6,7]

Update

In conclusion, the systematic review and meta-analysis undertaken should be reproducible. As more studies relevant to the clinical question are published, the meta-analysis can be periodically updated to include additional data.

Get started

To get started:

- Define the research question and identify relevant studies.
- Assemble a team of experts and develop a protocol.
- Conduct the meta-analysis, including searching for studies, critically appraising their quality, extracting and synthesizing data, and analyzing and interpreting results.
- Write and disseminate the results, following best practices for reporting meta-analyses.

Potential pitfalls

There are several potential pitfalls to consider when conducting a meta-analysis:

- Risk of bias: This can occur at various stages of the process and can be minimized by following a rigorous and systematic process.
- Risk of heterogeneity: This occurs when the studies included in the meta-analysis have different characteristics or measure different outcomes, leading to inconsistent results. This can be addressed by selecting studies that are similar in design, population, intervention, and outcome measures, and by using statistical techniques to adjust for differences between studies.
- Risk of publication bias: This occurs when the results of some studies are more likely to be published or disseminated than others, leading to an overestimation of the true effect of an intervention or exposure. To address this, it is important to conduct a comprehensive search for relevant studies.
- Risk of poor quality or relevance of studies: The validity and reliability of the results of a meta-analysis may be limited by the quality and relevance of the individual studies included in the analysis. It is important to critically appraise the quality and relevance of these studies and to consider the potential for confounding or other sources of bias.

Real-world example

Jongsma, Hidde, et al. "Angioplasty versus stenting for iliac artery lesions." *Cochrane Database of Systematic Reviews* 12 (2020).

Objective: To assess the effectiveness of percutaneous transluminal angioplasty (PTA) compared to primary stenting for the treatment of stenotic and occlusive lesions in the iliac artery.

Population- Patients with stenotic or occlusive lesions in the iliac artery
Intervention Percutaneous -transluminal angioplasty (PTA) with selective stenting
Comparison- Primary stenting (placing a stent without first attempting PTA)
Outcome-Technical success, complications, symptomatic improvement of peripheral arterial disease, patency, reinterventions, resolutions of symptoms and signs, and improvement in walking distance as reported by the patient.

Study Summary:

- Two randomized controlled trials (RCTs) involving a total of 397 participants were identified, but the evidence was of low certainty due to concerns about risk of bias and imprecision.
- The technical success rate was not significantly different between PTA with selective stenting and primary stenting in either the study involving stenotic lesions or the study involving iliac artery occlusions.
- In the study involving iliac artery occlusions, PTA was associated with a higher rate of major complications, including distal embolization.
- There was no significant difference between the two treatments in terms of symptomatic improvement of peripheral arterial disease, patency, or reinterventions.
- The available evidence does not support a clear advantage for either PTA with selective stenting or primary stenting as a treatment for stenotic and occlusive lesions in the iliac artery. Further research is needed to determine the most effective treatment approach.

Resources

Resources to refer to when writing a meta-analysis:

- Cochrane Collaboration: Provides guidelines and access to published reviews. Additionally, provides educational training resources.
- PRISMA: Guidelines for reporting systematic reviews and meta-analyses.
- Review Manager (RevMan): Software tool for conducting and reporting reviews and meta-analyses.

References

1. Moher D, et al. Preferred reporting items for systematic review and meta-analysis protocols (PRISMA-P) 2015 statement. *Syst Rev.* 2015;4:1.
2. Harrison H, et al. Software tools to support title and abstract screening for systematic reviews in healthcare: an evaluation. *BMC Med Res Methodol.* 2020;20(1):7.
3. Booth A, et al. The nuts and bolts of PROSPERO: an international prospective register of systematic reviews. *Syst Rev.* 2012;1:2.
4. Sterne JAC, et al. ROBINS-I: a tool for assessing risk of bias in non-randomised studies of interventions. *BMJ.* 2016;355, i4919. https://doi.org/10.1136/bmj.i4919.
5. Sterne JAC, et al. RoB 2: a revised tool for assessing risk of bias in randomised trials. *BMJ.* 2019;366, l4898. https://doi.org/10.1136/bmj.l4898.
6. Guyatt G, et al. GRADE guidelines: 1. Introduction-GRADE evidence profiles and summary of findings tables. *J Clin Epidemiol,* 2011;64(4):383−394.
7. Lasserson TJ, Thomas J, Higgins JP. Starting a review. In: *Cochrane Handbook for Systematic Reviews of Interventions*. 2019:1−12.

Chapter 43

Cost-effectiveness analysis: Design, measures, classic example

Babak Mohit[1] and Nariman Nezami[2,3]

[1]Sleep Disorders Center, Division of Pulmonary and Critical Care Medicine, Department of Medicine, University of Maryland School of Medicine, Baltimore, MD, United States; [2]Division of Vascular and Interventional Radiology, Department of Diagnostic Radiology and Nuclear Medicine, University of Maryland School of Medicine, Baltimore, MD, United States; [3]Experimental Therapeutics Program, University of Maryland Marlene and Stewart Greenebaum Comprehensive Cancer Center, Baltimore, MD, United States

Introduction

Cost-effectiveness is rooted in economics, which is the study of the value of limited resources and the allocation of resources. The origins of cost-effectiveness analysis (CEA) in healthcare can be traced to the first formal textbook published on the matter in 1996[1] (the updated textbook, from the Second Panel, was published in 2016[2,3]). However, the closely related Medical Subject Heading term "cost-benefit" was registered in Medline 20 years earlier.[4] From a health economics perspective, similar to radiology where the beauty of an image is in the eye of the beholder, the value of any health intervention is in the eye of the stakeholder.

In medicine, common stakeholders include patients, providers, payers, employers, health systems leaders, and others.[5] Each stakeholder values radiology, as a branch of medicine, for a different reason. For example, patients value radiology as a vehicle for reduced pain and suffering, enhanced quality of life, greater recreational engagement, and time with family and loved ones.[6] Payers value reduced healthcare spending.[3] Employers value increased workplace productivity.[3] Health systems value increased referrals, revenues, and retention.[7] Table 43.1 presents the perspectives of common stakeholders in radiology.

Definition of CEA

CEA is the study of methods to maximize the health outcomes of quality and quantity of life in a particular society that is constrained by a particular budget.[8] Researchers report the results of a cost-effectiveness or cost-utility analysis either as a cost-effectiveness ratio (CER) or, more frequently, as an incremental cost-effectiveness ratio (ICER). A CER is used to report the results of a single intervention and its unit is "dollars per quality-adjusted life year" (USD/QALY).[3,8] More frequently, the literature reports the results of cost-utility analyses as an ICER, where the costs and health outcome effects of one intervention are incrementally compared to the costs and health outcome effects of a competing intervention and reported with a USD/QALY unit[3,8]:

$$ICER = \frac{\text{Cost of Intervention 2} - \text{Cost of Intervention 1}}{\text{Effect of Intervention 2} - \text{Effect of Intervention 1}} = \frac{C_2 - C_2}{E_2 - E_1}(USD\,/\,QALY)$$

Table 43.2 outlines the various types of cost studies.

Indications of CEA

The most important guide toward a CEA is when well-defined competing alternatives exist. As an example, breast cancer can be diagnosed through self or medical examination, mammography, CT-Scan, or even an MRI. Furthermore, if we want to recommend one intervention (e.g., sonography for detecting breast cancer), should it be every year or 6 months? Each of these alternatives may lead to a different etiological/diagnostic pathway, where each has its own costs, and own effects and outcomes. Therefore, each of these alternatives could be a trigger for a CEA.

Goals of CEA

The objective of a CEA is to quantifiably identify the diagnosis or treatment alternative that either saves more

Translational Interventional Radiology. https://doi.org/10.1016/B978-0-12-823026-8.00005-5

TABLE 43.1 Important stakeholders of healthcare.

Stakeholder	Stake
Person/patient	Personal hygiene; reduced out of pocket costs; better quality of life; increased lifespan
Employer	More productivity; reduced insurance costs
Healthcare system	Increased referrals; increased efficiency through economies of scale; dominant market position
Government	Healthier population; satisfied population; increased tax revenue; lower healthcare costs
Societal	Healthier population; satisfied population; increased societal productivity; lower healthcare costs

quality-adjust life years (QALYs) for the same amount of money, or costs less money to save the same QALY. In technical terms, when interpreting an ICER if an intervention costs less than its comparator and produces more QALYs than its comparator, it has a higher value and is dominant to its comparator, suggesting that policymakers should adopt the intervention. If an intervention costs more than its comparator and produces less QALYs than its comparator, it has less value and is dominated by its comparator, and policymakers should not implement the intervention.[3,8] More frequently, however, interventions either cost less and produce less QALYs, or cost more and produce more QALYs. In these cases, the implementation of the intervention will depend on the stakeholder's willingness to pay.[3,8–10]

Required measures for CEA

The components of a CEA are implied in the name. To better understand them, they need to be broken down into their components. The cost elements of the economic burden of illness[11] and a CEA can be separated as direct and indirect costs.[3,8] Direct costs include costs directly associated with medical care for a given disease, such as medical visits, hospitalizations, diagnostic procedures, treatments, and other goods and services.[3] Indirect costs, on the other hand, include additional, nondisease-specific outcomes such as the cost of patient time, the cost of family (or nonfamily) caretakers, overall healthcare utilization (e.g., for comorbid diseases), accident risk, and workplace productivity, to name several.[3] Notably, each of these indirect costs is measurable in monetary terms and is borne primarily by one stakeholder (see Table 43.1—e.g., costs associated with workplace productivity are borne by employers).

The morbidity and mortality burden of disease are measured in QALYs. The mortality burden is measured in terms of life-years that the disease cuts short from the life expectancy of a population. The morbidity burden is measured as the product of the Health Related Quality of Life (HRQL) utility score of a health state, and the time (in years) the person spends in that state.[3,8,12] Therefore, if a person spends 1 year in the hypothetical state of perfect health, that person has lived one QALY. If a person spends 6 months (0.5 years) in perfect health, and 6 months in a diseased state with an HRQL utility score of 0.4 that person has lived $(0.5 \times 1) + (0.5 \times 0.4) = 0.7$ QALY, and has lost $(1-0.7) = 0.3$ QALYs due to disease. Also, if one patient loses one QALY from a bone fracture and one other patient loses one QALY from Cancer, from a health economics perspective the two losses are equivalent. Similarly, if one health intervention (e.g., a cast for bone fracture) can help one patient gain one QALY, and another intervention (e.g., chemotherapy for Cancer) can help another patient gain one QALY, the value of these two interventions is identical from a health economics perspective.[12]

The unit for morbidity, HRQL, itself is a term that indicates the breadth and diversity of characteristics of life that other traditional measures of "health" do not capture.[13,14] Notably, each component of HRQL is subjective and rated by the patient. In terms of assessment, HRQL is typically measured[15] in both general[16] and disease-specific domains.[14] For example, one of the most common general HRQL instruments used is the five-dimensional Euro Quality of Life (EQ-5D) developed by the Euro-QOL group.[17] The related EQ-5D-5L questionnaire has questions on the five dimensions of mobility, self care, usual activities, pain/discomfort, and anxiety/depression. Each of these dimensions has five levels, which include no problems, slight problems, moderate problems, severe problems, and extreme problems. Another common instrument to measure generic health utility scores is the 36-item Medical Outcomes Study Short-Form 36 (SF-36), which assesses HRQL in eight domains.[18] Subscale scores range from 0 to 100, based on community norms. Higher scores indicate better HRQL.

A standardized HRQL is often expressed as a utility score, which measures the utility preference weight of time and quality of life on a scale from zero to one.[3,8] An HRQL utility score of one represents the hypothetical "health state" of perfect health.[3,8] A utility score of zero corresponds with a "health state" equivalent in preference to being dead. HRQL utility scores between zero and one indicate greater or lesser morbidity, with weights closer to one corresponding to less severe, more preferred "health states," and weights closer to 0 corresponding to more severe, less preferred "health states."[3,8] The HRQL measure produces standardized utility scores that are both comparable across "health states," and can also be combined to estimate quality adjusted survival.[3,8]

Conduct of a CEA study

There are nine steps to conducting a successful cost-effectiveness research project (adopted from[8]). While reading these steps, it would help to follow them through an example of a cost-effectiveness study. Here the example will be an analysis of bedside placement of peripherally inserted central catheters.[19].

1. Thinking through and adopting a research question. The facts that need to become clear before drafting a question are:
   - **a.** What is the intervention? Does it have multiple levels? Multiple doses? The intervention in the reference paper is the placement of peripherally inserted central catheters (PICCs). In strategy A, all PICC placements are initially attempted at the patient's bedside, with failures referred to the interventional radiology service.
   - **b.** What is the comparator? In the reference paper in strategy B, all PICCs were placed by interventional radiologists under fluoroscopic and/or venographic guidance.[19]
   - **c.** What is the outcome and how is it measured? If the outcome is not in USD/QALY the term cost-effectiveness should not be used and other names introduced in Table 2 should be referenced. Since in the sample study, the outcome is the percentage of successful catheter placements, it should be considered a cost-consequence study.[19]
   - **d.** Who is the population? The population of the study is all inpatients in need of a catheter.[19]
   - **e.** Where is the study setting? The example study is performed at a hospital, Beth Israel Deaconess Medical Center in Boston, MA. USA.[19]
   - **f.** Who is the main stakeholder? What is the study perspective? Common perspectives are described in Table 43.1 of this chapter. In the sample study, the main stakeholders are inpatients and the health system. As Table 1 of the study indicates,[19] since there are no indirect costs, but all reasonable direct costs have been considered, the study is from the perspective of the health system.
   - **g.** How long is the analytical horizon? Why if it is different from the life expectancy of the cohort? In the example study, as the long-term consequences of the catheter placement are limited, the duration of the study is the analytical horizon, which is 3 months—from December 1996 to March 1997.[19]
2. Sketching out the map of the analysis in the form of a decision tree of various etiological, diagnostic, and treatment pathways on paper. The etiological pathways considered are depicted in Figure 1 of the reference paper.[19]
3. Collecting data on costs (direct and indirect), morbidity and mortality (HRQL life years, and QALY), and the probability of following varied etiological pathways. The footnote of Table 1 of the sample study hints toward the sources of the data.[19]
4. Adjusting the data. Table 1 of the sample study displays the data of the study after adjustment.[19]
5. Building the model in specialized software. Figure 1 of the sample study depicts the model as it is displayed in specialized software.[19] Complete model building also involves implementing the data on costs, effects, and probabilities of the pathways in the software.
6. Testing and running. Table 3 of the example study display the results in the form of the cost-consequence ratio of the intervention as a function of the costs of the comparator.[19]
7. Conducting sensitivity analysis to account for all the variabilities and interactions of the data. While the sample study does provide minimal sensitivity analysis based on successful catheter placement, an important factor of postcatheter placement infections is not analyzed.[19]
8. Writing and publishing findings. The Consolidated Health Economic Evaluation Reporting Standards is a checklist most often used by peer-reviewed journals for evaluating a cost-effectiveness study manuscript.[20]
9. Advocating evidence-based policies based on research findings. A list of advocacy methods is suggested in Table 43.3 of this chapter.

Final word

In summary, cost-effectiveness analysis is increasingly becoming a part of the evidence-based dossier of health interventions. As with all research, it is a specialized and nuanced field, offering various pitfalls for novice adopters of the practice. We, therefore, recommend appropriate training, familiarity, and collaboration with experts, before independent work. Furthermore, policy adoption is a

TABLE 43.2 Types of health economic analysis.

Name of study	Equation	Unit of reported results
Comparative effectiveness analysis	E_2-E_1	Standardized unit or QALY
Cost studies (also known as budget impact analysis)	C_2-C_1	Dollars (monetary unit)
Cost minimization studies	C_2-C_1	Dollars saved
Cost-effectiveness analysis[a]	$(C_2-C_1)/(E_2-E_1)$	Dollars per standardized unit
Cost-utility analysis[a]	$(C_2-C_1)/(U_2-U_1)$	Dollars per QALY
Cost-consequence analysis	$(C_2-C_1)/(Con_2-Con_1)$	Dollars per nonstandardized unit
Cost-benefit analysis (similar to return on iinvestment studies but also includes in-direct costs)	C_2/C_1	Dollars spent per dollars earned

In the equations presented, the effect is abbreviated as E, the cost is abbreviated as C, the utility is abbreviated as U, and the consequence is abbreviated as Con.

[a] *Although Cost-Effectiveness Analysis and Cost-Utility Analysis are used interchangeably in the healthcare literature, if the denominator is in QALYs, the technically more accurate name of the analysis is a cost-utility analysis. However, in this chapter the more frequently used name—cost-effectiveness analysis is used.*

TABLE 43.3 Guidelines and directions for future cost-effectiveness research.

Domain	Recommendation
Include health economic end points	Include measures of direct and indirect costs of target disorder in all future trials and studies of that disorder
Advance research on health utility scores for the disorder	Availability of HRQL utility score measures for the target disorder included in the international classification of disease (ICD)
Evaluate cost-effectiveness	Include measures of both general and disease-specific measures of health-related quality of life in all future disorder trials
Study-specific populations	Conduct health economic analyses among women; different ethnic groups; and patients with varying severities of the target disorder
Understand comorbid disorders	Study economic impact of the target disorder and its treatments in costly and chronic comorbid disease states such as heart failure, stroke, type 2 diabetes mellitus, Alzheimer's disease, and depression
Increase adherence	Study economic cost-benefit of interventions designed to increase treatment adherence, including telehealth and remote monitoring, automated approaches, and other interventions
Adopt employer perspective	Evaluate cost-benefit of target treatments from the perspective of the employers of adults: Impact on lost workplace productivity (i.e., absenteeism) and workplace accident and injury risk
Evaluate global impact	Evaluate cost-effectiveness of treating the target disorder worldwide in various health-care delivery systems
Compare economic effectiveness	Compare economic effectiveness between the treatments of the target disorder to empower stakeholders to make evidence-based decisions regarding allocation of scarce healthcare resources

complex field, in which cost-effectiveness presents only one aspect of the evidence. Therefore, the strategies presented in this chapter do not suffice as guidelines for policy change but are to be considered as a provoker of curiosity to the clinical practitioners, in search of improving the field of practice toward the better use of evidence.

References

1. Gold MR, Siegel JE, Russell LB, Weinstein MC. *Cost-Effectiveness in Health and Medicine*. New York, NY: Oxford University Press; 1996.
2. Neumann PJ, Sanders GD. Cost-effectiveness analysis 2.0. *N Engl J Med*. January 19, 2017;376(3):203–205. PubMed PMID: 28099837. Epub 2017/01/19. eng.
3. Neumann PJ, Sanders GD, Russell LB, Siegel JE, Ganiats TG. *Cost-Effectiveness in Health and Medicine*. 2nd ed. New York, NY: Oxford University Press; 2017.
4. National Library of Medicine (NLM). *Medical Subject Heading (MESH)—Cost Benefit Analysis*; 1976. Available from: https://www.ncbi.nlm.nih.gov/mesh/68003362.
5. Claxton K, Walker S, Palmer S, Sculpher M. *Appropriate Perspectives for Healthcare Decisions. CHE Research Paper 54*. York, UK: Centre for Health Economics: University of York; 2010.
6. Flowers CR, Garber AM, Bergen MR, Lenert LA. Willingness-to-pay utility assessment: feasibility of use in normative patient decision support systems. In: *Proceedings: A Conference of the American Medical Informatics Association AMIA Fall Symposium*. 1997:223–227. PubMed PMID: 9357621. PMCID: PMC2233361. Epub 1997/01/01. eng.
7. Neumann PJ, Jacobson PD, Palmer JA. Measuring the value of public health systems: the disconnect between health economists and public health practitioners. *Am J Publ Health*. December 2008;98(12):2173–2180. PubMed PMID: 18923123. PMCID: PMC2636521. Epub 2008/10/17. eng.
8. Muennig P, Bounthavong M. *Cost-Effectiveness Analysis in Health: A Practical Approach*. 3rd ed. San Francisco, CA: Jossey-Bass (Wiley); 2016.
9. Neumann PJ, Cohen JT, Weinstein MC. Updating cost-effectiveness–the curious resilience of the $50,000-per-QALY threshold. *N Engl J Med*. August 28, 2014;371(9):796–797. PubMed PMID: 25162885. Epub 2014/08/28. eng.
10. Tengs TO, Adams ME, Pliskin JS, et al. Five-hundred life-saving interventions and their cost-effectiveness. *Risk Anal*. June 1995;15(3):369–390. PubMed PMID: 7604170. Epub 1995/06/01. eng.
11. Lopez AD, Mathers CD, Ezzati M, Jamison DT, Murray CJL. *Global Burden of Disease and Risk Factors*. New York, NY: The World Bank & Oxford University Press; 2006.
12. Weinstein MC, Torrance G, McGuire A. QALYs: the basics. *Value Health*. March 2009;12(Suppl 1):S5–S9. PubMed PMID: 19250132.
13. Chapman RH, Berger M, Weinstein MC, Weeks JC, Goldie S, Neumann PJ. When does quality-adjusting life-years matter in cost-effectiveness analysis? *Health Econ*. May 2004;13(5):429–436. PubMed PMID: 15127423. Epub 2004/05/06. eng.
14. Patrick DL, Erickson P. *Health Status and Health Policy: Quality of Life in Health Care Evaluation and Resource Allocation*. New York, NY: Oxford University Press; 1993.
15. Guyatt GH, Feeny DH, Patrick DL. Measuring health-related quality of life. *Ann Intern Med*. 1993;118(8):622–629.
16. EuroQol Group. EuroQol—a new facility for the measurement of health related quality of life. *Health Pol*. 1990;16(3):199–208.
17. Herdman M, Gudex C, Lloyd A, et al. Development and preliminary testing of the new five-level version of EQ-5D (EQ-5D-5L). *Qual Life Res*. December 2011;20(10):1727–1736. PubMed PMID: 21479777. PMCID: PMC3220807.
18. Nichol MB, Sengupta N, Globe DR. Evaluating quality-adjusted life years: estimation of the health utility index (HUI2) from the SF-36. *Med Decis Making*. 2001;21(2):105–112.
19. Neuman ML, Murphy BD, Rosen MP. Bedside placement of peripherally inserted central catheters: a cost-effectiveness analysis. *Radiology*. February 1998;206(2):423–428. PubMed PMID: 9457195.
20. Husereau D, Drummond M, Petrou S, et al. Consolidated health economic evaluation reporting Standards (CHEERS) statement. *Value Health*. Mar–Apr 2013;16(2):e1–5. PubMed PMID: 23538200.

Chapter 44

Diagnostic test evaluation

Yechiel Mor
Apogee Physicians, Scottsdale, Arizona, United States

The objective of a diagnostic test is to provide additional clinical information to physicians to more effectively make a diagnosis, select an appropriate treatment, or monitor the efficacy of a specific treatment approach. The application of diagnostic tests in clinical practice is intended to facilitate and improve the collection of clinical information for the purpose of making a clinical decision. Clinicians today have a plethora of diagnostic modalities available to them including X-rays, CT scans, MRI scans, PET scans, ultrasounds, biopsies, pregnancy tests, endoscopies, EKGs, and EEGs among others. Although the majority of these tests are very informative, none of these tests are perfect. However, in today's clinical practice, the information generated by these tests is a critical component of the diagnostic and therapeutic decision-making process.

Before a new medical test can be introduced in clinical practice it needs to be assessed for diagnostic accuracy, which is defined as the ability to detect the presence or absence of a disease.

In this chapter, we will review measures used to evaluate the efficacy of a diagnostic test and examine an example. The classic parameters to determine the potential clinical value of a test include its sensitivity, specificity, positive predictive value, negative predictive value, likelihood ratio, and receiver operating characteristic (ROC) curve. All these parameters provide us with a measure of the accuracy of a test. Once this is established, one can evaluate the test's validity on clinical decisions, its impact on patient outcomes, and ultimately its utility in improving health care.

Sensitivity and specificity

The sensitivity of a clinical test indicates its ability to correctly identify individuals with the disease in question.[1] A test with 100% sensitivity correctly identifies all patients with the disease. As the sensitivity increases though it is prone to an increase in false-positive results, by incorrectly labeling people as having the disease when, in fact, they do not actually have it. The specificity of a test describes its ability to correctly identify people without the disease in question. A test with high specificity is unlikely to incorrectly label an individual as having a disease when in fact they do not have the disease. As the specificity decreases, the number of false-negative results increases by failing to identify the disease in individuals who actually have it.[2] A test with high sensitivity rules out a disease if negative and a test with high specificity rules in a disease when positive. Sensitivity and specificity are commonly believed not to vary with disease prevalence in the population of interest.[3] To calculate the sensitivity and specificity, it is important to take into consideration the following terms (Fig. 44.1):

- True positive (TP): the individual has the disease and the test is positive.
- False positive (FP): the individual does not have the disease but the test is positive.
- True negative (TN): the individual does not have the disease and the test is negative.
- False negative (FN): the individual has the disease but the test is negative.

With the above definitions,

$$\text{Sensitivity} = \text{TP}/(\text{TP} + \text{FN})$$

$$\text{Specificity} = \text{TN}/(\text{TN} + \text{FP})$$

		Gold Standard Test	
		Positive	Negative
Diagnostic Test being evaluated	Positive	TP	FP
	Negative	FN	TN

Sensitivity: TP/(TP+FN) Positive Predictive Value: TP/(TP+FP)
Specificity: TN/(TN+FP Negative Predictive Value: TN/(TN+FN)

FIGURE 44.1 Assessment of a new diagnostic test using sensitivity, specificity, PPV, and NPV.

Translational Interventional Radiology. https://doi.org/10.1016/B978-0-12-823026-8.00054-7

Positive and negative predictive value

Sensitivity and specificity describe a diagnostic test attributes relative to a gold standard. The positive and negative predictive values are depended on the prevalence of the disease in the population of interest.[4] The positive predictive value (PPV) is an attempt to establish whether people who tested positive on the diagnostic test do in fact have the disease of interest.

$$PPV = TP/(TP + FP)$$

A high PPV indicates that those people who tested positive are highly likely of truly having the disease. The negative predictive value (NPV) is an attempt to establish whether people who tested negative on the diagnostic test for the disease are truly disease free.

$$NPV = TN/(TN + FN)$$

A high NPV makes the disease quite unlikely in a subject with a negative test. PPVs and NPVs focus on the practical usefulness of a diagnostic test in clinical practice. A diagnostic test with high sensitivity and specificity is desired; however, the use of PPV and NPV are most appropriate when making clinical decisions about individual patients. PPV and NPV indicate the effectiveness of a test for categorizing people as having or not having a disease. They depend on the classification accuracy of the test as well as on the prevalence of the disease.[5]

Likelihood ratios

Another parameter useful to clinicians to evaluate diagnostic tests and that can help select one test over another is the use of positive likelihood ratios. A positive likelihood ratio is defined as the percent of patients who have a positive test result among those with the target disease divided by the percentage of patients who have a positive test result among those without the disease.[5] The numerator corresponds to the sensitivity and the denominator to 1 minus the specificity (1 − specificity). Because the likelihood ratio is calculated from sensitivity and specificity, it is stable with changes in the prevalence of the disease. A likelihood ratio of greater than 1 indicates the test result is associated with the disease. A likelihood ratio less than 1 indicates that the result is associated with the absence of the disease. Simply put, values greater than 1 increase the probability of the disease and the test is good at ruling in a disease.

Receiver operating characterisic curve

A receiver operating characteristic (ROC) curve is a way to compare diagnostic tests.[6] It is a plot of the true-positive rate against the false-positive rate and interestingly has its origins in World War II. ROC curves were developed by the British as a way to analyze radar data and differentiate signals from enemy aircraft from signal noise from a flock of geese for instance.[7] As the sensitivity of the receiver increased so did the number of false positives (1-specificity). A ROC curve shows this inverse relationship (tradeoff) between sensitivity (y-axis) and false positives (x-axis). The previously discussed measures of accuracy (sensitivity, specificity) were based on one specific cut-off point. The ROC curve plots sensitivity and specificity and tries out all of the cut-off points thereby allowing for comparison at different thresholds. The closer the curve is to the top-left corner the more accurate the test. A perfect test would go vertically from zero to the top-left corner and then go straight across the horizontal (Fig. 44.3, line A).[6] The point nearest to the left corner on a ROC curve is called the Youden Index and is one way of describing an optimal threshold by giving both sensitivity and specificity equal weight. An ROC curve lying on the diagonal line (Fig. 44.3, line C) reflects the performance of a diagnostic test that is no better than chance (e.g., from flipping a coin) and which yields a result unrelated to the true disease status.

The area under the ROC curve (AUC) is an overall summary of diagnostic accuracy.[5] A test with an AUC of 0.5 corresponds with a ROC curve that is purely by chance and has no discriminatory ability in diagnosing the absence or presence of a disease. A test with an AUC of 1.0 corresponds to perfect accuracy.[1] In general a test with a higher AUC has a better probability of providing better information with which to make a clinical diagnosis or decision.

Other commonly used metrics

In an era of evidence-based medicine, decision makers need high-quality data to support decisions about whether or not to use a diagnostic test in a specific clinical situation and thus warrant a discussion of a few more commonly used metrics. When making comparisons between tests, one can also use the diagnostic odds ratio, which is the ratio of the odds of positivity in disease relative to the odds of positivity in the nondiseased. Alternatively, it can be read as the ratio of the odds of disease in test positives relative to the odds of disease in test negatives.[8] With the earlier definitions,

$$DOR = (TN / FP)/(FN / TN)$$

Classification accuracy is the probability that an individual will be correctly classified by a test, that is, the number of correct predictions divided by the total number of individuals tested.

With the earlier definitions,

$$Accuracy = (TP / TN)/(TP + FP + TN + FN)$$

		Gold Standard RT-PCR		
		Positive (601)	Negative (413)	
CT chest	Positive (888)	580	308	PPV: 580/(580+308)= 65%
	Negative (126)	21	105	NPV: 105/(105+21)= 83%
		Sensitivity: 580/(580+21)= 96.5%	Specificity: 105/(308+105)= 74.6%	

FIGURE 44.2 Classical example.[10]

Another popular metric is the F1 score that rates the performance of a test or its accuracy based on two primary attributes: precision and recall. Recall being the ability to identify all the individuals with an attribute or its sensitivity and precision being the proportion of positive results that are truly positives also known as the PPV. The F1 score takes both these numbers and assigns a single number, easy for interpretation. A low F1 score is an indication of poor precision and recall. The higher the recall and precision, the higher the F-score.

Summary

Ultimately, we want to demonstrate that a diagnostic test has an impact on patient outcomes. By using the earlier criteria to assess diagnostic test performance, followed by randomized trials to measure clinical utility, we can choose a diagnostic test that will best help to solve a diagnostic problem. Proficiency in the earlier parameters will help clinicians evaluate the accuracy of a diagnostic test following which the effect on patient outcomes can be examined.

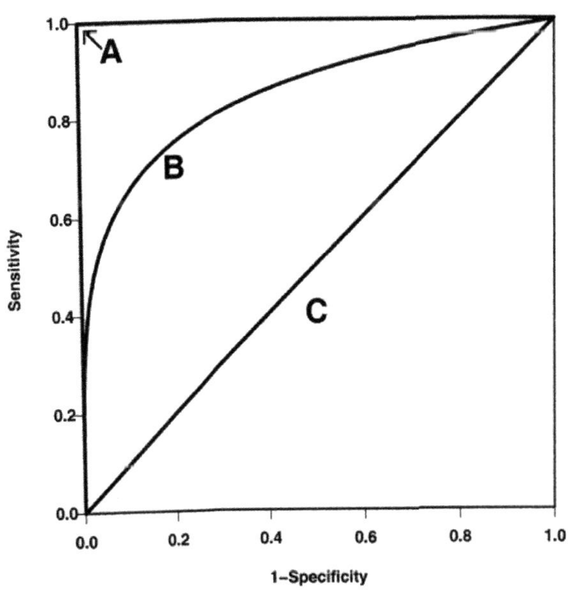

FIGURE 44.3 ROC curve.[6]

Classical example

As a classical example, we will consider the novel SARS-COV2 coronavirus responsible for COVID-19. SARS-COV2 was first identified in Wuhan, China in December 2019, which then went on to ravage through more than 188 countries.[9] On March 11, 2020, the World Health Organization characterized the spread of COVID-19 as a pandemic. Other notable pandemics include HIV and the 1918 Influenza pandemic. In the Journal *Radiology*, Liming Xia et al. evaluated chest CT as a diagnostic test for the diagnosis of COVID-19 infection.[10] From January 6 to February 6, 2020, 1014 patients in Wuhan, China who underwent both chest CT and RT-PCR tests were included. With RT-PCR as a reference standard test for the presence of SARS-COV2, the performance of chest CT in diagnosing COVID-19 was assessed. Of 1014 patients, 601 had positive RT-PCR results and 888 had positive chest CT scans (Fig. 44.2). The likelihood ratio is (580/601)/(308/413) = 1.29. The evaluation, as shown in Fig. 44.2, concludes that chest CT has a high sensitivity for diagnosing COVID-19 and consequently can be considered as a good test for COVID-19 screening. However, the likelihood ratio of 1.29, which is greater than 1 but only slightly so, suggests a small increase in the likelihood of the disease being present after the test.

References

1. Ransohoff D. Challenges and opportunities in evaluating diagnostic tests. *J Clin Epidemiol.* 2002;55:1178–1182.
2. Steurer Johann, Fischer Joachim, Bachmann Lucas, Koller Michael, Riet Gerben. Communicating accuracy of tests to general practitioners: a controlled study. *Br Med J.* 2002;324.
3. Kim Kyongjin, Visintin Irene, Alvero Ayesha, Mor Gil. Development and validation of a protein-based signature for the detection of ovarian cancer. *Clin Lab Med.* 2009;29:47–55.

4. Lord Sarah, Irwin Les, Simes John. When is measuring sensitivity and specificity sufficient to evaluate a diagnostic test, and when do we need randomized trials? *Ann Intern Med.* 2006;144:850–855.
5. Fardy J, Barrett B. Evaluation of diagnostic tests. Clinical epidemiology: practice and methods. *Methods Mol Biol.* 2015;1288:289–300.
6. Zou Kelly, O'Malley James, Mauri Laura. Receiver-operating characteristic analysis for evaluating diagnostic tests and predictive models. *Circulation.* 2007;115:654–657.
7. Lusted Lee. Signal detectability and medical decision-making. *Science.* March 26, 1971;171(3977):1217–1219.
8. Glas Afina, Lijmer Jeroen, Prins Martin, Bonsel Gouke, Bossuyt Patrick. The diagnostic odds ratio: a single indicator of test performance. *J Clin Epidemiol.* 2003;56:1129–1135.
9. COVID-19 Dashboard by the Center for Systems Science and Engineering (CSSE) at Johns Hopkins University (JHU). ArcGIS. Johns Hopkins University. Retrieved May 2020.
10. Tao Ai, Zhenlu Yang, Hongyan Hou, et al. Correlation of chest CT and RT-PCR testing in coronavirus disease 2019 (COVID-19) in China: a report of 1014 cases. *Radiology.* February 26, 2020; 296(2).

Chapter 45

Reliability study: Design, measures, and classic example

Arun Chockalingam[1] and Nariman Nezami[2,3]
[1]Department of Radiology, Mount Auburn Hospital/Harvard Medical School, Cambridge, MA, United States; [2]Division of Vascular and Interventional Radiology, Department of Diagnostic Radiology and Nuclear Medicine, University of Maryland School of Medicine, Baltimore, MD, United States; [3]Experimental Therapeutics Program, University of Maryland Marlene and Stewart Greenebaum Comprehensive Cancer Center, Baltimore, MD, United States

Introduction

Reliability of a particular test or instrument refers to the consistency of a measure, in that every time the test is completed the user should receive approximately the same response. This refers to the relationship between the error of measurement and the expected distribution of measurements over time and across circumstances. However, reliability is not equivalent to an agreement. For example, a new test that always gives a result of 100% regardless of patient or situation would have a perfect agreement but zero reliability, meaning that there is no new information provided to the clinician. Statistically speaking, reliability is the ratio of between-subject variation to the total variation (true differences between subjects divided by that plus measurement error).

$$\text{Reliability} = \frac{\text{"Between subject" variation}}{\text{Total variation}} = \frac{\text{True subject variation}}{\text{True subject variation} + \text{Measurement error}}$$

Reliability ranges from 0 (variation is due to error) to 1 (variation is due to differences between subjects). If the error of measurement is as large as the expected difference between measurements (or subjects) then the instrument would be useless. For example, if height among a group of people is measured and the expected range of measurements is 50–60 cm, errors of 4–5 cm might be acceptable to differentiate height between the subjects into "short," "average," or "tall" height. However, if the height being measured was of drinking glasses and the expected range of measurement is 7–10 cm, errors of 4–5 cm would be too high to adequately differentiate.

Measuring reliability falls into a set of techniques called clustered sampling, which is typically used in many types of sciences to investigate relationships between individual and group influences. Performance on a test between individuals within the same group, or clusters (i.e., employees in one organization, radiology residents in one classroom, or patients in one clinic), will most likely be similar than in individuals in different groups. Out of the many methods to describe reliability, the intraclass correlation coefficient (ICC) (form of equivalence) has been used extensively since as early as 1938 introducing the ICC index.[1] By 2013, many have detailed additional estimators of intraclass correlation coefficients with the best known still being ICC.[1] In practice, this concept is extremely important in evaluating the usefulness of a tool among different populations, with many considerations for the clinician-scientist.

Types of reliability measures

Reliability is assessed using three different criteria: homogeneity, stability, and equivalence. Homogeneity is also known as internal consistency, which is defined as the consistency of responses across the items on a multiple-item measure. Stability is also known as test–retest reliability, which is assessed when an instrument is given to the same set of participants more than once under similar circumstances and a statistical comparison is made between test scores during completion. Equivalence is generally measured using interrater reliability, involving a process to determine a level of agreement between two or more observers. These types of reliability measures will be discussed in further detail.

Translational Interventional Radiology. https://doi.org/10.1016/B978-0-12-823026-8.00060-2

Test–retest reliability

When researchers measure the reliability of a test that they assume to be consistent across time, the scores obtained should also be consistent across time, and this concept is validated by the use of test–retest reliability. Evaluating test–retest reliability requires using an instrument with a group of people once, and then again with the same group to assess test–retest correlation between the two sets of results. For an instrument to demonstrate stability, there needs to be a high correlation (>0.5) between the scores each time a participant completes the test.

For example, a person's intelligence is assumed to be consistent across time, so someone with high intelligence today will be highly intelligent next week. This means that a good test for intelligence should consistently produce the same scores for this individual next week as it would today. A measure that produces inconsistent scores over time cannot be a very good measure of a construct that is supposed to be consistent.

Test–retest reliability is useful in experiment designs that use a no-treatment control group. However, with this approach, there is no information about reliability until "after experiment" data are collected so this type of testing becomes useless if reliability is estimated to be low.

Parallel forms reliability

If a researcher is trying to validate the use of different tests to measure the same attribute, parallel forms of reliability can be used to measure the correlation between the tests. For example, in the assessment of tumor response to intraarterial therapies, if a researcher wants to create two different test sets to ensure the response is reflected persistently, they can use parallel forms of reliability to ensure that the tumor would show similar results across tests. The creation of such tests could prove to be difficult, as the creator would have to ensure that all markers evaluate the same tumor response or theories.

Interrater reliability

Interrater reliability measures the extent to which two or more "raters" obtain similar scores when evaluating the same subject. This type of reliability is the broadest and most clinically useful measure of reliability, especially when the error between evaluators is relevant. Interrater reliability can be assessed using Cronbach's α for quantitative data or Cohen's κ for categorical data. Interrater reliabilities tend to measure lower than other tests because this type of analysis compares across different evaluators and during possibly different time points, which is important to consider when deciding to use equivalence to measure reliability.

An example of interrater reliability would be in a dancing competition, where there is a panel of judges evaluating each contestant. The best way to assess the reliability of scoring across judges at one particular event would be with interrater reliability, measuring consistency in their scores across contestants.

An interesting caveat to this measure is that intrareader reliability would usually be equal or higher than interreader reliability, so if there is an acceptable level of interobserver reliability, no further reliability testing would be necessary. However, if there is poor interobserver reliability, knowledge of test–retest or intrareader reliability might help identify sources of error and modify accordingly. In the case of only one individual applying a test (i.e., self-reported quality of life questionnaires), test–retest reliability should be used instead.

Internal consistency reliability

Internal consistency reliability is defined as the consistency of people's responses across the items on a multiple-item measure and is typically assessed using split-half reliability, Kuder-Richardson coefficient and Cronbach's α. In split-half reliability, the results of a test are divided in half and correlations are calculated by comparing the halves. Strong correlations indicate high reliability, while weak correlations indicate low reliability. After items are split into two sets, a score is computed for each set of items and the relationship between the two sets of scores is examined, with a split-half correlation of 0.80 or greater indicating good internal consistency.

The Kuder–Richardson test is a more complex version of the split-half test, in that the average of all possible split-half combinations is determined and a correlation between 0 and -1 is generated. This test is more accurate than the split-half test but can only be completed on questions with a binary response. Cronbach's α is the most commonly used test to measure internal consistency. In this test, the average of all correlations in every combination of split-halves is determined. This measurement is more robust than Kuder–Richardson because nonbinary responses (more than two responses) can also be used in this test. The Cronbach's α result is a number between 0 and 1, with 0.7 and greater indicating an acceptable reliability score.

An example of internal consistency could be assessing a group of respondents presented with a set of statements designed to measure optimism or pessimism. Participants would have to rate the degree of agreement to each statement from 1 to 5. If this survey has high internal consistency, an optimistic participant would give generally higher ratings to the "optimistic" statements and lower ratings to the "pessimistic" statements and vice versa.

This type of reliability is important to assess the quality of each item in the survey in obtaining a certain homogenous result, technically assessing the construction of an instrument used to measure something in a group of subjects.

Designing a reliability study: get started

Keeping in mind what reliability means, and the different ways it can be measured, one must now decide how best to assess the instrument being designed. To do this, one must also design a study that assesses the most relevant reliability for the instrument. The best way to do this is by answering the series of questions:

What quality of the test or instrument is being assessed?

When designing an instrument and a study to assess the validity of the use of this instrument, the question that this instrument will be used in practice to answer is paramount. From the types of reliability discussed earlier, test–retest, intrareader, and interreader reliability all measure agreement between two or more test administrations. This is important as one should tread with caution when considering internal consistency, as this only requires a single administration of the test in question. Internal consistency as a result does not incorporate differences between settings, including times and users, indicating the weakest form of reliability.

Who will be applying the test or instrument in question?

The "raters" so to speak introduce their own bias that calls to question the validity of the instrument. To do this adequately, it is important to simulate real-life clinical environments for the users of the instrument to assess clinical reliability.

Who will be assessed using the test or instrument in question?: real-world examples

In research studies, diversity of samples is crucial, to strengthen conclusions and make observations seen more generalizable. When designing an instrument and determining the reliability of that instrument the same principle applies. For example, if there was a measurement for knee laxity and this was being piloted to a group of orthopedic surgeons, these clinicians would be more likely to agree on findings among a healthy group of volunteers. However, like discussed earlier, this would represent agreement, not reliability, as there is low between-subject variability. It is important to keep in mind that when the between-subject variability is larger relative to the error, reliability will be higher, and can be represented by testing the reliability of an instrument in a diverse subject population (i.e., one with variable pathologies in the discussed example).

Potential pitfalls

An important factor to consider in designing a reliability study is the sample of both raters and subjects. Increasing the number of either group will yield a more precise reliability estimate, reducing the size of the confidence interval, but a larger number of subjects would have a great impact than a larger number of raters. There are two approaches to establishing an adequate number of subjects. One method involves choosing a minimum acceptable reliability and selecting a sample size needed to prove a higher actual reliability. Another method involves calculating a sample size based on a desired precision of reliability (confidence interval), number of raters, and the expected intraclass correlation coefficient.

Generally, values of reliability close to 0 are agreed to represent poor measures while values close to 1 represent strong measures, and around 0.5 represent moderate measures. However, acceptable reliability depends on the clinical context as well as other instruments used in similar circumstance.

Resources

1. Price P, Jhangiani R, Chiang I. *Research Methods in Psychology*. BCOpenEducation; 2013.
2. Ghibes P, Partovi S, Grözinger G, et al. Reliability and accuracy of peri-interventional stenosis grading in peripheral artery disease using color-coded quantitative fluoroscopy: a phantom study comparing a clinical and scientific postprocessing software. *BioMed Res Int*. 2018;2018:1–11. https://doi.org/10.1155/2018/6180138.
3. Ben Hassen W, Malley C, Boulouis G, et al. Inter- and intraobserver reliability for angiographic leptomeningeal collateral flow assessment by the American Society of Interventional and Therapeutic Neuroradiology/Society of Interventional Radiology (ASITN/SIR) scale. *J Neurointerv Surg*. 2018;11(4):338–341. https://doi.org/10.1136/neurintsurg-2018-014185.
4. Rajebi M, Benenati M, Schernthaner M, et al. Reliability and accuracy of simple visual estimation in assessment of peripheral arterial stenosis. *J Vasc Intervent Radiol*. 2015;26(6):890–896. https://doi.org/10.1016/j.jvir.2015.02.018.
5. Hernaez R. Reliability and agreement studies: a guide for clinical investigators. *Gut*. 2015;64(7):1018–1027. https://doi.org/10.1136/gutjnl-2014-308619.
6. Shieh G. Optimal sample sizes for the design of reliability studies: power consideration. *Behav Res Methods*. 2013;46(3):772–785. https://doi.org/10.3758/s13428-013-0396-0.
7. Karanicolas P, Bhandari M, Kreder H, et al. Evaluating agreement: conducting a reliability study. *J Bone Jt Surg Am Vol*. 2009;91(Suppl 3):99–106. https://doi.org/10.2106/jbjs.h.01624.

Reference

1. Heale R, Twycross A. Validity and reliability in quantitative studies. *Evid Base Nurs*. 2015;18(3):66–67. https://doi.org/10.1136/eb-2015-102129.

Chapter 46

Surveys and questionnaires: Design, measures, and classic example

Maryam Boumezrag[1] **and Tao Liu**[2]
[1]*Medstar Georgetown University Hospital, Washington, DC, United States;* [2]*Department of Biostatistics, Brown University, Providence, RI, United States*

Key points

- Survey research is widely used in an increasingly broad range of disciplines including social, psychological, and clinical investigations.
- Survey research has the advantage of obtaining targeted information quickly with low cost. However, surveys require thoughtful planning to provide meaningful results.
- Successful design of survey requires the development of brief, relevant, unambiguous, specific, and objective (BRUSO) survey questionnaire (instrument).
- Once the study design process is finished, pilot testing of the instrument should be performed on a small number of responders. Quality surveys require thoughtful efforts and rigorous statistical analysis to provide valid and reliable results.

Potential pitfalls: what is survey research

Survey research is defined as an approach in which subjective data is obtained from a sample of subjects and analyzed to extrapolate information on the study population they represent.[1] Historically, this form of research was used to obtain information describing large populations, such as census data. In more recent times, survey research developed into an important asset in not only social and psychological research,[2] but also in an increasingly broad range of disciplines involving patients and other stakeholders.

Survey research has numerous advantages, including the ability to obtain limitless information in a relatively short period of time. The inherent versatility of surveys, however, can make this research method seem deceptively simple. To date, there is no consensus on a unified set of evidentiary rules on how surveys should be constructed and reported.[1] This methodological pitfall can significantly undermine the validity and reliability of study results.[3] As such, surveys require thoughtful planning to provide meaningful results. In this chapter, we aim to cover the main aspects of designing, conducting, and analyzing a survey as well as suggested minimums before manuscript submission.

The first step in study design is formulating a clear research question and identifying if a survey is the best instrument to answer it.[4] In general, survey research is best suited for questions that are (1) descriptive, (2) explanatory, or (3) explorative.[5] In clinical research, a potential use for a descriptive survey is to explore the distribution of a specific variable as opposed to why that distribution exists. For example, exploring the trends in interventional radiology procedures during the Covid-19 pandemic as opposed to understanding the factors driving those trends.

Real-world examples

Survey research can also be used for explanatory purposes, usually by investigating the relationship between two or more variables. An example would be a study investigating why certain referral patterns exist in some parts of the country and not others. Another use for survey research is in cases in which the goal is to gain general insight on a specific topic, usually with the goal of guiding future research. This can be achieved by using an exploratory survey.

Translational Interventional Radiology. https://doi.org/10.1016/B978-0-12-823026-8.00051-1

Survey design

One of the most important ingredients of high-quality survey research is the survey questionnaire (instrument). A literature review should be conducted to identify research already done in the field and to identify previously validated survey instruments that can be adapted for your purpose.[6] When possible, previously validated survey instruments should be used,[7] though keeping in mind that validated survey instruments still require piloting when being adapted to a different study.

When new questions are being developed, we recommend using the acronym created by Peterson,[8] BRUSO: brief, relevant, unambiguous, specific, and objective. Questions should be as brief as possible, both in number and in question length. The temptation to include questions not directly relevant to the study question should be avoided. Survey questions should be worded such that they do not leave room for interpretation, that is, unambiguous and specific. In addition, care must be taken to ensure that personal opinions do not bias answers, especially in situations in which inherent power differentials exist (such as the implicit power differential between patients and clinicians).

In addition to question content, an important consideration is the educational level and background of the intended respondents. In the United States, the average person reads at an eighth-grade level.[9] Surveys written for the general population should be within this level. Regardless of the intended audience, investigators should avoid technical words and jargon.

Choice of question and response options (scales) vary from survey to survey. Questions can be open or closed-ended, with both options providing advantages and disadvantages. Open-ended questions allow respondents to comprehensively convey their opinions without the constraints of predefined answers. However, they can be difficult to interpret and analyze and are at greater risk of response fatigue.[10] Closed-ended questions negate some of these pitfalls but pose a greater risk of excluding answer choices that otherwise would have been selected. As such, response options for closed-ended questions should be comprehensive.

Several additional design features should be tailored to best fit the study question including the target audience and the intended sample size. The respondent group should be representative of the population of interest. The number of respondent cohorts varies depending on the study question. Surveying only one cohort is common, but several cohorts can also be utilized to allow for comparisons based on differentiating characteristics.

Sample size estimates play a significant role in study quality and can be determined using one of several well-accepted formulas.[11] Of note, sample size estimates are based on completed surveys and must be adjusted to account for the response rate. While the relationship between sample size and confidence interval is not linear, a larger sample size usually correlates with a narrower confidence interval.

Piloting the study

Once the study design process is finished, pilot testing of the instrument should be performed on a small number of responders. This step ensures that questions are clear, and instructions are easy to follow and allows for any necessary modifications before full implementation.

Analytical method

Analysis of survey responses depends on the type of data collected. When possible, consultation with a statistician early in the study design process is advised. In cases in which a statistician is not accessible, a few general principles can be applied. For quantitative variables where the underlying population has a normal distribution, parametric tests such as the *t*-test or analysis of variance can be used. When the sample means are not normally distributed, parametric tests are not reliable and nonparametric tests (e.g., rank-based tests) should be used.[12] For qualitative variables, analysis predominantly involves coding or categorizing data. Significant advances have been made in the realm of software technology to simplify the process of data coding.

Reporting data

While there are no standardized guidelines for reporting survey research, a few key points to include are (1) information on the study instrument used, (2) sampling techniques, specifically with regards to inclusion and exclusion criteria and sample size, (3) response rate including analysis for nonresponse bias,[13] and (4) statistical analysis methods used.

Summary

Survey research is a unique asset that can be used to gather a large amount of population-based data. This chapter introduced general principles of surveys including when to conduct them, how to design them, and how to report their results. While seemingly simple to construct, high-quality surveys require thoughtful efforts to provide valid and reliable results.

Get started

1) Identify the research question.
2) Determine if a survey is the best research design to answer the question of interest.

3) Perform a literature survey and assess if any research instruments have been validated.
4) If no validated survey instruments exist, develop questions using the acronym BRUSO: brief, relevant, unambiguous, specific, and objective.
5) Define target audience and appropriate sample size.
6) Perform a pilot study and adjust survey as needed.
7) Conduct larger-scale data collection.
8) Analyze and interpret data.
9) Report or present results.

Resources

Practical guide to performing survey research available via JAMA Brasel K, Haider A, Haukoos J. Practical Guide to Survey Research. *JAMA Surg*. 2020;155(4):351–352. https://doi.org/10.1001/jamasurg.2019.4401.

References

1. Lau F. Chapter 13 methods for survey studies. In: Lau F, Kuziemsky C, eds. *Handbook of eHealth Evaluation: An Evidence-Based Approach*. Victoria (BC): University of Victoria; February 27, 2017.
2. Singleton RA, Straits BC. *Instructor's Manual to Accompany: Approaches to Social Research*. Oxford University Press; 1999.
3. Barbour RS. Checklists for improving rigour in qualitative research: a case of the tail wagging the dog? *BMJ*. 2001;322(7294):1115–1117. https://doi.org/10.1136/bmj.322.7294.1115.
4. Beauchamp GA, Perrone J. Making surveys count: enhancing the rigor of survey-based research in medical toxicology. *J Med Toxicol*. 2016;12(4):329–331. https://doi.org/10.1007/s13181-016-0565-5.
5. Babbie ER. *The Practice of Social Research*. Cengage Learning; 2021.
6. Rickards G, Magee C, Artino Jr AR. You can't fix by analysis what you've spoiled by design: developing survey instruments and collecting validity evidence. *J Grad Med Educ*. 2012;4(4):407–410. https://doi.org/10.4300/JGME-D-12-00239.1.
7. Tait AR, Voepel-Lewis T. Survey research: it's just a few questions, right? *Pediat Anesthesia*. 2015;25(7):656–662. https://doi.org/10.1111/pan.12680.
8. Peterson RA. *Constructing Effective Questionnaires*. Thousand Oaks, CA: SAGE Publications, Inc.; 2000. https://doi.org/10.4135/9781483349022.
9. Eltorai AE, Ghanian S, Adams Jr CA, Born CT, Daniels AH. Readability of patient education materials on the american association for surgery of trauma website. *Arch Trauma Res*. 2014;3(2):e18161. https://doi.org/10.5812/atr.18161. Published 2014 Apr 30.
10. Story DA, Tait AR. Survey research. *Anesthesiology*. 2019;130(2):192–202. https://doi.org/10.1097/ALN.0000000000002436.
11. Barlett JE, Kotrlik J, Higgins C. Organizational research: determining appropriate sample size in survey research. *Inf Technol Learn Perform J*. 2001;19.
12. Nahm FS. Nonparametric statistical tests for the continuous data: the basic concept and the practical use. *Korean J Anesthesiol*. 2016;69(1):8–14. https://doi.org/10.4097/kjae.2016.69.1.8.
13. Jones TL, Baxter MA, Khanduja V. A quick guide to survey research. *Ann R Coll Surg Engl*. 2013;95(1):5–7. https://doi.org/10.1308/003588413X13511609956372.

Chapter 47

Qualitative methods and mixed methods

Maryam Boumezrag
Medstar Georgetown University Hospital, Washington, DC, United States

Qualitative methods of research refer to the use of nonnumerical data to evaluate behaviors, perceptions, and experiences. Mixed methods research is an emergent method of research that combines quantitative and qualitative data within a single study.[1] Similar to survey research, mixed methods research gained popularity in the social sciences and most recently in other forms of health sciences. Due to its relative novelty, guidance regarding this methodology is limited. In this chapter, we present some of the advantages and disadvantages of mixed methods research and provide an approach to designing a mixed methods research study.

The limitations of single method studies have become increasingly apparent, especially when intended to help guide decision-making for practitioners and policymakers.[2] Traditionally, research methods focused on quantitative measures such as treatment efficacy. While these measures provide invaluable information, they can overlook the importance of qualitative metrics such as patient experience and satisfaction and the role those metrics play in patients' clinical outcomes.

Potential pitfalls

The most notable advantage of mixed methods research is the ability to address research questions more comprehensively than either quantitative or qualitative research alone.[1] On the other hand, mixed methods research studies can be difficult to implement. In cases in which discordant results must be resolved, data analysis can be extremely challenging. Additionally, high-quality mixed methods research requires a multidisciplinary team where researchers are willing and able to use methods outside of their usual scope (i.e., qualitative researchers who can perform quantitative analyses and vice versa).[3]

Mixed methods research can be designed and implemented in numerous ways. In fact, as many as 40 distinct types of mixed methods research design options have been described in the literature.[4] While this number of design options may seem overwhelming, it is largely a reflection of the various disciplines from which they originate as opposed to true differences in methodology. As such, they can be grouped into four major categories: (1) triangulation design, (2) embedded design, (3) explanatory design, and (4) exploratory design.[5]

Triangulation is a method that uses different but complementary information on the same topic. With this method, qualitative and quantitative data are gathered concurrently but separately. Several variations of this method exist, depending on the stage at which data are merged. While some studies merge quantitative and qualitative data during analysis, others perform separate analyses and merge results to formulate a final interpretation.

In the embedded design, one type of data is secondary to the other and is used to support the primary data's evidence. For example, a study published by Wade et al.[6] investigated the difference in efficacy between adaptive tutorials and web-based resources in engagement of junior radiology residents. The investigators obtained several quantitative metrics after each form of learning such as length of time to complete modules and mean quiz score after completing modules. In addition to those metrics, they embedded a qualitative survey to evaluate residents' perceived engagement. While it may be difficult to differentiate an embedded design from other forms of mixed methods research, the key distinguishing factor is whether the additional data is playing a secondary role to the primary data set.[4] In the example described earlier, the primary outcome was assessed using quantitative data but helpful additional information was obtained by including an embedded survey.

Explanatory design is a two-step method in which quantitative data is first obtained and analyzed. Then,

Translational Interventional Radiology. https://doi.org/10.1016/B978-0-12-823026-8.00047-X

qualitative data is collected to explain or expand on those results. This form of design is best suited for studies in which outliers or surprising results need to be explained.[7] It can also be used for studies in which quantitative characteristics are used to guide subsequent qualitative research.[8]

Real-world examples

Similarly, exploratory design involves two steps but begins by obtaining qualitative data first and quantitative data second. This design method is used for cases in which there is no guiding framework to explain a certain phenomenon. In such cases, the exploratory phase is used to help identify variables that can then be studied quantitatively. This model was used in a study conducted by Leets et al.[9] to investigate ways to improve caregiver perception of patient readiness to transition home. They began by assembling focus groups to explore clinical staff members' perspective on this issue. Analysis of information obtained from the focus groups guided the creation of a poster intervention that was then tested in a randomized controlled trial.

In summary, qualitative and mixed methods research are essential tools in answering research questions that cannot be answered by quantitative research alone. When conducting a mixed methods research study, investigators can choose between several design types depending on which design best addresses the research question.

Get started

1) Identify the research question.
2) Determine if a mixed methods study is the warranted to answer the question of interest.
3) Choose a study design based on which method (quantitative vs qualitative) will take priority and in what sequence data will be collected.
4) Gather data.
5) Analyze and interpret data.
6) Report or present results.

Resources

1. Practical guide to performing qualitative research available via JAMA, Schwarze ML, Kaji AH, Ghaferi AA. Practical Guide to Qualitative Analysis. *JAMA Surg*. 2020;155(3):252–253. https://doi.org/10.1001/jamasurg.2019.4385.
2. Practical guide to performing mixed methods research available via JAMA, Dossett LA, Kaji AH, Dimick JB. Practical Guide to Mixed Methods. *JAMA Surg*. 2020;155(3):254–255. https://doi.org/10.1001/jamasurg.2019.4388.

References

1. Creswell JW, Plano Clark VL. *Designing and Conducting Mixed Methods Research*. Sage publications; 2017.
2. Stern C, Lizarondo L, Carrier J, et al. Methodological guidance for the conduct of mixed methods systematic reviews. *JBI Evid Synth*. 2020;18(10):2108–2118. https://doi.org/10.11124/JBISRIR-D-19-00169.
3. Wisdom J, Creswell JW. *Mixed Methods: Integrating Quantitative and Qualitative Data Collection and Analysis while Studying Patient-Centered Medical Home Models*. Rockville, MD: Agency for Healthcare Research and Quality; February 2013. AHRQ Publication No. 13-0028-EF.
4. Tashakkori A, Teddlie C. *Handbook of Mixed Methods in Social & Behavioral Research*. Sage Publications; 2003.
5. Creswell J, Plano Clark V. *Designing and Conducting Mixed Methods Research*. SAGE Publications; 2011.
6. Wade SWT, Moscova M, Tedla N, et al. Adaptive tutorials versus web-based resources in radiology: a mixed methods analysis in junior doctors of efficacy and engagement. *BMC Med Educ*. 2020;20(1):303.
7. Morse JM. Approaches to qualitative-quantitative methodological triangulation. *Nurs Res*. 1991;40(2):120–123.
8. Teddlie CB, Tashakkori A. *Foundations of Mixed Methods Research: Integrating Quantitative and Qualitative Approaches in the Social and Behavioral Sciences*. Sage; 2009.
9. Leets LL, Cahill J, Sprenger AM, et al. Nudging discharge readiness with a poster: a sequential, exploratory mixed methods pilot study of patient caregivers. *J Patient Exp*. 2020;7(6):1341–1348. https://doi.org/10.1177/2374373520968976.

Part VI

Clinical trials

Chapter 48

Randomized control: Design, measures, and classic example

Tushar Garg[1] and Apurva Shrigiriwar[2]

[1]*Division of Vascular and Interventional Radiology, Russell H. Morgan Department of Radiology and Radiological Science, The Johns Hopkins University School of Medicine, Baltimore, MD, United States;* [2]*Division of Gastroenterology and Hepatology, The Johns Hopkins University School of Medicine, Baltimore, MD, United States*

Key points

- RCTs are the most rigorous and robust research tools for determining whether a cause—effect relationship exists between an intervention and an outcome.
- RCT is the preferred research method to generate evidence in basic, translational, and clinical research to improve the management of patients
- RCTs should be conducted only if they are ethically feasible, economically viable, and clinically worthwhile.

Why it matters?

Randomized controlled trial (RCT) is a prospective, comparative, quantitative study that is performed under controlled conditions with random allocation of participants to comparison groups. After this the intervention is performed, the experimental and control groups are followed to see if there is any significant difference in outcome between them. The results are then obtained and analyzed to assess the effectiveness of the treatment, procedure, or service. There are various types of randomized controlled trials, these types with their descriptions are shown in Fig. 48.1.

Why randomized control trials are the best?

The evidence based on RCTs is considered to be of the highest quality as compared to that gained from observational studies, as the latter are more prone to bias and confounding. To eliminate systematic differences like confounding between the experimental and control group in RCTs, the individuals are allocated randomly to experimental or control groups and this is called randomization. Therefore, the difference in observed outcome in the control and experimental group in RCT is due to the intervention being performed and not any other factors. However, even though randomization helps with minimizing confounding factors, sometimes the results obtained from RCTs can still be confounded. The advantages and disadvantages of RCTs compared to other methods are discussed in Table 48.1.

What can you test with a randomized control trial?

- Interventions that have sufficient uncertainty about their utility
- Interventions that are based on data obtained from in vitro laboratory studies, animal experiments, or observational studies
- Interventions that are safe for participants and do not expose them to undue harm/risk.
- When the research question is ethically appropriate to be tested in an RCT
- When we want to study the effectiveness and/or safety of diagnostic tests and screening tests

How to formulate a research question?

When one is conducting a clinical investigation, the first step is to develop a hypothesis that specifies an anticipated association between the predictor and outcome variable (or no association, i.e., null hypothesis), so that statistical tests of significance can be performed on the results. Key components of a sound research question are often abbreviated as PICOT (expanded later); after a research question

Translational Interventional Radiology. https://doi.org/10.1016/B978-0-12-823026-8.00100-0

Factorial	Participants randomly assigned to a group and specific intervention is delivered
Cluster	Participants from prespecified groups are randomly selected for intervention
Crossover	Participants receives one intervention and then another over time in a randomized sequence
Parallel Group	Participant is randomly assigned to a study group may or may not receive an interventon

FIGURE 48.1 Types of randomized controlled trials.

TABLE 48.1 Advantages and disadvantages of RCTs.

Advantages	Disadvantages
Rigorous experimental design	Challenging to conduct and time consuming
Randomization prevents confounding	Logistical and ethical restriction
Blinding helps with bias minimization	Setting and monitoring bias
Rigorous analysis method	Unsuitable for complex treatment studies
Useful in explaining cause and effect	Inappropriate for evaluation of side effects
	Limited follow-up necessitates surrogate endpoints
	Noncompliance of controls can threaten the validity

is converted into this format, we obtain a very well-defined, clear, feasible, specific, measurable, ethical, and clinically relevant question.

PICOT stands for:

- Population of Interest (P)
- Intervention to be studied (I)
- Comparator Intervention (C)
- Outcomes to be evaluated (O)
- Time duration for ascertainment of intervention outcome (T)

How to conduct an RCT?

Once a research question is generated, a well-defined protocol should then be prepared by clearly defining: the target population, inclusion and exclusion criteria, sample size, ethical requirements, consent, process of randomization, allocation, blinding of intervention, treatment and control delivery, definitions of outcomes, outcome assessment, and data management. The protocol should then be reviewed and approved by an independent ethics committee before starting a trial.

Random allocation

It is done to make sure each eligible participant has an equal chance to be allocated into the experimental or control group. Allocation should be done by the generation of a random sequence by a statistician who is not going to be involved in conducting the RCT.

Blinding

Blinding helps in the elimination of bias. Whenever possible, both participants and investigators of participants should be blinded to eliminate bias.

Conducting the trial

An RCT can be conducted at a single site or multiple sites. When several sites are included, it has the advantage of reaching the required sample size in a short period of time and it also helps to improve the generalizability of outcomes.

All the participants should be treated in the same way for procedures of treatment, diagnosis, investigations, and alternations in both experimental and control groups except for the intervention/control treatment.

Outcome ascertainment

Primary and secondary outcomes are must be collected by independent observers who are ideally unaware of the allocation and treatment arms of participants. Objective measures should be used as far as possible for ascertaining outcomes to prevent inclusion of a bias.

Power of a study

If significant differences in the primary outcomes between the two groups are seen, we can conclude that the difference is due to the intervention. The difference is thought to

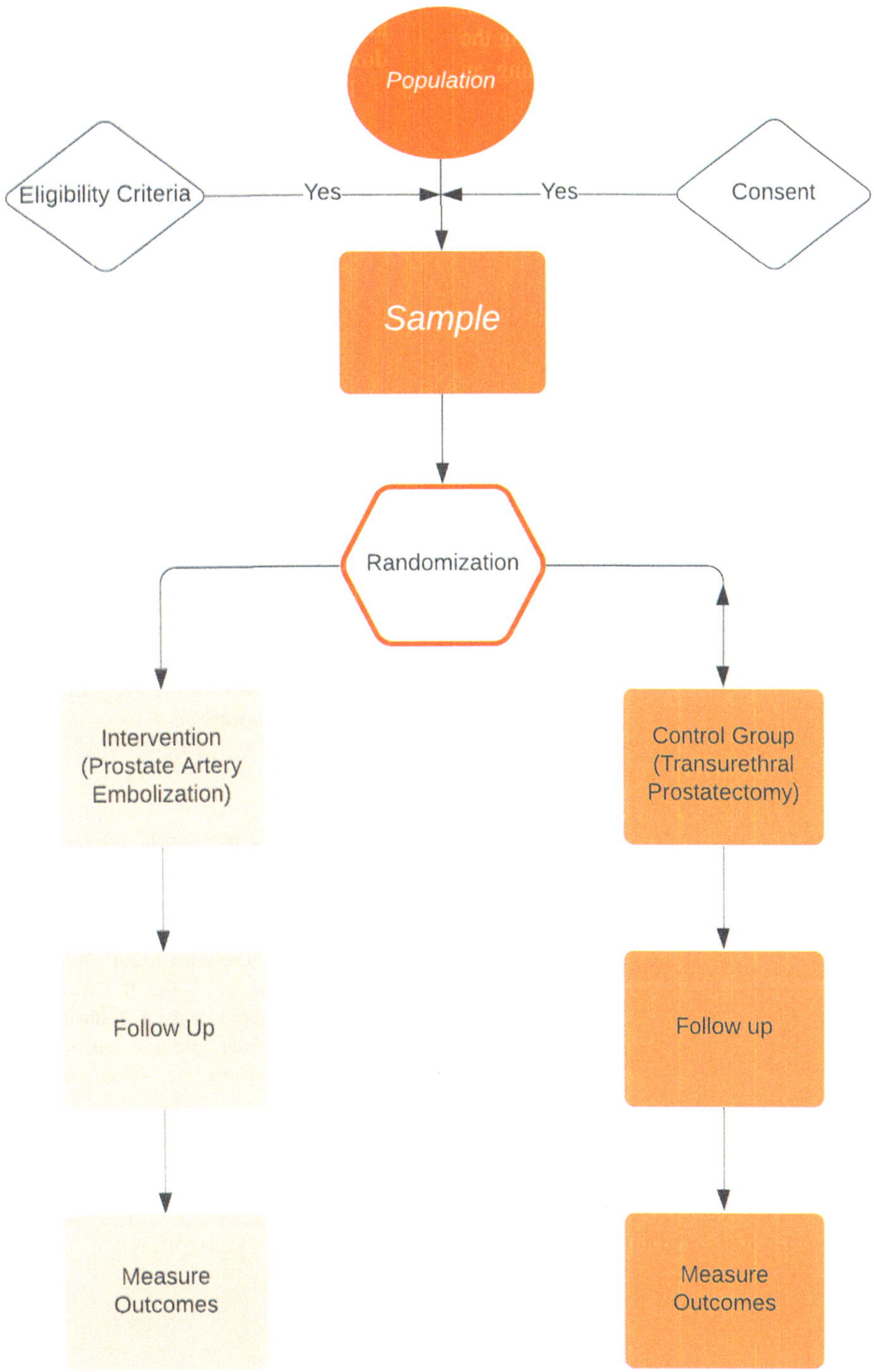

FIGURE 48.2 Simple example showing how to design a randomized controlled trial.

be statistically significant if the probability of this difference arising solely due to chance is <0.05 (*P*-value). The chance that the study will be able to demonstrate a significant difference if it is present is known as the power of the study; it is fixed at 0.8–0.95 (80%–95%) by convention.

Ethical considerations

Rigorous ethical principles must be applied to all RCTs involving experiments on animals, humans, or human biological material. Before the trial is started evaluating the risk and benefit to participants and society, obtaining an ethics approval and informed consent is essential.

Safety concerns

Participants should not be harmed when conducting an RCT, and therefore predefined, serious, adverse events have to be reported to the sponsor and trial monitoring committee. A very high frequency of adverse effects in the intervention group during the trial may lead to early termination of the RCT based on the decision of the data monitoring committee.

Other complications

It is critical that participants assigned to either the intervention or control group receive equal follow up and are all accounted for. Loss-to-follow-up, noncompliance, and contamination are potential problems that can reduce the power of an RCT and may introduce a bias if they occur in a nonrandom fashion.

How to report a randomized controlled trial?

CONSORT guidelines should be followed when reporting an RCT. Things that need to be reported in an RCT are as follows:

- Baseline characteristics of the two groups; no *P*-values reported according to CONSORT.
- Type of comparison made between the two groups: Superiority, noninferiority, or equivalence trial
- When the outcomes of all participants are reported by their randomization assignments, it is called intention-to-treat analysis (most commonly used)
- Effective size (along with its 95% confidence interval) should be reported for the primary outcome along with the *P*-value

A simple example demonstrating how to design a randomized controlled trial of your own (e.g., to compare outcomes of prostate artery embolization with Transurethral Prostatectomy) is shown in Fig. 48.2.

Examples of randomized controlled trials in interventional radiology:

1. Radiofrequency ablation duration per tumor volume may correlate with overall survival in solitary hepatocellular carcinoma patients treated with radiofrequency ablation plus lysothermosensitive liposomal doxorubicin[1]
   - Double-blind randomized controlled trial to study the burn time per tumor volume in radiofrequency (RF) ablation vs. RF ablation plus lysothermosensitive liposomal doxorubicin combination therapy for hepatocellular carcinoma
2. Randomized clinical trial of balloon occlusion versus conventional microcatheter prostatic artery embolization (PAE) for benign prostatic hyperplasia (BPH)[2]
   - A prospective, single-blind randomized controlled trial to compare the safety and efficacy of balloon occlusion (bPAE) versus conventional microcatheter PAE (cPAE)
3. Clinical and economic benefits of stent grafts in dysfunctional and thrombosed hemodialysis access graft circuits in the REVISE randomized trial[3]
   - A prospective randomized controlled trial to study the efficiency and cost-effectiveness to maintain an arteriovenous graft hemodialysis access circuits after rescue with percutaneous transluminal angioplasty with vs. without concurrent Viabhan stent grafts over 24 months.

References

1. Celik H, Wakim P, Pritchard WF, et al. Radiofrequency ablation duration per tumor volume may correlate with overall survival in solitary hepatocellular carcinoma patients treated with radiofrequency ablation plus lyso-thermosensitive liposomal doxorubicin. *J Vasc Intervent Radiol.* December 1, 2019;30(12):1908–1914.
2. Bilhim T, Costa NV, Torres D, Pisco J, Carmo S, Oliveira AG. Randomized clinical trial of balloon occlusion versus conventional microcatheter prostatic artery embolization for benign prostatic hyperplasia. *J Vasc Intervent Radiol.* 2019;30(11):1798–1806.
3. Mohr BA, Sheen AL, Roy-Chaudhury P, Schultz SR, Aruny JE, REVISE Investigators. Clinical and economic benefits of stent grafts in dysfunctional and thrombosed hemodialysis access graft circuits in the REVISE randomized trial. *J Vasc Intervent Radiol.* February 1, 2019;30(2):203–211.

Chapter 49

Nonrandomized controlled trials

Han G. Ngo[1], Chaitu Dandu[2], Girish B. Nair[3] and Sayf Al-Katib[4]

[1]*Oakland University William Beaumont School of Medicine, Rochester, MI, United States;* [2]*Wayne State University School of Medicine, Detroit, MI, United States;* [3]*Division of Pulmonary and Critical Care Medicine, Beaumont Health, Royal Oak, MI, United States;* [4]*Department of Diagnostic Radiology and Molecular Imaging, Beaumont Health, Royal Oak MI, United States*

This chapter aims to achieve the following:

- Explain nonrandomized controlled trials and discuss their strengths and weaknesses
- Demonstrate the concepts of nonrandomized controlled trials in three published studies
- Describe the general approach for designing a nonrandomized controlled trial

Introduction

Experimental studies can be divided into two broad categories: true experimental and quasi-experimental studies. An archetypical example of a true experimental study design is randomized controlled trials (RCTs), where participants are assigned to one of the comparison groups (e.g., either control or treatment group) randomly. Randomization yields comparison groups with minimal selection bias, enhancing the ability to assess the causal relationship between the intervention and the outcome. Although considered the gold standard for clinical research, RCTs are not always possible due to logistic and ethical concerns. Hence, nonrandomized controlled trials (NCTs)—a quasi-experimental study design that does not utilize random assignment—can be good alternatives when RCTs are not feasible.[1] It should be noted that NCTs can also fall under prospective studies in addition to experimental studies.

This chapter will focus on NCTs including basic principles, strengths and weaknesses, and applicability in interventional radiology research.

Nonrandomized controlled studies

NCT is a quasi-experimental study in which subjects are assigned to treatment arms deliberately. In some cases, researchers make participant assignments, while in others, participants effectively assign themselves. Ideally, the control and treatment groups should have similar baseline characteristics and differ only in the intervention evaluated in the study. This allows NCTs to assess the efficacy of an intervention by observing any differences between the control and treatment groups.[2] NCTs offer a good alternative to RCTs in several circumstances:

a. *Ethical reasons:* If an intervention has more risks than potential benefits, it would be unethical to randomly assign participants (or some of them) to the treatment group. Conversely, if there is a potentially effective intervention that can prolong life with minimal risks, it would be unethical to withhold treatment from the control group.[3]

b. *Less expensive and less time-consuming:* Since NCTs are not burdened with randomization or the ethical constraints associated with designing randomized trials, they often cost less and require less extensive planning (e.g., plans for randomization, blindness, and monitoring). This makes NCTs useful as preliminary studies before undertaking costly large-scale randomized trials.[3]

c. *Better real-world data simulation:* Unlike RCTs, NCTs are not limited to participants who consent to randomization, and thus can include a wider range of participants in their trials. This allows the samples of NCTs to represent key target populations and promote external validity—how well a study's results can be generalized to real-world settings.[4]

Despite the advantages discussed earlier, NCTs also have their limitations, most of which threaten the internal validity—how well a study outcome accurately reflects the true population. Below are three major limitations of NCTs (Table 49.1):

a. *Susceptible to bias:* Both NCTs and RCTs are susceptible to biases if they are not appropriately designed.

Translational Interventional Radiology. https://doi.org/10.1016/B978-0-12-823026-8.00092-4

TABLE 49.1 Advantages and disadvantages of nonrandomized controlled trials.

Advantages	Disadvantages
• Good alternatives when RCTs are not possible due to ethical/practical concerns • Cost less and require less planning than RCTs • Provide data that are closer to the real-world	• More susceptible to bias → a threat to internal validity • Harder to control for unmeasured confounders → a threat to internal validity • Association can be made, but not cause–effect relationship

However, inherently by its design, NCTs have a greater risk of selection bias, where the nonrandom assignment of study participants can affect the trial outcomes (e.g., sicker patients are more inclined to assign themselves to treatment group).[4]

b. *More difficult to control for confounding variables:* A confounder is a variable that was not accounted for during the trial and can weaken the internal validity of a study outcome. A confounder can affect the selection of treatment and/or the outcome. Unlike selection bias which can arise in the selection process, confounding factors can affect study results if they lurk in both control and experimental groups. Without randomization of treatment, confounding is inevitable. Although NCTs attempt to reduce confounding by creating two closely matched groups, it is impossible to know whether the control and treatment groups in NCTs are truly comparable with potential existence of unmeasured confounders.[2,5,6]

c. *NCTs cannot demonstrate causality*: As NCTs are highly susceptible to bias and confounding, only associations can be made between the interventions and outcomes. RCTs, on the other hand, minimize these limitations and are better suited to assess causality.

Real-world examples: interventional radiology studies that used nonrandomized controlled trials

#1 How nonrandomized controlled trials can help to resolve ethical challenges of randomization

A 2019 nonrandomized controlled study compared the effectiveness and safety of treating V2 trigeminal neuralgia using the percutaneous radiofrequency thermocoagulation at the foramen rotundum (FR) versus foramen ovale (FO). The 80 study subjects assigned themselves into two groups deliberately: the control group ($n = 40$) was treated at FO, while the treatment group ($n = 40$) was treated at FR. The study reported no significant difference in baseline characteristics between the two groups (age, gender, and preoperative facial pain scores). Follow-up data collected at 1 week, 6 months, and 1 year showed FR was more effective at treating V2 trigeminal neuralgia and associated with fewer postoperative complications.[7] In this study, the NCT design was particularly suitable due to the surgical risks associated with both FR and FO approaches. By allowing subjects to voluntarily choose their treatment based on an understanding of the risks and alternatives, the researchers were able to avoid the ethical problems associated with randomizing participants to one or the other treatment. Additionally, the study included patients who did not want to subject themselves to a randomization of treatment options, and thus was more representative of a real-world target population.

#2 An example of a well-designed nonrandomized controlled trial when randomization is infeasible or unethical

A 2002 study on the efficacy of transcatheter oily chemoembolization (TOCE) technique in treating inoperable hepatocellular carcinoma (HCC) illustrates the utility of NCTs in mitigating other limitations of RCTs. This study took place across five different hospitals, where 127 patients were selected to receive TOCE treatment and 127 patients were selected for the control (untreated) group. Patients in the control group qualified for TOCE but were not chosen to undergo this intervention for two reasons: (1) TOCE was not available at the time they were admitted; or (2) TOCE was not performed at their hospitals. At the study design phase, the control and treatment groups were matched as closely as possible on key clinical, anatomic, and biologic features that are well described as contributing to the development of HCC. Additionally, when multiple untreated patients matched a treated patient, the untreated patient with the longest survival potential was chosen to provide a valid parameter for comparison. Findings revealed that TOCE prolongs survival and improves the quality of life in patients with inoperable HCC.[8] This study overcame two limitations of an RCT design: first, random assignment would not have been possible due to the availability of the intervention; second, there would be major ethical issues associated with withholding life-improving treatment from patients in the control group. Additionally, this study serves as an excellent example of how a carefully designed NCT can promote internal validity and minimize selection bias through matching. By carefully creating matched groups and picking an untreated patient who had the longest survival potential, the study ensures that the two groups were comparable for meaningful conclusions to be made.

#3 An example of a nonrandomized controlled trial that fails to adequately control for confounding

Finally, a 2009 study assessing the safety and efficacy of electroacupuncture for pain reduction during and after

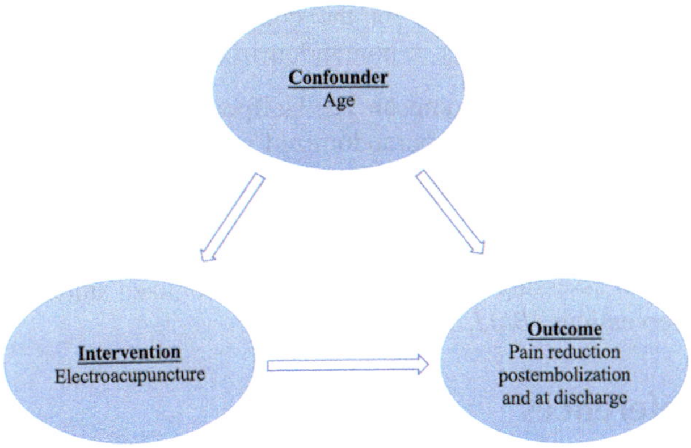

FIGURE 49.1 Diagram explaining how age could be a confounding factor affecting the study outcome of the electroacupuncture clinical trial. As the age difference between the control and treatment groups was statistically significant, potential differences in reporting of pain by age may have influenced the findings of this study.

uterine artery embolization for uterine leiomyomas illustrates the susceptibility of NCTs to various biases. The control group ($n = 37$) received the standard therapy of local anesthesia, while the treatment group ($n = 33$) consisted of patients volunteering to undergo electroacupuncture. Participants' baseline characteristics were not controlled for in this study and pain score was self-reported by the participants on a 10-point scale. Although the study concluded that the differences in mean pain scores between the two groups at postembolization and at hospital discharge were statistically significant, there are several limitations to consider.[9] First, the assignment to the treatment group was based on patient preference, and the study relied on self-reported method to measure pain. Thus, a placebo effect may have played a role in the reported pain scores. Second, the control and treatment groups significantly differed in age, which might represent a confounding factor among others (Fig. 49.1). Perhaps, younger subjects in the treatment group were generally more open to trying new interventions than older participants. If pain perception is presumptively decreased with age, then it is uncertain if differences in age may have influenced the reporting of pain independent of the treatment option.

Get started

Fig. 49.2 outlines the general steps of designing a nonrandomized controlled trial. Since NCTs and RCTs share many similar design features, readers are encouraged to read the chapter on *Randomized Controlled Trials* of this book for how to perform an experimental controlled study. This section will instead focus on discussing the key design features unique to NCTs and strategies that can minimize the limitations of these trials.

NCTs and RCTs mainly differ in how participants are allocated to groups. As allocation is not randomized in

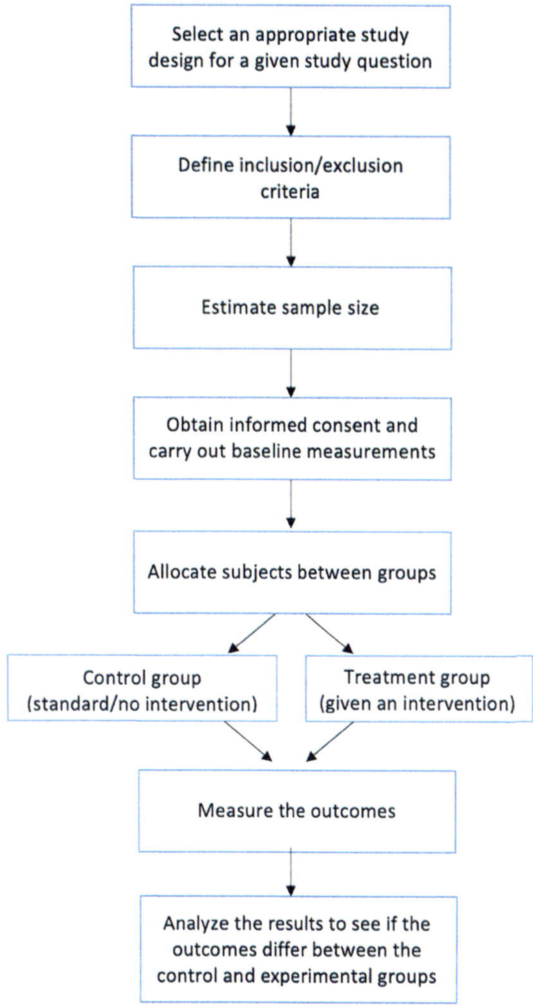

FIGURE 49.2 General step-by-step flow chart of how to design a nonrandomized controlled trial.

NCTs, the goal is to minimize the difference between control and treatment groups. This can occur at either the design stage or the analysis stage.

In the design stage, researchers should aim to make the control and treatment groups as similar in pretreatment characteristics as possible (e.g., demographic and clinical characteristics). One way to accomplish this is to use a propensity score, which is the likelihood an individual is assigned to a treatment group based on the baseline characteristics. Estimated by a logistic regression model, propensity score matching is a commonly used method in nonrandomized studies to balance the groups, control for selection bias, and minimize confounding factors. The method works by reducing each participant's baseline characteristics to a propensity score (ranges 0−1). Based on their propensity scores, participants in the untreated and treated groups are paired together. As long as the pair shares the exact same propensity score, it is likely that the baseline covariances between the untreated and treated participants are closely matched.[10] Furthermore, researchers should also collect adequate baseline covariates to capture all potential confounders. If researchers want to avoid removing subjects during matching due to an unequal number between the control and treatment group, they can use inverse probability weighting (IPW). IPW down weights the group with more participants and up weights the group with less participants, such that each participant regardless of the group has the same weight. In the analysis stage, researchers can use methods such as regression, matching, and reweighting to reduce bias from measured confounders.

Additionally, since NCTs are susceptible to more biases, researchers should try to anticipate for all potential bias before starting the trial. There exist quality assessment tools to evaluate the risks of bias in nonrandomized studies. Commonly used tools include Newcastle–Ottawa Scale, Methodological Index for Non-Randomized Studies (MINORS), bespoke appraisal tools, and Risk of Bias in Non-Randomized Studies (ROBINS-I). To learn more about these tools, readers are encouraged to read *Critical Appraisal of Nonrandomized Studies—A Review of Recommended and Commonly Used Tools.*[11]

Potential pitfalls

1. The potential for selection bias due to the lack of randomization.
2. The potential for confounding bias due to the lack of adjustment for differences in characteristics between control and treatment groups.
3. The potential to ascertain conclusions from NCTs are causal rather than correlational.

Resources

1. Additional readings on Inverse Probability Weighting (IPW):

Chesnaye NC, Stel VS, Tripepi G, et al. An introduction to inverse probability of treatment weighting in observational research. *Clin Kidney J*. 2021;15(1):14-20. Published 2021 Aug 26. https://doi.org/10.1093/ckj/sfab158

2. Additional readings on the commonly used tools to assess risk of bias in nonrandomized studies:

Quigley JM, Thompson JC, Halfpenny NJ, Scott DA. Critical appraisal of nonrandomized studies—A review of recommended and commonly used tools. *J Eval Clin Pract.* 2019; 25:44–52.

Sterne JA, Hernan MA, Reeves BC, et al. ROBINS-I: a tool for assessing risk of bias in non-randomized studies of interventions. *BMJ*. 2016; 355: i4919.

References

1. Penson DF, Wei JT, eds. *Clinical Research Methods for Surgeons*. 1st ed. Humana Press; 2007.
2. Balakrishnan N. *Methods and Applications of Statistics in Clinical Trials, Volume 1: Concepts, Principles, Trials, and Designs*. 1st ed. Incorporated: John Wiley & Sons; 2014:609–618.
3. Anderson S, Auquier A, Hauck WW, Oakes D, Vandaele W, Weisberg HI. *Statistical Methods for Comparative Studies: Techniques for Bias Reduction*. 1st ed. Wiley-Interscience; 1980: 31–45.
4. Sedgwick P. What is a non-randomised controlled trial? *BMJ*. 2014;348(g4115). https://doi.org/10.1136/bmj.g4115.
5. Gerstein HC, McMurray J, Holman RR. Real-world studies no substitute for RCTs in establishing efficacy | Enhanced Reader. *Lancet*. 2019;393(10168):210–211.
6. Pandis N, Tu Y-K, Fleming PS, Polychronopoulou A. Randomized and nonrandomized studies: complementary or competing? *Am J Orthod Dentofacial Orthop*. 2014;146(5):633–640.
7. Yan C-Y, Xue T-Q, Zhang Q-X, et al. *Non-Randomized Trial Radiofrequency Thermocoagulation through Foramen Rotundum versus Foramen Ovale for the Treatment of V2 Trigeminal Neuralgia*. Pain Physician J; 2019.
8. Bronowicki J, Vetter D, Dumas F, et al. Transcatheter oily chemoembolization for hepatocellular carcinoma. A 4-year Study of 127 French Patients. *Cancer*. 1994;74(1):16–24.
9. Pisco JM, Tsuchiya M, Bilhim T, Duarte M, Santos D, Oliveira AG. Uterine artery embolization under electroacupuncture for uterine leiomyomas- ClinicalKey. *J Vasc Intervent Radiol*. 2009;20(7): 863–870.
10. Linden A. Designing a prospective study when randomization is not feasible. *Eval Health Prof.* 2011;34:164–180.
11. Quigley JM, Scott DA. Critical appraisal of nonrandomized studies-A review of recommended and commonly used tools Present Address. *J Eval Clin Pract*. 2019;25.

Chapter 50

Historical control: Design, measures, and classic example

Mehdi Taghipour
Einstein Healthcare Network, Philadelphia, PA, United States

Definition of historical control

In any clinical trial, a novel treatment or procedure is compared to a control arm, which is a group with placebo or a standard of care. While the control arm can be recruited concurrently with the experimental group, there is an option to acquire a part or all of the control group from an old existing data, which is called historical control. In another word, a historical control is defined when a control group is borrowed from an old data to compare with the experimental group in a new trial. Therefore, in a study with a historical control group, all cases in the new trial will receive the new treatment or procedure and the control group will be retracted from historical data. The historical control is particularly helpful in rare diseases when recruiting enough number of both case and control group is difficult or in life-threatening diseases when there are ethical concerns in recruiting patients for control arm.

Sources of historical control

1. Real-world data
   a. Medical chart data from daily clinical care: A comprehensive real-world data, which contains medical notes, lab tests, and radiology studies along with other health-related information. Although there are some challenges such as frequent missing data, inaccuracies, and difficulties in extracting unstructured free text information in medical notes, medical chart data have the most accurate chronological laboratory, diagnostic and therapeutic procedures.
   b. Patient registry: A data collection tool for scientific, clinical, or policy purposes, which uses observational methods to collect uniform data of a particular disease. Keep in mind that this is not standard data and possibility of poor data quality, lack of consistent reporting, and missing data are always present. Medical claims and billings are another sources, which have more accurate diagnostic data.
2. Natural history trials: Tracking the natural course of a disease to identify demographics, genetics, environmental, and other variables, which correlate with the disease and its outcomes. Its difference from the patient registry is that there is a possibility of collecting control data more similar to the experimental group by a designed specific protocol. In a new trial, one of the limitations compared to the experimental group is that sometimes the patients had received the standard of care at the time of the natural history trial, which is different from the current standard of care.
3. Control group in previous randomized control trial (RCT): The control arm (placebo or standard of care) of prior RCTs can be treated as a source of control data. This is a very robust and rich source of high-quality data generated in a controlled environment. Also, some statistical methods can be used to decrease the biases.

Advantages of the historical control study

The historical data provide options to perform trials especially in rare diseases or in life-threatening conditions where enrolling a control group with placebo is nonethical. Moreover, using historical data will increase efficiency by decreasing the costs of study dramatically; therefore, more resources can be allocated to the experimental group. Single treatment design results in more patient recruitment for the experimental group, which results in increased statistical power and lower type I error rates.

Translational Interventional Radiology. https://doi.org/10.1016/B978-0-12-823026-8.00071-7

Disadvantages of the historical control study and how to overcome

In trials with historical control, there is no randomization or blindness and the control group is identified retrospectively. This potentially leads to a selection bias, which could affect the final results. Differences between the historical control group and the experimental group could come from a wide range of factors such as demographics (age, gender, ethnicity, and socioeconomic status), disease severity, duration of illness, prior treatment, disease definition in two groups, diagnostic technique, response assessment tools, comedications, and standard of care in different time points. For example, when a control arm is chosen from patients who were observed at some time in the past, the standard of care may have improved over the years; the same can happen with the improvement in accuracy of diagnostic tests over time. Potential confounding due to observed and unobserved differences between the two cohorts at different time is inevitable.

Moreover to be able to compare the control group with the experimental group, one of the most important factors is that the endpoint of both control data and experimental group should be the same. For example, if the study goal is assessing the survival benefits of a particular procedure, the control data should be derived from a prior trial that also reported survival data on the control group.

One solution to overcome the biases is to start with selecting a high-quality historical control group first then try to design an experimental group to mimic the control group. Another option is to do "dynamic borrowing," which is test-then-pool protocol. The scientists basically run statistical tests before the trial to figure out if the participants are equal enough to use historical data for the control. Several analysis methods have been offered to decrease the biases in studies with historical data. If designing a historical control study is planned, involving a statistician from the beginning of the study design is recommended (Fig. 50.1).

Types of historical controls

The historical control data can be incorporated into trials in two ways: single-arm trials or pooled trials. In single-arm trials, only the experimental group will be recruited and 100% of the control group will be borrowed from the historical data; while in pooled trials, the historical data will be added to the concurrent control data. The control group can be obtained using two different categories:

1. Well-documented population of patients with same diseases: Patients with similar disease entity without the specific intervention will be included as a control group. For example, to assess the value of middle meningeal artery (MMA) embolization in decreasing the rate of rebleeding in patients with chronic subdural hematoma (SDH), a similar group of patients with chronic SDH, who did not undergo MMA embolization, can be included as a historical control group. The rate of rebleeding in a certain follow-up time will be compared between these two groups.
2. Patients assigned as their own control: In this type of control data, patients' status before and after the new procedure/treatment will be compared. For example, the value of Y90 radioembolization in decreasing the rate of progression of metastatic hepatocellular carcinoma (HCC) can be evaluated by comparing the rate of progression of disease in each patient before and after treatment with Y90 radioembolization.

Samples of historical control study design

There are several published studies with historical control data. As an example, Takahiro Kinoshita et al. evaluated the mortality of patients with severe trauma who were managed in a new hybrid trauma suite, which is compatible with both interventional radiology and surgery procedures. The control group was recruited from historical data of those who were treated in conventional IR and surgery suites separately. In this study, the researchers only recruited the experimental group, and the whole control group was obtained from the historical data.

Below are a few research ideas, which can recruit a control group from historical data:

- To evaluate the value of a new catheter in the thoracic duct embolization procedure, the whole-control group can be obtained from historical data. It is extremely time consuming and may be practically impossible to enroll enough number of patients in both experimental and control groups; therefore, using a historical control group will be an alternative option.
- Proximal vessel thrombectomy is the standard of care in patients present with acute stroke, who fulfill the criteria for thrombectomy. It is not ethical to design a concurrent prospective case-control study to compare the group undergoing thrombectomy and a control group who is treated medically. In this situation, the control arm can be borrowed from historical data.
- The current standard of care for low-stage RCC is partial nephrectomy; however, many elderly patients with comorbidities cannot undergo surgery. These patients were followed by routine CT or MRI scans previously. Recently, minimally invasive ablation therapy was introduced, which demonstrated promising results. To design a research study to compare the effect of ablation therapy on overall survival, a control group with no treatment is needed. However, it is unethical to have a control group with no treatment while there is an alternative treatment

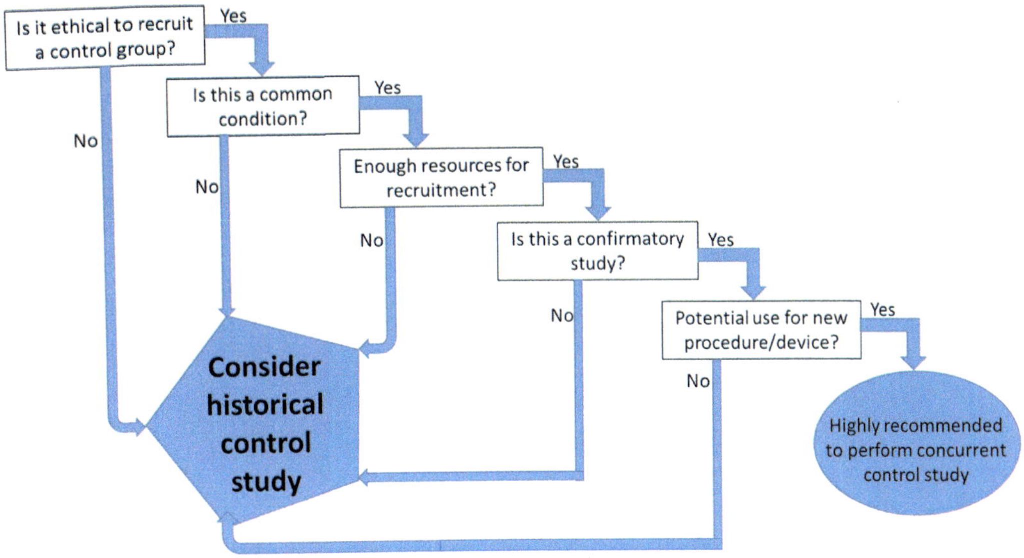

FIGURE 50.1 Decision-making diagram for using historical control data.

with promising results. Therefore, the control group for this study can be obtained from historical data.

On the other hand, using historical control is not always the best option. For example, it is possible to perform a study comparing prostate artery embolization and a historical control arm including patients with BPH who underwent medical management. However, using a concurrent prospective control arm with similar patient characteristics significantly increases the result reliability. Therefore, in such a situation, it is highly recommended to use a concurrent control group.

Further reading

1. Ghadessi M, Tang R, Zhou J, et al. A roadmap to using historical controls in clinical trials—by Drug Information Association Adaptive Design Scientific Working Group (DIA-ADSWG). *Orphanet J Rare Dis*. 2020;15(1):69.
2. Lim J, Walley R, Yuan J, et al. Minimizing patient burden through the use of historical subject-level data in innovative confirmatory clinical trials: review of methods and opportunities. *Ther Innov Regul Sci*. 2018;52(5):546–559.
3. Viele K, Berry S, Neuenschwander B, et al. Use of historical control data for assessing treatment effects in clinical trials. *Pharmaceut Stat*. 2014;13(1):41–54.
4. Kinoshita T, Yamakawa K, Matsuda H, et al. The survival benefit of a novel trauma workflow that includes immediate whole-body computed tomography, surgery, and interventional radiology, all in one trauma resuscitation room: a retrospective historical control study. *Ann Surg*. 2019;269(2):370–376.
5. Thomas N. Historical control. In: D'Agostino RB, Sullivan L, Massaro J, eds. *Wiley Encyclopedia of Clinical Trials*. 2008.
6. Baker SG, Lindeman KS. Rethinking historical controls. *Biostatistics*. 2001;2(4):383–396.

Chapter 51

Cross-over: Design and measures with classic example

Apurva Shrigiriwar[1] and Tushar Garg[2]

[1]Division of Gastroenterology and Hepatology, The Johns Hopkins University School of Medicine, Baltimore, MD, United States; [2]Division of Vascular and Interventional Radiology, Russell H. Morgan Department of Radiology and Radiological Science, The Johns Hopkins University School of Medicine, Baltimore, MD, United States

Goals- Definition of the study type. Benefits and limitations to such an investigation. Step by step for designing your own study of this type, using a simple, illustrative, fabricated example.

A cross-over design is a type of design in which an experimental unit (e.g., patients) receive different treatments during different time periods, and this is evaluated via repeated measurements of predefined outcomes, that is, the patients cross over from one treatment to another during the course of the trial. It is commonly compared to a parallel study design in which patients are randomized to a treatment group, and they remain on that group throughout the duration of the trial.

Study structure

The order of treatment administration in a cross-over clinical trial is called a sequence and the time of a treatment administration is called a period. Conventionally, the treatments are designated with capital letters, such as A, B, etc. The experimental units are allotted to sequences by randomization with adequate blinding.

A cross-over study can be labeled as—1. Uniform within sequences (Table 51.1)—each treatment appears the same number of times within each sequence. 2. Uniform within periods (Table 51.2)- each treatment appears the same number of times within each period. If a design is uniform within sequences and uniform within periods, then it is said to be a uniform cross-over. If the design is uniform across periods, we can remove the period effects. Period effects can occur when patients do better in a subsequent period because their state has changed, for example, their mental or health status has changed (independent of treatment). If the design is uniform across sequences, we can remove the sequence effects (sequence effects are defined as potential interactions among experimental units based on the sequences these treatments are presented).

The most frequently used cross-over design is the 2-sequence, 2-period, 2-treatment cross-over design, with sequences AB and BA, sometimes called the 2×2 cross-over design (Table 51.3).[1]

Some benefits of a cross-over study design includes the following:

1. Each experimental unit serves as his/her control, thereby reducing interparticipant variation.
2. As each patient serves as his/her control, fewer patients may be required in the cross-over design to attain the

TABLE 51.1 Cross-over study uniform within sequences.

	Period 1	Period 2	Period 3
Sequence ABB	A	B	B
Sequence BAB	B	A	B

TABLE 51.2 Cross-over study uniform within periods.

	Period 1	Period 2	Period 3	Period 4
Sequence ABCC	A	B	C	C
Sequence BCAA	B	C	A	A

Translational Interventional Radiology. https://doi.org/10.1016/B978-0-12-823026-8.00030-4

same level of statistical power or precision as a parallel design. According to various statistical estimates, using a cross-over design can reduce the sample size by 60% −70% compared to a parallel study design.

3. All participants are evaluated for all available treatments; this reduces the chance of encountering a confounding bias and makes the comparison between two or more interventions fairer.
4. Individual patient response to the intervention can be observed and evaluated in this study design.
5. Each patient receives an active intervention at some point in the trial; this is a unique feature that makes this trial an attractive option for patient recruitment.
6. Blinding can be maintained during the entire duration of the trial.

Disadvantages of a cross-over study design are as follows:

TABLE 51.3 2 × 2 cross-over design.

	Period 1	Period 2
Sequence AB	A	B
Sequence BA	B	A

1. The main disadvantage of a cross-over design is the carryover effect. The carryover effect is defined as the effect of the treatment from the previous time period on the response at the current time period. In simpler terms, if a patient receives intervention A during the first period and intervention B during the second period, then measurements taken during the second period could be a result of the direct effect of treatment B administered during the second period, and/or the carryover or residual effect of treatment A administered during the first period. The carryover period can be minimized by an interval washout period between two interventions.
2. This study design cannot be used for acute conditions as a significant washout period is required between two or more interventions for appropriate statistical analysis.
3. This study design is not applicable to diseases in which the first intervention is curative.
4. If a disease has a rapidly changing course, this study cannot be applied as it can result in falsified statistical significance.
5. There can be a significant drop out after the first intervention secondary to a longer trial duration, which can potentially make the study inaccurate. This disadvantage can be possibly overcome by the recruitment of new experimental units.

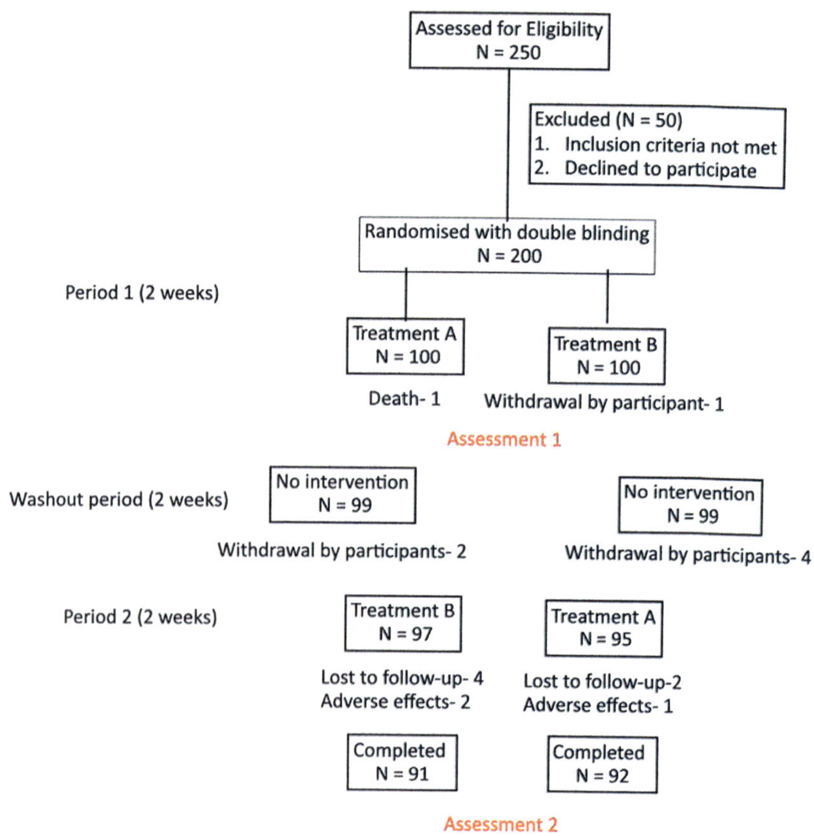

FIGURE 51.1 Step-by-step guide on conducting cross-over study.

6. Longer follow-up time compared to parallel clinical trials.
7. Other cons include potential period and sequence effects.[2]

Step-by-step guide for designing a cross-over study:

The following figure (Fig. 51.1) demonstrates a step-by-step guide on how to conduct a cross-over study. It also includes participants who withdrew from the trial, developed adverse effects, or are lost to follow-up. The choice of statistical test depends on the number of measurements analyzed at the end of each period.

References

1. Piantadosi Steven. Crossover designs. In: *Piantadosi Steven. Clinical Trials: A Methodologic Perspective*. 2nd ed. Hobaken, NJ: John Wiley and Sons, Inc; 2005.
2. Evans SR. Clinical trial structures. *J Exp Stroke Transl Med*. 2010;3(1):8–18. https://doi.org/10.6030/1939-067x-3.1.

Chapter 52

The withdrawal design

Tarek Almsaddi
Oakland University William Beaumont School of Medicine, Rochester, MI, United States

Key points

- The withdrawal design is a powerful tool for demonstrating the efficacy of a medical intervention.
- Also called the ABA design, the withdrawal study includes measuring the dependent variable at baseline, introducing an independent variable (i.e., clinical intervention) and then removing that independent variable.
- Results from the withdrawal design allow researchers and practitioners to make causal inferences about the intervention.

Why it matters?

The withdrawal study is an effective study design that further defines the relationship between clinical intervention and outcome. Also called the ***ABA design*, a withdrawal study** is defined as an experimental **single-case design** that compares outcomes after the presentation of an independent variable (clinical intervention) followed by subsequent removal of that variable. It is referred to as the ABA design because of its exclusive sequence of conditions—measuring the target response at baseline (A_1), adding an intervention (B) until the dependent variable has stabilized, and then removing that intervention and returning to baseline (A_2). This experimental analysis of behavior enhances the strength of the traditional A—B approach (baseline then intervention) because it allows researchers and practitioners to better understand the efficacy of the intervention. In other words, comparing the baseline before and after intervention represents an opportunity for drawing causal inferences about the intervention itself. If the withdrawal of the intervention results in a change in trend direction that is at a similar level to the first baseline phase, one can reasonably conclude that the intervention was responsible for the observed change. This design has historically been utilized in the context of basic and applied behavioral research, but it also represents the simplest single-subject research design for demonstrating cause—effect relation sand can therefore be effective in a multitude of clinical research subjects. The withdrawal design also includes several variations, such as ABAB and BAB designs, that serve to strengthen the observed relationships.

Fundamental components of the withdrawal study

To effectively carry out a withdrawal study, one must understand its key components. Throughout the breakdown of these fundamentals, the term ***behavior*** will be used to encompass a definition beyond the basic humanistic manner of conducting oneself. Behavior can also be the body's physiological response in the context that the independent variable is a drug, procedure, etc. One must also ensure that the administration and withdrawal of an intervention is practical and does not violate any ethical standards, see Chapter 45.

First, pretreatment target variables should take place at a baseline phase (A_1). The baseline phase should be stable, representative of the behavior of the targeted variable, and indicate a participant's "natural state" before intervention is introduced.[2] It is also critical that the dependent variable demonstrates **steady state,** meaning the baseline remains relatively consistent across data points. If baseline variables display signs of inconsistency (meaning upward or downward trends before an intervention is included), then the researcher's ability to interpret any intervention as efficacious is compromised. Additionally, the researcher must account for any **behavioral cyclicity**, as there may be slight fluctuations in baseline levels as a consequence of environmental or physiologic factors. This phenomenon should not be suggested as a confounding variable as long as the steady state is consistent and long enough to account for such variations.[2] The baseline phase and its consistent steady state are critical for the comparison of behaviors and/or responses once an intervention is presented.

Translational Interventional Radiology. https://doi.org/10.1016/B978-0-12-823026-8.00038-9

Once baseline observation has effectively been executed, the next fundamental feature of the withdrawal design is introduced: manipulation of the independent variable (aka intervention) phase (B). The purpose of this phase is to discover what specific impact does an independent variable have on the behavior/response, or whether an intervention produces a clinically useful response. The term **treatment integrity** is commonly used among researchers to describe how consistent is the independent variable in its intent and function. An example of this is seen in a study where researchers studied the efficacy of ultrasound-guided and fluoroscopy-guided percutaneous pericardial effusion drainage for patients with symptomatic pericardial effusion. In the study, pericardial drainage was performed via three routes: apical, subxiphoid, and transhepatic. To maintain the treatment integrity, these researchers had to define the anatomic landmarks of these routes and remain consistent in where these interventions were administered to draw conclusions on which route was preferable.[3] To make viable conclusions about the intervention, the independent variable's treatment integrity must be maintained. This means researchers are urged to describe how the specific intervention has been operationalized as well as explicitly mention the procedures for carrying out the intervention.[4]

The final feature of the withdrawal design is the return to baseline phase (A_2) after removing the intervention. The purpose of this phase is the essence of the whole design—to determine if a certain behavioral change/response to an intervention was indeed due to the intervention and not any other variable. Ideally, if there were no external variables affecting the data, the behavior should return to baseline levels similar to those before the intervention, allowing researchers to make causal inferences about the intervention.[2] A researcher's ability to show that returning to baseline was due only to an intervention establishes a study's **internal validity**. When a researcher can rule out the external variables contributing to the observed behavior, they can confidently support the efficacy of that intervention.[4]

Benefits and limitations

The benefits and limitations of the withdrawal design are outlined in Fig. 52.1.

Example 1

A 55-year-old male with a past medical history of obesity (body mass index 49 kg/m^2) presents to the emergency

FIGURE 52.1 Benefits and limitations of the withdrawal design.

Benefits	Limitations
• The withdrawal design allows for functional analysis of the behavior/response to intervention, rather than just a correlational analysis. • The withdrawal design also allows researchers and practitioners to conclude with greater certainty that the observed effect on behavior is due to the intervention. • The withdrawal design includes many variations, including the ABAB design which reintroduces the independent variable after the second baseline phase. Another is the BAB design, which starts with the intervention, followed by removing it, and then placing the subject back on the intervention. • The withdrawal design and its many variations makes it well equipped to offer a diverse assessment of the effects of different interventions and/or target responses.	• The withdrawal design may be susceptible to threats to internal and external validity. • The threat in internal validity may be seen in the case that introduction and withdrawal of the independent variable coincides with the naturally occurring behavioral cyclicity of the target response. This threat can be accounted for, however, by altering the number of observation periods in the second baseline condition, or even by reintroducing the independent variable a second time (i.e., creating an ABAB variation of the withdrawal design). • A_1 baseline values may not be fully retrieved in the A_2 values, even though these values should be approximated. • There may be practical and ethical complications with providing and withdrawing a potential effective intervention.[3]

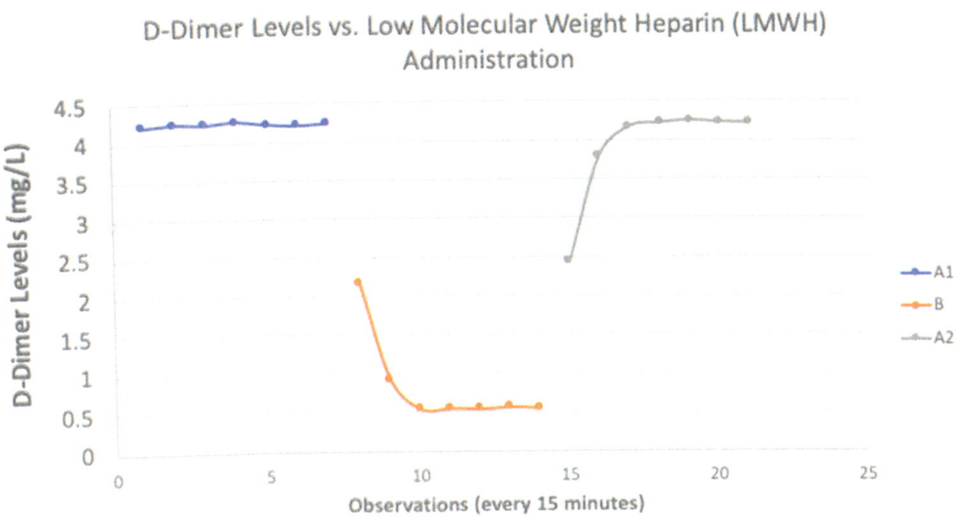

FIGURE 52.2 Example of withdrawal design graph. D-Dimer levels are observed in the first 7 observations (15 minute intervals) labeled A1. Then the low molecular weight Heparin (LMWH) is administered. Over these 7 observations, the D-Dimer levels dropped (labeled B). The LMWH is then removed, and the D-Dimer levels resume to levels similar to that before the intervention (labeled A2).

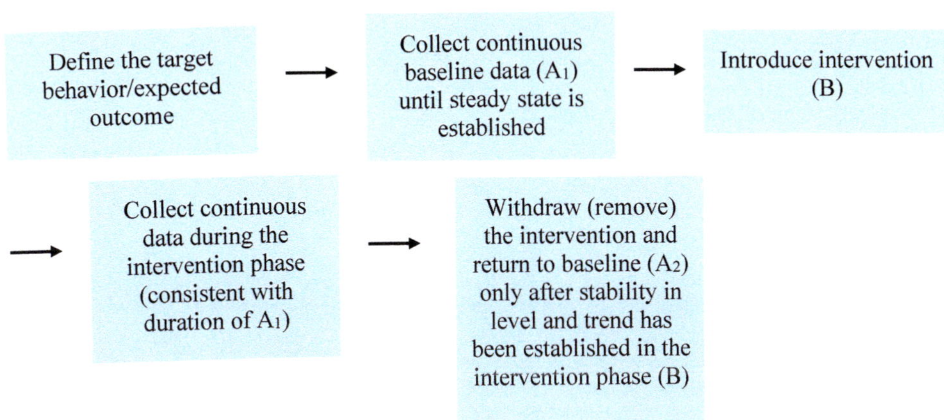

FIGURE 52.3 Simple step algorithm for performing a withdrawal study.

department with left calf pain and swelling after returning from a 13-h flight. The patient had no previous surgeries, denies smoking, and does not drink alcohol. The patient denied a family history of thromboembolism or clotting disorders and is not currently taking medications. His blood pressure was 132/93, heart rate of 92 bpm, and respiratory rate of 19 breaths per minute. Physical exam revealed obvious swelling of the left calf and moderate left calf tenderness with a +1 pitting edema. Diagnostic studies revealed an elevated D-dimer of 4.21 mg/L (normal <0.55 mg/L).[5] The patient was diagnosed with a left calf deep vein thrombosis, and the physician prescribed a low-molecular weight heparin (LMWH). To determine the efficacy of this LMWH, a withdrawal design was conducted to measure the drug's effects on lowering the D-dimer. D-Dimer levels were observed at 15 min intervals at baseline before the LMWH was administered (A_1), during the administration of LMWH (B), and after removal of LMWH (A_2). The results are depicted in Fig. 52.2.

Based on the results, one can reasonably conclude that the administration of LMWH significantly contributes to the drop in D-Dimer levels in a patient with a deep venous thrombosis of the lower extremity.

Get started

The simple steps required to carry out a withdrawal study can be summarized in Fig. 52.3. By maintaining steady state before including the intervention, maintaining treatment integrity during the intervention phase, and establishing internal validity after removing the intervention, researchers are better able to establish a clear cause-effect phenomenon from the intervention.

References

1. Gast D, Hammond D. Withdrawal and reverse designs. In: Ledford J, Gast D, eds. *Single Subject Research Methodology in Behavioral Sciences*. Taylor and Francis Group; 2009:234–275.
2. Barker J, McCarthy P, Jones M, Moran A. The withdrawal design in sport and exercise. In: Barker J, McCarthy P, Jones M, Moran A, eds. *Single-case Research Methods in Sport and Exercise Psychology*. Taylor and Francis Group; 2011:63–84.
3. Kim E, Won J, Kim J, Park J. Percutaneous pericardial effusion drainage under ultrasonographic and fluoroscopic guidance for symptomatic pericardial effusion: a single-center experience in 93 consecutive patients. *J Vasc Intervent Radiol*. 2015;26(10):1533–1538.
4. Morgan D, Morgan R. Single-case experimental designs: the withdrawal design. In: Morgan D, ed. *Single-Case Research Methods for the Behavioral and Health Sciences*. Thousand Oaks: SAGE Publications, Inc.; 2009:99–124.
5. *Labcorp.com*; 2020:115188. D-Dimer | Labcorp. [online] Available at: https://www.labcorp.com/tests/115188/d-dimer. Accessed June 17, 2020.

Chapter 53

Factorial design

Rebecca Tuan Le[1], Mauricio Hernandez[2] and Travis E. Meyer[3]
[1]*Diagnostic Radiology, Rochester General Hospital, Rochester, NY, United States;* [2]*Diagnostic Radiology, UF College of Medicine, Jacksonville, FL, United States;* [3]*Interventional Radiology, UF College of Medicine, Jacksonville, FL, United States*

Key points

- A factorial trial is a resource-saving study design that allows for concurrent evaluation of multiple interventions (i.e., factors) during one trial.
- Factorial trials are the only study design able to evaluate interactions between factors.
- Sample size must be evaluated before study initiation, as studies evaluating interactions require a larger target enrollment size compared to studies evaluating main effects to maintain the same power.

Why it matters: introduction

Originally utilized across medicine in prevention trials, factorial design is becoming more widespread when comparing interventions.[1,2] In factorial trials, investigators can evaluate two or more independent variables or factors within one trial. This contrasts with traditional randomized control trials (RCT), which require parallel evaluations of intervention, each with a placebo. Factorial designs can evaluate main effects or interaction effects. The main effect is the relationship between the factor (independent variable) and the clinical outcome, assuming the factors do not interact. An example of factors that do not interact includes treatments with different mechanisms of action that should yield similar outcomes.[1,3,4] Factorial trials can also study interactions between factors. Qualitative, or crossover,[5] interactions, indicate the reversal of one factor's effect in the presence of another. Quantitative, or spreading,[5] interactions, feature minor changes between factors that will not affect the direction of a treatment effect. Quantitative interactions can either be synergistic or antagonistic and either improve or detract from the projected effect, respectively.[4,6]

Benefits of factorial design

Factorial design may reduce research costs and time incurred. For instance, if there are two or more treatments without interactions, factorial design can evaluate their main effects with a reduced sample size while maintaining comparable precision as a parallel arms RCT. Factorial design is also the only design available to evaluate interactions between factors because it exposes groups to all available treatment combinations within the study and allows for responses to be directly compared.[1] Factorial design can complement a traditional RCT and be used to identify target candidates for further evaluation of interventions in a later RCT.[2] Investigators can also exclude certain treatment arms and design an incomplete, or partial, factorial trial. This usually occurs if there are infeasible treatment combinations or if the investigation lacks clinical equipoise.[1]

Limitations of factorial design

Factorial design is not suitable for evaluating interventions that cannot be used in conjunction.[3,4,7] Likewise, accessing an expanded number of factors may increase, rather than reduce, study costs, which may create challenges in study recruitment and protocol adherence. From an analysis standpoint, there is a risk of multiplicity or inflation of the type I error rate[8,9] as the design allows for comparisons across multiple treatment arms. Likewise, reducing treatment arms for a partial factorial trial will reduce the power to detect interactions and subsequently, complicate the analyses and interpretations of main effects and interactions (Fig. 53.1).[1]

Translational Interventional Radiology. https://doi.org/10.1016/B978-0-12-823026-8.00016-X

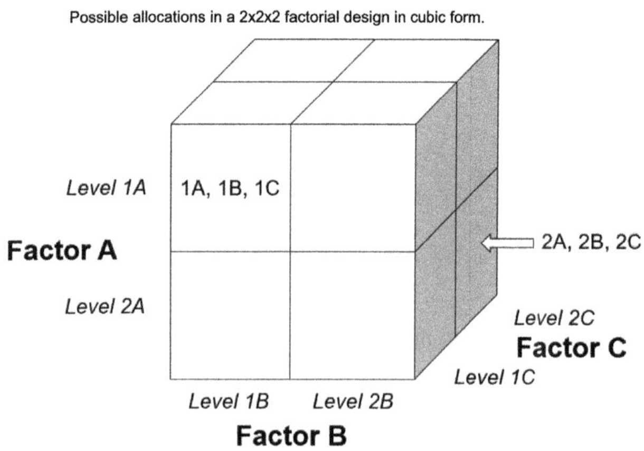

FIGURE 53.1 Possible allocations in a 2 × 2 × 2 factorial design in cubic form.

TABLE 53.2 Possible allocations in a 2 × 2 × 2 factorial design in list form.

Group	Treatment A	Treatment B	Treatment C
1	Level 1 A	Level 1 B	Level 1 C
2	Level 2 A	Level 1 B	Level 1 C
3	Level 1 A	Level 2 B	Level 1 C
4	Level 1 A	Level 1 B	Level 2 C
5	Level 2 A	Level 2 B	Level 1 C
6	Level 2 A	Level 1 B	Level 2 C
7	Level 1 A	Level 2 B	Level 2 C
8	Level 2 A	Level 2 B	Level 2 C

Factorial design—a step-by-step guide

The simplest and most common type of factorial design is a 2 × 2 design,[1,10] with two treatments and four allocation options (Table 53.1). Each value in the notation represents the factor, while the number of values represents the amount of levels per factor. The nomenclature becomes more complex as more factors or levels are added. For instance, if an investigator wanted to study three different medications (A, B, C) with two different dosage options, the study would be recorded as a 2 × 2 × 2 factorial design. It would feature eight allocation options, listed in the table (Table 53.2) and cubic (Fig. 53.2) format.

Investigators must first determine whether the primary goal of the study is to assess treatment interactions or to evaluate the effects of multiple noninteracting therapies. A common misconception is that a study can simultaneously investigate the main effect of factors and their interactions. However, this is not possible, because a trial evaluating interactions requires four times the subjects to maintain the same power as a trial only evaluating main effects.[1] A minimum sample size for statistical rigor/soundness can be calculated via a power analysis. Regardless of the study's goal, interaction dynamics should be assessed during the study design. Inadequate interaction assessment is a common pitfall when analyzing factorial trials, which produce misleading results.[11] There can be a degree of uncertainty in assessing the impact of interaction. One strategy is to evaluate a possible interaction clinically and gauge its influence within the trial.[12] Interactions could be statistically analyzed with a significance test (e.g., *P*-value). Furthermore, an investigator could utilize supplemental analyses, such as Bayesian methods, which produce results based upon previously known data, a measure of uncertainty (e.g., confidence interval), or a multiarm analysis as a test of sensitivity.[11,12]

Results from the factorial trial should include primary analysis from the margins, or a total number of participants allocated to each level of a factor, from the 2 × 2 table. Results entail an odds ratio or risk ratio and 95% confidence intervals that compare participants allocated to receive an intervention against participants receiving an alternative intervention, standard of care, or placebo. The study should also present descriptive statistics for outcome measures at follow-up for each intervention arm, to allow for interpretation of antagonism or synergism between the interventions.[12] A test and confidence interval for the interaction should also be provided, even if the trial is only powered for main effects. Because trials interested in evaluating main effects may be underpowered to study interactions, the confidence interval will be wide and should caution the reader of low validity.[12]

TABLE 53.1 Sample table for a 2 × 2 factorial design.

		Factor 1		
		Leve1 1A	Level 2A	Margin
Factor 2	Level 1B	1A + 1B	2A + 1B	Patients with 1B
	Level 2B	1A + 2B	2A + 2B	Patients with 2B
Margin		Patients with 1A	Patients with 2A	Total

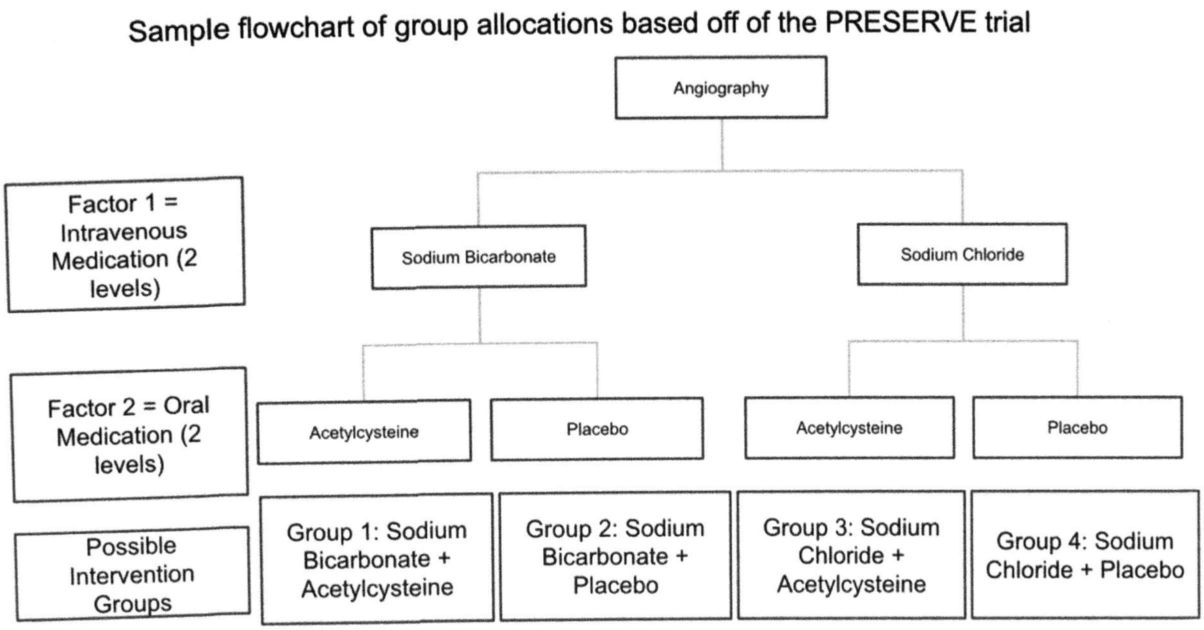

FIGURE 53.2 Sample flowchart of group allocations based off of the PRESERVE trial.

TABLE 53.3 Examples of published trials using factorial designs.

First author, reference number	Design	Cohort	Treatments	Outcomes
Lee[16]	2 × 2	Patients Requiring nonemergent Percutaneous coronary intervention	Factor 1: stent (everolimus-eluting stent or biolimus-eluting stent) Factor 2: implant Method (optical Coherence tomography or angiography)	Primary: difference in early strut coverage Secondary: cardiac death, myocardial infarction, stent thrombosis, major bleeding During 12-month follow-up
Kim[17] (PICASSO)	2 × 2	Patients with ischemic stroke and history of intracerebral hemorrhage	Factor 1: antiplatelet (cliostazol or aspirin) Factor 2: probucal arm (probucol or no probucol)	Primary: vascular events, cerebral hemorrhage Secondary: stroke, ischemic stroke, myocardial infarction, other vascular events, cardiovascular death, all-cause death
Nelson[18]	2 × 2 × 2	Patients with venous leg ulcer at least 1 cm in length for over 2 months	Factor 1: medication (oxpentifylline sustained-release 400 mg, three times per day vs. placebo) Factor 2: dressing (hydrocolloid dressing or knitted varicose dressing) Factor 3: layer amount (four-layer bandage or single-layer adhesive bandage)	Primary: time to complete healing Secondary: proportions, healed, withdrawals, adverse events

An example of a 2×2 factorial trial is the Prevention of Serious Adverse Events Following Angiography (PRESERVE) trial,[13] a multinational, randomized, double-blinded, 2×2 factorial clinical trial concurrently evaluating the effect of sodium bicarbonate and oral acetylcysteine on 90-day adverse outcomes (i.e., death, need for dialysis, a persistent increase of at least 50% from baseline in 3-month serum creatinine level) after angiography in patients with chronic kidney disease. The primary aim of this trial was to evaluate the main effects (i.e., sodium bicarbonate, oral acetylcysteine), and the investigators estimated 1920 patients per treatment arm for 90% power.[14] 5177 patients were allocated to one of four groups, illustrated in Fig. 53.2.

No interaction was noted between sodium bicarbonate and acetylcysteine ($P = .33$),[15] and there was no significant association between sodium bicarbonate use (OR, 0.93; 95% CI 0.72−1.22; $P = .62$) and oral acetylcysteine (OR, 1.02; 95% CI 0.78−1.33; $P = .88$).

Potential pitfalls

Potential pitfalls to avoid when using factorial design include choosing factors that cannot be used together in real life or choosing too many factors to evaluate, which would paradoxically increase study costs. Increasing the number of factors would also increase the complexity of the study protocol and provide additional challenges in study recruitment. From an analysis standpoint, there is a risk of multiplicity or inflating the type I error rate with factorial design. Moreover, reducing treatment arms in a partial factorial design can reduce the power in identifying interactions.

Real-world examples

There are many other examples of factorial design in the literature, in addition to the PRESERVE trial. Factorial design has been used to evaluate differences in early struct coverage based on type of coronary stent used and implantation method (16) or to evaluate incidence of vascular events in patients presenting with ischemic stroke and history of intracerebral hemorrhage by evaluating type of antiplatelet therapy or probucal administrations (17). Nelson et al developed a 2 x 2 x 2 factorial design study evaluating healing outcomes in patients with venous leg ulcers based on medication regiment, dressing, and layer amount of the dressing. Additional details of published factorial trials are available in Table 53.3.

Get started

1. Evaluate the study question and interventions of interest to gauge applicability in utilizing the factorial design. Are there multiple interventions that need to be evaluated that may be biologically unlikely to interact and may be expected to yield similar clinical outcomes? Would every proposed combination be ethically appropriate, and if not, could the study be modified into a partial factorial trial?
2. Determine whether the study needs to evaluate the main effects or interactions. Remember that a study should not be designed to simultaneously evaluate the main effects and qualitative interactions.
3. Consider time, costs, and recruitment requirements for the target sample size. Will the study have sufficient power to detect intended outcomes?
4. Analyze margins of tables, including primary analysis with an odds ratio and confidence intervals.
5. Further consultation with a team of experts or statisticians can be beneficial given the complexity behind the design, implementation, and analysis of factorial trials, so that interpretation of results is satisfactory.

Resources

For additional resources, consider purusing Piantadosi's chapter on factorial design (1).

References

1. Piantadosi S. Factorial designs. In: Piantadosi S, ed. *Clinical Trials: A Methodologic Perspective*. 3rd ed. John Wiley & Sons, Inc; 2017:671−683.
2. Collins LM, Dziak JJ, Kugler KC, et al. Factorial experiments: efficient tools for evaluation of intervention components. *Am J Prev Med*. 2015;47(4):498−504. https://doi.org/10.1016/j.amepre.2014.06.021.
3. Whelen DB, Dainty K, Chahal J. Efficient designs: factorial randomized trials. *J Bone Joint Surg Am*. 2012;94(Suppl 1 (E)):34−38. https://doi.org/10.2106/JBJS.L.00243.
4. Gurusamy KS, Gluud C, Nikolova D, et al. Design of surgical randomized controlled trials involving multiple interventions. *J Surg Res*. 2011;165(1):118−127. https://doi.org/10.1016/j.jss.2009.09.054.
5. Cuttler C. 9.2 interpreting the results OF a factorial experiment. In: Cuttler C, ed. *Research Methods in Psychology*. 3rd ed. Creative Commons; 2017.
6. McAlister FA, Straus SE, Sackett DL, et al. Analysis and reporting of factorial trials: a systematic review. *JAMA*. 2003;289(19):2545−2553. https://doi.org/10.1001/jama.289.19.2545.

7. Evans SR. Clinical trial structures. *J Exp Stroke Transl Med.* 2010;3(1):8–18. Available at: https://www.ncbi.nlm.nih.gov/pmc/articles/PMC3059315/. Accessed May 30, 2020.
8. Green S, Liu P, O'Sullivan J. Factorial design considerations. *J Clin Oncol.* 2002;20(16):3424–3430. https://doi.org/10.1200/JCO.2002.03.003.
9. Li G, Taljaard M, Wan den Heuvel ER, et al. An introduction to multiplicity issues in clinical trials: the what, why, when and how. *Int J Epidemiol.* 2017;46(2):746–755. https://doi.org/10.1093/ije/dyw320.
10. Factorial trials. In: Higgins JPT, Eldridge S, Li T, eds. *Cochrane Handbook for Systematic Reviews of Interventions*. 6th ed. Available from: www.training.cochrane.org/handbook.
11. Kahan BC, Tsui M, Jairath V, et al. Reporting of randomized factorial trials was frequently inadequate. *J Clin Epidemiol.* 2020;117:52–59. https://doi.org/10.1016/j.jclinepi.2019.09.018.
12. Montgomery AA, Peters TJ, Little P. Design, analysis and presentation of factorial randomised controlled trials. *BMC Med Res Methodol.* 2003;3:26. https://doi.org/10.1186/1471-2288-3-26.
13. Weisbord SD, Gallagher M, Jneid H, et al. Outcomes after angiography with sodium bicarbonate and acetylcysteine. *N Engl J Med.* 2018;378(7):603–614. https://doi.org/10.1056/NEJMoa1710933.
14. Weisbord SD, Gallagher M, Jneid H, et al. Outcomes after angiography with sodium bicarbonate and acetylcysteine. *N Engl J Med.* 2018;378(7):603–614. https://doi.org/10.1056/NEJMoa1710933. Protocol:.
15. Weisbord SD, Gallagher M, Jneid H, et al. Outcomes after angiography with sodium bicarbonate and acetylcysteine. *N Engl J Med.* 2018;378(7):603–614. https://doi.org/10.1056/NEJMoa1710933. Supplementary Appendix.
16. Lee S, Kim J, Yoon H, et al. Early strut coverage in patients receiving drug-eluting stents and its implications for dual antiplatelet therapy: a randomized trial. *JACC Cardiovasc Imag.* 2018;11(12):1810–1819. https://doi.org/10.1016/j.jcmg.2017.12.014.
17. Kim BJ, Lee E, Kwon SU, et al. Prevent of cardiovascular Events in Asian patients with ischaemic stroke at high risk of cerebral haemorrhage (PICASSO): a multicentre, randomised controlled trial. *Lancet Neurol.* 2018;17(6):509–518. https://doi.org/10.1016/S1474-4422(18)30128-5.
18. Nelson EA, Prescott RJ, Harper DR, et al. A factorial, randomized trial of pentoxifylline or placebo, four-layer or single-layer compression, and knitted viscose or hydrocolloid dressings for venous ulcers. *J Vasc Surg.* 2007;45(1):134–141. https://doi.org/10.1016/j.jvs.2006.09.043.

Chapter 54

Group allocation: Design, measures, classic examples

Priyam Choudhury[1] and Ashutosh Mohapatra[2]

[1]Department of Radiology, Mayo Clinic, Jacksonville, FL, United States; [2]Department of Orthopaedics, Mohapatra Hospital, Ulhasnagar, Maharashtra, India

Key points

1) Group allocation is an effective simple way to compare treatment outcomes among group/individuals.
2) It is important to prevent confounding bias, selection bias, and observer bias for this type of study.
3) Various measures such as randomization, blinding, and allocation concealment are best used to avoid bias in this type of study.

In a controlled clinical study (clinical trial) to compare two or more treatments for a disease, an objective method of allocating patients to treatment groups is needed. When it is not known in advance which treatment is going to be more effective, this process is called group allocation. Irrespective of whether the treatment is earlier established or new, a prospective, well-planned clinical study helps to achieve a meaningful comparison. The advantage of this study being—it decreases the difficulty of approaching people about the idea of randomization and have more acceptability, and the disadvantages being—may not be as effective as responses vary across clinics or groups. The various methods of group allocations and some simple modifications accompanying them:

Subjective procedure: While allotting treatment to a given patient of a certain group, clinicians may consciously or unconsciously choose the treatment for the patient which will have a better prognosis for them, beating the whole assumption behind a study that the better of the treatment is unknown. Hence methods like alternate allocation where treatment A is given to the first patient and treatment B is given to the second patient, treatment A to the third is being followed or other methods of allocation by month of birthday, first letter of last name, is being followed.[1]

Minimizing confounders is the core to conduct a proper research because they have the potential to introduce erroneous results because of unnoticed effects of variables.

A) "Randomization" is the process in which confounders can be reduced to minimum.

Benefits of randomization:

1. Eliminates confounding bias.
2. Eliminates selection bias.
3. Helps in forming the basis of statistical tests.[2]

Techniques for randomization:

Simple randomization: Randomization is done according to a single sequence of random assignment, which is not concerned with other variables. For example, rolling of dice.

Block randomization: It ensures that a number of participants in the study groups are nearly equal for a large sample size and longer follow-up time. Different blocks are created, which ensures the same number of people gets allocated in each block.[3] To ensure blindness, blocks of random block size are often used.

Stratified randomization: Stratification of the sample is done for a particular variable (age, gender, weight, etc.) to keep the sample as similar as possible in between groups as

Translational Interventional Radiology. https://doi.org/10.1016/B978-0-12-823026-8.00062-6

well as in each stratum to achieve a balance between baseline characteristics. For example: to assess treatment, results for immunotherapy for viral warts stratification should be done for the type of viral warts.

Cluster randomization: Also called as the "group randomization," this method randomizes groups of people instead of people. Group randomization is typically used when outcomes of participants from the same group tend to be correlated, for example, patients receiving care from the same clinics, vaccine trials, and nutritional interventions for family members.

B) Blinding is a process in which the participants or the assessing physicians are kept unaware of the treatment being received therefore eliminating bias. Types of blinding:

1. Unblinded or open-labeled: Every party in the study is aware of the treatment the participants are receiving because blinding is not possible sometimes.
2. Single-blind: The participants or the assessor interpreting the outcome in the study are unaware of the treatment modality being given.
3. Double-blind: Both the participants and the investigators are unaware of the treatment being given.
4. Triple-blind: Here, the participants, the investigators, and the data analysts are unaware of the treatment being given.

C) Concealment of allocation is done as the participant is entering the study, it prevents selection bias and conceals the randomization sequence.

D) Minimization: It is a process by which large numbers of strata can be balanced in a more effective way than simple and stratified randomization by the investigators. When there are more than one variable as prognostic factors, this method can be used to balance the numbers among all combinations of those variables, as well as between the remaining variables among themselves.[1] It helps in randomization while allocating subjects to the treatment groups while maintaining a balance of the multiple prognostics variables.

E) Optimal batchwise minimization (OBM): When evaluating closed group interventions, all the treatment allocations are done when the whole batch is enrolled. It is based on three steps based on algorithms to be carried out for each consecutive group of enrolled patients:

1. The sum of the squared factor class imbalances (SSFCI) is calculated for all the enrolled subjects in the study.[4]
2. Allocations with SSFCI are listed.
3. Randomly one of the allocations with minimum SSFCI is selected.

It was established that minimization and OBM have less risk of imbalance than other randomizations. Although it is difficult to measure imbalance it has been estimated that the composition, the number of prognostic factors, strength of relations between them, nature of the group, sample size, may have a subsequent impact on outcome.[5] To obtain overall balance, minimization is a better choice for methodology. OBM leads to better balance within batches and is better suited for studies with many prognostic factors with the lowest possible predictability.

Examples of group allocation: One of the earliest examples was a clinical trial by Celli in 1900, which was a step toward modern cluster-randomized trial design.[6] He studied whether mosquito net was an effective intervention in reducing malaria in households of Italian railway workers. Neighboring households were kept as controls and certain households were selected based on getting/not getting the intervention. In this study, members of the same household tended to have correlated outcomes, and hence were treated as one cluster. Clusters were used as the basic units for randomization.

Amberson et al. conducted a trial in 1931, challenging the use of gold-containing treatment (sanocrysin) for treating TB. Two matched groups were made by allocating through a coin toss, and the control was injected with distilled water. Both the patients and the investigators were blinded.[7] Sometimes contamination can be a problem in screening program studies between groups, a study by Tabar et al. in 1985 significantly reduced this problem by offering screening to communities selected randomly from matched communities separated by long distance (Fig. 54.1).

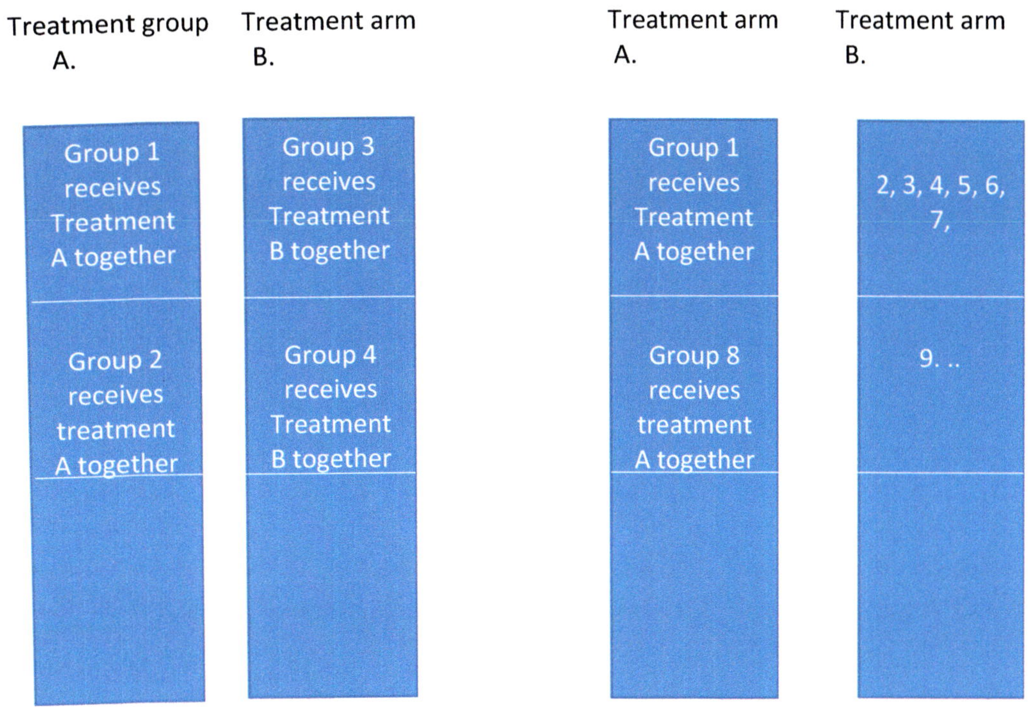

FIGURE 54.1 Design of a trial with group treatments in both arms and compares group treatment to individual treatment.

References

1. White SJ, Freedman LS. Allocation of patients to treatment groups in a controlled clinical study. *Br J Cancer*. 1978;37(5):849–857. https://doi.org/10.1038/bjc.1978.124.
2. Sil A, Kumar P, Kumar R, Das NK. Selection of control, randomization, blinding, and allocation concealment. *Indian Dermatol Online J*. 2019;10(5):601–605. https://doi.org/10.4103/idoj.IDOJ_149_19. Published 2019 Aug 28.
3. Parikh MN, Hazra A, Mukherjee J, Gogtay N. *Research Methodology Simplified Every Clinician A Researcher*. New Delhi: Jaypee Brothers Medical Publishers; 2010.
4. William CJ, Rosenberger F, Lachin JM. *Randomization in Clinical Trials: Theory and Practice*. 2nd ed. Vol 73. Hoboken: John Wiley & Sons. Biometrics; 2017:1054–1055. https://doi.org/10.1111/biom.12756 (3).
5. Faes MC, Reelick MF, Perry M, Olde Rikkert MG, Borm GF. Studies with group treatments required special power calculations, allocation methods, and statistical analyses. *J Clin Epidemiol*. 2012;65(2):138–146. https://doi.org/10.1016/j.jclinepi.2011.05.007.
6. Celli A. The new prophylaxis Against malaria: an account of experiments in latium. *Lancet*. 1900;156(4031):1603–1606. https://doi.org/10.1016/s0140-6736(01)89058-9.
7. Pask E. Sanocrysin in pulmonary tuberculosis. *Lancet*. 1931;218(5651):1346–1348. https://doi.org/10.1016/s0140-6736(00)99621-1.

Further reading

1. Tabár L, Gad A, Holmberg L, et al. Reduction in mortality from breast cancer after mass screening with mammography. *Lancet*. 1985;325(8433):829–832. https://doi.org/10.1016/s0140-6736(85)92204-4.

Chapter 55

Large, pragmatic: Design, measures, and classic examples

Rachel Drummey
The Russell H. Morgan Department of Radiology and Radiological Science, The Johns Hopkins Hospital, Baltimore, MD, United States

Key points

- Pragmatic trials aim to achieve greater external validity and generalizability by designing studies that closer resemble everyday clinical practice in contrast to explanatory trials that assess the effect of an intervention in ideal clinical practice.
- The intent of pragmatic studies is to be diverse, inclusive, practical, and patient-centered by testing interventions on representative populations mirroring those who receive the intervention in actual clinical care.
- Pragmatic and explanatory clinical trials exist on a continuum rather than a dichotomy.

Definition of a pragmatic clinical trial

The term pragmatic clinical trials was first identified by Schwarts and Lellouch in 1967 when they articulated a new approach to clinical trials that aimed to determine the effectiveness of interventions for the patient receiving care in the typical clinical setting.[1–3] The purpose behind this development was to expose discrepancies between outcomes observed in clinical trials and outcomes observed at the individual and population level.[4] Ultimately, PCTs aim to increase generalizability (or external validity) in exchange for an accepted drop in internal validity due to the intentionally more heterogeneous patient sample (Fig. 55.1).

Pragmatic and explanatory clinical trials are not binary, but rather exist on a spectrum. To illustrate this, the Pragmatic Explanatory Continuum Indicator Summary (PRECIS) was developed and revised (PRECIS-2 in 2015) to aid trial designers in quantifying the elements of a trial to assess where it lies on the explanatory-pragmatic continuum on a scale of 0–5 (Table 55.3).[5,6] The nine domains identified include eligibility criteria, recruitment, setting, organization, flexibility (delivery), flexibility (adherence), follow-up, primary outcome, and primary analysis (Table 55.1).[6]

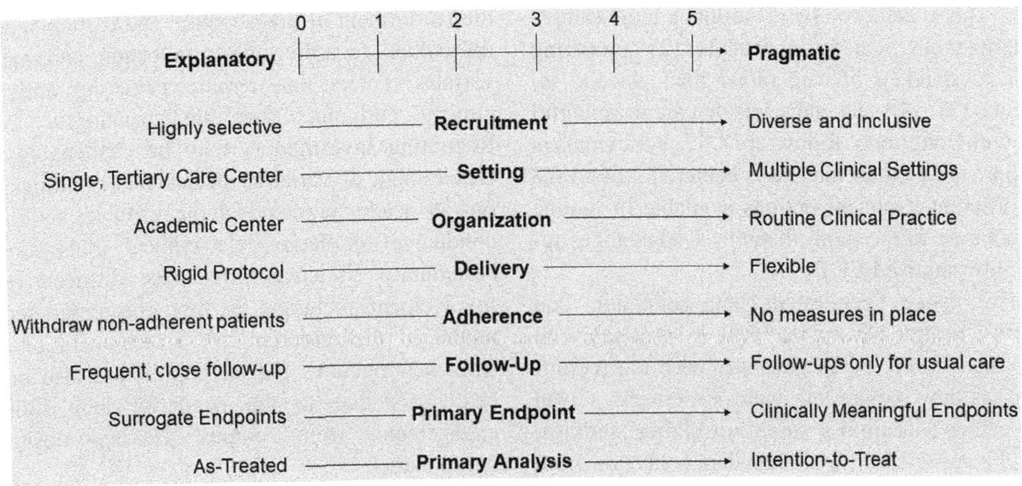

FIGURE 55.1 Pragmatic-explanatory continuum.

Translational Interventional Radiology. https://doi.org/10.1016/B978-0-12-823026-8.00075-4

TABLE 55.1 PRECIS-2 Domains to quantify trials on the pragmatic-explanatory continuum.[a]

Domains of pragmatism	Questions that quantify pragmatism	Pragmatic study characteristics
Eligibility	Who will be included in the trial?	No exclusion criteria.
Recruitment	How will participants be recruited to join the trial? Will more effort be put forth to recruit diverse participants compared to usual clinical trials?	Ample effort will be put into recruiting a heterogenous sample population.
Setting	In what clinical setting will the trial be completed?	A variety of practice settings, including community clinics and academic centers, in a mixture of urban, rural, and suburban areas.
Organization	What expertise and resources are necessary for this intervention to be performed?	Interventions performed in routine clinical care will be assessed.
Delivery	How should the intervention be delivered? How will the intervention be delivered?	Clinicians may implement the intervention without strict guidelines or oversight.
Adherence	Will participants be monitored for adherence? How will it be handled if participants do not adhere to the intervention?	Adherence is encouraged, but steps are not taken to ensure participants or clinicians adhere rigidly to protocol.
Follow-up	How closely will participants be followed?	Ideally long follow-up will be achieved, but with fewer formal, required visits.
Primary endpoint	Is this endpoint clinically meaningful to patient outcomes? Will this outcome help guide policy?	Clinically meaningful endpoints should be important to patients and easily assessed and obtained from EHRs.
Primary analysis	To what extent is all data included and analyzed?	All participants are included and analyzed with intention-to-treat methodology.

[a]*Loudon K, Treweek S, Sullivan F, Donnan P, Thorpe KE, Zwarenstein M. The PRECIS-2 tool: designing trials that are fit for purpose. Bmj. 2015;350:h2147. https://doi.org/10.1136/bmj.h2147 [published Online First: Epub Date].*
Adapted from Loudon K, Treweek S, Sullivan F, Donnan P, Thorpe KE, Zwarenstein M. The PRECIS-2 tool: designing trials that are fit for purpose. Bmj. 2015;350:h2147 https://doi.org/10.1136/bmj.h2147 [published Online First: Epub Date].

Benefits and limitations of PCTs

Benefits

A central strength of PCTs is the inclusive, practical, and patient-centered design, which operate by extrapolating data from trials and applying it directly to real-world clinical practice.[7] This is achieved by (1) utilizing large sample sizes of heterogenous, real-world patients, (2) comparing interventions to standard of care rather than placebo interventions, and (3) assessing endpoints that are meaningful to patients with long-term follow-up.[3,5,7,8] Policymakers favor this approach as their interest is centered around the efficacy of different treatment options available in routine practice in contrast to the more theoretical advantages reported in highly sanitized ECTs.[9]

In terms of patient recruitment, data collection, and analysis, PCTs design allows the trial to proceed with greater ease than ECTs. PCTs generally ease the recruitment burden, as they require less inclusion criteria. Given the need to gather data from a variety of clinical settings, trials should be designed to streamline data collection from electronic health records (EHRs) and consider linking data from multiple EHRs into a compiled database.

Limitations

Despite their many advantages, PCTs encounter barriers in the recruitment of investigators and participants, as well as difficulties regarding data collection and analysis from various centers that require gathering and quantifying clinical endpoints that are meaningful to patients.[7] Recruiting investigators may be challenging due to the requirement of variation in healthcare settings.[7,10] Clinics outside academia often lack the resources and infrastructure conducive for large-scale clinical trials. For example, community clinicians often face disincentives for trial involvement, including the time consuming and expensive nature of the research.[10–12] Recruitment of participants from heterogenous patient populations also poses unique challenges due to the need for long follow-up and engagement from clinics lacking onsite research coordinators.[10]

Regarding data collection and analysis, endpoints that are important to the patient may be more challenging to record and quantify, and data extraction from free-text EHRs can be laborious. Changes in private practice leadership or EHRs can lead to the loss of valuable patient data.[13] Additionally, a truly pragmatic trial cannot be blinded because the intervention is meant to be delivered by staff in a routine fashion. For these reasons, PCTs are more susceptible to bias and confounding variables than classic ECTs.[14]

Steps to design a pragmatic study

1. Selection and inclusion of sites where the intervention is typically delivered.
2. Participant selection, recruitment, and retainment.
3. Identification of a question that can be compared to the usual care for patients in everyday clinical practice. Potential challenge in variance of "standard clinical care" between institutions.
4. Selection of endpoint markers that are meaningful to patients, clinicians, and policymakers. Outcome measurements should not correct for imperfect decision-making in normal clinical settings, including medication nonadherence and medication interaction.
5. Utilizing existing systems, including EHR data, to streamline the data collection process and enable community clinician's participation.

Standard research team personnel and roles

See Table 55.2.

Budgetary considerations and funding pathways

At inception, pragmatic studies were very expensive and time-consuming due to the need for a large team of clinical staff to mine, compile, and analyze data. With the establishment of programs such as the NIH Health Care Systems Research Collaboratory, studies can now be performed at lower costs with collaborative efforts and resources.[15] PCTs are far more affordable in countries with national healthcare systems that enable easy and economical data-sharing, such as in the Netherlands.

Get started[5,16]

1. Build partnerships: Pragmatic trials require large, diverse patient populations, so collaboration with colleagues at different institutions and practice settings is essential.
2. Identify challenges at the start: Many PCTs struggle with participant recruitment and data collection and sharing.

TABLE 55.2 Research team for execution of PCTs and their roles.[a,b]

Research team member	Purpose	Action items
Clinician and healthcare staff	Contribute to study design and execute protocol	Integrate the needs of the study into existing clinic workflow
Researcher	Identify and answer questions that will improve clinical care and public health	Construct research studies investigating meaningful clinical questions that can be performed in a large-scale setting and integrated into existing clinical workflows.
Senior manager	Improve healthcare costs, healthcare efficiency, and patient outcomes.	Oversee work of all team members.
IT support	Enable swift and efficient data collection from existing clinical EHRs.	Optimize EHRs to streamline data collection. Advise trial designers in design trials to ensure data collection will be feasible.
Operational manager	Minimize interruptions to normal clinical workflow to ensure clinical success.	Identify interruptions to clinical workflow that may occur and proactively find solutions that will not interrupt normal workflow.

[a]*Johnson KE, Tachibana C, Coronado GD et al. A guide to research partnerships for pragmatic clinical trials. Bmj. 2014;349:g6826 https://doi.org/10.1136/bmj.g6826 [published Online First: Epub Date].*
[b]*Johnson KE, Neta G, Dember LM et al. Use of PRECIS ratings in the National Institutes of Health (NIH) Health Care Systems Research Collaboratory. Trials 2016;17(1):32 https://doi.org/10.1186/s13063-016-1158-y [published Online First: Epub Date].*
Adapted from Johnson KE, Tachibana C, Coronado GD et al. A guide to research partnerships for pragmatic clinical trials. Bmj. 2014;349:g6826. https://doi.org/10.1136/bmj.g6826 [published Online First: Epub Date].

3. Choose meaningful outcomes: Discuss with clinicians, patients, and policymakers to identify outcomes that are meaningful.
4. Ease study burden on clinicians: Ensure the study is designed such that data collection and statistical analysis will not be burdensome on individual healthcare providers or clinics. Select easily recordable data that is obtained from existing EHRs to make investigator and patient retention more successful.
5. Management: Include a senior manager in trial design to improve oversight and monitoring between and within sites to ensure patient and provider retention.

Potential pitfalls

1. Poor communication between partners: There must be clearly communicated inclusion criteria for patient recruitment and consistent understanding of necessary resources for the trial from the beginning. As demonstrated by Tables 55.4 and 55.5, the lack of patient enrollment can ultimately lead to trial suspension. Investigators must collaborate to overcome this challenge from the start.
2. Suboptimal system for data collection: Data collection must be integrated into existing workflows. If it is not, clinical investigators may drop out of the study due to clinician and staff burden.

Resources

1. https://getrealtrialtool.eu/
2. https://www.precis-2.org/

TABLE 55.3 Early versus deferred endovenous ablation of superficial venous reflux in patients with venous ulceration: the EVRA RCT.[a]

Inclusion criteria	Patients aged ≥18 years with a venous leg ulcer between 6 weeks' and 6 months' duration and an ankle-brachial pressure index of ≥0.8 who could tolerate compression and were deemed suitable for endovenous ablation of superficial venous reflux.
PRECIS-2 score in eligibility domain	Four—due to broad inclusion criteria with minimal exclusion criteria.

[a]*Gohel MS, Heatley F, Liu X, et al. Early versus deferred endovenous ablation of superficial venous reflux in patients with venous ulceration: the EVRA RCT. Health Technol Assess 2019; 23(24):1–96. https://doi.org/10.3310/hta23240 [published Online First: Epub Date].*

TABLE 55.4 Transarterial radioembolization versus chemoembolization for the treatment of hepatocellular carcinoma (TRACE): study protocol for a randomized controlled trial. Study has been suspended due to poor enrollment.[a]

Eligibility	Inclusion criteria: Intermediate stage hepatocellular carcinoma (BCLC class B; locally advanced, liver restricted disease, not eligible for curative treatment) with Eastern Cooperative Oncology Group (ECOG) performance status 0 to 1 and Child-Pugh A to B. Exclusion criteria: extrahepatic disease, abnormal organ or bone marrow function (as determined by bilirubin >45 µmol/L (or 2.6 mg/dL), serum albumin <28 g/L, AST/ALT > 5 × institutional upper limit of normal (ULN), creatinine >1.5 × institutional ULN, hemoglobulin <6.0 mmol/L, absolute neutrophil count <1.5 × 109/L, platelet count <60 × 109/L), compromised biliary system, hypersensivity to doxorubicin, pregnancy or breast feeding, >50% of liver involvement, main portal vein (right, left or common trunk) thrombosis, unmanageable intolerance for contrast medium, life expectancy less than 3 months or otherwise impossible follow-up.
Recruitment	Patients will be recruited after referral by their hepatologist or another treating physician. Treatment allocation will be performed by minimization, which is a valid alternative to standard randomization to ensure balance between groups for several prognostic factors.
Setting	This is a multicenter, randomized controlled trial.

[a]*Seinstra BA, Defreyne L, Lambert B, et al. Transarterial radioembolization versus chemoembolization for the treatment of hepatocellular carcinoma (TRACE): study protocol for a randomized controlled trial. Trials 2012; 13:144. https://doi.org/10.1186/1745-6215-13-144 [published Online First: Epub Date].*

TABLE 55.5 Liver resection surgery versus thermal ablation for colorectal LiVer MetAstases (LAVA): study protocol for a randomized controlled trial. Study has been suspended due to poor enrollment.[a,b]

Intervention	Thermal ablation conducted as per local policy (but centers were encouraged to recruit within Cardiovascular and Interventional Radiological Society of Europe guidelines) versus surgical liver resection performed as per center protocol.
Primary outcome	Disease-free survival (DFS) at 2 years postrandomization.
Secondary outcomes	Overall survival, timing and site of recurrence, additional therapy after treatment failure, quality of life, complications, length of hospital stay, costs, trial acceptability, DFS measured from end of intervention.
Control	Surgical liver resection performed as per center protocol.
Shortcomings	Study was suspended because only nine participants were randomized at 1 year. The key challenges included less eligible patients than anticipated for both treatments, misconceptions about eligibility criteria, surgeons' preference for one treatment with unconscious bias toward surgery, patients' preference of treatment, and lack of research nurses supporting the trial.

[a]*Davidson B, Gurusamy K, Corrigan N, et al. Liver resection surgery compared with thermal ablation in high surgical risk patients with colorectal liver metastases: the LAVA international RCT. Health Technol Assess 2020;24(21):1–38. https://doi.org/10.3310/hta24210 [published Online First: Epub Date].*
[b]*Gurusamy K, Corrigan N, Croft J, et al. Liver resection surgery versus thermal ablation for colorectal LiVer MetAstases (LAVA): study protocol for a randomized controlled trial. Trials 2018;19(1):105. https://doi.org/10.1186/s13063-018-2499-5 [published Online First: Epub Date].*

References

1. Ginexi EM, Hilton TF. What's next for translation research? *Eval Health Prof.* 2006;29(3):334–347. https://doi.org/10.1177/0163278706290409 [published Online First: Epub Date].
2. Rothwell PM. External validity of randomised controlled trials: "to whom do the results of this trial apply?". *Lancet.* 2005;365(9453):82–93. https://doi.org/10.1016/s0140-6736(04)17670-8 [published Online First: Epub Date].
3. Schwartz D, Lellouch J. Explanatory and pragmatic attitudes in therapeutical trials. *J Chron Dis.* 1967;20(8):637–648. https://doi.org/10.1016/0021-9681(67)90041-0 [published Online First: Epub Date].
4. Califf RM, Sugarman J. Exploring the ethical and regulatory issues in pragmatic clinical trials. *Clin Trials.* 2015;12(5):436–441. https://doi.org/10.1177/1740774515598334 [published Online First: Epub Date].
5. Thorpe KE, Zwarenstein M, Oxman AD, et al. A pragmatic-explanatory continuum indicator summary (PRECIS): a tool to help trial designers. *J Clin Epidemiol.* 2009;62(5):464–475. https://doi.org/10.1016/j.jclinepi.2008.12.011 [published Online First: Epub Date].
6. Loudon K, Treweek S, Sullivan F, Donnan P, Thorpe KE, Zwarenstein M. The PRECIS-2 tool: designing trials that are fit for purpose. *Bmj.* 2015;350:h2147. https://doi.org/10.1136/bmj.h2147 [published Online First: Epub Date].
7. Ford I, Norrie J. Pragmatic trials. *N Engl J Med.* 2016;375(5):454–463. https://doi.org/10.1056/NEJMra1510059 [published Online First: Epub Date].
8. Tunis SR, Stryer DB, Clancy CM. Practical clinical trials: increasing the value of clinical research for decision making in clinical and health policy. *JAMA.* 2003;290(12):1624–1632. https://doi.org/10.1001/jama.290.12.1624 [published Online First: Epub Date].
9. Patsopoulos NA. A pragmatic view on pragmatic trials. *Dialogues Clin Neurosci.* 2011;13(2):217–224. https://doi.org/10.31887/DCNS.2011.13.2/npatsopoulos [published Online First: Epub Date].
10. Garcia CJ, Haynes K, Pokorney SD, et al. Practical challenges in the conduct of pragmatic trials embedded in health plans: lessons of IMPACT-AFib, an FDA-catalyst trial. *Clin Trials.* 2020;17(4):360–367. https://doi.org/10.1177/1740774520928426 [published Online First: Epub Date].
11. Davidson B, Gurusamy K, Corrigan N, et al. Liver resection surgery compared with thermal ablation in high surgical risk patients with colorectal liver metastases: the LAVA international RCT. *Health Technol Assess.* 2020;24(21):1–38. https://doi.org/10.3310/hta24210 [published Online First: Epub Date].
12. Gurusamy K, Corrigan N, Croft J, et al. Liver resection surgery versus thermal ablation for colorectal LiVer MetAstases (LAVA): study protocol for a randomised controlled trial. *Trials.* 2018;19(1):105. https://doi.org/10.1186/s13063-018-2499-5 [published Online First: Epub Date].
13. Haff N, Choudhry NK. The promise and pitfalls of pragmatic clinical trials for improving health care quality. *JAMA Netw Open.* 2018;1(6):e183376. https://doi.org/10.1001/jamanetworkopen.2018.3376 [published Online First: Epub Date].
14. Foster N, Little P. Methodological issues in pragmatic trials of complex interventions in primary care. *Br J Gen Pract.* 2012;62(594):10–11. https://doi.org/10.3399/bjgp12X616238 [published Online First: Epub Date].
15. Johnson KE, Neta G, Dember LM, et al. Use of PRECIS ratings in the National Institutes of Health (NIH) health care systems research collaboratory. *Trials.* 2016;17(1):32. https://doi.org/10.1186/s13063-016-1158-y [published Online First: Epub Date].
16. O'Rourke PP, Carrithers J, Patrick-Lake B, et al. Harmonization and streamlining of research oversight for pragmatic clinical trials. *Clin Trials.* 2015;12(5):449–456. https://doi.org/10.1177/1740774515597685 [published Online First: Epub Date].

Chapter 56

Equivalence and noninferiority: Design, measures, and classic example

Ranjan Ragulojan
University of Minnesota, Minneapolis, MN, United States

Key points

- Equivalence and noninferiority trials are used when the objective is to establish that the current standard of treatment is equivalent or not superior to the proposed novel therapy (NT) with regards to the effect on the primary clinical endpoint (PCE).
- Such trials potentially allow for the validation of novel therapies where the benefits lie outside the PCE, such as cost, practicality, compliance, and adverse events.
- The primary limitations are the challenge in determining an acceptable difference in effect compared to the control that would validate the NT as equivalent or noninferior, the requirement for adequate effect of the control and sufficient historical data of such, the larger sample sizes required, and the requirement for high-quality trials to avoid the false establishment of noninferiority/equivalence.

Why it matters?

When it comes to evaluating a new therapeutic drug or intervention, superiority trials assessing the superiority of the novel therapy compared to the control or placebo are often considered the default approach for approval. However, equivalence and noninferiority trials can be more practical to support the validation of the novel therapy in clinical practice. This chapter sets out to expound on such trials.

Equivalence trials attempt to establish whether the effect of the novel therapy (NT) on primary clinical endpoint (PCE) is similar to the control therapy, within a predetermined acceptable difference (equivalence margin). Noninferiority trials evaluate only whether the effect of the control is not greater than the NT, by a quantity that exceeds a predetermined acceptable difference.[1] Thus if b is allowed to be the predetermined difference in effect on the PCE allowable between the control A and NT B, the null hypothesis in an equivalence trial dictates that the effect difference between A and B is $>b$ and $<-b$ (Fig. 56.1) with an alternative hypothesis of A = B. In a noninferiority trial, the null hypothesis only pertains to the upper limit of the effect difference and thus must be $> b$ (Fig. 56.2). Therefore rejection of the null hypothesis in noninferiority trials allows for the possibility that the NT is actually superior to the control, whereas in equivalence trials we can conclude that A and B are the same (equivalent).

There are several reasons why an equivalency or noninferiority trial may be utilized as opposed to a superiority trial when evaluating an NT. Among those include instances where the benefits of the NT lie outside differences in the primary clinical outcome such as with regards to cost, practicality, compliance, adverse side effects, etc. Additionally, in many instances, it would be unethical or illegal to compare the NT to a placebo.[2]

However, a primary challenge exists in determining the value of b, referred to as the boundary margin in trials, which is deemed reasonable for the consideration of B to be equivalent or noninferior to A. There is no consensus method to determine the boundary margin; however, regulators do strongly recommend it is established before carrying out the trial.[3] It is also advisable to use a boundary margin, which if not surpassed, would allow for clinical and statistical arguments that the novel intervention is not only noninferior to the control, but also likely superior to a placebo treatment and thus acceptable in light of other secondary benefits.[1] A typical method to determine the boundary margin in clinical noninferiority trials, known as the synthesis approach,[4] is to set it at a fraction (f) of the value of the effect demonstrated in a meta-analysis by the control therapy compared to the placebo (at the lower limit of the 95 confidence interval (CI)) (Fig. 56.3).[2] A smaller f

Translational Interventional Radiology. https://doi.org/10.1016/B978-0-12-823026-8.00052-3

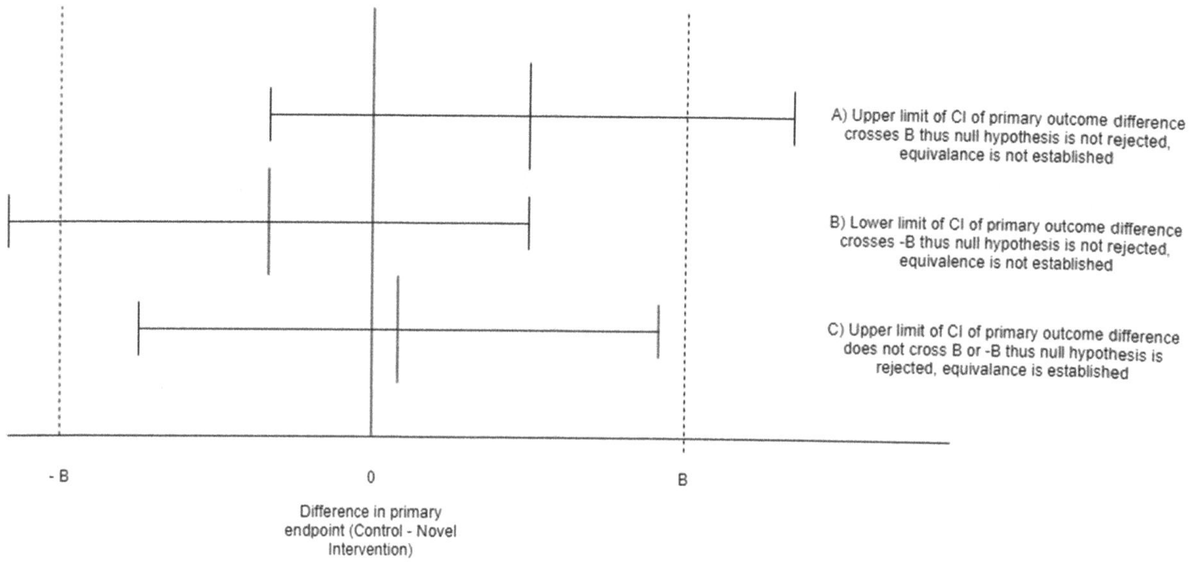

FIGURE 56.1 Equivalence trial. *B*, boundary value for difference in primary endpoint.

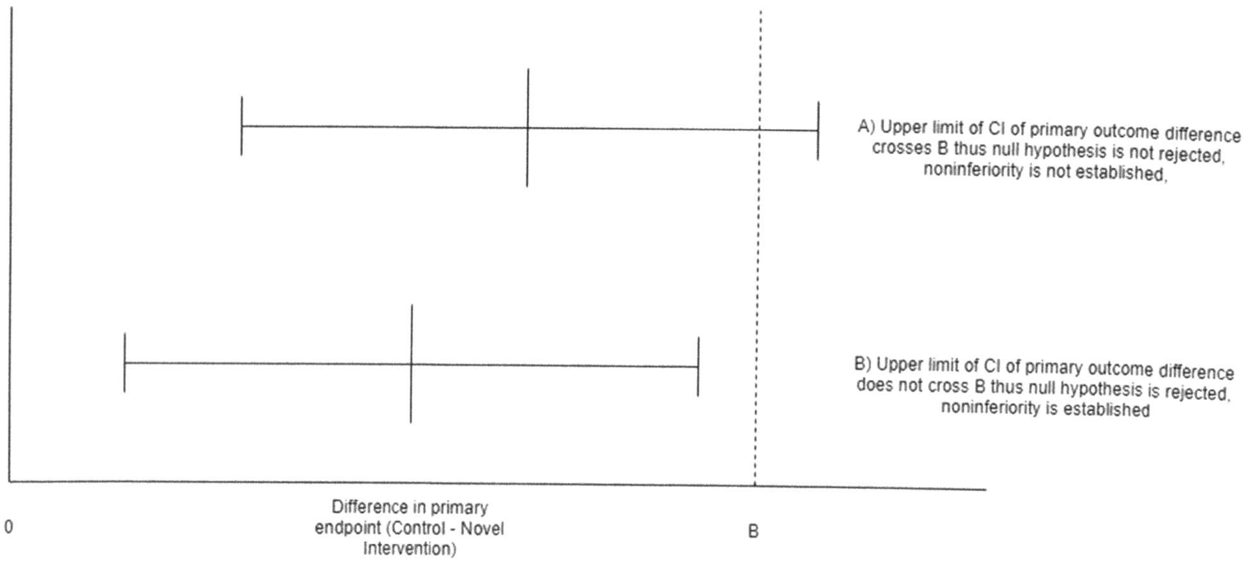

FIGURE 56.2 Noninferiority trial. *B*, boundary value for difference in primary endpoint.

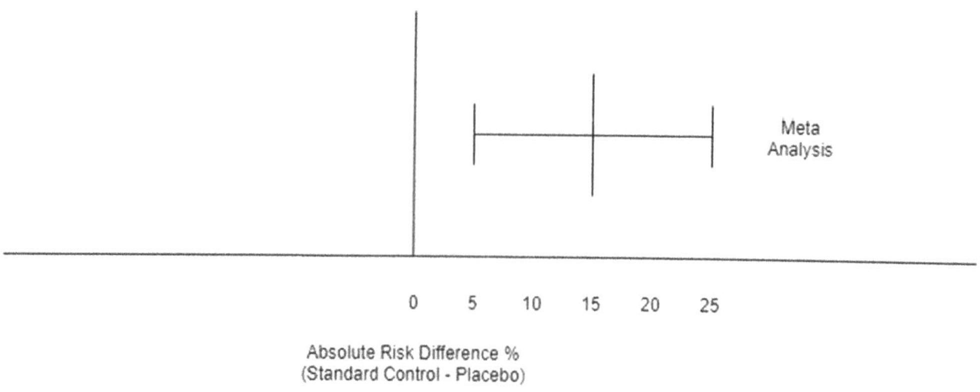

FIGURE 56.3 Determining boundary margin. The lower limit CI of the meta-analysis of the standard control compared to the placebo is a 5% absolute risk difference. Thus using a popular method, the formula for the boundary value would be B = 5f, where f = 0.33–0.5 as suggested by the CPMP.

would result in a higher threshold to establish equivalence/noninferiority and imply greater superiority of the NT compared to the placebo. The choice of f is made by clinical judgment and is influenced by the extent of willingness to sacrifice clinical efficacy toward the primary endpoint to reap the secondary benefits of the novel intervention. Of course, the synthesis approach requires the existence of multiple randomized control trials (RCTs) which compare the control to the placebo, as well as a relatively stable control and placebo over the span of the studies and which are similar to present day.[1,4]

The often increased sample sizes required for equivalence/noninferiority trials compared to superiority trials can also be a challenge; the sample size for the latter is generally inversely proportional to the boundary margin, which is in turn proportional to the effect of A as discussed above. Thus a very small effect of A demonstrated in historical meta-analysis compared to placebo may be a detrimental limitation that renders the sample size required nonfeasible.[4]

Another limitation of noninferior/equivalence trials compared to superiority trials arise from the fact that the poor quality trials and low efficacy of interventions actually increase the likelihood of the null hypothesis being rejected in the former as opposed to decreased likelihood in the latter. If there is overall low adherence and low treatment fidelity in a trial and/or both control and NT have low clinical efficacy, there is a greater chance of observing statistical noninferiority.[5] Thus ensuring study quality with rigorous monitoring throughout the duration of the trial is crucial to prevent false presumption of noninferiority/equivalence.[4]

Get started

Understand how noninferiority trials and equivalence trials differ from each other and RCTs.
Recognize potential situations in which to utilize noninferiority and equivalence trials.
Understand limitations/drawbacks of noninferiority and equivalence trials.
Understand accepted methods to determine the boundary value(s) for noninferiority/equivalence.
Understand the fundamental steps to carry out and analyze noninferiority/equivalence trials.

Potential pitfalls

With noninferiority/equivalence trials there is a primary challenge in determining the value of b, referred to as the boundary margin in trials, which is deemed reasonable for the consideration of B to be equivalent or non-inferior to A. There is no consensus method to determine the boundary margin, however regulators do strongly recommend it is established prior to carrying out the trial [3]. It is also advisable to use a boundary margin, which if not surpassed, would allow for clinical and statistical arguments that the novel intervention is not only noninferior to the control, but also likely superior to a placebo treatment and thus acceptable in light of other secondary benefits [1]. A typical method to determine the boundary margin in clinical noninferiority trials, known as the synthesis approach [4], is to set it at a fraction (*f*) of the value of the effect demonstrated in a meta-analysis by the control therapy compared to the placebo (at the lower limit of the 95 confidence interval(CI))(**figure 3**) [2]. A smaller f would result in a higher threshold to establish equivalence/noninferiority and imply greater superiority of the NT compared to the placebo. The choice of f is made by clinical judgement and is influenced by the extent of willingness to sacrifice clinical efficacy towards the primary endpoint in order to reap the secondary benefits of the novel intervention. Of course, the synthesis approach requires the existence of multiple RCTs which compare the control to the placebo, as well as a relatively stable control and placebo over the span of the studies and which are similar to present day [1, 4].

The often increased sample sizes required for equivalence/noninferiority trials compared to superiority trials can also be a challenge; the sample size for the latter is generally inversely proportional to the boundary margin, which is in turn proportional to the effect of A as discussed above. Thus a very small effect of A demonstrated in historical meta-analysis compared to placebo may be a detrimental limitation which renders the sample size required non-feasible [4].

Another limitation of non noninferior/equivalence trials compared to superiority trials arise from the fact that the poor quality trials and low efficacy of interventions actually increase the likelihood of the null hypothesis being rejected in the former as opposed to decreased likelihood in the latter. If there is overall low adherence and low treatment fidelity in a trial and/or both control and NT have low clinical efficacy, there is a greater chance of observing statistical noninferiority [5]. Thus ensuring study quality with rigorous monitoring throughout the duration of the trial is crucial to prevent false presumption of non-inferiority/equivalence [4] .

Example

We have designed an Inferior Vena Cava Filter (IVC) filter that spontaneously biodegrades after a time period of six months. The potential value of such a device exists, not in that it may be superior to the standard filter with regards to preventing pulmonary embolisms (PE), but that it eliminates the several risks associated with failure to retrieve the filter, such as filter migration, thrombosis and vessel penetration. Thus we want to perform a noninferiority study to establish whether our filter D is noninferior to filter C in respect to the PCE of PEs.

Establish the boundary margin(*b*): Identifying an appropriate value for the maximum allowable difference in outcome between our filter B and filter A is the most crucial and possibly ambiguous step. We will employ the commonly used synthesis approach. We first identify a recent relevant meta-analysis comparing the standard filter A to placebo and establish that the absolute risk reduction(ARR) of subsequent PE was 0.05 (95 % CI: 0.08 to 0.02) [10]. Thus our *b* should equate to the lower limit of the ARR, 0.02, multiplied by *f*. As stated above there is no generally applicable value of *f*, however guidance offered by the CPMP suggests a value of 0.33 to 0.5 [6]. If we are to choose 0.5 our *b* will be 0.01.

Determine sample size required: We must determine the sample size required to give our trial our desired power. In this example freely available online software [7], which utilizes an established formula [8], can be used to calculate our required sample size. In the calculator we set the significance level at 2.5% (see explanation on step 5), power at 80 percent, the expected percentage success of the control group at 98.1 % (which is determined from the same meta-analysis) and *b* (non-inferiority limit) at the predetermined 1 %. Because we believe our filter B is not inferior to filter A, we also set the percentage success of our experimental group to an identical 98.1%. The calculated sample size required is determined to be 5854.

Carry out Trial: The trial is carried out in accordance to the principles of an RCT with Filter D as the experimental intervention and Filter C as control. Strive to ensure quality of trail, including adherence to treatment protocols in both the Filter A and Filter B arms. Compliance to therapy is unlikely to be an issue in our trial consisting of just one initial procedure, however adherence to follow up timelines to assess for the clinical outcome of PE is a factor.

Analysis of Data: We must perform both Intention to treat (ITT) and Per Protocol(PP) data analysis to be most rigorous. In contrast to superiority trials, ITT is the less conservative approach in noninferiority trials because the inclusion of patients who do not undergo their intended therapies in the analysis generally leads to smaller differences between the therapy branches [4]. In PP, only patients who underwent their intended filter placement would be included in the data analysis. Agreement between ITT and PP analysis would lend heightened confidence to a noninferiority conclusion [2].

Testing Noninferiority: The absolute risk difference between the development of subsequent PEs in patients treated with our Filter B and control filter A is determined. The upper limit of the 95 % CI (corresponding to one-sided 97.5 % CI) is assessed to determine whether it exceeds the established noninferiority boundary; if not the null hypothesis is rejected and conclusion is made that Filter B is noninferior to Filter A. Since a one-sided 97.5 % CI is utilized to evaluate the null hypothesis the significance level of our trial is actually 2.5% as stated in step 2 [1].

Reporting of results: We must include specific details to in our reporting of our trial to ensure it may be assessed for quality [2]:

Justification for equivalence/noninferiority assessment rather than superiority assessment

Justification for why noninferiority would indicate the novel intervention be considered clinically beneficial through reference of historical efficacy results of the control intervention compared to placebo

Explanation of the reasoning behind the predetermined value of *b*

Method utilized to establish the sample size needed to achieve desired power

Reporting of both ITT and PP data

The process to perform an equivalence trial is similar to noninferior trials with a few key differences. There is now two boundary margins required, which simply equate to the positive and negative integer values of *b* via methods outlined in step 1. In step 2, an online sample size calculator software specific for equivalence trials is employed [9] using 5 % significance level rather than 2.5 %, as the boundary margins are compared against the upper and lower limits of a two sided CI [1]. In step 5, if the two-sided CI lies within the boundary margins the null hypothesis is rejected and Filter A and B are established as equivalent.

Examples of noninferiority and equivalence trials in the literature for further reference:

Kineflex lumbar artificial disc versus Charité lumbar total disc replacement for the treatment of degenerative disc disease: A randomized non-inferiority trial with minimum of 2 years' follow-up

(Pettine K, Hersh A. Kineflex lumbar artificial disc versus Charité lumbar total disc replacement for the treatment of degenerative disc disease: A randomized non-inferiority trial with minimum of 2 years' follow-up. *SAS J.* 2011;5(4):108-113. Published 2011 Dec 1. https://doi.org/10.1016/j.esas.2011.07.003)

Six months versus 12 months dual antiplatelet therapy after drug-eluting stent implantation in ST-elevation myocardial infarction (DAPT-STEMI): randomised, multicentre, non-inferiority trial

(Kedhi E, Fabris E, van der Ent M, et al. Six months versus 12 months dual antiplatelet therapy after drug-eluting stent implantation in ST-elevation myocardial infarction (DAPT-STEMI): randomised, multicentre, non-inferiority trial. *BMJ*. 2018;363:k3793. Published 2018 Oct 2. https://doi.org/10.1136/bmj.k3793)

Can Arthroplasty Stem INfluence Outcome? (CASINO): a randomized controlled equivalence trial of 125 mm versus 150 mm Exeter V40 stems in total hip arthroplasty

(Hamilton DF, Ohly NE, Gaston P. Can Arthroplasty Stem INfluence Outcome? (CASINO): a randomized controlled equivalence trial of 125 mm versus 150 mm Exeter V40 stems in total hip arthroplasty. *Trials*. 2018;19(1):226. Published 2018 Apr 16. https://doi.org/10.1186/s13063-018-2621-8)

Resoures

1. NYU Lesaffre E. Superiority, equivalence, and non-inferiority trials. *Bull NYU Hosp Jt Dis.* 2008;66(2): 150–154.
2. Walker E, Nowacki AS. Understanding equivalence and noninferiority testing. *J Gen Intern Med.* 2011; 26(2):192–196. https://doi.org/10.1007/s11606-010-1513-8.

References

1. NYU Lesaffre E. Superiority, equivalence, and non-inferiority trials. *Bull NYU Hosp Jt Dis*. 2008;66(2):150–154.
2. Walker E, Nowacki AS. Understanding equivalence and non-inferiority testing. *J Gen Intern Med*. 2011;26(2):192–196. https://doi.org/10.1007/s11606-010-1513-8.
3. Greene CJ, Morland LA, Durkalski VL, Frueh BC. Noninferiority and equivalence designs: issues and implications for mental health research. *J Trauma Stress*. 2008;21(5):433–439. https://doi.org/10.1002/jts.20367.
4. https://www.fda.gov/regulatory-information/search-fda-guidance-documents/non-inferiority-clinical-trials. (Accessed 24 June 2020).
5. Rief W, Hofmann SG. Some problems with non-inferiority tests in psychotherapy research: psychodynamic therapies as an example. *Psychol Med*. 2018;48(8):1392–1394. https://doi.org/10.1017/S0033291718000247.
6. Bikdeli B, Chatterjee S, Desai NR, et al. Inferior vena cava filters to prevent pulmonary embolism: systematic review and meta-analysis. *J Am Coll Cardiol*. 2017;70(13):1587–1597. https://doi.org/10.1016/j.jacc.2017.07.775.
7. CPMP Garrett AD. Therapeutic equivalence: fallacies and falsification. *Stat Med*. 2003;22(5):741–762. https://doi.org/10.1002/sim.1360.
8. Sealed Envelope Ltd. *Power Calculator for Binary Outcome Non-inferiority Trial*; 2012 [Online] Available from: https://www.sealedenvelope.com/power/binary-noninferior/. Accessed June 24, 2020.
9. Blackwelder WC. Proving the null hypothesis in clinical trials. *Control Clin Trials*. 1982;3:345–353.
10. Sealed Envelope Ltd. *Power Calculator for Binary Outcome Equivalence Trial*; 2012 [Online] Available from: https://www.sealedenvelope.com/power/binary-equivalence/. [Accessed 25 June 2020].

Chapter 57

Adaptive: Design, measures, and classic example

Christopher Ovanez

Russell H. Morgan Department of Radiology and Radiological Science, Johns Hopkins Hospital, Baltimore, MD, United States

Key points

- In a manner of speaking, adaptive study design developed as an alternative approach to randomized control trial study design. Whereas randomized control trials are traditionally run in a manner by which there are subjects who are randomized into either an intervention group or a control group, in adaptive design, clinical outcomes are measured at certain points, and one can apply modifications to the study design depending on the outcomes.
- There are several different types of adaptive study design strategies, each with their own individual merits. One must also be aware of potential drawbacks in utilizing these strategies.
- Adaptive study design strategies appear to be unfortunately underutilized, possibly in part due to unfamiliarity with these types of strategies by researchers.

Introduction

What is adaptive study design?

In terms of a definition, the Pharmaceutical Research Manufacturer Association Working Group on Adaptive Design came up with one, and they defined it essentially as "a clinical trial design that uses accumulating data to decide how to modify aspects of the study as it continues, without undermining the validity and integrity of the trial."[1]

In a manner of speaking, adaptive study design developed as an alternative approach to randomized control trial study design.[2] Why would you want to do something like that? Randomized control trials are considered the highest standard when it comes to research study design, and are usually hailed as the gold standard when it comes to testing the merit of a new medication or product, but, it inherently requires a large amount of time and resources to proceed properly.[2] Whereas randomized control trials are traditionally run in a manner by which there are subjects who are randomized into either an intervention group or a control group, in adaptive design, clinical outcomes are measured at certain points, and one can apply modifications to the study design depending on the outcomes.[2]

Adaptive designs have been around for a while, and according to Bothwell et al., the biostatistics literature demonstrated growth on the topic of adaptive design in the 1960s and 70s, but were not utilized much at all. The individuals who attempted adaptive trials encountered issues, some of which included: "increased risks of falsely detecting treatment effects, prematurely discarding promising therapies and causing statistical and operational bias."[2] Probably, a reason why these types of trials had not skyrocketed in popularity early on was that there was insufficient information and a paucity of experience from prior successfully completed trials to lay a proper foundation for future researchers and as to what to expect, so more than likely, many did not embrace this type of study design as much they did randomized control trial study design.

Despite these types of issues, adaptive design gained traction, and there has been a renewed interest in adaptive study design within the United States and internationally.[2] As an example, the pharmaceutical community embraced adaptive design for instance to increase the efficiency or likelihood to show drug efficacy in phase 2 or 3 registration trials.[2]

Get started

This chapter discusses adaptive study design, its merits and several examples. The examples of adaptive study design strategies provided within this chapter are not meant to be exhaustive, but a starting point for someone who plans to design research studies who would like to learn more about

Translational Interventional Radiology. https://doi.org/10.1016/B978-0-12-823026-8.00057-2

this underappreciated approach, and may inspire them to consider these types of strategies in the future.

Potential pitfalls

While adaptive study design strategies may seem appealing and useful with regard to integration within research studies, it should also be noted that there are also drawbacks and consequences to utilizing these types of strategies. These strategies should be used carefully in order not to potentially jeopardize study efficacy or patient safety. For instance, an adaptive design strategy could lead to the early termination of a clinical trial when data indicated efficacy or futility, and which this may be considered a positive effect, it may also lead to negative consequences. For instance, as pointed out by Weson et al, premature termination of a trial may lead to inability to provide accurate information on secondary outcomes, safety, or subgroup effects.[5] Additionally, Weson et al point out logistical limitations with regard to adaptive study design methods, namely caused by delays in performing interim analyses or implementation of changes.[5] Additional drawbacks may be encountered with the utilization of adaptive study design methods depending upon the strategy utilized and manner of study being pursued.

Examples

There are several different types of adaptive designs, and we will touch upon some examples.

For instance, there is adaptive randomization, whereby the manner in which randomization itself occurs is adjusted after monitoring study results over time, and patients who are subsequently enrolled in the trial are more likely to be placed into the treatment group(s) that were deemed more effective among subjects earlier on during the course of the study.[1,2] This type of study design is often referred to as enrichment design.

Another example is an adaptive hypothesis, where a study hypothesis is altered depending upon results that are garnered from the study.[1,2]

Having a study in which phase 2 and phase 3 blend with one another is another example of adaptive design, whereby there is a continuation between the two phases, without pausing.[1,2] The Indacaterol to Help Achieve New COPD Treatment Excellence (INHANCE) trial, was a randomized clinical trial of inhaled indacaterol, utilizing an adaptive two-stage strategy.[4]

Altering sample size based on study results is another manner of adaptive design.[1,2] In the Cangrelor versus Standard Therapy to Achieve Optimal Management of Platelet Inhibition (CHAMPION) PHOENIX trial, which was a double-blind placebo-controlled trial, the sample size was increased only after careful review of interim study results, in order for the researchers to ensure that the results were promising before proceeding.[4]

Another type of adaptive design would be to lose the treatment arm that is found to be inferior during the course of the study, changing the treatment arm, or perhaps even adding treatment arms depending upon study results.

As you can imagine, there are several other types of adaptive study designs, and I encourage you to explore them should you wish to learn more, and perhaps utilize adaptive design methods in your own research pursuits.

Conclusion

Adaptive trial design truly has several merits, and it is probably fair to say that it is unfortunately an underutilized type of study design. I am willing to bet that many people, even among clinicians, have never even heard of it, and among those who have heard of it, they probably are not able to provide a great elaboration as to what it entails.

As the name implies, the adaptive design allows a researcher to be adaptive and flexible when it comes to their research pursuits, allowing for tailoring of numerous variables and components of a study to allow for increased efficiency and probably a more cost-effective approach in terms of resource utilization, for instance, given the ability and freedom to alter study hypotheses, patient allocation and randomization, and several other mechanisms.[1–3]

Within this chapter, we touched upon some examples of adaptive study design, and by no means an exhaustive review. There are several excellent resources that go into greater detail regarding adaptive design methods and applications, and I encourage you to explore these, as it might pique your interest and you might find yourself utilizing these methods within your own studies.

References

1. Chow SC. Adaptive clinical trial design. *Annu Rev Med.* 2014;65:405–415.
2. Bothwell LE, Avorn J, Khan NF, Kesselheim AS. Adaptive design clinical trials: a review of the literature and ClinicalTrials.gov. *BMJ Open*. February 10, 2018;8(2):e018320.
3. Li Q, Lin J, Lin Y. Adaptive design implementation in confirmatory trials: methods, practical considerations and case studies. *Contemp Clin Trials*. November 2020;98:106096.
4. Bhatt DL, Mehta C. Adaptive Designs for Clinical Trials. *N Engl J Med*. July 7, 2016;375(1):65–74.
5. Wason JMS, Brocklehurst P, Yap C. When to keep it simple - adaptive designs are not always useful. *BMC Med*. August 2, 2019;17(1):152.

Chapter 58

Randomization

Christopher Ovanez
Russell H. Morgan Department of Radiology and Radiological Science, Johns Hopkins Hospital, Baltimore, MD, United States

Key points

- Approaching randomization in a systematic and well-thought manner is very important when designing a research study. If the approach to randomization is not well-executed, it may lead to potentially dire consequences in several respects, for instance – unbalanced study groups, inaccurate data and conclusions, and possible introduction of unconscious bias.
- There are several methods or means by which randomization could be accomplished in research study design. The best approach to randomization is usually dependent upon the specific needs and goals of your study.
- When designing a study, a researcher must pay close attention to ensure that randomization is ethical, feasible, and executed in a proper, methodical, and conscientious manner.

Introduction

Randomization is undoubtedly a key and crucial element that must be carefully planned when designing a research study, and if not handled and executed in a proper manner, could lead to potentially devastating issues down the road—for instance, yielding data that are not truly valid, misinterpretation of these data, and at the end of a study ultimately arriving at conclusions that are not truly accurate, which in turn will lead to suboptimal patient outcomes if these conclusions potentially relate to the performance of medical devices or medications.

So, what does "randomization" truly mean when it comes to designing a research study?

In essence, it entails randomly assigning study participants or entities into two or more treatment (or exposure) groups in a particular research study. If this randomization of study participants is performed in a proper and thorough manner, the study will have groups of participants who possess theoretically equally distributed factors associated with a particular outcome except for the treatment status (thus allowing for establishing a causal relationship between treatments and outcome), and avoids "harmful" factors that may be introduced into a study, for instance, potential confounding factors (i.e., unknown or known factors which are associated with both treatments and outcomes and could potentially limit the ability to see a true causal effect of a particular exposure upon a particular outcome), bias, and other sorts of issues that could hypothetically lead to serious consequences down the road in a study.

Get started

Approaching randomization in a systematic and well-thought manner is very important when designing a research study. If the approach to randomization is not well-executed, it may lead to potentially dire consequences in several respects, for instance – unbalanced study groups, inaccurate data and conclusions, and possible introduction of unconscious bias. This chapter will discuss the importance of randomization in study design, as well as various methods/approaches to randomization.

Potential pitfalls

A researcher should also always be mindful from an ethical standpoint, as there are instances when the process of randomization for study groups is not always a feasible option, for instance, if you have an exposure (i.e., a particular treatment or device) that is known to be either undoubtedly beneficial or undoubtedly harmful and hence would pose inherent harm to patients.

Hence, when designing a study, the researcher must pay close attention to ensure that randomization is ethical, feasible, and executed in a proper, methodical, and con scientious manner.

Randomization methods in which the researcher who designs the study is privy to the information of which treatment is being allocated to specific participants would be considered less than ideal. This opens up the possibility

Translational Interventional Radiology. https://doi.org/10.1016/B978-0-12-823026-8.00003-1

to potential unconscious or conscious bias or manipulation and could potentially taint the truth when it comes to the validity of study results.

Examples

There are several methods or means by which randomization could be accomplished in research study design. A wide variety of methods are available, which could be drawn upon, with some examples including random number generators, random number lists, tossing a coin, sealed and envelope method.[1]

A more ideal type of randomization is that performed by a third party, which is completely independent of the research team. This third-party distributes a particular assignment to study participants in a random manner and does not disclose the truth with regard to the assignments until the study is completed. As an example, you could have a study in which an independent pharmacist randomly assigns numbers to vials of placebo and medication(s) whose contents are not disclosed to the researchers or the study participants, and only after the study is completed will the list of true identities be disclosed.[1] This type of randomization accomplished by a third party is considered ideal, as it eliminates bias and other potentially confounding factors from the study design.

There are alternative types of randomization that may be pursued depending on the specific situation surrounding a particular type of study design—some methods potentially being safeguards against unexpected situations.

Block randomization is a form of randomization that tries to keep a steady balance of study participants in different treatment arms throughout the period of enrollment, with the goal of allocating participants into treatment arms in a fixed ratio (e.g., 1:1 for a two arms design) within small blocks. A reason to utilize this type of randomization would be to safeguard against the potential situation in which enrollment is terminated earlier than expected, and despite this situation, there is continued maintenance of a relative balance with regard to the number of participants in the treatment arms.[1] Block size can be fixed or random. Treatment arm assignment by block randomization is not fully random as, within each block, there is a fixed number of patients in each treatment arm; so, the treatment assignment of the last or last few patients can be deducted given the treatment assignments of the other patients in the block. For this reason, block randomization with random block sizes is often preferred.

There are also stratified and non-stratified methods of randomization. For stratified randomization, an example might be having a research study that assesses a new treatment and you have two treatment groups, and you want to make sure that both of these groups are the same with regard to prognosis, and this method entails the creation of randomization lists, such that those patients with a better prognosis are randomized using one particular list, and those with a worse prognosis are randomized utilizing another list,[1] to ensure that the prognosis is balanced between two treatment groups. On the other hand, non-stratified randomization essentially means that random chance determines the assignment of study participants.

Minimization is another strategy that could be used. With this particular strategy, the idea is that "the next patient" to be assigned to a particular treatment arm takes into consideration the arms as a whole, for instance, going back to our discussion of stratified randomization—the example of wanting to have treatment arms that are similar with regard to the prognosis of study participants; with minimization, if you have a large number of study participants in one treatment arm with an overall worse prognosis, the next participant with a relatively worse prognosis would not be assigned to this particular arm but would be assigned to the second treatment arm, with the intention being to create a sense of balance among treatment arms with regard to prognosis.

Randomization in research study design is a vast topic that one could spend a great deal of time exploring, and there are several nuances and other methods that have not been discussed within the scope of this chapter, but the goal was to touch upon some key points and ideas.

Conclusion

The main point to convey is that randomization is a critical component of study design and should be taken very seriously when designing a research project, be it within the realm of interventional radiology, or any other field. How one goes about randomization needs to be tailored to fit a researcher's specific needs, the needs of the patients, and the needs of the study itself—a researcher must take into consideration the specific goals of one's study.

I cannot emphasize enough the importance of properly utilizing randomization. The consequences of not taking randomization seriously are quite dire and can unintentionally distort data, the validity of the study, and the conclusions that are made at the end of the study—essentially rendering the study useless, and a large waste of time, effort, and resources. Randomization is an important tool that a researcher must feel comfortable utilizing and must be properly adapted to the specific type of study that is being designed.

Resources

Porta M. *A Dictionary of Epidemiology*. 5th ed. Oxford University Press; 2014.

Kestenbaum B. *Epidemiology and Biostatistics: An Introduction to Clinical Research*. Springer; 2009.

Foundations of Evidence-Based Medicine: Clinical Epidemiology and Beyond. CRC Press; 2020. 2nd ed.
Sashegyi A., Felli J., Noel R. *Benefit-Risk Assessment in Pharmaceutical Research and Development*. CRC Press; 2014.

References

1. Van Den Broek J, Brestoff J. *Epidemiology: Principles and Practical Guidelines*. Springer; 2013.

Further reading

1. Porta M. *A Dictionary of Epidemiology*. 5th ed. Oxford University Press; 2014.
2. Kestenbaum B. *Epidemiology and Biostatistics: An Introduction to Clinical Research*. Springer; 2009.
3. *Foundations of Evidence-Based Medicine: Clinical Epidemiology and beyond*. 2nd ed. CRC Press; 2020.
4. Sashegyi A, Felli J, Noel R. *Benefit-Risk Assessment in Pharmaceutical Research and Development*. CRC Press; 2014.

Chapter 59

Blinding: Who and how?

Jonas Kruse[1], Conner D. Reynolds[2,3], Ahmed Farag[4] and Alvin Anene[1,5]

[1]School of Medicine, Texas Christian University, Fort Worth, TX, United States; [2]Transitional Year Residency, John Peter Smith Health Network, Fort Worth, TX, United States; [3]Diagnostic Radiology Residency, Department of Medical Imaging, University of Arizona College of Medicine, Tucson, AZ, United States; [4]Interventional Radiology Fellowship, Baylor University Medical Center, Dallas, TX, United States; [5]Radiology Associates of North Texas, Fort Worth, TX, United States

Blinding in trials of interventional procedures is possible and many creative methods have been used to maintain the blinding. Interventional procedures are challenging to blind, but the effort is worthwhile because of the obvious benefits, such as avoiding bias, as well as the less evident benefits, such as avoiding patient drop-out in the control arm.

Wartolowska et al. (2017).

Key points

- Blinding is a pivotal part of bias reduction in interventional procedure trials.
- This chapter outlines the importance of blinding, identifies which trial actors should be blinded, outlines potential blinding strategies for interventional trials, and provides a standardized reporting template.

Introduction: importance of blinding and impact on findings

The ideal trial would be a trial devoid of bias. Investigators strive to approach this ideal through a multitude of mechanisms. The gold standard for evidence-based medicine is randomized clinical trials (RCTs), in which blinding is a foundational element.[1–3] Blinding is the process of masking information that may bias parties involved in the trial and confound study results.[4] Such bias may be introduced into a randomized sample either by disparate treatment or by inconsistent assessment of treatment groups.[5] Trial participants may also change their expectations, adherence, behavior, and subjective response if made aware of group designation.[6] Explicitly stated, blinding is a crucial component of ensuring differential outcomes are due to the intervention alone.

Despite efforts to maintain a strictly objective approach, results and processes in clinical studies are often influenced by a subjective component. Numerous studies have demonstrated distinct discrepancies among nonblinded studies. For example, Kaptchuk et al. found that patient encouragement from a nonblinded healthcare team yielded a statistically significant improvement in patient's symptoms when treated with a known placebo.[7] In another study investigating outcomes of focal cryoablation for prostatic lumpectomy, the author reported investigator bias, where the nonblinded interventional radiologist performing the procedure was involved in follow-up assessments, as a study limitation.[8] In this example, it is simple to envision how an investigator removed from the cases may interpret the outcomes differently than if they had also performed the procedures. Furthermore, one may be less inclined to publish their own adverse outcomes, leading to a reporting or publication bias. In a meta-analysis of procedural outcomes of specialty devices in interventional radiology for the treatment of infrapopliteal atherosclerosis, Razavi et al. reported publication bias favored percutaneous transarterial atherectomy and drug-eluting stents.[9]

Not only do nonblinded studies have the potential to skew the direction of an outcome, but magnitude as well. Hróbjartsson et al. found that nonblinded participants have exaggerated reported outcomes compared to blinded trials, observing a Pygmalion-like effect. This is of particular importance to interventional radiology, as this phenomenon was found to be magnified in trials with invasive procedures.[10] One observational study of 33 meta-analyses found a 17% statistically significant increase in effect estimates in trials that were not double-blinded.[11] Additionally, while the effects of open-label bias can be most easily conceptualized within subjective metrics, a recent literature review has suggested that lack of blinding can also play a

Translational Interventional Radiology. https://doi.org/10.1016/B978-0-12-823026-8.00025-0

significant role in objective outcomes.[4] To mitigate these biases, investigators should strive to mask the treatment to both the participant and the assessors/investigators whenever possible.

Deciding who to blind

When determining which trial actors should be blinded, recent literature suggests that at least 11 distinct groups should be considered: patients, clinicians (interventional radiologists), data collectors, trial managers, pharmacists, technicians, outcome assessors, outcome adjudicators, statisticians, safety and data monitoring committees, and manuscript writers.[12] When planning a blinded study, researchers will readily identify treatment assignment as the most important blinding measure, there are many other factors to consider. Concealing additional information such as patient demographics, clinical characteristics, and study purpose will yield greater bias minimization, treatment effect isolation, and real-world generalizability.

Hierarchy of blinding studies for interventional trials

When determining which blinding measures will be implemented, it is important to remember that blinding is not an "all-or-none" phenomenon, but a spectrum. The natural hierarchy of blinding ranges from nonblinded open-label studies to completely blinded studies. The feasibility of blinding trial actors varies considerably with the intervention being tested. Here, we will present the primary approaches to blinding, their advantages and limitations, and several examples.

Double-blinded studies

In double-blinded studies, treatment assignment is concealed from both participants and investigators. Among all patients, this blinding method reduces all patients' cognitive tendency to exaggerate subjective effects of the intervention (response bias).[13] It also improves the likelihood of adherence to intervention protocols in the control group (compliance bias). Among investigators, it prevents them from inadvertently conveying high expectations to intervention-group patients, thus altering their subjective reporting (experimenter-expectancy bias).[14] Their high degree of variable control leads to superb minimization of bias and replicability. As such, double-blind studies are considered the gold standard for interventional trials and should be implemented wherever feasible and ethically permissible.

There are several disadvantages to double-blind studies. First, they do not always reflect real-world circumstances. While variable control is beneficial for isolating treatment effects, clinical practice has innumerable confounders and practice patterns that cannot be isolated. Second, some double-blinded studies are too small to reflect a representative sample size. While the size needed is typically derived from effect size, Beth Israel's Winchester Hospital reports that a good double-blind study should enroll 100–300 patients.[15] Third, to achieve such a sample size, there is typically a high monetary cost. These three factors can limit a double-blind study's generalizability and should be addressed before undertaking.

In a classic double-blind study, Yoon et al. evaluated whether superior hypogastric nerve blockade provided adequate analgesia after uterine artery embolization.[16] The authors utilized two separate interventionist and nursing teams to achieve the blinding of interventionalists. The first team was responsible for delivering for performing the procedure but had no knowledge of whether the patient received true blockade or sham treatment. The second team was responsible for managing patients who received the blockade. This clever split-team model for blinding works particularly well in interventional trials. Another such study evaluating the value of antibiotic prophylaxis for infection reduction after percutaneous gastrostomy tube placement effectively blinded all patients, operators, nursing staff, and study staff.[17]

Complete blinded studies

A complete blinded study is an extension of the double-blinded study described previously. It expands the scope of blinding to include all personnel who evaluate the patient or analyze data, such as the data management and statistical staff. While the analytic process is known for its inherent objectivity, blinding these staff members reduces their likelihood of inadvertently revealing participants to another blinded staff member. Moreover, because statisticians must sometimes make post hoc decisions during analysis, blinding ensures that results are not skewed toward a particular conclusion.

The advantages and disadvantages of complete blinded studies are also magnified when compared to double-blinded studies. The high degree of control achieved by complete blinded studies outshines all others in isolating treatment effects. However, they are the most artificial and cannot reflect real-world confounders. Their relative complexity makes them difficult to orchestrate and financially tenuous.

There are very few studies able to achieve true complete blinding. One that comes close is the VAPOUR Trial, which compared the safety and efficacy of vertebroplasty for analgesia in patients with refractory pain after osteoporotic spinal fractures.[18] Once the patient was in the trial room, immediately before the procedure, an automated phone service would inform the interventionalists whether they should receive vertebroplasty or placebo treatment. After the procedure, neither interventionalists nor other staff had any additional patient contact during the trial. All other staff members, investigators, and outcome assessors remained blinded to treatment assignment.

Single-blinded studies

In single-blinded studies, either the patient or investigator is blinded. These studies are usually employed when specific knowledge of treatment assignment would significantly bias the data collected or its analysis. Single-blinded studies facilitate bias reduction in studies where blinding both parties is not feasible or ethically permissible. For example, in studies comparing different rehabilitation, chronic disease management, or lifestyle modification programs on functional outcomes after stroke, it would not be feasible for blind patients, as they must actively participate in the programs. However, interventionalists or other clinical staff involved in data collection can easily be blinded. Overall, these can reduce design complexity and cost, especially for smaller institutions or practices that do not regularly host trials.

Several disadvantages also exist. In single-blinded studies where investigators are nonblinded, they may inadvertently convey their expectations for certain outcomes to blinded patients. In studies where patients are nonblinded, knowledge of treatment assignment introduces the cognitive tendency to exaggerate the overall treatment effects. It can also lead to loss of adherence and higher attrition among control group patients.

A key study where only patients were blinded was performed by the COLLISION Trial Group, which evaluated the noninferiority of thermal ablation compared to resection in patients with colorectal liver metastasis.[19] Another single-blinded study evaluated whether endovascular embolization before surgical resection of spinal metastases reduced intraoperative blood loss. In this study, nonblinded interventional radiologists completed the embolization, while spine surgeons, anesthesiologists, and staff members obtaining outcome data were kept blinded to the allocation.[20]

Open-label studies

In open-label studies, neither the patient nor investigator are blinded. This method may improve patient recruitment and retention, as they have knowledge of being in the treatment group. The utility of these studies is often limited to early investigative phases, particularly when establishing a safety profile.[21] Patient recruitment and retention for open-label studies can also often be easier in the previously described blinded studies. However, the risk of bias is highest in this study type. A great open-label study can be seen in the Mulcahy et al. [22] study, which evaluated the cotreatment of radioembolization with chemotherapy for colorectal liver metastases.[22]

How to blind effectively

Blinding in interventional trials presents unique challenges when compared to pharmaceutical trials. As a result, blinding is less prevalent in studies investigating interventional procedures.[23] Still, if the goal is to produce rigorous evidence for the establishment and improvement of treatment guidelines, blinding is of equal importance in the presence of said challenges.[23] In this effort, Wartolowska et al. described strategies for blinding that can be utilized in interventional placebo-controlled trials. Of note, the placebo in this context is an imitation procedure meant to mimic the procedure of interest without the component believed to be vital to effective intervention.[24]

To blind the participant in procedures requiring general anesthesia, only the access wound needs to be indistinguishable between the two study groups.[24] As most interventional radiology procedures utilize conscious sedation as well as mobile imaging techniques that render visually occlusive drapes nonfeasible, interventional radiology trials must employ alternative creative measures to achieve adequate participant blinding. Previous studies of minimally invasive procedures have utilized the concept of structure access sans intervention as placebo, such as diagnostic laparoscopy without laparoscopic surgery[25] or endoscopy without radiofrequency energy delivery.[26] Within interventional radiology, such "sham" procedures could include vascular or percutaneous access without device delivery (e.g., stent, tube, embolic, cement, etc).

Additionally, the placebo procedure should mimic the active arm procedure as closely as possible, including visual, auditory, and physical cues.[24] Wartolowska et al. emphasize the importance of this standardization as interventionalists vary in experience and will gain experience as the study progresses. To this end, previous interventional trials instructed surgeons to verbalize procedure steps,[27] obscure the participants' view,[28] manipulate the procedure site with tactile simulation and imitation irrigation,[29,30] and even recreate distinct procedural smells.[31] Other considerations include matching the duration of the procedure and ensuring the placebo has no therapeutic effect, such as local analgesic/anesthesia in procedures with pain reduction as a primary endpoint (geniculate artery embolization, kyphoplasty, etc.).[32] In an ideal trial, every step of the preprocedure, intraprocedure, and postprocedure process would be identical in the two arms except the key interventional component under investigation.

Following the procedure, additional components such as postoperative medications, imaging, and follow-up care should also be standardized. Participants may exhibit changes secondary to protocol lifestyle alterations such as self-monitoring or avoiding alcohol, rather than the intervention itself.[24]

While less common, there are trials that have blinded the interventionist, either by assigning the delivery of experimental treatment or placebo to a technician[33] or having a preloaded device with contents blinded to the proceduralist.[34] In the absence of blinding the interventionist, there is a priority in blinding the remaining members of the care team and assessment mentioned in Lang and Stroup (2020).[12] Measures may include providing postoperative care at a separate site[35] or by providers blinded to the protocol.[36] Boutron et al. summarized a subset of blinding methods as found below (Fig. 59.1)[37]:

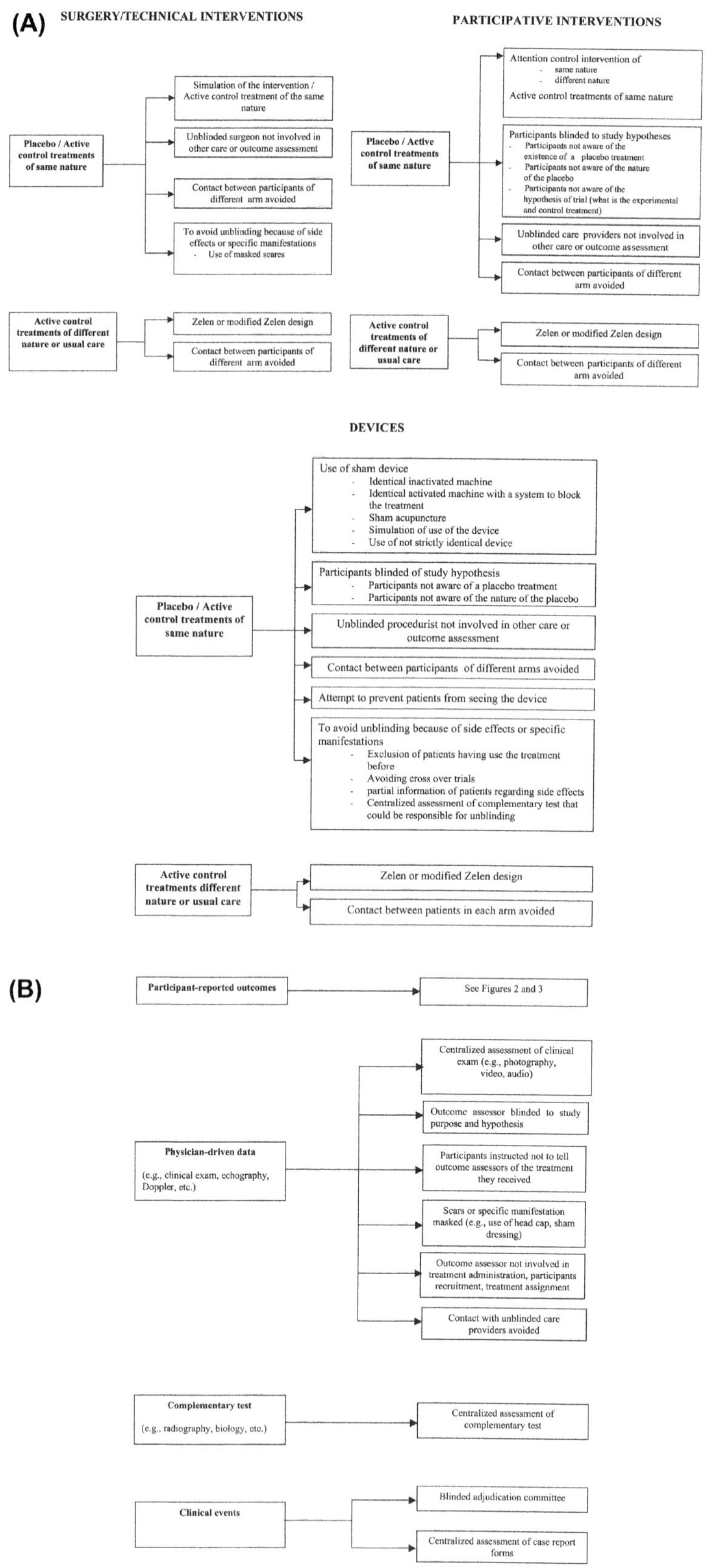

FIGURE 59.1 Methods of blinding (A) study participants and investigators; and (B) outcome assessors.

TABLE 59.1 A standard table for reporting the use of blinding in randomized trials of pharmaceutical interventions.

Group or individual blinded[a]	Information withheld[b]	Method of blinding[c]	Blinding compromised
Required fields to be completed for all trials described as blinded			
Person assigning participants to groups	Group assignment	Concealed allocation schedule	No
Participants	Group assignment	Placebo medications; sham surgeries	No
Care providers	Group assignment	Not told of group assignment	No
Data collectors and managers	Group assignment	Not told of group assignment	No
Outcome assessors	Purpose of study; group assignment; participant characteristics	Participants given numerical identifiers	No
Statisticians	Participant and group identities	Participants and groups given numerical identifiers	No
Supplemental fields for all blinded groups or individuals not mentioned earlier			
Trial manager	Not applicable		
Pharmacists	Not applicable		
Laboratory technicians	Participant identities	Participants given numerical identifiers	
Outcome adjudicators	Group assignment	Groups given numerical identifiers	Yes (put details in text)
Data monitoring and safety committees	Not applicable		
Manuscript writers	Not blinded		

[a]*Other groups or individuals in a trial that were capable of being blinded should be listed in the table, and whether or not they were blinded in the study should be indicated. Individuals with dual responsibilities, such as caregiving and data collecting, should be identified by combining the entries in the same row heading.*
[b]*Although group assignment is the information most commonly withheld in a blinded trial, data assessors, such as pathologists and radiologists, are often blinded to the purpose of the trial, group assignment, and the demographic and clinical characteristics of participants whose biopsy samples or images they are interpreting.*
[c]*If blinding has been compromised, authors should report the fact and indicate the potential implications the loss of blinding might have for interpreting the results. Reprinted with permission from Ref. 35. Copyright 2020 Lang et al. Full text available from https://trialsjournal.biomedcentral.com/articles/10.1186/s13063-020-04607-5.*

Detailed reporting

Descriptive reporting of blinding methods has not been common practice in previous trials, be it pharmaceutical or nonpharmaceutical.[6] A review of 200 RCTs found that 56% of articles failed to report the blinding status of any person in the trial and 19% of "double-blind" trials did not blind participants, care team members, or data collectors.[38] Further, surveying the authors of these trials resulted in 15 varying interpretations of "double blind" with the majority of respondents believing their own definition as the most common.[38] To avoid confusion, the 2010 Current Consolidated Standards of Reporting Trials (CONSORT) guidelines recommend explicit specification of blinding methodology, blinded and not blinded trial persons, intervention similarities, and concealed information in place of terms such as "double blind."[39] In 2013, the Standard Protocol Items: Recommendations for Interventional Trials (SPIRIT) guidelines further affirmed this concept, advising authors to detail blinded individuals and methodology.[40]

Monaghan et al. recently noted continued insufficient adherence to the previously stated CONSORT and SPIRIT recommendations.[4,41,42] A uniform table (Table 59.1) was proposed by Lang and Stroup in 2020 that offers a detailed model to ensure clarity of blinding protocol.[12]

Conclusion

During any interventional trial, researchers must minimize bias to isolate true treatment effects that will inform future best practices in healthcare. Well-designed and executed blinding measures are paramount. This chapter has introduced the importance of blinding, identified which trial actors can be blinded, outlined potential blinding strategies unique to interventional trials, discussed limitations and disadvantages of each blinding technique, and provided a standardized reporting template to optimize reproducibility.

Potential pitfalls

1. Make sure the type of blinded study used is commiserated with the research stage
2. Failure to implement and maintain blinding throughout the trial risks the validity of results

References

1. Murad MH, Asi N, Alsawas M, Alahdab F. New evidence pyramid. *Evid Base Med.* 2016;21:125.
2. Psaty BM, Prentice RL. Minimizing bias in randomized trials: the importance of blinding. *JAMA.* 2010;304:793–794.
3. Bhide A, Shah PS, Acharya G. A simplified guide to randomized controlled trials. *Acta Obstet Gynecol Scand.* 2018;97:380–387.
4. Monaghan TF, Agudelo CW, Rahman SN, et al. Blinding in clinical trials: seeing the big picture. *Medicina.* 2021;57:647.
5. Karanicolas PJ, Farrokhyar F, Bhandari M. Practical tips for surgical research: blinding: who, what, when, why, how? *Can J Surg.* 2010;53:345–348.
6. Schulz KF, Chalmers I, Altman DG. The landscape and lexicon of blinding in randomized trials. *Ann Intern Med.* 2002;136:254–259.
7. Kaptchuk TJ, Friedlander E, Kelley JM, et al. Placebos without deception: a randomized controlled trial in irritable bowel syndrome. *PLoS One.* 2010;5:e15591.
8. Onik G. Rationale for a "male lumpectomy," a prostate cancer targeted approach using cryoablation: results in 21 patients with at least 2 Years of follow-up. *Cardiovasc Intervent Radiol.* 2008;31:98–106.
9. Razavi MK, Mustapha JA, Miller LE. Contemporary systematic review and meta-analysis of early outcomes with percutaneous treatment for infrapopliteal atherosclerotic disease. *J Vasc Intervent Radiol.* 2014;25:1489–1496. e3.
10. Hróbjartsson A, Emanuelsson F, Skou Thomsen AS, Hilden J, Brorson S. Bias due to lack of patient blinding in clinical trials. A systematic review of trials randomizing patients to blind and nonblind sub-studies. *Int J Epidemiol.* 2014;43:1272–1283.
11. Schulz KF, Chalmers I, Hayes RJ, Altman DG. Empirical evidence of bias: dimensions of methodological quality associated with estimates of treatment effects in controlled trials. *JAMA.* 1995;273:408–412.
12. Lang TA, Stroup DF. Who knew? The misleading specificity of "double blind" and what to do about it. *Trials.* 2020,21:697.
13. Response bias. In: Salkind NJ, ed. *Encyclopedia of Research Design.* SAGE Publications, Inc.; 2010.
14. Experimenter-expectancy bias. In: Salkind NJ, ed. *Encyclopedia of Research Design.* SAGE Publications, Inc.; 2010.
15. *WinchesterHospital. Double-Blind Study.* 2022.
16. Yoon J, Valenti D, Muchantef K, et al. Superior hypogastric nerve block as post–uterine artery embolization analgesia: a randomized and double-blind clinical trial. *Radiology.* 2018;289:248–254.
17. Ingraham CR, Johnson GE, Albrecht EL, et al. Value of antibiotic prophylaxis for percutaneous gastrostomy: a double-blind randomized trial. *J Vasc Intervent Radiol.* 2018;29:55–61. e2.
18. Clark W, Bird P, Gonski P, et al. Safety and efficacy of vertebroplasty for acute painful osteoporotic fractures (VAPOUR): a multicentre, randomised, double-blind, placebo-controlled trial. *Lancet.* 2016;388:1408–1416.
19. Puijk RS, Ruarus AH, Vroomen LGPH, et al. Colorectal liver metastases: surgery versus thermal ablation (COLLISION) – a phase III single-blind prospective randomized controlled trial. *BMC Canc.* 2018;18:821.
20. Clausen C, Dahl B, Frevert SC, Hansen LV, Nielsen MB, Lönn L. Preoperative embolization in surgical treatment of spinal metastases: single-blind, randomized controlled clinical trial of efficacy in decreasing intraoperative blood loss. *J Vasc Intervent Radiol.* 2015;26:402–412.e1.
21. Day RO, Williams KM. Open-label extension studies. *Drug Saf.* 2007;30:93–105.
22. Mulcahy MF, Mahvash A, Pracht M, et al. Radioembolization with chemotherapy for colorectal liver metastases: a randomized, open-label, international, multicenter, phase III trial. *J Clin Oncol.* 2021;39:3897–3907.
23. Cook JA. The challenges faced in the design, conduct and analysis of surgical randomised controlled trials. *Trials.* 2009;10:9.
24. Wartolowska K, Beard D, Carr A. Blinding in trials of interventional procedures is possible and worthwhile [version 2; peer review: 3 approved]. *F1000Research.* 2018;6.
25. Sutton CJG, Ewen SP, Whitelaw N, Haines P. Prospective, randomized, double-blind, controlled trial of laser laparoscopy in the treatment of pelvic pain associated with minimal, mild, and moderate endometriosis**Supported by Birthright Research Grant (Ref:S3/90), Royal College of Obstetricians and Gynaecologists, London and Laserscope, Cwmbran, United Kingdom. *Fertil Steril.* 1994;62:696–700.
26. Arts J, Bisschops R, Blondeau K, et al. A double-blind sham-controlled study of the effect of radiofrequency energy on symptoms and distensibility of the gastro-esophageal junction in GERD. *Offic J Am Coll Gastroenterol.* 2012;107.
27. Rothstein R, Filipi C, Caca K, et al. Endoscopic full-thickness plication for the treatment of gastroesophageal reflux disease: a randomized, sham-controlled trial. *Gastroenterology.* 2006;131:704–712.
28. Salem M, Rotevatn S, Stavnes S, Brekke M, Vollset SE, Nordrehaug JE. Usefulness and safety of percutaneous myocardial laser revascularization for refractory angina pectoris. *Am J Cardiol.* 2004;93:1086–1091.
29. Bradley JD, Heilman DK, Katz BP, G'Sell P, Wallick JE, Brandt KD. Tidal irrigation as treatment for knee osteoarthritis: a sham-controlled, randomized, double-blinded evaluation. *Arthritis Rheum.* 2002;46:100–108.
30. Sihvonen R, Paavola M, Malmivaara A, et al. Arthroscopic partial meniscectomy versus sham surgery for a degenerative meniscal tear. *N Engl J Med.* 2013;369:2515–2524.
31. Kallmes DF, Comstock BA, Heagerty PJ, et al. A randomized trial of vertebroplasty for osteoporotic spinal fractures. *N Engl J Med.* 2009;361:569–579.
32. Bono CM, Heggeness M, Mick C, Resnick D, Watters III WC. North American Spine Society: newly released vertebroplasty randomized controlled trials: a tale of two trials. *Spine J.* 2010;10:238–240.

33. Freeman BJC, Fraser RD, Cain CMJ, Hall DJ, Chapple DCL. A randomized, double-blind, controlled trial: intradiscal electrothermal therapy: versus: placebo for the treatment of chronic discogenic low back pain. *Spine*. 2005;30:2369–2377.
34. Gillespie MB, Wylie PE, Lee-Chiong T, Rapoport DM. Effect of palatal implants on continuous positive airway pressure and compliance. *Otolaryngology-Head Neck Surg*. 2011;144:230–236.
35. Thomsen J, Bretlau P, Tos M, Johnsen NJ. Meniere's disease: endolymphatic sac decompression compared with sham (placebo) decompression. *Ann N Y Acad Sci*. 1981;374:820–830.
36. Cotton PB, Durkalski V, Romagnuolo J, et al. Effect of endoscopic sphincterotomy for suspected sphincter of oddi dysfunction on pain-related disability following cholecystectomy: the EPISOD randomized clinical trial. *JAMA*. 2014;311:2101–2109.
37. Boutron I, Guittet L, Estellat C, Moher D, Hróbjartsson A, Ravaud P. Reporting methods of blinding in randomized trials assessing non-pharmacological treatments. *PLoS Med*. 2007;4:e61.
38. Haahr MT, Hróbjartsson A. Who is blinded in randomized clinical trials? A study of 200 trials and a survey of authors. *Clin Trials*. 2006;3:360–365.
39. Schulz KF, Altman DG, Moher D, the CG. CONSORT 2010 Statement: updated guidelines for reporting parallel group randomised trials. *BMC Med*. 2010;8:18.
40. Chan A-W, Tetzlaff JM, Altman DG, et al. SPIRIT 2013 statement: defining standard protocol Items for clinical trials. *Ann Intern Med*. 2013;158:200–207.
41. Probst P, Zaschke S, Heger P, et al. Evidence-based recommendations for blinding in surgical trials. *Langenbeck's Arch Surg*. 2019;404:273–284.
42. Penić A, Begić D, Balajić K, Kowalski M, Marušić A, Puljak L. Definitions of blinding in randomised controlled trials of interventions published in high-impact anaesthesiology journals: a methodological study and survey of authors. *BMJ Open*. 2020;10:e035168.

Chapter 60

Multicenter considerations

William Wagstaff

The University of Chicago Pritzker School of Medicine, Chicago, IL, United States

Key points

- Completing a successful multicenter clinical trial is a substantial endeavor that requires an immense amount of resources, planning, and teamwork
- The five fundamental resources are patients, clinical staff, clinical care centers, funding, and time
- Contingency planning is key to overcoming inevitable obstacles

Introduction

A multicenter clinical trial is a collaborative study in human subjects involving multiple, independent care centers that investigates the risks and benefits of a novel clinical intervention with the goal of maximizing the generalizability of the results. Then known as cooperative investigations, multicenter clinical trials were first conducted in the early 20th century. They were designed to overcome errors attributed to individual investigators working in relative isolation (i.e., single-center trials) by standardizing the evaluation of therapeutic interventions in cohorts with hundreds of patients (Fig. 60.1). The Patulin Clinical Trials were the very first multicenter clinical trials on record. A collaboration between the Therapeutic Research Corporation and the Medical Research Council, the Patulin Clinical Trials ultimately demonstrated that patulin, a derivative of *Penicillium patulum*, showed no clinical benefit against the common cold. Although the treatment was unsuccessful, the study design served as a model for future multicenter clinical trials. One particularly helpful outcome was the increase of government regulation in multicenter clinical trials. Passed in 1938, the US Federal Food, Drug, and Cosmetic Act subjected preclinical and clinical trials to regulation by the FDA. Future legislation would further standardize the process of running trials and implementing the interventions in clinical settings.

The five fundamental resources of multicenter clinical trials are the patients, the clinical staff (physicians, epidemiologists, biostatisticians, etc.), clinical care centers (academic and community hospitals), funding (government and private industry), and time. Those resources are deployed over four distinct phases (planning phase, project development phase, study execution phase, and the dissemination phase) for a complete multicenter clinical trial (Fig. 60.2). In this chapter, each phase of a multicenter clinical trial will be defined and practical considerations will be highlighted. At the end of the chapter, potential pitfalls in multicenter clinical trials will be summarized along with potential solutions to those challenges.

Planning phase

The first stage of the planning phase is to establish an executive planning committee. The executive planning committee is responsible for oversight of the multicenter clinical trial and all of its subcommittees. It comprises clinical staff members with professional backgrounds ranging from physician to biostatistician to clinical ethicists. This group should also include members of sponsoring organizations such as government agencies and private industry partners. Together, members of this diverse body will complete a systematic literature review to develop a central research question that is feasible, interesting, novel, ethical, and relevant. Once the research question has been finalized, the next stage is to draft an Institutional Review Board (IRB) application; keeping in mind that multiple care centers will have to adopt and adhere to the standards set forth by the executive planning committee. Eventually, the primary investigators will be responsible for holding virtual or face-to-face meetings to discuss the details of the protocol such as the submission and review process. Doing so will surely save time, as obtaining IRB approval is often a rate-limiting step in multicenter clinical trials. Once the framework for the IRB has been drafted, the next step is to identify the outcome measures. The outcome measures are the data that are used to answer the central hypothesis of the trial. Beginning with

Translational Interventional Radiology. https://doi.org/10.1016/B978-0-12-823026-8.00095-X

FIGURE 60.1 A table comparing single center clincical trials to multicenter clinical trials in four categories: number of patients, statistical power, generalizability, and cost.

	Single Center Clinical Trial	Multicenter Clinical Trial
Number of Patients	Small N	Large N
Statistical Power	Low	High
Generalizability	Low	High
Cost	Lower	Higher

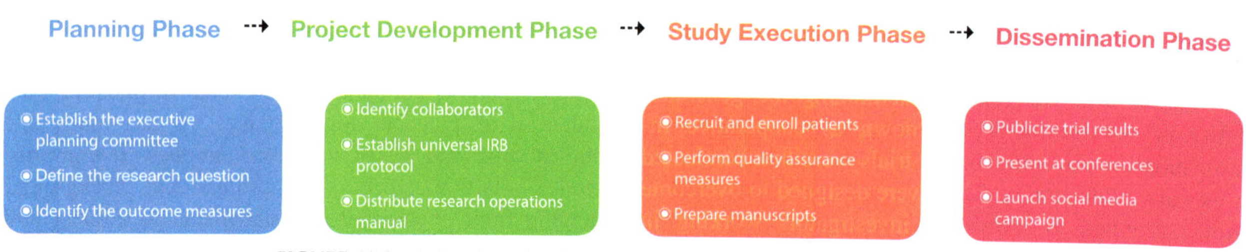

FIGURE 60.2 A flowchart showing the four phases of a multicenter clinical trial.

the end in mind facilitates the conception of a focused trial. The final stage of the planning phase is to conduct pilot studies to ensure feasibility and to estimate statistical power. There are two types of pilot studies. External pilot studies are small-scale test runs of the larger study that are designed to troubleshoot the mechanics of the study protocol and design. Internal pilot studies are designed with two goals in mind: (1) to help adjust sample size and statistical power calculations using realistic recruitment rates and (2) to help set realistic expectations for how long the study will take to complete. The data gathered during the internal pilot study are typically integrated into the study at large. Together, data from external and internal pilot studies contribute to strong funding applications, and—in some cases—they help refine the central hypothesis and outcome measures.

Project development stage

The project development phase begins with the identification of key collaborators. Collaborators with relevant basic science expertise and clinical trial experience are important members of the trial. However, the most important characteristic of any collaborator is their ability to devote sufficient time to the trial. Availability is of paramount importance because collaborators will be responsible for executing IRB contracts at their respective sites and staying in constant communication with members of the executive planning committee. The next step is for the collaborators to select a coordinating center. For some trials, that means one sole coordinating center. For others, there can be clinical coordinating centers, data collection centers, and data analysis centers. When the sites have settled on IRB standards, investigators from the coordinating center should submit the multicenter trial to a federal registry. Doing so provides researchers, journal editors, and reviewers with detailed study information. The delineation of responsibility highlights another important stage of the project development phase: creating a research operations manual. One key component of the research operations manual is the organizational chart. The organizational chart should clearly outline the roles and responsibilities of each subcommittee in the trial. The operations manual should include a step-by-step policy guide and standard operating procedures for study execution. It should include information regarding study flow, inclusion and exclusion criteria, study methods, training, and details regarding data collection, sharing, and analysis. The overall goal is to have a research operations manual that can be clearly interpreted by any individual associated with the trial regardless of their level of engagement or location. To increase buy-in—thereby increasing the likelihood of a successful trial—input from each collaborating center should be solicited. Feasibility testing during the project development stage allows for a functional test of the data management system using sham data. Doing this decreases the likelihood of data uploading errors. Lastly, the formation of a manuscript committee ensures academic rigor, introduces transparency to the authorship assignment process, manages conflicts of interest, and creates a space to discuss data visualization.

Execution phase

The execution phase is marked by the recruitment and enrollment of patients in the trial via screening, case finding, or referral. It is crucial to establish standardized consent and randomization processes early in the recruitment and enrollment stage so that the data are not compromised in any way. As the data collection process begins, quality assurance measures must be put in place. Members of the data collection subcommittee must ensure that no clinical care center begins patient enrollment until a sufficient number of clinical staff has demonstrated proficiency in the procedure. This requires intermittent data monitoring and review in addition to in-person site visits by members of the executive planning committee to each of the sites. Centralization of data collection can really benefit a multicenter trial. When appropriate, performing laboratory tests in a central location allows for unbiased assessment of the specimens, standardization of analysis, reduction in variability, and ease of quality control. Multicenter trials manage large data sets that are both viewed and accessed by investigators across institutions, so it is important to maintain adequate data security and privacy with the help of information technology specialists. Later in the execution phase, data organization becomes a priority. Proper data visualization—with guidance from the manuscript committee—allows the executive steering committee to terminate the trial in the case of en masse, unanticipated, and undue harm or high likelihood of indifferent results.

Dissemination phase

The last, and often overlooked, phase of a multicenter clinical trial is the dissemination phase. The trial is complete, but it is incumbent upon the executive planning committee to market the results. During this phase, investigators work together to publicize trial results and publications via traditional media, social media, and conference presentations. The goal of the dissemination phase is to ultimately translate the results into improvements in patient health outcomes. One strategy for optimizing the impact of the trail—if guidelines permit—is to submit abstracts and presentations to several different journals and conferences. The social media campaign should be designed ahead of time such that diverse subgroups of major stakeholders are made aware of the results. Medical Subject Headings (MeSH) terms should also be selected in this phase so that trial results and publications rise to the top of search engine queries.

Challenges

Multicenter clinical trials involve several moving parts developing over several years. In this section, challenges faced by multicenter trials will be presented vis-a-vis the Keynote 937 multicenter trial.

First posted to www.clinicaltrials.gov in March of 2019, the Keynote 937 trial is a phase 3 trial that will evaluate the safety and efficacy of pembrolizumab versus placebo as adjuvant therapy in patients with hepatocellular carcinoma and complete radiological response after surgical resection or local ablation. The estimated completion date in 2025 highlights that time is a challenge that multicenter trials face. Not only do patients in the treatment arm of the study require at least 17 cycles of a 21-day treatment, but some of the primary outcome measures require tracking patients for up to 6 years. Working efficiently helps to protect the novelty of the trial results that is important for publishing in high-impact journals. Sponsored by Merck, this trial shows that funding is also a major obstacle to completing a multicenter trial. Without funding none of this work would be possible. A legislative solution in the form of the American Recovery and Reinvestment Act of 2009 resulted in the allocation of over $1 billion from the federal government to support clinical trials research. The target enrollment of Keynote 937 is around 950 patients. It is important that the 193 study locations have adequate geographical, racial, socioeconomic, gender, and lifestyle diversity to optimize generalizability. The National Institutes of Health (NIH) Revitalization Act of 1993 ensured

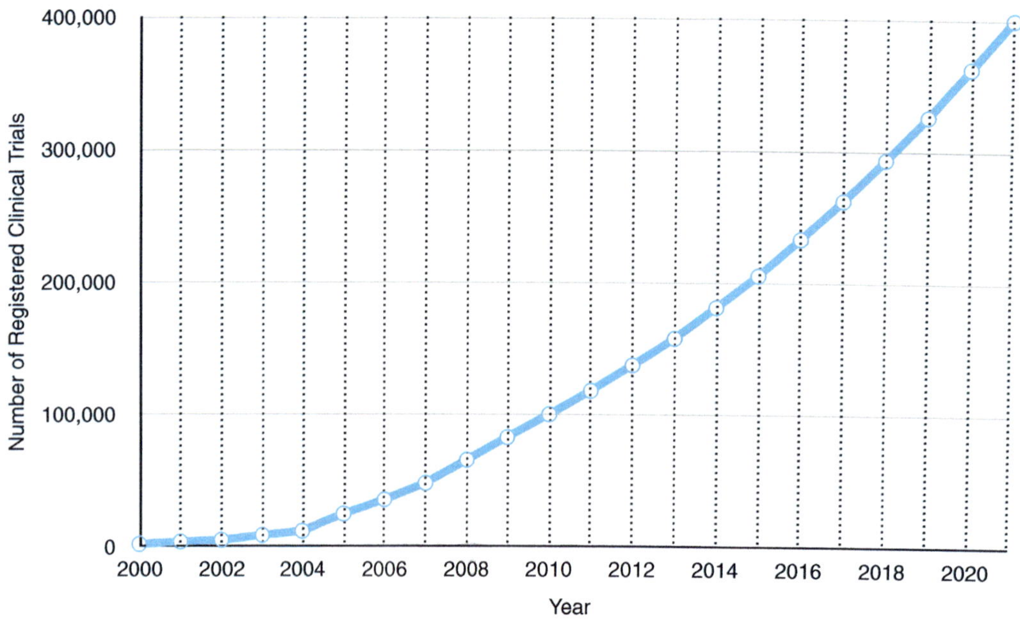

FIGURE 60.3 A bar graph showing the number of registered clinical trials by year from 2000 to 2021 according to www.clinicaltrials.gov. *(https://www.clinicaltrials.gov/ct2/resources/trends)*

that women and minorities are included in clinical trials such that members of those underrepresented communities can benefit from multicenter clinical trial results as well. Additionally, it is important to have a reasonable proportion of patients who received surgical intervention versus tumor ablation therapy. As those 193 locations are located all over the world, another challenge, globalization, is presented. Culturally competent materials in several different languages must have been developed to reduce bias and to have standardized IRB and operation protocols. It is ethically important that patients—regardless of their country of origin—can benefit from the results of the trial. With 193 sites, there are several authors to consider when it comes to maintaining morale, offering protected research time, managing conflicts of interest, and ensuring fair acknowledgment at publication. To guarantee that credit is fairly distributed junior faculty and trainees should be appointed to subcommittees. Additionally, they should be encouraged to take on more substantive leadership roles in ancillary studies.

The number of multicenter clinical trials is growing year after year (Fig. 60.3). This means that groups are both preparing for anticipated challenges and successfully navigating unanticipated ones. Contingency planning with diverse, experienced teams is crucial to overcoming challenges both foreseen and unforeseen.

Get started

- Review landmark clinical trials in interventional radiology at http://rfs.sirweb.org/landmark-clinical-trials/
- Form a preliminary executive planning committee and begin a thorough literature review
- Identify a central research question that is feasible, interesting, novel, ethical, and relevant

References

1. Cheng A, Kessler D, Mackinnon R, et al. Conducting multicenter research in healthcare simulation: lessons learned from the INSPIRE network. *Adv Simul*. 2017;2(1). https://doi.org/10.1186/s41077-017-0039-0.
2. Junod SW. *FDA and clinical drug trials: a short history. A Quick Guide to Clinical Trials*. 2008:25–55.
3. Friedman L, Furberg C, DeMets D, Reboussin D, Granger C. *Multicenter Trials. Fundamentals of Clinical Trials*. 2015:501–518. https://doi.org/10.1007/978-3-319-18539-2_21.
4. *Implementation Tools. Clinical Trials Transformation Initiative*; 2020. Published. https://www.ctti-clinicaltrials.org/briefing-room/tools. Accessed July 1, 2020.
5. Chung K, Song J. A guide to organizing a multicenter clinical trial. *Plast Reconstr Surg*. 2010;126(2):515–523. https://doi.org/10.1097/prs.0b013e3181df64fa.
6. Cheng A, Kessler D, Mackinnon R, et al. Conducting multicenter research in healthcare simulation: lessons learned from the INSPIRE network. *Adv Simul*. 2017;2(1). https://doi.org/10.1186/s41077-017-0039-0.
7. Dashevsky B, Bercu Z, Bhosale P, et al. Multicenter research studies in radiology. *Acad Radiol*. 2018;25(1):18–25. https://doi.org/10.1016/j.acra.2017.05.019.
8. Greenberg G. Logistics and management of clinical trials. *Br J Clin Pharmacol*. 1982;14(1):25–30. https://doi.org/10.1111/j.1365-2125.1982.tb04930.x.

Chapter 61

Registries

Stephanie S. Dybicz[1], **Kristina M. Chapple**[2] **and Santiago M. Cornejo**[3]

[1]*Department of Radiology, University of Minnesota, Minneapolis, MN, United States;* [2]*Trauma and Acute Care Surgery Department, St. Joseph's Hospital and Medical Center, Phoenix, AZ, United States;* [3]*Department of Radiology, St. Joseph's Hospital and Medical Center, Phoenix, AZ, United States*

Summary

1. A registry is a standardized collection of data that can be used for observational studies investigating a wide range of clinical questions, from a disease's natural course, to treatment outcomes, product safety, and cost comparisons.
2. Registries can offer investigators of all backgrounds and resources the ability to extract "real-world" data without the extensive involvement demanded of a controlled, randomized trial.
3. Efforts to centralize and standardize a registry in the interventional radiology community can revolutionize the field's capacity to demonstrate its value in providing high-quality, evidence-based, first-line solutions compared to that of traditional approaches.

Introduction

The manner in which medicine is administered has trended to include the patient in the decision-making of their medical care, otherwise known as *patient-centered care.*[1] Patient-centered care is providing care that is respectful of and responsive to individual patient preferences, needs, and values to ensure that the patient's values guide all clinical decisions. For this reason, various stakeholders like patient advocacy groups, hospital administration, and clinical peers have demanded that clinical investigators pursue patient-centered outcomes research (PCOR) that seeks to address the questions and concerns most relevant to patients.[2] PCOR delivers two desired clinical outcomes, not only to increase the satisfaction of patients and healthcare providers with medical care, but also to improve health outcomes that are most meaningful to patients and clinicians.

In addition to patient-centric medical practices, the technological advances and ever-changing protocols that are particularly unique to the field of interventional radiology challenge the physician to demonstrate "real-world" evidence to its stakeholders that the care provided is high quality, cost-effective, and patient-centered. With that in mind, *registries* afford physicians of all institutional capacities and caliber an easily accessible source of patient data to support (or dispute) years of experience and intuition that have guided their medical decision-making. Ultimately, a registry is well suited for patient-centered research, thus increasing the likelihood that the evidence produced can both support and guide clinical and health policy decisions.

Definition

A *registry* is defined by the Agency for Healthcare Research and Quality (AHRQ) as an "organized system that uses observational study methods to collect uniform data (clinical and other) to evaluate specified outcomes for a population defined by a particular disease, condition, or exposure, and that serves one or more predetermined scientific, clinical or policy purposes."[3] In other words, registries use retrospective or prospective observational study methods to collect and harmonize data about the treatment, outcomes, and well-being of patients who receive care over time. They permit the investigator to aggregate large data sets and analyze trends or patterns in treatments and outcomes.

The common types of registries pertinent to clinical management include patient (i.e., disease), health service, product, and pregnancy exposure registries. To elaborate, a patient registry often gathers data pertaining to the health

Translational Interventional Radiology. https://doi.org/10.1016/B978-0-12-823026-8.00090-0

status of patients and the care they receive for a specific disease or condition. A product registry will focus instead on the effectiveness, safety, and value of pharmacologic or medical device products in patient care.

Benefits and limitations

Benefits of a registry are numerous and vary depending on the stakeholders involved.[4–6] For the physician, one is able to witness real-world evidence of the disease course, effective treatment options and alternatives, as well as desired or aversive outcomes. Physician organizations and societies can gain valuable insight from registry data to assess whether providers are indeed practicing evidence-based medicine. Patients and advocacy groups benefit from a deeper understanding of the natural course of disease and the success of treatment options for an improved quality of life.

In the era of patient-centered observational research, stakeholders rely on a rich source of data to evaluate the course of a disease and the effectiveness of new treatments. Few will challenge the merits of a prospective, randomized, controlled trial (RCTs). RCTs control for confounding variables, blind subjects, and clinicians, as well as optimize data quality control that pulls away the noise cluttering around the control and experimental groups or interventions. This will inevitably lead to breakthrough conclusions between groups that are both trusted and reputed.[7]

Nevertheless, one must consider the extensive organization, recruitment, and funding demanded of a randomized controlled clinical trial and admit that it is not always feasible to take on such a study design. Smaller caliber facilities face barriers of lower patient volumes, less funding, time, or even a lack of desire to participate in clinical research. To not include those data, however, would be to bias patient outcomes toward those which are produced by larger, academic, resource-rich institutions.[7] Comparatively, registries are accessible and inclusive of community hospitals and clinics, thus producing a realistic snapshot of a disease's progression, patient outcomes, technical success, product safety, and so forth. Additionally, observational patient registries supply "real-world" data through its inherent inclusion of noncompliance, variabilities in clinician decision-making, and operator techniques, unlike the somewhat artificial and restrictive nature of RCTs.

Despite its many contributions to clinical research, registries are still an underused form of data collection. Studies designed around a patient registry face many limitations: funding, patient engagement, minority participation, transparency of data collection, and opt-out ability, to name a few.[1] Moreover, the greatest challenge is the lack of centralized data sourcing across national borders, specialties, diseases, professional societies, and other entities. The Registry of Patient Registries effort developed by the AHRQ, in conjunction with the National Library of Medicine, has taken to improve data centralization in registry-based research with the goal to include all types of patient registries.[8] Even with these efforts, the accessibility and visibility of registries have been impeded by the lack of standardized data reporting, centralized data sharing, and limited national and international data collaborations.[9] Additionally, attempts to ensure complete data collection (e.g., medications and products used) and to standardize common core data sets and coding terminologies have been laborious.

As with any study design mentioned in earlier chapters, it is of utmost importance for the investigator to appraise the study's goals and select the most effective study design in the context of its advantages, disadvantages, costs, and patient population.

Get started

The following guide can assist the interventional radiologist in initiating a registry to support his or her study adapted from the Effective Health Care Program user's guide developed by the AHRQ.[3]

State the objectives

An investigator must critically evaluate whether a registry is appropriate for a given clinical question in the planning phases of a study. For example, a proposed investigation may benefit from a registry if the objectives of the study are to *observe* the course of a disease, variations in treatment and outcomes, factors influencing prognosis and quality of life, care patterns across providers, and institutions; the safety, harm, effectiveness, and quality of care; and quality improvement.[10]

Is a registry appropriate?

As a general rule, the successful development of a registry relies on at least one variable being considered "active," wherein the data are said to be collected directly for the registry and not extrapolated indirectly from other forms (e.g., billing, intake forms, etc). That is not to say that registries cannot incorporate other sources such as the electronic medical record (EMR) or other external databases for completeness and breadth. Moreover, registries depend on consistent and clear data definitions with easily submitted and accessible data core sets. Lastly, it is important to decide whether the data needed for the study exists elsewhere as to avoid the labors of recollection. An important distinction from other study designs is that the patient management reflected in a registry is not determined by the protocol; rather, it takes on a natural, observational form of data collection.

Identify the stakeholders

The type of data collected will matter greatly for whom the data matters, so it is crucial to identify those stakeholders before the development of the registry. The primary stakeholder typically includes the funding or supervising authority (e.g., National Institute of Health (NIH), AHRQ, hospital administration). Secondary stakeholders are those that would primarily gain knowledge from the registry's data and practically apply it to a patient's medical management (e.g., physician, patient, societies). Other stakeholders could include public health or regulatory authorities, product manufacturers, healthcare service providers, patients/advocacy, clinician groups, academic institutions, and professional societies (SIR).

Create a project plan

Once a study is appraised as a good fit for a registry, the investigator must create a thoughtful, organized project plan. One must identify the scope of the registry, which is commonly limited to available funding and number of stakeholders involved. Other planning milestones should include a timeline, schedule, cost and budget, staff and roles, procurement of products, medications or devices, and expected risks.

Define the standard research team personnel and roles

Personnel to include for the development and management of a registry depends entirely on the size, scope, and purpose of the study. A registry' team can range from only one interventional radiologist with the informal assistance of multiple trusted consultants or experts to a number of individuals, including but not limited to, the following roles:

1. *Project manager*: develop timelines, milestones, deadlines, and overall goals and budgets met.
2. *Executive lead*: absorbs the grand financial, scientific, and legal role to steer the direction and purpose of the registry.
3. *External review*: provide independent oversight for safety of patients and quality of data acquisition. This can include core laboratory physicians.
4. *Liaison*: manage day-to-day operations between stakeholders
5. Database manager
6. Lead experts in the subject matter
7. Statistician
8. Epidemiologist

As a general rule, the larger and more involved the registry, the more diversified the stakeholder involvement should be.

Choose your registry software product

The number of registries has grown over the past several decades as healthcare information has become digitized. Yet despite their increase in use and significance, registries face real challenges in establishing the participation, engagement, and utility needed to drive their sustainability. Modern clinical data registries address these limitations by going beyond data and storage. They rely on advanced analytics and data science to transform data into meaningful insights that are useable for a variety of stakeholders to achieve a desired outcome.

Populate initial patient list

There are a few different strategies in acquiring an initial patient list for a registry. Commonly, patient lists are generated from the institution's EMR with billing codes specific to a diagnosis, product, or medication, depending on the study's objectives. A sophisticated registry software can offer a robust extraction of patient lists tailored to the needs of the study.

Financing a registry

Funding required to build and maintain a registry is directly proportional to the number of sites, patients, and data points collected. Registries have historically been funded by a number of sources such as nonprofit organizations, professional societies, and government or regulatory departments. These include the NIH, Center for Disease Control, Centers for Medicare and Medicaid services, the AHRQ, the Food and Drug Administration, health insurance providers, patient advocacy groups, private funding, product manufacturers, and professional societies. Funding can occur through any combination of these sources depending on the primary stakeholders' needs and the purpose of the study.

Patient-Centered Outcomes Research Institute (PCORI) finances those investigators specifically interested in PCOR, with strict guidelines to maintain constant patient engagement throughout the research project. To qualify for PCORI grants, the objectives of the study should center around the outcomes most important to patients to ultimately assist them, their caregivers, clinicians, insurers, and policymakers in making better-informed decisions about their health and healthcare options.

Potential pitfalls

There are best practices that will mitigate the common pitfalls in creating registries. Namely, *selection bias* can be most troublesome during the patient and physician enrollment process.[3] Creating a standardized inclusion and exclusion criteria for the registry will combat any biases related to the recruitment of institutions, physicians, and patients alike.

To avoid institution bias, one must not be too inclusive of academic institutions with little involvement of private groups or community-based practices. By only enrolling patients from one type of institution, the data that follow will inevitably not be able to reflect "real-world" medicine or a disease process across all departments and patient populations. Bias can also be introduced during the recruitment of interventional radiologists. Here, it is essential to include a breadth of experience levels and technical skills, treatment of diseases, and avoiding only deemed experts in the field or subspecialties of IR. Selection bias can also occur when enrolling patients into the registry. Physicians should be aware of their biases toward recruiting stable patients and less so older, sicker patients. Understandably, the additional barriers of informed consent, follow-up, and time available to educate make it burdensome to recruit unstable patients. Nevertheless, not including them will hurt a registry's capacity to present "real-world" data.

Similarly, confusing terminology can hinder patient recruitment of all abilities and background, which can risk unattended exclusion of minority patient populations. Lee and colleagues remarked that their registry design may have hindered Spanish-speaking patients from participating in the registry due to a language barrier and limited understanding of the study's purpose; thus, it is important to have a clear, concise, correctly translated, and easily understood protocol.[11] Whether from a language barrier or limited experience in research (or both), patient education is of utmost importance during the recruitment phase to avoid the pitfall of incomplete patient data from minority populations.

VIRTEX SIR data registry

In the fall of 2020, the Society of Interventional Radiology (SIR) launched the VIRTEX data registry supported by ArborMetrix software. With the implementation of standardized reporting templates available on the SIR website, IR departments are able to uniformly submit their patient data to the centralized registry. The VIRTEX registry system is then able to extract and code data specific to IR procedures from the reports, eliminating the need for manual chart abstraction or other information technologies. At its inception, there are standardized reports for over 70 procedures, including biopsies, central venous catheter use, vertebroplasty, inferior vena cava filters, prostatic artery embolization, uterine artery embolization, peripheral artery disease, venous thromboembolism, and stroke. Future plans of the VIRTEX SIR data registry are to implement automatic surveys delivered to patients undergoing IR procedures and to expand the standardized reports to most, if not all, IR procedures. The enrollment of institutions and practices is currently underway.

These efforts hope to promote the collection of dedicated clinical data specific to IR and obtain the depth of evidence needed to demonstrate the abilities and potential of IR. Additionally, VIRTEX shows true promise in offering a standardized IR registry that will support PCOR. Ultimately, a central registry gives investigators an opportunity to produce evidence that IR therapies are high-quality, evidence-based, first-line solutions and are often less invasive and more efficient than traditional approaches, all while delivering equivalent or better outcomes.

Real-world examples

SCVIR transluminal angioplasty and revascularization registry[12]

Registry details: A total of 1802 patients were recruited from 7 participating sites in the northeast ranging in private practice, community, and academic hospitals between January 1, 1992 and December 31, 1994. The registry was sponsored by the Society of Cardiovascular and Interventional Radiology (SCVIR) concomitantly with Cardiovascular and Interventional Radiology Research and Education Foundation (CIRREF) designed to retrospectively record the angiographic and clinical outcomes of patients who have undergone percutaneous transluminal revascularization procedures involving the abdominal aorta and lower extremities.

Objectives: Westcott and colleagues retrospectively studied 25 patients from the SCVIR transluminal angioplasty and revascularization registry with infrarenal aortic atherosclerotic disease receiving conventional balloon percutaneous transluminal angioplasty (PTA) versus PTA plus the Palmaz intravascular stent to evaluate for differences in safety and short-term efficacy. The investigation yielded no statistical difference between the two interventions when measuring the end outcomes of stenosis reduction, transtenotic pressure gradient, and initial clinical improvement postoperatively.

The authors noted an important limitation of the registry that was likely confounding their data. The registry made no distinction of whether an interventionalist first chose PTA alone followed by subsequent addition of the Palmaz stent during the case, or if it was always planned to perform PTA with Palmaz from the start. Was the stent placed because the first pass of PTA was unsuccessful? Thus, it may be necessary to add postprocedural decision-making rationale to a registry's core data set.

Takeaway: Registries can introduce selection bias.

The unprotected left main trunk intervention multicenter assessment registry[13]

Registry details: The unprotected left main trunk intervention multicenter assessment (ULTIMA) registry was a multicenter, international patient list consisting of 277 patients from 25 different sites from 1994 to 1996.

Objectives: Marso and colleagues extracted 40 patients from the ULTIMA registry suffering cardiogenic shock secondary to acute myocardial infarction of an unprotected left main stenosis. They then characterized the immediate, 2- and 12-month outcomes following either percutaneous transluminal angioplasty or a primary stent for emergency percutaneous revascularization and whether either would result in improved survival rates and/or no indication for Coronary artery bypass graft (CABG) at 1 year.

Marso and colleagues mitigated the possibility of selection bias by limiting the number of patients inputted into the registry to no more than 30 patients per institution. That way, larger centers with the volume to attain greater than 30 patients were prevented from skewing the data to represent one single center.[14]

Takeaway: There are simple and practical strategies to reduce selection bias in registries.

Pulmonary embolism response to fragmentation, embolectomy, and catheter thrombolysis registry[15]

Registry details: The registry included 101 patients with massive and submassive pulmonary embolisms (PEs) treated with catheter-directed therapy (CDT) from 6 sites in the United States and 1 international site from 2011 to 2014.

Objectives: Kuo and colleagues investigated whether CDT was a safe and effective alternative method to the treatment of massive and submassive PE compared to traditional systemic thrombolysis. Incidentally, the authors discovered that there was no difference in outcomes with the use of ultrasound-assisted thrombolysis (USAT) versus standard infusion catheters during CDT in the management pulmonary embolisms.

The ULTIMA (not to be confused with the aforementioned registry) and SEATTLE II trials were two randomized, controlled clinical trials studying CDTs in the management of massive and submassive PEs.[16,17] Both trials' protocols included the use of the high-cost USAT in lieu of the standard infusion catheters. The present study compared patients in the registry who were treated with USAT versus standard CDT and found no statistically significant difference in pulmonary artery pressure changes between pre- and postprocedure. Thus, the study supports the use of low-dose thrombolytic infusions through standard catheters for patients with submassive PEs without the additional unnecessary cost.

The conclusions generated from the pulmonary embolism response to fragmentation, embolectomy, and catheter thrombolysis registry supported the refinement of protocols to reflect the most cost-effective options for the patient and all stakeholders involved. Moreover, the registry's data also yielded 0% rate of major complications, lower than the previously reported 2.4% rate. The authors concluded that the reduction was likely due to eliminating the AngioJet device from the present study's protocol. This was yet another way in which the availability of registries supported real-time protocol modification and identified procedure-related complications, as these studies were done in mere years of one another.[18]

Takeaways: Registries can bring costs down.

The registry of AngioVac procedures in detail[19]

Registry details: 234 patients with caval thromboemoboli (CTEs), right heart masses, catheter-related thrombi, and pulmonary emboli were enrolled from 21 institutions from March 2016 to August 2019.

Objectives: Moriarty and colleagues examined the procedural safety and technical success of the AngioVac device, which was defined by at least 70% tissue removal of emboli, thrombus, or mass removal using the AngioVac suction system.

The registry of AngioVac procedures in detail study evaluated the efficacy of the AngioVac system across multiple patient populations and pathologies. The authors cautioned that readers not apply any one conclusion made by the study toward pathologies in which limited registry data was available, and in their case, pulmonary emboli. A centralized IR registry can improve sample sizes available to the investigator and, in the same breath, the validity of a study's conclusions.

Takeaway: A standard IR registry can improve sample size and study validity.

References

1. Daugherty SE, Lee SB, Nowell B, et al. The increasing focus on the patient in patient registries. In: Gliklich RE, Dreyer NA, Leavy MB, et al., eds. *21st Century Patient Registries: Registries for Evaluating Patient Outcomes: A User's Guide*. 3rd ed. Rockville (MD): Agency for Healthcare Research and Quality (US); March 2018. Addendum [Internet].
2. PCORI About Us. http://www.pcori.org/about-us. Accessed March 27, 2020.
3. Gliklich R, Dreyer N. *Registries for Evaluating Patient Outcomes: A User's Guide*. Rockville, MD: Agency for Healthcare Research and Quality; 2010.
4. Kennedy L, Craig AM. Global registries for measuring pharmacoeconomic and quality-of-life outcomes: focus on design and data

collection, analysis and interpretation. *Pharmacoeconomics.* 2004;22(9):551–568.
5. Charrow J, Esplin JA, Gribble TJ, et al. Gaucher disease: recommendations on diagnosis, evaluation, and monitoring. *Arch Intern Med.* September 14, 1998;158(16):1754–1760.
6. Bushby K, Lynn S, Straub T, et al. Collaborating to bring new therapies to the patient–the TREAT- NMD model. *Acta Myol.* July 2009;28(1):12–15.
7. Trotter JP. Patient registries: a new gold standard for "real world" research. *Ochsner J.* 2002;4(4):211–214.
8. *Developing a Registry of Patient Registries (RoPR). Project Overview. Agency for Healthcare Research and Quality.* May 31, 2012.
9. McGettigan P, Alonso Olmo C, Plueschke K, et al. Patient registries: an underused resource for medicines evaluation: operational proposals for increasing the use of patient registries in regulatory assessments [published correction appears in Drug Saf. 2019 Sep 5]. *Drug Saf.* 2019;42(11):1343–1351.
10. LaBresh KA, Gliklich R, Liljestrand J, et al. Using "get with the guidelines" to improve cardiovascular secondary prevention. *Joint Comm J Qual Saf.* October 2003;29(10):539–550.
11. Lee SB, Zak A, Iversen MD, Polletta VL, Shadick NA, Solomon DH. Participation in clinical research registries: a focus group study examining views from patients with arthritis and other chronic illnesses. *Arthritis Care Res.* 2016;68(7):974–980.
12. Westcott MA, Bonn J. Comparison of conventional angioplasty with the Palmaz stent in the treatment of abdominal aortic stenoses from the STAR registry. SCVIR Transluminal Angioplasty and Revascularization. *J Vasc Interv Radiol.* 1998;9(2):225–231.
13. Marso SP, Steg G, Plokker T, et al. Catheter-based reperfusion of unprotected left main stenosis during an acute myocardial infarction (the ULTIMA experience). Unprotected Left Main Trunk Intervention Multi-center Assessment. *Am J Cardiol.* 1999;83(11): 1513–1517.
14. Ellis SG, Tamai H, Nobuyoshi M, et al. Contemporary percutaneous treatment of unprotected left main coronary stenoses: initial results from a multicenter registry analysis 1994-1996. *Circulation.* 1997;96(11):3867–3872.
15. Kuo WT, Gould MK, Louie JD, Rosenberg JK, Sze DY, Hofmann LV. Catheter-directed therapy for the treatment of massive pulmonary embolism: systematic review and meta-analysis of modern techniques. *J Vasc Interv Radiol.* 2009;20(11):1431–1440.
16. Kucher N, Boekstegers P, Müller OJ, et al. Randomized, controlled trial of ultrasound-assisted catheter-directed thrombolysis for acute intermediate-risk pulmonary embolism. *Circulation.* 2014;129(4): 479–486.
17. Piazza G, Hohlfelder B, Jaff MR, et al. A prospective, single-arm, multicenter trial of ultrasound-facilitated, catheter-directed, low-dose fibrinolysis for acute massive and submassive pulmonary embolism: the SEATTLE II study. *JACC Cardiovasc Interv.* 2015;8(10): 1382–1392.
18. Kuo WT, Gould MK, Louie JD, Rosenberg JK, Sze DY, Hofmann LV. Catheter- directed therapy for the treatment of massive pulmonary embolism: sys- tematic review and meta-analysis of modern techniques. *J Vasc Interv Radiol.* 2009;20(11): 1431–1440.
19. Moriarty JM, Rueda V, Liao M, et al. Endovascular removal of thrombus and right heart masses using the AngioVac system: results of 234 patients from the prospective, multicenter registry of AngioVac procedures in Detail (RAPID). *J Vasc Interv Radiol.* 2021;32(4):549–557.e3.

Chapter 62

Phases of clinical trials

Matthew Henry[1,3,4] and Daniel Kirkpatrick[2]

[1]*Wayne State University School of Medicine, Detroit, MI, United States;* [2]*Department of Radiology, Michigan Medicine, University of Michigan, Ann Arbor, MI, United States;* [3]*Rackham Graduate School, University of Michigan, Ann Arbor, MI, United States;* [4]*Department of Radiology, Medical College of Wisconsin, Milwaukee, WI, United States*

Key points

1. Each clinical trial phase seeks to address a specific question and must meet certain criteria before the intervention can be investigated in the next phase.
2. Investigators must follow established ethical guidelines when leading a clinical research trial.
3. Designing and conducting methodologically robust trials is crucial for successful research.

A clinical trial is a prospective study that compares the effects of an intervention in humans against a control group.[1] Contemporary clinical trials have undergone a drastic evolution compared to their first iterations (Table 62.1).[1–3] King Nebuchadnezzar, a Babylonian ruler, directed the first clinical trial when he commanded his subjects to eat a diet consisting of legumes and water instead of meat and wine.[2] Following the 10-day experiment, the king concluded that the vegetable consuming group appeared better nourished. It was not until the mid-18th century that a physician conducted the first planned clinical trial with a control group when Dr. James Lind investigated treatment for scurvy among sailors aboard the Salisbury.[2] In the following century, Dr. Austin Flint carried out the first placebo-controlled clinical trial in which he investigated rheumatism treatments. The 1900s saw the first contemporary clinical trials with the introduction of blinding and randomization.[2]

Clinical trials are a fundamental and vital component of modern medicine and research. Data produced from methodologically robust clinical trials are considered the highest level of evidence because they are prospective and incorporate features that limit or reduce bias. As a result, there is increasing emphasis on conducting clinical trials to ensure the highest quality data are obtained. Clinicaltrials.gov is a website for patients, researchers, and the public providing access to information about ongoing and completed clinical trials. It launched in February of 2000

TABLE 62.1 Evolution of clinical trials.

Year	First known documentation of a	Description
~550 BCE	Clinical trial	A Babylonian king tests two different diets in his people
1537	Clinical trial of novel therapy	Battlefield surgeon Ambroise Pare compares a novel concoction versus oil to treat battle wounds
1747	Controlled clinical trial	Dr James Lind investigates different treatment options for scurvy in sailors
1863	Placebo-controlled clinical trial	Dr Austin Flint designed a study to investigate herbal extract versus placebo for the treatment of rheumatism
1931	Randomization and blinded clinical trial	Dr James Amberson conducted a trial utilizing both randomization and single-blinding to compare sanocrysin versus distilled water for the treatment of tuberculosis

Adapted from Friedman LM, Furberg CD, DeMets D, Reboussin DM, Granger CB. *Fundamentals of Clinical Trials.* 5th ed. Springer International Publishing; 2015. doi:10.1007/978-3-319-18539-2; Bhatt A. Evolution of clinical research: a history before and beyond James Lind. *Perspect Clin Res.* 2010;1(1):6–10. Collier R. Legumes, lemons and streptomycin: a short history of the clinical trial. *CMAJ.* 2009;180(1):23-24. doi:10.1503/cmaj.081879

Translational Interventional Radiology. https://doi.org/10.1016/B978-0-12-823026-8.00105-X

with slightly over 2000 registered clinical trials. Twenty years later, newly registered studies neared 37,000 per year (Fig. 62.1).[4]

When a researcher decides to conduct a clinical trial, it is paramount to adhere to certain ethical guidelines.[5] First, the investigators must establish clinical equipoise and assure the intervention has the potential to provide either social or scientific value. The design and procedures of the proposed trial should be reviewed by an independent party. Furthermore, the intervention should be performed and investigated in a manner that produces reliable results. It is crucial to minimize risks to the patient while also maximizing the benefit. Informed consent is important to select patients fairly to not exploit the vulnerable and favor the privileged. Once subjects are enrolled, the investigators must protect their rights and health.

Before one treatment can be compared to another, it must pass through the five phases of a clinical trial. Each phase answers a specific question about the intervention and must meet certain milestones before it can progress to the next stage (Table 62.2).

In 2006, phase 0 studies were introduced by the United States Food and Drug Administration (FDA) to accelerate the development of novel drugs.[6] These studies do not evaluate therapeutic potential. Instead, phase 0 studies evaluate pharmacokinetic and pharmacodynamic information or assay methodology and modulation.[7] Phase 0 studies are very small, and investigators often enroll fewer than 25 patients. Either healthy volunteers or patients are enrolled in phase 0 studies. Because the drugs are administered at subtherapeutic doses, there are ethical challenges if the subjects have terminal illnesses. The drug or device they are receiving is known to provide no therapeutic benefit, and the patient's enrollment in the study could prevent participation in other clinical research that could potentially treat their disease.

Phase I trials are relatively small, usually enrolling fewer than 100 subjects and lasting less than 1 year. Subjects may be healthy volunteers or patients with disease, especially if their disease is the target of the intervention. The primary objective of phase I clinical trials is safety and dosage, specifically the maximum tolerated dose. If the intervention is a drug, the dose is escalated, and the corresponding adverse effects are monitored. Controlling with a placebo group, randomization, and blinding are not typical features of these studies because there is usually no control group.

Mimura et al. performed phase I and phase II studies investigating the safety of radiofrequency ablation (RFA) for renal cell carcinoma.[8] A total of nine patients were enrolled in phase I. Three subjects underwent RFA and were monitored for severe adverse events over 4 weeks. The next three patients were treated only if severe events were seen in one patient or less. Four weeks is insufficient to evaluate the clinical efficacy of RFA. However, the goal of the phase I component was to ascertain the intervention's safety and the investigators successfully addressed this objective by reporting no grade 4 or grade 5 adverse events.

Phase II studies may begin after the intervention's safety profile is deemed acceptable. The goal of this phase is to assess the therapeutic effects as well as monitor for toxicity and adverse events. Studies can include several

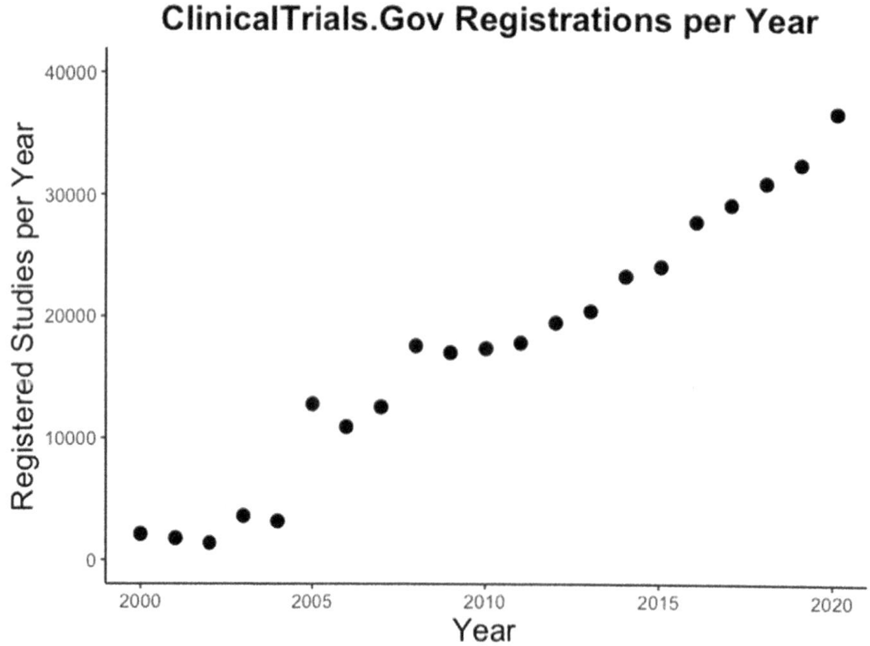

FIGURE 62.1 Number of clinical trials registered at ClinicalTrials.Gov per year since the launch of the website in February of 2000.[4]

TABLE 62.2 Different phases of clinical trials, their purpose, design features, types and number of participants, and typically duration.

Phase	Purpose	Design	Type of participants	Number of participants	Duration
Phase 0	Pharmacokinetics Pharmacodynamics Assay methodology Target modulation	Single arm No blinding No randomization	Healthy subjects or subjects with specified disease	Less than 25	Months
Phase I	Safety Dosage	Single arm No blinding No randomization	Healthy subjects or subjects with specified disease	Less than 100	Months
Phase II	Evidence of efficacy Adverse events	Single or multiple arms Can be controlled and randomized	Subjects with specified disease	Hundreds	Months or years
Phase III	Clinical efficacy Adverse events	Controlled Blinded Randomized	Subjects with specified disease	Hundreds or thousands	Years
Phase IV	Postmarketing clinical efficacy Adverse events	Controlled Blinded Randomized	Subjects with specified disease	Hundreds or thousands	Years

hundred subjects with the targeted condition, although most procedural studies have fewer than 100 participants. Because of their large size, these studies can last a year or longer. Importantly, phase II studies are not powered to detect a clinically significant effect. They may have single or multiple arms, in which case randomization and blinding are common study characteristics. Controlling with a placebo is not often utilized.

Gonsalves et al. performed a phase II clinical trial investigating the effect of radioembolization for treatment of uveal melanoma hepatic metastases.[9] They enrolled 48 subjects and included a treatment-naïve group, however they did not randomize or blind. All patients underwent the intervention and were evaluated for adverse effects every 3 months for at least 2 years. They found that overall survival was prolonged, and the toxicity was less than other available therapies. They concluded that radioembolization was safe, effective, and had an acceptable toxicity profile for those with hepatic metastasis secondary to uveal melanoma.

After an intervention has demonstrated potential clinical effect and an acceptable toxicity profile, it moves to a phase III study. The primary objective of phase III is to demonstrate a clinically meaningful treatment benefit of the intervention. As a result, randomization and blinding are necessary components of the study design. If an effective treatment is already available for the disease, then an active-treatment controlled trial is preferred to a placebo-controlled trial. When the intervention being studied is a procedure, placebo or sham groups are more common. Phase III clinical trials often enroll hundreds or thousands of patients, and they can take several years to complete. Because many more subjects are enrolled compared to prior phases, the majority of safety data are collected during phase III. These are commonly multiinstitutional to maximize patient enrollment that requires substantial coordination and communication between sites. The success of the trial is dependent on strict adherence to the study protocol.

Compared to prior phases, phase III clinical trials face unique challenges. First, recruiting patients to clinical trials within a reasonable timeframe is challenging. Over 80% of Americans say that they are likely to participate in a clinical trial if recommended by their doctor.[10] However, only 18% of people surveyed had participated or knew a family member who participated in a clinical trial. Many cite lack of trust or availability of information as a major barrier preventing their involvement.[10] Therefore, establishing robust subject recruitment procedures is vital for success. Another challenge with phase III trials is blinding subjects and investigators. Blinding of subjects can be addressed by performing a sham procedure when the study is investigating a procedure. Blinding of investigators is more difficult but can be done depending on the type of intervention. A treatment may be administered by a third party after a catheter has been placed by the investigator. When the intervention is a drug, placebo tablets designed to replicate the shape, size, and color of the medication facilitate appropriate and successful blinding.

Wei et al. performed a phase III trial investigating combination microwave ablation and chemotherapy compared to chemotherapy monotherapy in patients with advanced nonsmall-cell lung cancer.[11] Fourteen sites in China enrolled patients over the course of 2 years and randomized them in a 1:1 ratio into each group. They did not utilize blinding; otherwise, this study was methodologically sound. Participants were followed for over a year and were found to have longer overall survival and progression-free survival with a combination of microwave ablation (MWA) and chemotherapy. There was no statistically significant difference in adverse events between the groups. The researchers concluded that longer progression-free and overall survival can be achieved via treating advanced nonsmall cell lung cancer with microwave ablation plus chemotherapy compared to chemotherapy alone.

Following phase III, the drug or device is approved by the FDA and made available to the public. Phase IV involves monitoring the patient population for adverse events and evaluating the efficacy of treatment outcomes. These are long trials with the largest sample size and follow-up. As a result, they play an important role in evaluating slowly manifesting or very rare adverse events.

The modern era of clinical trials began less than a century ago. Today, clinical trials are the foundation of evidence-based medicine. Novel medications, procedures, and new indications for preexisting interventions are tested in different phases of clinical trials, with each phase answering a specific question. Planning and executing clinical trials necessitate meticulous planning and attention to detail. In doing so, the clinical trial is the gold standard to establish the clinical effectiveness of intervention while also identifying potential adverse effects.

Get started

- Familiarize yourself with clinicaltrials.gov
- Identify important clinical trials of all phases in your area of expertise
- Critically evaluate often cited clinical trials for the strengths and weaknesses and how its findings are implemented in everyday clinical practice
- Brainstorm clinical trials that would help answer important clinical questions and ultimately improve patient care

Potential pitfalls

- Be careful interpreting the clinical results of phase II studies, as they are not powered to detect a clinically significant effect.
- Recruitment of subjects is perhaps the most challenging aspect of completing a clinical trial. Spend ample time designing a comprehensive subject recruitment strategy.
- Adhering to ethical guidelines is paramount. Ensure that the protocol does not violate any ethical standards such as using a placebo control group when a disease has proven treatment options.

Real-world examples

- Mimura et al. performed phase I and phase II studies investigating the safety of radiofrequency ablation for renal cell carcinoma.[8]
- Gonsalves et al. published a phase II clinical trial investigating the effect of radioembolization for treatment of uveal melanoma hepatic metastatses.[9]
- Wei et al. conducted a phase III trial investigating the combination of microwave ablation and chemotherapy compared to chemotherapy monotherapy in patients with advanced non-small cell lunger cancer citation.[11]

Resources

1. Friedman LM, Furberg CD, DeMets D, Reboussin DM, Granger CB. *Fundamentals of Clinical Trials.* 5th ed. Springer International Publishing; 2015. https://doi.org/10.1007/978-3-319-18539-2.
2. Emanuel EJ, Wendler D, Grady C. What makes clinical research ethical? *JAMA.* 2000;283(20):2701–2711. https://doi.org/10.1001/jama.283.20.2701.

References

1. Friedman LM, Furberg CD, DeMets D, Reboussin DM, Granger CB. *Fundamentals of Clinical Trials.* 5th ed. Springer International Publishing; 2015. https://doi.org/10.1007/978-3-319-18539-2.
2. Bhatt A. Evolution of clinical research: a history before and beyond James Lind. *Perspect Clin Res.* 2010;1(1):6–10.
3. Collier R. *Legumes, lemons and streptomycin: a short history of the clinical trial.* 1st 180. Can. Med. Assoc. J.; 2009:23–24. In press.
4. Trends, Charts, and Maps, ClinicalTrials.gov. Accessed March 3, 2021. https://www.clinicaltrials.gov/ct2/resources/trends.
5. Emanuel EJ, Wendler D, Grady C. What makes clinical research ethical? *JAMA.* 2000;283(20):2701–2711. https://doi.org/10.1001/jama.283.20.2701.
6. Kummar S, Rubinstein L, Kinders R, et al. Phase 0 clinical trials: conceptions and misconceptions. *Cancer J.* 2008;14(3):133–137. https://doi.org/10.1097/PPO.0b013e318172d6f3.
7. Kummar S, Kinders R, Rubinstein L, et al. Compressing drug development timelines in oncology using phase "0" trials. *Nat Rev Cancer.* 2007;7(2):131–139. https://doi.org/10.1038/nrc2066.
8. Mimura H, Arai Y, Yamakado K, et al. Phase I/II study of radiofrequency ablation for malignant renal tumors: Japan interventional radiology in oncology study group 0701. *Cardiovasc Intervent Radiol.* 2016;39(5):717–723. https://doi.org/10.1007/s00270-015-1275-0.
9. Gonsalves CF, Eschelman DJ, Adamo RD, et al. A prospective phase II trial of radioembolization for treatment of uveal melanoma hepatic metastasis. *Radiology.* 2019;293(1):223–231. https://doi.org/10.1148/radiol.2019190199.

10. America Speaks: Poll Data Summary. vol. 18. Research!America. Published April 4, 2018. Accessed March 3, 2021. https://www.researchamerica.org/polls-publications/publications/america-speaks-poll-data-summary-volume-18.

11. Wei Z, Yang X, Ye X, et al. Microwave ablation plus chemotherapy versus chemotherapy in advanced non-small cell lung cancer: a multicenter, randomized, controlled, phase III clinical trial. *Eur Radiol.* 2020;30(5):2692–2702. https://doi.org/10.1007/s00330-019-06613-x.

Chapter 63

The IDEAL framework

Matthew Henry[1,2,4] **and Daniel Kirkpatrick**[3]
[1]*Wayne State University School of Medicine, Detroit, MI, United States;* [2]*Rackham Graduate School, University of Michigan, Ann Arbor, MI, United States;* [3]*Department of Radiology, Michigan Medicine, University of Michigan, Ann Arbor, MI, United States;* [4]*Department of Radiology, Medical College of Wisconsin, Milwaukee, WI, United States*

Key points

- Evaluation of surgical and interventional procedures is not standardized like the evaluation of medications.
- The IDEAL framework and recommendations aim to establish standards and guidelines for the evaluation of innovative surgical and interventional procedures.
- The IDEAL framework still requires more acceptance and implementation within the procedural community to effectively and efficiently promote the development of novel procedures.

Interventional radiologists have a strong history of innovation. The specialty was born in the mid-20th century with catheter-based treatments pioneered by Charles Dotter and Melvin Judkins.[1–3] A short time later, Dotter and his colleagues designed catheter-directed thrombolysis and intravascular stent placement. Soon thereafter, interventional radiologists were tackling aortic and liver disease with the introduction of stent grafts, transhepatic embolization, and transhepatic portosystemic shunts. Today, the interventionists' procedural repertoire continues to grow. Procedures such as bariatric and geniculate artery embolization are actively under research and may be in widespread use in the coming years.[4,5]

Despite the propensity to develop new procedures, interventional radiologists do not have a strong track record of conducting randomized controlled trials. Thankfully this is changing, as the number of these studies published per year is increasing.[6] Akin to surgery, interventional radiology procedures are complex and present unique challenges during evaluation.[7] First, while the surgeon or interventionalist performs the procedure, there is a large team also involved in the care of the patient, such as the nursing staff and anesthesiologists. Additionally, each procedure is composed of intricately linked steps that are challenging to isolate. Because of this intertwined construction, steps of procedures are difficult to study independently.

First proposed in 2009, the innovation, development, exploration, assessment, and long-term (IDEAL) framework addresses these issues and structures the process based on the principles of evidence-based medicine.[8] The IDEAL recommendations describe the stages through which a new and innovative interventional procedure is developed (Table 63.1).[9,10]

These recommendations also promote the use of prospective databases and registries to increase the overall quality of research regarding interventional procedures. Throughout this chapter, we use prostate artery embolization research studies as an example to evaluate the IDEAL stages (Table 63.2).

Stage 0, or Pre-IDEAL studies, involve prehuman development and investigation.[8,10] Not all procedures start at this stage, particularly if the procedure or device improves upon an already established procedure. The first manuscript describing prostate artery embolization (PAE) was published in 1980.[11] In this study, Darcwicz et al. investigated the utility of internal iliac artery embolization to treat hemorrhage following prostate surgery. Twenty-nine years later, researchers investigated treating benign prostatic hyperplasia (BPH) with PAE in a canine model.[12] They concluded that it was a feasible procedure. These two studies represent stage 0 in the development of PAE for the treatment of BPH, because they both occurred in animal models.

Stage 1 is the development and investigation of a new procedure, but unlike stage 0, it is implemented in humans.[8–10] This is usually the initial report of a procedure performed in a human for the first time and can be either accidental or planned. The goal of this phase is to prove that the idea has the potential to evolve into a practical procedure. During the first stage, there is one or a small group of physicians developing the procedure in a select few patients. Investigators explain the rationale behind the procedure and describe it in detail. They report their successes and failures, especially adverse events, in online

Translational Interventional Radiology. https://doi.org/10.1016/B978-0-12-823026-8.00007-9

TABLE 63.1 Stages and characteristics of the IDEAL framework.

IDEAL framework	Goal	Number of subjects	Number of surgeons/ interventionists	Preferred study - design	Outcomes
Stage 0: Preclinical	Feasibility and technical consistency	N/A	One or very few	Animal and cadaver models	Demonstration of technical feasibility, potential safety risks
Stage 1: Idea	Proof of concept	Very few	One or very few	Structured case reports	Technical success, clinical success, adverse events
Stage 2a: Development	Procedural development and description	Few	Few	Prospective case series	Acceptable safety profile and procedural feasibility
Stage 2b: Exploration	Establishing consensus	Many	Many	Pilot randomized controlled trials or multi-institutional cohort studies	Safety, clinical, feasibility, and patient-reported outcomes
Stage 3: Assessment	Comparison to current standards of care	Many	Majority	Randomized controlled trials	Clinical and patient-reported outcomes, short and middle-term outcomes, and cost-effectiveness
Stage 4: Long-term study	Surveillance	All eligible	Most or all	Registries, long-term databases, rare event case reports	Long-term outcomes, rare adverse events, quality control

Adapted from McCulloch P, Altman DG, Campbell WB, et al. No surgical innovation without evaluation: the IDEAL recommendations. *Lancet.* 2009;374(9695):1105–1112. https://dx.doi.org/10.1016/S0140-6736(09)61116-8; Hirst A, Philippou Y, Blazeby J, et al. No surgical innovation without evaluation: evolution and further development of the IDEAL framework and recommendations. *Ann Surg.* 2019;269(2):211–220. https://dx.doi.org/10.1097/SLA.0000000000002794; McCulloch P, Cook JA, Altman DG, Heneghan C, Diener MK, IDEAL Group. IDEAL framework for surgical innovation 1: the idea and development stages. *BMJ.* 2013;346:f3012. https://dx.doi.org/10.1136/bmj.f3012.

TABLE 63.2 Prostate artery embolization as it progresses through the different stages of the IDEAL framework.

IDEAL framework stage	Prostate artery embolization	Year(s) published
Stage 0	Treatment of BPH via PAE in nine dogs	2009
Stage 1	Two case series demonstrating that PAE could potentially serve as the primary treatment modality for BPH	2010
Stage 2a	Eleven patients with acute urinary retention secondary to BPH, who also failed medical therapy and were waiting for further intervention, receive PAE.	2012
Stage 2b	Several cohort studies are published investigating the clinical utility of PAE in the treatment of BPH in slightly different patient populations	2014–18
Stage 3	Trials comparing the efficacy of PAE for the treatment of BPH compared to both transurethral resection of the prostate (TURP) and sham procedures	2016–ongoing

registries to allow others to learn. There is an ethical obligation to report these cases in sufficient detail to allow others to replicate the procedure and prevent failures.[13]

The first report of PAE improving symptomatic BPH involved a patient who developed gross hematuria and was treated with embolization of the right inferior vesicle artery.[14] Following the embolization, the patient experienced an improvement in both subjective and objective measures of BPH. A small case series involving two patients was subsequently published and demonstrated that PAE could be a possible treatment modality for BPH.[15,16] The clinical improvement in BPH following PAE in these three patients demonstrated proof of concept and allowed the investigation of PAE to move to stage 2.

Stage 2 of the IDEAL framework is split into two substages: development (2a) and exploration (2b).[8–10] During the development phase, the technique of the procedure undergoes rapid, iterative modification and testing. The goal is to improve the safety and success of the procedure. This stage raises several ethical and logistical challenges because of the rapid evolution in both the procedural steps and eligibility.[13] Due to the limited quality of retrospective studies, the IDEAL framework recommends prospective analysis. Prospective protocols must be unambiguous and registered a priori. Reports include eligibility criteria, description of procedural steps, outcomes, and complications. This phase only includes a small number of patients and investigators. Additionally, local ethics approval is necessary.

Carnevale et al. published the first prospective stage 2a study of PAE for the treatment of BPH.[17] The investigators recruited 11 patients with acute urinary retention secondary to BPH who failed medical therapy and were on the waiting list for transurethral resection of the prostate. The manuscript provides a detailed description of the procedure such that other providers can understand the steps. The investigators also report their complications and outcomes of the procedure. However, the manuscript provided only limited data about the inclusion and exclusion criteria, but the investigators subsequently published a detailed article on patient selection, imaging protocol, technique, and postprocedural management.[18]

Stage 2b (exploration) is the second phase of stage 2 and expands to involve the early adopters alongside the initial investigators.[8–10] The purpose is to establish a consensus on the procedural steps, indications, and contraindications through prospective data collection. Research studies are prospective cohort designs or pilot studies to assess the feasibility of future randomized controlled trials. Measured outcomes include short-term clinical results and the safety profile, both of which are documented using standardized measurements and terminology to avoid ambiguity. For example, reporting outcomes in common terminology, such as the Dindo-Clavien classification of complications,[19] facilitates easy and confident interpretation of the data. The practicality and achievability of the intervention are also studied, including determining the learning curve and ease of implementation. Similar to stage 2a, protocols should be published before the study begins.

There are several stage 2b studies investigating PAE for the treatment of BPH.[20–29] They involve relatively small cohorts of patients and each have different inclusion and exclusion criteria. This is typical during this phase as the ideal patient population has yet to be established. These early studies demonstrated that PAE had a potential role in the treatment of BPH with a tolerable safety profile.

With more data, the procedure is considered less experimental. This transition marks the start of stage 3: assessment. The primary goal of stage 3 is to compare the procedure against current standards of care.[8–10] The technique of the intervention may still be improved upon, but the procedural steps are standardized enough to allow comparison against the gold standard treatment. In this stage, randomized controlled trials are the preferred study design due to their prospective nature and design that limits bias. Per the IDEAL framework, procedures in stage 3 are widely disseminated with most interventionalists able to offer the procedure to patients. There is generally enough data at this point for patients to be aware of the procedure.

Like the previous stages, protocols should be published a priori. Investigators should follow protocol reporting guidelines such as the Standard Protocol Items: Recommendations for Interventional Trials (SPIRIT).[30] SPIRIT provides recommendations for the minimum detail that should be included in the protocol of a clinical trial. Researchers should publish their protocols a priori because it improves peer-review evaluation of the methodology of the study and promotes transparency of research. Additionally, investigators should adhere to guidelines such as the Consolidated Standards of Reporting Trials (CONSORT). CONSORT aims to improve the reporting of randomized controlled trials, facilitating peer-review and appropriate interpretation of the data.[31] By adhering to standards such as SPIRIT and CONSORT, researchers publish high-quality data that allows a honest evaluation of a procedure.

PAE is currently in stage 3. Randomized controlled trials comparing PAE to sham procedures and current standards of care are being published.[32–34] The next step is to evaluate the procedure against the standard of care rather than controlling with a placebo group. As mentioned earlier, the procedural technique may still be refined in this stage. There are several randomized studies investigating improvements to the procedure such as arterial access location,[35] embolic agents,[36,37] and other procedural techniques such as balloon occlusion versus microcatheter PAE.[38]

Stage 4 is the final stage in the assessment of an interventional procedure.[8–10] Similar to phase IV clinical trials, stage 4 pertains to long-term scrutinization and surveillance. Most if not all interventionalists offer the procedure, and the eligible patient population is established. Monitoring for rare adverse events, risk adjustment, and quality assurance continues. In this stage, registries are the typical study design with a focus on key outcomes and relevant information to facilitate complete and accurate data entry.

Since its introduction in 2009, the IDEAL framework has been adopted into the surgical and interventional research communities. However, there remains room for improvement. The current level of understanding and implementation of the framework in clinical research is low.[39] To increase the utilization and implementation of the framework, checklists and tips for researchers have been published, and there is evidence that an increasing number of investigators are following the guidelines.[40,41]

New treatments undergo a standardized regimen from inception to approval via clinical trial phases. However, surgical and interventional procedures have historically undergone little formal assessment. As evidence-based medicine continues to be emphasized, it is necessary to establish and follow guidelines for procedural innovation. The IDEAL framework outlines the appropriate process for new invasive techniques and improvements through a standardized process. Using the IDEAL framework will enhance and accelerate high-quality research in the interventional radiology community.

Get started

Familiarize yourself with the IDEAL framework and its steps In procedures you commonly perform, identify how the steps were studied and if it follows the IDEAL framework guidelines Brainstorm different research protocols that would follow the IDEAL framework and could be implemented to better understand steps, patient populations, and outcomes for procedures you commonly perform

Potential pitfalls

- Do not progress through the stages too fast. A procedure that is poorly described and established will risk negative clinical trials or heterogenous implementation.
- By following established guidelines and standards such as SPIRIT and CONSORT, the quality of the research is substantially higher, easier to interpret, and more efficiently implemented in clinical practice.
- Not publishing protocols a priori can result in biased and ambiguous research. All protocols should be published before the research commences to obtain the highest quality data possible.

Real-world examples

Jeon et al. studied the utility of prostate artery embolization in nine beagle dogs with benign prostatic hyperplasia, representing stage 0 of the IDEAL framework.[12] For stage 1 of the IDEAL framework, Carnevale et al. published two studies on prostate artery embolization as the primary treatment for benign prostatic hyperplasia.[15–16] Carvanela et al. also published two studies that represent the second stage of the IDEAL framework.[17–18] In these manuscripts, the authors described the procedural steps in detail, in addition to reporting their outcomes and complications.

Resources

1. Barkun JS, Aronson JK, Feldman LS, et al. Evaluation and stages of surgical innovations. *Lancet.* 2009;374(9695):1089–1096. https://doi.org/10.1016/S0140-6736(09)61083-7.
2. McCulloch P, Altman DG, Campbell WB, et al. No surgical innovation without evaluation: the IDEAL recommendations. *Lancet.* 2009;374(9695):1105–1112. https://doi.org/10.1016/S0140-6736(09)61116-8.
3. Hirst A, Philippou Y, Blazeby J, et al. No surgical innovation without evaluation: evolution and further development of the IDEAL framework and recommendations. *Ann Surg.*

References

1. Baum RA, Baum S. Interventional radiology: a half century of innovation. *Radiology.* 2014;273(2S):S75–S91. https://doi.org/10.1148/radiol.14140534.
2. Midulla M, Pescatori L, Chevallier O, et al. Future of IR: emerging techniques, looking to the Future…and learning from the past. *J Belg Soc Radiol.* 2019;103(1). https://doi.org/10.5334/jbsr.1727.
3. Rösch J, Keller FS, Kaufman JA. The birth, early years, and future of interventional radiology. *J Vasc Intervent Radiol.* 2003;14(7):841–853. https://doi.org/10.1097/01.RVI.0000083840.97061.5b.
4. Weiss CR, Abiola GO, Fischman AM, et al. Bariatric embolization of arteries for the treatment of obesity (BEAT obesity) trial: results at 1 year. *Radiology.* 2019;291(3):792–800. https://doi.org/10.1148/radiol.2019182354.
5. Bagla S, Piechowiak R, Hartman T, Orlando J, Gaizo DD, Isaacson A. Genicular artery embolization for the treatment of knee pain secondary to osteoarthritis. *J Vasc Intervent Radiol.* 2020;31(7):1096–1102. https://doi.org/10.1016/j.jvir.2019.09.018.
6. Hong SJ, Yoon DY, Cho YK, et al. Characteristics and quality of radiologic randomized controlled trials: a bibliometric analysis between 1995 and 2014. *AJR Am J Roentgenol.* 2016;206(5):917–923. https://doi.org/10.2214/AJR.15.15640.
7. Ergina PL, Cook JA, Blazeby JM, et al. Challenges in evaluating surgical innovation. *Lancet.* 2009;374(9695):1097–1104. https://doi.org/10.1016/S0140-6736(09)61086-2.
8. Barkun JS, Aronson JK, Feldman LS, et al. Evaluation and stages of surgical innovations. *Lancet.* 2009;374(9695):1089–1096. https://doi.org/10.1016/S0140-6736(09)61083-7.
9. McCulloch P, Altman DG, Campbell WB, et al. No surgical innovation without evaluation: the IDEAL recommendations. *Lancet.* 2009;374(9695):1105–1112. https://doi.org/10.1016/S0140-6736(09)61116-8.
10. Hirst A, Philippou Y, Blazeby J, et al. No surgical innovation without evaluation: evolution and further development of the IDEAL framework and recommendations. *Ann Surg.* 2019;269(2):211–220. https://doi.org/10.1097/SLA.0000000000002794.
11. Darewicz J, Cylwik B, Musierowicz A, Boczon S. The effect of internal iliac artery embolization of the prostatic gland in the dog. *Int Urol Nephrol.* 1980;12(1):37–41. https://doi.org/10.1007/BF02085379.
12. Jeon GS, Won JH, Lee BM, et al. The effect of transarterial prostate embolization in hormone-induced benign prostatic hyperplasia in

dogs: a pilot study. *J Vasc Intervent Radiol.* 2009;20(3):384–390. https://doi.org/10.1016/j.jvir.2008.11.014.
13. McCulloch P, Cook JA, Altman DG, Heneghan C, Diener MK, IDEAL Group. IDEAL framework for surgical innovation 1: the idea and development stages. *BMJ.* 2013;346:f3012. https://doi.org/10.1136/bmj.f3012.
14. DeMeritt JS, Elmasri FF, Esposito MP, Rosenberg GS. Relief of benign prostatic hyperplasia-related bladder outlet obstruction after trans-arterial polyvinyl alcohol prostate embolization. *J Vasc Intervent Radiol.* 2000;11(6):767–770. https://doi.org/10.1016/s1051-0443(07)61638-8.
15. Carnevale FC, Antunes AA, da Motta Leal Filho JM, et al. Prostatic artery embolization as a primary treatment for benign prostatic hyperplasia: preliminary results in two patients. *Cardiovasc Intervent Radiol.* 2010;33(2):355–361. https://doi.org/10.1007/s00270-009-9727-z.
16. Carnevale FC, da Motta-Leal-Filho JM, Antunes AA, et al. Midterm follow-up after prostate embolization in two patients with benign prostatic hyperplasia. *Cardiovasc Intervent Radiol.* 2011;34(6):1330–1333. https://doi.org/10.1007/s00270-011-0136-8.
17. Carnevale FC, da Motta-Leal-Filho JM, Antunes AA, et al. Quality of life and clinical symptom improvement support prostatic artery embolization for patients with acute urinary retention caused by benign prostatic hyperplasia. *J Vasc Intervent Radiol.* 2013;24(4):535–542. https://doi.org/10.1016/j.jvir.2012.12.019.
18. Carnevale FC, Antunes AA. Prostatic artery embolization for enlarged prostates due to benign prostatic hyperplasia. How I do it. *Cardiovasc Intervent Radiol.* 2013;36(6):1452–1463. https://doi.org/10.1007/s00270-013-0680-5.
19. Dindo D, Demartines N, Clavien P-A. Classification of surgical complications: a new proposal with evaluation in a cohort of 6336 patients and results of a survey. *Ann Surg.* 2004;240(2):205–213. https://doi.org/10.1097/01.sla.0000133083.54934.ae.
20. Kurbatov D, Russo GI, Lepetukhin A, et al. Prostatic artery embolization for prostate volume greater than 80 cm^3: results from a single-center prospective study. *Urology.* 2014;84(2):400–404. https://doi.org/10.1016/j.urology.2014.04.028.
21. Somani BK, Hacking N, Bryant T, et al. Prostate artery embolization (PAE) for benign prostatic hyperplasia (BPH). *BJU Int.* 2014;114(5):639–640. https://doi.org/10.1111/bju.12672.
22. de Assis AM, Moreira AM, de Paula Rodrigues VC, et al. Prostatic artery embolization for treatment of benign prostatic hyperplasia in patients with prostates > 90 g: a prospective single-center study. *J Vasc Intervent Radiol.* 2015;26(1):87–93. https://doi.org/10.1016/j.jvir.2014.10.012.
23. Bagla S, Smirniotopoulos JB, Orlando JC, van Breda A, Vadlamudi V. Comparative analysis of prostate volume as a predictor of outcome in prostate artery embolization. *J Vasc Intervent Radiol.* 2015;26(12):1832–1838. https://doi.org/10.1016/j.jvir.2015.08.018.
24. Amouyal G, Thiounn N, Pellerin O, et al. Clinical results after prostatic artery embolization using the PErFecTED technique: a single-center study. *Cardiovasc Intervent Radiol.* 2016;39(3):367–375. https://doi.org/10.1007/s00270-015-1267-0.
25. Wang MQ, Wang Y, Yan JY, et al. Prostatic artery embolization for the treatment of symptomatic benign prostatic hyperplasia in men ≥75 years: a prospective single-center study. *World J Urol.* 2016;34(9):1275–1283. https://doi.org/10.1007/s00345-016-1771-0.
26. Isaacson AJ, Raynor MC, Yu H, Burke CT. Prostatic artery embolization using embosphere microspheres for prostates measuring 80-150 cm(3): early results from a US trial. *J Vasc Intervent Radiol.* 2016;27(5):709–714. https://doi.org/10.1016/j.jvir.2016.01.146.
27. Yu SCH, Cho CCM, Hung EHY, Chiu PKF, Yee CH, Ng CF. Prostate artery embolization for complete urinary outflow obstruction due to benign prostatic hypertrophy. *Cardiovasc Intervent Radiol.* 2017;40(1):33–40. https://doi.org/10.1007/s00270-016-1502-3.
28. Rampoldi A, Barbosa F, Secco S, et al. Prostatic artery embolization as an alternative to indwelling bladder catheterization to manage benign prostatic hyperplasia in poor surgical candidates. *Cardiovasc Intervent Radiol.* 2017;40(4):530–536. https://doi.org/10.1007/s00270-017-1582-8.
29. Ray AF, Powell J, Speakman MJ, et al. Efficacy and safety of prostate artery embolization for benign prostatic hyperplasia: an observational study and propensity-matched comparison with transurethral resection of the prostate (the UK-ROPE study). *BJU Int.* 2018;122(2):270–282. https://doi.org/10.1111/bju.14249.
30. Chan A-W, Tetzlaff JM, Altman DG, et al. SPIRIT 2013 statement: defining standard protocol Items for clinical trials. *Ann Intern Med.* 2013;158(3):200–207. https://doi.org/10.7326/0003-4819-158-3-201302050-00583.
31. Schulz KF, Altman DG, Moher D, CONSORT Group. CONSORT 2010 Statement: updated guidelines for reporting parallel group randomized trials. *Open Med.* 2010;4(1):e60–68.
32. Carnevale FC, Iscaife A, Yoshinaga EM, Moreira AM, Antunes AA, Srougi M. Transurethral resection of the prostate (TURP) versus original and PErFecTED prostate artery embolization (PAE) due to benign prostatic hyperplasia (BPH): preliminary results of a single center, prospective, urodynamic-controlled analysis. *Cardiovasc Intervent Radiol.* 2016;39(1):44–52. https://doi.org/10.1007/s00270-015-1202-4.
33. Pisco JM, Bilhim T, Costa NV, et al. Randomised clinical trial of prostatic artery embolisation versus a sham procedure for benign prostatic hyperplasia. *Eur Urol.* 2020;77(3):354–362. https://doi.org/10.1016/j.eururo.2019.11.010.
34. Insausti I, Sáez de Ocáriz A, Galbete A, et al. Randomized comparison of prostatic artery embolization versus transurethral resection of the prostate for treatment of benign prostatic hyperplasia. *J Vasc Intervent Radiol.* 2020;31(6):882–890. https://doi.org/10.1016/j.jvir.2019.12.810.
35. Bhatia S, Harward SH, Sinha VK, Narayanan G. Prostate artery embolization via transradial or transulnar versus transfemoral arterial access: technical results. *J Vasc Intervent Radiol.* 2017;28(6):898–905. https://doi.org/10.1016/j.jvir.2017.02.029.
36. Wang MQ, Zhang JL, Xin HN, et al. Comparison of clinical outcomes of prostatic artery embolization with 50-μm plus 100-μm polyvinyl alcohol (PVA) particles versus 100-μm PVA particles alone: a prospective randomized trial. *J Vasc Intervent Radiol.* 2018;29(12):1694–1702. https://doi.org/10.1016/j.jvir.2018.06.019.
37. Torres D, Costa NV, Pisco J, Pinheiro LC, Oliveira AG, Bilhim T. Prostatic artery embolization for benign prostatic hyperplasia: prospective randomized trial of 100-300 μm versus 300-500 μm versus 100- to 300-μm + 300- to 500-μm embospheres. *J Vasc Intervent Radiol.* 2019;30(5):638–644. https://doi.org/10.1016/j.jvir.2019.02.014.

38. Bilhim T, Costa NV, Torres D, Pisco J, Carmo S, Oliveira AG. Randomized clinical trial of balloon occlusion versus conventional microcatheter prostatic artery embolization for benign prostatic hyperplasia. *J Vasc Intervent Radiol.* 2019;30(11):1798–1806. https://doi.org/10.1016/j.jvir.2019.06.019.
39. Khachane A, Philippou Y, Hirst A, McCulloch P. Appraising the uptake and use of the IDEAL Framework and Recommendations: a review of the literature. *Int J Surg*. 2018;57:84–90. https://doi.org/10.1016/j.ijsu.2018.07.008.
40. Pennell CP, Hirst AD, Campbell WB, et al. Practical guide to the idea, development and exploration stages of the IDEAL framework and recommendations. *Br J Surg*. 2016;103(5):607–615. https://doi.org/10.1002/bjs.10115.
41. Bilbro NA, Hirst A, Paez A, et al. The ideal reporting guidelines: a Delphi consensus statement stage specific recommendations for reporting the evaluation of surgical innovation. *Ann Surg*. 2021;273(1):82–85. https://doi.org/10.1097/SLA.0000000000004180.

Chapter 64

Studies to assess AI methodology in clinical research

Benedict Thomson[1] and Gregory C. Makris[2]

[1]*Department of Interventional Radiology, Guy's and St. Thomas' NHS Foundation Trust, London, United Kingdom;* [2]*Department of Endovascular and Interventional Radiology, Guys and St. Thomas' Hospitals NHS Foundation Trust, London, United Kingdom*

Key points

- AI is set to benefit clinical practice in IR but current research is on its practical application is lacking.
- This chapter seeks to highlight several practical examples of the use of AI methodology in clinical research and how that may be applicable to IR in particular in providing advanced procedural support, improved clinical decision tools and enhancing the trainee experience.
- We also discuss potential pitfalls and challenges when undertaking AI research.

Introduction

Artificial intelligence (AI) is an umbrella term pertaining to the ability of a machine to imitate intelligent human behaviour. Machine learning (ML) is subset of AI devoted to algorithms that can improve at performing tasks and making predictions through exposure to data.[1] Deep learning algorithms differ from traditional models of machine learning in that they are built from multilayers, or deep, artificial neural networks (ANNs) (Fig. 64.1.). This allows them to automatically identify relevant features from

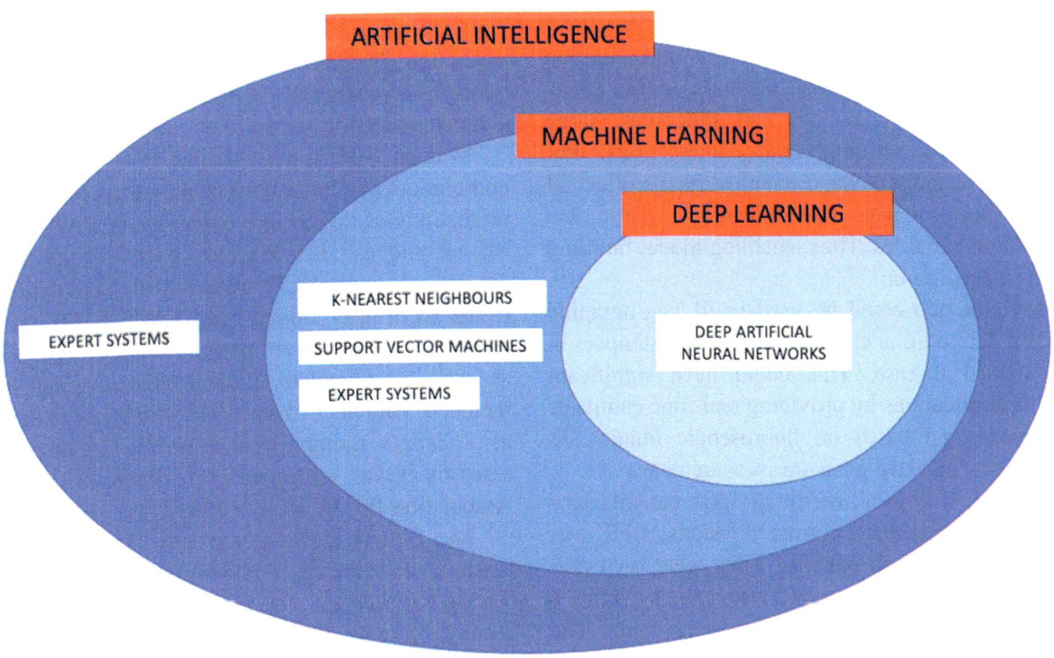

FIGURE 64.1 Venn diagram shows relation between artificial intelligence (AI), machine learning, and deep learning. Expert systems are an example of AI that historically do not use machine learning to emulate human problem-solving ability. K-nearest neighbors, support vector machines, and decision trees are examples of commonly encountered algorithms in traditional machine learning. Deep learning (subset of machine learning) employs deep artificial neural networks to learn and make predictions.

Translational Interventional Radiology. https://doi.org/10.1016/B978-0-12-823026-8.00023-7

data and excel at learning from large, raw data sets. ANNs are a computational framework inspired by the organisation and function of the brain. Information that flows through the ANN causes network changes which allow it to learn based on the input and output data. ANNs are deep learning models where complex relationships between inputs and outputs are modelled or patterns are found.[1]

The potential for this technology in the clinical setting of interventional radiology (IR) practice has been suggested to include aiding tasks related to image analysis, patient selection modeling, predictive tools for treatment modeling, and trainee education.[1] However, the use of AI in IR research has been relatively underreported thus far.[2] In the following chapter, we discuss three recently published examples of AI methodology used in clinical research and how this may be applicable to future IR research. In particular, we will focus on three areas: imaging, clinical modeling, and trainee education.

Angiography-based machine learning

In 2019, Cho et al. published their study looking at the role of angiography-based machine learning for predicting fractional flow reserve (FFR) in intermediate coronary artery disease (CAD).[3] FFR is an invasive procedure requiring direct pressure measurements across indeterminate coronary vessel lesions. FFR is a physiological metric considered the gold standard for identifying ischemia-related stenosis in CAD.[4] The accuracy of quantitative coronary angiography diameter stenosis, as a single morphological parameter for predicting FFR, has been reported to be only 60%.[5] The development of prediction models, using traditional statistical methods, is limited by the nonlinearity between factors and outcomes, large numbers of predictors, and intervariable interactions.[6] For this reason, ML techniques have emerged as an effectual computer algorithm for pattern identification in large data sets with multiple variables. Thus, enabling model building for data-driven prediction.[7]

A similar approach could be used in IR angiographic research when determining clinically significant stenoses in peripheral arterial disease. This could have significant future clinical applications by providing real-time estimates of physiological data based on fluoroscopic images, as opposed to direct, invasive pressure measurements.

In their study, Cho et al. recruited 1501 patients prospectively with intermediate coronary lesions. FFR was invasively measured for all patients. They were then randomized to training versus test sets. During angiography, a series of lumen diameter measurements were made. The angiographic features and four clinical features (age, sex, body surface area, and) were then used for ML by XGBoost. The model was then independently trained and tested by 2000 bootstrap iterations. Including all 28 features, the ML model with fivefold cross-validation in the 1204 training samples predicted fractional flow reserve $\leq$0.80 with an overall diagnostic accuracy of 78% (averaged area under the curve: 0.84 0.03). The 12 high-ranking features selected by scatter search were involved segment; body surface area; distal lumen diameter; minimal lumen diameter; length of a lumen diameter <2.0 mm, <1.5 mm, and <1.25 mm; mean lumen diameter within the worst segment; sex; diameter stenosis; distal 5 mm reference lumen diameter; and length of diameter stenosis >70%. Using those 12 features, the ML predicted fractional flow reserve $\leq$0.80 in the test set with sensitivity of 84%, specificity of 80%, and overall accuracy of 82% (area under the curve: 0.87). The methodology is summarized in Fig. 64.2.

Clinical modeling

Transarterial chemoembolization (TACE) is a well-established therapy to treat patients with unresectable hepatocellular carcinoma (HCC). Response to treatment is assessed by measuring the reduction in tumor volume on follow-up cross-sectional imaging.[8] However, it remains clinically challenging to predict which patients will respond to TACE therapy and who should be offered this treatment. No single clinical or imaging feature is predictive of response.[8] Supervised ML can be used to overcome this problem. For example, patient baseline imaging and clinical data, treatment characteristics, and treatment outcomes can be applied from a retrospective cohort to teach a model to learn the relationship between these variables and treatment outcomes. The model predicts treatment response in patients preprocedurally, provided the planned treatment characteristics are specified.

In their 2018 paper, Abajlan et al. sought to use MRI and clinical patient data to create an AI framework for the predictions of therapeutic outcomes of TACE by applying ML techniques.[9] It was a retrospective study that included 36 patients with HCC treated with conventional TACE (cTACE) or drug eluting bead-TACE (DEB-TACE). Only patients with follow-up imaging within 30 days of TACE, as well as complete clinical and imaging data, were included. Patients had an MRI 1 week before treatment and at 1 and 3 months posttreatment. Treatment response assessment was performed using the quantitative European Association for the Study of the Liver (qEASL) method.[8]

The clinical and imaging variables were then used to teach two learning modules (referred to as features). Twenty-five clinical, laboratory, demographic, and imaging features were assembled in the retrospective patient cohort. These 25 features were then sorted into binary categories to standardize the scale of each variable. A variance threshold was then applied to remove features that were all negative or positive more than 80% of the time. A univariate chi-

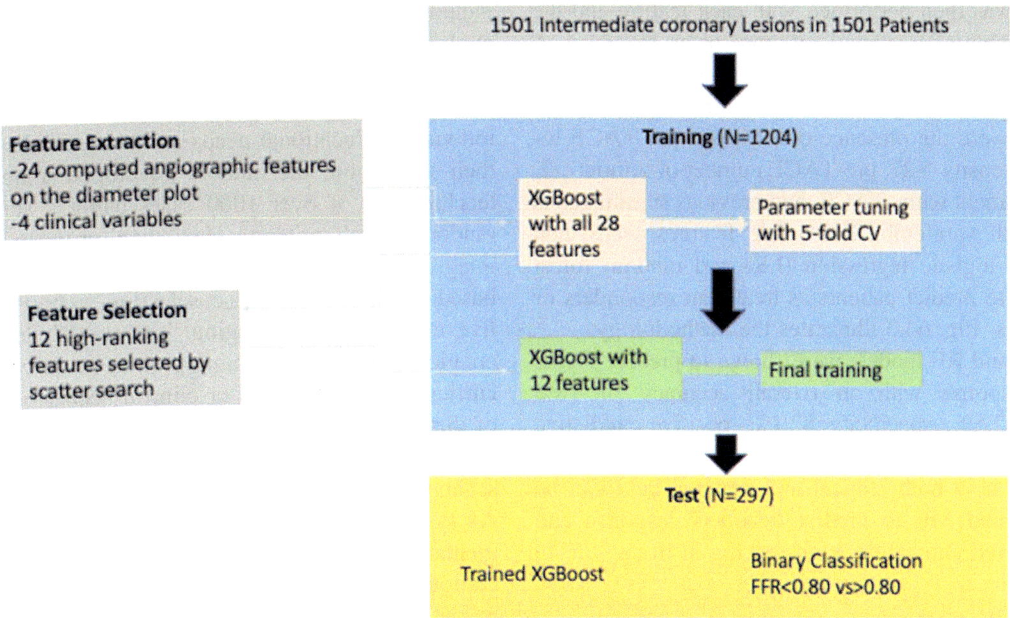

FIGURE 64.2 Workflow for developing the machine learning model. The 12 high-ranking features selected by scatter search were involved segment; body surface area; distal lumen diameter; minimal lumen diameter; length of a lumen diameter of <2.0 mm, <1.5 mm, and <1.25 mm; mean lumen diameter within the worst segment; sex; diameter stenosis; distal 5-mm reference lumen diameter; and length of diameter stenosis >70%. CV indicates cross-validation; FFR, fractional flow reserve.[3]

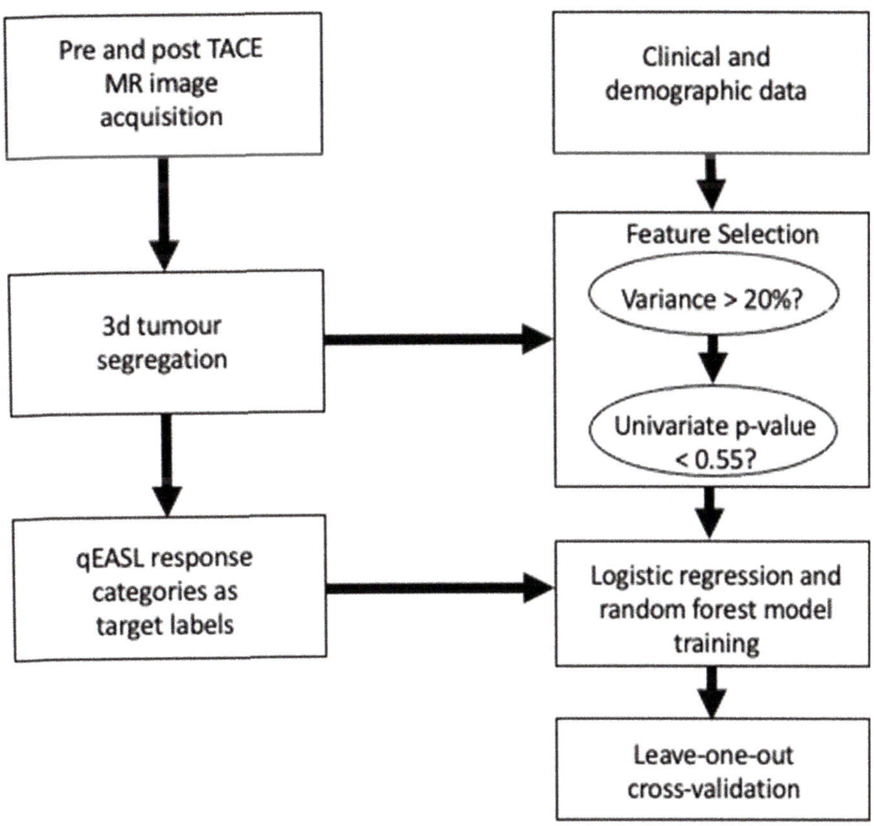

FIGURE 64.3 Workflow describing MR image preprocessing and feature selection steps. The qEASL response outcomes were combined with the MRI and clinical and imaging features, passing selection criteria, to train logistic regression and random forest plot models to predict treatment outcomes. The accuracy of the models compared to qEASL assessment of treatment response was evaluated using leave-one-out cross-validation.

squared test was then performed with each feature and the outcome variable of treatment response or no-response. A P-value <0.55 was selected after variance thresholding with the aim of reducing the number of features to 5. These five features were the presence of cirrhosis, pre-TACE tumor signal intensity >27, pre-TACE number of tumors >2, previous treatment with cTACE, and previous simultaneous treatment with sorafenib. These five features were then used to train logistic regression (LR) and random forest (RF) models to predict patients as treatment responders or nonresponders. Fig. 64.3 illustrates the methodology.

Both LR and RF models were shown to predict TACE treatment response with an overall accuracy of 78% (sensitivity 62.5%, specificity 82.1%, positive predictive value 50.0%, negative predictive value 88.5%). This study demonstrates how both clinical and imaging data can be used by AI and ML to predict treatment response and enable improved clinical decision making in IR.

Trainee education

pi Recently there has been a movement in postgraduate medical education to understand AI and its potential applications to clinical practice and research. However, the application of AI to medical and radiological education remains very limited.[10] In this section, we will discuss a worked example of how an AI-integrated framework may be used to augment radiology education. We then discuss how this could be used in medical educational research to demonstrate the benefits of "AI precision education."

In 2019 Duong et al. published a paper assessing models of AI-based education used in their radiology institute.[10] Case teaching files, in the form of picture archiving and communication systems (PACS), are commonplace in postgraduate radiology education.[11] However, the curation of these files is labor intensive. Also deciding which cases are deemed "valuable" and to be included in the file rests with the trainer or trainee. Duong et al. proposed an AI-led system where large collections of automatically curated cases are presented to trainees based on their own "learning profile" accounting for both their individual educational needs and subspecialty interests. At their institution, the authors developed a neuroradiology teaching file of over 1000 cases using a semiautomated, random search criteria. They then propose, through the integration of teaching files with Bayesian networks, AI-based supervision of case selection, report drafting, and live teaching file cataloging. The model depends on the creation of a trainee learning profile, based on the standard curriculum of diagnoses or clinical scenarios. A summary of the process is illustrated in Fig. 64.4.

The authors propose a scenario where a patient is scanned, and the imaging study is added to the task list. The AI system then performs an analysis of the lesion segmentation and identifies a potential diagnosis. It then compares the diagnosis with the profile of a trainee unfamiliar with that diagnosis and assigns the case to the trainee it predicts to benefit most. The trainee then identifies important clinical and radiologic features from the study and formulates a tentative differential diagnosis within a draft report. The diagnosis can either be generated completely unassisted, with supplementation from AI quantitation tools, or with AI-guided decision making and putative diagnosis sharing. The trainee and supervising radiologist will collaborate on the final report and sign off. The AI utilizes its neurolinguistic programming to anonymize the case, add it to a teaching file, extract the diagnosis and categorize the case, and update the trainee's case experiential learning profile.

This integrated, AI-augmented radiological education promotes greater independence under attending supervision, high fidelity simulation, and real-time situational feedback. The development of platforms to track trainees, their activity, and case logs demonstrates the utility of virtual tools for postgraduate education. This can be

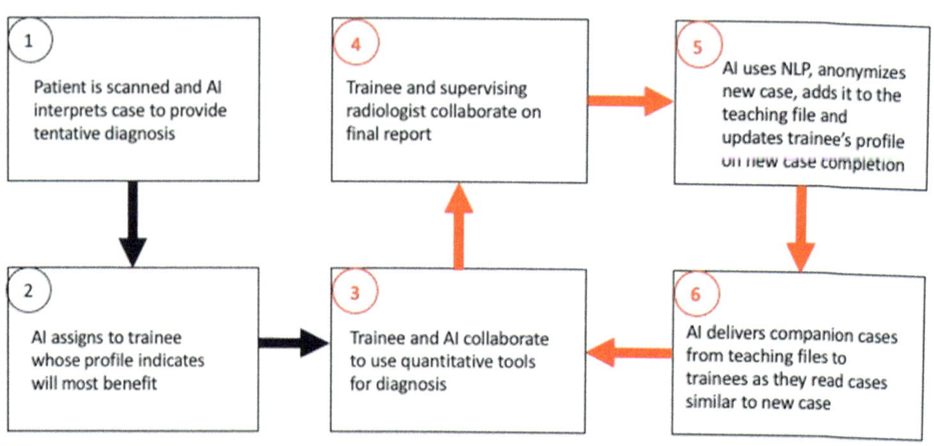

FIGURE 64.4 Summary of a future model for radiology precision education where AI identifies diagnoses and assigns the case to the trainee most likely to benefit.

integrated with AI as a tool for supervisors to assess individual strengths and weaknesses. Such approaches have the potential to improve efficiency in radiology education and practice by targeting specific trainee needs. The authors then go on to propose various research strategies to prove the utility of their educational model. This could either be through qualitative methods such as trainee feedback and satisfaction surveys or through quantitative methods such as examination scores in those who have received AI integrated education and those who have not.

Budgetary considerations and how to fund this stage

So far in this chapter, we have sought to demonstrate the potential applications of AI research in IR. However, undertaking AI research has several financial, ethical, and organizational challenges. In the following sections, we seek to identify these challenges and outline how a research group may overcome them.

Data sharing between different research groups is common in computer science but not in radiology where clinical and imaging data are closely stored. This is due to the ethical and legal implications of patient confidentiality. Research will require large datasets made available for collaboration. Stanford's Medical ImageNet is an example of such a system with permissive rules of access for research.[12]

For many applications, expert annotations (measurements, contouring, and descriptions) for training AI algorithms are very expensive to obtain.[13] A direction of research should be to make this process more efficient, which can be done with AI. The medical image Computing and Computer-Assisted Intervention Workshop on Large-scale Annotation of Biomedical data and Expert Label Synthesis is an example of an academic group with this research focus.[14]

Collaboration between different AI research groups requires a common computing framework and standardization software framework. There are several deep learning frameworks that are publicly available (TensorFlow, Microsoft Cognitive Toolkit).[15] However, the cost of these systems is high. Specialized hardware capable of highly parallel processing such as graphics processing units and more recently tensor processing units is required. Shared public resources are available and are particularly beneficial for large-scale projects.[16] It is generally not cost-effective for individual research groups to build large, specialized servers as they can quickly become obsolete. Resource allocation grants are available from various national computing bodies to help fund such computing hardware and software.[16]

Current AI frameworks are not accessible by noncomputer scientists. Therefore, research groups require computer scientist collaboration. Future developments may circumvent this by implementing AI applications integrated into PACS systems and improving ease of use. However, presently academic and industrial collaborations are necessary for future developments but create several questions. First, who owns the intellectual property on the models developed jointly? Second, should data sharing agreements be signed with individual sites or with the research consortium in the case of multicentre studies? Third, is there sufficient synergy between publicly funded research groups and private industry?

Implementation of AI research

As has been illustrated, the potential applications of IR AI research are far reaching. However, one of the future challenges will be how to implement this new data into everyday clinical practice. In this section, we outline how this might be achieved successfully.

First, we recommend training and educating the current and future IR workforce and leadership to become proficient adopters of AI-based clinical decision support systems.[17] IR professional societies should develop AI education modules for radiologists, and formal training should be included in the fellowship. Future closer collaborations between IRs in training and nonclinical experts in the development of AI applications will be important.[18] Second, efforts to standardize IR terminology should continue. This would allow for reports to consist largely of structured data elements. Thus far patient-reported outcomes integration and analysis by AI have been hindered by lack of data standardization and electronic health record integration.[19] Third, formalizing standards for external and continuous validation of AI models and increasing research on algorithm fairness are needed to minimize AI bias in IR. Using nonrandomized, real-world data could introduce bias into the algorithm.[17] This could affect the validity of predictive modeling.[20] Therefore, standards and guidelines for the validation of AI systems are needed to promote clear and uniform measures of their accuracy and correctness.[17] Fourth, establishing consensus around a robust yet sensible regulatory framework that fairly assigns liability due to AI-related error while ensuring an acceptable level of quality and safety for AI tools is necessary to foster trust with both IRs and patients. However, updating preexisting legislative frameworks to regulate AI in healthcare will likely be insufficient. Lawmakers should obtain multidisciplinary counsel from ethicists, computer scientists and informaticians, clinicians, patients, professional societies, and health technology companies to construct a new regulatory framework that takes into consideration AI's self-learning characteristics.[17] Finally, standardizing research reporting and conducting prospective randomised controlled trials that demonstrate improvement in traditional patient outcomes are essential for AI adoption. Because AI systems are costly, healthcare leadership will be more willing to adopt AI systems when there is evidence of improved patient outcomes (e.g., survival, quality of life, and cost-effectiveness).[17] Researchers have recently published consensus-based guidelines for evaluating and reporting clinical trials for AI interventions.[21]

Conclusions

As has been highlighted, there are several areas where AI is set to benefit clinical practice in IR but current research on its practical application is lacking. This chapter has sought to highlight several practical examples of the use of AI in methodology in clinical research and how that may be applicable to IR in particular in providing advanced procedural support, improved clinical decision tools, and enhancing the trainee experience.

References

1. Meek R, Lungren M, Gichoya J. Machine learning for the interventional radiologist. *Am J Roentgenol.* 2019;213:782–784.
2. Letzen B, Wang CJ, Chapiro J. The role of artificial intelligence in interventional oncology: a primer. *J Vasc Intervent Radiol.* 2019;30(1):38–41.e1.
3. Cho H, Lee JG, Kang SJ, et al. Angiography-based machine learning for predicting fractional flow reserve in intermediate coronary artery lesions. *J Am Heart Assoc.* 2019;8:e011685.
4. Pijls NH, Van Gelder B, Van der Voort P, et al. Fractional flow reserve. A useful index to evaluate the influence of an epicardial coronary stenosis on myocardial blood flow. *Circulation.* 1995;92:3183–3193.
5. Park SJ, Kang SJ, Ahn JM, et al. Visual-functional mismatch between coronary angiography and fractional flow reserve. *JACC Cardiovasc Interv.* 2012;5:1029–1036.
6. Goldstein BA, Navar AM, Carter RE. Moving beyond regression techniques in cardiovascular risk prediction: applying machine learning to address analytic challenges. *Eur Heart J.* 2017;38:1805–1814.
7. Krittanawong C, Zhang H, Wang Z, Aydar M, Kitai T. Artificial intelligence in precision cardiovascular medicine. *J Am Coll Cardiol.* 2017;69:2657–2664.
8. Tacher V, Lin M, Duran R, et al. Comparison of existing response criteria in patients with hepatocellular carcinoma treated with transarterial chemoembolisation using a 3D quantitative approach. *Radiology.* 2016;278(1):275–284.
9. Abajian A, Murali N, Savic L, et al. Predicting treatment response to intra-arterial therapies for hepatocellular carcinoma with the use of supervised machine learning—an artificial intelligence concept. *J Vasc Intervent Radiol.* 2018;29:850–857.
10. Duong MT, Rauschecker AM, Rudie JD, et al. Artificial intelligence for precision education in radiology. *Br J Radiol.* 2019;92:20190389.
11. Rosset A, Ratib O, Geissbuhler A, Vall e J-P. Integration of a multimedia teaching and reference database in a PACS environment. *Radiographics.* 2002;22:1567–1577.
12. *Radiology Informatics Lab at Stanford Medicine. Medical ImageNet*; 2016. Available at: http://langlotzlab.stanford.edu/projects/medicalimage-net/. Accessed July 1, 2021.
13. Channin DS, Mongkolwat P, Kleper V, Rubin DL. The annotation and image mark-up project. *Radiology.* 2009;253:590e2.
14. Cardoso MJ, Arbel T, Lee SL, et al. *Intravascular Imaging and Computer Assisted Stenting, and Large-Scale Annotation of Biomedical Data and Expert Label Synthesis.* New York: Springer; 2017.
15. Varangaonkar A. *Top 10 Deep Learning Frameworks.* Packt Hub; 2017. Available at: https://datahub.packtpub.com/deep-learning/top-10-deeplearning-frameworks/. Accessed July 1, 2021.
16. Compute Canada. Available at: https://www.computecanada.ca/research-showcase/. Accessed July 1, 2021.
17. Chua IS, Gaziel-Yablowitz M, Korach ZT, et al. Artificial intelligence in oncology: path to implementation. *Cancer Med.* 2021;10:4138–4149.
18. American Medical Association. AMA Adopt Policy, Integrate Augmented Intelligence in Physician Training. https://www.ama-assn.org/press-center/press-releases/ama-adopt-policy-integrate-augmented-intelligence-physician-training. Accessed July 4, 2021.
19. Howell D, Molloy S, Wilkinson K, et al. Patient-reported outcomes in routine cancer clinical practice: a scoping review of use, impact on health outcomes, and implementation factors. *Ann Oncol.* 2015;26:1846–1858.
20. Sherman RE, Anderson SA, Dal Pan GJ, et al. Real-world evidence—what is it and what can it tell us. *N Engl J Med.* 2016;375:2293–2297.
21. Cruz Rivera S, Liu X, Chan A-W, et al. Guidelines for clinical trial protocols for interventions involving artificial intelligence: the SPIRIT-AI extension. *Nat Med.* 2020;26:1351–1363.

Chapter 65

Patient perspectives

Jim Zhong and Alexandra Gilbert

Leeds Institute of Medical Research at St. James's, University of Leeds, Leeds, United Kingdom

Key points

- Appreciate that patient perspectives are an important outcome measure in IR
- Learn about the current state of patient-reported outcomes (PROs) in IR
- Understand how to incorporate PROs into future IR research

Definitions

Patient-reported outcome (PRO)—a health outcome directly reported by the patient without interpretation by the clinician or another individual, that relates to the patient's health, quality of life, or functional status associated with health care or treatment.[1]

Why it matters?

Patient perspectives are central to what interventional radiology (IR) offers. Despite being the new kid on the block, IR is changing how medicine is practiced through the eyes of the patient who is able to feel less pain, go home quicker, or avoid debilitating side effects thanks to the minimally invasive nature of IR treatments. It is vital for interventional radiologists to understand the patient perspectives to incorporate this into their own practice to optimize patient care.

Introduction

Patient perspectives are important considerations for medical care and central to evaluating the quality of care interventional radiologists provide. The British National Health Service outlines eight domains defined as critical to a "good" patient experience, which include respect, information and communication, physical comfort, emotional support, and access to care.[2] All these components should be honored with every patient interaction.

Patient-centered care leads to better patient satisfaction and perceived quality of care.[3] Focusing on the diagnosis and treatment of physical pathologies can result in the psychological impact of therapies being overlooked particularly when patients may not always volunteer this information. Interventional radiologists perform many palliative procedures and the effect of these therapies on quality of life (QoL) is vitally important, particularly when therapies may offer only marginal survival benefits but improve QoL. IR must engage with the patient and their caregivers as partners to help identify common goals, weighing up their options, and helping them to make informed decisions based on what is important to them.

In this chapter we consider the patient pathway in IR, in what setting the measurement of patient perspectives could be valuable, how it can be measured with a focus on patient-reported outcomes (PROs), and how data on patient perspectives can be used and interpreted in both the research and clinical setting.

The patient pathway in IR

Evaluating the patient's perspective is multidimensional and complex. Breaking down the patient pathway into different stages, with each having unique issues to address, could allow for a more tailored and effective approach in gauging patients' views. Fig. 65.1 shows the potential patient concerns at various stages of their IR treatment pathway.

At the beginning of the patient's IR journey is the initial consultation on the procedure being undertaken. The issues at this stage are around gaining informed consent from the patient, developing a rapport and also discussing the patient's own expectations of the treatment. The patient's

Translational Interventional Radiology. https://doi.org/10.1016/B978-0-12-823026-8.00012-2

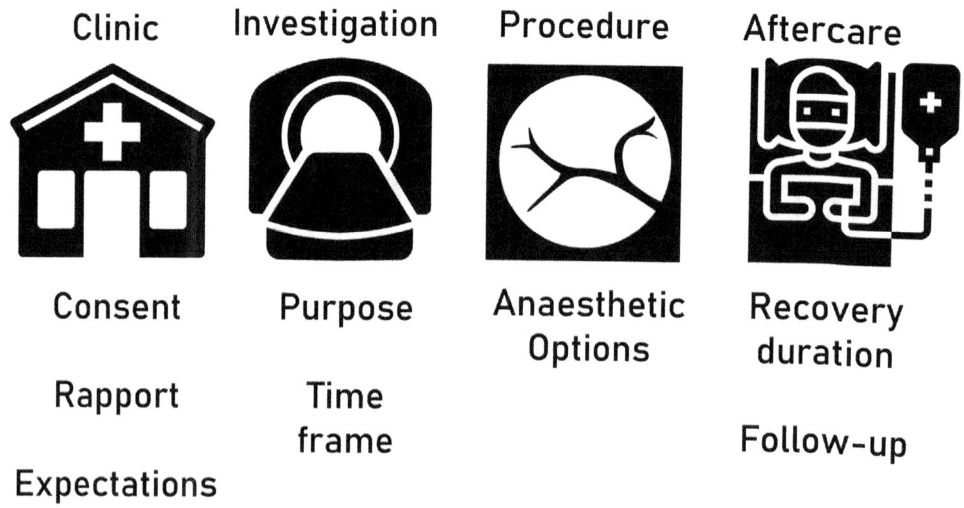

FIGURE 65.1 Patient perspectives and concerns throughout the IR pathway from the clinic, to pre-procedural investigations, to the procedure and aftercare.

decision should be free of coercion and reflective of their values. Although the legal part of consent is easily achievable with a signed document, from the patient's perspective and the need to respect patient autonomy, the ethical component is more complex. This requires taking into account the health literacy of the patient, psychosocial issues, cultural aspects, and the environment in which the consent discussion takes place in an attempt to help the patient fully understand.[4] It is uncertain whether patients and their families can be fully informed due to the complexities of the information being shared, and it has been reported that trust in their clinician may often bridge this gap in understanding.[5] Taking the time to understand the patient's perspective and preferences will also increase patient–doctor trust.

Traditionally, patients were referred to an interventional radiologist once they had already agreed in principle to a procedure that they may have already discussed with a surgeon or oncologist managing their condition. They would fleetingly meet with the radiologist, who may not know the extensive past medical history of the patient, to consent to a life-changing procedure and the next time they would meet would be the day of the procedure. After this, they might go back to seeing their "own doctor." The lack of clinical presence however from the part of IR has issues with continuity of care and patient responsibility, both of which likely impact on the patients' perspective or experience of their treatment. In the future, IR must be involved throughout the entire patient pathway, from the multidisciplinary team decision to the treatment and follow-up of patients as advocated by IR societies such as the Society of Interventional Radiology and British Society of Interventional Radiology. Only with a more integrated approach can we ensure the service provided to patients is optimal. Maintaining a patient-centered approach throughout this journey is more likely to result in a positive experience for both the patient and their family.[6,7]

Measuring patient perspectives

Measuring patient perspectives highlights opportunities to improve care. Keeping the patient central to this process, a variety of methods are worth exploring including validated patient-reported outcomes, feedback surveys, diaries, and other qualitative methods through interviews, workshops, or focus groups. PROs (often in the form of a questionnaire) are health outcomes directly reported by the patient without interpretation by the clinician or another individual, that relates to the patient's health, quality of life, or functional status associated with health care or treatment.[1] PROs are increasingly used in clinical practice and research to evaluate treatment outcomes.

When selecting PRO tools, minimum standards are set out by the International Society of Quality of Life Research, including evidence describing the reliability and validity of the tool; meaning the content of the PRO is deemed to be relevant and

comprehensive for the clinical scenario and patient population.[8] The style of reporting the PRO results, to patients and clinicians, should also be considered to help with data interpretation and maximize patients understanding.[9] A number of PRO consensus documents, summarising the optimal implementation of patient-reported outcomes in clinical trials and routine care, have been published over the past 10 years and may be easily accessed via the PROTEUS website.[10]

Additional considerations for PRO tools, also known as patient recorded outcome measures (PROMs), include the number of questions involved and language versions available, as this is likely to impact on how patients understand and engage with it. Costs of implementation and running costs must also be considered in the current financial climate to ensure longevity for such tools. In the digital age, PROs can also be collected electronically with a case study demonstrating the successful integration of an electronic QoL reporting platform into the follow-up assessment of deep vein thrombosis and lymphedema patients in a tertiary care IR practice.[11]

Using PROs for selecting treatments

PROs can be used to help inform clinicians on treatment selection and facilitate shared decision-making between patients and clinicians. For example, a study using QoL endpoints to compare embolotherapy techniques for patients with hepatocellular carcinoma found radioembolization had greater improvement in QoL domains, compared to transaterial chemoembolization, despite radioembolization being used to treat patients with more advanced disease.[12] In the SARAH trial (*S*or*A*fenib vs. *R*adioembolization in *A*dvanced *H*epatocellular Carcinoma), radioembolization was found to have significantly better QoL scores despite no statistical difference in overall survival, the primary endpoint.

As well as QOL measurement, much research has highlighted the differences in patient and clinician reporting of adverse events, with more subjective issues often underreported by clinicians.[13] To fully understand the impact of treatments on patients, we have to ask the patients themselves as some symptoms or adverse effects may not be easily observable. The FDA and the Critical Path Institute explored the use of PROs and QoL measures to complement existing clinical safety assessments, better inform patients on treatment tolerability, and help select the most appropriate treatment.[14] Many instruments are in common use across oncology, including the National Cancer Institute PRO version of the CTCAE (PRO-CTCAE),[15] the Functional Assessment of Chronic Illness Therapy (FACIT) Searchable Library, and the EORTC Quality of Life Group Item Library.[16,17] These online PRO systems allow researchers to use existing questionnaires (in the case of FACIT and EORTC) alongside the opportunity to develop a customized questionnaire for a specific research study. This flexible use of PRO instruments is useful for IR procedures where patients are treated in several organ systems through embolization or ablative techniques with a wide variety of symptom areas, systemic effects, and potential treatment-related toxicity, alongside those of their disease process. It is important for IR researchers to be familiar with commonly used validated tools that can be integrated into IR practice and future research should consider the validation of commonly used PROs in IR cohorts and developing extensions to existing PROs for use in IR.

Future PRO considerations in IR

The use of PROs in IR has been mainly confined to the research setting despite other specialties succeeding in routinely collecting PROs in clinical practice, with positive feedback from both patients and healthcare professionals.[6,18] A major shift in mindset is required to promote PROs use in IR and this involves engagement with all key stakeholder groups and leverage of existing expertise from other specialties such as oncology through collaborative working. IR procedures used to treat cancer such as tumor ablation are still not widely available and the availability of PRO data would further support the implementation of these therapies nationwide.[19,20] An IR consensus panel document found that patient experience during and following minimally invasive IR procedures was vastly different from traditional treatments such as open surgery or medical options, therefore routinely collecting PROs data is vital to highlight the advantages of IR therapies.[21]

Conclusions

Patient perspectives must be considered at each step in the patient's journey and IR has a key role to play by recognizing what concerns the patient may have and how to use PROs to evaluate the patient experience and the impact of treatment. The use of tailored PROs and QoL assessments will not only expand the evidence base in support of minimally invasive image-guided procedures but help educate the public on the meaningful impact on QoL that interventional radiologists offer patients.

Get started

1) Think about what are the factors that are important to a patient when they go for an IR procedure.
2) At what points in the patient's journey through the IR pathway can we make a difference?
3) What makes a good patient reported outcome tool?
4) Understand the important of validated tools for assessing quality of life.

Potential pitfalls

1) Focusing on procedural technical outcomes without considering patient perspectives and impact on quality of life.
2) Only addressing patient perspectives after procedure rather than throughout the patient journey from first clinic visit to aftercare.
3) Selecting patient reported outcome tools that are not appropriate for the patient.

Real-world examples

1) Understanding the minimum standards for patient reported outcome measures – Reeve BB, Wyrwich KW, Wu AW, Velikova G, Terwee CB, Snyder CF, et al. ISOQOL recommends minimum standards for patientreported outcome measures used in patient-centered outcomes and comparative effectiveness research. Qual Life Res. 2013;22:1889–1905.
2) An example of an electronic patient reported outcome tool – Hoang NS, Hwang W, Katz DA, Mackey SC, Hofmann LV. Electronic patient-reported outcomes: semi-automated data collection in the interventional radiology clinic. J Am Coll Radiol. 2019;16:472e477.
3) Examples of patient groups who may benefit specifically from using patient reported outcomes measures – Kwan SW, Charalel RA, Stover AM, et al. Development of national research and clinical agendas for patient-reported outcomes in IR: proceedings from a multidisciplinary consensus panel. J Vasc Intervent Radiol. 2018;29:1e8.

Resources

PROTEUS Consortium. Patient-Reported Outcomes Tool: Engaging Users and Stakeholders (PROTEUS). https://more.bham.ac.uk/proteus/about-proteus/.

References

1. Weldring T, Smith SMS. Article commentary: patient-reported outcomes (PROs) and patient-reported outcome measures (PROMs). *Health Serv Insights*. 2013;6. HSI.S11093.
2. Department of Health. *NHS Patient Experience Framework*; 2012. https://www.gov.uk/government/uploads/system/uploads/attachment_data/file/215159/dh_132788.pdf.
3. McMillan SS, Kendall E, Sav A, et al. Patient-Centered approaches to health care. *Med Care Res Rev*. 2013;70:567–596.
4. Grady C. Enduring and emerging challenges of informed consent. *N Engl J Med*. 2015;372:855–862.
5. Keller EJ, Perez AW, Makary MS. Informed consent: beating a dead horse or an opportunity for quality improvement? *J Vasc Intervent Radiol*. 2020;31:139–140.
6. Valderas JM, Kotzeva A, Espallargues M, et al. The impact of measuring patient-reported outcomes in clinical practice: a systematic review of the literature. *Qual Life Res*. 2008;17:179–193.
7. Kotronoulas G, Kearney N, Maguire R, et al. What is the value of the routine use of patient reported outcome measures toward improvement of patient outcomes, processes of care, and health service outcomes in cancer care? A systematic review of controlled trials. *J Clin Oncol*. 2014;32:1480–1501.
8. Reeve BB, Wyrwich KW, Wu AW, Velikova G, Terwee CB, Snyder CF, et al. ISOQOL recommends minimum standards for patient-reported outcome measures used in patient-centered outcomes and comparative effectiveness research. *Qual Life Res*. 2013;22:1889–1905.
9. Snyder C, Smith K, Holzner B, Rivera YM, Bantug E, Brundage M, et al. Making a picture worth a thousand numbers: recommendations for graphically displaying patient-reported outcomes data. *Qual Life Res*. 2019;28:345–356.
10. PROTEUS Consortium. Patient-Reported Outcomes Tool: Engaging Users and Stakeholders (PROTEUS). https://more.bham.ac.uk/proteus/about-proteus/.
11. Hoang NS, Hwang W, Katz DA, Mackey SC, Hofmann LV. Electronic patient-reported outcomes: semi-automated data collection in the interventional radiology clinic. *J Am Coll Radiol*. 2019;16:472–477.
12. Salem R, Gilbertsen M, Butt Z, et al. Increased quality of life among hepatocellular carcinoma patients treated with radioembolization, compared with chemoembolization. *Clin Gastroenterol Hepatol*. 2013;11:1358–1365. e1.
13. Atkinson TM, Ryan SJ, Bennett AV, et al. The association between clinician-based common terminology criteria for adverse events (CTCAE) and patient-reported outcomes (PRO): a systematic review. *Support Care Cancer*. 2016;24:3669–3676.
14. Kluetz PG, Kanapuru B, Lemery S, et al. Informing the tolerability of cancer treatments using patient-reported outcome measures: summary of an FDA and critical Path Institute workshop. *Value Health*. 2018;21:742–747.
15. Basch E, Reeve BB, Mitchell SA, et al. Development of the national cancer institute's patient-reported outcomes version of the common terminology criteria for adverse events (PRO-CTCAE). *JNCI J Natl Cancer Inst*. 2014;106. dju244–dju244.
16. Webster K, Cella D, Yost K. The functional assessment of chronic illness Therapy (FACIT) measurement system: properties, applications, and interpretation. *Health Qual Life Outcome*. 2003;1:79.
17. European Organization for Research and Treatment of Cancer. *EORTC Quality of Life*; 2021. https://qol.eortc.org/item-library/.
18. Chen J, Ou L, Hollis SJ. A systematic review of the impact of routine collection of patient reported outcome measures on patients, providers and health organisations in an oncologic setting. *BMC Health Serv Res*. 2013;13:211.
19. Zhong J, Atiiga P, Alcorn DJ, et al. Cross-sectional study of the provision of interventional oncology services in the UK. *BMJ Open*. 2017;7:e016631.
20. Yennu Nandan A, Mukundu Nagesh N, Lee MJ. Patient reported outcomes in interventional radiology: time to measure what we do. *Cardiovasc Intervent Radiol*. 2019;42:931–932.
21. Kwan SW, Charalel RA, Stover AM, et al. Development of national research and clinical agendas for patient-reported outcomes in IR: proceedings from a multidisciplinary consensus panel. *J Vasc Intervent Radiol*. 2018;29:1–8.

Part VII

Clinical: preparation

Chapter 66

Meaningful outcome measurements

Travis William McCain Pebror
Interventional and Diagnostic Radiology, Department of Radiology and Imaging Sciences, Indiana University School of Medicine Indianapolis, IN, United States

Key points

- The value proposition of an intervention is best characterized through outcome measurements that directly assess how an individual feels, functions, or survives following the intervention.
- Indirect outcome measures such as biomarkers and vital signs offer a more practical yet less accurate means of predicting the likelihood an intervention will ultimately have a meaningful effect.
- Patients are an important source of direct feedback regarding the potential utility of a novel intervention, however very few patient-reported outcomes are validated and readily available for interventional radiology (IR) procedures.

Why it matters?

Translation of an innovation through a clinical strategy can be thought of as a collection of outcomes more so than a collection of studies. A carefully designed study is only as insightful as the underlying outcomes measured. The value of any study is therefore somewhat predetermined during the planning process when the final research question is articulated and outcome measures selected. There are several outcome measures relevant to clinical research that offer tradeoffs in terms of three important characteristics; validity, reliability, and responsiveness. Together these characteristics describe how readily an outcome can be measured and how applicable the outcome measure is toward describing the potential clinical benefit of an innovation.

Characteristics of outcome measures

Several characteristics describe the ability of an outcome measure or endpoint to convey the benefit and/or risk of an intervention. **Validity** describes the degree to which an outcome is able to measure what it intends to. Another word for validity is accuracy. Certain measures are more valid than others. For example, death is clearly a valid endpoint for measuring harm whereas a poorly constructed patient questionnaire might not accurately reflect postoperative pain. Another characteristic often confused with validity is reliability (see Fig. 66.1). **Reliability** describes how well a measure will demonstrate the same outcome when repeated in similar clinical circumstances. Reliability is critical for attributing an observed change in an outcome to the intervention versus sampling variation inherent in the measure itself. Therefore, reliability plays an important role when determining the significance of an observed outcome.

Finally, an outcome measure must change appreciably when an intervention exerts a true effect. **Responsiveness**

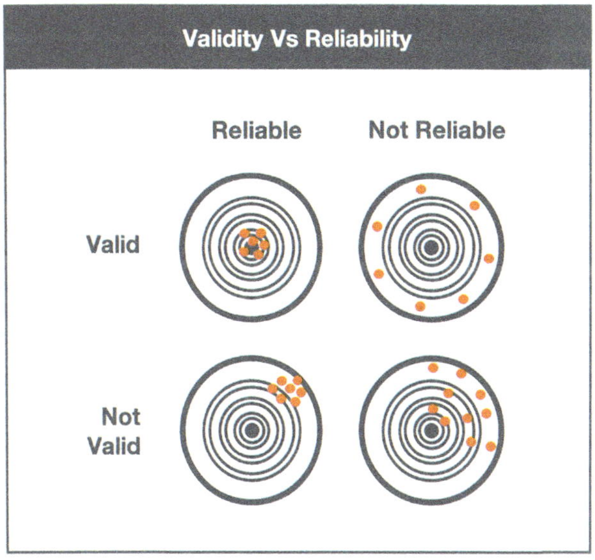

FIGURE 66.1 The bullseye diagrams illustrate the different sampling distributions of a measure based on validity and reliability. The upper left bullseye shows a measure centered around the bullseye that is both valid and reliable. The other three show measures that lack either reliability (top right), validity (bottom left) or both (bottom right).

Translational Interventional Radiology. https://doi.org/10.1016/B978-0-12-823026-8.00049-3

describes how sensitive an outcome measure is toward detecting the benefit of an intervention. It is important to note that the responsiveness of an outcome is dramatically affected by how feasible it is to measure the outcome. The time between intervention and outcome; rarity of outcome; and difficulty of measuring the outcome all contribute to a make an outcome less responsive. It is important to have a firm understanding of these characteristics because trade-offs between each characteristic are often made when choosing which types of outcome measures to pursue.

Real-world examples: clinically meaningful outcome measures

A clinical outcome that directly measures how an individual feels, functions, or survives is a **clinically meaningful endpoint**.[1] This definition of clinical significance aligns how a patient perceives their illness or functions in daily life following an intervention with the value of that intervention rather than endpoints that merely describe technical "success" of the intervention. While the latter are of often easier to measure (i.e., ablation zone diameter) and of great focus during the development of a technology into a novel intervention, the former are ultimately most important.

Common clinically meaningful endpoints include the frequency of or time to events such as death, adverse events, local progression of disease, permanent disability, myocardial infarction, and stroke as well as disease-specific measures such as recurrence rate and disease-free survival in cancer. While these direct measures are inherently valid and reliable, many appear less responsive in the context of a clinical trial limited by time, enrollment, sampling, and resources to observe the endpoints in human subjects, especially when the event of interest occurs infrequently.

Under these circumstances, a composite endpoint can allow a study to be well-powered and sensitive despite a reduced sample size and duration of follow-up that increases feasibility.[2] A **composite endpoint** is reached when a subject experiences any one of a collection of distinct but related clinical endpoints. A well-known example is the major cardiovascular event (MACE) composite endpoint, generally defined as the composite of death, stroke, or myocardial infarction. While collecting separate endpoints together makes the composite endpoint increasingly more responsive, doing so decreases the validity of the measure by muddling the effects of several outcomes together. For example, Kip et al.[3] demonstrated how slight differences in published definitions of MACE result in widely differing estimations of cardiovascular disease progression at 1 year follow-up among nearly 7000 patients who received a drug-eluting coronary stent. By including several endpoints together, it can be difficult to interpret what the observed difference in the composite endpoint reflects about the intervention (i.e., safety vs. effectiveness) thus rendering composite endpoints a more responsive though sometimes less valid measure.[4]

Real-world examples: surrogate outcome measures: often more feasible, sometimes less valid

An alternative to directly measuring how a patient feels, functions, or survives following a therapy is to instead measure an endpoint that readily predicts these clinically meaningful outcomes. This is especially useful if the intervention under investigation aims to affect future consequences of a slowly progressive disease. **Surrogate endpoints** are responsive and easily measured laboratory results and physical exam signs that reflect a biological response to therapy to predict future clinical benefits. The goal of surrogate endpoints is to shorten the length of time required to gather sufficient data for regulatory approval.

The preamble to the FDA's accelerated approval rule states that surrogate endpoints must be "reasonably likely" to predict future clinical benefits.[5] To varying degrees of success, several radiographic features, vital signs, laboratory results, functional assessments, and other endpoints have been proposed as candidate surrogate endpoints. The validity of a surrogate endpoint rests squarely on the body of evidence that correlates the surrogate endpoint with future clinical outcomes of a specific disease. To this effect, the FDA recognizes three levels of validity for surrogate endpoints that have bearing on postapproval study of therapies approved under the accelerated rule (see Table 66.1). Even when a surrogate marker is seemingly well correlated to the outcome of interest, the surrogate endpoints may not provide reliable evidence of the intervention's efficacy or safety.[6]

Consider blood pressure as a surrogate endpoint for cardiovascular disease in patients with renal artery stenosis undergoing revascularization with percutaneous transluminal renal angioplasty and stenting. Several early studies (DRASTIC,[7] STAR[8], and ASTRAL[9]) all demonstrated decreased blood pressure in patients following revascularization. However, in the most recent major trial comparing renal artery stenting in addition to antihypertensive medical therapy, blood pressure was a poor predictor of cardiovascular risk. The CORAL trial found that patients who received renal artery stenting in addition to medical therapy experienced significantly lower blood pressure during the trial, but similar rates of cardiovascular-related death, stroke, and myocardial infarction (MI) as those receiving medical therapy alone.[10] These results demonstrate how even a surrogate marker with considerable evidence of predictive ability in one context must be interpreted with skepticism in another.

TABLE 66.1 There are three types of surrogated endpoints recognized by the FDA based on the body of evidence demonstrating how well the surrogate endpoint predicts a future clinical benefit. Both validated and reasonably likely surrogate endpoints.

	FDA BEST definition	Evidence	Use in FDA approval/clearance	Additional studies
Validated	*An endpoint supported by a clear mechanistic rationale and clinical data providing strong evidence that an effect on the surrogate endpoint predicts a specific clinical benefit.*	High	Yes	None
Reasonably likely	*An endpoint supported by strong mechanistic and/or epidemiologic rationale such that an effect on the surrogate endpoint is expected to be correlated with an endpoint intended to assess clinical benefit in clinical trials, but without sufficient clinical data to show that it is a validated surrogate endpoint.*	Medium	Yes, in accelerated approval process	Postmarket confirmatory trials required
Candidate	*An endpoint still under evaluation for its ability to predict clinical benefit.*	Low	No	NA

Adapted from FDA-NIH Biomarker Working Group. *BEST (Biomarkers, EndpointS, and Other Tools) Resource*; 2016.

Real-world examples: patient-reported outcome measures

An obvious but often neglected method to directly measure the clinical benefit of an intervention is simply by asking patients. **Patient-reported outcomes** (PROs) capture how patients feel or experience their health status without edit or interpretation by a clinician. Examples include standardized ratings scales and questionnaires regarding physical symptoms and psychological stress. While seemingly simple, it is often difficult to construct PROs that are both valid and reliable. Given the dearth of PROs in IR, the Society of Interventional Radiology Foundation sponsored an expert panel in 2017 to outline a research agenda for the development of PROs. Of three treatment areas with significant PROs available (see Table 66.2) uterine fibroid embolization had the most well-developed measure, the 37-item Uterine Fibroid Symptom and Quality of Life (UFS-QOL) questionnaire.[11,12] In a single-arm, prospective trial, UFS-QOL was able to demonstrate the effectiveness of UAE by measuring a reduction in UFS-QOL score in 1287 patients 3 years after UAE.[13] UFS-QOL demonstrates how PROs are especially important for understanding the value of an innovation that modifies a disease with difficult to measure functional impairment.

Choosing the right outcome measures to support translation

There are important tradeoffs to consider when deciding which outcome measures to pursue in the context of a clinical strategy. Ultimately, the outcome measures should convey whether the innovation is able to deliver clinically meaningful benefits and offer a unique value proposition. Direct measures including patient-reported outcomes are critical, however many come with important feasibility limitations. In well-validated circumstances, indirect outcome measurements can offer compelling evidence of clinical benefit while being substantially more feasible.

Get started

- **Begin by formulating a research question in the context of a clinical strategy.** Consider the overall clinical strategy tied to the value proposition of the innovation (Chapter 41) and decide which uncertainty needs to be resolved through a clinical study. Formulate several iterations of the research question following guidance in Chapter 11.
- **Define outcome measures capable of answering the research question.** For each iteration of the research question, define a set of possible outcome measures capable of addressing the research question. These outcome measures are specific to disease processes and available within the literature. Below are several additional general inventories of outcome measures:

Potential pitfalls

- Choosing an outcome measure that relates well to the value proposition, but is too burdensome or infrequent to measure.

TABLE 66.2 Three disease areas of IR practice with significant development of PROs as outlined by SIR working group on PROs.

Disease	Publications for review
Uterine leiomyomata	• The UFS-QOL, a new disease-specific symptom and health-related quality of life questionnaire for leiomyomata[12] • Uterine artery embolization for treatment of leiomyomata: long-term outcomes from the FIBROID registry[13] • Uterine fibroid embolization: measurement of health-related quality of life before and after therapy[14] • The FIBROID Registry: symptom and quality-of-life status 1 year after therapy[15]
Superficial and deep venous disease	• Recommended reporting standards for endovenous ablation for the treatment of venous insufficiency: joint statement of The American Venous Forum and The Society of Interventional Radiology[16] • Evaluation of outcomes in chronic venous disorders of the leg: development of a scientifically rigorous, patient-reported measure of symptoms and quality of life[17] • VEINES-QOL/Sym questionnaire was a reliable and valid disease-specific quality of life measure for deep venous thrombosis[18] • Rationale and design of the ATTRACT Study: a multicenter randomized trial to evaluate pharmacomechanical catheter-directed thrombolysis for the prevention of postthrombotic syndrome in patients with proximal deep vein thrombosis[19]
Hepatocellular carcinoma	• Longitudinal quality of life assessment of patients with hepatocellular carcinoma after primary transarterial chemoembolization[20] • Quality of life and its associated factors in patients with hepatocellular carcinoma receiving one course of transarterial chemoembolization treatment: a longitudinal study[21] • Health-related quality of life in patients with hepatocellular carcinoma after hepatic resection, transcatheter arterial chemoembolization, radiofrequency ablation, or no treatment[22] • Increased quality of life among hepatocellular carcinoma patients treated with radioembolization, compared with chemoembolization[23]

Table adapted from Kwan SW, et al. Development of National research and clinical agendas for patient-reported outcomes in IR: proceedings from a multidisciplinary consensus panel. *J Vasc Intervent Radiol.* 2018;29(1):1.

- Selecting a surrogate endpoint that is responsive but a poor predictor of the clinical outcome of interest.
- Focusing exclusively on physiologic outcomes and failing to incorporate patient reported outcomes.

Resources

- ClinicalTrials.gov
- BEST (Biomarkers, EndpointS, and other Tools) by FDA-NIH
- FDA "Guidance for Industry" documents
- Centers for Medicare and Medicaid Measures Inventory (cmit.cms.gov)
- National Quality Form (qualityform.org)
  - Patient-Reported Outcome measures
    - IR disease-specific Patient-Reported Outcomes (Table 66.2)
    - Patient-Reported Outcomes Measurement Information Systems PROMIS (healthmeasures.net/PROMIS)
    - FACIT Measurement System, including Functional Assessment of Cancer Therapy FACT (facit.org/facitorg/questionnaires).
- **Select an outcome measure by comparing the relative merits of possible outcome measures.** List the strengths and limitations of each outcome measure to address the research question in terms of validity, reliability, and responsiveness. Choose one outcome measure to serve as the primary endpoint.
- **Select a clinic study type capable of producing the desired outcome measure.** Review clinical study designs in prior chapters to see which study designs are capable of producing the desired outcome measure. If the study design type is infeasible to implement,

consider the next-best outcome measure and review which study design type might be.

References

1. FDA-NIH Biomarker Working Group. *BEST (Biomarkers, EndpointS, and Other Tools) Resource*. 2016.
2. Fleming TR, Powers JH. Biomarkers and surrogate endpoints in clinical trials. *Stat Med*. 2012;31.25:2973–2984.
3. Kip KE, Hollabaugh K, Marroquin OC, Williams DO. The problem with composite end points in cardiovascular studies: the story of major adverse cardiac events and percutaneous coronary intervention. *J Am Coll Cardiol*. 2008;51.7:701–707.
4. Gad, Cox S, eds. *Clinical Trials Handbook*. Vol. 8. John Wiley & Sons; 2009.
5. *57 FR 58942-44*. December 11, 1992.
6. Fleming TR. Surrogate endpoints and FDA's accelerated approval process. *Health Affairs*. 2005;24(1):67–78.
7. van Jaarsveld BC, Krijnen P, Pieterman H, et al. The effect of balloon angioplasty on hypertension in atherosclerotic renal-artery stenosis. *N Engl J Med*. 2000;342(14):1007–1014.
8. Bax L, Woittiez AJ, Kouwenberg HJ, et al. Stent placement in patients with atherosclerotic renal artery stenosis and impaired renal function: a randomized trial. *Ann Intern Med*. 2009;150(12):840–848.
9. Investigators A. Revascularization versus medical therapy for renal-artery stenosis. *N Engl J Med*. 2009;361(20):1953–1962.
10. Cooper CJ, Murphy TP, Cutlip DE, et al. Stenting and medical therapy for atherosclerotic renal-artery stenosis. *N Engl J Med*. 2014;370(1):13–22.
11. Kwan SW, Charalel RA, Stover AM, et al. Development of National Research and Clinical Agendas for patient-reported outcomes in IR: proceedings from a multidisciplinary consensus panel. *J Vasc Intervent Radiol*. 2018;29(1):1.
12. Spies JB, Coyne K, Guaou Guaou N, et al. The UFS-QOL, a new disease-specific symptom and health-related quality of life questionnaire for leiomyomata. *Obstet Gynecol*. 2002;99(2):290–300.
13. Goodwin SC, Spies JB, Worthington-Kirsch R, et al. Uterine artery embolization for treatment of leiomyomata: long-term outcomes from the FIBROID Registry. *Obstet Gynecol*. 2008;111:22–33.
14. Spies JB, Warren EH, Mathias SD, Walsh SM, Roth AR, Pentecost MJ. Uterine fibroid embolization: measurement of health-related quality of life before and after therapy. *J Vasc Intervent Radiol*. 1999;10(10):1293–1303.
15. Spies JB, Myers ER, Worthington-Kirsch R, Mulgund J, Goodwin S, Mauro M. The FIBROID Registry: symptom and quality-of-life status 1 year after therapy. *Obstet Gynecol*. 2005;106(6):1309–1318.
16. Kundu S, Lurie F, Millward SF, et al. Recommended reporting standards for endovenous ablation for the treatment of venous insufficiency: joint statement of the American Venous Forum and the Society of Interventional Radiology. *J Vasc Surg*. 2007;46(3):582–589.
17. Lamping DL, Schroter S, Kurz X, Kahn SR, Abenhaim L. Evaluation of outcomes in chronic venous disorders of the leg: development of a scientifically rigorous, patient-reported measure of symptoms and quality of life. *J Vasc Surg*. 2003;37(2):410–419.
18. Kahn SR, Lamping DL, Ducruet T, et al. VEINES-QOL/Sym questionnaire was a reliable and valid disease-specific quality of life measure for deep venous thrombosis. *J Clin Epidemiol*. 2006;59(10):1056–e1.
19. Vedantham S, Goldhaber SZ, Kahn SR, et al. Rationale and design of the ATTRACT Study: a multicenter randomized trial to evaluate pharmacomechanical catheter-directed thrombolysis for the prevention of postthrombotic syndrome in patients with proximal deep vein thrombosis. *Am Heart J*. 2013;165(4):523–530.
20. Wible BC, Rilling WS, Drescher P, et al. Longitudinal quality of life assessment of patients with hepatocellular carcinoma after primary transarterial chemoembolization. *J Vasc Intervent Radiol*. 2010;21(7):1024–1030.
21. Shun SC, Chen CH, Sheu JC, Liang JD, Yang JC, Lai YH. Quality of life and its associated factors in patients with hepatocellular carcinoma receiving one course of transarterial chemoembolization treatment: a longitudinal study. *Oncol*. 2012;17(5):732.
22. Toro A, Pulvirenti E, Palermo F, Di Carlo I. Health-related quality of life in patients with hepatocellular carcinoma after hepatic resection, transcatheter arterial chemoembolization, radiofrequency ablation or no treatment. *Surgl Oncol*. 2012;21(1):e23–e30.
23. Salem R, Gilbertsen M, Butt Z, et al. Increased quality of life among hepatocellular carcinoma patients treated with radioembolization, compared with chemoembolization. *Clin Gastroenterol Hepatol*. 2013;11(10):1358–1365.

Chapter 67

Sample size

Christopher Ovanez
Russell H. Morgan Department of Radiology and Radiological Science, Johns Hopkins Hospital, Baltimore, MD, United States

Key points

- Consideration of sample size is quite important when setting out to design a research study. Several factors must be carefully taken into consideration when deciding upon the most optimal or appropriate sample size for a particular research study.
- While there are various ways to calculate sample size, one must take into consideration and be mindful of the fact that there is no truly perfect way to determine the most optimal sample size by only taking statistical calculations into account, as there are several non-statistical factors that will undoubtedly affect a study (directly or indirectly).
- The power of a statistical study increases as the sample size increases. If a study possesses too small a sample size, it could lead to the study being inadequate with respect to being capable of detecting a true difference when it actually exists, hence the study would be deemed underpowered. If a study possesses a sample size that is too large for the research question being posed, there are a number of potential problems which may arise.

Introduction

Consideration of sample size is quite important when setting out to design a research study. The power of a statistical study increases as the sample size increases. If a study possesses too small a sample size, it could lead to the study being inadequate with respect to being capable of detecting a true difference when it actually exists, hence the study would be deemed underpowered. If a study possesses a sample size which is too large for the research question being posed, there are a number of problems that may arise, for instance, if studying the effects of a particular intervention or medication upon a patient population, there are ethical issues that could potentially arise with regard to exposing potential risk to a larger group than is needed, in addition to needlessly wasting resources, effort, and time. There must be careful consideration of the research question being asked, the type of study which is most optimal to answer that question, and the tailoring of further study design accordingly, to determine the optimal number of participants that should be included in a study.

Get started

Several factors must be carefully taken into consideration when deciding upon the most optimal or appropriate sample size for a particular research study. For instance, examples of such factors may include[1] the following: the type of study design that is being selected (i.e., prospective, retrospective, observational, randomized clinical trial, etc.),[2] the research question(s) posed (i.e., toxicity, efficacy, etc),[3] taking into account potential issues that may occur during the course of the undertaken study (i.e., loss to follow-up, study population heterogeneity, rarity of a particular factor to be examined—outcome, disease, or patient characteristic, etc.),[4] and acceptable margin of error or significance level such as *P*-value. These are but a few of the factors that may be taken into consideration when selecting an appropriate sample size.

Potential pitfalls

There must be careful consideration of the research question being asked, the type of study that is most suitable to answer that question, and the tailoring of further study design accordingly, to determine the optimal number of participants that should be included in a study.[3]

The power of a statistical study increases as the sample size increases. If a study possesses too small a sample size, it could lead to the study being inadequate with respect to being capable of detecting a true difference when it actually exists, hence the study would be deemed underpowered. If a study possesses a sample size that is too large for the research question being posed, there are a number of problems which may arise, for instance—if studying the effects of a particular intervention or medication upon a

Translational Interventional Radiology. https://doi.org/10.1016/B978-0-12-823026-8.00010-9

patient population, there are ethical issues that could potentially arise with regard to exposing potential risk to a larger group than is needed especially when the intervention or medication is ineffective, in addition to needlessly wasting resources, effort, and time.

So, how does one go about figuring out the most optimal number of participants for a particular study? In general, sample size is calculated mathematically based upon certain considerations and desired parameters for the specific study being designed, for instance—factors that may be taken into account include acceptable margin of error, desired confidence level, population size, and incidence/prevalence of a particular disease process, among other factors.[1–4] There are statistical formulae that may be utilized to calculate study size, as well as several online sample size calculators and a variety of statistical programs that could also calculate the number best suited for your study based upon your desired parameters. Theoretically, given a particular study design, a larger sample size will result in a narrower confidence interval, a larger analysis power, and will more closely reflect the true population.[4]

While there are various ways to calculate sample size, one must take into consideration and be mindful of the fact that there is no truly perfect way to determine the most optimal sample size by only taking statistical calculations into account, as there are several nonstatistical factors that will undoubtedly affect a study (directly or indirectly), for instance—issues involving ethics, resources, internal validity, and information loss[3]—this is beyond the scope of this chapter, but many excellent resources are available if you wish to further explore this topic. For instance, if someone is designing a phase-1 type of trial where a new medication or intervention is being examined, and the risks for which are largely unknown and potentially highly adverse, the study power may not be the primary focus, and the sample size will likely be lower than that which would be most "optimal" given the potential risk being posed to study participants.[3] Thus, depending upon the type of study being designed, the research question(s) being posed, and several other factors that must be taken into consideration, the most "optimal" study size will change.

Hypothetically, let us that say there are two studies (both with the same type of study design)—one with a sample size of 100,000, and another with a sample size of 100. The study with the larger sample size will inherently have an increased likelihood to detect something that is statistically significant, that is of course, if the result actually exists within the study population.[2] Another key point is that studies with larger sample sizes may highlight results or associations that are statistically significant, but may possess no clinical significance of any kind. So, when examining/interpreting data that are collected, an investigator should be critical of several things—*P*-value certainly being important, but also the confidence interval, and if there is a violation of aspects that would deem data/results not statistically significant, for instance if the confidence interval crosses a value of 0 (the null value).

While designing a study, as well as while interpreting data, it is important to not only have an appropriately large sample size to sift out what you are looking for, but you must also keep in mind several other critical factors.

Examples

Let us consider an example of a 5-year randomized multicenter clinical trial study that finds that there is a statistically significant difference of 0.01 with regard to ankle-brachial index in two groups of patients treated by two different revascularization techniques, with a *P*-value of 0.03, and a 95% confidence interval [0.005, 0.02]. While this result may have been "statistically significant" by the number crunching, is this clinically significant? No.

Let us consider a different example. Let us say that there is single-center study looking at 10 patients, which finds that a new revascularization technique increases ankle-brachial index by 0.8 compared to a routinely used revascularization technique, with *P*-value 0.05 and 95% confidence interval [−0.5, 1.1]. In this example, as the confidence interval crosses 0 (the null value), this raises concern that there is no actual statistically significant difference between the two techniques. There are some reasons that may explain why there seems to be a clinically significant difference, one possibility being that this study's sample size is too low, and that if the study were carried out for a longer period of time with a larger sample size, the results may demonstrate statistical significance.

Another point to keep in mind is that if you have a very large sample size, let us say 50,000 patients in a multi-institutional study, this large group will likely include a wider variety of patients with regard to races/ethnicities, ages, socioeconomic backgrounds, genetic predisposition/makeup, etc. Now, let us hypothetically say that there is a rare complication associated with a certain brand of drug-coated stent that is seen in approximately 0.01%, or 1 out of every 1000 patients. In our example of the study with 50,000 participants, this complication would theoretically occur in approximately 50 people. So, if the goal of a study was to evaluate this rare complication, it would be very difficult to do so in a single-center study at a small hospital that may only utilize this certain type of drug-coated stent infrequently and let us say it might have 175 patients where this specific drug-coated stent was utilized. This small center may never see this rare complication occur, compared to the aforementioned larger study. Hence, the larger study with the larger sample size is clearly better suited to examine this rare complication, and try to figure out why those 50 patients developed the complication, be it something specific with regard to a certain gene, perhaps predisposition to a particular race/ethnicity, geographic location, environmental factors, etc, and hence, further studies could be designed to examine that particular subgroup of patients to elucidate the reason for the complication. In summation, a larger sample size is better for studying a rare complication compared to a smaller sample size.[2]

Conclusion

In summary, sample size determination is of great importance when designing a research study, and as discussed within this chapter, there are several factors and nuances that must be kept in mind and carefully considered.

Resources

Sashegyi A, Felli J, Noel R. *Benefit-Risk Assessment in Pharmaceutical Research and Development*. CRC Press; 2014.

References

1. Van Den Broek J, Brestoff J. *Epidemiology: Principles and Practical Guidelines*. Springer; 2013.
2. Porta M. *A Dictionary of Epidemiology*. 5th ed. Oxford University Press; 2014.
3. Kestenbaum B. *Epidemiology and Biostatistics: An Introduction to Clinical Research*. Springer; 2009.
4. Jenicek M. *Foundations of Evidence-Based Medicine: Clinical Epidemiology and Beyond*. 2nd ed. CRC Press; 2020.

Further reading

1. Sashegyi A, Felli J, Noel R. *Benefit-Risk Assessment in Pharmaceutical Research and Development*. CRC Press; 2014.

Chapter 68

Budgeting

Ryan Bitar and Geoffrey D. Clarke
University of Texas Health Science Center San Antonio, San Antonio, TX, United States

Naturally, translational research requires funding. This reality is not exclusive to IR translational research, but rather nearly all facets of research and in all fields of medicine. Most research projects receive funding in the form of grants, a defined sum of money allocated for an exact purpose and particular length of time, although some can be supported by contracts to perform specific tasks. Magnitude and allotted duration of grant funds vary based on the awarding body and grant category; however, the acquisition of grant funding is dependent on a scientifically strong grant application supported by a justified budget. The awarding body must be assured that the award recipient will use the funds properly. Thus, it is important to construct a budget that clearly demonstrates the intended utilization of funds by enumerating the categories/types of expenditures involved. Once awarded, the recipient, or principle investigator (PI), is responsible for ensuring that funds are expended in a timely and appropriate manner.[1] This chapter discusses the intricacies behind constructing a grant budget and some general policies regarding proposals and management.

Given the investigator has an estimate of the overall cost and duration of a project, it is important to select an organization whose mission coincides with the researcher's proposal/interests and select a mechanism for which the proposal is eligible. Additionally, the investigator should determine whether the budget limits of the funding mechanism would be sufficient to complete the project.[2] With a budget of over $40B, the National Institutes of Health (NIH) serves as the most important funding source for biomedical research in the United States. The NIH consists of over 20 different institutes and centers, including the National Centers for Advancing Translational Sciences ($833M budget authority in 2020), although translational research proposals are accepted by most of the Institutes.[3] Within the NIH, a multitude of grant types exists with different amounts of funding, based on the scale and expenditures of the project. Investigator-initiated translational research projects typically utilize the R series (Research Grant) mechanisms listed in Table 68.1.[4] The National Science Foundation is another important federal agency (>$7B in 2020) but it funds fundamental biomedical research, rather than applications of scientific knowledge like clinical research.[3] While the Department of Defense presents opportunities for generous financial support for biomedical research, the project must be deeply related to a military objective (i.e., Gulf War illness, skin burns) or "hot" medical topics such as epilepsy and cancer.[5] Other federal agencies, including the Health Resources and Services Administration, the Centers for Disease Control, and the National Aeronautics and Space Administration also fund biomedical projects. Private foundations with specialized interests, such as the American Cancer Society and the American Heart Association, may also provide support for translational research projects, but their grant mechanisms usually have shorter durations and lower maximum budgets than those of federal agencies.[3]

Generally, funding agencies request a detailed budget regarding the expenditures of the project. In constructing a budget, one must follow two fundamental principles: The budget must pose a reasonable request for funds required for completing the project in a suitable amount of time, and it must itemize funds into subdivided categories to justify the necessity for the funds. An exception is the NIH's modular budget format for totals of <$250,000 of direct costs per year (in modules of $25,000) for some applications that do not require a full detailed budget.[6] Most often funding agencies will clearly state a budget limit for grant programs; however, in cases where they do not, the PI should directly contact the program administrator to determine an appropriate budget limit for the specific project.[7] Grant budgets are divided into two distinct categories: direct and indirect costs. Direct cost budget categories include personnel salaries, fringe benefits, equipment, materials and supplies, publication costs, consultant services, computer services, participant incentives, shipping, travel, and trainee costs

Translational Interventional Radiology. https://doi.org/10.1016/B978-0-12-823026-8.00076-6

TABLE 68.1 Listing of Research Grant Programs for which investigators may consider to pursue funding.

R01	Research Project Grant Program
R03	Small Grant Program
R15	Academic Research Enhancement Award
R21	Exploratory/Developmental Research Grant Award
R41/42	Small Business Technology Transfer
R43/44	Small Business Innovative Research
U01	Research Project Cooperative Agreement

(tuition/fees/health insurance). Indirect costs, which cover infrastructure expenses, are not directly related to the project itself, but are required to conduct research as a whole, such as overhead fees, maintenance, and rent if applicable. Indirect costs are typically calculated as a percentage of the direct costs on a basis that is negotiated between the funding agency and the institution submitting the application.[6] Based on these principles, it can be difficult to begin the process of generating an initial draft of a budget proposal. One valid approach is to prepare budget calculations following the completion of the first draft of the methodology. Methods that are cost-effective and have an appropriate sensitivity and specificity for measuring the results and answering the research question should be selected. Essentially, after the scope of the project is defined in regards to an essential set of equipment, personnel, and facilities, appropriate budget calculations can be conducted.[7]

It is important for applicants to establish a strong relationship with their department's grant preparation office. The office will have staff members experienced with the regulatory, budgetary, and documentation requirements of the granting agency and are also familiar with the institution's policies regarding research grant applications. The department also will generally have an individual who will interact with the institution's grant management office, which includes the individual designated by the institution to approve all grant submissions. If a grant includes collaborations with investigators at other institutions, the subcontracts will require approval from each institution's grant office, so the PI should allow plenty of time for this process ahead of the grant deadline.[7] PIs also must ensure their requests are allowed, as some funding agencies will not allow funds for faculty salaries or cover indirect costs.[6] Costs for contracted personnel should be determined on a pro rata basis, considering position, role, and proportion of time they will devote. Benefits such as insurance and housing should be included. Consultants and technical advisors should only include service stipend and be categorized separately from associated facility or equipment fees.[7] Materials and supplies should not only be included, but also subcategorized into areas such as office, lab, or animal. A subcategory of supplies should be created if total costs for the category exceed $1000; otherwise, categories costing less than $1000 may be combined as miscellaneous supplies. Equipment greater than $500 should be listed along with any shipping costs; the necessity for the unit should be specified in the budget justification.[8] Transportation modality, destination, and quantity should be explained. The standard charges for related patient care costs should be acquired from the host clinic/hospital.[9] Once the budget is constructed, a budget justification must follow, as reviewers look favorable upon explicitly detailed budgets to understand the necessity of each requested expenditure. For example, the justification should describe the unique role of each personnel, explain why certain equipment is essential for the project, and indicate the purpose for any travel associated with the project.[8]

After obtaining the funds to conduct Translational Research, four key principles guide the financial management of project funds: The use of funding solely for its intended purpose, regular monitoring of expenditures against the constructed budget, maintaining accurate accounting records, and adequate filing/backing up of financial information. While the PI constructs an initial budget and should maintain continuous awareness of expenditure, generally, accounting staff of the host institution serve a critical role regarding the surveillance of funds related to a project. Purchasing agents will evaluate whether requested purchases follow administrative requirements, such as whether adequate purchase bids have been documented. Accounting personnel usually conduct critical tasks such as monetary and bank handling, charting and cost coding, budget preparation and monitoring, storekeeping and tendering, manage exchange rate variations, dealing with irregularities/fraud, and provide standard forms and protocol for fund expenditures.[3] PIs are required to certify on a monthly basis that all expenditures that occur for a grant are appropriate. A valid strategy is to maintain a balance sheet that not only summarized the current financial position of the project and expenditures, but also includes a cash flow

TABLE 68.2 An example template for which investigators may allocate funding to different components of a proposed project.

Direct costs	Year 1	Year 2	Etc.
Personnel			
Consultant costs			
Equipment (further subdivided)			
Supplies (further subdivided)			
Travel			
Patient care costs			
Other expenses			
Total direct costs			

forecast. To avoid overexpenditures, the PI should compare expenditures against the initial budget near the end of the grant's funding. In the event of overexpenditure of funds, it is generally unfavorable to request additional funding, and deficits must be made up from other accounts. By proceeding with a project with these principles in mind, a strong budget proposal helps to frame the execution of a project while also maximizing the chances of successfully completing a project (Table 68.2).

References

1. Visovsky C. Writing a successful grant: tips and tools. *J Adv Pract Oncol.* 2015;6(3):279–280.
2. NIH. Develop Your Budget. National Institutes of Health Grants & Funding. https://grants.nih.gov/grants/how-to-apply-application-guide/format-and-write/develop-your-budget.htm#personnel. Published March 7, 2019. Accessed October 14, 2020.
3. Ogden T, Goldberg I. The NIH and other sources of research support. In: Ogden T, Goldberg I, eds. *Research Proposals.* 3rd ed. Academic Press; 2002:3–19.
4. NIH. Types of Grant Programs. National Institutes of Health Grants & Funding. https://grants.nih.gov/grants/funding/funding_program.htm#RSeries. Published February 27, 2019. Accessed October 15, 2020.
5. Dept. Of Defense. Program Funded Opportunities. CDMRP. Published September 25, 2020. Accessed October 5, 2020. https://cdmrp.army.mil/funding/archive/archive.
6. Smith PG, Morrow RH, Ross DA, eds. *Field Trials of Health Interventions: A Toolbox.* 3rd ed. Oxford (UK): OUP Oxford; June 1, 2015 (Chapter 8), Preparing grant applications. Available from: https://www.ncbi.nlm.nih.gov/books/NBK305513/.
7. Higdon J, Topp R. How to develop a budget for a research proposal. *West J Nurs Res.* December 2004;26(8):922–929. https://doi.org/10.1177/0193945904269291. PMID: 15539536.
8. Reif-Lehrer L. *Gtan Application Writer's Handbook.* 4th ed. Sudbury, MA: Jones & Bartlett Learning; 2005. V.
9. NIH. Understanding Indirect costs. National Institutes of Health NIAID. https://www.niaid.nih.gov/grants-contracts/understanding-indirect-costs. Published April 19, 2017. Accessed October 6, 2020.

Chapter 69

Ethics and review boards

Brianna L. Gibney
Oakland University William Beaumont School of Medicine, Rochester, MI, United States

- This chapter provides an ethical framework for designing and conducting translational research.
- Ethical considerations for institutional review board approval are explained.
- Unique ethical questions related to IR research are examined.

Why it matters?

Translational research aims to apply findings from human studies to improve clinical decision-making. Each phase of translational research involves ethical considerations. This chapter outlines the basic ethical principles for human subjects research applied to interventional radiology (IR), including securing institutional review board (IRB) approval, getting informed consent, and navigating ethical concerns about emerging technologies.

Ethical principles

The Belmont Report published by the National Commission for the Protection of Human Subjects of Biomedical and Behavioral Research in 1979 has identified the three basic ethical principles of respect for persons, beneficence, and justice as the framework for addressing ethical issues arising from the conduct of research with human subjects.[1]

Respect for persons protects the autonomy of research participants and requires investigators to obtain informed and voluntary consent from all participants involved in a research study. Respect for persons recognizes that all people have the right to voluntarily participate in research after being adequately informed about the nature of the study and reserve the right to discontinue research participation at any time.[1]

Beneficence encompasses the dual obligation of human subjects research to do no harm and to maximize benefits and minimize harms involved in performing the study. Beneficence requires that the benefits of scientific knowledge discovered from the research outweigh the inconvenience and risk experienced by research participants. This involves minimizing the risk of possible physical harm that may occur from the research intervention and psychosocial harm that could occur such as breaches in confidentiality. All individually identifiable health information must be kept confidential in accordance with the Federal Health Privacy Rule (HIPAA).[1]

Justice requires that the benefits and burdens brought about by research are distributed fairly. The selection of research subjects must be scrutinized to ensure that disadvantaged and vulnerable populations are not systematically targeted to bear the burdens of a study and that the development of therapeutic interventions does not provide advantages only to those who can afford them.[1] It must be ensured that the human subjects of a study are from groups likely to be among the beneficiaries of the subsequent applications of the research.

The ethical principles laid out in the Belmont Report provide an important framework that informs each step of the translational research process. These include the development of a research question, establishment of clinical endpoints, design of methodology, study implementation, and analysis. Respect for persons, beneficence, and justice must be protected throughout each phase of the translational study. In the next section of the chapter, the application of these principles will be demonstrated using an example from translational research in IR.

Review boards

Before the onset of a research investigation, the ethical acceptability of a study must be examined. To ensure ethical principles are followed, regulations by the US Food and Drug Administration (FDA) and US Department of Health and Human Services (HHS) require that research with human participants is approved by an IRB.[2] An IRB is a formally designated committee knowledgeable in matters of legal and

Translational Interventional Radiology. https://doi.org/10.1016/B978-0-12-823026-8.00065-1

ethical issues pertaining to research who have the responsibility of reviewing and monitoring human subjects research.[2] IRBs closely evaluate research design, benefits and risks, subject selection, informed consent, confidentiality, security, and privacy.[3] For an IRB to approve a research study, it must determine that risk is minimized, subject selection is just, informed consent is obtained, and confidentiality is adequately maintained in accordance with HIPAA.[3] Fig. 69.1 highlights helpful questions to consider when designing and conducting a translational research study.

For details regarding specific IRB requirements including which studies may be exempt from the full IRB review process, the US Office for Human Research Protections and each institution's IRB should be consulted.

Special considerations for IR research

IR is a newly emerging medical specialty at the forefront of the latest technological innovations in medicine and science.[7] The establishment of IR as a medical specialty has relied on the development of cutting-edge imaging technology using artificial intelligence (AI) and novel medical devices for image guidance. Emerging technologies present unique ethical concerns. Consider for example Table 69.1:

A 55-year-old man is referred for ablation of a renal cell carcinoma abutting the renal pelvis and right hepatic lobe. The plan is to use the practice's standard cryoablation system. The resident consents the patient, listing off the same general risks. A device representative happens to be present for the case and suggests trying their new ablation system, which is FDA approved for lung ablations but has some limited data showing better outcomes for kidneys. The new device is used but malfunctions (unclear if device or user error), causing damage to the renal pelvis and right hepatic lobe. The patient requires a short hospital stay and drain for a urinoma but does well overall. The patient is told this is an expected complication by the clinician as advised by risk management.[8]

This case sparks ethical questions about disclosure when a new approach or device is used, when complications arise, and what constitutes research. How big does a complication have to be for it to be attributed to something beyond an expected complication of the procedure? If an IR uses a newly approved off-label device or performs an untested technique to gain vessel access midprocedure, the IR may want to study these cases retrospectively.[8] How would informed consent be obtained? Is it even necessary?

Another important ethical consideration related to IR research is the increasing use of AI. With emerging technology such as AI, there are extrinsic concerns of justice, access and equality, individual rights and liberties, autonomy, health, safety, and the environment.[9] There are also intrinsic concerns of "playing God," hubris, and respecting nature that should be considered.[9] A study that investigated how an AI framework using magnetic resonance imaging and clinical patient data to predict therapeutic outcomes of transarterial chemoembolization poses unique ethical questions.[10] Is it ethical to rely on AI algorithms to predict

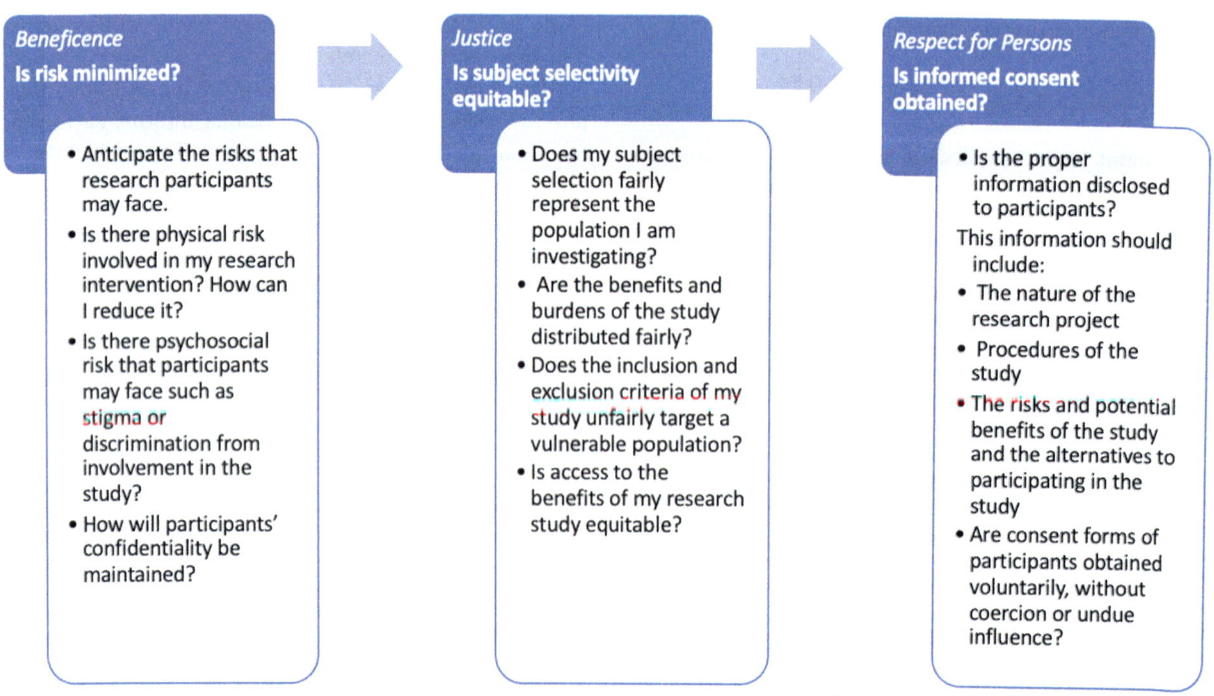

FIGURE 69.1 Ethical considerations in research design[3].

TABLE 69.1 **Informed and voluntary consent**[3,5].

Informed and voluntary consent
Consent forms: Consent forms must disclose the nature of the research project, procedures involved, potential risks and benefits of the study, and the alternatives to participating in the study. IRBs may have template consent forms they prefer investigators to use. IRBs should be consulted on specific institutional requirements. Consent forms used in a research study must be IRB-approved.
Participant comprehension: Technical jargon and complicated language should be avoided in consent forms to avoid confusing research participants. It is recommended that consent forms are written at an 8th grade reading level or below. Participants who lack decision-making capacity and are not able to give informed consent may have a legally authorized representative grant permission of study participation; however, the protocol should be scrutinized to ensure the research study cannot be conducted in a population capable of giving consent.
Voluntary consent: Participant coercion or undue influence such as excessive payments must be avoided. Participants should understand their medical care will not be compromised if they decline participation in the study and that they can withdraw from the study at any time.
Exceptions to informed consent: Informed consent may not be required in the case of scientifically important studies in which carrying out informed consent would be highly difficult or impossible. Consult the IRB for specific details regarding exceptions to informed consent.

patient therapy? Who is liable when the system makes mistakes? To program AI algorithms, large collections of medical data must be used. How will patient confidentiality be maintained using AI? It is impossible to know what emerging ethical dilemmas will arise as translational research pushes the field of IR forward. When designing and conducting translational research in IR, the three basic ethical principles of respect for persons, beneficence, and justice should guide investigators, but challenging dilemmas will nonetheless require thoughtful planning.

Get started

The ethics of designing and conducting translational research ultimately relies on the ethical judgment of the investigator. Fig. 69.2 provides a basic framework of ethical factors to consider when starting the translational research process.

Potential pitfalls

This chapter provided an overview of ethical considerations in translational research specifically related to IR. The potential pitfalls that apply to the ethical integrity of translational research in general, apply to IR. In the early phases of translational research, the primary ethical pitfalls are related to traditional scientific research ethics, such as study design and animal bioethics. Once the investigation advances to the clinical world, potential pitfalls include unfair resource allocation, breaches in data confidentiality and research misconduct in carrying out the study.

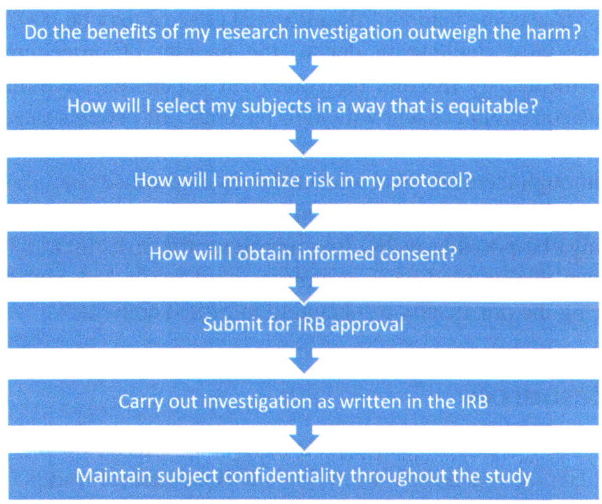

FIGURE 69.2 Basic steps to ethical study design.

Real-world examples

To highlight important considerations of the IRB process, an IR investigation comparing postoperative ovarian function following uterine artery embolization (UAE), myomectomy, and hysterectomy will be examined.[4] Fiftyfive patients with symptomatic uterine leiomyomas were prospectively enrolled in the study. Thirty-three patients underwent UAE, 7 underwent myomectomy, and 15 underwent hysterectomy.[4] Patients had serum follicestimulating hormone (FSH) and estradiol levels measured preprocedure and at regular follow-up visits for 6 months.[4]

Considering this study involved invasive surgical procedures, the physical risk was an important consideration for the investigators. Care needed to be taken to ensure the benefit of scientific knowledge gained from the study and the benefit to the patient of carrying out the procedure outweighed the risk of not operating at all. Risks of the study were minimized by having each study participant consult with a gynecologist before the study who determined solely based on clinical evaluation which intervention was the best option for each patient. Study participants were women between the ages of 18e50 years old with symptomatic uterine leiomyomas.[4] Selection of study participants was from the population most likely to be among the beneficiaries of subsequent applications of the research, and the benefits and burdens of the study were distributed fairly, thus the criterion of justice was met. The respect for person standard was met by utilizing an informed consent process that adequately informed participants about the nature of the procedure, its risks and potential benefits, and alternative forms of treatment, and assessed their understanding of each of these.

Due to the nature of IR, informed consent may be p0075 especially challenging due to the breadth of procedures performed by IRs and the context in which IRs have discussion about consent with their patients.[6] IR patients range from healthy outpatients to very sick patients in the ICU and although there are ideal informed consent practices based on medical literature, carrying out these practices may not always be possible in IR. It is important for investigators to keep this in mind when writing consent forms and considering the practicality of obtaining informed consent.

Resources

The US Office for Human Research Participants' Website at HHS.gov and each institution's IRB are helpful resources to refer to when considering the ethical acceptability of a translational research study.

References

1. Ryan KJ, Brady JV, Cooke RE, et al. *The Belmont Report*. 1979.
2. *Institutional Review Boards Frequently Asked Questions*. FDA; 2019. Published https://www.fda.gov/regulatory-information/search-fda-guidance-documents/institutional-review-boards-frequently-asked-questions. Accessed June 8, 2020.
3. Hulley SB, Cummings SR, Browner WS, Grady DG, Newman TB. *Designing Clinical Research*. 4th ed. 2013.
4. Hovsepian DM, Ratts VS, Rodriguez M, Huang JS, Aubuchon MG, Pilgram TK. A prospective comparison of the impact of uterine artery embolization, myomectomy, and hysterectomy on ovarian function. *J Vasc Intervent Radiol*. 2006;17(7):1111−1115. https://doi.org/10.1097/01.RVI.0000228338.11178.C8.
5. Informed Consent Guidance - How to Prepare a Readable Consent Form. https://www.hopkinsmedicine.org/institutional_review_board/guidelines_policies/guidelines/informed_consent_ii.html. Accessed June 10, 2020.
6. Keller EJ, Perez AW, Makary MS. Informed consent: beating a dead horse or an opportunity for quality improvement? *J Vasc Intervent Radiol*. 2020;31:139−140. https://doi.org/10.1016/j.jvir.2019.08.010.
7. Marx MV. Evolution of Interventional Radiology and ABR Certification: Birth of the ABR Certificate in Interventional Radiology and Diagnostic Radiology.
8. Bozorghadad S, Newton IG, Perez AW, Makary MS, Keller EJ. Research ethics in IR: the intersection between care and progress. *J Vasc Intervent Radiol*. 2020;31(5):846−848. https://doi.org/10.1016/j.jvir.2020.02.014.
9. Sandler R. *Ethics and Emerging Technologies*. New York: Palgrave Macmillan; 2014.
10. Abajian A, Murali N, Jeanette Savic L, et al. Predicting treatment response to intra-arterial therapies for hepatocellular carcinoma with the use of supervised machine learning-an artificial intelligence concept. *J Vasc Intervent Radiol*. 2018;29:850−857. https://doi.org/10.1016/j.jvir.2018.01.769. e1.

Chapter 70

Regulatory considerations for drugs and devices seeking FDA approval or off-label uses

David Bradley Money[1], Katherine Shin-Ying Fu[2] and Brandon Olivieri[3]

[1]*University of Southern Florida Morsani College of Medicine, Tampa, FL, United States;* [2]*University of Florida College of Medicine, Gainesville, FL, United States;* [3]*Department of Vascular and Interventional Radiology, Mount Sinai Medical Center, Miami Beach, FL, United States*

Key points

1. Highlight the areas of interface between the FDA and interventional radiologists as it relates to getting medical device approval.
2. Give an overview of the three main processes by which a device manufacturer can gain FDA approval (510(k), PMA, HDE).
3. Discuss off-label use of medical devices and what data can be collected when using a device "off-label."

Medical device approval

The understanding of the device development process is imperative to interventional radiology where innovation is at the cornerstone of the specialty. In the United States, medical device development falls under federal regulation, headed by the Center for Devices and Radiological Health (CDRH) at the FDA. When a device is approved, it means that the FDA considers the product to be safe (i.e., the benefits outweigh the risks) and effective (i.e., the device consistently does what it is supposed to) for a particular indication. The FDA lays out a three-tiered classification system for medical devices, found in section 513 of the FDA guidelines (Fig. 70.1). The system is based on the level of risk posed for the patient and the vitality to sustaining human life. As a rule of thumb, increasing the class number from I to III means that the risk to humans is increased and as a result, the regulatory controls increase. By extension, the obliged preclinical data is less when involving class I or II when compared to some class II and all class III medical devices.[1]

As per the FDA, a medical device needs to meet this criterion: "an instrument, apparatus, implement, machine, contrivance, implant or an in vitro reagent that is either recognized in the official National Formulary or the US Pharmacopeia, is intended for use in the diagnosis of disease or other conditions, or the cure, mitigation, treatment or prevention of disease or is intended to affect the structure or function of the body of humans."[2] The FDA is required to use the "least burdensome" approach to approving medical devices, meaning that manufacturers are required to provide only data that are necessary to demonstrate the safety and effectiveness of the device.[3] There are currently three pathways to obtain device approval: the 501(k) clearance process, the premarket approval (PMA) process, and the humanitarian device exemption process. The pathway an entity chooses essentially depends on the device's class in the three-tiered classification system.

Premarket notification, also known as a 510(k), is a type of submission made to the FDA demonstrating that a device is as safe and effective while proving that it is substantially equivalent to a preexisting ("predicate") legally marketed device.[4] That is, for a device to be submitted under a 510(k), there must be a demonstration that the device has the same intended use and technologic characteristics as a predicate device.[5] For the FDA to evaluate the characteristics of a device, manufacturers must provide an overall description of the device design, materials used in construction of the device, and other key technologic features.[6] Interestingly, once a device has been cleared, it can itself serve as a predicate. Devices cleared through the 510(k) process generally do not require a clinical trial, though manufacturers of certain 510(k) devices are now being

Translational Interventional Radiology. https://doi.org/10.1016/B978-0-12-823026-8.00093-6

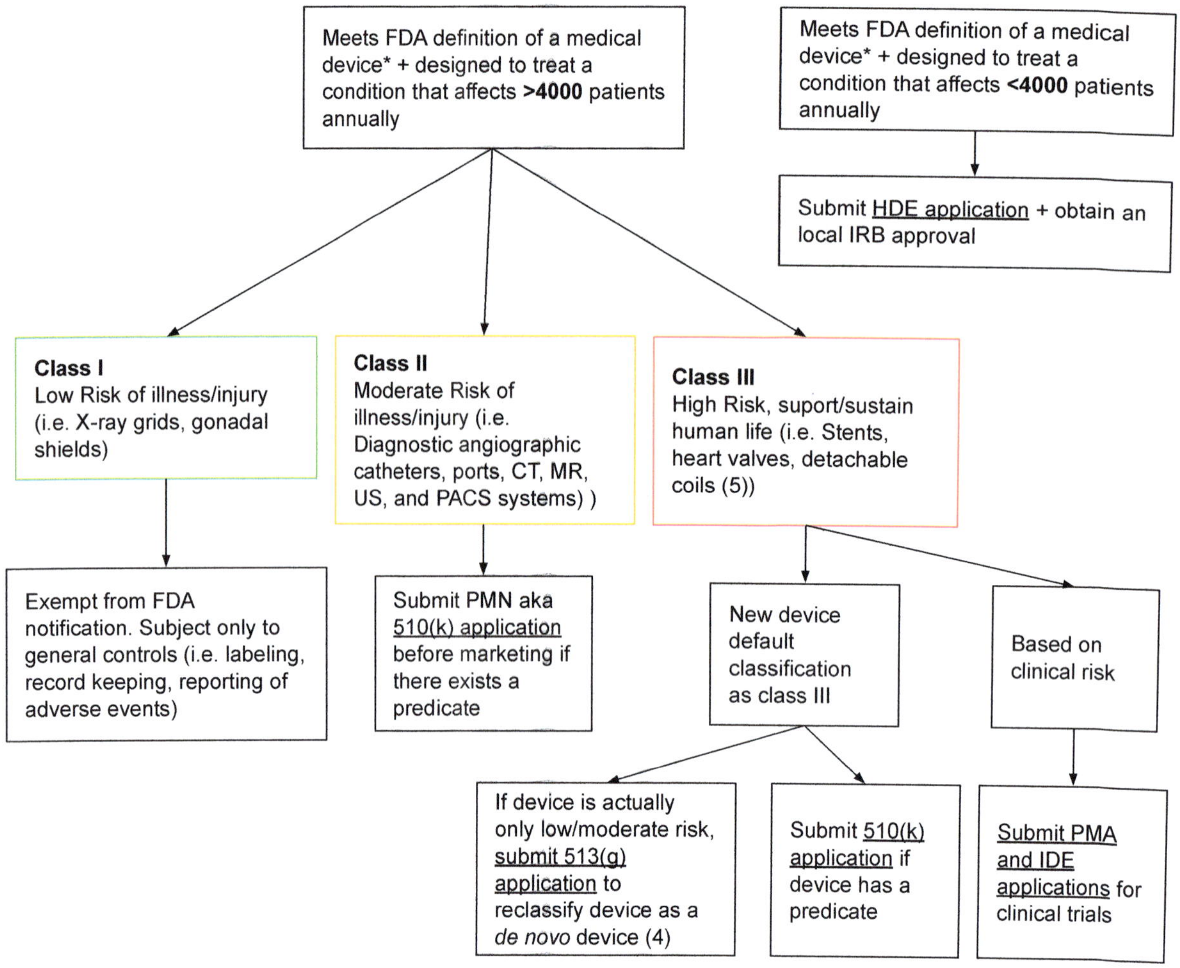

FIGURE 70.1 Process selection is based on FDA three-tiered medical device classification system.

required to provide advanced clinical trial data to support clearance, though still less than the PMA process. Devices that fall into class II, meaning they have moderate risk of illness/injury to the patient, such as diagnostic angiographic catheters are eligible for approval through the 510(k) process. Key advantages of this pathway are that it is generally faster and less expensive than the PMA process.

The PMA process includes rigorous clinical trials to test the safety and efficacy of the device and usually requires several years of invested time and millions of dollars in funding.[7] All class III (i.e., stents) and some class II devices (if they have changed significantly from predicate devices or are seeking a new indication for use) are obliged to undergo this process. Information contained in a PMA submission typically includes a device description and indications for use, manufacturing information, reference to any performance standard or voluntary standard, results of nonclinical laboratory studies, and results of human clinical studies. A PMA submission must demonstrate the safety and efficacy of a device with a thorough performance analysis of bench and clinical evidence data that demonstrate the ability to achieve the intended use of the device. Multiple phases of clinical trials provide the necessary data to support device manufacturing, approval, and adoption.[8] Fig. 70.2 highlights the PMA process, which is derived from the four-step breakdown seen in Kaplan et al.[17]

To initiate the PMA process, the manufacturer may choose to meet with the FDA in a preinvestigational device exemption (pre-IDE) meeting where the preclinical testing data is reviewed and plans for clinical trials are discussed. Following this, a formal IDE application that includes the intended use of the device, preclinical test results, any available clinical data from outside the US, instructions for use, and a proposed trial protocol can be filed, which starts a formal review process will generate data capable of supporting a PMA.[9] The protocol must include a description of the procedure, a standard of care, the steps taken to monitor the effects of the device and minimize risk, proposed endpoints, and the statistical rationale for the proposed sample size.[10] A completed PMA includes the

Stage 1: Pre-investigational Device Exemption/Pre-clinical Evaluation
- Medical device company has series of meetings with FDA/CDRH focused on preclinical testing (bench and animal), as well as plans for clinical trial

↓

Stage 2: IDE/Clinical Testing
- Ethical review of protocols by the IRB
- Centers for Medicare/Medicaid Services (CMS) approval required when treating Medicare patients (approximately 6-12 months timeline)
- Clinical trial performed according to Good Clinical Practices headed by the FDA, including formal site initiation/auditing, core lab data analysis, and independent data analysis

↓

Stage 3: Data Review: Premarket Approval (PMA)
- PMA application is prepared by sponsor/submitted to FDA/CDRH
- FDA/CDRH review process can take as long as 12 months
- FDA/CDRH will stipulate specifics of the indications statement

↓

Stage 4: Postmarket Approval
- After approval, the sponsor is subject to control regulations required under Good Manufacturing Practices guidelines
- FDA Surveillance includes regular inspets to ensure compliance

CDRH= Center for Devices and Radiological Health

FIGURE 70.2 Overview of the PMA process. *Headed by the FDA

analysis of data from the IDE clinical trial as well as the trade name of the device, the registration number of the owner, the device classification, labeling or proposed labels to describe the device and its intended use, directions for its use, and a statement of disclosure.

The FDA recognizes there may not be sufficient market returns for certain diseases or conditions affecting smaller patient populations. For this reason, it endorses the regulation of the Safe Medical Devices Act of 1990 regarding the humanitarian use of devices to encourage research investigations in the treatment of these less represented populations.[11] A device that treats fewer than 4000 patients per year is known as a humanitarian use device (HUD). A HUD is eligible to apply for a humanitarian device exemption (HDE), which is comparable to a PMA. An HDE authorizes the marketing of a HUD and requires a similar application to the PMA, but it is exempt from demonstrating results of scientific efficacy for its intended purpose. HUDs require the approval of both the FDA and the Institutional Review Board (IRB) before use. Labeling for the device must state it is a humanitarian use device, which has been authorized by Federal Law but whose effectiveness has not been demonstrated.[12]

During the approval process, the FDA will inspect the manufacturing laboratories and facilities where the device will be made to check for good manufacturing practices. If appropriate, the FDA will consult an advisory committee that consists of a group of experts who provide the FDA with independent advice on an issue. The panels recommend whether a product should be approved or not.

Off-label use of medical devices

A trademark of interventional radiology is the innovative solutions that arise while troubleshooting complications during existing procedures. Amid these efforts, the use of nonapproved or off-label devices is common. "Off-label use" of a device or drug signifies it is being applied outside the realm of its FDA-approved indication.[13] Importantly, the FDA does not prohibit the use of off-label devices, and while it governs the marketing of medical products, it does not govern the use of them in clinical practice. The January 1998 FDA Guidance Document for "Off-Label and Investigational Use of Marketed Drugs, Biologics, and Medical Devices" acknowledges the value of off-label device applications in day-to-day practices.[14] When a physician uses a marketed device outside of its intended use in the absence of research efforts, an IRB review is not required. In the interest of documentation, physicians may choose to retrospectively report and publish the results of their off-label experience, so long as manufacturers do not attempt to alter the way the product is marketed or promoted.[15]

Emergency situations and compassionate use

Aside from investigational use, a nonapproved FDA device may be used for emergency situations and "compassionate use." For emergency use, a physician must conclude a patient has a life-threatening condition that requires immediate treatment with no existing, alternative treatment. Under these terms, an investigational device may be applied outside of the originally approved IDE protocol and before FDA approval. The term "compassionate use" applies to situations in which the patient has no other option but an investigational device for the management of a serious or life-threatening condition. In contrast to emergency situations, compassionate use warrants FDA approval of an IDE supplement before proceeding.[16] For either circumstance, the experience can be retrospectively documented, but not used for research purposes (Fig. 70.3).

Get started

The process of getting a medical device FDA-approved is tedious, but understanding the basic ins and outs of the FDA processes to approval can make the experience go much smoother. It is vital to find out which class the proposed device falls under in the three-tiered classification system to know which approval process will be best suited. If there exists a predicate device, the 510(k)-application process can save considerable time and money. Other devices that do not have a predicate will need to follow the PMA process outlined, which can be time consuming.

Understanding that the FDA does not prohibit the use of off-label devices on patients is key. While the FDA governs

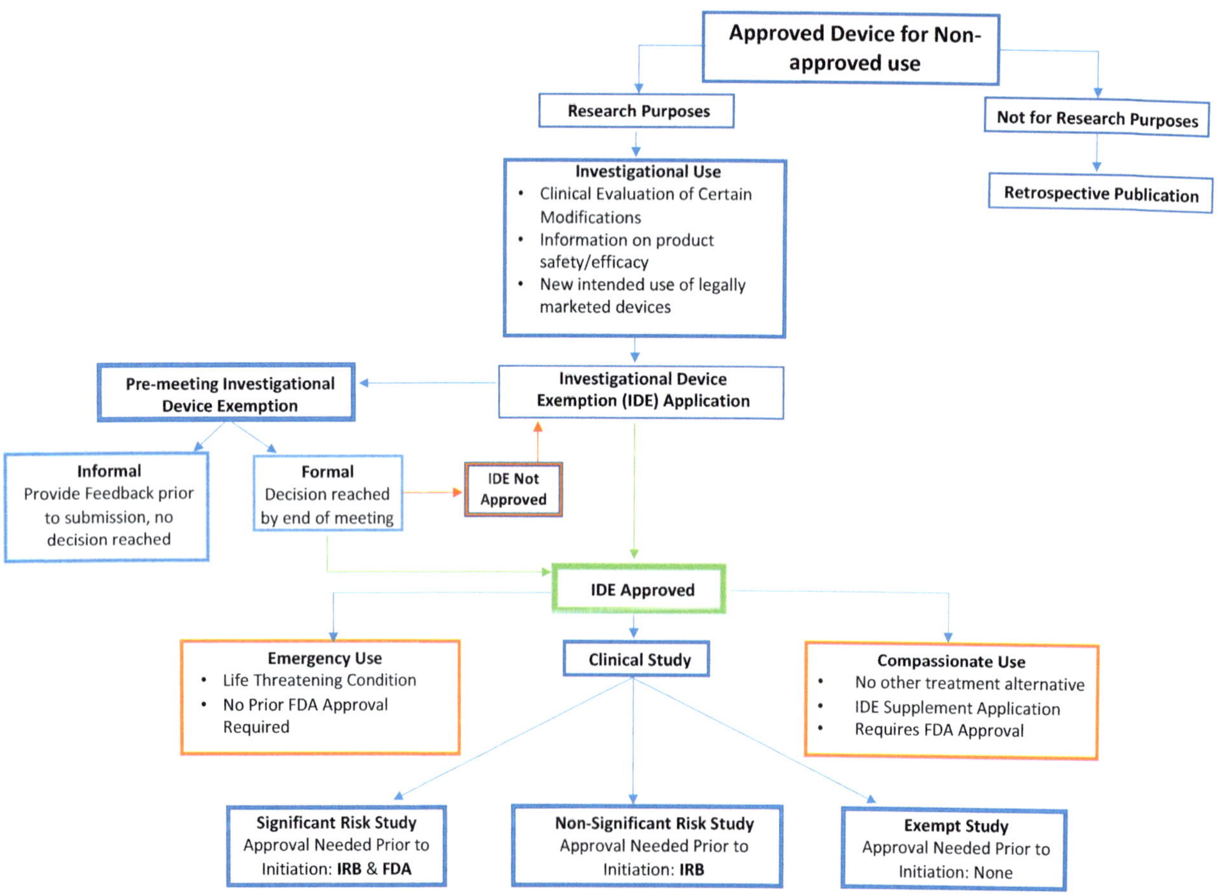

FIGURE 70.3 Guide for conducting research for a non-approved FDA device.

the marketing of medical products, it does not govern the use of them in clinical practice, which can be useful in emergent-type situations.

Potential pitfalls

The process to getting a medical device FDA-approved is tedious but understanding the basic ins and outs of the FDA processes to approval can make the experience go much smoother. It is vital to find out which class the proposed device falls under in the three-tiered classification system to know which approval process will be best suited. If there exists a predicate device, the 510(k)-application process can save considerable time and money. Other devices which do not have a predicate will need to follow the PMA process outlined which can be time-consuming. Understanding that the FDA does not prohibit the use of off-label devices on patients is key. While the FDA governs the marketing of medical products, it does not govern the use of them in clinical practice which can be useful in emergent-type situations.

Real-world examples

PMA Process: The Gore Viabahn VBX Balloon Expandable Endoprosthesis, falling under a class III medical device, submitted its PMA (PMA#P160021) on 01/05/2016 and W.L Gore & Associates, Inc. received its FDA notice of approval on 01/27/2017. It indicated for the treatment of de novo or restenotic lesions found in iliac arteries with reference vessel diameters ranging from 5mm−13mm and lesion lengths up to 110mm, including lesions at the aortic bifurcation. In the PMA, it showed *In vitro* bench testing, sterilization, packaging, product shelf life, animal, and biocompatibility testing, along with the clinical studies discussed above, provided reasonable assurance that the GOREÒ VIABAHNÒ VBX Balloon Expandable Endoprosthesis is safe and acceptable for clinical use, in accordance with regulatory and industry standards and guidance documents. The VBX FLEX prospective, multicenter clinical study demonstrated safety and effectiveness of the GOREÒ VIABAHN VBXÒ Balloon Expandable Endoprosthesis since the composite MAE of 2.3%, with the upper limit of the 95% confidence interval at 6.5%, was less than the performance goal of 17.0% ($p<0.001$). This performance goal was established based on previous iliac stenting study outcomes. Similar iliac stenting studies reported composite MAEs ranging from 0−15% at 9 months (18)

HDE Process: A target demographic for utilizing the HDE pathway in IR is pediatric medical devices. The limited sample size and individual heterogeneity makes obtaining safety data and regulatory approval difficult for this patient population. Reports in pediatric interventional cardiology estimate greater than 60% of the cardiac catheterization cases are exposed to off-label use of adult devices. (19) In attempt to appease this issue, the US legislation passed the "Pediatric Medical Device Improvement and Safety Act" in 2007, which encouraged data-sharing registries for medical manufacturers to increase development as well as permit pediatric device makers to generate a profit for HDE devices. A cohort study analyzing the approved FDA pediatric devices between January 1, 2008 and December 31st, 2011 found three approved devices via the HDE program, including but not limited to Medtronic Melody® Transcatheter Pulmonary Valve in 2010 and the Berlin Heart EXCOR® Pediatric VAD for bridge to transplant in 2011 (20,21) A more recent example, in 2018 the FDA granted HDE approval in adults for the PK Papyrus Covered Coronary Stent System indicated for the treatment of acute perforations of native coronary arteries and coronary bypass grafts in vessels 2.5 to 5.0mm in diameter. (22)

Off-Label Use: A study published in 2016 compared the outcomes of Endurant II (Medtronic) stent graft used under instructions for use (IFU) versus off-label use in high-risk patients who were deemed unfit for conventional surgery. The study retrospectively analyzed data from patients who underwent endovascular aneurysm repair (EVAR) with Endurant II stent-graft from December 2012 to March 2015. The study concluded the device to be safe in an off-label setting and was justified in patients considered unfit for conventional surgery. This published data now serves as scientific evidence and support for future off-label applications of Endurant II (Medtronic) stent graft and similar devices for patients presenting with aortic abdominal aneurysms that are considered non-ideal surgical candidates.

Resources

1. Norman GAV. Drugs, devices, and the FDA: Part 2. *JACC: basic to Transl Sci.* 2016;1:277−287.
2. Federal Food and Drug and Cosmetics Act 21 U.S.C. 321(h).
3. Smith JJ, Shyjan AM. Defining "least burdensome means" under the Food and Drug Administration Modernization Act of 1997". *Food Drug Law J.* 2000;55:435−447.
4. Frank RA, Rucker DW, Ferguson MA, Sweeney TJ. Evidence requirements for innovative imaging devices: from concept to adoption. *J Am Coll Radiol.* 2011;8:124−141.
5. Maisel WH. Medical device regulation: an introduction for the practicing physician. *Ann Intern Med.* 2004;140:296−302.
6. Center for Devices and Radiological Health. U.S. Food and Drug Administration. The510(k)Program:

Evaluating Substantial Equivalence in Premarket Notifications (510(k)): guidance for Industry and Food and Drug Administration Staff. July28,2014. White Oak, MD: US Food and Drug Administration; 2014.

7. Adamovich, A, Park S, Siskin G, *et al.* The ABCs of the FDA: a primer on the role of the United States Food and Drug Administration in Medical Device Approvals and IR Research. *J Vasc Intervent Radiol.* 2015;26:1324–1330.

8. Frank RA, Rucker DW, Ferguson MA, Sweeney TJ. Evidence requirements for innovative imaging devices: from conceptto adoption. *J Am Coll Radiol.* 2011;8:124–141.

9. Abel DB, Farb A, Moynahan M. FDA response to medical deice regulatory challenges. *Endovascular Today.* March 2013:57–61

10. Chen EA, Patel-Raman SM, O'Callaghan K, Hillebrenner MG. FDA's perspectives on cardiovascular disease. *J Cardiovasc Trans Res.* 2009;2:143–146.

11. https://www.fda.gov/medical-devices/device-approvals-denials-and-clearances/hde-approvals. Accessed June 07, 2020.

12. Fda.gov. 2006. *Information Sheet Guidance For Irbs, Clinical Investigators, And Sponsors.* https://www.fda.gov/files/about%20fda/published/Frequently-Asked-Questions-About-Medical-Devices—Information-Sheet.pdf. Accessed June 4, 2020.

13. Hillman BJ. The FDA and radiologists: momentum for change. *J Am Coll Radiol.* 2005;26:459–460.

14. https://www.fda.gov/regulatory-information/search-fda-guidance-documents/label-and-investigational-use-marketed-drugs-biologics-and-medical-devices.

15. Abel DB, Farb A, Moynahan M. FDA response to medical deice regulatory challenges. *Endovasc Today.* March 2013;57–61.

16. Fda.gov. 2006. *Information Sheet Guidance For Irbs, Clinical Investigators, and Sponsors.* https://www.fda.gov/files/about%20fda/published/Frequently-Asked-Questions-About-Medical-Devices—Information-Sheet.pdf. Accessed June 4, 2020.

17. Kaplan AV, Williams D O. Medical device regulatory landscape. *Circul Cardiovas Intervent.* 2012;5:2–5.

18. https://www.accessdata.fda.gov/cdrh_docs/pdf16/P160021B.pdf. Accessed June 06, 2020.

19. Sutherell JS, Hirsch R, Beekman RH, III. Pediatric interventional cardiology in the United States is dependent on the off-label use of medical devices. *Congenit Heart Dis.* 2010;5(1):2–7.

20. Yousufi U, Levi DS. Utilizing devices from interventional radiology in congenital heart disease. *Intervent Cardiol.* 2013;5:113–129.

21. Almond CS. The FDA review process for cardiac medical devices in children: a review for the clinician. *Prog Pediatr Cardiol.* 2012;33(2):105–109. https://doi.org/10.1016/j.ppedcard.2012.02.002.

22. https://www.accessdata.fda.gov/scripts/cdrh/cfdocs/cfhde/hde.cfm?id=H170004. Accessed June 16, 2020.

23. Pecoraro F, Corte G, Dinoto E, Badalamenti G, Bruno S, Bajardi G. Cinical outcomes of Endurant II stent-graft for infrarenal aortic aneurysm repair: comparison of on-label versus off-label use. *Diagn Interv Radiol.* 2016;22(5):450–454. https://doi.org/10.5152/dir.2016.15418.

References

1. Norman GAV. Drugs, devices, and the FDA: Part 2. *J Am Coll Cardiol.* 2016;1:277–287.
2. Federal Food and Drug and Cosmetics Act 21 U.S.C. 321(h).
3. Smith JJ, Shyjan AM. Defining "least burdensome means" under the food and drug administration modernization Act of 1997". *Food Drug Law J.* 2000;55:435–447.
4. Frank RA, Rucker DW, Ferguson MA, Sweeney TJ. Evidence requirements for innovative imaging devices: from concept to adoption. *J Am Coll Radiol.* 2011;8:124–141.
5. Maisel WH. Medical device regulation: an introduction for the practicing physician. *Ann Intern Med.* 2004;140:296–302.
6. Center for Devices and Radiological Health. U.S. Food and Drug Administration. The510(k)Program: evaluating substantial equivalence in premarket notifications (510(k)). In: *Guidance for Industry and Food and Drug Administration Staff. July28,2014.* White Oak, MD: US Food and Drug Administration; 2014.
7. Adamovich A, Park S, Siskin G, et al. The ABCs of the FDA: a primer on the role of the United States food and drug administration in medical device approvals and IR research. *J Vasc Intervent Radiol.* 2015;26:1324–1330.
8. Frank RA, Rucker DW, Ferguson MA, Sweeney TJ. Evidence requirements for innovative imaging devices: from conceptto adoption. *J Am Coll Radiol.* 2011;8:124–141.
9. Abel DB, Farb A, Moynahan M. FDA response to medical deice regulatory challenges. *Endovascular Today. March.* 2013: 57–61.
10. Chen EA, Patel-Raman SM, O'Callaghan K, Hillebrenner MG. FDA's perspectives on cardiovascular disease. *J Cardiovasc Trans Res.* 2009;2:143–146.
11. https://www.fda.gov/medical-devices/device-approvals-denials-and-clearances/hde-approvals. Accessed June 07, 2020.
12. Fdagov. *Information Sheet Guidance For Irbs, Clinical Investigators, And Sponsors*; 2006. https://www.fda.gov/files/about%20fda/published/Frequently-Asked-Questions-About-Medical-Devices—Information-Sheet.pdf. Accessed June 4, 2020.
13. Hillman BJ. The FDA and radiologists: momentum for change. *J Am Coll Radiol.* 2005;26:459–460.
14. https://www.fda.gov/regulatory-information/search-fda-guidance-documents/label-and-investigational-use-marketed-drugs-biologics-and-medical-devices.
15. Abel DB, Farb A, Moynahan M. FDA response to medical deice regulatory challenges. *Endovasc Today.* March 2013:57–61.
16. Fdagov. *Information Sheet Guidance for Irbs, Clinical Investigators, and Sponsors*; 2006. https://www.fda.gov/files/about%20fda/published/Frequently-Asked-Questions-About-Medical-Devices—Information-Sheet.pdf. Accessed June 4, 2020.
17. Kaplan AV, Williams DO. Medical device regulatory landscape. *Circul Cardiovas Intervent.* 2012;5:2–5.

Chapter 71

Funding approaches

Pranav Moudgil and Jatin Sharma
Oakland University William Beaumont School of Medicine, Rochester, MI, United States

Key points

- The three major sources of funding are: government, private organizations/foundations, and industry
- Strong grant applicants know their audience, explain why their idea is testable and important, and are persistent
- Weak grant applicants lack awareness of the project's greater significance, have overambitious research design, and fail to recognize potential pitfalls and their remedies

Why it matters?

Research funding is the proverbial oil that lubricates the cogs of scientific innovation and acquiring funding for interventional radiology studies is no different than for other academic disciplines. In this chapter, we discuss funding sources and how they present applicants with unique advantages and disadvantages, what makes for a good grant application and what makes for a poor grant application, and how you can position yourself to be in the former rather than the latter. Ultimately, most research will not just require an abundance of time, effort, and dedication but also the capital to invest in raw materials, devices, hardware, and software, and other design components to take an idea from the bench to the bedside.

For the purpose of this chapter, we will focus on public agency funding from sources within the United States. The major players when it comes to this form of grant funding are federal agencies, including and most notably, the **Department of Health and Human Services,** which is the encompassing agency for the **NIH**.[1] For nonmedical fields, including engineering, an independent agency exists, which is the **NSF**.[2] The NSF alone has an annual budget of $8.3B (FY2020) and provides 25% of the funding for basic science research in the United States.[2] The NIH, in turn, provides $41.7B annually in medical research for the US population through approximately 50,000 competitive grants.[3] These grants have specific "activity code" identifiers that are commonly used terms thrown around by researchers. Research project grants, also known as **RO1**, are the most commonly used grant program and are generally awarded for discrete projects that encompass 3–5 years worth of time. **RO3**, or NIH's Small Grant Program, provides short-term funding for smaller projects such as feasibility studies or secondary analysis of an existing study and is generally for $50,000 a year and is not renewable.[4] Another notable series of grants awarded by the NIH are the **K-award** series, also known as the career development awards, which include, for example, the K08 Mentored Clinical Scientist Research Career Development Award.[4] These types of awards provide funding and protected time for intensive, supervised research career development experience to promising clinicians with demonstrated aptitude. Knowing about the various grant programs and what their stated purpose and mission are will help the grant applicant in tailoring his or her grant toward the activity code best suited to provide funding and support for his or her specific project. As an example, a landmark IR paper from a group at Northwestern that demonstrated Y90-radioembolization prolonged time to progression versus conventional TACE in patients with HCC was funded, in part, by an NIH RO1 (CA126809).[5]

Numerous private agencies exist to also provide research funding and oftentimes these entities are nonprofit organizations. Private agencies vary widely in mission and in scope when it comes to granting research funding. However, as a researcher, it is in your best interest to maximize your chances to find funding. It is also important to keep in mind that though grant applications open and close during various times of the year, it also takes lengthy periods of time to hear back from prospective funders. Private agencies, though often smaller in scale compared to

Translational Interventional Radiology. https://doi.org/10.1016/B978-0-12-823026-8.00048-1

their public counterpart, are a common source for seed funding for pilot studies that can eventually be brought to scale to apply for a larger public grant. Because these agencies tend to be smaller and more focused on a particular field of medicine, one can better tailor the grant application to suit the mission of the organization. Also, because the funding source is by nature smaller, the number of potential grant applications tends to be smaller, therefore raising the likelihood of being granted funding. A resource very beneficial to IR research is the SIR Foundation, which supports its members with various grant programs aimed at advancing IR clinical practice.[6] The added benefit of receiving funding from an organization such as this, is the added networking benefits of having your grant application reviewed by leaders in your field of subfield of interest, thereby advancing your research and career aspirations in other ways besides just funding support. A real-life example would be the PRESERVE IVC Filter trial that is funded by a joint collaboration of the Society of Interventional Radiology and the Society of Vascular surgery called the IVC Filter Group Foundation.[7]

The field of interventional radiology is very fortunate to have a staunch industry backing for both its clinical and research-based missions. As you know, interventional radiology is a field of innovation and is constantly changing. This is by no means an accident, and much of our field's advancement is due to the direct relationship between our physicians and scientists and the private industry. Industry support comes in various shapes and sizes. Several of our field's annual and national meetings are heavily attended by industry device companies, and they help provide funding for those meetings. Additionally, device companies have helped provide funding for various clinical trials and also donate heavily to the aforementioned SIR Foundation. Many industry representatives are very friendly with physicians and scrub into our cases to help us use their latest technology. Additionally, they have various programs to support residents and fellows in training with dedicated workshops. When it comes to applying for research funding from private industry, it is important to remember that there can be an inherent conflict of interest. Many of the companies provide funding in an intervention or subfield where they have a preexisting product or a potential product that they would like to bring to market. Generally speaking, this is not problematic and companies are adept at minimizing this conflict with their grant recipients. Unlike the standardized processes of a public or private agency, each industry source will have its own grant application process but generally follows a similar format. Numerous IR research has been funded through industry sponsors and this relationship of physicians and industry is likely to only grow further with time and should be taken advantage of when appropriate and possible. As an example, the ZILVERPASS trial that compared the use of a paclitaxel-coated stent versus surgical bypass in patients with femoropopliteal lesions was funded largely by Cook Medical.[8]

Writing a grant application is a time-consuming and involved process but it is also a learnable skill. There are discrete criteria that reviewers use to evaluate grants with a raw number score assigned to federal grant applications.[9] Many times a percentile score is attached to the grant as well.[10]

The most important first step you can take to writing a successful grant application is to find people or organizations within your institution that have successfully secured grant funding in the past.[11] They can act as experienced mentors to guide you through the process, proofread your work, and offer their own insight as to what has worked well for them in the past.

Another component of a successful grant application is a clear, testable idea that asks a scientific question that is deserving of funding.[12] Reviewers see hundreds, if not thousands, of applications and are looking to fund innovative ideas that advance knowledge within a field of science or remove a barrier to progress within that field. Questions or ideas that stand to advance knowledge incrementally are less likely to receive funding. Once an innovative project idea is selected it is helpful to frame the research objectives within the **SMART** framework to clearly identify how your scientific question is specific, measurable, achievable, realistic, and time sensitive.[11]

A large percentage of grants are not immediately accepted but instead returned to applicants with reviewer comments.[12] A successful grant application uses these comments as constructive criticism and addresses all of them during the resubmission process. The changes mentioned in the review comments should either be implemented within the grant or addressed in a manner that explains how that change has already been enacted.

Weak grant applications have distinguishing features that reviewers are able to identify. Evidence of a poor literature review is one such weakness.[13] When formulating a research question, it is vital to conduct a thorough literature review to establish whether the question has already been answered. For example, applicants may use the **NIH RePORTER** or **HHS TAGGS** search tools to query existing funded research projects.[14,15] Reviewers are unlikely to approve a grant in which the primary question has already been answered. Even if the scientific question is unique, proper integration of the existing literature into the grant is necessary to provide context for the question being investigated (Table 71.1).

Another weakness that is apparent to reviewers is the lack of a focused problem or hypothesis. It is necessary to provide both a specific and measurable question as well as a scientific hypothesis. Reviewers like to see experimental designs informed by hypotheses as opposed to access to population, method, or technology being the driver for the

TABLE 71.1 Table of public agencies.

Public agency	Location	Website
National Science Foundation (NSF)	4201 Wilson Boulevard Arlington, Virginia 22230	www.nsf.gov
National Institutes of Health	9000 Rockville Pike Bethesda Maryland 20892	www.nih.gov
National Cancer Institute	6130 Executive Boulevard Rockville, Maryland 20852	www.cancer.gov/
National Heart, Lung, and Blood Institute (NHLBI)	6701 Rockledge Drive, MSC 7956 Bethesda, Maryland 20892-7956	https://www.nhlbi.nih.gov/
National Institute of Diabetes and Digestive Kidney Diseases (NIDDK)	31 Center Drive, MSC 2560 Bethesda, Maryland 20892-2560	www.niddk.nih.gov
Agency for Healthcare Research and Quality	2101 East Jefferson Street, Suite 501 Rockville, Maryland 20852	https://www.ahrq.gov/funding/index.html

grant.[13] Hypothesis-driven projects are seen as more likely to result in peer-reviewed manuscripts (Table 71.2).

Grant reviewers will also critically examine whether your research team has the expertise to conduct the project that you are proposing.[13] Applications with a strong idea or hypothesis can be weakened by a research team with a lack of expertise in the experimental field. Proposals should highlight why their teams are uniquely positioned and qualified in their fields to answer the proposed question (Table 71.3).

Get started

In summary, the take-home action items the reader should focus on after reading this section are as follows:

- Review your current research project and take a step back from a bird's eye view and recapture the greater significance of your work and why it is important to be funded
- Ensure your idea has a testable hypothesis and importantly, can be tested in the allotted time for which you are approaching the agency for funding
- Research potential funders and find one which aligns with your goals and which you can see yourself realistically working with on your path to finding the answer to your clinical/laboratory question

TABLE 71.2 Table of private agencies.

Private agency	Location	Website
Society of Interventional Radiology Foundation	3975 Fair Ridge Drive Fairfax, Virginia 22030	www.sirfoundation.org/grants_awards/
American College of Radiology (ACR) Innovation Fund	1891 Preston White Dr. Reston, VA 20191	www.acr.org/Member-Resources/Innovation/FAQ
Radiological Society of North America (RSNA)	820 Jorie Boulevard Oak Brook, Illinois 60521	www.rsna.org/en/research/funding-opportunities
American Roentgen Ray Society (ARRS)	44211 Slatestone Court Leesburg, VA 20176-5109	www.arrs.org/scholarships/
Radiology Alliance for Health Services Research (RAHSR)	820 Jorie Boulevard, Suite 200 Oak Brook, IL 60523	www.aur.org/en/affinity-groups/rahsr

TABLE 71.3 Table of industry grant donors.

Industry company	Location	Website
MedTronic	710 Medtronic Parkway Minneapolis, Minnesota	https://www.medtronic.com/us-en/about/corporate governance/medtronic-charitable-donations.html
Cook Medical	P.O. Box 489, 750 Daniels Way Bloomington, IN 47402-0489, USA	https://www.cookmedical.com/charitable-donations-grants/
Boston Scientific	300 Boston Scientific Way Marlborough, MA 01752-1234	https://www.bostonscientific.com/content/gwc/en-US/corporate-social-responsibility/giving/grants-donations-exhibits.html
BD Bard	1 Becton Drive Franklin Lakes, NJ 07417-1880	https://www.crbard.com/en-US/healthcare-professionals/research-grants

- Make your application reflect that of a successful grant applicant: be persistent, avoid common pitfalls, and know your audience!

References

1. *Grant-making Agencies | GRANTS.GOV*. Grants.Gov; 2020. https://www.grants.gov/learn-grants/grant-making-agencies.html. Accessed June 17, 2020.
2. *About NSF - Overview | NSF - National Science Foundation*. Nsf.Gov; 2020. https://www.nsf.gov/about/. Accessed June 16, 2020.
3. *Budget*. National Institutes Of Health (NIH); 2020. https://www.nih.gov/about-nih/what-we-do/budget. Accessed June 16, 2020.
4. *Types of Grant Programs | Grants.Nih.Gov*. Grants.Nih.Gov; 2020. https://grants.nih.gov/grants/funding/funding_program.htm. Accessed June 16, 2020.
5. Salem R, Gordan AC, Mouli S, et al. Y90 radioembolization significantly prolongs time to progression compared with chemoembolization in patients with hepatocellular carcinoma. In: *Gastroenterology*. Vol 151. Elsevier BV; 2016:1155−1163.e2. no. 6. doi:10.1053/j.gastro.2016.08.029. [Accessed 16 June 2020].
6. *SIR Foundation - Grant Opportunities*. Sirfoundation.Org; 2020. https://www.sirfoundation.org/grants_awards/current-opportunities/. Accessed June 16, 2020.
7. *SIR Foundation - Research*. Sirfoundation.Org; 2020. https://www.sirfoundation.org/clinical-research2/clinical-trials/preserve-trial/. Accessed June 16, 2020.
8. Bosiers M, Setaci S, Donato GD, et al. ZILVERPASS study: ZILVER PTX stent vs bypass surgery in femoropopliteal lesions. In: *J Endovasc Therapy*. 27(2). SAGE Publications; 2020:287−295. https://doi.org/10.1177/1526602820902014. Accessed 16 June 2020.
9. *Scoring & Summary Statements*. National Institute of Allergy and Infectious Disease; 2020. https://www.niaid.nih.gov/grants-contracts/scoring-summary-statements. Accessed June 24, 2020.
10. *R01 Applications by Percentile Score*. National Institutes of Health (NIH); 2010. https://www.nidcr.nih.gov/grants-funding/funded-research/funding-outcomes/r01-applications-percentile-score. Accessed June 24, 2020.
11. *How to Write A Grant that Will Get Funded | AJE*. Aje.Com; 2020. https://www.aje.com/arc/how-to-write-grant-that-will-get-funded/. Accessed June 16, 2020.
12. Lewis Jr , Neil A. Help funders help you: five tips for writing effective funding applications. In: *Science*. American Association For The Advancement Of Science (AAAS); 2019. https://doi.org/10.1126/science.caredit.aaz6040. . Accessed June 16, 2020.
13. Moore C. Common strengths and weaknesses in grant applications. In: *The Clinical Research Education Library*. Vol 4. American Speech Language Hearing Association; 2015:1−50. https://doi.org/10.1044/cred-pvd-path019. no. 1 . Accessed June 16, 2020.
14. *"NIH RePORTER" Research Portfolio Online Reporting Tools*; 2020. https://projectreporter.nih.gov/reporter.cfm. Accessed June 24, 2020.
15. *"TAGGS Site Search" Tracking Accountability in Governments Grant System*; 2020. https://taggs.hhs.gov/SearchSite. Accessed June 24, 2020.

Chapter 72

Subject recruitment: Where, when, and how

Zain Badar[1] and Prakash Muthusami[2]
[1]*Interventional Radiology Department, University Health Network, University of Toronto, Toronto, ON, Canada;* [2]*Department of Diagnostic Imaging, Hospital for Sick Children, Toronto, ON, Canada*

- Subject recruitment and retention are critical for timely and successful completion of a trial. A low power study can result in a type II error.
- Subject recruitment can be categorized into protocol phase, conduct phase, and follow-up period.
- Participants' convenience and individual circumstances must be considered to minimize challenges with follow-up.

Real-world examples

Subject recruitment is a challenging aspect of translational research, specifically in randomized controlled trials (RCTs), where it serves as one of the main reasons for the failure of the study. Investigators often overestimate the available number of patients who meet inclusion criteria for their study (the eponymous "Lasagna's Law" of research).[1] Insufficient patient recruitment can result in delays with a resultant increase in costs and difficulty in maintaining funding, leading to study closure. Additionally, a small sample size compromises the integrity and validity of the trial/study, as it increases the chance of a type II error. Abraham et al. found that patients who meet eligibility criteria for a study often may not want to participate in RCTs either due to patient-treatment preference, complexity of the trial, fear of random assignment, or fear of negative outcome.[2] The results of poor participation in a trial can result in selection bias, or worse, skewed results with implications for translational applicability. Thus, as a researcher, the development of a robust recruitment strategy is essential, keeping in mind the major barriers for recruitment: subject-related, investigator-related, and protocol-related.

Thoma et al. categorizes subject recruitment into the following[3]:

1. Study protocol phase
2. Study conduct phase
3. Study follow-up period

Study protocol phase

This stage outlines the protocol development that can establish the base of patient recruitment. In this stage, researchers categorize their trial based on efficacy (explanatory trial) or effectiveness (management/pragmatic trial).[3] Overall, it is easier to recruit patients for the latter.

An adequate sample size must be achieved for subject recruitment. Thus, it is crucial to account for subject drop out or subjects lost to follow-up while calculating the appropriate sample size. Often, researchers will increase the sample size to account for such loss, with inflation of sample sizes ranging from 10% to 40%. Another strategy to minimize recruitment issues at this stage is to *know the patient population and simplify the study protocol.* In translational research, it is crucial to understand the minimum clinically important difference (MCID) between interventions that would likely change practice. This can be determined via conducting a pilot study before trial initiation or via analysis of previous literature. MCID is also important as it is a component of power analysis and thus has a major impact on sample size and patient recruitment.[3]

Real-world examples and potential pitfalls: recruitment strategies

The cost of the recruitment method and study population should be considered before initiating the trial. Different recruitment methods include media (e.g., television, radio, and newspaper), fliers, mail, telemarketing calls, internet, and physician referrals.[4] Chin Feman et al. stated that

Translational Interventional Radiology. https://doi.org/10.1016/B978-0-12-823026-8.00046-8

referral and flyer recruitment methods were the most effective in their study, yielding the highest enrollment.[4] Furthermore, clinical trial recruitment companies can aid in recruiting patients for a specific trial. These companies can aid in developing a screener, identifying adequate locations, and helping with digital advertising. However, before recruiting such companies, one must consider a budget model since these companies usually work under either a "flat fee" or a "pay for performance" program. Often times, while considering such vendors, researchers can look for additional services provided, including whether the companies provide an in-house marketing team, or if they have worked in a similar therapeutic area, whether the company specializes in a particular phase of trial, or how quickly the company can develop a recruitment strategy.[5]

Thoma et al. identify four different recruitment strategies—all participants recruited with simultaneous initiation of the trial; participants enter the trial in a "batch" at a predetermined time; participants recruited continuously throughout the trial until sample size is achieved; and participants recruited until a specific date.[3,6]

Real-world examples and potential pitfalls: study conduct phase: recruitment and retention

This phase evaluates the inclusion and exclusion criteria as part of the patient screening process. Reevaluating inclusion and exclusion criteria can aid in minimizing recruitment issues. The researcher should perform a thorough analysis on sites with low recruitment and problem-solve to fix pertinent issues, including protocol or trial-specific recruitment issues. Recruitment quotas can be set to avoid low recruitment, with HIPAA-compliant portals to identify underperforming sites. The option of adding new investigators can aid in addressing staff and/or site-specific recruitment issues. Participant retention can be maximized by spending time with patients to answer any detailed questions and addressing any concerns they have about the study. Education of participants before trial initiation, including providing them with written information and an opportunity to ask questions, results in a lower chance of subjects being lost to follow-up.[3,7]

Real-world examples and potential pitfalls: study follow-up period

Participant follow-up is crucial in identifying patterns and for collecting a complete dataset of results. Follow-up visits should ideally coincide with routine visits to the clinic to facilitate patient convenience. Patients unlikely to comply with follow-up should be identified at the outset. These include patients who are unable to complete outcome assessments, need to travel for a lengthy period of time, patients who do not have a permanent home address, or who plan on moving to a different state, city, or country over the course of the follow-up period. The most important point as a researcher is to be able to make every effort to locate a patient lost to follow-up. Techniques to minimize such instances include having physician's office provide reminders before appointments, obtaining next of kin information before enrollment, and reducing the frequency of visits.[3,8]

In conclusion, it is important for a researcher to consider the various factors associated with recruitment before trial initiation. Techniques and strategies mentioned earlier can avoid difficulties encountered during the subject recruitment process.

Get started

Trial initiation can be a cumbersome process. It is essential for a researcher to consider patient recruitment without falling in the trap of over or under-estimation.

Resources

Additional reading

Journals:

- Contemporary Clinical trials.

Articles:

Brahams D. Medicine and the law. Randomised trials and informed consent. Lancet 1988;ii:1033–1034.

Baum M, Zilkha K,Houghton J. Ethics of clinical research: lessons for the future. BrMedj 1989;299:251–253.

Brewin CR, Bradlev C. Patient preferences and randomised clinical trials. BrMed 1989;299:313–315.

References

1. Lasagna L. Problems in publication of clinical trial methodology. *Clin Pharmacol Ther.* 1979;25(5 Pt 2):751–753. https://doi.org/10.1002/cpt1979255part2751.
2. Abraham NS, Young JM, Solomon MJ. A systematic review of reasons for nonentry of eligible patients into surgical randomized controlled trials. *Surgery.* 2006;139(4):469–483. https://doi.org/10.1016/j.surg.2005.08.014.
3. Thoma A, Farrokhyar F, McKnight L, Bhandari M. Practical tips for surgical research: how to optimize patient recruitment. *Can J Surg.* 2010;53(3):205–210.
4. Chin Feman SP, Nguyen LT, Quilty MT, et al. Effectiveness of recruitment in clinical trials: an analysis of methods used in a trial for irritable bowel syndrome patients. *Contemp Clin Trials.* 2008;29(2):241–251. https://doi.org/10.1016/j.cct.2007.08.004.

5. Ryerson N. 15 clinical trial patient recruitment companies (and how to choose one). *Clin Trial Pat Recruit*. https://www.antidote.me/blog/15-clinical-trial-patient-recruitment-companies. Accessed August 9, 2020.
6. Jack WJ, Chetty U, Rodger A. Recruitment to a prospective breast conservation trial: why are so few patients randomised? *BMJ*. 1990;301(6743):83–85. https://doi.org/10.1136/bmj.301.6743.83.
7. Csimma C, Swiontkowski MF. Large clinical trials in musculoskeletal trauma: are they possible? Lessons learned from the international study of the use of rhBMP-2 in open tibial fractures. *J Bone Joint Surg Am*. 2005;87(1):218–222. https://doi.org/10.2106/JBJS.D.01938.
8. Sprague S, Leece P, Bhandari M, et al. Limiting loss to follow-up in a multicenter randomized trial in orthopedic surgery. *Contr Clin Trials*. 2003;24(6):719–725. https://doi.org/10.1016/j.cct.2003.08.012.

Chapter 73

Data management

Christopher Ovanez
Russell H. Morgan Department of Radiology and Radiological Science, Johns Hopkins Hospital, Baltimore, MD, United States

Key points

- In the context of research study design, data management refers to the process of collecting data, what one does with that data once it has been collected, and what happens to this data after it has been sorted, interpreted, and placed into concise graphs, tables, etc.
- Careful and meticulous data management is critical to the success of a research study.
- An institutional review board (IRB) is a group of individuals who review and monitor research studies which involve human subjects. The role of an IRB is to ensure that research endeavors take the necessary steps in order to protect the welfare and rights of human research subjects. An IRB possesses the ability to approve, deny, or require modification of proposed research endeavors.

Introduction

So, what exactly is "data management" in the context of research study design?

The definition is somewhat pliable, but in essence, this refers to the process of collecting data, what one does with that data once it has been collected, and what happens to this data after it has been sorted, interpreted, and placed into concise graphs, tables, etc.

Get started

There are several steps which should be taken to ensure that data is handled appropriately - for instance during the process of gathering, organizing, and interpreting data, as well as safeguarding data to ensure that it is protected.

Discussion/potential pitfalls/ examples

As a critical first step before actually collecting data involving human or animal subjects, the appropriate protocols must be followed at your respective institution and/or governing body to be allowed to access and collect certain data. For researchers at academic institutions, this body will be the institutional review board (IRB), for whom permission must be sought usually in the manner of a proposal that must detail the specifics regarding the type of project you wish to pursue, be it prospective and requiring access to clinical records, research involving human or animal subjects, retrospective-type studies, etc. In my experience, even if you are planning to pursue a project that you might believe will not require IRB approval, it is usually a safe bet to send a submission through the IRB anyway, just so that they can look it over to make sure that you are not missing something that might be an issue during the course of the study and possibly get you into trouble.

Once the appropriate approval has been granted (if applicable to your particular situation), a data collection/ extraction plan is a great idea before the collection of data. A plan should enumerate the data needed to address each research aim. Time should be taken to define variable names, types (numeric, logical, count, etc.) with labels. For longitudinal studies, the plan should include how repeated measurements will be stored. This process will generate a data dictionary that will be critical and affect the quality of research. There also should be a plan as to how to sort subject identifiers (if applicable). This plan should be put together and set into motion as a cooperative effort by

Translational Interventional Radiology. https://doi.org/10.1016/B978-0-12-823026-8.00070-5

members of the research team, for instance, the principal investigator, statistician, and data manager.

Data should be collected and handled in a conscientious and careful manner, and every effort should be made to make sure that it is completely free of error, bias, or manipulation, in order to produce results and interpretations that in essence reflect "the truth," and will ultimately be of significance in the medical/scientific field, and potentially positively influence patient care.

To take a step back, data gathering is not as simple a task as one might think. Of course, you might think it easy to just sit down one day and jot down some numbers into a spreadsheet, but in my opinion, this is a common pitfall. One might be quite zealous in beginning a study, with the process of data gathering being treated as a grueling task that must be accomplished before one can truly "begin" a study. I feel that many people unfortunately have this sort of mindset, and believe that the analysis/interpretation phases and making the pretty-looking graphs, tables, etc. are the most important parts.

A pearl that I can bestow, and one which has served me well, is to take your time before even thinking about starting to gather a single piece of data, and thinking long and hard about what your short-term and long-term goals are. What do I mean by this? Let us say you are planning to embark on a new clinical research project, and you have a fantastic idea and you just cannot wait to start. Hypothetically, let us say that this project is a retrospective study that wants to look at the long-term clinical outcomes of a variety of patient groups treated with a particular type of drug-coated stent versus another. You have come up with a list of patients who have been treated and that you have been following in clinic, and you are excited to start gathering some data to see what you will find. Of course, you might be excited to get that data gathering process started, but, have you stopped to think what type of information you need to be extracting? Sure, you might have a vague idea, but every study is different, and the goals of every study are different.

Many start the process of data gathering without a well-thought solid plan, and in essence, this is something that is quite detrimental, and perhaps something that many people do not really think about, but really should. What do I mean by this? I am trying to say that you need to come up with a concise battle plan of what you plan to accomplish with the data that will be amassed, and what type of data you will be wanting to gather to answer the question(s) that you are posing by embarking on this research project.

A great way would be to sit down and think about all of the nuances of your hypothesis/hypotheses, and start listing them down, so that the data gathering process will go as smoothly as possible. A common issue in data gathering is to start the process of gathering prematurely, and after combing through hundreds of patient charts, realizing that you wanted to gather another piece of data that you think is important, and needing to once again go back through those hundreds of charts—this is inefficient and truly a waste of time. If you had just sat down and made a great plan before gathering the data, this would go a lot more smoothly.

Ok, so that sounds great and all—but you might be asking yourself—how do I come up with a great plan before gathering data to make the most efficient use of my time, efforts, and resources? Glad you asked. My advice is this—you have a study idea (let us use the hypothetical example I listed earlier); you have a list of patients; now, think of all the questions you want to pose, and what you plan to accomplish with this study—in terms of the hypothetical example, to start with the basics, you might want to stratify your study by various patient factors, for instance—age, gender, comorbidities, ethnicity, race, socioeconomic factors, etc., and then think—what would be additional specific pieces of information that I would want to grab from the charts and place into my spreadsheet? You might want to look at additional medications patients are on, any complications from treatment, potential confounders for treatment response, anything specific to your particular study, perhaps in terms of the specific treatment (i.e., stent, medication, etc.) that you are looking at.

This planning before the data gathering will surely be helpful. Now, once the appropriate data has been gathered, it must be organized in a manner that will be useful for your project needs. Data must be kept confidential and treated very carefully to ensure that it is not lost or accessed by parties who are not supposed to. This could lead to large issues for yourself and your institution with regard to lawsuits. Data should only be accessed by those who are part of the project and are authorized to do so (and are listed as such in an IRB proposal). Great ideas would be to encrypt data, and if utilizing spreadsheets to input data, these files should be password-protected. If data is chosen to be kept on a hard-drive or USB-type device, the data files should be encrypted, and password-protected, and the computer should not be publicly accessible, and it too should be password-protected. Many institutions utilize internal applications on their networks that are protected and could be utilized. One should definitely not send, view, or utilize sensitive patient data with personal email services, as this is a manner by which data could potentially be compromised, accessed, or stolen.

Also, other issues to keep in mind with regard to data gathering—one must be conscientious to double-check data entries to ensure that there have been no errors with regard to data entry (for instance, key-stroke errors, double entry of data, placement of data into the wrong column or group, etc.) because this will certainly cause issues down the line when it comes to analysis and interpretation of the data.

A master database should be maintained where all data is input in an organized manner, and access is restricted to team members who are listed within the study's IRB proposal.

The master database should be regularly maintained for accuracy and organization, and ideally, the data that is entered into the database should be readily identifiable with regard to whom entered a piece of data. There should be a designated team member who oversees the data inputting to ensure that there are no mistakes, for instance, duplicated data entry or entry of data into an incorrect section of the database. There should be regularly scheduled team meetings where the objective is to discuss data gathering progress and database organization/modification. Sometimes by gathering a certain amount of data, patterns might emerge in the data that may necessitate modification of the database.

Once data has been gathered and organized in the appropriate manner, it must be analyzed and interpreted. Using a dedicated biostatistician is a great idea, even if you are well-versed in statistics and are fully capable of accomplishing the tasks of a biostatistician, in order to decrease the chances of unconscious bias or manipulation of data.

Once the data is analyzed, you can start making some nice-looking graphs and tables and sit down to see what interpretations you could come up with from the data. You might find that you might need to go back and gather more data if certain findings are equivocal. This is something that might happen even if you are careful in the pre-planning phase before data gathering, but you never know what the cards will show after data interpretation.

There should also be a plan in place as to how to share data, be it within a research team or with investigators that are not part of the immediate research team. Another good idea would be to have a plan in place for data deposition so that it could be accessed for public use, of course after deidentifying all data in the proper manner, as well as gaining the appropriate approval to do this. Sharing of data within the scientific/medical community is very important for continued research progress in a particular area and is also commonly required for NIH-funded studies.

Conclusion

In conclusion, data management is a crucial component of designing and conducting a study, and something that should be pursued and accomplished with the utmost care to ensure the success of your project. Your goal should be to provide the highest quality of data by ensuring that you reduce the chance of potential errors, bias, and manipulation throughout the phases of data gathering, management, and interpretation.

Resources

1. Gerritsen MG, Sartorius OE, vd Veen FM, Meester GT. Data management in multi-center clinical trials and the role of a nation-wide computer network. A 5 year evaluation. *Proc Annu Symp Comput Appl Med Care*. 1993:659–662.
2. Krishnankutty B, Bellary S, Kumar NBR, Moodahadu LS. Data management in clinical research: an overview. *Indian J Pharmacol*. 2012 Mar-Apr;44(2):168–172.
3. Porta Miguel. *Dictionary of Epidemiology*. 6th ed. Oxford University Press; 2014.
4. *Epidemiology and Biostatistics: An Introduction to Clinical Research by Byran Kestenbaum*. 2nd ed. Springer; 2019.
5. *Epidemiology: Principles and Practical Concerns by Jan Van Den Broek, Jonathan Brestoff*. Springer; 2013.
6. *Epidemiology and Biostatistics: An Introduction to Clinical Research by Bryan Kestenbaum*. Springer; 2009.
7. *FDA website – Information Sheet, Institutional Review Boards Frequently Asked Questions, Guidance for Institutional Review Boards and Clinical Investigators*; Janurary 1998. https://www.fda.gov/regulatory-information/search-fda-guidance-documents/institutional-review-boards-frequently-asked-questions.

Chapter 74

Quality control in IR translational research

Ishmael Raheem
Wake Forest School of Medicine, Winston-Salem, NC, United States

Key points

- Quality control is the concept and action of analyzing and verifying the suitability of a product according to predetermined criteria.
- Quality control allows interventional radiology (IR) to be conducted and refined to minimize errors and adverse outcomes.
- Good quality management (QM) practices should include the initial integration of QM in the study plan and a team specifically designated to handle all aspects and problems regarding QM.

Why does quality management matter?

The ability to monitor, assess, and revise is the key to maintaining a standard of excellence in any field. Quality management brings this ability to processes, tools, and outcomes in research. To establish confidence and high-caliber performance, Quality management must be a primary consideration for translational research in interventional radiology (IR).

Who deals with quality management?

In the United States, the FDA issues overall QM system regulatory guidance for medical devices in the United States through the form of "Quality System Regulations" (Code of Federal Regulations Title 21, Part 820, aka **21 CFR 820**). The *International Organization for Standardization* issues overall QMS regulatory guidance for medical devices in the European Union and Canada through "**ISO 13485.**"[1] In both cases, there are largely general requirements with no specific guidance. This vagueness allows flexibility for the wide range of individual products, but leaves a significant creative burden on the sponsor of the trial and product to ensure the satisfaction of the guidelines and appropriate QM criteria.

When a sponsor initiates an investigation, they ensure the study protocol is within the regulatory guidelines. Next, the sponsor makes a protocol that, in good practice, is rigorously inclusive of the criteria necessary to satisfy the desires of the consumer. In terms of translational research, this means that the sponsor's protocol is meant to realize a product that addresses a specific clinical need or problem of the consumer. Upon establishment of the protocol, the sponsor selects the investigator to conduct the investigation and the QT to monitor adherence to the protocol(s). Fig. 74.1 provides an illustration of QM responsibilities.

Where and when is quality control applicable?

Quality control is applicable whenever a product is designated for use from testing to clinical implementation. If the product under investigation is used outside of protocol guidelines, or undocumented, that is an opportunity for QC to determine the causes leading to the misuse and placing strategies to prevent further misuse of the product.

How is quality control conducted?

As mentioned earlier, there are not many mandatory guidelines for QC to bring a product to market. There is not a one-size-fits-all approach to QM for clinical trials; however, there are excellent considerations and examples that can guide anyone to success.

Quality management systems

A QMS has several major "controls" that are necessary for successful implementation: management controls; design

Translational Interventional Radiology. https://doi.org/10.1016/B978-0-12-823026-8.00072-9

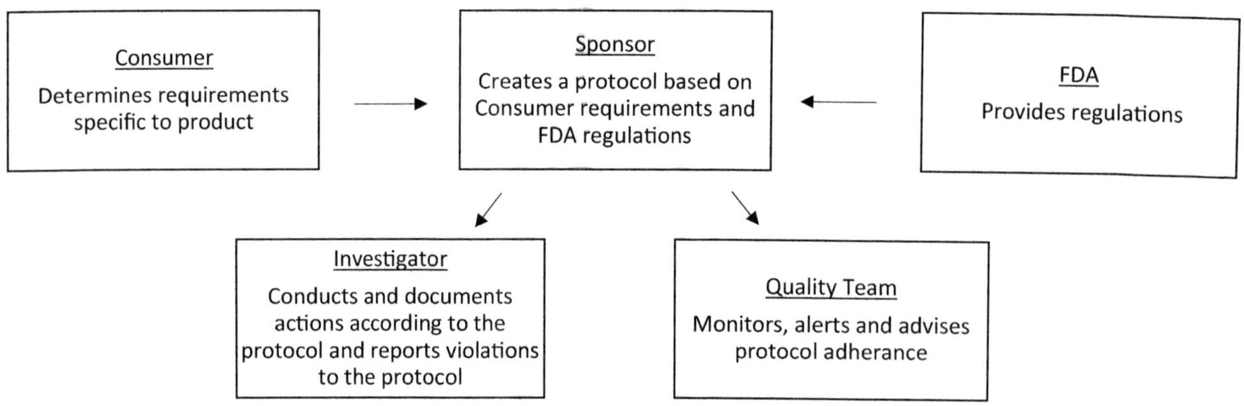

FIGURE 74.1 Individual responsibilities and contributions of each entity involved in quality management.

controls; corrective and preventative actions (RDCC); records, documents, and change controls (RDCC); facilities and equipment controls (FEC).[2] These components are part of a multifaceted approach to QM.

Management controls are the overarching guidelines and authorized processes for ensuring successful QM. Every other component of a QMS has its roots in management controls. Table 74.1 further illustrates this concept.

Key components of management controls:

1. Objectives and protocols that are thoroughly defined and documented.
2. Designated individual to lead the quality team.
3. Efficient and accessible resources to enable the achievement of objectives and protocols.
4. Monitoring systems with regular auditing to ensure inadequacies are addressed and remediated.

Design controls are specific processes used to manage design specifications and their modifications. They are also foundational to any good QMS and are distinguished by being strictly applicable to the product alone. The FDA requires a complete description of design controls before Class III (high risk, implantable, or life sustaining), Class II, and a select few Class I (minimally harmful) medical devices are cleared for use.[1]

Key components of design controls:

1. Product design is strictly determined by the needs of the consumer.
2. Thorough description of the plan for procedures, responsibilities, and interactions relevant to the design of the product.
3. Risk management for likelihood and potential extent of harm to any individuals or property involved in the investigation.
4. Verification and validation of the product satisfying design and consumer requirements.

Corrective and Preventative Actions (CAPA) are put in place to protect against failures of protocol and study

TABLE 74.1 Demonstration of nested objectives between management controls and other QMS controls.

	Redundancies of QMS components with management controls		
Objectives and protocols are thoroughly defined and documented.	Design control: Risk management for likelihood and potential extent of harm to any individuals or property involved in the investigation.	CAPA: Collect, analyze, and document information regarding problems with the product and protocol violations.	RDCC: Establishes a protocol to cover evaluating, verifying, and documenting changes to policies, processes, and products
Efficient and accessible resources to enable the achievement of objectives and protocols.	Design control: Thorough description of the plan for procedures, responsibilities, and interactions relevant to the design of the product.	RDCC: Ensure unanimous agreement regarding implementing and tracking protocols.	
Monitoring systems with regular auditing to ensure inadequacies are addressed and remediated	Design control: Verify and validate that product satisfies design and consumer requirements	CAPA: Implement and record processes to address QA and QC violations and concerns.	FEC: Verification of installation, maintenance, and standard of use for equipment.

design. CAPA is closely related to medical device reporting. The first step in the successful use of CAPA is to collect, analyze, and document information regarding problems with the product and protocol violations. The second step is to implement and record processes to address QA and QC violations and concerns.

Records, documents, and change controls are meant to verify and document any changes regarding protocols and objectives. RDCC ensures a secure, comprehensive, and centralized approach to record management for the QMS. Most importantly, the RDCCs ensure unanimous agreement regarding implementing and tracking protocols and establish a protocol to cover evaluating, verifying, and documenting changes to policies, processes, and products.

Finally, the last component of a QMS is facilities and equipment controls. This is meant to ensure a standard qualification of all facilities and equipment involved in the investigation. The relevant components of this set of controls are verification of installation, maintenance, and standard of use for equipment.

A comprehensive approach to QM will follow the guidelines and components of QMS described earlier. Good study design includes proper and thorough QM measures from the beginning to ensure they are integrated through the entire investigative process. Despite the information earlier, implementation of the components of QMS can be challenging. Two process-oriented ideologies that provide useful insight for thorough QMS are described here.

Lean

The "Lean" methodology is meant to design, improve, and manage processes by disposing of "useless" efforts that use resources but provide no additional value.[3] Its methodology is summarized as follows:

1. Specify consumer definition and extent of value
2. Identify and formalize the processes involved with developing the product
3. Conduct value stream mapping
4. Eliminate waste
5. Integrate results into future iterations

One of the most important, yet likely unfamiliar, aspects of Lean is "Value Stream Mapping." This process can be described as a thorough graphical depiction of the key people, resources, activities, and information flows involved in the product design, development, and delivery.

In summary, Lean is a formal implementation of continuous reduction of waste and improving workflow. Lean may be of particular use with QM in translational research by providing a structured process to analyze breakdowns in protocols, and empowering the QT to devise solutions that encompass the cause and inhibit future instances of protocol breakdown. This is especially valuable in regards to translational research that is dependent on process-driven outcomes, such as investigations of new techniques or procedures.

Six Sigma

The second methodology to consider in the application of QM is called "Six Sigma." Six Sigma is a management strategy that uses statistical analysis methods to achieve efficiency, uniformity, and precision of processes.[4] The methodology of Six Sigma is called the DMAIC improvement model and is composed of the following five steps:

1. Define the project goals and consumer requirements
2. Measure the process to determine current performance
3. Statistical analysis to determine the root cause(s) of relevant defects
4. Improve the process by eliminating or mitigating the root causes of defects
5. Controlling future process performance by building checks into the process

In summary, Six Sigma is a statistical methodology to approach a "Six Sigma" rate of defects. The usefulness of this methodology in translational research can be seen particularly well in applications of new or modified devices, where deviations in the protocol regarding their use may be better suited for a statistical comparison and approach to uncover new considerations. While both methodologies have individual benefits, they both possess ideological overlap that is illustrated in Table 74.2. An understanding of the similarities may help guide a QMS without needing to adhere to Lean or Six Sigma.

Real-world examples

Quality management has proven to be a vital tool for the successful implementation of cutting-edge technologies and procedures in Interventional Radiology. While the information addressed earlier applies very well to interventional radiology, the following part of this text will specifically focus on real-world examples of problems and opportunities for improvement with QM in IR.

An extremely valuable application of QM is fostering a culture of discipline in reporting and standardization. While unacceptable performance and adherence to guidelines reflect a lack of discipline, it also suggests poor study design and oversight. Foundational flaws in QM and study design may yield incorrect procedures, improper documentation, and unachieved goals of the investigation.

With considerable foresight, the Society of Interventional Radiology (SIR) released guidelines specifically regarding the implementation of IR practices to establish a

TABLE 74.2 Common themes between Lean and Six Sigma.

Lean versus Six sigma		
Steps and themes	**Lean**	**Six sigma**
Initiation	Specify consumer definition and extent of value	Define the project goals and consumer requirements
Quantification	Identify and formalize the processes involved with developing the product	Measure the process to determine current performance
Evaluation	Conduct value stream mapping	Statistical analysis to determine the root cause(s) of relevant defects
Revision	Eliminate waste	Improve the process by eliminating or mitigating the root causes of defects
Application	Integrate results into future iterations	Controlling future process performance by building checks into the process

strong culture of standardization and discipline. The essential pillars of SIR's QC guidelines are to: "form a quality committee; identify QA standards; understand the system and define processes; establish metrics and set threshold values; collect and analyze data; and initiate QI (Quality Improvement) projects."[5] These pillars highlight the necessity of QC being integral to the study design process, and also serve as an excellent example of institutional efforts to establish QC as an integral aspect to the practice of IR. Further elaborating on this idea, Statler et al. used these ideas to develop guidelines specifically regarding the use and labeling of medication.[6]

A valuable consideration in implementing QC includes difficulties in integrating data collection. Where data must be manually entered, workflow is disrupted. In an insightful effort to further integrate QC into practice, Gupta et al. noted that the use of dictation software and electronic medical records improved reporting compliance and minimized workflow hindrances.[7] In a similar vein, Burch et al. analyzed an adverse event database of "simple" procedures such as biopsies, aspirations, drainages, and therapeutic injections. The database used a billing record to cross-reference with a database and sent reminders to log the procedures. Between the reminders and audit of the overall database, the compliance was overall 100%.[8] In both of these examples, the use of management controls yielded excellent results. The use of dictation software and Electronic Medical Records (EMRs) is along the lines of efficient and accessible resources, and auditing is a directly stated management control to verify compliance with protocols.

These examples are proof of the strong value QM has in the IR and translational research setting. Not only is it necessary to provide quality care and results, but, given enough consideration and foresight, it is possible to implement QM in a way that improves the efficiency of the entire investigation. QM is a powerful tool that allows excellent results that would not be possible otherwise.

References

1. Regulatory Compliance Associates® Inc. *Correspondence between ISO 13485:2016 and 21 CFR Part 820 QMS Requirements*. WI: Pleasant Prarie; 2019. https://www.rcainc.com/wp-content/uploads/2017/06/ISO-Comparison-Matrix-jw-mp.pdf.
2. Molitor SC. *Biodesign: The Process of Innovating Medical Technologies*. Vol 303. 2010. https://doi.org/10.1001/jama.2010.680.
3. Womack JP, Jones DT. Lean thinking—banish waste and create wealth in your corporation. *J Oper Res Soc*. 1997;48(11). https://doi.org/10.1038/sj.jors.2600967, 1148–1148.
4. Kwak YH, Anbari FT. Benefits, obstacles, and future of six sigma approach. *Technovation*. 2006;26(5–6):708–715. https://doi.org/10.1016/j.technovation.2004.10.003.
5. Steele JR, Wallace MJ, Hovsepian DM, et al. Guidelines for establishing a quality improvement program in interventional radiology. *J Vasc Intervent Radiol*. 2010;21(5):617–625. https://doi.org/10.1016/j.jvir.2010.01.010.
6. Statler JD, Towbin RB, Ronsivalle JA, et al. Recommendations for the implementation of joint commission guidelines for labeling medications. *J Vasc Intervent Radiol*. 2009;20(7 Suppl):S251–S255. https://doi.org/10.1016/j.jvir.2009.04.024.
7. Gupta S, Patel J, McEnery K, et al. Early experience using an online reporting system for interventional radiology procedure-related complications integrated with a digital dictation system. *J Digit Imag*. 2011;24(4):672–679. https://doi.org/10.1007/s10278-010-9319-0.
8. Burch EA, Shyn PB, Chick JF, Chauhan NR. Auditing an online self-reported interventional radiology adverse event database for compliance and accuracy. *J Am Coll Radiol*. 2017;14:494–498. https://doi.org/10.1016/j.jacr.2016.09.038.

Chapter 75

Statistical software

Shakthi Kumaran Ramasamy[1] and Rahul A. Sheth[2]
[1]Department of Radiology, Stanford University School of Medicine, Stanford, CA, United States; [2]Department of Interventional Radiology, The University of Texas MD Anderson Cancer Center, Houston, TX, United States

Key points

- This chapter will be a guide for Quantitative Statistical Analysis software for interventional radiologists.
- Which statistical software is effective? SPSS or R or SAS or ...
- Frequency and distribution of various statistical software's in JVIR.

Why it matters?

Statistical software plays a vital role in the analysis of medical research and a tool to communicate the results of the research effectively. Statistical software application programs provide a set of statistical toolboxes that may be used to analyze data and improve the expression of data into beautiful plots and graphs. As data sets for clinical and preclinical studies become increasingly complex, it has become increasingly important for researchers to develop familiarity and facility with statistical software packages. While each package has unique capabilities, ultimately it is individual experience and familiarity that often drives software selection. In this chapter, we summarize the most commonly used statistical packages and provide a summary of their relative advantages and disadvantages. We also present a survey of the frequency of each software package's use in Interventional Radiology research.

Overview of common statistical software packages

SPSS

SPSS is one of the most popular statistical software packages and is regarded to be exceptionally user-friendly.

The main advantage of SPSS is the user interface as the menus and dialogue boxes are simple and straightforward to use, with a similar feel to Excel. For many users, it is easier to learn when compared to other packages and is compatible to use with Windows and MacOS. Despite the syntax and script language, automating and integrating into other applications is more challenging than with other alternatives. There are no reliable techniques available for Least Absolute Deviation Regression and Quantile Regression. Also, the yearly update cycle affects stability. Despite its wide popularity, SPSS is not open source. The license costs $99 for a single user per month.

R

R is an open-source statistical program that allows users to extend its capabilities and functionalities through user-generated packages (about 16,000), making it suitable for a wide range of data analyses. R is frequently used in conjunction with RStudio for a more user-friendly interface.

The main strengths of using R are that it is open source, and the active development company makes it future proof. R is a scripting language with a lot of power and flexibility, for example, supports object-oriented programming. The program's flexibility enables it to handle large amounts of data with hundreds of millions of data points in a highly effective manner. R interacts well with other software such as Python, Bioconductor, WinBUGS, JAGS, Git, LaTeX, ODBC, Oracle R Enterprise, teradataR, Apache Hadoop, and Microstrategy. Community assistance is excellent, as is fee-based help from third-party suppliers. The open-source nature of R, coupled with the active developer community, has extended R's capabilities to analyze not only conventional quantitative or qualitative data sets but also genomic data, spatial data, and imaging data. New users can teach themselves using a variety of online instructions, manuals, and tutorials. It is quite simple to create attractive plots that are suitable for publication and extend your capabilities in a big way. Users can quickly knowledge from a certain sort of analysis, which will inspire them to pursue a specific

Translational Interventional Radiology. https://doi.org/10.1016/B978-0-12-823026-8.00013-4

project. The main disadvantage is its steep learning curve. An initial effort in learning how to write simple scripts and visualize R data can pay out well in the long term (Figs. 75.1 and 75.2).

Excel

Excel is a very commonly used statistical software, in particular for small data sets. There is a broad familiarity with its use. The advantages of Excel are ease of use and familiarity. The disadvantages are that, for beyond basic analyses, it is very cumbersome to do statistical tests, that is, T-test is easy, and logistic regression is complicated. Statistical tests that can be done in a single command line in other packages takes a lot of manipulation in Excel. Also, it is limited by data set size; for many applications, that size limit is not prohibitive, but for applications like "big data" or certainly any genomic analysis, Excel is not a feasible option.

SAS

In 1966, Anthony Bar of North Carolina State University started developing SAS (Statistical Analysis System) and later was joined by James Goodnight. This initiative was financed by the National Institute of Health with the objective of studying agricultural data to increase crop yields. The first stable release of SAS (SAS Institute in Cary, North Carolina) was in 1972. Since then, SAS has grown to applications of data management well beyond statistics.

The advantages of SAS are that new statistical approaches can be quickly integrated, and routines are extremely robust and dependable. They provide professional assistance and excellent documentation. It is well suited to dealing with big data collections. The disadvantage is that it is a complex system with an old interface. The financial factor weighs in as it costs approximately $9000 for licensing for the first year alone. A user will be able to run SAS on a MacOS, but only if they have parallels or another virtual machine technology running that hosts the compatible Windows operating system for the SAS version. Furthermore, extending the capability of the base software package often requires purchasing additional extensions that add to the financial limitations of this software.

STATA

Bill Gould and Sean Becketti wrote a regression and data management program with 44 commands, which was the first stable release of STATA (StataCorp Inc, College Station, Texas) in January 1985. In 2003, the graphical user interface (menus and dialogue boxes) was made available.

One of the advantages of STATA is that it works well with data from panel, survey, and time series. There are multiple packages that extend the capabilities of STATA generated by the software development community that are available. One of the highlights of STATA is the availability of matrix programming in Mata. Most of the time, the syntax is utilized, although menus are also available. The major limitation is the ability to store only one dataset in memory at a time. Another limiting factor is the specific STATA packages (Stata/IC, Stata/SE, and Stata/MP), which limit the size of datasets. Depending on the package, the number of variables may have to be sacrificed in favor of the number of observations, or vice versa. Previous

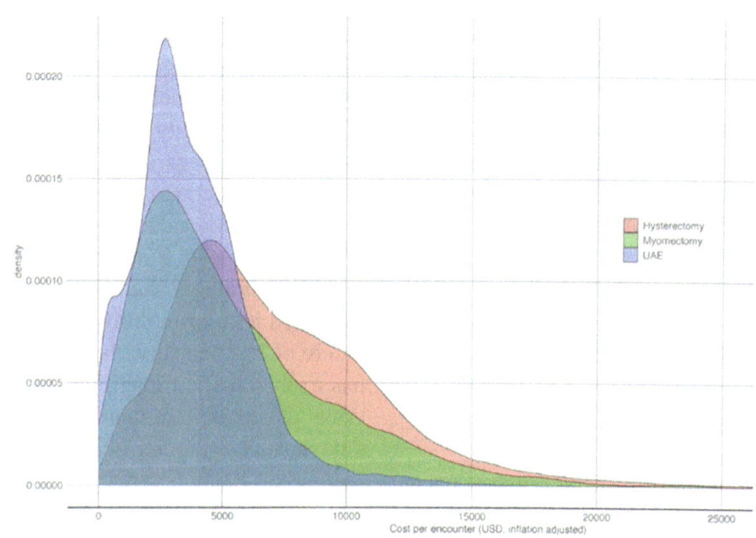

FIGURE 75.1 Histogram of cost per encounter for uterine leiomyomata interventions. Although the most frequent charges for all 3 interventions were similar, laparoscopic hysterectomy and myomectomy had a broader overall distribution of charges, resulting in a significantly greater median charge than UAE. Data are inflation-adjusted to 2015 US dollars.[1]

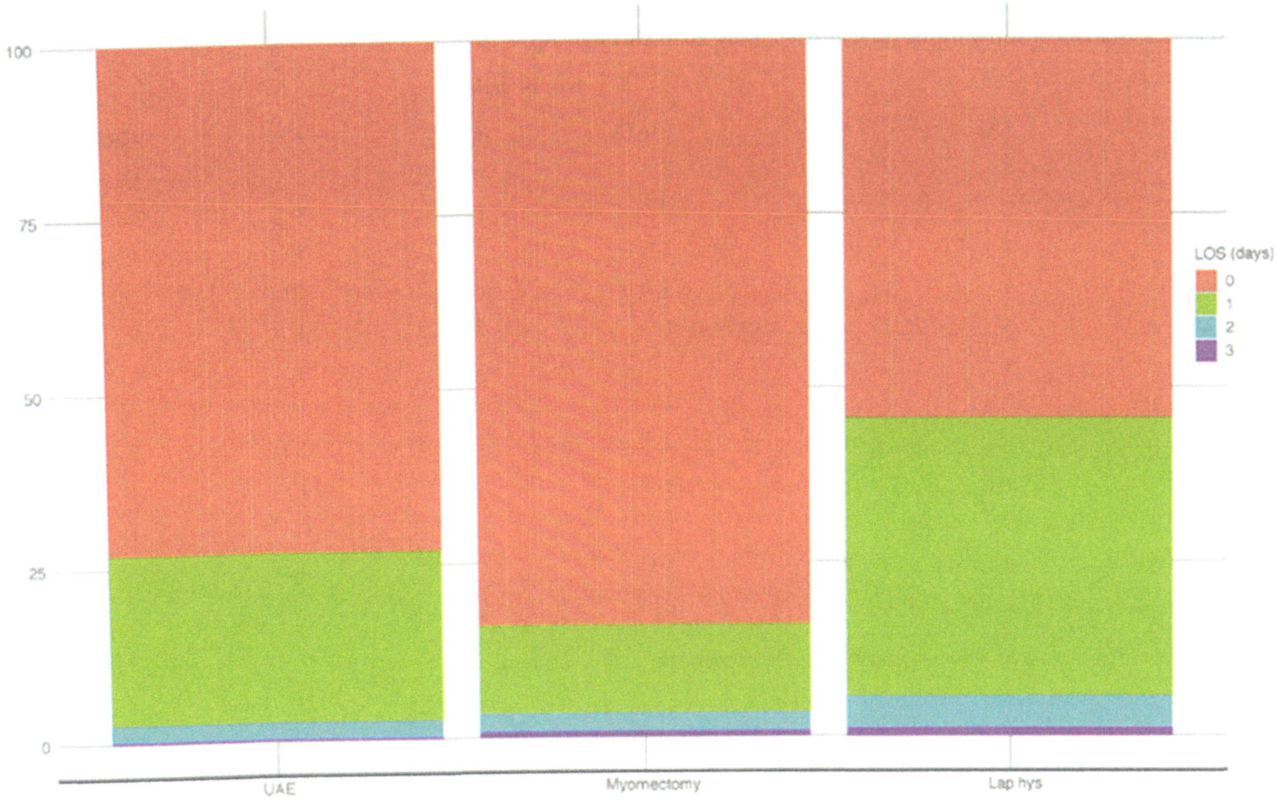

FIGURE 75.2 Distribution of durations of hospital stay for patients undergoing uterine leiomyomata interventions. Although the length of stay for the majority of patients in all three interventions was 0 days, this proportion was the lowest for patients undergoing hysterectomy. Likewise, the proportions of patients with lengths of stay >1 day were greatest in the hysterectomy arm.[1]

versions of Stata are unable to read datasets created by later versions, while newer versions may save in the format of older versions. The license costs $125–425 for a single user per year.

The articles in JVIR were surveyed to identify the statistical software used in the research. The authors retrospectively reviewed 405 articles in JVIR published from May 2020 to April 2021. The statistical software was identified and characterized according to the type of statistical software used, version, whether multiple software was used for a clinical study, and articles other than clinical studies using statistical software. As shown in Table 75.1, a total of 174 out of 405 articles (43%) used at least one statistical software with 19 (4.7%) articles using a combination of two or more software. The most used software is SPSS version 25 with a total of 19 (4.7%) articles. A variety of 18 different statistical software were used with the most frequent being SPSS with 65 articles (16%), followed by SAS with 24 (6%) and R with 22 (5.4%). The three most used software are SPSS version 25, SAS version 9.4, and SPSS version 22. A total of 30 (7.4%) clinical studies have no mention of the statistical software that was used. The other less commonly used statistical software among Interventional radiologists are GraphPad, MedCalc, REDCap, Julia, JMP, EasyR, SigmaPlot, MATLAB, PASW, MedDRA, Review manager software, XLSTAT, and Google Sheets, in the order of decreasing usage frequency (Fig. 75.3).

Comparison of statistical software (Tables 75.2 and 75.3)

Get started

With R

- Install R and RStudio.
- ggplot2 package, grammar of graphics.
- Check out The R Graph Gallery https://r-graph-gallery.com/.

Potential pitfalls

- STATISTICS is an important part of a study. Statistical software will be of help only if the statistical principles are applied and the results are interpreted correctly.

TABLE 75.1 Statistical software of JVIR articles, May 2020 through April 2021.

Month	No. of articles					
	SAS	SPSS	STATA	Excel	R	GRAPHPAD
May-20	2	3(2)	2	(1)	–	(1)
June-20	3	6	(1)	–	1	(2)
July-20	1	6(1)	2	1	3(1)	1
August-20	2(1)	7(3)	5(1)	(1)	(1)	1
September-20	5	2	1	2	1	1
October-20	1	6	–	1	4	1
November-20	2	3	–	1	1	2
December-20	2	12(2)	–	–	2(1)	2(1)
January-21	2(1)	6(1)	3	–	1	(1)
February-21	1	7	1	(1)	(1)	(1)
March-21	3(2)	3(2)	2(1)	–	4(2)	2
April-21	–	4	(1)	3	3	–

Note: Numbers in parentheses are articles using more than one software.

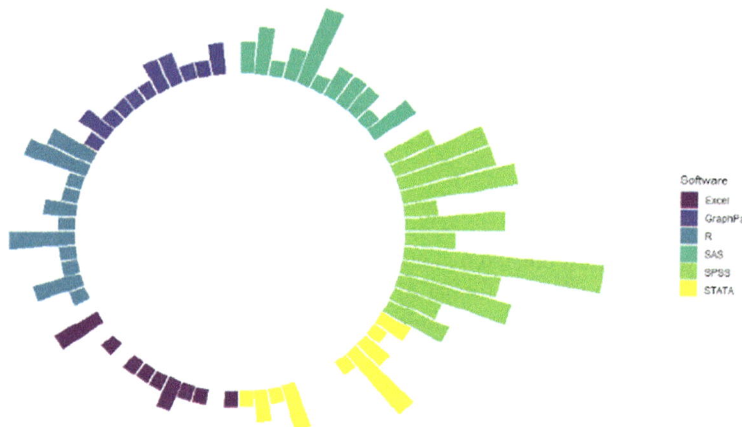

FIGURE 75.3 Circular bar plot of the frequency and distribution of five most used statistical software in articles published in JVIR from May 2020 to April 2021.

Real-world examples

1. R: R is widely used in academia and research for data analysis and visualization. It is also used in industries such as finance, healthcare, and marketing for data mining, machine learning, and predictive modeling.
2. Python: Python is a general-purpose programming language that is also popular for data analysis and visualization. It has a large and active community, and many libraries and frameworks are available for machine learning, natural language processing, and data visualization.
3. SAS: SAS is a commercial software suite that is commonly used in businesses and organizations for data management, data visualization, and advanced analytics.
4. SPSS: IBM SPSS is a data analysis and statistics software. It is widely used by researchers in various fields, such as social sciences, health sciences, and market research, for data management and analysis.
5. STATA: Stata is a general-purpose statistical software package.It is widely used in the social sciences, economics, and political sciences, for data management and analysis.

TABLE 75.2 Basic comparison of the most commonly used statistical software by interventional radiologists.

Criterion	Excel	R	SPSS	SAS	STATA	Python
First version	1985	1993	1968	1972	1985	1991
Mac/ Windows	Both	Both	Both	Windows	Both	Both
Data formats	xls, xlsx	xls, xlsx txt, dat, csv	sav, sps dta	xls, xlsx txt, dat, csv	dta, do, txt, dat, csv	xls, csv txt, tsv
		sav, dta sas7bdat	xls, xlsx txt, dat, .csv	sav, dta jmp, xml	xls, xlsx XPT, xml	
		xml, json	sas7bdat			
Graphics	Yes	Yes	Yes	Yes	Yes	
License and cost	Free	Free Open source	$99 single user/per month	$9000 approx. for first year	$125–425 single user/per year	Free

TABLE 75.3 Comparison of statistical abilities of the most commonly used statistical software by interventional radiologists.

Criterion	Excel	R	SPSS	SAS	STATA	Python
Descriptive statistics						
Base statistics	+	+	+	+	+	+
Normality tests	+	+	+	+	+	+
Nonparametric statistics						
CTA	+	+	+	+	+	+
Nonparametric comparison, ANOVA	+	+	+	+	+	+
Quality control	+	+	+	+	+	+
Survival analysis	+	+	+	+	+	+
Cluster analysis	+	+	+	+	+	+
Discriminant analysis	+	+	+	+	+	+
Data processing						
BDP	+	+	+	+	+	+
Extended	+	+	+	+	+	+

ANOVA, Analysis of Variance; *BDP*, Base Data processing; *CTA*, Contigency table analysis.

6. Matlab: Matlab is a programming language and environment for numerical computation and visualization. It is used in fields such as engineering, finance, and signal processing, for data analysis and algorithm development.
7. Minitab: Minitab is a statistical software package that is used for data analysis and statistical process control. It is commonly used in manufacturing and quality control to improve products and processes
8. Eviews: Eviews is a statistical software package designed for econometrics and time series analysis. It is widely used in economics, finance, and business for forecasting and statistical analysis of time series data

9. JMP: JMP is a software for statistical discovery that is used in fields such as engineering, manufacturing, and life sciences. It allows users to explore data and build models using interactive graphical interface.
10. RStudio: RStudio is an integrated development environment (IDE) for R. It is popular among R users because it provides a user-friendly interface and features that make working with R easier and more efficient.

It is always important to choose the right software based on the needs of your project, data and the question that you want to address.

Reference

1. Balabhadra S, Kuban JD, Lee S, et al. Association of inferior vena cava filter placement with rates of pulmonary embolism in patients with cancer and acute lower extremity deep venous thrombosis. *JAMA Netw Open*. 2020;3(7):e2011079. https://doi.org/10.1001/jamanetworkopen.2020.11079.

Additional reading

1. R for Data Science - Book by Garrett Grolemund and Hadley Wickham.

Chapter 76

Report forms: Harm and quality of life

Jim Zhong and Alexandra Gilbert
Leeds Institute of Medical Research at St. James's, University of Leeds, Leeds, United Kingdom

Key points

- *Understand error, adverse events, and complications within interventional radiology (IR)*
- *Learn how to mitigate harm in IR*
- *Understand the value and importance of reporting quality of life*
- *Learn about the role of patient recorded outcome measures in IR*

Background

Harm to a patient can be defined as any negative effect of clinical care, even if the patient is not aware of this. Such events or complications can arise from medical errors and may encompass physical or mental effects along with the potential to cause serious physical harm, such as near misses, for example, prescription errors that were picked up before the patient came to any physical harm.[2] Adverse events may be expected side effects of particular treatments such as post-embolization syndrome[3] but may also be unexpected, where unintended harm takes place that could be modified or reduced by good practice. Unintended harm in healthcare also has significant financial implications, costing billions of pounds in additional hospital stays alone from United Kingdom data.[4] Research suggests up to a quarter of patients come to some harm in acute care with half of these thoughts to be preventable.[5–7]

Why do errors happen

Several factors have been identified as contributing to healthcare-related errors including human factors (teamwork, communication, stress/burnout), structural factors (reporting systems, infrastructure, workforce loads, equipment, and the work environment), and clinical factors (complex care and length of hospital stay).

In surgery, an even higher number at 54% of complications are thought to be preventable.[8] Due to the invasive nature of interventional radiology, factors that affect patient safety are likely comparable to surgery. Measures to prevent, identify and manage adverse events are vital including the World Health Organisation (WHO) surgical safety checklist and other IR-specific checklists that have helped to reduce preventable errors.[9,10]

Error classification systems have been developed to identify types and sources of systems failures. One widely used system in pediatric cardiac surgery found non-technical errors (e.g., coordination or communication failures) to be more prevalent than technical ones (e.g., expertise or skill failures) along with task-specific threats (e.g., patient-specific procedural difficulties such as variant anatomy or equipment failures) being the most prevalent systemic source of error.[11] Similar errors were identified in vascular IR procedures that were largely attributed to ineffective planning, communication errors, and equipment difficulties, rather than the result of technical or patient-related issues.[12] Over half of these failures were also preventable. Patient safety is a key priority in healthcare and efforts to identify and overcome errors in IR will improve patient outcomes.

Adverse event classification systems in interventional radiology

An ideal classification system for adverse events should be accurate, consistent, and practical for both clinical and research purposes. Several systems exist for reporting adverse events in surgery, with the most commonly used one being the Clavien-Dindo system.[13,14]

In 2017, the Society of Interventional Radiology (SIR) updated its adverse event (AE) classification system to ensure reporting was more consistent between practitioners, with less ambiguity.[15] The adaptations made to this new

Translational Interventional Radiology. https://doi.org/10.1016/B978-0-12-823026-8.00087-0

system also reflect the transition of IR to become a more clinical specialty as all AEs occurring within 30 days of the IR procedure are captured, which highlights the greater responsibility that interventional radiologists now have in following up the patient after the procedure, as is the case with surgical procedures.

How to mitigate harm in IR

As IR is a relatively new specialty compared to surgery, it has benefited from observing the methods used to reduce patient safety failures, such as the WHO surgical safety checklist.[9] IR-specific checklists have also been designed and implemented, leading to less deviation from optimal procedural processes and fewer case cancellations.[16] The Cardiovascular and Interventional Radiological Society of Europe adapted the WHO checklist[17] and Corso et al.[10] showed that using a version of this eliminated AEs associated with IR procedures in the first year of use. A key factor was the involvement and awareness of patient safety among the entire multidisciplinary IR team that requires engagement with the training required. The SIR has also published its own preprocedure checklist, with emphasis on keeping the design simple to ensure adequate compliance.[18]

Additional steps such as a pre-procedural team rehearsal can also help to reduce errors and compliments the preprocedural checklist. Equipment errors, largely due to unavailability are frequent in IR due to a large number of equipment and kit required. Locating the right size or type of desired equipment can take several minutes and can be stressful in the intraoperative setting therefore a rehearsal will familiarise staff members with where such kit is located to speed this process up. This has been shown to be successful when implemented in a busy IR environment with the time taken to complete a rehearsal also decreasing during the study period, likely reflecting a better understanding of the process and also a change in staff culture who prepared thoroughly before the case.[12]

Additional measures to relieve the burden caused by errors are required. Duncan[19] describes a three-part process to tackle errors including reducing the frequency of errors, reducing the burden/severity resulting from the error when it inevitably occurs, and finally detecting the error before it causes any harm. The example of a simple arterial puncture for obtaining vascular access can be used as an example to illustrate this approach (Fig. 76.1).

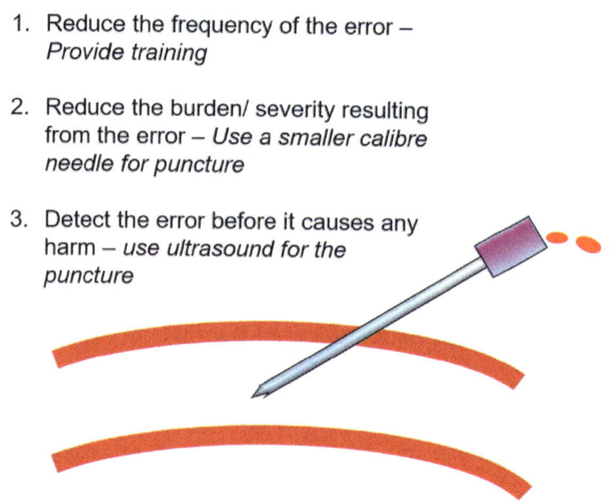

FIGURE 76.1 Example of how to tackle errors arising from an arterial puncture using a three-part approach by Duncan.[19]

The value of quality of life

Tools called PROMs can be used to measure quality of life (QoL), and patient-reported symptomatic AEs, an essential part of measuring health outcomes. Revicki et al. 20 defined QoL as a "broad range of human experiences related to one's overall well-being." QoL is a multidimensional, subjective, and personal concept that reflects the patient's satisfaction with their own life based on specific domains that they consider important. Measuring these metrics will show levels of side effects from IR procedures that are acceptable to the patient. For example, following tumor ablation procedures, patients may experience a constellation of symptoms such as fever, fatigue, and discomfort that are encompassed by the term postablation syndrome; however, it is relevant to determine how these symptoms impact the patient's QoL to know where to focus for the management.[21]

In IR, QoL considerations are important to help patients and clinicians decide on the most appropriate treatment. For patients with cancer, their symptoms may be caused by the underlying cancer, disease progression, or side effects of treatment; therefore, it is vital to measure and monitor this. Many IR procedures are performed with palliative intent, for example, to relieve abdominal discomfort with ascitic drainage or relieve malignant abdominal pain with a celiac plexus nerve block. Although these procedures do not extend life, they are done with an aim to improve QoL.

The United States Food and Drug Administration has published a roadmap on how to create patient-focused outcome measures within a clinical trial setting.[22] The first part is understanding the effects of the treatment, taking into account the disease of interest and identifying the desired outcomes from the patient perspective. This knowledge is then incorporated into the selection, adaptation, and/or development of appropriate endpoints meaningful and important to patients. This highlights the value of involving patient groups in IR research in order to identify the outcome measures important to them. Cross-speciality collaboration is also likely to help formulate more patient relevant questions as it offers a more holistic insight.

Different types of PROMs

PROMs, usually in the form of a questionnaire, may be used to measure patient experience, including symptomatic adverse events and more complex multidimensional domains such as physical functioning and QoL. PROMs may be defined as generic or specific. Generic PROMs are not disease or treatment specific and are designed to cover a range of different aspects of health outcomes including overall QoL, function, and cost-effectiveness of interventions. The disease or treatment-specific PROMs allow evaluation of particular aspects of a condition or side effects and their impact on patient experience. Often generic and disease/treatment-specific PROMs are used in combination.

When selecting a questionnaire, it should be relevant to the patient group, be straightforward to use, and be both validated and reliable. For example, pain is multidimensional and can be assessed in different ways, from using a basic numerical scale for pain severity to evaluating the reactive dimension of pain using more advanced tools such as the brief pain inventory (BPI) that assesses how pain affects the patient's mood, general activity, normal work, relations with other people, sleep, and enjoyment of life. The BPI is a validated tool that has been used as endpoints in oncology trials assessing cancer bone pain that have involved IR procedures.[23]

Other tools such as the European Organization for Research and Treatment of Cancer Quality of Life Questionnaire core 30 (EORTC-QLQ-C30), one of the most commonly used QoL tools for oncology trials, are more generic.[24] This tool consists of five functional scales (physical, role, emotional, cognitive, and social), three symptom scales (fatigue, nausea and vomiting, and pain), and a global health status/QoL scale. The EORTC-QLQ-C30 remains a core set of questions designed for general use in oncology patients however cancer-specific questionnaires may be used in conjunction with these core questions to capture additional aspects of QoL important to specific patient subgroups. The EORTC QLQ development process is well documented; starting with a systematic review followed by interviews with clinicians and patients to identifying QoL issues. Question development and translations are then completed before initial questionnaire pilot testing and then international validation of the questionnaire is carried out.[25,26]

The collection of PROMs has historically required paper records however with the advancement of technology services and the development of more robust electronic platforms, this has enabled larger-scale data collection across many conditions with readily available outcome data that can be analyzed.[27] Electronic PROMs (ePROMs) would also give real-time results for clinicians to act on and for patients to receive their results along with self-management feedback.[28] Beyond the clear advantages of physical storage, ease of retrieval, and access in using ePROMs, the transition from paper to electronic requires additional infrastructure to set up and ongoing checks to ensure the data generated from different formats are comparable, likely to come with an additional financial cost. Good practice recommendations should be followed when migrating PROMs to electronic platforms.[29]

The role of PROMs in IR research

As IR procedures encroach upon existing therapies, PROMs may help clinicians decide between therapies where clinical outcomes are comparable but greater improvement in QoL or other aspects of improved patient experience are observed. For example, in the SARAH trial (*S*or*A*fenib vs. Radioembolization in Advanced Hepatocellular Carcinoma), QoL was preserved for a longer duration with radioembolization compared to Sorafenib despite no statistical difference in overall survival, the primary endpoint.[30] The impact of minimally invasive IR treatments such as embolization and ablation on QoL, therefore, offers great potential for patients with cancer.

QoL results should also be reported in the appropriate context and clinically meaningful thresholds or a minimally important difference (MID) should be identified. An MID puts QoL measurements into the necessary clinical context for determining when changes in symptom burden are significant. An example where this has been shown is with the Expanded Prostate Cancer Index-Composite QoL tool used for prostate cancer patients where MIDs were established.[31–33] This allows for direct comparison between prostate IR therapies such as prostate artery embolization and other surgical or focal therapies when it comes to assessing urinary and bowel symptoms.

The international multidisciplinary consortium—Setting International Standards of Patient-Reported Outcomes and Quality of Life Endpoints in Cancer Clinical Trials—IMI (SISAQOL-IMI) was set up to address how PROMs should be analyzed.[34] To optimize the way PROMs are collected and analyzed in clinical trials, there should be agreed appropriate statistical methods by which PROMs data are analyzed and accepted ways to deal with missing data. This will ensure the comparability of results between research studies. Future academic interventional radiologists should refer to these recommendations to make sure any prospective PROMs research is comparable across different specialties and limit interpretational barriers.

Conclusions

Understanding the healthcare-related factors that lead to adverse events will help clinicians to better identify and mitigate patient harm. Using PROMs will allow interventional radiologists to determine the most important patient-centered domains that are relevant to IR procedures to demonstrate comparative effectiveness and improve patient care. It will also help interventional radiologists to better appreciate their impact on the patient's QoL and reaffirm our expanding role in the palliation of oncology patients.

Get started

1) Write down all the potential causes for patient's coming to harm in IR.
2) Think about which ones are preventable and how these can be mitigated.
3) Using any IR procedure you can think of, apply Duncan's (see 'How to mitigate harm in IR' section) three part process to tackle potential errors.
4) Think about how IR procedures help improve a patient's quality of life.

Potential pitfalls

1) Underestimating the importance of good planning and communication in mitigating healthcare related errors in IR.
2) Not involving all team members in the safety checklist.
3) Placing less weight on quality of life and patient reported outcomes when discussing treatment plans.

Real-world examples

1) An example of surgical safety checklist - https://www.who.int/teams/integrated-health-services/patient-safety/research/safe-surgery/tool-and-resources.
2) An example of an IR-specific Adverse Event Classification System — Khalilzadeh, O. et al. Proposal of a new adverse event classification by the society of interventional radiology standards of practice committee. J Vasc Interv Radiol. 2017;28:1432–1437.e3.
3) An example of a research study showing the quality of life impact that an IR procedure has for patients — Pereira, H, et al. Health-related quality of life in locally advanced hepatocellular carcinoma treated by either radioembolisation or sorafenib (SARAH trial). Eur J Cancer. 2021;154:46–56.

Resources

1). Rodziewicz TL, Houseman B, Hipskind JE. Medical error reduction and prevention. [Updated 2022 May 1]. In: StatPearls [Internet]. Treasure Island (FL): StatPearls Publishing; 2022 Jan-. Available from: https://www.ncbi.nlm.nih.gov/books/NBK499956/.
2). How to select the most suitable outcome measurement instrument — https://www.cosmin.nl/.

References

1. Clavien PA, Barkun J, de Oliveira ML, et al. The Clavien-Dindo classification of surgical complications. *Ann Surg*. 2009;250(2):187–196.
2. Health and Safety Executive. *Accidents and Investigations*. 2021.
3. Piscaglia F, Tovoli F, Pini P, Salvatore V. A new horizon in the prevention of the postembolization syndrome after transcatheter arterial chemoembolization for hepatocellular carcinoma. *Hepatology*. 2018;67(2):467–469.
4. Department of Health. *An Organisation with a Memory: Report of an Expert Group on Learning from Adverse Events in the NHS*. 2000.
5. The Health Foundation. *Levels of Harm*. 2011.
6. Landrigan CP, Parry GJ, Bones CB, Hackbarth AD, Goldmann DA, Sharek PJ. Temporal trends in rates of patient harm resulting from medical care. *N Engl J Med*. 2010;363(22):2124–2134.
7. Vincent C. Adverse events in British hospitals: preliminary retrospective record review. *BMJ*. 2001;322(7285):517–519.
8. Gawande AA, Thomas EJ, Zinner MJ, Brennan TA. The incidence and nature of surgical adverse events in Colorado and Utah in 1992. *Surgery*. 1999;126(1):66–75.
9. Haynes AB, Weiser TG, Berry WR, et al. A surgical safety checklist to reduce morbidity and mortality in a global population. *N Engl J Med*. 2009;360(5):491–499.
10. Corso R, Vacirca F, Patelli C, Leni D. Use of "Time-Out" checklist in interventional radiology procedures as a tool to enhance patient safety. *Radiol Med*. 2014;119(11):828–834.
11. Catchpole KR, Giddings AEB, de Leval MR, et al. Identification of systems failures in successful paediatric cardiac surgery. *Ergonomics*. 2006;49(5–6):567–588.
12. Morbi AHM, Hamady MS, Riga CV, et al. Reducing error and improving efficiency during vascular interventional radiology: implementation of a preprocedural team rehearsal. *Radiology*. 2012;264(2):473–483.
13. Dindo D, Demartines N, Clavien P-A. Classification of surgical complications. *Ann Surg*. 2004;240(2):205–213.
14. Clavien PA, Sanabria JR, Strasberg SM. Proposed classification of complications of surgery with examples of utility in cholecystectomy. *Surgery*. 1992;111(5):518–526.
15. Khalilzadeh O, Baerlocher MO, Shyn PB, et al. Proposal of a new adverse Event classification by the Society of Interventional Radiology Standards of Practice Committee. *J Vasc Interv Radiol*. 2017;28(10):1432–1437.e3.
16. Koetser ICJ, de Vries EN, van Delden OM, Smorenburg SM, Boermeester MA, van Lienden KP. A checklist to improve patient safety in interventional radiology. *Cardiovasc Intervent Radiol*. 2013;36(2):312–319.
17. Gillams A, Khan Z, Osborn P, Lees W. Survival after radiofrequency ablation in 122 patients with inoperable colorectal lung metastases. *Cardiovasc Intervent Radiol*. 2013;36:724–730.
18. Rafiei P, Walser EM, Duncan JR, et al. Society of interventional radiology IR pre-procedure patient safety checklist by the Safety and Health Committee. *J Vasc Interv Radiol*. 2016;27(5):695–699.
19. Duncan JR. Strategies for improving safety and quality in interventional radiology. *J Vasc Interv Radiol*. 2008;19(1):3–7.
20. Revicki DA, Osoba D, Fairclough D, et al. Recommendations on health-related quality of life research to support labeling and promotional claims in the United States. *Qual Life Res*. 2000;9(8):887–900.

21. Zhong J, Bambrook J, Bhambra B, et al. Incidence of post-ablation syndrome following image-guided percutaneous cryoablation of renal cell carcinoma: a prospective study. *Cardiovasc Intervent Radiol.* 2018;41(2):270–276.
22. Food and Drug Administration. Roadmap to Patient-Focused Outcome Measurement in Clinical Trials.
23. Li D, Madoff D. Incorporating quality of life metrics in interventional oncology practice. *Semin Intervent Radiol.* 2017;34(04):313–321.
24. European Organization for Research and Treatment of Cancer. EORTC QLQ-C30. 2021
25. Kavadas V, Blazeby JM, Conroy T, et al. Development of an EORTC disease-specific quality of life questionnaire for use in patients with liver metastases from colorectal cancer. *Eur J Cancer.* 2003;39(9):1259–1263.
26. *European Organization for Research and Treatment of Cancer.* EORTC Quality of Life; 2021.
27. O'Connell S, Palmer R, Withers K, Saha N, Puntoni S, Carolan-Rees G. Requirements for the collection of electronic PROMS either "in clinic" or "at home" as part of the PROMs, PREMs and Effectiveness Programme (PPEP) in Wales: a feasibility study using a generic PROM tool. *Pilot Feasibility Stud.* 2018;4(1):90.
28. Absolom K, Warrington L, Hudson E, et al. Phase III randomized controlled trial of eRAPID: eHealth intervention during chemotherapy. *J Clin Oncol.* 2021;39(7):734–747.
29. Coons SJ, Gwaltney CJ, Hays RD, et al. Recommendations on evidence needed to support measurement equivalence between electronic and paper-based patient-reported outcome (PRO) measures: ISPOR ePRO good research practices task force report. *Value Heal.* 2009;12 (4):419–429.
30. Pereira H, Bouattour M, Dioguardi Burgio M, et al. Health-related quality of life in locally advanced hepatocellular carcinoma treated by either radioembolisation or sorafenib (SARAH trial). *Eur J Cancer.* 2021;154:46–56.
31. Szymanski KM, Wei JT, Dunn RL, Sanda MG. Development and validation of an abbreviated version of the expanded prostate cancer index composite instrument for measuring health-related quality of life among prostate cancer survivors. *Urology.* 2010;76(5):1245–1250.
32. Wei JT, Dunn RL, Litwin MS, Sandler HM, Sanda MG. Development and validation of the expanded prostate cancer index composite (EPIC) for comprehensive assessment of health-related quality of life in men with prostate cancer. *Urology.* 2000;56(6):899–905.
33. Skolarus TA, Dunn RL, Sanda MG, et al. Minimally important difference for the expanded prostate cancer index composite short form. *Urology.* 2015;85(1):101–106.
34. Coens C, Pe M, Dueck AC, et al. International standards for the analysis of quality-of-life and patient-reported outcome endpoints in cancer randomised controlled trials: recommendations of the SISAQOL Consortium. *Lancet Oncol.* 2020;21(2):e83–96.

Chapter 77

Subject adherence

Kimberly Coffman
School of Public Health, The University of Texas Health Science Center, Houston, TX, United States

Key points

- For studies that involve human subjects, nonadherence to study protocol is inevitable. - Nonadherence can cause bias when evaluating an intervention or treatment.
- Many causes can lead to study nonadherence and should be handled differently.
- Some measures can be taken to prevent or reduce the impact of nonadherence.

Introduction

In any study, there will inevitably be a subset of research subjects who do not comply with the protocol in question. Subject adherence is one of the most important aspects of an experimental study design. Yet, it has been shown that protocol deviations can be as high as 40% in some studies.[1] Nonadherence can introduce effects into a study that have nothing to do with the tested intervention. This introduces artifact into the study which will artificially modify effect size. Not only can it affect statistical analysis and therefore measured differences between treatment and control groups, but it can be incredibly costly due to the need to recruit additional subjects to maintain statistical power.

In this chapter, we will review the types of subject nonadherence, as well as how to minimize and address it when it inevitably arises.

Causes of nonadherence

The term "nonadherence" can be misleading. From an epidemiological standpoint, it is essentially a deviation from the research protocol to which the patient was assigned by randomization. There are two main categories of nonadherence, real-world nonadherence, and artifactual nonadherence. The causes of real-world nonadherence are many and apply not only to clinical trials but to general clinical practice as well. Examples include disruption of the subjects' lifestyle, duration of a study, miscommunication or inadequate explanation of the therapy instructions, subjects' perceived risk, and side effects. Artifactual nonadherence is a phenomenon unique to clinical trials and is due to professional subjects, which are discussed further on in this chapter. There are multiple subtypes of nonadherence (Fig. 77.1), which are outlined later.

- The subject does not receive treatment to which they were randomized.
  - In studies involving oral medications, some subjects may intentionally or unintentionally fail to take their assigned dose. In rare cases, one may unknowingly enroll a "professional subject" or a person who enrolls in concurrent or serially monogamous studies (one immediately after another, without washout) for monetary gain, with no intention of taking the prescribed medication. This is likely the largest source of artifactual nonadherence. Professional subjects are more likely to respond to placebo, which

FIGURE 77.1 Contributors to nonadherence.

Translational Interventional Radiology. https://doi.org/10.1016/B978-0-12-823026-8.00064-X

can lessen the observed effect of the studied therapy and the sample size requirement for statistical power while decreasing the observed effect size of the medication of interest.[1] As such, these subjects must be discovered and removed from the study. This issue is probably exceedingly rare in the field of interventional radiology as many of the interventions are procedurally based and there is objective evidence of disease with an observable intervention.

- The subject receives the treatment assigned to another arm of the study.
  - Occasionally a patient may become acutely ill and require an unplanned intervention or may withdraw consent for the study intervention in favor of more conventional treatment. Example: A 42-year-old male subject is enrolled in a study comparing intravenous heparin and heparin plus thrombectomy. He is randomized to the heparin-only arm of the study, but due to the rapid worsening of his condition thrombectomy is performed. The patient has therefore unexpectedly crossed over to the other arm of the study.

Alternatively, professional subjects can also contribute to this subtype of nonadherence by enrolling in the same study multiple times, separate studies at the same time, or one after another without sufficient washout time.

- Partial adherence
  - Partial adherence is likely the most common type of nonadherence, especially in studies that require regular administration of therapy by the patient in an unobserved fashion, such as taking a medication each day at home. The patient may take the medication most days, but occasionally forgets or is unable to take the medication for one reason or another. In the case of lifestyle interventions, the patient may not be able to comply with the medication or intervention protocol 100% of the time due to weather changes or travel.
- The subject receives a therapy that has been outlined as "prohibited" by the study.
  - Some studies may outline therapies that a subject should not participate in during the observation time of the study due to potential interactions they may have with the assigned therapy. Example: A 56-year-old female patient is prescribed warfarin as a part of a clinical trial, but takes vitamin K regularly. Example: A 56-year-old female patient is prescribed warfarin as a part of a clinical trial, but takes vitamin K regularly.
- Lost to follow-up
  - Inevitably, most studies will experience some degree of patient dropout or attrition. This can be due to various circumstances, from a patient moving and merely losing contact, to a patient dying during the observation period. Patients can be lost to follow-up after randomization, but before the intervention, or after the intervention, but before the termination of the observation period.

Detection and prevention of nonadherence

Many clinical trials in various fields rely on the integrity of patient self-reporting for the determination of subject treatment compliance. This is particularly true of studies of nutrition, behavior, and oral medications. Studies in which patients must self-administer oral medications may traditionally rely on pill counts, though these are unreliable when compared to random urine or blood screening for the target drug or related metabolites.[1] The testing of bodily fluids only allows for the qualification of medication adherence at one point in time. However, most patients may only be partially noncompliant, forgetting to take a pill only occasionally. Because of readily available and relatively inexpensive testing, Riboflavin has been used in the past as an addition to study drugs and placebo to monitor medication adherence. However, dietary intake and supplementation can interfere with measurements.

Loss to follow-up is a difficult and complex issue to address due to the many, at times uncertain, reasons that subjects fail to follow up. Studies that are authorized to monetarily compensate subjects may choose to space out compensation and only distribute a portion at each follow-up or stage of the trial, rather than in a lump sum at the initiation or conclusion of the follow-up period.

Detection of artifactual nonadherence, as it pertains to professional or duplicate subjects, is essential to maintaining the integrity of the statistical analysis. Such subjects are more likely to respond to placebo and can be detrimental to the outcomes of a comparison analysis.[1] Although many subject questionnaires may ask about recent participation in other trials, professional subjects may withhold this information for the opportunity of monetary gain. Large subject databases exist, which if used can allow for early detection of these types of subjects so that they can be excluded from the study.

Addressing nonadherence

Upon learning of unplanned crossover or dropout in a particular study, it is the researcher's greatest temptation to recategorize these patients according to the intervention that they received, for example moving a patient that withdrew consent and did not receive an intervention to the control arm or excluding a patient who did not take the medication exactly as directed. However, to do so would completely undermine the purpose and effort of the initial randomization.

There are several options for how to handle a nonadherent subject. An investigator may choose to categorize a nonadherent subject depending on the type of nonadherence (real world vs. artifactual as defined previously).

The decision of how to address these scenarios is best made before the start of the trial to avoid bias. For example, an investigator may reduce the number of professional subjects in their trial by excluding applicants who have participated in another clinical trial within the last year, and confirm this by cross-referencing the list of applicants against a subject database. They also may outline in the protocol that all patients who were randomized to get the intervention will be considered as a part of the "exposed" group, meaning that some patients who drop out, are lost to follow up, or do not take the medication as directed will still be considered as having received the intervention for statistical purposes.

Randomization is performed as an attempt to uniformly distribute known and unknown characteristics of a study population into the different treatment groups. To perform an intention-to-treat analysis means that once a participant is randomized into a group, they are part of that group for statistical analysis, regardless of whether they received the intervention to which they were randomized.[2] In other words, "once randomized, always randomized." For example, a patient is randomly assigned to a treatment group. However, after enrolling in the study they unexpectedly become unstable and require hospitalization, precluding them from receiving the study drug. In an intention-to-treat analysis, this patient would still be counted in the "treated" group, even though they never received the medication.

An explanatory analysis, also referred to as a "per-protocol" analysis, is one in which participants are randomized initially, but recategorized based on the intervention that they received. In an explanatory analysis, the same subject introduced above would be recategorized as a control or removed from the analysis entirely. If an intention-to-treat analysis is not performed, and instead an explanatory analysis is implemented, it essentially defeats the purpose of the initial randomization. When an investigator chooses to recategorize or remove subjects, the known and unknown characteristics may no longer be evenly distributed, undermining the study in the process. On the other hand, intention-to-treat analysis can underestimate the efficacy of a studied intervention or exposure if a significant number of patients unexpectedly crossover to another treatment group.[3] Fig. 77.2 shows how the

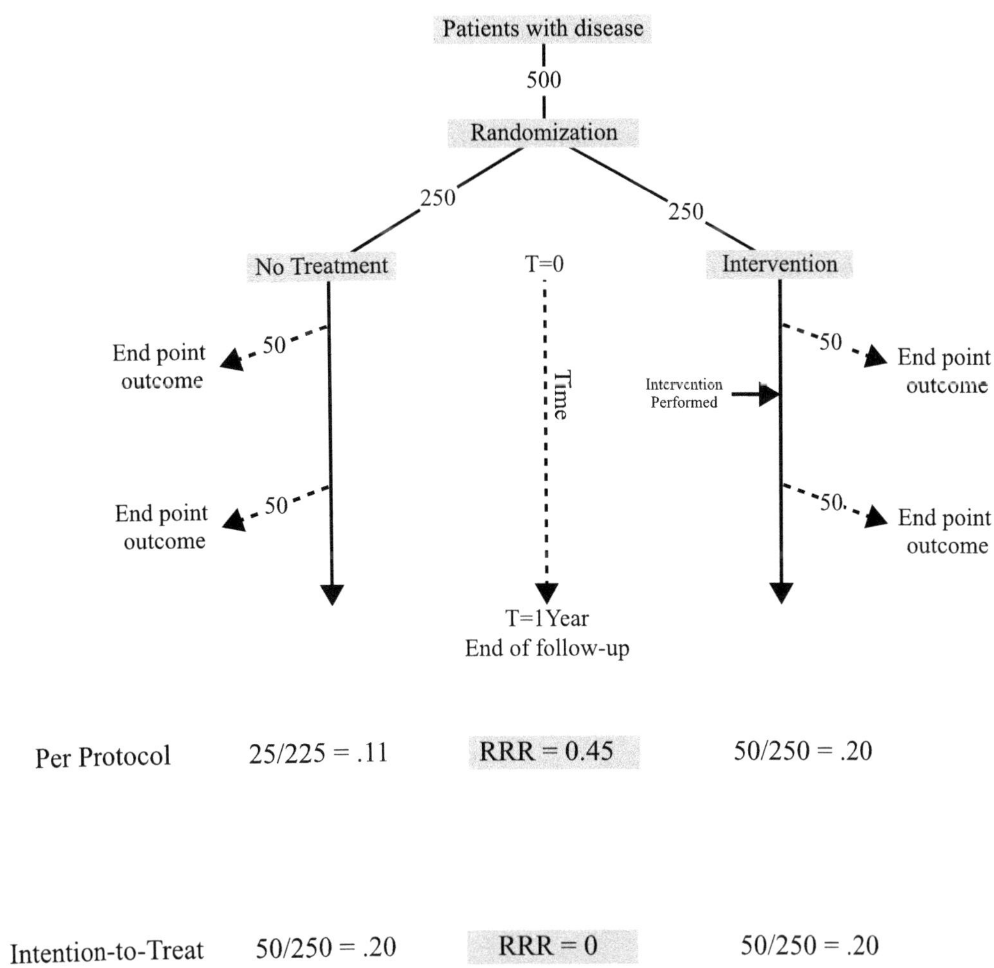

FIGURE 77.2 Intent-to-treat versus per-protocol analysis of treatment.

statistical outcomes of a study can be altered based on how crossover and nonadherence are addressed.

There are pros and cons to different methods of addressing noncompliance. It is a decision best made in the planning period when possible. However, if there is a significant amount of unexpected nonadherence your statistician may suggest an explanatory analysis. Whatever method you employ, it is important to critically evaluate the level of compliance in your study population, detail the reasoning behind your statistical approach in the methods, as well as address the effect that it may have had on outcomes in the discussion section of your manuscript.

Get started

- Review types of nonadherence and brainstorm potential sources of nonadherence in your study.
- Consider utilizing a communal subject database to identify and prevent potential sources of artifactual nonadherence.
- Determine, in advance, guidelines for how noncompliant patients will be categorized.

The book *A Dictionary of Epidemiology* (Porta M, 2014) provides a great resource on this topic from an epidemiological perspective [cite the book].

Potential pitfalls

- Study nonadherence is inevitable. Do not plan a study with an assumption of perfect adherence.
- Understanding the causes and planning a study properly, nonadherence can potentially be reduced to improve the study quality.
- Statistical analyses should account for nonadherence. Its impact should be discussed when publishing study findings.

References

1. McCann D, Petry N, Bresell A, Isacsson E, Wilson E, Alexander R. Medication nonadherence, "Professional subjects," and apparent placebo responders: overlapping challenges for medications development. *J Clin Psychopharmacol*. 2015;35(5):566–573.
2. Ranganathan P, Pramesh C, Aggarwal R. Common pitfalls in statistical analysis: intention-to-treat versus per-protocol analysis. *Perspect Clin Res*. 2016;7(3):144–146.
3. Hernán MA, Robins JM. Per-protocol analyses of pragmatic trials. *N Engl J Med*. 2017;377(14):1391–1398.

Further reading

1. Porta M, ed. *A Dictionary of Epidemiology*. 6th ed. Oxford University Press; 2014 [Accessed 25 June 2020].

Chapter 78

Monitoring committee in clinical trials

Sydney Cooper[1] and Nathan D. Kauffman[2]
[1]*Michigan State University College of Osteopathic Medicine, MI, United States;* [2]*Michigan State University College of Human Medicine, Grand Rapids, MI, United States*

Key points

- Data monitoring committee (DMC) is used as an unbiased committee to periodically review unmasked trial data and make interim suggestions to the trial sponsor
- DMC should only be used in long-term, large, potentially unsafe, or morbidity/mortality endpoint trials
- DMC enhances the ceiling for the success of trials when implemented properly and can actually hurt a trial when they are not

Who serves on monitoring committees

Data monitoring committees (DMCs) are composed of individuals with various professional backgrounds and have relevant experience. Clinicians with experience in specialties relevant to the clinical trial should be appointed to the DMC. Statisticians with expertise about the design of the clinical or with previous experience should also be considered for positions on a DMC.[1] Depending on the clinical trial, a DMC should also include experts within the fields of ethics, pharmacology, and epidemiology (Table 78.1). It is suggested that members have a fundamental understanding of clinical trials and previous participation on a DMC.[1] It is not required that members of the DMC have prior experience or that trial sponsors provide training, although it is highly sought. In the instance that training resources are provided, there are numerous options including but not limited to lectures, articles, and textbooks.[2] It is important to note that individuals appointed to serve on a DMC should not have significant conflicts of interest.[1,3] This includes investigators in the trial and those with financial, intellectual, and personal conflicts of interest. The trial sponsor or members of the trial steering committee hold the responsibility of selecting individuals to serve on the DMC.[4]

When to use a monitoring committee

Use of a DMC is only required when informed consent is not used, therefore most trials require the sponsor to make the decision. There are many variables that need to be considered when a sponsor is deciding to include a DMC. The major variables are study length, size, safety, and endpoint (Table 78.2).[1] Only studies that are long term should include a DMC. DMCs take time to review data and introduce potential interventions, so short-term studies would not be able to reap any benefits of a DMC and would only complicate the trial. Long-term studies can also typically accept major changes to trial criteria, giving the DMC an avenue for eliciting adjustments. Trials of significant size, particularly multicenter trials, become increasingly complex and warrant unbiased data review expertise to properly manage the study. DMCs provide important safety recommendations during the trial and therefore are critical in any trials with fragile patients (e.g., pregnant women), patients with increased risk of death, particularly invasive procedures, or potential toxicities.[1] Endpoints that deal with morbidity and mortality also warrant the use of a DMC. Any study with an endpoint that is significant enough to consider early study termination after mid-trial analysis showing favorable or unfavorable results strongly benefits from a DMC.[1] The DMC is able to use interim analysis to protect patients by either pushing the intervention forward more quickly or ending the study before more futile treatments.

Data monitoring committee responsibilities

The Data monitoring committee has varied responsibilities that focus on protecting the individuals participating in a clinical trial. Once members of the committee have been

Translational Interventional Radiology. https://doi.org/10.1016/B978-0-12-823026-8.00106-1

TABLE 78.1 Summary of potential professionals to include on a data monitoring committee depending on needs and focus of the clinical trial.

Member:	Potential role of member:
Clinician	Possesses expertise about specialties relevant to the clinical trial
Statistician	Extensive knowledge of the clinical trial method and statistical analysis
Ethicist	Understanding of ethical codes in the setting of clinical trials
Pharmacologist	Knowledge about pharmacologic agents relevant to a clinical trial

TABLE 78.2 Summary of which major trial attributes result in recommendation of a data monitoring committee and why trials with those attributes benefit from use of a data monitoring committee.

Trial attribute:	Relationship of attribute to DMC:
Length	Long-term studies allow time and space for DMC review and intervention
Size	Large or multicenter studies are complex and benefit from DMC's
Safety	DMC's improve safety in trials with fragile subjects, invasiveness, or toxicity
Endpoint	Critical endpoints are more likely to warrant early trial termination via DMC

appointed, the roles and expectations of each individual should be communicated. An initial meeting should be held to establish committee and individual responsibilities. Following this, it is expected that regularly scheduled meetings will be held. These meetings can be open, allowing individuals directly involved in conducting the clinical trial to participate, or closed, for DMC members only. An executive session is usually held following a closed session to discuss recommendations from the DMC with the study sponsor (Fig. 78.1). Both meeting types offer benefits in achieving the ultimate goal of overseeing and evaluating the progress of clinical trials.[5] Based upon the data assessed by the DMC, it is necessary that they provide recommendations, such as continuation with or without conditional approval, or termination of the trial. Assessments comparing the potential harm and outstanding benefit of ongoing clinical trials are conducted by DMCs (Fig. 78.2). Less data is necessary when evaluating for potential harm compared to benefit. For example, if statistical analysis suggests that the clinical trial is leading to harm to trial subjects, less proof would be necessary before a recommendation is made when compared to a recommendation for trial termination potential benefit.

The DMC is also responsible for monitoring the effectiveness of the clinical trial. This is accomplished by periodic assessment of data trends and results. Interim data should be assessed to evaluate whether the proposed benefit of the study is achievable.[1] The responsibility of reviewing data related to the conduct of a clinical trial is also held by the DMC. Monitoring supplementary information for potentially impactful data is necessary throughout ongoing trials. This is especially important in regard to information relating to safety.[1] A process of communication between the DMC and sponsor should be established and the suggested format is written. Recommendations on how to proceed with a clinical trial in regard to modification, continuation, or termination, are provided by the DMC, but the responsibility of acting on them is left up to the sponsor.[5]

Real-world examples

Broadly, these committees are important because both large and invasive trials have unique challenges that require an unbiased group to make data-centered decisions. These decisions can allow for slight iterations in endpoints or design to achieve efficacious results or for pausing futile studies before they cost the patients and sponsors additional time and money. Below are three examples of studies that successfully utilized DMC's:

The radial versus femoral access for coronary angiography and intervention in patients with acute coronary syndromes (RIVAL) trial was a randomized clinical trial across 158 hospitals and 32 countries. The trial found that both radial and femoral access are safe for percutaneous coronary intervention, but that radial access had significantly less severe outcomes. This trial used an independent DMC to evaluate their unmasked data.[6] The large size of the trial encouraged the sponsor to include a DMC to properly assess large quantities of data and make any

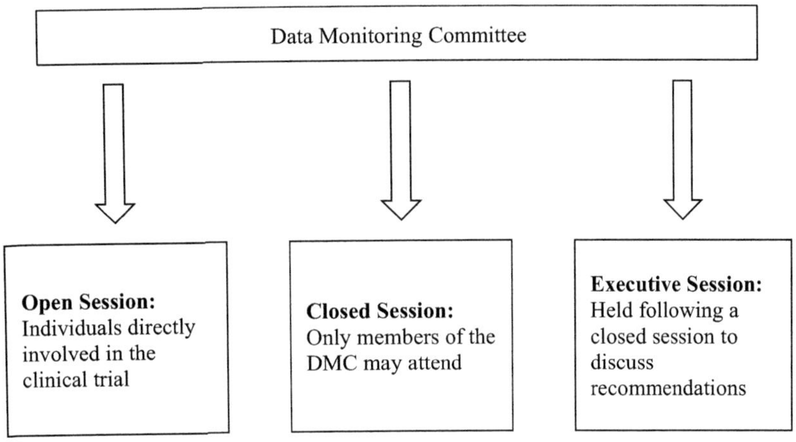

FIGURE 78.1 Flow chart demonstrating the types of meetings held by a clinical trial DMC.

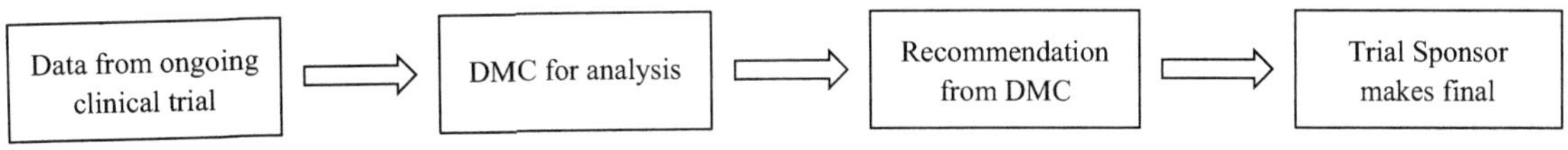

FIGURE 78.2 Flow chart demonstrating the process of developing a recommendation for an ongoing clinical trial.

necessary adjustments throughout the study duration. Dealing with data from such a large number of research centers would have not only been a difficult task for one of the other committees associated with the trial but would have been impossible to appraise the data in an unbiased way.

The Heidelberg liver cancer (HeiLivCa) trial was a phase III randomized control trial with one treatment arm receiving chemoembolization with sorafenib and the other receiving chemoembolization with placebo. The trial used a data and safety monitoring board (DSMB) composed of three physicians and one statistician to review their study periodically. The responsibilities of the DSMB included monitoring of serious adverse events stratified by body system, severity, and causality.[7] No adverse events or specific DSMB interventions were reported, likely meaning the close monitoring of these events allowed the trial to continue in a safe manner. Utilization of a DSMB in this trial by the sponsor was warranted, as the trial was multicenter. Furthermore, patients with advanced hepatocellular carcinoma are very sick meaning any indication of significant benefit or futility would encourage early termination of the trial.

Celsion's Phase III OPTIMA study was advised by their DMC to discontinue monitoring of their subjects as the study was unlikely to meet its endpoints. The global, double-blind trial was investigating radiofrequency ablation treatment of liver cancer with or without the use of ThermaDox. The treatments had already been delivered to patients and the study was 2 years into the 5-year posttreatment follow-up when the second interim analysis report from the trial's DMC advised terminating the study as the endpoints were not being achieved and there was high futility. Celsion ultimately decided to continue following up with patients as the expensive treatments were already completed and there was no risk to patient safety by continuing the study.[8] In this scenario, the DMC did not advise changes to the trial during the treatment phase but rather the patient outcomes analysis, which was still costing Celsion money to perform. Celsion could have chosen to terminate the trial to save money based on the unbiased analysis showing treatment futility, which is information they would not have if they did not have a dedicated DMC to review unmasked data and make a recommendation.

Get started

- Fully understand the characteristics of your trial conditions and subjects
- Make a decision as to whether a DMC is necessary for your clinical trial
- Decide what professionals should be included on a potential DMC
- Contact professionals with prior DMC experience

Potential pitfalls

- Avoid choosing to include a DMC when it is not needed as it will only increase the complexity of the study
- Avoid choosing members for the DMC that are not qualified or fit their particular role as they will only make poor judgments that could derail the project

- Avoid conflicts of interest when choosing a DMC, particularly if any member of the DMC is advising other trials, as this can lead to many complications
- Avoid having the sponsor gain access to unmasked interim data through DMC as this leads to unmeasurable biases throughout the remainder of the study

Defined terms:

- Data monitoring committee: An independent data review entity, comprised of a panel of experts, that provides advice on trial continuation based on unmasked data available only to the committee
- Data and safety monitoring board: See data monitoring committee
- Data and safety monitoring committee: See data monitoring committee
- Serious adverse event: Death or life-threatening illness warranting hospitalization after medical intervention
- Other adverse events: Any unwanted or unintended side effect associated with recent medical treatment that may or may not be directly related to the treatment
- Endpoint: The clinical study objective that is being measured
- Futility: The inability of a trial to achieve its objectives

Resources

- Training for DMC members: Coursera, Open CourseWare, and iTunes U[2]

References

1. *Guidance for Clinical Trial Sponsors, Establishment and Operation of Clinical Data Monitoring Committees*. U.S. Department of Health and Human Services Food and Drug Administration; 2006. https://www.fda.gov/media/75398/download. Accessed March 22, 2021.
2. Fleming TR, DeMets DL, Roe MT, et al. Data monitoring committees: promoting best practices to address emerging challenges. *Clin Trials*. 2017;14(2):115–123.
3. Fleming TR, Ellenberg SS, DeMets DL. Data monitoring committees: current issues. *Clin Trials*. August 2018;15(4):321–328.
4. Bierer BE, Li R, Seltzer J, et al. Responsibilities of data monitoring committees: consensus recommendations. *Ther Innov Regul Sci*. September 2016;50(5):648–659.
5. Calis KA, Archdeacon P, Bain R, et al. Recommendations for data monitoring committees from the clinical trials transformation initiative. *Clin Trials*. 2017;14:342–348.
6. Jolly SS, Yusuf S, Cairns J, et al. Radial versus femoral access for coronary angiography and intervention in patients with acute coronary syndromes (RIVAL): a randomised, parallel group, multicentre trial. *Lancet*. April 23, 2011;377(9775):1409–1420.
7. Hoffmann K, Glimm H, Radeleff B, et al. Prospective, randomized, double-blind, multi-center, Phase III clinical study on transarterial chemoembolization (TACE) combined with Sorafenib versus TACE plus placebo in patients with hepatocellular cancer before liver transplantation – HeiLivCa [ISRCTN24081794]. *BMC Cancer*. November 26, 2008;8:349.
8. *Celsion Corporation Receives Recommendation from Independent Data Monitoring Committee to Consider Stopping the Phase III OPTIMA Study*. Celsion Corp; 2020. https://investor.celsion.com/news-releases/news-release-details/celsion-corporation-receives-recommendation-independent-data. Accessed March 22, 2021.

Part VIII

Regulatory

Chapter 79

The FDA: An overview of their role

Shashidhara Murthy[1] and Sandeep Murthy[2]
[1]Mount Sinai Health System, New York, NY, United States; [2]University of Virginia School of Medicine, Charlottesville, VA, United States

Key points

- The Food and Drug Administration (FDA) is a government agency that regulates many products to ensure public safety and consumer health in the United States.
- The FDA oversees clinical trials and approves marketing and sale of applicable drugs and medical devices.
- Organizationally, the FDA is divided into many different branches allowing for each branch to specialize in a particular niche.

Why it matters?

The Food and Drug Administration (FDA) is the key organization that directly oversees the approval process for drugs, biologicals, and medical devices. As a clinical researcher, it is vital to have knowledge of the FDA's role in clinical research and how to navigate the FDA's process of successfully bringing a drug or device to market. The FDA's responsibilities are not exclusive to clinical research, and it is useful to understand the broad range of consumer and medical products the FDA regulates. As the United States expanded, our desire to study human health led to the production of various drugs and medicinals. The nation quickly realized that there needed to be an organization to ensure the products entering the market were safe for the public. Today, the FDA is the go-to regulatory agency for a growing list of products.

FDA history

Throughout both the consumer and medical history of the United States, there have been numerous issues of food and drug safety that have impacted public health. These issues were initially unsuccessfully managed by state and local governments. As these issues continued, a division called the Bureau of Chemistry, under the Department of Agriculture, was created to enforce the Pure Food and Drugs Act of 1906. This department eventually became the FDA, which found a home under the division of the US Department of Health and Human Services. Over time, the FDA gained more responsibilities and now regulates food, drugs, medical devices, radiation-emitting products, vaccines, blood and biologicals, veterinary products, cosmetics, and tobacco (Fig. 79.1).[1]

The FDA is currently responsible for the 50 states in the United States, the District of Columbia, Puerto Rico, Guam, the Virgin Islands American Samoa, and other US territories and possessions.[1]

What does the FDA regulate?

The FDA regulates numerous product categories. In this section, we will overview the different categories and products that may fall within each category. A condensed example list can be found in Fig. 79.2.

- Food: This category includes all food and food products for sale through interstate commerce except meat from livestock, poultry, and some egg products. The latter products are regulated by the Department of Agriculture. Other notable products in this category include dietary supplements, bottled water, food additives, infant formulas, and other food products. Although bottled water is regulated by the FDA, national standards for drinking water from municipal water supplies are set by the Environmental Protection Agency.[2]
- Drugs: This category includes both prescription and over-the-counter medicine. Ensuring safety and efficacy of approved drugs as well as the approval process are regulated by the FDA. Abuse and misuse of controlled substances, such as prescription medication, are monitored by the Drug Enforcement Agency. It is important to notice that alcohol is not regulated by the FDA, rather by the department of the Treasury's Alcohol and Tobacco Tax and Trade Bureau[2]

Translational Interventional Radiology. https://doi.org/10.1016/B978-0-12-823026-8.00086-9

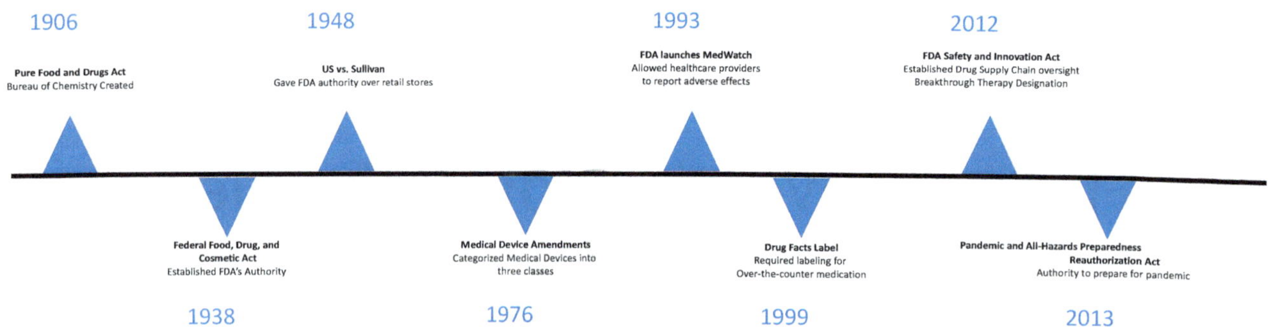

FIGURE 79.1 FDA timeline.

- Biologics: Common biologics the FDA regulates include human vaccines, blood and blood products, cellular and gene therapy products, and allergenics.[2]
- Medical devices: Medical devices are broken into three classes based on intended use and potential adverse effects. Typically Class I and II devices are exempt from FDA regulation but sometimes may require a 510(k) notification made to the FDA before device marketing efforts. Class III devices are highly regulated and require Premarket approval before market availability. Examples of medical devices include dental floss (class I), pregnancy tests (class II), and stents (class III).[2]
- Electronic products that give off radiation: It is commonly known that radiation-producing products are potentially harmful to health. These devices require some sort of monitoring and regulation. This category includes products such as microwave ovens, X-ray equipment, laser products, ultrasonic therapy equipment, mercury vapor lamps, and sunlamps.[2]
- Cosmetics: Although makeup is not regulated by the FDA, the color additives in makeup typically are. This category also includes personal care products, skin moisturizers and cleansers, nail polish, and perfume. The FDA does not have the legal authority to approve cosmetics before they go on the market; however, they do have the ability to take legal action against a manufacturer if they have evidence that the cosmetic does not abide by their legal requirements for safety.[2]
- Veterinary products: Similar to human food and drugs, animal-specific feeds and veterinary drugs are regulated by the FDA. Vaccines for animals are not regulated by the FDA; however, they are regulated by the US Department of Agriculture's Animal and Plant Health Inspection Service.[2]
- Tobacco products: This category includes cigarettes, roll-your-own tobacco, and smokeless tobacco. Recently, electronic cigarettes were added to this category for regulatory purposes.[2]

How is the FDA organized?

The FDA is organized into divisions focused on a central task. A general breakdown of the major offices and centers is listed in the following and outlined in Fig. 79.3.[3]

The Office of the Commissioner[4]:

- Includes the Commissioner of Food and Drugs, the Principal Deputy Commissioner, and Chief of Staff.

The Office of the Chief Counsel[5]:

- Reports directly to the general counsel of Health and Human Services
- Advises the Commissioner of Food and Drugs.

The Office of the Executive Secretariat[6]:

- Reports directly to the Commissioner of Food and Drugs.

The Office of the Counselor to the Commissioner[7]:

- Leads advocacy and advancement of Commissioner's priorities
- Advises the Commissioner on issues that are not well defined enough to be managed by one particular center or office.

Center for Biologics Evaluation and Research (CBER)[8]:

- Reports directly to the Commissioner of Food and Drugs.

Center for Devices and Radiological Health (CDRH)[9]:

- Oversees medical devices and radiation-emitting products.
- Aids in medical-device innovation.

Center for Drug Evaluation and Research (CDER)[10]:

- Regulates over-the-counter and prescription drugs.
- Includes other over-the-counter products such as sunscreen and medicated shampoos.

Center for Food Safety and Applied Nutrition[11]:

Product Category	Examples	Key Exclusions
Food	• Dietary supplements • Food Additives • Infant Formulas • Bottled Water	• Meat • Poultry • Tap Water • Catfish
Drugs	• Prescription drugs (antibiotics, opiates, etc...) • Many Over-The-Counter Drugs	• Alcohol • Fitness supplements
Biologicals	• Vaccines • Blood and blood products • Cellular and Gene therapy products • Tissue and tissue products • Allergenics	• Household chemicals • Pesticides
Medical Devices	• Floss • Surgical implants • Prosthetics	• Gloves • Shoe covers
Electronic Products that give off Radiation	• Microwave ovens • X-ray equipment • Laser products • Ultrasonic therapy equipment • Mercury vapor lamps • Sunlamps.	• Laptops • Power tools
Cosmetics	• color additives • skin moisturizers and cleansers • nail polish and perfume.	• Soap
Veterinary Products	• livestock feeds • pet foods • veterinary drugs	• Animal vaccines • Animal devices
Tobacco Products	• Cigarettes • Roll-your-own tobacco • Smokeless tobacco • E-cigarettes	• Smoking paraphernalia

FIGURE 79.2 FDA regulatory product categories.[27]

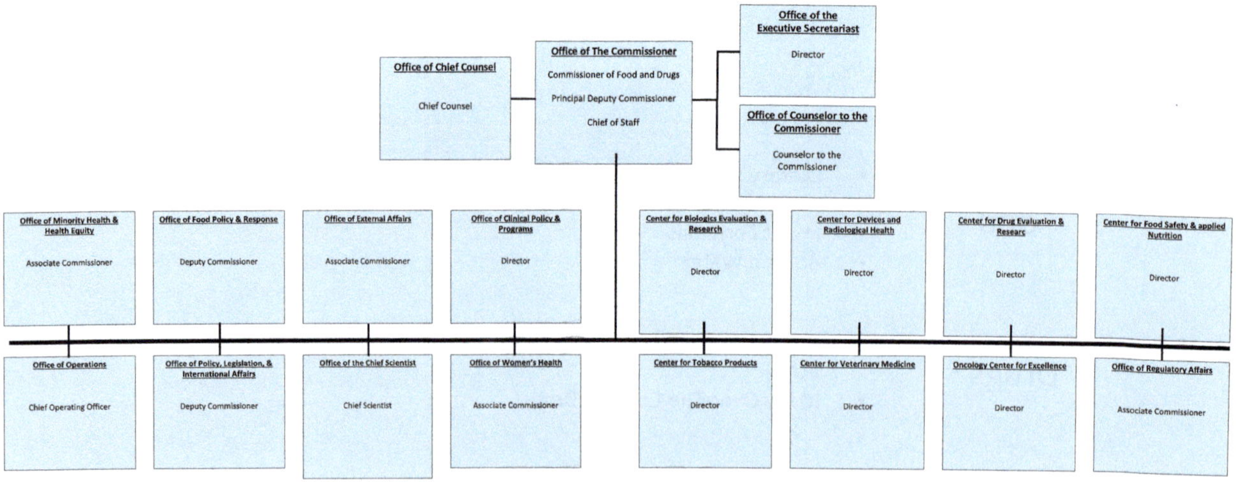

FIGURE 79.3 Organizational schema of the FDA.

- Determines the safety of new ingredients/additives for food.
- Ensures proper labeling of products.
- Responsible for investigating causes of foodborne illness outbreaks.

Center for Tobacco Products[12]:

- Oversees implementation of the Tobacco Control Act passed by Congress in 2009.
- Responsible for regulating advertising and promotion of tobacco products.

Center for Veterinary Medicine[13]:

- Ensures safe and effective drugs for pets and food-producing animals.
- Ensures safety of pet food and proper labeling.

Oncology Center of Excellence[14]:

- Involved in expediting development of medical products used for oncologic and hematologic malignancies.
- Provides clinical review to ensure safe and effective oncologic products.

Office of Regulatory Affairs (ORA)[15]:

- Lead office for all agency field activities.
- Responsible for continuous monitoring of regulated products.
- Ensures safety of imported products.

Office of Clinical Policy and Programs[16]:

- Involved in both creating and implementing policies related to medical product development and evaluation.
- Composed of the Clinical Policy Program and the Office of Good Clinical Practice.

Office of External Affairs[17]:

- Responsible for communicating and educating the news media, health professionals, and the general public.
- Maintains agency-wide communication to maintain uniform practices.
- Liaison between the FDA and advocacy groups.

Office of Food Policy and Response[18]:

- Involved in a cross-center collaboration that helps advance important food policy.

Office of Minority Health and Health Equity[19]:

- Protects the health of minority health groups.
- Promotes increased clinical trials involving racial and ethnic minorities.

Office of Operations[20]:

- Coordinates emergency response to incidents that involve FDA-regulated products.

Office of Policy Legislation, International Affairs[21]:

- Provides Congress with the most updated and accurate domestic and international information regarding medical research that can impact policy and legislation.
- Determines how pending or recently passed legislation impacts current FDA-approved products.

Office of the Chief Scientist[22]:

- Involved in supporting scientific research and development of drugs and technology that will target public health needs.
- Protects and enforces scientific integrity when training FDA scientists.

National Center for Toxicological Research[23]:

- Facilitates toxicology research and training

- Oversees national and international collaborations between government, academia, and industry scientists.

Office of Women's Health[24]:

- Established by a Congressional mandate in 1994
- Responsible for scientific, ethical, and policy issues that pertain to women's health
- Promotes inclusion of women in clinical trials

Each of the above offices has branches within them that consist of teams focused on compliance, statistics, management, and much more. The FDA has clinical investigators leading their own research as well as teams designated to monitoring and reporting on health outcomes. For example, "MRI-related FDA Adverse Event Reports: A 10-yr Review," authored by investigators in the FDA's Center for Devices and Radiological Health, illustrates reported adverse events over a 10-year span.[25] Reviews such as these contribute to the scientific and medical community by increasing awareness of long-term adverse effects in patients.

How this relates

The FDA is the main regulatory agency responsible for the oversight and review of clinical trials.

Many times, drugs and medical devices require multiple reviews before approval. These multiple reviews seem to play a vital role in ensuring products on the market are both safe and effective.[26] Four of the relevant branches of the FDA that play a direct role in clinical research are the CBER, CDER, CDRH, and the ORA. It is important to understand the branch that your research falls under so that questions can be directed to the correct teams.

Get started

- Go to the FDA's website at www.FDA.gov and click on drugs, then proceed to the "tools to keep you informed" page. Browse through the app descriptions and download any that may be of value.
- Visit the FDA Fast Facts page and read some of the headlines. This should give a better understanding of the vast reach of the FDA.

References

1. *What Does FDA Do?*; 2018. Retrieved from https://www.fda.gov/about-fda/fda-basics/what-does-fda-do.
2. *What Does FDA Regulate*; 2018. Retrieved from https://www.fda.gov/about-fda/fda-basics/what-does-fda-regulate.
3. FDA Organization. Retrieved from https://www.fda.gov/about-fda/fda-organization; 2020.
4. *Office of the Commissioner*; 2020. Retrieved from https://www.fda.gov/about-fda/fda-organization/office-commissioner.
5. Office of the Chief Counsel. Retrieved from https://www.fda.gov/about-fda/office-commissioner/office-chief-counsel; 2019.
6. Office of the Executive Secretariat. Retrieved from https://www.fda.gov/about-fda/office-commissioner/office-executive-secretariat; 2018.
7. FDA Staff Manual Guides. *Office of the Counselor to the Commissioner*; 2018. Retrieved from https://www.fda.gov/media/81977/download#: ~ :text=1.,which%20is%20not%20well%20defined.
8. Center for Biologics Evaluation and Research (CBER). Retrieved from https://www.fda.gov/about-fda/fda-organization/center-biologics-evaluation-and-research-cber; 2019.
9. Center for Devices and Radiological Health. Retrieved from https://www.fda.gov/about-fda/fda-organization/center-devices-and-radiological-health; 2020.
10. Center for Drug Evaluation and Research I CDER. Retrieved from https://www.fda.gov/about-fda/fda-organization/center-drug-evaluation-and-research-cder; 2019.
11. Center for Food Safety and Applied Nutrition (CFSAN). Retrieved from https://www.fda.gov/about-fda/fda-organization/center-food-safety-and-applied-nutrition-cfsan; 2018.
12. *Center for Tobacco Products*; 2018. Retrieved from https://www.fda.gov/about-fda/fda-organization/center-tobacco-products.
13. Center for Veterinary Medicine. Retrieved from https://www.fda.gov/about-fda/fda-organization/center-veterinary-medicine; 2019.
14. Oncology Center of Excellence. Retrieved from https://www.fda.gov/about-fda/fda-organization/oncology-center-excellence; 2020.
15. Office of Regulatory Affairs. Retrieved from https://www.fda.gov/about-fda/fda-organization/office-regulatory-affairs; 2020.
16. Office of Clinical Policy and Programs. Retrieved from https://www.fda.gov/about-fda/office-commissioner/office-clinical-policy-and-programs; 2020.
17. Office of External Affairs. Retrieved from https://www.fda.gov/about-fda/office-commissioner/office-external-affairs; 2018.
18. Office of Food Policy and Response. Retrieved from https://www.fda.gov/about-fda/office-commissioner/office-food-policy-and-response; 2020.
19. Office of Minority Health and Health Equity. Retrieved from https://www.fda.gov/about-fda/office-commissioner/office-minority-health-and-health-equity#: ~ :text=Mission,science%20that%20addresses%20health%20disparities; 2020.
20. *Office of Operations*; 2019. Retrieved from https://www.fda.gov/about-fda/office-commissioner/office-operations.
21. Office of Policy, Legislation, and International Affairs. Retrieved from https://www.fda.gov/about-fda/office-commissioner/office-policy-legislation-and-international-affairs; 2019.
22. Office of the Chief Scientist. Retrieved from https://www.fda.gov/about-fda/office-commissioner/office-chief-scientist; 2018.
23. National Center for Toxicological Research. Retrieved from https://www.fda.gov/about-fda/office-chief-scientist/national-center-toxicological-research; 2019.
24. Office of Women's Health. Retrieved from https://www.fda.gov/about-fda/office-commissioner/office-womens-health; 2019.
25. Delfino JG, Krainak DM, Flesher SA, Miller DL. MRI-related FDA adverse event reports: a 10-yr review. *Med Phys*. 2019;46(12):5562–5571. https://doi.org/10.1002/mp.13768.
26. Ross JS, Dzara K, Downing NS. Efficacy and safety concerns are important reasons why the FDA requires multiple reviews before approval of new drugs. *Health Aff*. 2015;34(4):681–688. https://doi.org/10.1377/hlthaff.2014.1160.
27. *Regulated Products*; 2020. Retrieved from https://www.fda.gov/industry/import-basics/regulated-products#: ~ :text=The%20FDA%20regulates%20a%20wide,human%20use%3B%20radiation%2Demitting%20electronic.

Chapter 80

Investigational new drugs

Jennifer J. Dennison and Thaddeus Maguire
Vascular and Interventional Radiology, Medical College of Wisconsin, Milwaukee, WI, United States

Key points

- The investigational new drug (IND) application is the process in which a sponsor receives permission from the Food and Drug Administration (FDA) to distribute a new drug to investigators across state lines for clinical study, emergency use, or treatment despite Federal law requiring an approved marketing application beforehand.
- After a sponsor deems a molecule to have sufficient pharmacological potential to qualify for further development and to be reasonably safe for diagnostic and therapeutic testing in humans, the molecule's legal status changes to a new drug under the Federal Food, Drug, and Cosmetic Act and it becomes accountable to the requirements of drug regulation.
- Resources are available to researchers who aim to simplify what has been considered an intimidating and drawn-out application process.

Why it matters?

Advancement of a pharmaceutical agent from preclinical development by a commercial sponsor (pharmaceutical manufacturer or potential marketer) to clinical development by a sponsor-investigator (physician) places the molecule under the jurisdiction of the Food and Drug Administration (FDA) and requires an investigational new drug (IND) application. There are three types of IND applications that fall into two general categories: commercial or research (non-commercial). Commercial sponsors generally employ individuals with specific training to oversee the development process. This chapter is meant to aid the individual investigator in navigating the IND application process. It will discuss the components of each IND application in the research category and guide the reader through the IND application process for the independent sponsor-investigator, utilizing drugs often encountered in interventional radiology procedures as examples. Common challenges and budgetary considerations will also be discussed.

Overview[1,2]

Clinical trials involving human subjects that utilize drugs either not yet approved by the FDA or else in an off-label fashion require the completion and certification of an IND application.[1] These studies must adhere to strict regulations regarding the investigational agent itself, in addition to the standard guidelines set forth for the protection of human research participants. Failing to do so could result in legal and monetary penalties for the investigator conducting the research and the associated institutions.

Despite requiring only three sets of forms, the IND application process can be daunting to navigate. Paperwork errors may result in a clinical hold being placed by the FDA that can lead to frustrating delays in research. Adequate support can prevent delays and ensure efficient completion of all steps. The FDA website provides ample resources such as downloadable forms, explanatory guidance documents, lists of specific requirements for each type of IND, manuals for policies and procedures, as well as contact information for FDA center officials to answer more individualized questions.[2] These resources along with this book chapter should help ease researchers through the application process (Fig. 80.1).

Process/pathways[1,3–12]

The first step is to identify the study sponsor, determine whether the proposed trial requires an IND and, if so, the type of IND application required. If the potential study uses an FDA-approved drug for its on-label indication, then an IND is not necessary. In certain cases, a trial may also qualify for an IND exemption. Exemptions may be granted to trials employing a drug that is

Translational Interventional Radiology. https://doi.org/10.1016/B978-0-12-823026-8.00050-X

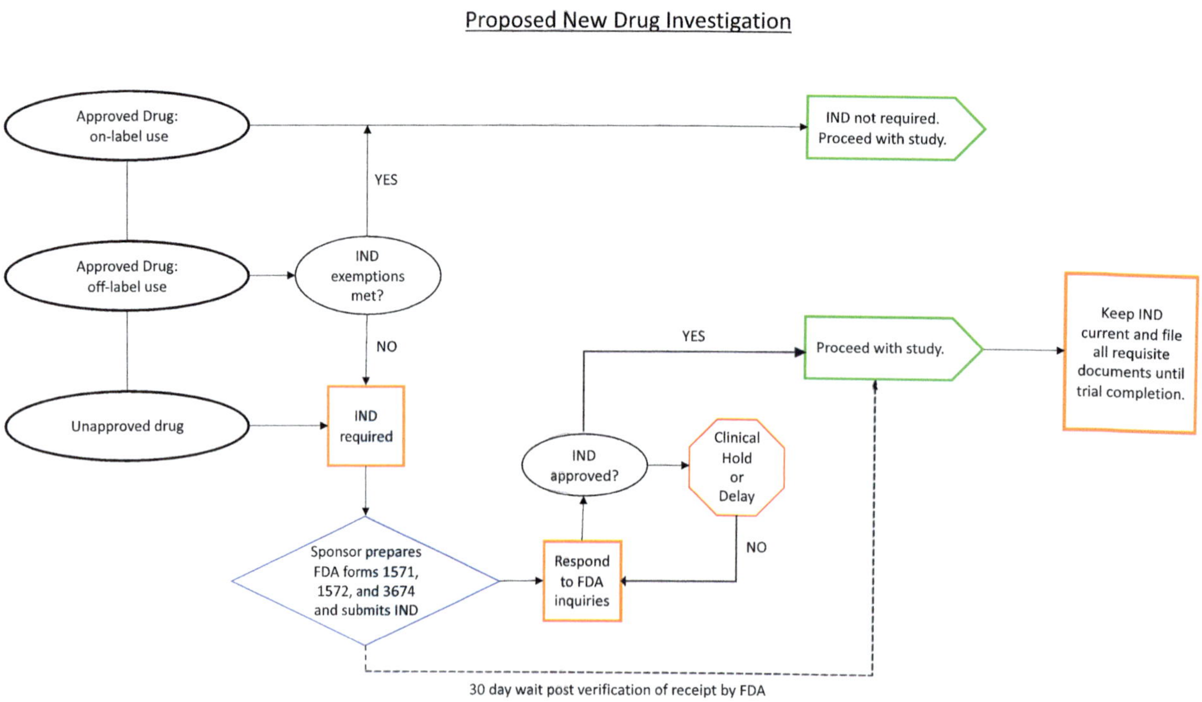

FIGURE 80.1 IND application process map.

marketed and commercially available but in the evaluation of a new route of administration, dosage, proportion, or patient population.[3] Further information regarding the regulations and exemptions can be found on the FDA website.[4]

The sponsor (initiator) type determines the category of the research trial.[5] Studies initiated by corporations or those intended to develop pharmaceuticals require a Commercial IND. A Research IND application, on the other hand, is intended for the individual researcher, academic institution, or nonprofit organization.[1] An investigator IND is for the investigator who will initiate, conduct, and supervise the study and is the focus of this chapter.[1] Less common applications include the Emergency IND, for permission to use an experimental drug in an emergent situation, and the Treatment IND to enable the delivery of investigational drugs for the treatment of serious, life-threatening illnesses that have no other alternative treatment.[1]

Once the appropriate category has been determined, the investigator can begin the application process. Although not required, the FDA encourages the initiation of contact early in the clinical plan development process as a part of the pre-IND consultation program.[6] The intention of this program is to promote early communication between the agency and the sponsor; however, sponsors can request advice at any point during IND development by submitting a pre-IND meeting request to the appropriate office.[7]

Once the clinical investigational plan is developed the IND application can be completed, and the investigator can submit the application and cover letter. The completed application can be submitted in triplicate either electronically (preferred) or on paper to the appropriate division that typically would be one of either the Center for Drug Evaluation and Research (CDER) (most common), the Center for Biologics Evaluation and Research, or the Center for Devices and Radiological Health.[8] A letter will be sent to the investigator to acknowledge the date of receipt of the application. The letter will contain contact information for the FDA manager tasked with reviewing the application as well as the IND number assigned to the study. In the absence of further correspondence from the FDA, the investigator may begin the trial 30 days from the date the FDA received the application.[1] The only exception to this is if the FDA notifies the investigator that they may begin earlier than the 30 days,[5] or a partial or complete hold has been placed until further information is provided by the investigator. In the latter case, the study cannot begin until the FDA explicitly states that the hold has been lifted.[9]

Acceptable completion of an IND marks the beginning of the investigators' responsibility to keep the IND current, complete annual reports, and notify the FDA of changes in protocol, safety breeches or that the study is terminated for any reason.[10] Accurate and complete recording of data and strict adherence to ethical standards is imperative.[11] The investigator must correct any issues or terminate a study and notify all parties involved if these requirements cannot be met.[12]

TABLE 80.1 IND Application Components. There are three necessary forms for an IND application: FDA Form 1571, FDA Form 1572, and FDA Form 3674. A cover letter must accompany an IND upon submission. Additionally, other forms for correspondence and monitoring of a study are required.[18]

IND application components	Description
IND application, FDA Form 1571	Many sections will not be required for the sponsor-investigator. Additionally, other responses are not required or require only brief answers if the drug is already approved by the FDA such as Sections 6, 8, 13, 18, and 19. Enter "0000" for the initial serial number and increase that value by 1 with each successive amendment. In studies that will utilize an approved drug according to the labeling, all drug names, dosage information, and lot number should be included.[14]
IND application, FDA Form 1572	Referred to as the "Investigator's statement" the information requested by this form comprises the qualifications and contact information of the principal investigator. These requirements can generally be fulfilled by supplying an academic curriculum vitae and noting its inclusion on the form.[15]
IND application, FDA Form 3674	Certifies that the study is registered through ClinicalTrials.gov in accordance with section 402(j) of the Public Health Service Act. Investigator must register and obtain an NCT number by filing trial information with the Protocol Registration System of clinical trials run by the US National Library of Medicine at the National Institutes of Health. If the overseeing institution has a registration account, a separate account will not need to be made.[16]
Cover letter	Must be signed and addressed to the Director of the appropriate Review Division in the Office of New Drugs. It should identify the submission as an IND, include the study title and type, the IND product's name and proposed formulation, the manufacturer's name and contact information, the disease or condition under investigation, and reference to any existing IND application.[17]
Table of contents	Information in the application should be indexed by volume and page with enough detail to facilitate effortless location of information.[18]
Introductory statement and general investigational Plan	A 2–3-page document to aid in the understanding of the clinical development plan and facilitate the anticipation of future program needs by the FDA. If a detailed development plan is incomplete, the reasons should be clearly stated and an explanation for future plans should be briefly explained.[18]
Investigator's brochure (IB)	Investigators that have right of reference to an existing manufacturer's IND application do not require and IB. However, the IB is then updated. Content expectations and format suggestions are provided on the FDA website.[12]
Nonclinical components	Chemistry, manufacturing, and control information as well as pharmacology and toxicology information are required to a lesser extent than that required for a commercial IND.[19]
Clinical Protocol	Thoroughly detail all elements pertaining to study safety, enrollment criteria, and toxicity assessment. Considerations vary according to trial phase and are detailed on the FDA website.[20]
Response to clinical hold	Address all cited deficiencies and submit in writing within 30 days along with a cover letter identifying the correspondence as such and marked with the IND number.[21]
Protocol amendments (21 CFR 312.30)	Submitted for changes in the study protocol or the addition of a new investigator beforehand unless the change is to avoid immediate harm to study participants.[22]
Information amendments (21 CFR 312.31)	Filed for changes to crucial IND information that does not apply to safety, annual or protocol reports. Restricted to one per 30 days. Contents must be

Continued

TABLE 80.1 IND Application Components. There are three necessary forms for an IND application: FDA Form 1571, FDA Form 1572, and FDA Form 3674. A cover letter must accompany an IND upon submission. Additionally, other forms for correspondence and monitoring of a study are required.[18]—cont'd

IND application components	Description
	clearly identified in the title and include clear reasoning, data, and request for FDA comment (if desired).[10]
Safety reports (21 CFR 312.32)	Any real or suspected adverse event or reaction regardless of expectation is required to be reported. Reporting time parameters are outlined according to level of severity and the report should be labeled accordingly. Submit form 3500A accompanied by Form 1571 to the appropriate review division. Follow-up reporting should be done to update the initial report with relevant additional information no later than 15 days from the time the sponsor is made aware. Submit electronically.[23]
Annual reports (21 CFR 312.33)	Submit no later than 60 days after the effective anniversary date. Should summarize study status and progress and include individual study information such as clinical and nonclinical information on adverse experiences, safety reports, subject fatality, or withdrawal, gained understanding of drug actions, a list of preclinical study status, and any significant manufacturing or microbiological changes made. Other information expected include updates to the general investigational plan, IB, protocol, foreign marketing developments, and a log of outstanding business.[24]

Components: See Table 80.1.

Examples: See Table 80.2.

Common challenges[13]

Completing an IND application may seem overwhelming to the novice investigator. However, the majority of IND applications are submitted by sponsor-investigators, which offers proof that regulatory obligations can successfully be met not only by corporate sponsors.[13] Common obstacles that result in delaying approval can be avoided with a little effort by:

1. Starting with an end goal and working backwards to identify possible issues.
2. Staying well organized and fostering a working relationship with the FDA.
3. Knowing the investigational molecule/drug well.

Budgetary considerations and how to fund

Securing funding can also be a challenge. Search engines, federal and state databases, and internal grants provided by the home institution may provide crucial resources. When creating the budget, it is important to consider the cost related to storage and handling of the investigational agent. Chapters 71, 74, and 115 provide a more in-depth discussion on funding and budget considerations.

Get started

- Become familiar with the process by visiting the FDA website page "Investigator-Initiated Investigational New Drug (IND) Applications."

TABLE 80.2 Relevant clinical trial examples.

	Clinical trial	Version date	NCT number	Phase
1	Lipiodol as an imaging biomarker of tumor necrosis after transcatheter chemoembolization therapy in patients with primary and metastatic liver cancer[25]	December 14, 2015	NCT01877187	II
2	A phase II trial of systemic chemotherapy gemcitabine and cisplatin) in combination with conventional transarterial Chemoembolization (cTACE) in Patients with advanced intrahepatic cholangiocarcinoma (ICC)[26]	July 19, 2018	NCT02994251	II
3	Pharmacokinetics of Doxorubicin in cTACE of liver cancer[27]	August 3, 2020	NCT02753881	Completed

- Initiate early contact with the FDA CDER's pre-IND Application Consultation Program.
- Begin developing or complete as much of the clinical investigation plan as possible.

Potential pitfalls

- Underestimating the time and resources required for IND approval.
- Disorganization of the application or overall development plan.
- Failing to initiate communication with the FDA through pre-IND consultations.

References

1. U.S. Food and Drug Administration. Investigational New Drug (IND) Application. https://www.fda.gov/drugs/types-applications/investigational-new-drug-ind-application#Introduction. Updated February 24, 2021. Accessed May 19, 2021.
2. U.S. Food and Drug Administration. Investigator-Initiated Investigational New Drug (IND) Applications | FDA. https://www.fda.gov/drugs/investigational-new-drug-ind-application/investigator-initiated-investigational-new-drug-ind-applications. Updated March 26, 2020. Accessed May 20, 2021.
3. U.S. Food and Drug Administration. Guidance for Clinical Investigators, Sponsors, and IRBs Investigational New Drug Applications (INDs) — Determining Whether Human Research Studies Can Be Conducted Without an IND. U.S. Department of Health and Human Services,. https://www.fda.gov/media/79386/download. Updated Sept 2013. Accessed May 20, 2021.
4. U.S. Food and Drug Administration. IND Application Procedures: Exemptions from IND Requirements. https://www.fda.gov/drugs/investigational-new-drug-ind-application/ind-application-procedures-exemptions-ind-requirements. Updated October 9, 2015. Accessed August 1, 2021.
5. U.S. Food and Drug Administration. IND Application Procedures: Overview | FDA. https://www.fda.gov/drugs/investigational-new-drug-ind-application/ind-application-procedures-overview. Updated Oct 9, 2015. Accessed.
6. U.S. Food and Drug Administration. IND Application Procedures: Interactions with FDA. https://www.fda.gov/drugs/investigational-new-drug-ind-application/ind-application-procedures-interactions-fda. Updated October 9, 2015. Accessed August 8, 2021.
7. U.S. Food and Drug Administration. Center for Drug Evaluation and Research - Pre-IND Consultation Contacts.
8. U.S. Food and Drug Administration. Office of New Drugs | FDA. https://www.fda.gov/about-fda/center-drug-evaluation-and-research-cder/office-new-drugs. Updated March 15, 2021. Accessed May 20, 2021.
9. U.S. Food and Drug Administration. IND Application Procedures: Clinical Hold | FDA. https://www.fda.gov/drugs/investigational-new-drug-ind-application/ind-application-procedures-clinical-hold. Updated Oct 9, 2015. Accessed May 20, 2021.
10. U.S. Food and Drug Administration. IND Application Reporting: Information Amendments | FDA. https://www.fda.gov/drugs/investigational-new-drug-ind-application/ind-application-reporting-information-amendments. Updated Oct 8, 2015. Accessed May 20, 2021.
11. U.S. Food and Drug Administration. IND Application Procedures: Investigator's Responsibilities | FDA. https://www.fda.gov/drugs/investigational-new-drug-ind-application/ind-application-procedures-investigators-responsibilities. Updated Oct 9, 2015. Accessed May 20, 2021.
12. U.S. Food and Drug Administration. CFR - Code of Federal Regulations Title 21 Section 312.56. https://www.accessdata.fda.gov/scripts/cdrh/cfdocs/cfcfr/CFRSearch.cfm?fr=312.56. Updated April 1, 2020. Accessed May 20, 2021.
13. U.S. Food and Drug Administration. CDER ORIGINAL INDs RECEIVED - CALENDAR YEARS 1986 - 2008. https://www.fda.gov/media/76869/download. Updated 2008. Accessed May 20, 2021.
14. U.S. Food and Drug Administration. FORM FDA 1571. Updated April 2019. Accessed May 20, 2021.
15. U.S. Food and Drug Administration. FORM FDA 1572. https://www.fda.gov/media/71816/download. Updated March 2019. Accessed May 20, 2021.
16. U.S. Food and Drug Administration. FORM FDA 3674. https://www.fda.gov/media/69901/download. Updated March 2019. Accessed May 20, 2021.
17. U.S. Food and Drug Administration. CFR - Code of Federal Regulations Title 21 Section 312.23. https://www.accessdata.fda.gov/scripts/cdrh/cfdocs/cfcfr/CFRSearch.cfm?fr=312.23. Updated April 1, 2020. Accessed May 20, 2021.
18. U.S. Food and Drug Administration. IND Applications for Clinical Investigations: Regulatory and Administrative Components | FDA. https://www.fda.gov/drugs/investigational-new-drug-ind-application/ind-applications-clinical-investigations-regulatory-and-administrative-components. Updated Feb 21, 2018. Accessed May 20, 2021.
19. U.S. Food and Drug Administration. IND Applications for Clinical Investigations: Chemistry, Manufacturing, and Control (CMC) Information | FDA. https://www.fda.gov/drugs/investigational-new-drug-ind-application/ind-applications-clinical-investigations-chemistry-manufacturing-and-control-cmc-information. Updated Oct 9, 2015. Accessed May 20, 2021.
20. U.S. Food and Drug Administration. IND Applications for Clinical Investigations: Pharmacology and Toxicology (PT) Information | FDA. https://www.fda.gov/drugs/investigational-new-drug-ind-application/ind-applications-clinical-investigations-pharmacology-and-toxicology-pt-information. Updated February 16, 2016. Accessed May 20, 2021.
21. U.S. Food and Drug Administration. Submitting and Reviewing Complete Responses to Clinical Holds (Revised) | FDA. https://www.fda.gov/regulatory-information/search-fda-guidance-documents/submitting-and-reviewing-complete-responses-clinical-holds-revised. Updated May 5, 2020. Accessed May 20, 2021, 2021.
22. U.S. Food and Drug Administration. IND Application Reporting: Protocol Amendments | FDA. https://www.fda.gov/drugs/investigational-new-drug-ind-application/ind-application-reporting-protocol-amendments. Updated Dec 23, 2015. Accessed May 20, 2021.
23. U.S. Food and Drug Administration. IND Application Reporting: Safety Reports | FDA. https://www.fda.gov/drugs/investigational-

new-drug-ind-application/ind-application-reporting-safety-reports. Updated Oct 9, 2015. Accessed May 20, 2021.

24. U.S. Food and Drug Administration. IND Application Reporting: Annual Reports | FDA. https://www.fda.gov/drugs/investigational-new-drug-ind-application/ind-application-reporting-annual-reports. Updated Oct 9, 2015. Accessed May 20, 2021.
25. Todd Schlachter M. Lipiodol as an imaging biomarker of tumor necrosis after transcatheter chemoembolization therapy in patients with primary and metastatic liver cancer. NIH U.S. National Library of Medicine,. https://clinicaltrials.gov/ProvidedDocs/87/NCT01877187/Prot_SAP_000.pdf. Updated Dec 14, 2015. Accessed May 20, 2021.
26. Todd Schlachter M. A Phase II Trial of Systemic Chemotherapy (Gemcitabine and Cisplatin) in Combination with Conventional Transarterial Chemoembolization (cTACE) in Patients with Advanced Intra-Hepatic Cholangiocarcinoma (ICC). NIH U.S. National Library of Medicine,. https://clinicaltrials.gov/ProvidedDocs/51/NCT02994251/Prot_SAP_000.pdf. Updated July 19, 2018. Accessed May 20, 2021, 2021.
27. Todd Schlachter M. Pharmacokinetics of Doxorubicin in cTACE of Liver Cancer - Full Text View - ClinicalTrials.gov. NIH U.S. National Library of Medicine,. https://clinicaltrials.gov/ProvidedDocs/81/NCT02753881/Prot_SAP_000.pdf. Updated Aug 3, 2020. Accessed May 20, 2021, 2021.

Chapter 81

New drug application process

Sandeep Murthy[1] and Shashidhara Murthy[2]
[1]University of Virginia School of Medicine, Charlottesville, VA, United States; [2]Mount Sinai Health System, New York, NY, United States

Key points

- The FDA's drug approval process is complex but can be successfully navigated by preparing and using the appropriate resources.
- A new drug application (NDA) is debatably the most vital part of the drug approval process.
- Submitted NDAs are either approved, approved with changes, or denied by the FDA.

Why it matters?

During 2019, the FDA received 618 investigational new drug (IND) applications and approved 48 NDAs.[1] IND applications consist of preclinical data, human subject experimentation design, and other components that require analysis and approval by the FDA before Phase I testing. NDAs consist of all the information in an IND plus phase I–III clinical trial data and are required for drug approval. The approval process is complex, and the NDA is the last major hurdle to overcome before the drug enters the market. Throughout this chapter, we will discuss the importance of the NDA and provide a foundation as well as resources on this vital step in drug approval.

History of the NDA

Before 1938, the American Medical Association had a voluntary program of drug approval. This process was largely ineffective and resulted in the sulfanilamide tragedy. Sulfanilamide was a drug used to treat streptococcal infections, which was safe and efficacious in its tablet and powder forms. Wanting to sell the drug in liquid form, Sulfanilamide was dissolved in diethylene glycol and without further testing, the drug was shipped for use immediately. Although the initial drug was safe, the combination with diethylene glycol was poisonous and resulted in the death of more than 100 people in 15 states.[2]

As a result, in 1938 Congress passed The Federal Food, Drug, and Cosmetic act that required new drugs to demonstrate safety and provided the framework for what we have today. Many amendments have been passed since then and the process has been refined to become the current system. Today, all new drugs have had to go through the process of submitting an NDA to the FDA to have them formally review and approve a drug for sale and marketing in the United States. The branch of the FDA that reviews this application is the Center for Drug Evaluation and Research (CDER).[3]

Investigational new drug submission

Once safety standards are met in animal trials, an IND can be filed to start clinical trials in humans. A pre-IND meeting can be arranged with the FDA to discuss the design of the animal studies, the intended protocol for conducting the clinical trial, and the chemistry, manufacturing, and control of the investigational drug. This will help the applicant organize their research and submit a strong application. Once an IND is submitted, the applicant must wait 30 calendar days before starting any clinical trials.[4]

Next, phase I clinical trials start, lasting approximately 1–3 years and include human subjects in the tens. This phase is largely to assess the safety of this drug on humans. If all goes well in phase I trials, you can proceed to phase II, which involves further studies to evaluate the safety of varying dosages and the efficacy of the drug. If phase II studies can demonstrate efficacy, then phase III studies begin. This phase involves enrolling thousands of human subjects to ensure safety, efficacy, and assess for side effects in a diverse patient population that should be representative of the target population (Fig. 81.1).[4]

Components of an IND application

An IND application must contain material from the following three categories[4]:

(1) Animal pharmacology and toxicology studies, which includes both safety testing carried out in animals as well as any studies related to drug use in foreign countries.

Translational Interventional Radiology. https://doi.org/10.1016/B978-0-12-823026-8.00081-X

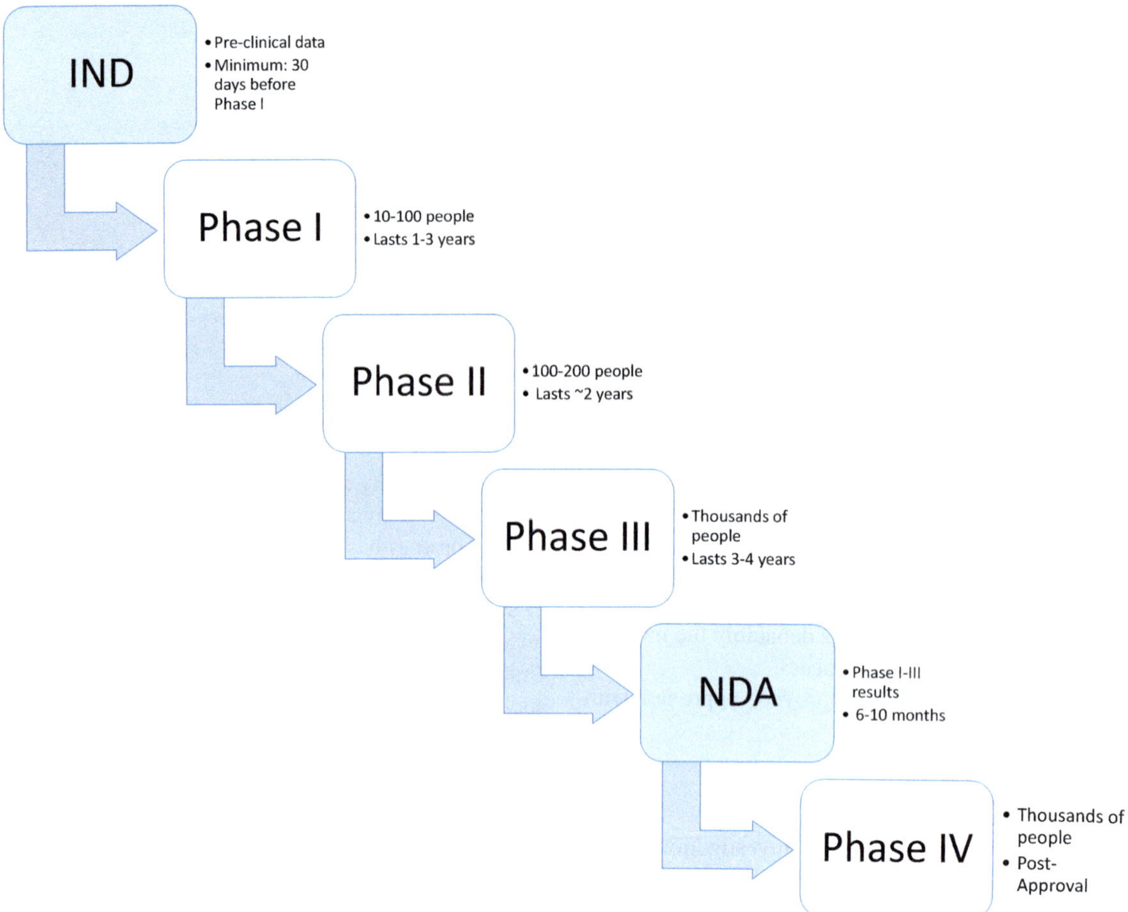

FIGURE 81.1 An overview of the approval process, beginning with the Investigational New Drug application and ending with an NDA and Phase IV trial.

(2) Manufacturing information, which is all the information regarding the chemical composition, how it is manufactured, and the expected shelf-life and stability.

(3) Clinical protocols and investigator information category includes the proposals for clinical studies in humans and commitments to obtain approval from an institutional review board (IRB). This category also includes information about the individuals who will be overseeing these studies.

Types of IND[4]:

(1) Investigator IND is submitted by a physician who will directly oversee the administration and distribution of the drug. This needs to be submitted to research a new drug and/or to research a new application for an already approved drug.

(2) Emergency Use IND allows for the use of an experimental drug in an emergency situation. During the SARS-CoV-2 pandemic, the FDA issued an emergency use authorization for chloroquine phosphate and hydroxychloroquine sulfate to be used in certain hospitalized patients while the drug was studied for this use. After early data showed that there was no significant improvement in patients who were treated with these drugs, the FDA revoked this emergency use authorization.[5]

(3) Treatment IND is submitted before final studies are completed for drugs that are showing significant improvements in early clinical testing for serious or immediately life-threatening conditions.

NDA submission

After all three phases are successfully passed, an NDA can be submitted with all the data from the animal and human studies for review. The actual application requires the submission of the FDA-356h form, which is the actual application to market a new drug, biologic, or an antibiotic; the FDA-3397, which is a cover sheet; FDA-3331 form, which is a new drug application field report. This will be reviewed by the CDER, which is composed of a team of scientists. Approval of an NDA can range from 6 to 10 months. After the NDA is granted, the company can market and sell the approved drug. They will also enter phase IV clinical trials, which will focus on understanding the long-term effects of the drug.[3]

After review of the submitted NDA, the FDA can take the three actions. The first is that they can not approve the application if they determine that the material submitted does not meet their requirements. The second outcome is that they approve the application with a contingency that minor changes are made. The final outcome is that they approve the application in its submitted state allowing immediate marketing and sale of the new drug. If the decision is either not approved or approvable, the applicant can meet with the FDA and discuss what changes need to be made (Fig. 81.2).[6]

The standard review process can take up to 10 months. However, once a drug application is submitted, it can also get a designation that can help accelerate the review process if it meets certain criteria. The Fast Track designation is for drugs that treat serious conditions and fill a medical need that has not been solved yet. The Breakthrough Therapy designation is given when the initial studies for a drug demonstrate that there is a substantial improvement when compared to currently available therapy. The Priority Review designation gives priority review to drugs that would significantly improve the treatment, diagnosis, or prevention of serious illnesses (Fig. 81.3).[6]

Abbreviated new drug application submission

Abbreviated new drug application is the process for generic drugs that must be pharmaceutically equivalent to the pioneer drug. This is an abbreviated application since it does not require preclinical or clinical data because as long as the generic is pharmaceutically equivalent, these requirements have already been met by the predecessor. What is required is data that demonstrates that this drug is identical to the reference drug.[7]

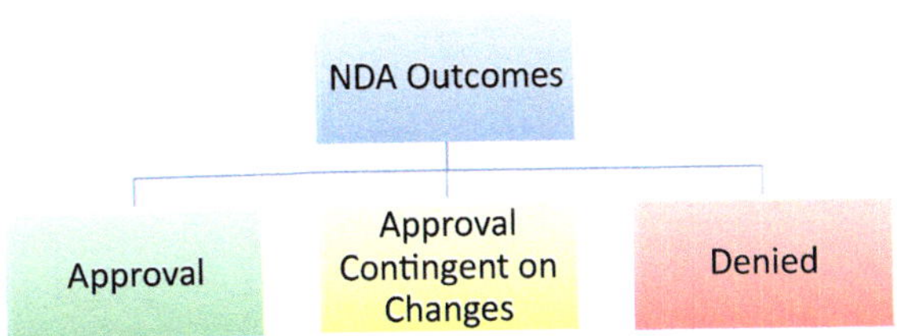

FIGURE 81.2 An overview of potential outcomes following submission of a New Drug Application (NDA).

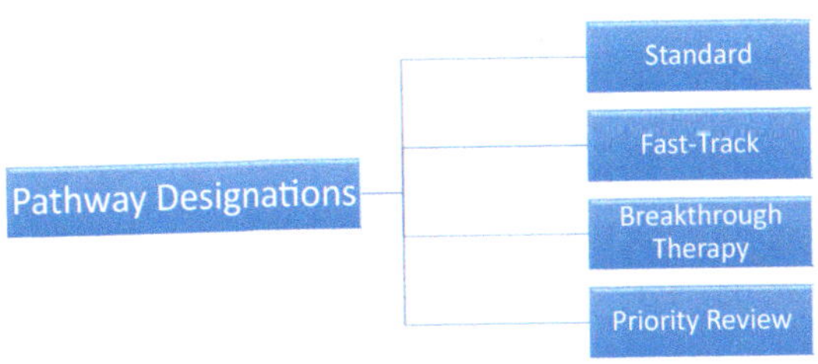

FIGURE 81.3 The pathway options that an application may fall under.

Cost of filing an NDA

In 1992, Congress passed the Prescription Drug User Fee Act (PDUFA) that allowed the FDA to collect fees to review an NDA and is required to be paid before the application is considered complete. These fees can be substantial, ranging from approximately $1.4 million if no clinical data is required to $2.9 million if clinical data is required.[8] Although these fees are substantial, there are certain waivers and reductions that an applicant can apply for. For instance, a waiver or reduction can be granted if the drug is necessary to protect public health or if the applicant is a small business submitting their first NDA.[8] If a drug was not approved or withdrawn and the same drug product is resubmitted by the same entity, a second payment is not required.

Get started

- Visit https://www.fda.gov/drugs/and click on the "Guidances" section under priority areas and initiatives. Documents under this tab can help provide information on FDA's current guidelines on a range of subjects.
- Go to https://www.fda.gov/drugs/types-applications/new-drug-application-nda and click on "guideline for submitting supporting documentation in drug applications for the manufacture of drug products." Review this document to better prepare supporting documents needed for a successful NDA.

References

1. *Novel Drug Approvals for 2019*; 2020. Retrieved from https://www.fda.gov/drugs/new-drugs-fda-cders-new-molecular-entities-and-new-therapeutic-biological-products/novel-drug-approvals 2019.
2. Ballentine C. *Sulfanilamide Disaster*. FDA Consumer Magazine; 1981. Retrieved from https://www.fda.gov/files/about%20fda/published/The-Sulfanilamide-Disaster.pdf.
3. New Drug Application (NDA). Retrieved from https://www.fda.gov/drugs/types-applications/new-drug-application-nda; 2019.
4. *Investigational New Drug (IND) Application*; 2020. Retrieved from https://www.fda.gov/drugs/types-applications/investigational-new-drug-ind-application.
5. Emergency Use Authorization. Retrieved from https://www.fda.gov/emergency-preparedness-and-response/mcm-legal-regulatory-and-policy-framework/emergency-use-authorization; 2020.
6. The Drug Development Process. Retrieved from https://www.fda.gov/patients/learn-about-drug-and-device-approvals/drug-development-process; 2018.
7. Abbreviated New Drug Application (ANDA). Retrieved from https://www.fda.gov/drugs/types-applications/abbreviated-new-drug-application-anda; 2019.
8. The Drug Development Process. Retrieved from https://www.fda.gov/patients/learn-about-drug-and-device-approvals/drug-development-process; 2018.

Chapter 82

Medical device pathways

Sakshum Chadha[1], Akanksha Acharya[2] and Ron C. Gaba[3]
[1]Rutgers New Jersey Medical School, Newark, NJ, United States; [2]W.L. Gore and Associates, Inc, Elkton, MD, United States; [3]Department of Radiology, College of Medicine University of Illinois at Chicago, Chicago, IL, United States

Key points

1. The FDA separates medical devices into three classes: the class of medical device helps determine the best method for device approval.
2. Submitting a device application for approval can require extensive research and technical specifications depending on the type of device, its classification, and the intended use.
3. Compliance with FDA regulations for device approval is essential to commercial viability, success of the device, and patient safety.

Why it matters?

Medical device approval in the United States (US) is controlled by the Food and Drug Administration (FDA). To market a medical device in the US, the FDA must approve the device and many of its marketing materials. Device manufacturers and entrepreneurs should be aware of the various classifications of medical devices delineated by the FDA, as well as the appropriate approval pathways and their respective characteristics. The approval process ranges from 3 to 7 years in total and can require significant logistical involvement, including preparation of technical specifications, requirement of both lab and clinical data, and sufficient documentation of FDA requirements including potential labeling and marketing materials.[1] This documentation is considered very seriously by the FDA and the US criminal justice system. As an example, in the court case of US v. Synthes, the bone cement company Norian, then acquired by Synthes, was alleged with noncompliance with FDA lawful functions related to marketing their medical device for unapproved indications and the misrepresentation of device safety through clinical research.[2] Due to adverse events that resulted in patient deaths, the company was criminally charged, and company officials were held personally responsible.[2] This case demonstrates that compliance with FDA regulations is not only vital to the commercial success of a medical device but also to patient safety, and noncompliance with FDA regulations can result in patient harm and criminal liability.

Device classification

The Food, Drug, and Cosmetic Act (FDCA) defines a **medical device** as "an instrument intended for use in diagnosis or in cure, mitigation, treatment, or prevention of disease in man and which does not achieve intended purpose through chemical action within or on the body."[2,3] The FDA Center for Devices and Radiological Health is the regulatory body responsible for companies that manufacture new devices, relabel or repackage existing devices, and/or import devices to be sold in the US.[4] Devices are categorized into three classes based on the level of risk and level of control and oversight the FDA holds (Table 82.1). Additionally, classification is determined by **intended use** of the device and specific **indications for use**.[5] The best pathway for device approval depends on the class of devices outlined in Table 82.1.

Of note, devices with significant risk (class III devices) require an **Investigational Device Exemption (IDE)** to be used in a clinical study intended to obtain safety and efficacy data for FDA approval.[2] These studies have a substantial amount of FDA and Institutional Review Board (IRB) control over them (Fig. 82.1).[2]

Device pathways

FDA exempt

The FDA has published a list of class I and class II devices that are exempt from 510(k) Premarket Notification (see the following section) and FDA clearance before marketing the

Translational Interventional Radiology. https://doi.org/10.1016/B978-0-12-823026-8.00006-7

TABLE 82.1 Device classifications and pathways.[1,2]

Class	Level of risk of injury/illness	Level of control by FDA	Device pathway*	Examples
I	Low risk	*General control* (adherence to good manufacturing practices) to provide reasonable assurance of safety and efficacy	FDA exempt OR Premarket notification (510(k))	Gloves, gauze
II	Moderate risk	*General control + special control* (standards and postmarket surveillance) to provide reasonable assurance of safety and efficacy	FDA exempt OR Premarket notification (510(k))	Suture, infusion pumps, needles
III	Significant risk	*General control + premarket approval*: a Higher level of control for devices where just general and special controls are insufficient or there is insufficient information to determine safety/effectiveness	Premarket approval	Heart valves, ventilators pacemakers

*These are general guidelines, and exceptions/limitations may exist.

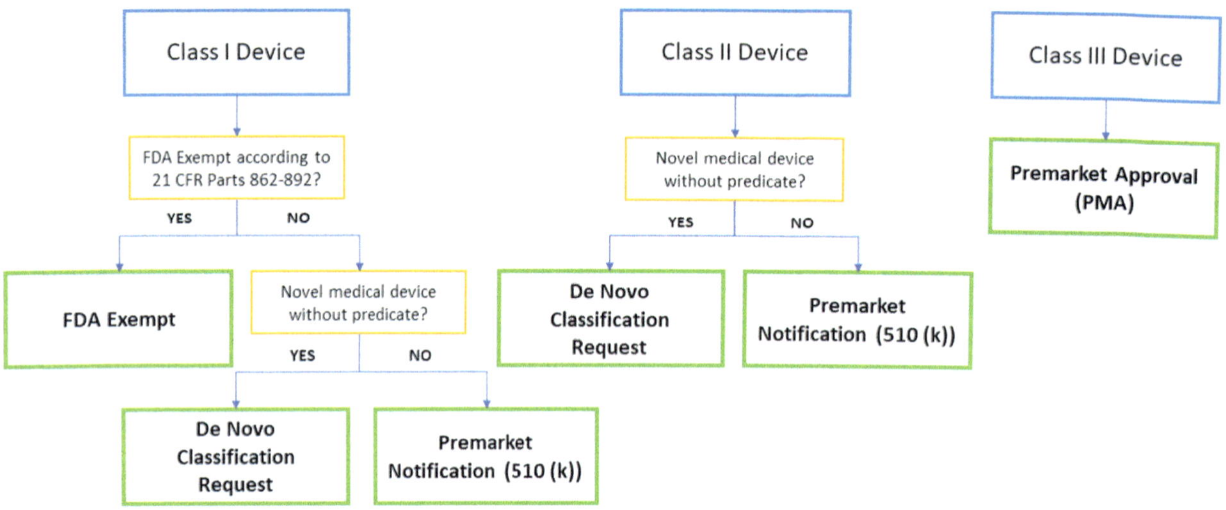

FIGURE 82.1 Device pathways flowchart.[1]

device in the US.[6] These exempt devices range from stethoscopes to guidewires used for intravascular procedures (e.g., coronary stenting).[6] Certain generic class I devices and a few class II devices are also exempt from **Good Manufacturing Practice** requirements.[7] Manufacturers are still required to register with the FDA and provide the generic category and classification name of their device using the online FDA United Registration and Listing System.[6] Such generic categories of devices are outlined in title 21 Code of Federal Regulations (CFR), parts 862–892. The **FDA exempt** process requires three key components: (1) payment of an annual registration fee, (2) submission of registration and listing information electronically, and (3) electronic mail (e-mail) notification from the FDA verifying the completion of requirements.[8]

FDA 510(k) (premarket clearance/premarket notification)

Section 510(k) of the FDCA creates a pathway for approval of devices that can be shown as substantially equivalent to any **predicate** medical device. This can include (1) a medical device approved before 1976, (2) a medical device previously found to be substantially equivalent through the 510(k) process, (3) a medical device approved for marketing through the de novo classification method (see the following section), or (4) a medical device reclassified from class III to class II or class I[9]. Most class II devices fall into this category of clearance. As an example, the GORE® VIABIL® device (W.L. Gore and Associates, Newark DE) is a biliary stent intended to increase bile flow through malignant bile duct strictures.[9,10] In 2017, the manufacturer—W.L. Gore and Associates—demonstrated substantial equivalence of their biliary stent to a predicate device, a previous version of their biliary stent approved in 2008.[9,10] FDA documentation outlines the difference between the two devices as the inclusion of a guidewire exit port at the distal end of the catheter component of the endoscopic delivery system to allow for single user control of a shorter guidewire with no other changes.[10] Despite no change in intended use, due to the change that allowed for a single user to control implantation of the stent, a new 510(k) application was submitted and approved.[10]

510(k) approval limits the use of the new medical device to the intended use on the label of the predicate device and the manufacturer's proposed labeling. As described earlier, if a previously approved medical device is modified in any way that could alter its safety or effectiveness, a new 510(k) submission is required as the device has been altered but remains similar to a previously approved predicate device. Because the goal of this approval process is to demonstrate substantial equivalence to a predicate device, no clinical trials are required if safety and effectiveness can be otherwise demonstrated.[2] The 510(k) pathway is considered a relatively fast and inexpensive approval process compared to premarket approval.

Submission of a 510(k) application requires that the manufacturer pay a medical device review fee to the FDA.[4] After submission of the application electronically, the FDA has 90 days to decide whether to approve the medical device.[2] During this time period, the FDA goes through three separate review stages for the application before submitting a decision letter.[11] If the FDA deems the medical device is not substantially equivalent to a predicate device, the manufacturer should resubmit the 510(k) with new data that demonstrate substantial equivalence, file a reclassification petition, or pursue another approval pathway such as de novo review or premarket approval.[11] A special 510(k) approval process exists for expedited clearance by the FDA (30 days) when the device is a modified version of a previously approved device submitted by the manufacturer, and the change to the device does not alter its intended use. The majority of medical devices on the market today are cleared through a 510(k) approval process.

Premarket approval

Under the FDCA, **premarket approval** (PMA) is an approval process typically used for class III devices. Few exceptions outlined in the CFR exist for certain class III devices that may be better suited for a 510(k) submission.[12] This process is the most stringent and requires the manufacturer to notify the FDA 90 days before marketing with a 180-day total review timeline.[1,12]

Though premarket approval does not require a randomized controlled trial, data from clinical trials are still required to support device safety and efficacy, and this approval process tends to require more time and financial investment compared to the 510(k) approval process. Data are required in three distinct categories: (1) technical information related to the device, (2) nonclinical laboratory studies, and (3) clinical investigations.[12] Manufacturers should submit an IDE application to use the device in clinical studies intended for collecting data to facilitate device approval.[13] The manufacturer must demonstrate the safety and efficacy of the device and provide information about the conducted clinical trials with the application; this can include information delineated in title 21 CFR part 812 such as a protocol, risk analysis, labeling, and consent procedures.[13] These trials are overseen by an IRB to ensure patient safety and protection.

Due to stringent requirements and a lengthy review process, the FDA recommends manufacturers remain in close contact with the FDA before submission to ensure compliance with all regulatory requirements. A presubmission process exists where manufacturers may obtain feedback from the FDA before application submission.[12] Some medical devices intended for life-threatening conditions may be eligible for the **Expedited Access Pathway Program** (now **Breakthrough Devices Program**), which allows for a quicker PMA or de novo review process.[12]

De novo review

A **de novo classification request** is utilized for new devices that cannot be shown to have a predicate device and for which general and/or specific controls can demonstrate reasonable assurance of safety and effectiveness.[14] These devices must be shown to have low or moderate risk and to be poorly suited for other approval methods whether due to lack of a predicate device for 510(k) or class I or II designation as opposed to class III for premarket approval, especially as most novel medical devices without predicate devices are automatically considered to be class III.[1] Similar to other device pathways, an electronic submission is made to the FDA containing information related to proposed marketing, clinical and nonclinical research data, discussion of safety and efficacy, and technical information.[14]

The review process for a de novo application is less stringent than premarket approval and involves stages of review, including (1) an FDA acceptance review, which is an administrative stage that ensures application completeness, as well as (2) an FDA substantive review to verify the correct classification of the device and ensure adequate information regarding the device has been provided. If additional information is required, the manufacturer has 180 days to provide such information before the application is removed from the system.[14] At the end of the review process, the FDA provides a final decision granting or denying the de novo request.

Humanitarian device exemption

The humanitarian device exemption (HDE) pertains to devices indicated for conditions that afflict fewer than 8000 patients annually in the US and that have no other treatment alternatives.[2] This approval mechanism is typically reserved as a marketing application for class III devices, and the manufacturer is still required to demonstrate that the benefits of the device exceed the risks. For example, the TheraSphere yttrium-90 glass microspheres (Boston Scientific, Marlborough MA) have an HDE to be employed for radiation therapy of primary liver cancer (hepatocellular carcinoma) through targeted delivery via a fluoroscopically positioned transarterial catheter.[15,16] The particular subset of patients stricken with hepatocellular carcinoma that have no other treatment options (e.g., surgical resection or liver transplantation) is less than 8000 annually in the US.[17,18] As such, this device has been approved by the FDA as an HDE.

The HDE has notable restrictions relating to profit: a humanitarian use device can only be sold for profit limited to a certain quantity determined as the annual distribution number if it is intended for the treatment of a disease that fulfills one of two conditions: (1) occurs in pediatric populations or (2) occurs almost exclusively in adult populations (and not pediatric populations).[19] The purpose of this restriction is to incentivize the development of medical devices for pediatric populations.[19]

Special considerations

Section 3051

Section 3051 of the FDCA is a part of the 21st Century Cures Act. It created a program in the FDA that allows priority review and management attention for devices that can be seen as "breakthrough" devices in that they provide a "more effective treatment" or diagnose a "life threatening or irreversibly debilitating human disease or condition."[2,3] For such devices, there is no cleared or approved alternative.

Section 3054

Section 3054 of the FDCA is also a part of the 21st Century Cures Act. It requires that the FDA review class I and class II devices once every 5 years beginning in the year 2017 to consider if the devices can bypass the 510(k) approval process with reasonable assurance of safety and effectiveness.[2,3] Such devices are declared 510(k) exempt.

Get started

The FDA recommends four key stepsfor device approval. As a device manufacturer or entrepreneur, the first step in bringing a medical device to market involves (1) classifying the device based on existing federal regulations and identifying the regulatory controls the FDA will hold over the device (Table 82.1).[20] Based on device characteristics and classification, it is important to next (2) identify the best device pathway of those outlined in this chapter and to complete the premarket submission application.[20] Important considerations at this stage include the types of lab-based and clinical testing/research that may be required to demonstrate the safety and effectiveness of the device, as well as the specific indications and intended use that will be included on device labeling.[20] Supplying the necessary information may become a part of the (3) FDA review process that can involve various stages of review relating to submission fees, additional information required, or electronic submissions with a back and forth between the FDA and device manufacturer.[20] Finally, essential to the review process is (4) compliance with all FDA-recommended regulatory controls including device labeling, device registration/listing, and adherence to good manufacturing practices (Fig. 82.2).[20]

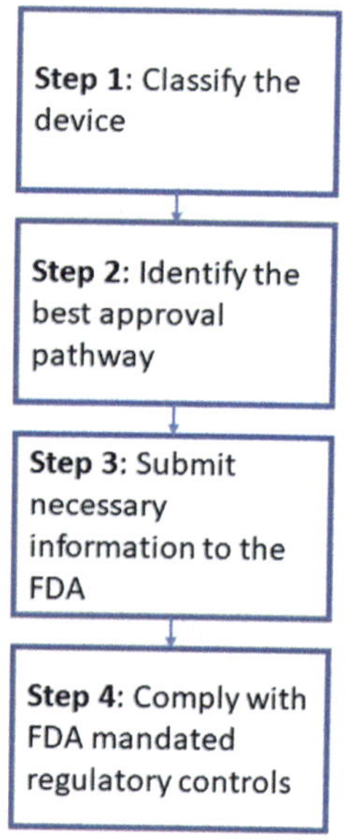

FIGURE 82.2 Flowchart of FDA recommended steps.[20]

References

1. Van Norman GA. Drugs, devices, and the FDA: Part 2: an overview of approval processes: FDA approval of medical devices. *JACC Basic Transl Sci*. 2016;1(4):277–287.
2. Sullivan GB. *Federal Medical Device Enforcement: A Case Study of u.s. V. Synthes & Norian*. Penn Law School; 2017:81.
3. Congress US. Food, Drug, and Cosmetics Act. In. 211934.
4. Administration FaD. Overview of Device Regulation. FDA.gov. https://www.fda.gov/medical-devices/device-advice-comprehensive-

regulatory-assistance/overview-device-regulation. Updated 8/31/2018. Accessed.

5. Administration FaD. Classify Your Medical Device. FDA.gov. https://www.fda.gov/medical-devices/overview-device-regulation/classify-your-medical-device. Updated 02/07/2020. Accessed.
6. Administration USFaD. Food and drugs. In: FDA, ed. 21. Vol Parts Vols 862−892. CFR- Code of Federal Regulations: Office of the Federal Register.
7. Administration FaD. Class I/II Exemptions. https://www.fda.gov/medical-devices/classify-your-medical-device/class-i-ii-exemptions. Updated 07/01/2019. Accessed.
8. Administration FaD. How to Register and List. Food and Drug Administration. https://www.fda.gov/medical-devices/device-registration-and-listing/how-register-and-list. Updated 8/31/2018. Accessed.
9. Tillman D-B. In: Services DoHaH, ed. *Re: K081184 GORE® VIABIL® Biliary Endoprosthesis*. Food and Drug Administration; 2008.
10. Maisel WH. In: Services DoHaH, ed. *Re: K170740 : GORE® VIABIL® Short Wire Biliary Endoprosthesis*. Food and Drug Administration; 2017.
11. Administration FaD. Premarket Notification 510(k). https://www.fda.gov/medical-devices/premarket-submissions/premarket-notification-510k. Updated 03/13/2020. Accessed.
12. Administration FaD. Premarket Approval (PMA). https://www.fda.gov/medical-devices/premarket-submissions/premarket-approval-pma. Updated 05/16/2019. Accessed.
13. Administration USFaD. Food and drugs. In: FDA, ed. 21. Vol Part 812. CFR- Code of Federal Regulations: Office of the Federal Register.
14. Administration FaD. De Novo Classification Request. FDA.gov. https://www.fda.gov/medical-devices/premarket-submissions/de-novo-classification-request. Updated 11/20/2019. Accessed.
15. Corporation BS. Summary of safety and probable benefit. In: *Humanitarian Device Exemption Database*. 1999.
16. Richter KC. *Re: H980006 Therasphere TM*. Services DoHaH; 1999.
17. White DL, Thrift AP, Kanwal F, Davila J, El-Serag HB. Incidence of hepatocellular carcinoma in all 50 United States, from 2000 through 2012. *Gastroenterology*. 2017;152(4):812−820. e815.
18. Gholam PM, Iyer R, Johnson MS. Multidisciplinary management of patients with unresectable hepatocellular carcinoma: a critical appraisal of current evidence. *Cancers*. 2019;11(6).
19. Administration FaD. Humanitarian Device Exemption. FDA.gov. https://www.fda.gov/medical-devices/premarket-submissions/humanitarian-device-exemption. Updated 09/05/2019. Accessed.
20. Administration FaD. How to Study and Market Your Device. FDA.gov. Updated 12/13/2019. Accessed.

Chapter 83

Radiation-emitting electronic products

Nicholas Vollano
College of Osteopathic Medicine, Kansas City University, Joplin, MO, United States

Radiation-emitting electronic devices are widely used throughout our daily lives. They range from cell phones and common household appliances like the microwave oven to the detection of major medical problems via mammography and computerized tomography. Despite its appearance in everyday life, not everyone thinks about the impact these devices have on our society. This chapter discusses radiation-emitting electronic devices, different forms of radiation, and how to safely use devices that emit ionizing radiation and concludes with a discussion about general regulations with the resultant budgetary implications.

Radiation-emitting electronic products are defined as electronically operated devices capable of emitting any form of radiation.[1] Radiation is simply energy in motion and can be categorized as electromagnetic or particulate radiation.[2] Examples of radiation-emitting electronic devices include the following:

- Nonmedical devices—cellular telephones, compact disc players, microwave ovens, laser products, police radar, tanning beds
- Medical devices—magnetic resonance imaging, computed tomography (CT), mammography, diagnostic X-ray device

Electromagnetic radiation involves energy moving in the form of varying wavelengths at the speed of light. This type of radiation has no mass or charge. The varying wavelengths constitute the electromagnetic spectrum, demonstrated in Fig. 83.1. This includes, but is not limited to, radiowaves, γ rays, microwaves, X-rays, infrared waves, and ultraviolet waves. Energy and wavelength are inversely related with shorter wavelengths displaying high energy and long wavelengths having low energy levels.

Particulate radiation involves fast-moving, charged (e.g., α particles, β particles, positrons) or uncharged (e.g., neutrons) subatomic particles capable of producing ionizing radiation.[1] Ionizing radiation can be caused by particulate or short-wavelength electromagnetic radiation, as illustrated in Fig. 83.2.

Ionizing radiation

Ionizing radiation can cause cellular damage in varying ways. It can directly damage cellular organelles leading to cell death or disruption of deoxyribonucleic acid (DNA). Ionizing radiation can directly impact DNA by causing single- or double-strand breaks in the double-helical structure.[3] Indirect effects can occur when the ionizing radiation displaces orbiting electron(s) around an atom and results in the secondary formation of ion pairs with the displaced electron and residual atom (ionization).[4] These ion pairs are reactive and capable of interacting with the DNA to cause breaks in the DNA strand(s).[2] The human defense mechanisms used to repair DNA damage are complex and outside the scope of this chapter.

The amount of damage done by ionizing radiation depends on many factors including the type of ionizing radiation, exposure dose, and amount of time exposure ensued. It is important to have a basic understanding of the varying types of ionizing radiation to determine how to protect against their potentially dangerous effects.

- α particles are heavy, positively charged particles with a large linear energy transfer (LET) capability. Due to their large mass (approximating the size of a Helium atom) and charge, these molecules cause intense ionization and significant secondary damage to local tissues but can only travel a few cell diameters in distance.[5] Therefore, α particles transfer a massive amount of energy to a small region.
- β particles are negatively charged electrons. The mass of these particles makes up only a small fraction of an α particle's mass and, therefore, travels much farther. β particles travel several feet in the air and can penetrate

Translational Interventional Radiology. https://doi.org/10.1016/B978-0-12-823026-8.00021-3

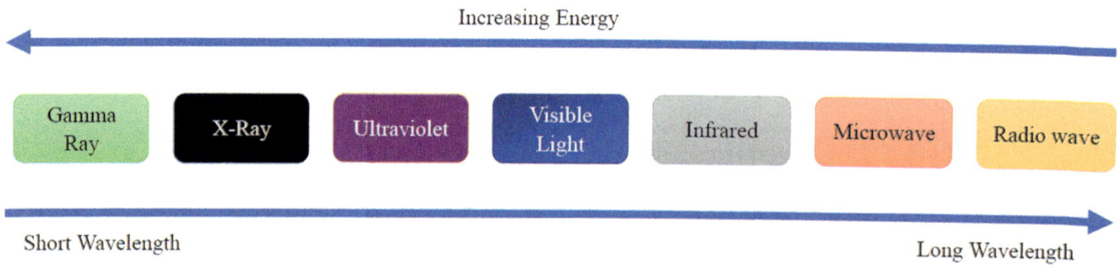

FIGURE 83.1 Electromagnetic spectrum.

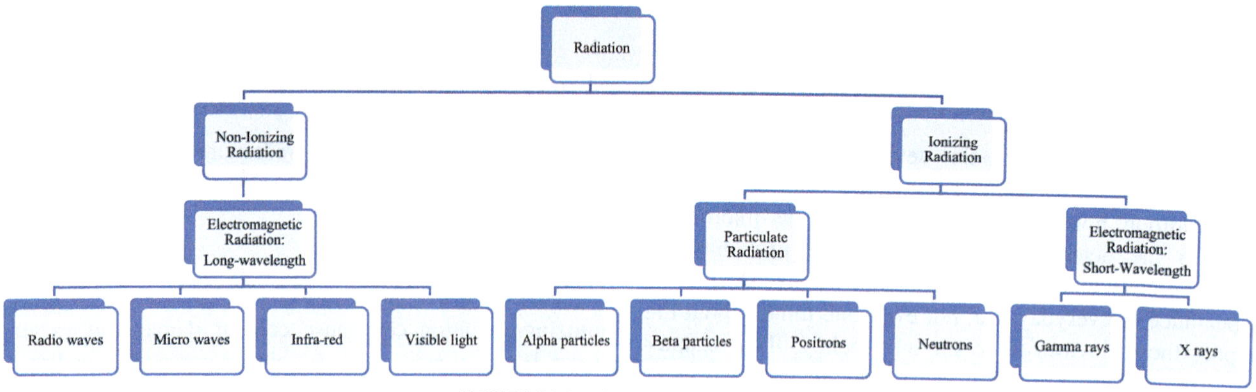

FIGURE 83.2 Categories of radiation.

up to several millimeters into the skin but are unable to cause as much damage due to a relatively lesser LET.[5]

- Positrons display similar mass to β particles but have a single positive charge. Due to these similarities, positrons share a similar range and LET as β particles in matter. Positron emission tomography is widely used in combination with CT or magnetic resonance imaging as a highly sensitive diagnostic tool.[6]
- Neutrons are chargeless particles with the approximate mass of a proton. They are emitted from nuclei of heavy elements in response to high-speed particle bombardment and have a longer range than the prior three particles mentioned.
- γ waves are high-energy, short wavelength, electromagnetic waves that are spontaneously emitted from a radioactive atomic nucleus. They do not possess a charge. γ waves can travel several kilometers in the air and even penetrate through the human body.
- X-rays share similar characteristics to γ waves except for their origin. X-rays are emitted secondary to orbital shell transitions by an atom's electron(s). Diagnostic X-rays and CT scans produce images by directing a surge of electrons toward a target material within a vacuum tube. When the electrons interact with the target's atoms, X-rays are produced. Both γ- and X-rays penetrate much farther than α and β particles, so protective shielding is important in limiting accidental radiation exposures.

Radiation safety

Safe use of radiation-emitting devices involves limiting exposure to ionizing radiation above levels typically encountered as background radiation.[7] All humans are exposed to a small amount of background radiation emanating from natural and man-made sources daily. Proper ionizing radiation safety focuses on three basic principles: distance, shielding, and time.[8]

1. **Distance.** Whenever operating a radiation-emitting device, maximize the distance from the source of radiation. The intensity of γ and X-ray radiation follows the inverse square law, which means radiation intensity decreases inversely proportional to the square distance from the source.
2. **Shielding.** Proper shielding materials placed between a radiation source and a worker/patient protect an individual from unnecessary exposure. Proper shields are created from materials capable of partially or completely absorbing different types of radiation and can decrease or eliminate the absorbed radiation dose. Fig. 83.3 illustrates shielding examples that block various forms of ionizing radiation. A common example is using a lead apron or lead gonad shield to protect a patient's germinal tissue during a diagnostic X-ray procedure.
3. **Time.** When using a radiation-emitting device, minimize the amount of time spent near the radiation source.

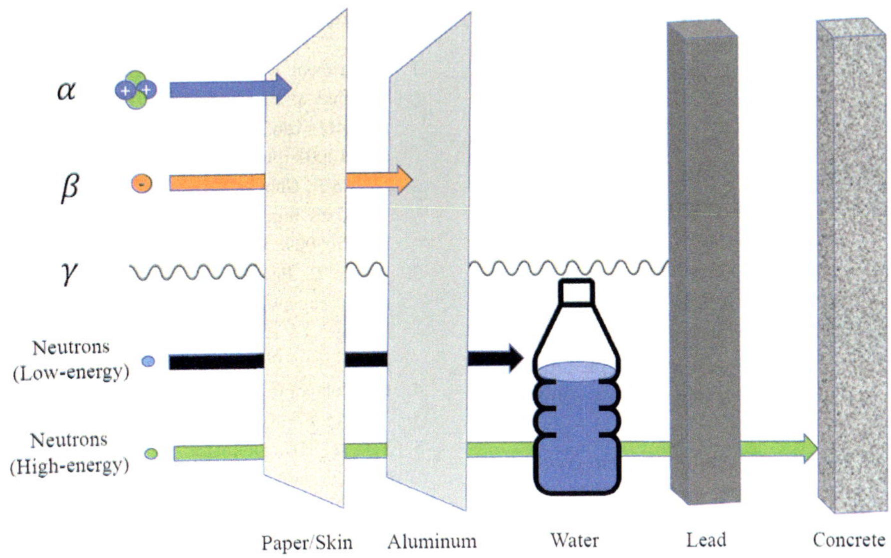

FIGURE 83.3 Comparison of travel distances and protective shielding against different forms of ionizing radiation.

Procedures that require such devices should plan to operate with the shortest amount of exposure time as possible.

Monitoring radiation exposure

Monitoring radiation exposure is an integral part of radiation safety programs and requires special devices to measure these levels. Introducing a device capable of emitting ionizing radiation to a work environment requires an area survey to demarcate its boundaries and establish a controlled exposure area of operation. Additionally, personnel who are at high risk of exposure use wearable detectors such as the thermoluminescent dosimeter badge to monitor individual exposure levels to radiation. These are best suited for measuring the cumulative absorbed dose from long-term, low-dose radiation exposure.[2] Facilities operating radiation-emitting devices specify how radiation levels are measured and monitored in their radiation safety manuals.

Regulation

All radiation-emitting devices must be registered and used in accordance with the regulations of the United States (US) Food and Drug Administration (FDA) and the individual State regulatory requirements. The US FDA authority to regulate radiation-emitting electronic devices was granted through the EPRC program (an expansion of the original Radiation Control for Health and Safety Act of 1968). The EPRC program is outlined in the Federal Food, Drug, and Cosmetic Act, Chapter 5, subchapter C and was created to limit the emissions of and the public's exposure to electronic device radiation.[1]

All radiation-emitting electronic devices produced must abide by the radiation safety requirements in the Code of Federal Regulations, title 21, subchapter J, parts 1000–1050. If the product is also a medical device, it must comply with additional medical device regulations (e.g., premarket notification and 510(k) clearance) within the provisions of the Federal Food, Drug and Cosmetic Act. An outlier among the radiation-emitting, medical devices is mammography. Mammography is uniquely regulated under the Mammography Quality Standards Act of 1992 (MQSA).

Additional rules, regulations, and radiation dose limits have been developed by national and international regulatory agencies and scientific organizations. Some examples include the US Environmental Protection Agency, the International Council on Radiological Protection, the National Academy of Sciences, the Occupational Safety and Health Administration, and the National Council on Radiation Protection and Measurements, among many others.

Effective radiation safety requires the use of radiation-emitting electronic devices in compliance with these regulations and by limiting exposures to as low as reasonably achievable.

Budgetary considerations

Many of the current regulations were created using the linear no-threshold (LNT) model in conjunction with a precautionary approach. This model states that cancer risk increases linearly with increased doses of ionizing radiation, even at the smallest levels. It also infers that there is no minimum level before ionizing radiation increases cancer risk, meaning that no levels are safe and even the smallest levels are damaging.[9–13] Although it is widely agreed that high levels of ionizing radiation follow this pattern, current debate exists regarding the validity of this relationship at levels less than a short-term exposure to low doses (e.g., 100 mSv).[9–13]

There is no surprise that government regulators use a precautionary perspective when drafting policies because the goal is to protect the general population and workers who handle radiation-emitting devices. It is important to note that epidemiological data studying low doses of ionizing radiation used to support and to disprove the use of the LNT model is unlikely to demonstrate definitive evidence for either side due to uncertainties in epidemiologic data. Therefore, we must follow the conservative thresholds as outlined by current regulations.

Understanding it is not reasonable to remove every potential radiation exposure, a limit of acceptable risk was established. For the public, manufactured radiation below 1 mSv per year is acceptable; whereas, the acceptable risk for people who work with radiation-emitting devices is below 20 mSv per year.[13] However, evidence suggests that dose rates of up to 500 mSv per year should be considered safe, and regulating radiation levels to much lower levels is expensive.[13] One study estimates that US$100 million spent to control low-level radiation emissions may save one human life-year. If that same amount was reallocated for screening and treatment of cervical cancer, an estimated 2000 human life-years could be saved.[14] Most would argue that the excess money spent to protect from such low doses could be better used for protecting against known harms.[9] Complete reallocation, however, is not the answer to this excessive economic burden but suggestions to adopt a graded approach regarding low radiation dose regulation are making its way into international agencies and will eventually play a role in the future.

Conclusion

Having a basic understanding of the different types of radiation, how to safely protect against harmful effects, general regulations, and some of the budgetary considerations is important for carefully operating and studying radiation-emitting electronic devices. Whether it be studying the public health impacts of fifth-generation, 5G, radiofrequency radiation, developing α particle therapy for the treatment of tumors, or using a microwave oven to reheat old food, radiation-emitting electronic devices play a significant part in our everyday lives for many years to come.[5,15]

References

1. Department of Health and Human Services. Radiological Health. Codified at 21 CFR §1000-1050. https://www.ecfr.gov/cgi-bin/text-idx?SID=1db3ea87be53f1c686a910cee5f184e3&mc=true&node=pt.21.8.1000&rgn=div5. Updated April 1, 2020. Accessed June 21, 2021.
2. Akram S, Chowdhury YS. Radiation Exposure of Medical Imaging. StatPearls. https://www-ncbi-nlm-nih-gov.proxy.kansascity.edu/books/NBK565909/. Updated December 5, 2020. Accessed June 23, 2021.
3. Burgio E, Piscitelli P, Migliore L. Ionizing radiation and human health: reviewing models of exposure and mechanisms of cellular damage. An epigenetic perspective. *Int J Environ Res Publ Health.* 2018;15(9):1971. https://doi.org/10.3390/ijerph15091971.
4. Havránková R. Biological effects of ionizing radiation. *Cas Lek Cesk.* 2020;159(7–8):258–260.
5. Tafreshi NK, Doligalski ML, Tichacek CJ, et al. Development of targeted alpha particle therapy for solid tumors. *Molecules.* 2019;24:4314. https://doi.org/10.3390/molecules24234314.
6. Santos BS, Ferreira MJ. Positron emission tomography in ischemic heart disease. *Rev Port Cardiol.* 2019;38(8):599–608. https://doi.org/10.1016/j.repc.2019.02.011.
7. Frane N, Bitterman A. Radiation Safety and Protection. StatPearls. https://pubmed-ncbi-nlm-nih-gov.proxy.kansascity.edu/32491431/. Updated May 30, 2020. Accessed June 23, 2021.
8. Occupational Safety and Health Administration. Ionizing Radiation - Control and Prevention. Osha.gov. https://www.osha.gov/ionizing-radiation/control-prevention. Accessed June 21, 2021.
9. Sykes P. Until there is a resolution of the Pro-LNT/Anti-LNT debate, we should head toward a more sensible graded approach for protection from low-dose ionizing radiation. *Dose Resp.* 2020;18(2). https://doi.org/10.1177/1559325820921651, 1559325820921651.
10. Cardarelli JJ, Ulsh B. It is time to move beyond the linear no-threshold theory for low-dose radiation protection. *Dose Resp.* 2018;16(3). https://doi.org/10.1177/1559325818779651, 1559325818779651.
11. Kaminski CY, Dattoli M, Kaminski JM. Replacing LNT: the integrated LNT-Hormesis model. *Dose Resp.* 2020;18(2). https://doi.org/10.1177/1559325820913788, 1559325820913788.
12. Ricci PF, Tharmalingam S. Ionizing radiations epidemiology does not support the LNT model. *Chem Biol Interact.* 2019;301:128–140. https://doi.org/10.1016/j.cbi.2018.11.014.
13. Vaiserman A, Koliada A, Zabuga O, Socol Y. Health impacts of low-dose ionizing radiation: current scientific debates and regulatory issues. *Dose Resp.* 2018;16(3). https://doi.org/10.1177/1559325818796331, 1559325818796331.
14. Graham JD. *Comparing Opportunities to Reduce Health Risks: Toxin Control, Medicine, and Injury Prevention. NCPA Policy Report No. 192.* Dallas, TX: National Center for Policy Analysis; 1995. Updated July 1, 1995 http://www.ncpathinktank.org/pdfs/st192.pdf. Accessed June 27, 2021.
15. Hardell L, Carlberg M. Health risks from radiofrequency radiation, including 5G, should be assessed by experts with no conflicts of interest. *Oncol Lett.* 2020;20(4). https://doi.org/10.3892/ol.2020.11876.

Chapter 84

Combination products

Tulasi Talluri[1] and Capt. Tej Ishaan Mehta[2,3]

[1]*Oakland University William Beaumont School of Medicine, Rochester, MI, United States;* [2]*Department of Interventional Radiology, The Johns Hopkins Hospital, Baltimore, MD, United States;* [3]*United States Air Force Medical Corps, Falls Church, VA, United States*

Key points

- Combination products are any combination of two or more drugs, devices, and/or biological products generally designed to allow for a more precise, efficacious, and safer delivery of treatment to patients than using independent drug products.
- Common interventional radiological uses of combination products include transarterial embolization therapies, hydrogel-coated embolization coils, and drug-eluting stents, which are all drug/device combinations.
- Challenges to combination products include proving the efficacy and necessity of using the combination of products as opposed to using products alone.

Why does it matter?

Combination products are defined as therapeutic and diagnostic products that combine drugs, devices, and/or biological products. They are tightly regulated by the US Food and Drug Administration (FDA)[1] due to potential product interactions. Advancements in combination products have allowed for precise, targeted, and safe delivery of therapies to patients suffering from a litany of diseases.

The processes and pathways of combination products vary greatly, depending on the particular product. Most often, combinations can include a drug/device, biological product/device, or drug/biological product. Table 84.1[2] outlines a list of several common combination products that are approved by the FDA. In combinations involving a drug and a device, for example, an instrument or apparatus that is not intended for metabolism or chemical alteration within or on the body, such as an inhaler, is combined with a drug that is intended for a condition, such as albuterol inhalers for asthma.[3] This combination mechanism allows albuterol to be delivered immediately and directly to the lungs, where it can perform its primary action of bronchodilation. Additionally, combination therapies may allow for synergistic therapeutic mechanisms resulting in a supra-additive therapeutic effect, such as in the case of combination sacubitril-valsartan for heart failure with reduced ejection fraction.[4]

Due to the innovative nature and expanding use of combination products as technology advances greatly, there are many regulatory processes and pathways in place to approve these combinations. Therefore, this chapter will discuss common interventional radiology combination products and provide an overview of the FDA approval process for novel combination products.

Examples of common combination products used in radiologic interventions

Transarterial chemoembolization

Transarterial chemoembolization (TACE) is a procedure that combines a drug and a device. TACE is used in a variety of oncologic applications, such as in patients with unresectable hepatocellular carcinoma (HCC). TACE uses a combination of a chemotherapeutic agent with an embolic agent to deliver chemotherapy and block off blood supply to the tumor simultaneously.[5] Using the right chemotherapeutic and embolic agent may cause a beneficial effect to the patient compared to using bland transarterial embolization (TAE) alone. For example, one study conducted a randomized phase II/III trial investigating the use of cisplatin 50 mg as the chemotherapeutic agent and polyvinyl alcohol particles as the embolic agent.[6] Patients either received TACE or TAE, and the tumor progression in patients who received TACE was significantly less than in those who received TAE. Fig. 84.1[6] illustrates these results that using the combination of chemotherapy and embolization is more effective than using bland transarterial embolization alone.

Translational Interventional Radiology. https://doi.org/10.1016/B978-0-12-823026-8.00088-2

TABLE 84.1 List of common combination products approved by the FDA.[2]

Type	Description	Common example(s)
1	Convenience kit or copackage *Drug and device are provided as individual constituent parts within the same package*	Drug or biological product vials packaged with device(s) or accessory kits (empty syringes, autoinjectors, transfer sets), first aid or surgical kits containing devices and drugs
2	Prefilled drug delivery device/system *Drug is filled into or otherwise combined with the device AND the sole purpose of the device is to deliver drug*	Prefilled drug syringe, autoinjectors, metered-dose inhalers, dry powder inhalers, nasal spray, pumps, transdermal systems, prefilled iontophoresis system or microneedle "patch"
3	Prefilled biologic delivery device/system *Biological product is filled into or otherwise combined with the device AND the sole purpose of the device is to deliver biological product*	Vaccine or other biological product in a prefilled syringe, autoinjector, nasal spray, transdermal systems, or microneedle patch preloaded with biological product
4	Device coated/impregnated/otherwise combined with drug *Device has an additional function in addition to delivering the drug*	Drug pills embedded with sensors, contact lens coated with a drug, drug-eluting stents, drug-eluting leads, condoms with spermicide, dental floss with fluoride, antimicrobial coated catheters/sutures, bone cements with antibiotics
5	Device coated or otherwise combined with biologic *Device has an additional function in addition to delivering the drug*	Live cells seeded on or in a device scaffold, extracorporeal column with column-bound protein
6	Drug/biologic combination	Antibody-drug conjugates, progenitor cells combined with a drug to promote homing
7	Separate products requiring cross labeling	Light-activated drugs or biological products not copackaged but labeled for use with a specific light source device
8	Possible combination based on cross labeling of separate products	Drug/biological product under development utilizes a device, but unclear whether the final product will require that the two be cross-labeled
9	Other type of part 3 combination product (e.g., drug/device/biological product) *Combination product not otherwise described*	All three articles are combined in a single product (e.g., a prefilled syringe containing an antibody-drug conjugate), device to manufacture a biologic also includes a drug or biologic in the kit, or the product contains two different combination product types (e.g., Type 1 and Type 2 are provided together

Hydrogel-coated embolization coils

Hydrogel-coated embolization coils are a drug/device combination product using a platinum coiling wire surrounded by an expandable hydrogel polymer.[7] The purpose of this combination product is to offer a method for vascular occlusion in the management of states such as visceral intracranial and peripheral aneurysms, high flow arteriovenous fistulae, portosystemic shunts, and endoleak management after aneurysm repair. This drug/device combination allows for an effective method for mechanical vascular occlusion as opposed to placing an uncoated coil, as the hydrogel-coating gives these embolization coils the ability to expand up to four times their original size once deployed into the vasculature. The hydrogel-coil system is delivered via a delivery catheter and battery-operated detachment controller, making this a multiple drug/device combination product. The hydrogel polymer acts as a therapeutic drug, because once it comes into contact with blood, it is activated, starting a process of volume expansion. The hydrogel coating expands by 80% in the first 3 min of deployment, and after 20 min, reaches maximal expansion (Fig. 84.2)[7].

Drug-eluting stents

Drug-eluting stents (DESs) are an incredibly useful combination product used in the treatment of peripheral and coronary artery disease. This is the combination of a therapeutic drug and a metal stent device. Oftentimes, an antiinflammatory or antiproliferative drug is coated onto the metal stent. Before DESs were developed, bare metal stents were already a powerful device in the treatment of vascular occlusion. However, the risk of neointimal hyperplasia leading to in-stent restenosis remained. With the advancement of combination products today, DESs have shown to be a potent drug/device combination that reduces angiographic restenosis and target vessel revascularization when

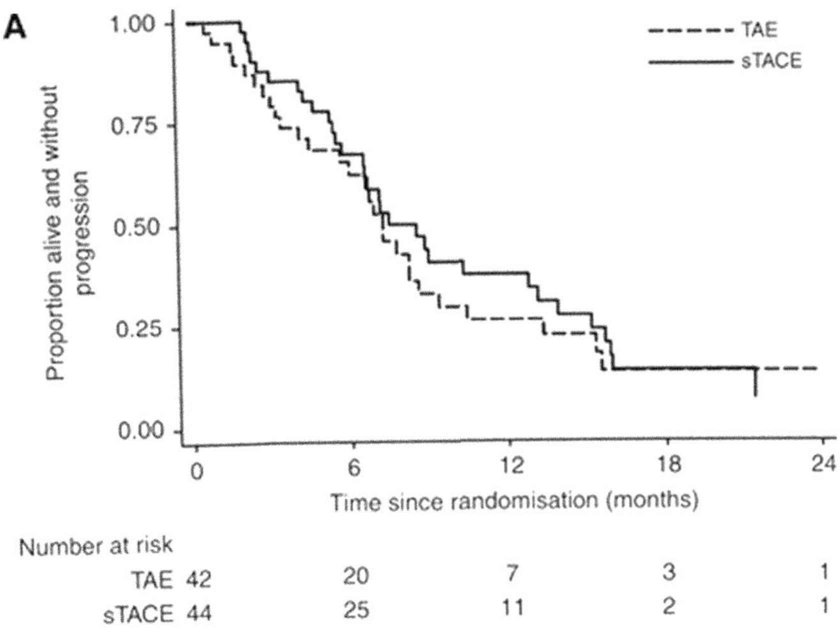

FIGURE 84.1 Progression-free survival is greater in HCC patients who received TACE as compared to TAE.[6]

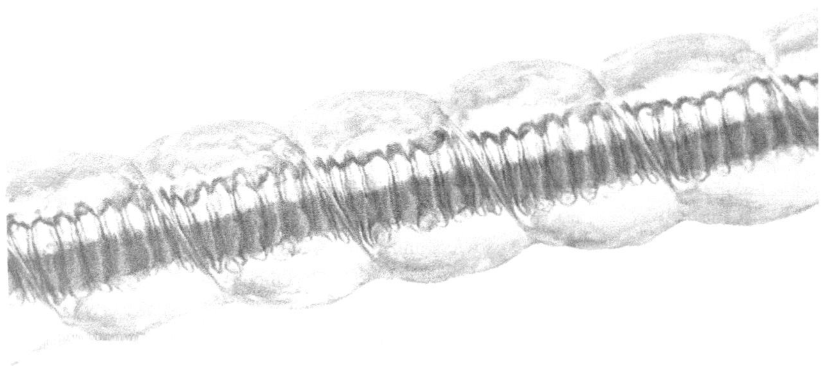

FIGURE 84.2 A magnified view of the hydrogel-coated embolization coil. The platinum wire is covered with the expandable hydrogel polymer.[7]

compared to bare metal stents.[8] One such example of a DES that is commercially available today is the sirolimus-eluting stent.[9] The stent is coated with a nonerodable polymer incorporated with sirolimus (rapamycin), and a topcoat is placed on top of that to function as a diffusion barrier, for the controlled release of the drug. Another recent development is triformulation paclitaxel-coated stents. This stent is coated with paclitaxel and has three different drug-release formulations (slow, moderate, and fast). For these drugs to be released so timely, they must be part of a precise drug/device combination mechanism. Alone, the drug would not be able to reach the target site with precision, nor would it be able to release in a timed manner but combining the drug and the stent device together allows the DES to accomplish this.

Overview of FDA approval of combination products

The FDA approval process for combination products generally begins at the FDA's Office of Combination Products (OCP). Based on the primary mode of action, the OCP will assign the combination product to one of the FDA's three human medical product centers: Center for Biologics Evaluation and Research, Center for Drug Evaluation and Research, or Center for Devices and Radiological Health.[4] The lead center will have oversight responsibility for the review and regulation of the combination product. The FDA has also developed a Standard Operating Procedure for the Intercenter Consultative and Collaborative Review Process, to ensure the timely and effective communication of consultations and collaborations.

Combination products may be subject to investigational new drug applications or investigational device exemption applications. Relevant to many applications in interventional radiology, a product may qualify for investigational device exemption if the combination product has a device-based primary mechanism of action. Therefore, many interventional radiology-related products may have an expedited FDA approval process owing to the heavily device-dependent nature of interventional radiology practice.[10]

The cost of getting a combination product approved is significantly less than getting a novel drug approved. The median cost of a new drug's development is estimated to be about $2-3 billion,[11] while combination products are less than that. Funding for novel combination products in their early stages of development primarily comes from the government and philanthropic organizations. Later, it comes from pharmaceutical companies.[12]

Get started

This chapter discusses the role of combination products in interventional radiology (IR). Herein, we will define combination products, illustrate examples of them in the practice of IR, and finally outline potential pitfalls to avoid.

Potential pitfalls

The primary pitfalls regarding FDA approval of combination products are regarding the FDA-defined primary mechanism of action and GxP practices of any given combination product. The primary mechanism of action of the combination product in question may determine if the product requires consideration as a new investigational drug therapy or may receive a device exemption. If the primary mechanism of action appears to be primary of a "drug" or "biologic" basis, an investigational new drug application may be required. Conversely, if a combination product appears to have a primarily "device" mechanism of action, a device exemption may be granted, generally significantly expediting the approval process. The second major pitfall is regarding GxP practices. Specific GxP practices are outlined in 78 FR 4307.[13] As an overview, any combination products are subject to GxP regulations and oversight for all components in conjunction. If different parts of combination products are manufactured at separate facilities, GxP regulations apply to those parts both when they are separated and when they are combined.

Proving the efficacy of a drug/device/biologic product being used in combination as opposed to being used alone may be arduous and expensive, and several studies with large sample sizes may be needed to observe any significant benefit. Additionally, when developing a new combination product that is similar to an already approved product, the directions for its use and population that uses it must be considered. For example, EpiPen is a drug/device combination that allows patients at risk for anaphylaxis to self-inject epinephrine. A similar device was also approved much later in 2010 called Adrenaclick, which is also an epinephrine self-injector, but has different directions of administration than the widely used EpiPen. This may create added confusion, particularly in acute scenarios and lead to inappropriate use of drugs and/or devices. Conversely, expanded combination products may lead to increased market competition, translating to improved products and reduced costs for consumers as well as expanding therapeutic applications. Thus, while there are numerous therapeutic benefits to combination products, there are also hurdles in the process of developing a novel combination product that must be considered.[14]

References

1. Office of the Commissioner. Combination products. U.S. Food and Drug Administration. https://www.fda.gov/combination-products. Accessed September 1, 2021.
2. Office of the Commissioner. Combination product definition combination product types. U.S. Food and Drug Administration. https://www.fda.gov/combination-products/about-combination-products/combination-product-definition-combination-product-types. Accessed September 1, 2021.
3. Office of the Commissioner. Classification of products as drugs and devices and additional issues. U.S. Food and Drug Administration. https://www.fda.gov/regulatory-information/search-fda-guidance-documents/classification-products-drugs-and-devices-and-additional-product-classification-issues#drug. Accessed September 1, 2021.
4. Office of the Commissioner. Early development considerations for innovative combination products. U.S. Food and Drug Administration. https://www.fda.gov/regulatory-information/search-fda-guidance-documents/early-development-considerations-innovative-combination-products. Accessed September 1, 2021.
5. Raoul J-L, Forner A, Bolondi L, Cheung TT, Kloeckner R, de Baere T. *Updated Use of TACE for Hepatocellular Carcinoma Treatment: How and when to Use it Based on Clinical Evidence. Cancer Treatment Reviews, 72*. 2019. https://doi.org/10.1016/j.ctrv.2018.11.002.
6. Meyer T, Kirkwood A, Roughton M, et al. A randomised phase II/III trial of 3-weekly cisplatin-based sequential transarterial chemoembolisation vs embolisation alone for hepatocellular carcinoma. *Br J Cancer*. 2013;108(6). https://doi.org/10.1038/bjc.2013.85.

7. Ferral H. Hydrogel-coated coils: product description and clinical applications. *Semin Intervent Radiol.* 2015;32(04). https://doi.org/10.1055/s-0035-1564809.
8. Htay T, Liu MW. Drug-eluting stent: a review and update. *Vasc Health Risk Manag.* 2005;1(4). https://doi.org/10.2147/vhrm.2005.1.4.263.
9. Muni NI, Gross TP. Problems with drug-eluting coronary stents — the FDA perspective. *N Engl J Med.* 2004;351(16). https://doi.org/10.1056/NEJMp048262.
10. FDA. Combination Products FAQs. https://www.fda.gov/media/106799/download.
11. Institute of Medicine (US) Forum on Drug Discovery, Development, and Translation. *Breakthrough Business Models: Drug Development for Rare and Neglected Diseases and Individualized Therapies: Workshop Summary*. Vol. 2. Washington (DC): National Academies Press (US); 2009 (Current Model for Financing Drug Development: From Concept Through Approval).
12. DiMasi JA, Grabowski HG, Hansen RW. Innovation in the pharmaceutical industry: new estimates of R& D costs. *J Health Econ.* 2016;47. https://doi.org/10.1016/j.jhealeco.2016.01.012.
13. FDA. Biological Product Definitions. https://www.fda.gov/files/drugs/published/Biological-Product-Definitions.pdf.
14. Current good manufacturing practice requirements for combination products. Federal Register. https://www.federalregister.gov/documents/2013/01/22/2013-01068/current-good-manufacturing-practice-requirements-for-combination-products. Published January 22, 2013. Accessed September 1, 2021.

Chapter 85

Cosmetics

Satya K. Morar
Integrated Medical Student, Case Western Reserve University School of Medicine, Cleveland, OH, United States

Overview of cosmetics in IR regulatory basics

The regulatory basics of conducting translational research in cosmetic interventional radiology (IR) is relatively new. There is demand for IR treatments in cosmetics, specifically in varicose veins care, botulinum toxin type A (Botox) injections under ultrasound guidance, and the management of vascular irregularities of the soft tissue.[3] Similar to other IR regulatory basics, cosmetics in IR translational research is indicated by more specific indicators for pathological processes, regularly scheduled clinical practice performance overview, and further investigation of current practice techniques.[1]

There are processes and pathways in the regulation of translational IR research unique to the subsidiary of IR. Similarly, the regulation of research components for cosmetic IR is as specific and divergent as the cosmetic procedures themselves. Lastly, although there are common challenges to all translational research and all IR translational research, there are also common challenges to regulating cosmetics in IR that are specific to the subject. Financial challenges are among the most prominent, and specific budgetary considerations must be made for cosmetics in IR, per the funding source and action plan.

Process/pathways

There are a few prominent pathways for the propagation of cosmetic IR translational research. As part of a growing market of cosmetic medicine, there is a defined pathway for researching academic institutions looking to increase the cash-based procedures where medical coding and insurance reimbursement do not even enter the equation. However, the most prominent is independent Office-Based Labs, which were already focusing on or marketing their services as vein specialists or clinicians. This is because they have a captive audience and an active referral base. There are minimal overhead costs due to similar equipment. There is a monetary incentive to show utility for cosmetics procedures.[2] Finally, practice-based and performance strategies can be implemented by the researching and practicing clinicians to build trust in this sector with patients and the general public to initiate the demand for the services.[4]

Components

There are multiple components of regulation for cosmetics in IR. The initial steps are proof of concept for the technique and any pharmaceutical in a laboratory setting. Treatments like varicose vein management have already been shown to be effective with an indication of pain, which means that the procedure and the equipment used are safe and effective. The next component is the FDA approval to continue with the research story. Moving ahead with the project now involves various components of equipment, patients, medications, and funding.

After proof of concept, the procedure can be translated to the real world via an FDA-approved Clinical trial where the equipment such as the ultrasound, office space, medical billing and coding, physician extenders, or support personnel will be needed. On top of this, a patient population will need to be established or accused in which the trial can still show unbiased efficacy of the treatment. Any disposable components such as pharmaceuticals would need to be established and stored, and funding for the entire operation would need to be procured. This is not an exhaustive list.

Examples

A prime example of translational research and the regulation of cosmetics in IR is the use of Botox in ultrasound-guided injections to improve wrinkles. Botox has initially been FDA-approved for medical use within the United States in 2002 with an action mechanism that blocked the release of acetylcholine from the motor neurons in the injected muscle, thus preventing contractions via paralysis.[5]

Additionally, the laser removal of dermal impurities has been a subset of medicine that has also translated to the IR

Translational Interventional Radiology. https://doi.org/10.1016/B978-0-12-823026-8.00044-4

space. The techniques used by IR physicians to carry out these types of procedures have already been established for more pathological indications. Both the technique and the outcomes can be translated to the IR space due to the success and prior approval of each.

Common challenges

One of the biggest and most crucial obstacles for the triumphant promotion of translational research is the significant and front-loaded capital and opportunity cost. The most considerable concern with the opportunity cost is the time of the IR Physician overseeing the research.[1] There is trouble with proof of concept and the importance of some of the research to acquire matters and funding for cosmetic IR specifically. Relatively recently, in 2007, IR Physicians around the world realized that the skills they acquired for minimally invasive treatments could be used to do Botox injection, collagen injections, and laser-based blemish removals.[2]

Besides these significant more traditional barriers, all significant cultural barriers within radiology inhibit translational research efforts on the cosmetics side. Chief of which is the inherent lack of intrinsic research beyond medical devices within IR. The specialty does not lend itself well to traditional Physician researchers such as M.D. and Ph.D. holders, and the residency program plus fellowship addition is almost all clinically oriented.[1]

Lastly, a significant inhibitor to the success and validation of translation research in IR is the lack of general public awareness surrounding the industry or the medical specialty. For example, the layperson knows the difference between neurosurgery and orthopedic surgery yet can tell the difference between diagnostic and IR. Nevertheless, they might not even know what IR is or never heard of it before. In addition to the general lack of public awareness toward IR, there is also a culture all to its own that surrounds the medical cosmetics space. In the public's eye, this culture can lead to a negative association with the clinical field. These two factors coupled together make it very difficult for funding grant requests to have much value or for there even to be a strong patient referral base to study.

Budgetary considerations

Holistic and groundbreaking projects take ample time and energy to activate and maintain. Extensive budgetary consideration is equipment use and time.[1] For cosmetics in IR, this equates to ultrasound scanner purchase, storage, maintenance, and funding. Materials such as Botox or Collagen also have their own storing requirements and considerations. The equipment has an overhead cost and may be a burden on the project. Additionally, the equipment is usually single faceted, meaning that it is only used for research and cannot be an asset to other aspects of the associated practice. Botox has little utility in a pathological-based IR practice where it is predominantly being used for a translation trial for cosmetics. Funding can also be a considerable concern with new equipment as the overhead cost can leave tiny behind for the rest of the research.

On the positive side, the often elective nature and cash-based model for cosmetics procedures can increase revenue from research and integration of associated procedures into established practices. However, some IR practices face lower and lower reimbursements from their standard procedure and patient acquisition.

How to fund

One of the greatest assets that the regulation of cosmetics in IR has is that the cosmetic medicine industry represents a billion-dollar business. As a result, a common way to fund cosmetic IR ventures is to pivot from one of two previously established capital bases. First, those in the medical cosmetics space can horizontally integrate IR into their practice and use the cash flow from their established practices to fund the project. Second, an IR group can expand their services to include cosmetic treatments such as Botox or collagen injections and use the revenue from their "bread-and-butter" IR services that have established reimbursements to finance the venture.

The other more traditional funding strategies come from a third-party origination that grants teams willing to dedicate their time to the project. This method is less popular due to the variety of strings that often come attached to the money. The bearer of these grants is professional and disease-focused societies or companies who often require a proposal, slide deck, interview, or more to make their allocation decision that has both a financial and opportunity cost. Additionally, some grants even go so far as to stipulate that the named grant awardee must spend a specific amount of time on their research-related activities. This is often 50%–75%.[3]

Get started

Getting started doing cosmetics in Interventional Radiology is not difficult if a private Office-based lab (OBL) model already exists. Outside of normal operations, there are new budgetary considerations for conducting cosmetics in IR, predominantly for equipment use and opportunity cost. Funding new equipment and training requires a capital investment from large funding sources. Regarding training and time, the opportunity cost lies in implementing new practices and not doing standard procedures. Interventional Radiologists can easily offer cosmetic vein work, BOTOX, and filler but must be trained.[6]

Potential pitfalls

Practitioners should be aware of a few potential pitfalls regarding implementing translation cosmetic interventional radiology. While financial incentives are a big bonus for implementing cosmetic procedures, they require a high front-end cost for the institution. Equipment and training have to be procured before starting.[2] In addition, there may be cultural barriers making it difficult to introduce potential patients to such procedures. Transitioning from generally medically-necessary procedures to elective cosmetic cases brings different conversations in the clinic.[2]

Other pitfalls include a lack of knowledge or demand from the patient population. While external marketing is necessary, patients are not acutely aware that IR offers cosmetics services. Surveys of patient demand should be conducted to understand if offering cosmetics services is a viable practice in a specific market.[3] If there is demand, then current practice patents may be the easiest to recruit for procedures to start. Lastly, an important pitfall involves 'biting off more than you can chew.' When implementing, a staged approach may be beneficial in which certain procedures are added in phases and not all at once. This can ensure that adequate demand exists before investment and can also allow a quicker start to practice.

Real-world examples

[1] Shah J, Kantharia S, Greif EF, Honig S, Sobolevsky S, Mobley D. Exploring the landscape of outpatient cosmetic interventional radiology: A new market for a new era. *J Vasc Interv Radiol.* 2015;26(2). https://doi.org/10.1016/j.jvir.2014.12.559.

[2] Bennett JD, Miller TA, Richards RS. The Use of Botox® in Interventional Radiology. *Tech Vasc Interv Radiol.* 2006;9(1):36–39. https://doi.org/10.1053/j.tvir.2006.08.008.

[3] Huq S, Bee E. Interventional Radiology: Giving Cosmetic Medicine a Makeover. *J Vasc Interv Radiol.* 2017;1(3). https://doi.org/10.1055/s-0041-1729795.

Resources

[1] Carrafiello G, Laganà D, Fontana F, et al. Cosmetic treatments: An emerging field of interest for interventional radiologists. *Int J Surg.* 2008;6. https://doi.org/10.1016/j.ijsu.2008.12.016

[2] Masuda E, Ozsvath K, Vossler J, et al. APPROPRIATE USE CRITERIA. *J Vasc Surg Venous Lymphat Disord* 2020;8(4). https://doi.org/10.1016/j.jvsv.2020.02.001

[3] Haskal Z. The Cosmetic Frontier: BOTOX and Beyond. *Radiol Bus.* Published online 2010.

[4] Sabharwal T, Fotiadis N, Adam A. Modern trends in interventional radiology. *Br Med Bull.* 2007;81–82(1): 167–182.

References

1. Petrou M, Foerster BR, Reich DS. Translational research in radiology. *Acad Radiol.* 2009;16(5):593–596. https://doi.org/10.1016/j.acra.2009.01.017.
2. Cosmetic practice offers makeover for cash-poor irs. Diagn Imag. https://www.diagnosticimaging.com/view/cosmetic-practice-offers-makeover-cash-poor-irs. Published July 1, 2007. Accessed August 2, 2021.
3. Carrafiello G, Laganà D, Fontana F, et al. Cosmetic treatments: an emerging field of interest for interventional radiologists. *Int J Surg.* 2008;6. https://doi.org/10.1016/j.ijsu.2008.12.016.
4. Shah J, Kantharia S, Greif EF, Honig S, Sobolevsky S, Mobley D. Exploring the landscape of outpatient cosmetic interventional radiology: a new market for a new era. *J Vasc Intervent Radiol.* 2015;26(2). https://doi.org/10.1016/j.jvir.2014.12.559.
5. FDA approves Cosmetic use of botulinum toxin. Center for Infectious Disease Research and Policy. https://www.cidrap.umn.edu/news-perspective/2002/04/fda-approves-cosmetic-use-botulinum-toxin. Published April 18, 2002. Accessed August 2, 2021.
6. Haskal Z. The Cosmetic Frontier: BOTOX and Beyond. *Radiol Bus.* Published online 2010.

Chapter 86

CMC and GxP

Marvee Espiritu[1], Capt. Tej Ishaan Mehta[2] and Purushottam K. Dixit[3]

[1]*Oakland University William Beaumont School of Medicine, Rochester, MI, United States;* [2]*Department of Interventional Radiology, The Johns Hopkins Hospital, Baltimore, MD, United States;* [3]*Department of Radiology, Division of Interventional Radiology, Beaumont Health, Royal Oak, MI, United States*

Key points

- Chemistry, manufacturing, and controls (CMC) is an integral aspect of an investigational new drug (IND) application to ensure characterization, safety, and stability of the drug product.
- CMC processes maintain the stability and reliability of pharmaceutical products to ensure those products are biologically equivalent to the same products used in clinical trials.
- Good "x" Practices (GxP) policies ensure compliance of drug products with FDA guidelines.

Chemistry, manufacturing, and controls

Overview

The CMC process is a critical part of IND applications (Fig. 86.1). CMC practices provide the reliability of drug product identity and stability. A critical component of long-term drug manufacturing is fidelity to drugs used in clinical trials. CMC practices help ensure the effectiveness and safety of drug products, maintaining this fidelity (Fig. 86.2).

The following components of the CMC are required in the summary report of the product.

Components of CMC: *section 312.12*[1]

- Description of the investigational drug to ensure "proper identification, quality, purity, and strength of the investigational drug." (*Section 312.12 (a) (7) (i)*).
- *Drug substance*—Description of the physical and chemical properties of the drug product, the manufacturer, the method of preparation, methods used to ensure identity, purity, and quality of the product, and information supporting the stability of the drug product during toxicological and clinical studies (*Section 312.12 (a) (7) (iv) (a)*).
- *Drug product*—A section listing the components used in the manufacture of investigational drug products including alternatives used in the manufacturing (*Section 312.12 (a) (7) (iv) (b)*).
- Description of the composition, manufacture, and control of placebo used in controlled clinical trial. This should include tests showing the absence of the active ingredient as well as stability testing throughout the duration of the study (*Section 312.12 (a) (7) (iv) (c)*).
- *Labeling*—"Copy of all labels and labeling to be provided to each investigator" (*Section 312.12 (a) (7) (iv) (d)*).
- Environmental assessments of the drug substance (*Section 312.12 (a) (7) (iv) (e)*).
- Information regarding the dosage and route of administration of the drug product.

Common challenges: CFR 312.23[1]

- CMC lacks sufficient information to assess the risks to subjects.
- Subjects would be exposed to significant risks or injury.
- Changes in scale production from clinical trials to a larger scale for market production may inevitably lead to changes in product manufacturing, which in turn may impact drug quality, its physical and chemical properties, as well as its purity.[2]
- Insufficient data provided to assure stability of the drug product.
- Insufficient data supporting "comparability of toxicology and proposed clinical lots."
- Lacks sufficient information regarding the raw materials of animal origin.
- Impurity profile indicates risks or safety concerns.
- Lack of information regarding product transfer protocols between manufacturing or packaging sites may delay approval of the application. For instance, in the packaging of Apixaban, multiple manufacturing sites

Translational Interventional Radiology. https://doi.org/10.1016/B978-0-12-823026-8.00061-4

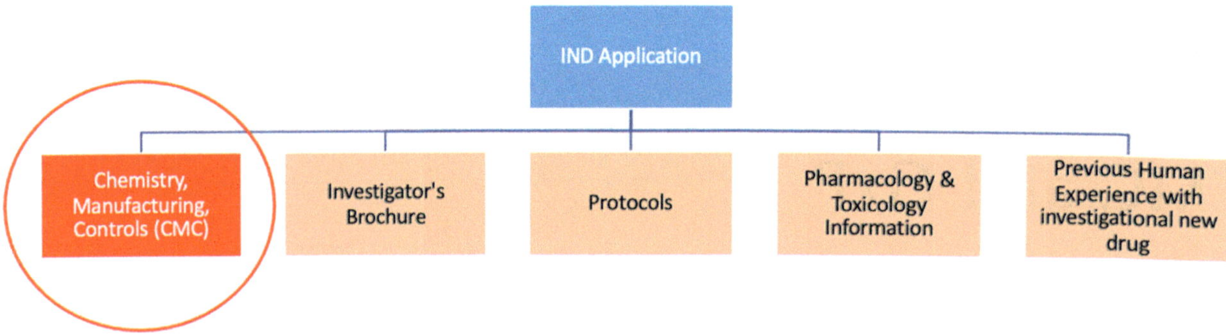

FIGURE 86.1 Schematic diagram demonstrating the role of the CMC as a critical aspect of an IND application.

FIGURE 86.2 A schematic diagram demonstrating the role of CMC practices in maintaining the fidelity of manufactured drugs to clinical trial drugs, ensuring the effectiveness and safety of the drug products.

were used—Bristol-Myers Squibb (BMS) sites at Mount Vernon Illinois and Humacao, Puerto Rico. Protocol regarding the transfer of drug products between sites was not provided leading to a delay in the CMC approval.

Real-world examples in interventional radiology

Apixaban (*Eliquis*) is an anticoagulant often encountered by interventional radiologists in the periprocedural management of patients with coagulopathy requiring image-guided interventions. Apixaban is a factor Xa inhibitor indicated for the treatment of Deep Vein Thrombosis (DVT) and Pulmonary Embolism (PE) as well as a therapeutic agent used to reduce the risk of stroke or systemic emboli in patients with atrial fibrillation. Below is an example of the components of Apixaban's CMC information that was provided in the IND application with clarifications provided as needed.[3]

Drug substance: A key way of providing evidence of the drug's identification is a description of its physical and chemical properties. CMC information on Apixaban Tablets 2.5 and 5.0 mg included a description of its physical properties (chemical name, structural formula, molecular formula, molecular weight) and chemical properties characterized by studies such as elemental analysis, UV and visible absorption spectral analysis, and mass spectral analysis to show that the molecule has no chiral centers. Chirality plays a critical role in drug development as one enantiomer may be beneficial whereas another enantiomer may be toxic. Therefore, it is essential to ensure that the drug has no chiral center that could potentially alter its biochemical actions. Throughout the process of drug development, confirmation of the structure is critical to prove that the drug used in clinical trials has the same identity and properties as the drug used in preclinical animal studies.[2]

Information on batch analyses produced by different processes was provided to show the consistent quality of Apixaban in the commercial process. In addition, adequate stability studies were provided to support the integrity and lack of degradation of the drug product under long-term or harsh environmental conditions. Stability data of the drugs at each manufacturing site was also provided to ensure the integrity of the drug product. Data regarding microbial limits testing showed no chemical degradation or microbial growth in the batches, storage, and packaging conditions. Impurities were addressed from the chemistry and safety perspectives. Chemistry aspects included identifying the impurities and providing protocols set in place to control their levels. From a safety perspective, information was provided regarding impurities that were not found in batches used in clinical studies or impurities present but in insignificant levels.[4] Convincing data that control strategies for potential impurities and routine drug batch testing for impurities were provided.

Drug product: The active ingredients used in the manufacturing of Apixaban were listed addressing safety concerns from the ingredients including raw materials and their animal origin. For instance, one of the active ingredients of Apixaban was lactose. Further safety information was provided to address concerns regarding Transmissible Spongiform Encephalopathies. Analysis of drug bioavailability with the use of excipients must be provided.[5] In the case of Apixaban, detailed information regarding the excipients and film coating materials was submitted to ensure the safety of the drug product including certificates of analysis for the excipient lots.

Labeling: Copies of the labels were provided in the CMC section of the IND. Lack of labels may delay the approval of the process as was the case with Apixaban's CMC application.

Any changes to the manufacturing or packaging sites must be filed as a CMC amendment as part of compliance. For example, a supplemental new drug application was filed to add Bristol-Myers Squibb facility in Humacao, Puerto Rico to package Apixaban. In the review of this amendment, a container closure system in the amended facility as well as product stability was investigated before approval.

Another applicable example in the realm of interventional radiology is Doxorubicin. Doxorubicin is a cytotoxic chemotherapeutic drug used in the treatment of cancers such as acute lymphoblastic leukemia, breast, ovarian cancer, and bladder cancer. In interventional radiology, Doxorubicin has been used in transcatheter arterial chemoembolization (TACE) as a treatment for Hepatocellular Carcinoma.[6] In the CMC application of Doxorubicin, failure to provide a complete batch record and stability data from a batch that used an alternate manufacturing process delayed approval of the CMC application.[7] In addition, an amendment to the CMC application was completed when the manufacturing and packaging site shifted from Columbus, Ohio to Albuquerque, New Mexico. In review of this amendment, the site was inspected to assure compliance with GxP as well as assessed for container closure system and product stability before approval.

Further relevant example in interventional radiology is the use of Bevacizumab in the treatment of colorectal cancer liver metastases during TACE.[8] In the Drug Product and Drug Substance sections of the CMC application, the quality, potency, and safety of Bevacizumab were assessed for fidelity across clinical trial stages.[9] In the drug development of Bevacizumab, changes to the manufacturing process including modifications in the formulation and vial configuration were submitted as an amendment to the CMC application. The amendment was accompanied by stability data documenting that the changes did not affect the quality or safety of the product.

Strategies for CMC application

- Submit amendments to the IND related to manufacturing changes as soon as reasonably possible to allow for sufficient time for the FDA to review and assure the safety of any manufacturing changes.
- CMC amendments may include:
  - Changes in manufacturing site or processing.
  - Changes in the packaging or container closure system of the drug product that may affect the quality of the drug.
  - Changes in sterilization methods.
- Stability Testing:
  - Description of tests used to confirm the stability of the drug product during toxicology and clinical studies.
  - Provide data supporting the stability of the drug product during handling and administration and collect these data throughout the "clinical development."
- Include primary data for the toxicology and clinical material.
- Provide stability data updates in the annual CMC report.

Good "x" practices

GxP, an acronym for "Good" "x" "Practices," is a collection of regulations set forth by the FDA to ensure the safety of drug products throughout their manufacturing, storage, and distribution. The "x" may represent manufacturing "M" as in Good Manufacturing Practices (GMP), clinical "C" as in GCP, laboratory "L" as in GLP, storage "S" as in GSP, distribution "D" as in GDP, or review "R" as in GRP. For instance, GMP are guidelines that ensure consistent manufacturing processes and product quality. While the goals of CMC and GxP may overlap, CMC is specifically related to pharmaceutical drug products themselves whereas GxP is related to the manufacturer's practices (GMP), the distribution of the drug (GDP), and storage of the drug (GSP) (Fig. 86.3).

Specific regulations regarding GxP may be referenced in 21CF211.[10] In brief, the following are critical components of GxP:

- High fidelity processes to ensure pharmaceutical quality control.
- Accurate, up-to-date records of pharmaceutical manufacturing processes.
- Verification of purity, strength, and quality of each component in the pharmaceutical manufacturing process.
- Reliable storage methods for pharmaceuticals.

GxP practices are particularly relevant when manufacturing processes shift from small to large scale. Strict adherence to the GMP regulations set forth by the FDA aids in assuring that the drug batches made in the large-scale clinical setting are the same drugs used in smaller batches during preclinical settings.

FIGURE 86.3 Schematic diagram of the overlap between CMC and GxP with both components promoting the safe, consistent, and quality production of the drug.

Furthermore, if multiple facilities are utilized in the manufacturing, testing, or packaging of the drug product, each manufacturing site must undergo GMP inspection.[5] Review of the manufacturer's compliance with GMP regulations is needed before approval of the CMC application. For example, the CMC approval of Apixaban was delayed before the thorough review of the manufacturer's facilities and equipment to ensure compliance of the manufacturing process with GMP.

A number of challenges often arise in GxP manufacturing that is not fully elucidated by 21CF211. Namely, GxP manufacturers must verify both the product quality of their own products as well as the raw materials used in their manufacturing processes. This may be particularly challenging when manufacturers change or upgrade fabrication methods as unforeseen errors or interactions may occur.

In summary, CMC and the GxP processes are crucial components of a successful IND application. Detailed reports and data regarding the drug product's identity, safety, stability, and manufacturing compliance are critical to ensure safe and consistent production. Any changes in the manufacturing process of the drug must be submitted immediately to avoid further delays in the approval of the CMC application. Furthermore, it is critical to incorporate all aspects of GxP (GCP, GDP, GLP, GMP, and GSP) to ensure approval of the IND application.

Get started

1. Review all the required components of the CMC application.
2. Gather all documentations and data supporting the identity, safety, and stability of the drug product.
3. Ensure that all proper documentations, such as a copy of the FDA's GxP inspection (GMP inspection documents for all manufacturing or packaging sites used), are in place.
4. Submit the CMC and GxP information as part of the IND application.
5. Respond to amendment requests promptly to increase the likelihood of application approval.

Potential pitfalls

- Failure to provide sufficient information to delineate the drug's safety and stability.
- Failure to submit or provide sufficient information regarding changes in the manufacturing process and manufacturing or packaging sites as an amendment to the IND in a timely manner.
- Failure to provide a complete batch record and stability data after using an alternate manufacturing process.
- Failure to ensure that GxP inspections are up-to-date on all sites used in the manufacturing process of the drug.

Resources

Additional reading

1. US Food and Drug Administration: CFR—Code of Federal Regulations Title 21—Food and Drugs; Part 211—Current Good Manufacturing Practice for Finished Pharmaceuticals
2. US Food and Drug Administration: CF—Code of Federal Regulations Title 21—Food and Drugs; Part 312—Investigational New Drug Application

References

1. FDA. *FDA US Food & Drug Administration CFR - Code of Federal Regulations Title 21*. 2020.
2. DiFeo TJ. Safety and efficacy: the role of chemistry, manufacturing, and controls in pharmaceutical drug development. *Drug Dev Ind Pharm*. 2004;30(3):247–257. https://doi.org/10.1081/DDC-120030428.

3. FDA. *US Food and Drug Administration Drug Approval Package - Eliquis*. 2013.
4. Sheinin E, Williams R. Chemistry, manufacturing, and controls information in NDAs and ANDAs, supplements, annual reports, and other regulatory filings. *Pharm Res (N Y)*. 2002;19(3):217–226. https://doi.org/10.1023/A:1014437513188.
5. DiFeo TJ. Drug product development: a technical review of chemistry, manufacturing, and controls information for the support of pharmaceutical compound licensing activities. *Drug Dev Ind Pharm*. 2003;29(9):939–958. https://doi.org/10.1081/DDC-120025452.
6. Lencioni R, Petruzzi P, Crocetti L. Chemoembolization of hepatocellular carcinoma. *Semin Intervent Radiol*. 2013;30(1):3–11. https://doi.org/10.1055/s-0033-1333648.
7. FDA. *US Food & Drug Administration FDA Application - Adriamycin*. 2003.
8. Fiorentini G, Sarti D, Nardella M, et al. Transarterial chemoembolization alone or followed by bevacizumab for treatment of colorectal liver metastases. *Hepatic Oncol*. 2021:HEP40. https://doi.org/10.2217/hep-2020-0031. Published online.
9. FDA. *US FOOD & Drug Administration Drug Approval Package - Avastin*. 2005.
10. FDA. *US Food & Drug Administration CFR - Code of Regulations Title 21 Part 211 - Current Good Manufacturing Practice for Finished Pharmaceuticals*. 2020.

Chapter 87

Non-US regulatory

Daniel Reyes and Daniel Braga
Department of Vascular and Interventional Radiology, Mount Sinai Medical Center, Miami Beach, FL, United States

Broadly speaking, the design and execution of any research endeavor in the field of interventional radiology is developed through much the same avenues as other medical subspecialties as a whole in the era of evidence-based medicine; however, clinical and translational research efforts regarding IR and other predominantly proceduralist subspecialties are invariably tied not only to written theories on the pathophysiology of various disease states and their corresponding medical management, but also the added responsibility of being able to pioneer, develop, and evaluate numerous different kinds of medical devices and the underlying biologic mechanisms that contribute to their efficacy as well as any potential risks to the patient. It is therefore essential for the IR operator to understand the various regulatory bodies responsible for ensuring the safety of patients when it comes to future research as well as new medications and medical devices on a more broadened and international scale, especially since a growing body of lesser-known novel transcatheter techniques often may originate in various international research groups of interventionalists, which then gradually assimilate into clinical practice in the United States (US) over a number of years. Furthermore, although the US is traditionally viewed as the undisputed global leader in medical innovation and research, recent evidence has questioned this over concerns of reduced investments in early-stage basic science biomedical research in favor of later-stage clinical research, underfunding, disproportionate resource allocation with respect to individual disease burden, and the globalization of medical R&D.

The regulation of medical drugs and devices outside the United States is incredibly varied and complex, ranging from comprehensive oversight to none at all. Regulatory bodies must ensure that medical drugs and devices are safe and effective while they transit through investigative processes and into public use. The most prominent regulatory system outside of the United States Federal Food and Drug Administration (FDA) is the European Union's European Commission (EC). While the FDA is a single centralized agency based on consumer protection, the EC's multifaceted regulatory system is based on a need to synchronize the commercial interests and autonomy of its 28 member states through both centralized and decentralized networks within individual member states.[1] Critics have long drawn comparisons between the US and EU approval processes, with the topics of concept to market time and transparency of drug approval data being the most heavily scrutinized. It is important to note these two important topics, as they are often the target of updated regulatory measures; however, a thorough discussion of these topics falls outside the scope of this chapter and will not be discussed in the great detail that they deserve.

Before the formation of the European Union, various regulatory processes for medical drug approval within its now member states stifled the progression of drug concepts to market and led to limited availability of certain drugs throughout Europe. In 1965, the EC Directive 65/65/EEC was introduced to curb the heterogeneity of drug approval processes throughout Europe and to establish consistent guidelines regarding the information that must be submitted for drug approval.[1] After the formation of the European Union in 1993, further measures were introduced to streamline and centralize the drug approval process. Many of these drug approval processes are similar to those of the FDA.

An investigator must first gain preauthorization for use of the proposed drug in clinical trials, which will be guided by Regulation No. 536/2014 of the European Parliament, replacing its predecessor, Clinical Trials Directive of the European Commission (2001/20/EC).[1] Regulation No. 536/2014 has been adopted by the EC and is managed primarily by the European Medicines Agency (EMA)'s Management Board and is due for an independent audit by a third party in December 2020 with plans to go live shortly thereafter.[2] This measure will ensure a more harmonious environment for conducting clinical trials within the EU while setting a

Translational Interventional Radiology. https://doi.org/10.1016/B978-0-12-823026-8.00077-8

higher standard for interstate collaboration and public ease of access to clinical trial information, including results and safety data. These updates aim to increase the efficiency of all trials within the EU while avoiding trial duplication or repetition of unsuccessful trials.[3] Additionally, it introduces an "authorization process based on a single submission via a single EU portal, an assessment procedure leading to a single decision, rules on the protection of subjects and informed consent, and transparency requirements."[3] Lastly, Regulation No. 536/2014 aims to facilitate the conduction of multinational clinical trials by pharmaceutical companies, which should increase the number of studies conducted within the EU.[3]

Before initiating a clinical trial, approval must be obtained from an independent ethics committee (IEC), which is the equivalent of the FDA's institutional review board (IRB). An IEC is a body that reviews and subsequently approves or rejects clinical trial protocols based on their risk/benefit assessment. The structure of IECs and IRBs differ greatly between the US and EU. In the EU, individual IECs are created and regulated by each individual member state, which are approved and regulated by a local regulatory body known as a competent authority.[4] IRBs in the US are regulated centrally by the Office for Human Research Protections within the Department of Health and Human Services.[5] The biggest difference between IEC and IRB approval is that IRB approval in the US is pursued by the investigator and IEC approval in the EU is obtained by the sponsor of the clinical trial.

Once a proposed pharmaceutical obtains preauthorization and IEC approval for clinical trials, the next steps are similar to the drug approval process in the United States. The proposed pharmaceutical advances through phase 1 trials, where the drug is tested in a small number of healthy subjects to determine pharmacology and dose range. Next, Phase II trials investigate and establish the dose-response relationship in several hundred patients. Phase III trials are then conducted in hundreds to thousands of subjects to evaluate and confirm the safety and efficacy of the drug of interest. Analogous to the FDA's process, special approval processes exist for drugs to be used in emergency conditions and for "orphan drugs', which are proposed for the treatment of rare conditions which would make randomized clinical trials nearly impossible.[1]

The EMA was formed in 1995 to synchronize the regulation processes of the EU's individual member states; however, the EMA does not parallel the FDA as the EU's single approval body. In stark contrast to the FDA's unilateral approval process, there are four routes in which pharmaceuticals may be approved in Europe, depending on the drug class and manufacturer preference.[1] The centralized process is governed by the EMA, where each of the 28 EU member states is represented. The EMA's Committee for Medicinal Products issues a single license that is valid in all EU member states. Some classes of drugs including those used in HIV/AIDS, diabetes, oncology, neurodegenerative disorders, and others are mandated to go through this centralized process.[1] The second and third routes of approval are known as the national process and mutual recognition. The national process allows individual member states to oversee their own approval process for drugs that do not meet the mandate of the centralized process while the mutual recognition process allows for drugs approved via an EU member state's own process to obtain marketing authorization in another EU member state.[1] The fourth route for approval is known as the decentralized process, in which manufacturers may apply for approval simultaneously in more than one EU state for drugs that have not yet been approved in any EU state and do not fall under the mandatory centralized process.[1] The majority of applications are submitted through this process.

Aside from drug approval, European device regulation is similar to the aforementioned processes, but is less well defined than European drug regulatory processes with a number of stark differences when compared to the device approval process in the US. Legislative directives, applicable to all EU member states, set forth by the EC regarding medical devices include directive 90/385/EC (implantable devices, 93/42/EC (other medical devices), and 98/79/EC (in vitro diagnostic devices), although some of these directives and others have been reviewed and subsequently repealed recently in favor of Medical device regulation 2017/745 to be implemented in the near future. New medical devices, in theory, would need to coincide to these directives, and in addition undergo risk stratification and a complex decentralized legal process of obtaining a Conformité Europeenne (CE) mark to officially be considered directive-compliant and hence allowed to be marketed within any EU member state.[1]

Preapproval stratification into regulatory device classes in Europe is similar to that of the FDA in that the individual class of device (low, intermediate, and high risk) often will determine the rigor and quality of evidence needed to apply for and obtain approval for that device, as well as the specific directive(s) with which it must comply. There are four different device classes outlined by the EC, as follows. Class I is considered low risk (e.g., gloves) and requires the involved party "self-declare" their conformity with a checklist of safety and performance requirements such as the Essential Requirements to the respective Competent Authority in the country of origin. Class I devices do not require a clinical trial, similar to obtaining premarket approval from the FDA in the US. Class IIa is considered a low-medium risk (e.g., suctioning equipment), Class IIb is a medium-to-high risk (e.g., ventilators), and Class III is a high risk (e.g., cardiac pacemaker). These classes require evidence of safety and performance in the absence of a substantially similar "predicate" device, as is generally the case in the US with Class III (high risk) devices per the FDA.[1] Depending on the classification, typically higher risk devices with the required adjunctive data, approval may then be sought from any of the member EU states with a device application that undergoes a further decentralized review process consisting of over 70 primarily state-authorized private companies, each of which is termed a "Notifying Body," which is abbreviated as "NB". For a fee, the NB then issues a CE mark if a device meets the specified requirements, ultimately allowing the device to be freely marked throughout the EU member states.

In general, the system of NBs currently in place arose out of prior socioeconomic and medical industrial concerns and represents a multinational effort to streamline or "harmonize" commercial processes in the EU with respect to biomedical innovation. This is fundamentally different from the principles with which the FDA comprises, which is primarily that of consumer health protection. It is, therefore, crucial to recognize unfortunately that the CE mark applied to European medical devices, according to EU regulations, does not strictly require prior clinical efficacy data for approval as it would in the US, reducing its clinical value to that of basic safety and on-paper legal compliance with various European regulatory bodies. The large number of decentralized NBs, which in theory would help expedite approval in contrast to the more centralized FDA, may have varying individual regulatory standards and fees, introduces an enormous conflict of interest from a financial and regulatory standpoint, superimposed on top of an already problematic CE mark NBs may or may not grant to medical devices whose clinical efficacy may largely be undetermined.[1] Perhaps most concerning is the fact that safety and clinical data submitted to various NB's also fall victim to decentralized commercialization and are not routinely available to the public for fully informed risk versus benefit decision-making by both patient and physician, especially with respect to newer devices.

The approval of medical drugs and devices outside of the United States and EU is widely variable and ranges from country-specific and regional regulatory bodies to a complete absence of a regulatory process. For example, in Japan, a proposed pharmaceutical initially follows a process similar to that of the US and EU, with the completion of Phase I, II, and III clinical trials followed by a manufacturer-filed application with the Pharmaceuticals and Medical Devices Agency (PMDA).[6] The application then undergoes review and the PMDA provides a recommendation to the Ministry of Health, Labor, and Welfare, which will then serve as the regulatory authority to approve or reject the pharmaceutical of interest.[6] China's regulatory body is based on the FDA but exercises more flexibility to increase the global competitiveness of its drugs and devices. Southeast Asian Nations have complex regulatory processes, which despite most nations adopting EU regulations, are highly country specific. For example, halal compliance is required in Malaysia and Indonesia. The drug regulatory bodies of South American countries are also highly variable with the regulatory bodies of Argentina and Brazil functioning autonomously, while those of Chile and Peru are overseen by the ministries of health.[7] The regulation of medical drugs and devices in Russia is unique in its complexity and process, with oversight and issuance of certificates directed by multiple state agencies. The approval process in the Middle East possesses both country-specific and collaborative international approval bodies, influenced by past and present geopolitical ties. In Africa, several international organizations have been established to promote continental standardization for the approval and regulation of medical drugs and devices; however, in 2005, the WHO reported that only 7% of the 46 sub-Saharan African countries had national regulatory bodies in place, with 30% of these countries having no regulation.[2]

Get started

Study planning must include a working understanding of the requirements set forth by the various regulatory bodies specific to a country and/or region in which approval is sought. As the medical device and drug industry will not infrequently aim to maximize commercial market share internationally, either in succession or simultaneously, it is also essential for researchers to familiarize themselves with regulatory processes outside of where approval was originally granted, particularly with concerns regarding patient safety.

Potential pitfalls

When involved in research regarding a previously approved device or medication, safety databases that are offered by some regulatory bodies should be referenced for any issues that have been encountered by other clinicians, especially as there can be a lag between reported adverse events and when the subsequent regulatory action is taken against it. Furthermore, user-reported safety databases may unfortunately not be as well organized or regulated as the processes in place that conferred regulatory approval to a particular device in the first place, and it is of course not known how many adverse events go underreported or even unreported altogether.

Real-world examples

Three real world literature examples regarding our topic:

1. Hwang TJ, Sokolov E, Franklin JM, Kesselheim AS. Comparison of rates of safety issues and reporting of

trial outcomes for medical devices approved in the European Union and United States: cohort study. *BMJ.* June 28, 2016;353:i3323. https://doi.org/10.1136/bmj.i3323. PMID: 27352914; PMCID: PMC4925918.
2. Wild C, Erdös J, Zechmeister I. Contrasting clinical evidence for market authorisation of cardio-vascular devices in Europe and the USA: a systematic analysis of 10 devices based on Austrian pre-reimbursement assessments. *BMC Cardiovasc Disord.* November 4, 2014;14:154. https://doi.org/10.1186/1471-2261-14-154. PMID: 25366498; PMCID: PMC4232631.
3. Kramer DB, Xu S, Kesselheim AS. How does medical device regulation perform in the United States and the European union? A systematic review. *PLoS Med.* 2012;9(7):e1001276. https://doi.org/10.1371/journal.pmed.1001276. Epub 2012 Jul 31. PMID: 22912563; PMCID: PMC3418047.

Resources

1. U.S. FDA Medical Devices Section. https://www.fda.gov/medical-devices
2. U.S. FDA Manufacturer and User Facility Device Experience (MAUDE) Database. https://www.accessdata.fda.gov/scripts/cdrh/cfdocs/cfmaude/search.cfm
3. EURAMED – European Database on Medical Devices. https://ec.europa.eu/tools/eudamed/#/screen/home
4. European Commission Website. https://commission.europa.eu/index_en

References

1. Gail A, Norman Van. Drugs and devices: comparison of European and U.S. Approval processes. *J Am Coll Cardiol.* 2016;1(5):399–412. https://doi.org/10.1016/j.jacbts.2016.06.003. ISSN 2452-302X.
2. Clinical Trials - Regulation EU No 536/2014." European Commission. Published 13 May 2020. https://ec.europa.eu/health/human-use/clinical-trials/regulation_en.
3. *Clinical Trial Regulation.* European Medicines Agency; June 12, 2020. https://www.ema.europa.eu/en/human-regulatory/research-development/clinical-trials/clinical-trial-regulation.
4. Regulation (EU) No 536/2014 of the European Parliament and of the Council of 16 April 2014, on Clinical Trials on Medicinal Products for Human Use, and Repealing Directive 2001/20/EC. European Commission. https://ec.europa.eu/health/sites/health/files/files/eudralex/vol1/reg_2014_536/reg_2014_536_en.pdf.
5. Electronic Code of Federal Regulations. Published July 19, 2018. https://www.ecfr.gov/cgi-bin/retrieveECFR?gp=&SID=83cd09e1c0f5c6937cd9d7513160fc3f&pitd=20180719&n=pt.45.1.46&r=PART&ty=HTML.
6. Information On Japanese Regulatory Affairs. Regulatory Information Task Force. Japan Pharmaceutical Manufacturers Association. Published 2018. http://www.jpma.or.jp/english/parj/pdf/2018.pdf.
7. Sandri MM, De Seta MH, Luiza VL. Drug regulatory authorities in South America: an analysis based on organizational rules. *Rev Panam Salud Pub.* 2013;34(3):169–175.

Chapter 88

Postmarketing drug safety monitoring

Shrey Patel
Interventional Radiologist, Baylor Radiologists, Houston, TX, United States

Key points

- Clinical trials provide important information regarding a drug's safety and efficacy; however, continued surveillance is required to overcome their inherent limitations.
- The most well-known task in postmarketing surveillance is monitoring drug safety, which ranges from manufacturing inspections to guidelines for generic drugs to active and passive surveillance.
- The Food and Drug Administration (FDA) also regulates drug advertising and whether an already marketed drug can be used in additional ways.

Although drugs are approved on the basis of clinical trials, postmarketing surveillance is an important regulatory practice that allows the FDA to more accurately monitor the safety and efficacy of drugs. The most widely known component of postmarketing surveillance is safety monitoring. The FDA is also tasked with overseeing additional regulations that impact consumers, manufacturers, and healthcare providers.[1]

Supplemental applications

Supplemental applications are petitions submitted to the FDA for an already approved new drug application (NDA).[2] The FDA must approve all important changes to ensure the conditions originally set for the drug are maintained. If approved, these supplemental applications allow companies to change labels, market new dosages, or change manufacturing practices. For example, cefiderocol, an antibiotic, was initially approved to treat complicated urinary tract infections in adults. Later, the FDA approved the pharmaceutical company's supplemental application to expand the treatment indication to certain hospital-acquired pneumonia.[3]

INDs for marketed drugs

Investigational new drugs (INDs) are defined by the FDA as new drugs to be used in clinical investigations in humans. These may already be developed drugs that use a "new" aspect of its use. In the most general terms, an IND application is needed any time an unapproved drug is to be used in a clinical trial. This can be used if the sponsor aims to change the indication, route of administration, dose, patient population, or other factors that significantly alter the risk profile of the drug. The drug sponsor does not have to file an IND if certain exemption criteria are met (Table 88.1). Drug sponsors are encouraged to consult with the FDA for guidance and advice during any stage of the IND application. Once the application is filed and approved, investigators may administer the drug to human subjects.[4–6]

Manufacturer inspections

The FDA conducts routine inspections and assessments of drug manufacturing sites to ensure compliance with the applicable laws and regulations. Inspections may be routine, part of a survey, or in response to a complaint. The FDA has the authority to shut down a facility if minimum standards are not being met or maintained. The FDA usually discloses a portion of their inspections to increase the public's awareness of the FDA's practices in maintaining public safety.[1,7]

Drug advertising

The FDA's Office of Prescription Drug Promotion (OPDP) regulates prescription drug advertising and promotional labeling. All advertisements must be accurate, not misleading, and contain information regarding drug effectiveness, side effects, and prescription information. Additionally, all promotional items must promote only "on label" or FDA-approved indications for the drug.[8,9] Drug advertisements generally appear in journals, magazines, newspapers, radio, and television.[8]

Promotional labeling is another method of drug promotion aimed at selling prescription drugs. The principal difference from advertising is the way promotional

Translational Interventional Radiology. https://doi.org/10.1016/B978-0-12-823026-8.00088-2

TABLE 88.1 Exemption criteria for IND applications of marketed drugs.

- Drug is lawfully marketed in the United States.
- Study does not aim to change the indication or labeling of the drug.
- Study does not aim to change advertising.
- Investigation does not involve a change in route of administration, dosage level, or use in a patient population that significantly increases (or decreases the acceptability of) risks associated with the drug.
- Investigation is conducted in compliance with an investigational review board requirements.

labeling is distributed. Pharmaceutical companies may give out brochures, mailed letters, or other promotional items (such as magnets, pens, or cups) either directly to patients or physicians to promote a drug. A drug's prescribing information must accompany the promotional labeling.[8]

The OPDP has the authority to issue reprimands if a drug promotion is violative of its guidelines (Table 88.2). This can be done by requiring the company to cease the faulty promotions. The OPDP can also take more severe measures that can require a company to perform corrective advertising.[10] For example, a drug company producing dextroamphetamine had online advertisements presenting only the benefits without *any* of the risks. As dextroamphetamine is a schedule II controlled substance with a significant risk profile, the OPDP requested that the manufacturer immediately cease the advertisement and issue a corrected, nonpromotional statement addressing each violation.[11]

Generic drugs

New drugs are generally patent protected to allow the company sole rights of manufacturing and distribution for a specific period of time. This allows the company to be compensated for the resources associated with bringing the drug to market. Once these patents have expired, other manufacturers can produce and sell the drug. The FDA ensures that generic drugs have the same dosage, safety, strength, effectiveness, quality, and intended uses as their branded counterparts.

TABLE 88.2 Common violations of drug advertisements.

- Excluding risk information
- Minimizing risks
- Misrepresenting scientific results
- Exaggerating efficacy
- Using deceiving imagery to give a false impression

Generic drug manufacturers do not have to repeat clinical studies to demonstrate safety and efficacy since this has already been done. This helps confer the most widely known benefit of generic drugs—lower price. Generic drug manufacturers file abbreviated new drug applications to the FDA that ensure the generic drug is equivalent to its branded counterpart. Monitoring safety and efficacy of generic drugs once they are marketed also falls within the purview of the FDA.[12]

Reporting problems

The FDA has several programs that allow consumers, healthcare providers, and the industry to report problems with approved drugs. MedWatch is the principal gateway to allow for voluntary reporting of adverse or sentinel events associated with a drug. Important safety information directed at consumers and healthcare professionals is also disseminated by the MedWatch platforms.[13] For example, the MedWatch website reported the possible carcinogenic properties of lorcaserin, a weight-loss drug. The FDA requested the drug manufacturer voluntarily withdraw the drug from the market and recommended that patients cease ingestion.[14] This illustrates the importance of continued surveillance for a drug that was initially approved by the FDA as an NDA.

Medical Product Safety Network (MedSun) is an adverse event reporting program that monitors the safety of medical devices.[15] The program is a collaborative effort with the clinical community to identify and solve problems with medical devices. It comprises 350 hospitals and healthcare facilities.[16] Each month the FDA releases a MedSun newsletter that highlights the summaries submitted from the various clinical sites.[15]

Active surveillance

The FDA complements its passive safety monitoring with the Sentinel System (Fig 88.1), an active surveillance measure. The Sentinel System is a component of the broader Sentinel Initiative, which aims to improve the ability to identify and assess the safety of medical drugs and devices.[17,18] This system uses large electronic databases including electronic health record systems, insurance claims, and registries to provide active safety monitoring. This initiative requires the FDA to work with private, public, and academic institutions to create a network that uses various databases to monitor the safety of approved medicinal products.[19] By using these large grids of information, the FDA can more quickly and efficiently track patient results and outcomes on a larger scale. For example,

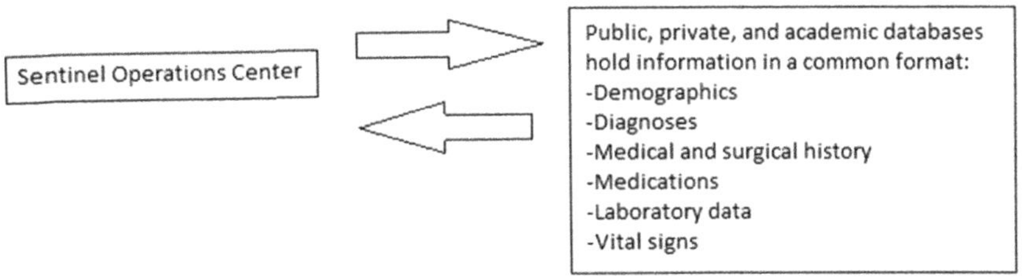

FIGURE 88.1 **Sentinel system.** The Sentinel Operations Center sends queries to the network of databases. The results are returned to the operations center.

this network can potentially answer the following types of questions[20]:

- How many tablets of a specific drug were prescribed to outpatients?
- Which patients are taking a combination of certain medications?
- What is the risk of a side effect in patients on two medications rather than just one?

Given the rapid flux of information sources, data mining, and databases, the full potential of the Sentinel Initiative remains unknown. To this end, the FDA continues to evolve and refine data sources to advance real-time safety surveillance under this program.[21]

Get started

The FDA has broad oversight in regulating drugs after they enter the marketplace. This chapter aimed to clarify and review the FDA's role in the monitoring of the safety and efficacy of drugs. The reader may now be better able to navigate through when to file certain types of applications (i.e., supplemental vs. IND) to perhaps find a new use for an already marketed drug. Additionally, the reader may be able to identify the components of an accurate drug advertisement that meets the FDA's regulations. Lastly, this chapter emphasizes the importance of active and passive surveillance. Readers may now be able to identify the various surveillance measures, report safety violations, and gather safety data put out by these programs.

References

1. Step 5: FDA Post-Market Drug Safety Monitoring. https://www.fda.gov/patients/drug-development-process/step-5-fda-post-market-drug-safety-monitoring. Accessed June 14, 2020.
2. Drugs@FDA Glossary of Terms. https://www.fda.gov/drugs/drug-approvals-and-databases/drugsfda-glossary-terms. Accessed June 14, 2020.
3. Wunderink RG, Matsunaga Y, Ariyasu M, Echols R, Menon A, Nagata TD. LB4. Efficacy and safety of cefiderocol vs. High-dose meropenem in patients with nosocomial pneumonia—results of a phase 3, randomized, multicenter, double-blind, non-inferiority study. *Open Forum Infect Dis*. 2019;6(suppl 2). https://doi.org/10.1093/ofid/otz415.2487.
4. *Guidance for Clinical Investigators, Sponsors, and IRBs Investigational New Drug Applications (INDs) — Determining whether Human Research Studies Can Be Conducted without an IND*. U.S. Department of Health and Human Services, Food and Drug Administration; 2013. https://www.fda.gov/media/79386/download.
5. Holbein MEB. Understanding FDA regulatory requirements for investigational new drug applications for sponsor-investigators. *J Invest Med*. 2009;57(6):688–694. https://doi.org/10.2310/jim.0b013e3181afdb26.
6. Investigational New Drugs. NC TraCS Institute. https://tracs.unc.edu/index.php/services/regulatory/ind-and-ide-application-support/investigational-new-drugs. Accessed June 14, 2020.
7. Inspections Classification Database. https://www.fda.gov/inspections-compliance-enforcement-and-criminal-investigations/inspection-classification-database. Accessed June 14, 2020.
8. Drug Advertising: A Glossary of Terms. https://www.fda.gov/drugs/prescription-drug-advertising/drug-advertising-glossary-terms. Accessed June 14, 2020.
9. The Office of Prescription Drug Promotion (OPDP). https://www.fda.gov/about-fda/center-drug-evaluation-and-research-cder/office-prescription-drug-promotion-opdp. Accessed June 14, 2020.
10. Marwah U, Huettenmoser D, Patel S. Prescription Drug Advertising and Promotion Regulations and Enforcement in Select Global Markets. Food and Drug Law Institute (FDLI). https://www.fdli.org/2017/08/prescription-drug-advertising-promotion-regulations-enforcement-select-global-markets/. Accessed June 14, 2020.
11. Warning Letter: Outlook Pharmaceuticals, Inc. - 605071. https://www.fda.gov/inspections-compliance-enforcement-and-criminal-investigations/warning-letters/outlook-pharmaceuticals-inc-605071-02212020. Accessed June 14, 2020.
12. Generic Drugs: Questions & Answers. https://www.fda.gov/drugs/questions-answers/generic-drugs-questions-answers. Accessed June 14, 2020.
13. MedWatch: FDA Safety Information & Adverse Event Reporting Program. U.S. and Drug Administration. https://www.fda.gov/safety/medwatch-fda-safety-information-and-adverse-event-reporting-program. Accessed June 14, 2020.
14. Belviq, Belviq XR (lorcaserin) by Eisai: Drug Safety Communication - F. https://www.fda.gov/safety/medical-product-safety-information/belviq-belviq-xr-lorcaserin-eisai-drug-safety-communication-fda-requests-withdrawal-weight-loss-drug. Accessed June 14, 2020.
15. MedSun: Medical Product Safety Network. https://www.fda.gov/medical-devices/medical-device-safety/medsun-medical-product-safety-network. Accessed June 14, 2020.
16. Engleman D, Rich S, Powell T, et al. Medical product safety network (MedSun) collaborates with medical product users to create specialty subnetworks. Assessment. In: Henriksen K, Battles JB, Keyes MA, et al., eds. *Advances in Patient Safety: New Directions and*

Alternative Approaches. Vol 1. Rockville (MD): Agency for Healthcare Research and Quality; 2008. Available from: https://www.ncbi.nlm.nih.gov/books/NBK43633/.

17. FDA's Sentinel Initiative - Background. https://www.fda.gov/safety/fdas-sentinel-initiative/fdas-sentinel-initiative-background. Accessed June 14, 2020.
18. Findlay S. The FDAs Sentinel Initiative. Health Affairs. https://www.healthaffairs.org/do/10.1377/hpb20150604.936915/full/. Published June 4, 2015. Accessed June 14, 2020.
19. Sentinel Initiative. https://www.sentinelinitiative.org/. Accessed June 14, 2020.
20. Coyle, D. Sentinel System Overview. FDA. https://www.Fda.gov/media/110795/download.
21. Gottlieb S, Dal Pan G. The Future of FDAs Electronic Safety Surveillance. https://www.fda.gov/news-events/fda-voices/future-fdas-electronic-safety-surveillance. Accessed June 14, 2020.

Additional references

Ball R, Robb M, Anderson SA, Dal Pan G. The FDA's sentinel initiative–A comprehensive approach to medical product surveillance. *Clin Pharmacol Ther*. 2016;99(3):265–268. https://doi.org/10.1002/cpt.320.

Craigle V. MedWatch: the FDA safety information and adverse event reporting program. *J Med Libr Assoc*. 2007;95(2):224–225. https://doi.org/10.3163/1536-5050.95.2.224.

FDA's Bad Ad Program. *Empowering HCPs to Recognize and Report False or Misleading Drug Promotion*. Office of Prescription Drug Monitoring. United States Food and Drug Administration; 2016. https://www.fda.gov/media/100294/download.

Guarino RA, ed. *New Drug Approval Process*. Boca Raton: CRC Press; 2009.

Mantus D, Pisano DJ, eds. *FDA Regulatory Affairs*. Boca Raton: Taylor & Francis; 2014.

Sharrar RG, Dieck GS. Monitoring product safety in the postmarketing environment. *Therap Adv Drug Saf*. 2013;4(5):211–219. https://doi.org/10.1177/2042098613490780.

Van Norman G. Drugs, devices, and the FDA: Part 1: an overview of approval processes for drugs. *J Am Coll Cardiol*. 2016;1(3):170–179. https://doi.org/10.1016/j.jacbts.2016.03.002.

Chapter 89

Postmarket device safety monitoring

Christopher Ovanez
Russell H. Morgan Department of Radiology and Radiological Science, Johns Hopkins Hospital, Baltimore, MD, United States

Key points

- Postmarket device safety monitoring is an important patient safety measure, which involves the close monitoring of a medical device once it has been approved and released into the market for usage. It is important to closely monitor for adverse outcomes in patients, as well as to ensure that a particular device functions appropriately and as clinically intended.
- The United States Food and Drug Administration (FDA) possesses many functions, among which includes ensuring the safety and efficacy of foods, drugs, vaccines, biologic agents, medical devices which are to be consumed/used by people.
- The Sentinel Initiative was created by the FDA to figure out new and improved methods to assess the safety of medical products approved in the United States.

Introduction

Postmarket device safety monitoring, as the name implies, is the close monitoring of a medical device once it has been approved and released into the market for usage. This type of monitoring is of critical importance to ensure that a particular device operates as it is intended, and to monitor for any unexpected complications that may arise that could potentially compromise the safety of patients in which the device is being utilized. If there is a complication that arises due to a particular device, this type of monitoring would theoretically detect such an occurrence, flag it, and have mechanisms in place to have the appropriate actions taken to address the issue.

Get started

Postmarket device safety monitoring is very important patient safety measure. It is useful to be acquainted with the different methods of monitoring, as well as the organizations/entities which are involved.

Potential pitfalls

A researcher or clinician should be conscious of the potential for pitfalls when it comes to postmarket device safety monitoring. When a device is approved for usage, care should be taken to ensure that this particular device continues to be safe after it is implemented for use in patients. One should closely monitor patients after a device is utilized in order to ensure that it is functioning appropriately and produces the expected clinical results, with ongoing periodic check-ups with patients to ensure that they are doing well, and to see if they are reporting any adverse effects which may be related to the device. If any unexpected adverse consequences arise, this should be documented with as much detail as possible, and reported immediately, with concise information available for agencies which collect this type of information. If patients are not closely monitored, there is the potential to miss significant adverse effects or consequences, which could potentially place additional patients at risk. If conducting a postmarket prospective clinical research study, care should be taken to concisely detail not only if adverse effects arise and what they entail, but ideally, detail should also be provided with regard to patient demographics, to determine whether there is a particular inclination of these adverse effects occurring in a particular segment of a patient population (i.e. comorbidities, ethnicity, race, age, gender, etc.) in order to allow for further investigation with increased efficacy.

Discussion/examples

Postmarket device safety monitoring is accomplished by various means, including the gathering and interpretation of

Translational Interventional Radiology. https://doi.org/10.1016/B978-0-12-823026-8.00089-4

various types of data from a particular medical device once it is released into the market.

For instance, "passive surveillance" is a term that denotes a process by which data is reported to the Food and Drug Administration (FDA) by individuals and organizations (for instance hospital systems), and this data is collected by the FDA, who then analyzes it and determines whether there are any irregularities which might necessitate that action be taken—for instance, if there is a complication which leads to increased morbidity or mortality within the patient population upon whom a particular medical device is being used.[2]

In contrast to passive surveillance methods, there are also more active types of postmarket surveillance for a particular medical device, for instance—postmarket prospective clinical research studies which could potentially be designed and led by members of an interventional radiology department within a medical center, or perhaps interventional radiologists within a private practice setting. Pursuit of this type of endeavor could potentially lead to a multi-institutional level study, which could monitor a particular device across a variety of clinical settings (i.e., urban vs. rural; large hospital vs. small medical center; etc.), as well as across a wide variety of patients (i.e., races, ethnicities, ages, socioeconomic status, comorbidities, etc.), and potentially could provide a large amount of beneficial data regarding the safety and performance of a particular medical device, especially if it better reflects the diverse population of the country, as opposed to a more isolated study that may not well reflect the population as a whole, which in turn, could potentially lead to bias and findings that may not extrapolate well to a more diverse population with a wide variety of demographics/factors as noted above.

Multi-institutional research endeavors could venture an additional step further, for instance into the realm of molecular epidemiology, which as the name implies, combines the world of an epidemiological approach, methods, and standards, with that of molecular biology. This is not an avenue that is commonly pursued in the world of interventional radiology, but should be considered as a potential way to elevate research potential on a multi-institutional level, and potentially improve patient care and safety by identifying factors on a genetic level that may influence the way treatments are administered, and identifying which patient populations would benefit more or be at increased risk from the effects of a particular treatment. For instance, let us consider a hypothetical example of a drug-coated vascular stent—let us say that this particular stent, which is coated with drug X has been recently approved for usage in the United States, and after analyzing patient data, someone makes the realization that there is increased risk of morbidity/mortality in certain patients. Upon further exploration of the data, it is demonstrated that a large majority of these patients happen to belong to a particular ethnicity. Analysis of genetic information from these patients demonstrates that there happens to be a particular mutation common among this particular ethnicity, and after some postdocs churn out some amazing bench research, it is found that drug X happens to interact adversely with a particular cardiac myocyte potassium channel subunit in this particular patient population, and this mutation leads to increased morbidity/mortality by drug X, perhaps sudden cardiac death. This is a hypothetical example, but you can see where I am going with this, and how this type of research has limitless potential and could uncover answers to many questions and potentially lead to a future of treatments that could be tailored to suit the needs of particular patient populations and prevent morbidity/mortality in others.

In addition to the aforementioned method of potential active postmarket surveillance, another method of active-type surveillance could be the setting up of device registries, whose role includes the active gathering and reporting of data with regard to a particular medical device's safety and function/performance within a database,[3] which would be compiled and readily accessible to monitor for any potential aberrancies such as critical complications or patient safety issues.

The FDA is a government organization in the United States which has numerous roles and functions, one of which includes monitoring the safety and performance of devices once they have been approved and released into the market.[1] The FDA possesses officials that conduct in-person inspections of facilities that manufacture devices, and these inspections may be ones that are scheduled in advance, or they may be prompted by a particular issue or concern which may arise regarding a device.[1] The FDA possesses the authority to actually close down a manufacturer if they ascertain that specific standards are not being fulfilled.

The FDA has several ways by which device manufacturers, medical professionals, as well as patients/consumers can report issues that arise with a particular device. One method by which to report issues is MedWatch—which is an adverse event reporting program set up by the FDA for both devices and medications, which also provides updated safety information regarding a particular device or medication. A second method is via the Medical Product Safety Network (MedSun), whose role is to monitor device safety and effectiveness, and a place to report adverse events.

The Sentinel Initiative was created by the FDA to figure out new and improved methods by which to assess the safety of medical products approved in the United States, which includes drugs, vaccines, and medical devices.[2] There are several components involved in and crucial to the Sentinel Initiative, some of which include: (1) The Sentinel

System, (2) Active Postmarket Risk Identification and Analysis (ARIA) System, and (3) FDA-Catalyst.[2] The Sentinel System essentially creates programs that analyze electronic healthcare data, which includes analyzing information within medical billing as well as electronic health records, and in turn, these programs help to answer questions posed by the FDA regarding approved medical products. The ARIA System is part of the Sentinel System and has programs that allow the FDA to analyze healthcare data that is processed by the Sentinel System. The FDA-Catalyst essentially provides data of patient–provider interactions. The Sentinel Initiative is of critical importance to the FDA and its ability to provide quality postmarket device safety monitoring as well as other key tasks related to medical products.[2] Please refer to the websites of the FDA and The Sentinel Initiative for more helpful information that will surely prove useful for medical professionals and researchers.

Conclusion

In conclusion, postmarket device safety monitoring is accomplished by various methods, and is of critical importance to ensure the most optimal of standards for performance of all approved medical devices, quickly identify and act upon any potential complications that might arise from the usage of a particular medical device, and most importantly, allow active surveillance to ensure the ongoing safety of our patients.

References

1. FDA Website [www.fda.gov/patients/device-development-process/step-5-fda-post-markey-device-safety-monitoring].
2. Sentinel Initiative Website [www.sentinelinitiative.org].
3. Rajan PV, Kramer DB, Kesselheim AS. Medical device post-approval safety monitoring: where does the United States stand? *Circ Cardiovasc Qual Outcomes*. January 2015;8(1):124–131. https://doi.org/10.1161/CIRCOUTCOMES.114.001460. Epub 2015 Jan 6. PMID: 25563662; PMCID: PMC4385582.

Further reading

1. Mehran R, Leon MB, Feigal DA, et al. Post-market approval surveillance: a call for a more integrated and comprehensive approach. *Circulation*. June 29, 2004;109(25):3073–3077. https://doi.org/10.1161/01.CIR.0000134694.78653.B6. PMID: 15226222.
2. Van Den Broek J, Brestoff J. *Epidemiology: Principles and Practical Guidelines*. Springer; 2013.
3. Kestenbaum B. *Epidemiology and Biostatistics: An Introduction to Clinical Research*. Springer; 2009.

Part IX

Clinical implementation

Chapter 90

Implementation research in interventional radiology

Ravi Tyagi[1], Omair Ali[2] and Raj Shah[2]

[1]*Department of Radiology, Section of Interventional Radiology, Rush University Medical Center, Chicago, IL, United States;* [2]*Department of Radiology, Section of Interventional Radiology, University of Illinois at Chicago College of Medicine, Chicago, IL, United States*

Key points

- Common obstacles encountered on the journey from research to practice include: access to adequate funding, time constraints for early-career investigators, difficulty in identifying clinically relevant endpoints in the early stages of translational research, and lack of infrastructure or staff to support translational research.
- Once the efficacy of a product/procedure has been verified, randomized control trials (RCTs) assessing the added benefit compared to conventional treatments facilitate increased implementation.
- Budgetary considerations while initiating the implementation process include: Seeking funding through public and corporate avenues, cost-minimization through multidisciplinary collaboration, and prioritizing the use of materials that are immediately necessary for completing a translational research study.

Why it matters?

Understanding the iterative and structured nature of biodesign is crucial in efficiently translating research into practice. In this chapter, we discuss common barriers encountered in this process while describing the steps needed to effectively make the transition. We further explore areas of active research in interventional radiology with examples of products and procedures that successfully transitioned from the experimental to clinical space. Finally, we discuss budgetary considerations to anticipate while scaling research from benchtop to practice.

Translational science: background

Translational science builds on basic science research to develop new therapies and medical interventions to treat disease. Generally speaking, basic science can involve animal research and studying of molecular pathways at a cellular level. These pathways help identify areas for intervention that can be exploited. Once an area of intervention is identified, therapeutics/diagnostics can be developed and translated to humans. These new therapies are observed in humans to determine their efficacy, their safety, and whether they provide an advantage over the current treatment. Any new therapy must be safe, must be efficacious, and preferably provides an advantage over the current treatment. These qualities are initially tested through small studies to determine their safety in humans.

Once the therapeutic is determined to be safe, more studies can be conducted to determine its efficacy. If deemed efficacious, large scale-clinical trials can be performed to directly compare the new therapy versus the standard of treatment. The next step is awareness. After proving the new therapeutic or intervention is superior to the previous solution, it is important to spread the message. This can be a time-consuming process. While many academic physicians keep up with novel therapies, some physicians in the community could be harder to reach. Nevertheless, translational research plays a vital role in the development of new therapeutic and the progress of medicine.

Barriers to implementation

Implementing biomedical translational research presents many challenges to investigators given the extensive effort required to prove that a novel process can be applied safely and effectively in patients while eventually establishing evidence that suggests an overall benefit for society. Investigators must also meet rigorous regulatory, ethical, and legal criteria, have adequate funding, and maintain a dedicated staff that can manage the complexities of translational research.

Translational Interventional Radiology. https://doi.org/10.1016/B978-0-12-823026-8.00085-7

Implementing translational research is challenging due to limited opportunities for funding, despite the expensive and time-consuming nature of the translational research pipeline.[1,2] Available public funding for translational research has historically been limited, and grant applications are competitive and sometimes inaccessible due to focused areas of priority, which further bottlenecks the implementation of interventional radiology translational research.[3]

Translational research implementation is also challenging because it requires an appropriately trained research staff to perform complicated tasks and an infrastructure within an investigator's institution that supports a translational research lab.[4] Early career physician investigators may be unable to fully participate in certain translational research projects given that they often have heavy clinical responsibilities and limited protected research time.[3]

Another barrier to translational research is the challenge of identifying clinically relevant and cost-effective endpoints. Promising data seen in preclinical studies must later be demonstrated in human trials through clinically pertinent endpoints, but there is a high failure rate in the early stages of translational research.[1,2] Further, some desirable endpoints are unobtainable due to high costs associated with acquisition or storage, such as the use of magnetic resonance imaging.[3]

A key facilitator of translational research is having a multidisciplinary team to address the wide scope of research questions posed by translational research.[2,3,5,6] Translational research should involve scientists, clinicians, public health experts, and other disciplines represented in healthcare. Having a multidisciplinary team providing oversight would enhance the research process by improving the data collection process, information exchange, and troubleshooting when challenges arise.[2]

Areas of current research

Major areas of research in interventional radiology (IR) include the use of image-guided regional therapy for liver disease and liver malignancy. Interventional oncology is a growing area of research and development focused on by many interventional radiologists. Many labs develop animal models, including rabbits and pigs, to facilitate instrumentation for the development and application of various interventional techniques and better understand the genomics of hepatocellular carcinoma (HCC). Through a better understanding of HCC genomics, more targeted approaches and treatments can be ascertained to treat HCC.

Other research focuses on the postprocedural outcomes of transjugular intrahepatic portosystemic shunts (TIPS). While the efficacy of TIPS is not in question, established guidelines do not exist regarding TIPS reductions/occlusions in the setting of TIPS complications or side effects. The decision to occlude or reduce the TIPS can largely depend on the operator or institutional preference, depending on the clinical situation. Furthermore, other areas of interest include novel techniques to treat disease. One such example is geniculate artery embolization (GAE) for osteoarthritis of the knee. While the efficacy of GAE is not well established, early small studies do suggest a correlation between GAE and symptom relief from knee osteoarthritis. While these are just some examples, the full gamut of potential IR research is endless.

Real-world examples

Yttrium-90 transarterial radioembolization

Yttrium-90 (Y90) was initially developed in the 1960s as a nonsystemic treatment for HCC.[7] As previously described in this text, needs identification is the first step in biomedical design, and in this case, HCC was an ideal application given the poor outcomes with other treatment modalities at the time. Anatomically, the hepatic artery supplies up to 80% of liver tumors while the portal vein supplies the remaining parenchyma.[8] In light of this differential perfusion, Y90 transarterial radioembolization was successfully tested in multiple small patient cohorts in the mid-1960s followed by the first phase 1 clinical trial in the late 1980s−early 1990s.[7] Since then, Y90 has been validated as a viable treatment for HCC in a number of trials and remains an area of investigation for other novel applications.

Of note, the development of Y90 follows a typical progression for other IR products and devices; that is, beginning as an experimental treatment in small patient cohorts, further tested through RCTs, and ultimately approved by the FDA for a number of applications (Fig. 90.1).

Prostatic artery embolization

Prostatic artery embolization (PAE) was first conceptualized in the 1980s as a potential treatment for prostate hemorrhage.[9] Initially tested in canine models with promising results,[10] PAE was again tested in a pig model almost 30 years later with decreased postembolization prostate volume noted on exam.[11] In 2010, PAE was tested in two patients resulting in sustained reduction of benign prostate hyperplasia (BPH) symptoms,[12] then validated through multiple RCTs in the last 10 years highlighting its safety and efficacy for this application. While certainly an effective treatment for BPH, PAE has not yet proven itself to be more efficacious than transurethral resection of the prostate.[13] From a translational perspective, this demonstrates the importance of *minimally* establishing noninferiority in increasing procedure/device implementation. Furthermore,

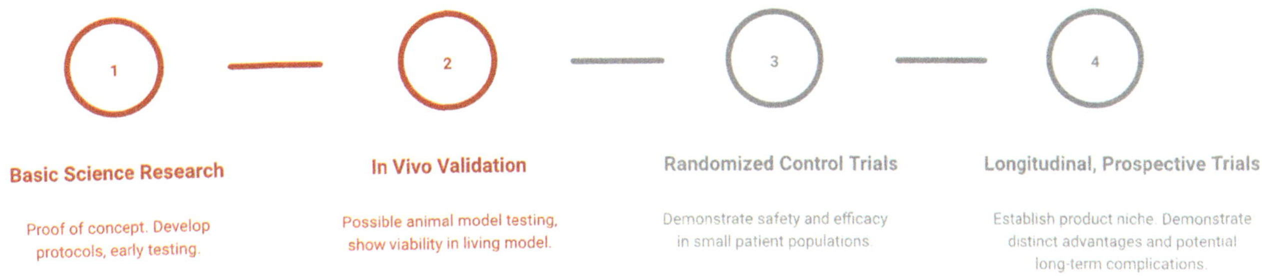

FIGURE 90.1 Study design process for product implementation. Anticipating these steps early on can be valuable in expediting the implementation process.

while short-term RCTs may be useful in demonstrating efficacy, long-term studies are often required to establish sustained benefit, potential complications, and even unique advantages compared to conventional techniques.

Endovascular arteriovenous fistula creation

Surgical fistula creation technique for hemodialysis has remained relatively unchanged since the 1960s. However, in the last 10 years, there has been an increased interest in endovascular arteriovenous fistula (endoAVF) creation, a minimally invasive technique that relies on the proximity of the radial artery and vein or the proximal radial artery (PRA) and deep communicating vein (DCV). Currently, two such devices exist, each unique in its fistulization technology.[14,15] As these devices are relatively new, ongoing investigation highlighting the advantages and disadvantages of endoAVF versus surgical fistula creation is still required.

EndoAVF creation, like other minimally invasive interventional techniques, requires longitudinal data to establish its benefit over current techniques. IR is a relatively young field with a growing number of interventions becoming available for numerous applications. As such, long-term RCTs establishing the efficacy of these interventions over conventional treatments will be required to increase implementation and to move the field forward.

Budgetary considerations

Obtaining funding for translational research is difficult due to limited and competitive funding mechanisms.[1] Currently, there are research programs sponsored by various radiology-friendly organizations, including the following: the National Institutes of Health's (NIH) National Center for Advancing Translational Sciences (NCATS); the Radiological Society of North America; and the Society of Interventional Radiology.

For example, the NIH's KL2 and TL1 programs subsidize translational research training and NCATS's Clinical and Translational Science Awards Program sponsors translational research.[3] To be competitive applicants, investigators must explain the following: their research objectives; the research's significance; a cost-effective design of study methods; and assurance that data is attainable, valid, and meaningful. Entrepreneurially inclined investigators can seek funding through corporate partners, whose primary aims may involve information exchange or translating an experimental product into commercialization.[5]

Given the time-consuming and expensive nature of translational research that often coincides with budgetary constraints, efforts should be made to structure the budget's ability to support a research study along with subsequent experimentation to complete the desired trajectory within the research pipeline. The budget will need to allocate funds for an appropriately trained staff, cost-effective materials for experimentation, financially feasible data acquisition and storage methods, and the flexibility to pivot research priorities should a study fail to meet its primary endpoints. The research cost may also be reduced with multidisciplinary collaboration, which may ease the process of data acquisition, experimentation, or other crucial components of translational research (Figs. 90.2 and 90.3).

Potential pitfalls

- Anticipate establishing the added advantage of a product early in the design process; waiting too long may make the product obsolete or prolong the implementation process.
- A well-designed translational research study is often expensive, time consuming, and difficult to complete, which means that seeking funding and resources from public (i.e., NIH) or private entities (i.e., industry) is almost always a necessity along with having a dedicated research staff.

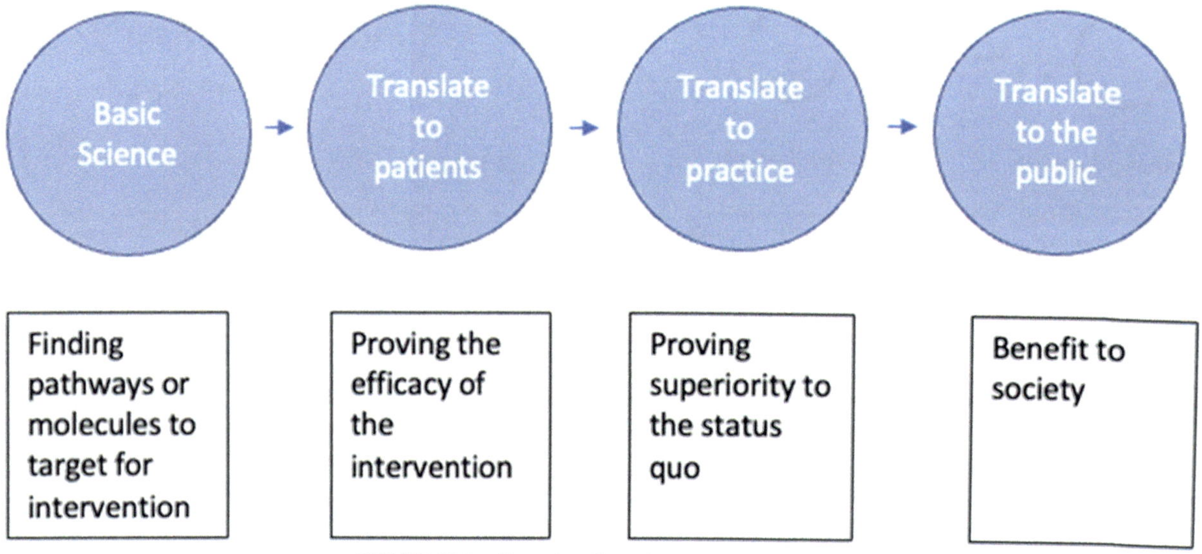

FIGURE 90.2 General outline of translational research.

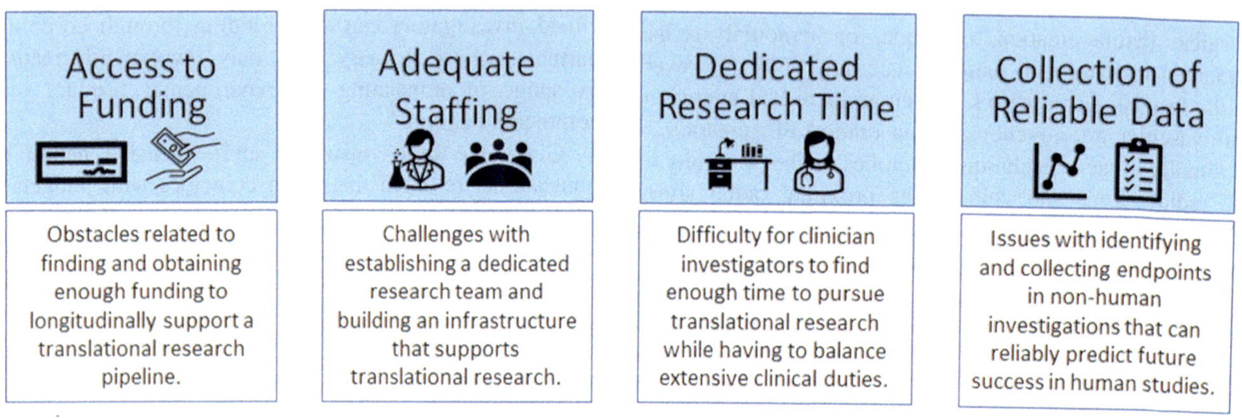

FIGURE 90.3 Common barriers to implementing translational research.

Get started

- Consider approaches to navigate systemic barriers like the following: initiate collaboration in a multidisciplinary fashion before starting a new project in the pipeline; ensure that an available budget adequately addresses all needs to carry out a research study to completion before starting.
- Plan the implementation process early on so key steps can be performed in parallel
- Identify stakeholders and potential investors; consider these groups in the design process to successfully secure funding in the implementation phase

References

1. Hörig H, Marincola E, Marincola F. Obstacles and opportunities in translational research. *Nat Med.* 2005;11:705–708. https://doi.org/10.1038/nm0705-705.
2. Seyhan AA. Lost in translation: the valley of death across preclinical and clinical divide – identification of problems and overcoming obstacles. *Transl Med Commun.* 2019;4:18. https://doi.org/10.1186/s41231-019-0050-7.
3. Petrou M, Foerster BR, Reich DS. Translational research in radiology: challenges and role in a patient-based practice. *Acad Radiol.* 2009;16(5):593–596. https://doi.org/10.1016/j.acra.2009.01.017.
4. Fudge N, Sadler E, Fisher HR, Maher J, Wolfe CD, McKevitt C. Optimising translational research opportunities: a systematic review and narrative synthesis of basic and clinician scientists' perspectives of factors which enable or hinder translational research. *PLoS One.* 2016;11(8):e0160475. https://doi.org/10.1371/journal.pone.0160475. Published 2016 Aug 4.
5. Rubio DM, Schoenbaum EE, Lee LS, et al. Defining translational research: implications for training. *Acad Med.* 2010;85(3):470–475. https://doi.org/10.1097/ACM.0b013e3181ccd618.
6. Fernandez-Moure JS. Lost in translation: the gap in scientific advancements and clinical application. *Front Bioeng Biotechnol.* 2016;4:43. https://doi.org/10.3389/fbioe.2016.00043. Published 2016 Jun 3.

7. Saini A, Wallace A, Alzubaidi S, et al. History and evolution of yttrium-90 radioembolization for hepatocellular carcinoma. *J Clin Med.* 2019;8(1):55. https://doi.org/10.3390/jcm8010055.
8. Breedis C, Young G. The blood supply of neoplasms in the liver. *Am J Pathol.* 1954;30(5):969–977.
9. Darewicz J, Cylwik B, Musierowicz A, Boczon S. The effect of internal iliac artery embolization of the prostatic gland in the dog. *Int Urol Nephrol.* 1980;12(1):37–41. https://doi.org/10.1007/BF02085379.
10. Yu H, Isaacson AJ, Burke CT. Review of current literature for prostatic artery embolization. *Semin Intervent Radiol.* 2016;33(3):231–235. https://doi.org/10.1055/s-0036-1586141.
11. Sun F, Sánchez FM, Crisóstomo V, et al. In: *Radiology*. Vol 246. 2008:783–789 (3).
12. Abt D, Hechelhammer L, Müllhaupt G, et al. Comparison of prostatic artery embolisation (PAE) versus transurethral resection of the prostate (TURP) for benign prostatic hyperplasia: randomised, open label. *Non-Inferiority Trial BMJ.* 2018;361:k2338. https://doi.org/10.1136/bmj.k2338.
13. Carnevale FC, Antunes AA, da Motta Leal Filho JM, et al. Prostatic artery embolization as a primary treatment for benign prostatic hyperplasia: preliminary results in two patients. *Cardiovasc Intervent Radiol.* 2010;33(2):355–361. https://doi.org/10.1007/s00270-009-9727-z.
14. Rajan DK, Ebner A, Desai SB, Rios JM, Cohn WE. Percutaneous creation of an arteriovenous fistula for hemodialysis access. *J Vasc Intervent Radiol.* 2015;26(4):484–490. https://doi.org/10.1016/j.jvir.2014.12.018.
15. Mallios A, Jennings WC, Boura B, Costanzo A, Bourquelot P, Combes M. Early results of percutaneous arteriovenous fistula creation with the ellipsys vascular access system. *J Vasc Surg.* 2018;68(4):1150–1156. https://doi.org/10.1016/j.jvs.2018.01.036.

Chapter 91

Implementation research: Design and analysis

Conner D. Reynolds[1,2], Scott Bittle[3], Jonas Kruse[4] and Gregory J. Woodhead[2]

[1]Transitional Year Residency, John Peter Smith Health Network, Fort Worth, TX, United States; [2]Diagnostic Radiology Residency, Department of Medical Imaging, University of Arizona College of Medicine, Tucson, AZ, United States; [3]Diagnostic Radiology Residency, Baylor University Medical Center, Dallas, TX, United States; [4]School of Medicine, Texas Christian University, Fort Worth, TX, United States

Summary

- Dissemination and implementation research has rapidly evolved to address the considerable time lag between research discovery and frontline implementation.
- This chapter outlines a five-step process for designing implementation trials.
- This process involves defining the problem, identifying viable implementation strategies, developing appropriate research questions, defining measurable outcomes, and utilizing a study design that suits the implementation.

Introduction

In 2018, the United States invested $194.2 billion in clinical and health-related research and development (R&D), with the lion's share coming from private industry.[1] While the total R&D investment has risen over the past several years, there is a considerable lag between research and implementation. Previous work suggests that in 17–20 years, less than 50% of innovations from this research will ever reach clinical practice.[2–5] Further, nearly 80% of investments in medical research yield no public health impact at all.[6] This represents a logistical and economic problem for translational research.

The time lag between clinical research and implementation is a complex, multifactorial problem.[7,8] Dissemination and implementation (D&I) research has rapidly developed to address this problem, emerging as the field of Implementation Science. D&I research employs lessons from healthcare, public health, business, and others to identify methodologies for deploying new evidence-based practices on the frontline.[7,9–11]

Most D&I methodologies divide the implementation process into several phases but do not necessarily move linearly. Aarons et al.[12] developed the EPIS model to characterize four phases of D&I research: exploration, preparation/adoption, implementation, and sustainment. In the exploratory phase, researchers identify an organizational problem and evaluate one or more interventions to address it. This is closely followed by the preparation/adoption phase, where researchers strategize how to deploy programs in their organization. The implementation phase evaluates how well these programs were deployed and strategizes how to optimize fidelity. Finally, the sustainment phase evaluates strategies to maintain programs over time and extend their use in new organizations and communities. D&I research completes the tripartite translational research pipeline previously described by the National Institute of Health (NIH) and others, accelerating the translation of biomedical discovery to frontline clinical practice.[13]

Planning effective D&I research requires the same scientific rigor as biomedical research. Accomplishing each step necessitates detailed planning, interdisciplinary collaboration, pilot testing, and gradual refinement throughout intervention development. D&I has been applied to multiple diagnostic and interventional radiology areas, including establishing integrated IR residency,[14] improving workflow,[15,16] optimizing electronic records,[17] reporting time,[18,19] and communication with consultants,[20] reducing errors and enhancing the quality of healthcare delivery.[21–23] The following section will discuss critical steps for the implementation of research design.

Translational Interventional Radiology. https://doi.org/10.1016/B978-0-12-823026-8.00082-1

Step 1: Define the problem and where it exists

Solving any problem begins with identifying the components that define it. Given the variety of moving parts within an organization, D&I research planning should start with a **situational analysis**. This analysis should engage interdisciplinary groups of stakeholders and future end-users of the intervention to understand the problem at their level within the organization. Groups should also discuss expected outcomes for implementation programs. Engaging these groups early and often will improve the program's fidelity and adoption by organizational units. One study, which found reductions in pulmonary CT angiography in the ED after implementing a clinical decision support tool, credited its success to engaging multidisciplinary groups, educating faculty, and changing culture before rolling out the tool.[24]

Once a preliminary framework has been established, researchers should conduct a **knowledge synthesis**. This process contextualizes and integrates multiple studies to solidify the problem targeted by researchers. There are many types of knowledge synthesis, including systematic reviews,[25] scoping reviews,[26] network meta-analyses,[27] rapid reviews,[28] diagnostic reviews,[29] prognostic reviews,[30] and economic reviews.[31] Each of these types has its use case and can be tailored to researchers' overall needs. Failure to utilize some form of knowledge synthesis can lead to significant time delays, reduce implementation fidelity, and introduce bias. Basing decisions solely on expert opinion can introduce bias. One previous investigation compared the results of meta-analyses of RCTs to clinical expert recommendations for the treatment of myocardial infarction. These authors found experts can be up to 10 years behind on the latest research, failing to mention important advances, exhibiting delayed recommendation of effective preventive measures, and continuing recommendations for ineffective therapies.[32] Basing decisions on a single clinical study can also be problematic, as up to 30% of highly cited clinical studies may contradict or have reduced effect size in subsequent evaluation.[33]

Step 2: Identify viable implementation strategies

With a firm understanding of the problem and its drivers, researchers can begin identifying strategies for implementation. Given the complex, dynamic nature of healthcare systems, most implementation strategies target change at multiple levels. The Expert Recommendations for Implementing Change (ERIC) study previously identified 73 discrete components of D&I research strategies, which can be categorized into nine discrete domains: engage consumers, use evaluative and iterative strategies, change infrastructure, adapt and tailor to the context, develop stakeholder interrelationships, utilize financial strategies, support clinicians, provide interactive assistance, and train and educate stakeholders.[34] Once researchers identify the most salient strategies for their situation, they can begin building the intervention program itself.

Step 3: Develop appropriate research questions

As discussed in the introduction, the D&I research process can be subdivided into four key phases (e.g., exploration, preparation/adoption, implementation, and sustainment). Each of these phases corresponds to its own unique research questions.

Exploration phase research aims to determine which organizations or units are currently interested in implementing the new program. While some may only identify the total number of interested organizations or units, these data may reveal key demographic insights about those interested.[35]

Preparation/adoption phase research aims to assess barriers and facilitators to implementing the new program. These may be identified through implementation readiness and capacity assessments.[35–37] Leveraging facilitators and utilizing training, coaching, audit, and feedback systems to overcome barriers will be critical to the success of D&I research in future phases. These will serve to align stakeholders and leadership, cultivating a climate where program use is expected, supported, and rewarded at all levels.[38,39]

Implementation research questions aim to compare various implementation strategies and assess the speed of implementation and **implementation fidelity**. Implementation fidelity can be defined by how closely the program was implemented in the clinic, compared to preimplementation design.[35]

Sustainment phase research aims to evaluate ongoing processes and support structures that will maintain the new program beyond the researchers' initial involvement.[35,38]

Step 4: Define measurable outcomes

High-quality measurement of D&I research outcomes is vital for making evidence-based decisions within individual organizations and advancing the field. Proctor et al. provide one of the most comprehensive taxonomies of implementation outcomes, placing them into eight distinct subcategories: acceptability, adoption, appropriateness, cost, feasibility, fidelity, reach or penetration, and sustainment. There are a variety of existing tools to measure each of these outcomes. Interestingly, the authors provide a conceptual framework wherein implementation outcomes impact future service and client-based outcomes.[40,41] Overall efficiency and success of implementation deployment have also been measured via the Stages of Implementation Completion (SIC) tool.[42,43] Well-defined outcomes will yield high dividends during statistical analysis and interpretation of results.

Step 5: Designing the study

Study design in D&I research is fundamentally different from that of traditional biomedical research. Biomedical studies focus primarily on efficacy (i.e., *does the intervention provide substantial clinical benefit to patients when all other factors are held constant?*). This leads them to rely heavily on gold standard, randomized control trials that maximize internal validity. D&I studies focus on effectiveness in the context of real-world patient care (i.e., *does the intervention provide substantial benefit to patient care and the systems that support it?*). This leads them to utilize study designs that maximize external validity. While randomization is still important in D&I studies, there are some circumstances where it is not feasible, ethical, or acceptable.[44]

There are a plethora of study designs available for different use cases in D&I research. Brown et al.[13] provide a practical framework for D&I study designs by placing them into three distinct categories: within-site designs, between-site designs, and within- and between-site designs.

Within-site designs

Postdesign

The postdesign is one of the simplest and most common study designs in D&I research. It is a quasi-experimental design, which describes healthcare processes, utilization, and outputs after implementing a new evidence-based program within single sites or units. Although they have no preimplementation baseline for comparison, the results of postdesign studies can be compared to standards established in previous literature.[13,35,45] This study design is advantageous in the early phases of D&I research when researchers are evaluating interest, barriers, and facilitators to new programs. It is also helpful when implementing new health policies or guidelines.[13]

Prepost design

In contrast to poststudies, the prepost study evaluates program outcomes compared to a preimplementation baseline. This quasiexperimental design is useful for assessing the impact of complex implementation strategies within a single organization or across multiple sites representative of a particular population.[13,35,45] These designs are also useful for evaluating the adoption of guidelines, black box warnings, or other new programs within a healthcare system.[13]

Pre-post studies are utilized throughout interventional radiology literature. One previous study employed a retrospective design to evaluate the complication rates of percutaneous gastronomy tube placement after initiating a dedicated preprocedure clinic. These authors saw a statistically significant reduction in major complications.[46] Another study found that failures during VIR procedures in their facility were primarily due to ineffective planning, communication errors, and equipment issues. However, these were mitigated with preprocedural team rehearsals.[21]

Alternatives to randomization in quasi-experimental within-site study designs

Given the nature of quasi-experimental studies, there are several valuable alternatives to randomization to improve internal validity in within-site studies. The most common of these alternatives is an interrupted time-series design, which involves multiple observations before and after implementing new programs.[47] These designs employ additional analyses to account for trend, serial dependence, and other sequence characteristics.[45,48] This adds flexibility by allowing multiple sites at multiple baselines, replicating single cases, and increasing internal validity.[45] Other alternatives to randomization not discussed here include nonequivalent group,[49] propensity score matching,[50] synthetic control group,[51] and regression-discontinuity designs[52]

Between-site designs

New program versus usual practice design

The New versus Implementation As Usual (IAU) design is one of the most straightforward studies in D&I research. In these study designs, some sites implement a new program while others maintain usual conditions, then outcomes are compared. New and IAU sites can be assigned through random and nonrandom paradigms. Randomized New versus IAU trials aim to evaluate whether a new program produces better, more efficient processes and outputs than standard conditions. When conducting randomization, researchers should seek to randomize only at the level of the implementation.[53] This maximizes internal validity exactly where the new program is designed to occur. Nonrandomized New and IAU trials may utilize self-selection processes, leveraging adoption by sites that opt-in to implementing new programs. While more convenient to researchers, those analyzing this outcome data should be careful to control assignment bias via propensity scoring or another equivalent method.[13,50,54]

One example of this study design in radiology comes from Stiell et al. who tested an implementation of the

Canadian C-Spine rule using a matched pair cluster randomized trial. When 12 emergency departments in Canada were randomly assigned to use the Canadian C-Spine rule or usual protocols, those utilizing the rule saw a 12.8% reduction in cervical spine imaging without missing injuries or increasing morbidity.[55] Another stepped-wedge, cluster-randomized implementation trial found that a clinical decision support tool improved appropriateness scores for ordering imaging studies in primary and specialty care ambulatory clinics.[56]

Head-to-head randomized implementation design

Head-to-head randomized implementation trials are designed to compare two different implementation strategies to determine which is more successful at deploying a new program. The two strategies are mainly compared on the quality, quantity, and speed of implementing the new program.[13]

Van deer Leest et al. recently utilized this study design to perform a head-to-head implementation comparison of multiparametric magnetic resonance imaging (mpMRI) versus MR-guided biopsy for biopsy-naïve men with suspected prostate cancer. This prospective, multi-center trial of 626 patients found the mpMRI had identical detection rates of clinically significant prostate cancer when compared with biopsy.[57]

Factorial design

Factorial designs evaluate a combination of two or more implementation strategies, each with two or more levels (e.g., presence or absence). In a 2 × 2 factorial design, units are randomly assigned to one of four possible experimental conditions. This allows researchers to estimate the effect of each factor, as well as the interactions between them.[13] Incomplete factorial designs may be utilized when it is deemed unethical to withhold both implementation strategies. Factorial designs can also be used to assess the contribution of numerous components of an implementation strategy, as is seen in Multiphase Optimization Strategy Implementation Trials (MOST)[58,59] and Sequential Multiple Assignment Randomized Implementation Trial (SMART) implementation designs.[13,60]

Ong et al. recently utilized a factorial implementation trial to improve adherence to infection control precautions during inpatient transfers between radiology and wards. Their intervention involved both color-coded cues and pretransfer checklist forms. In groups that received no intervention, adherence to infection control precautions was 38%. Groups utilizing color-coded cues had compliance rates of 73%, and those utilizing checklists had compliance rates of 71%. When both strategies were implemented, compliance rates of 74% were attained.[61] This example demonstrates the power of factorial designs to evaluate both main and interaction effects in implementation trials.

Within- and between-site designs

Roll-out randomized implementation design

Making comparisons within- and between-site designs requires that study designs utilize a crossover mechanism, wherein they begin with one implementation strategy before transitioning to another. The most common of these are termed roll-out randomized implementation trials. To accomplish this, each unit starts in the standard practice setting before implementing the new program at a randomly assigned interval. This facilitates pre-post comparisons within individual units and comparisons of quality, quantity, and speed of implementing the new program between organizations. The roll-out paradigm has been utilized in dynamic wait-listed and stepped-wedge design effectiveness trials for several decades.[13,62,63] The most significant advantage to this paradigm is that all units eventually receive the program and maximally benefit.[13]

Conclusion

The United States invests billions into clinical and health-related R&D, but there is still a considerable time lag between research discovery and implementation. D&I research has rapidly developed to address this dilemma. This chapter has provided a five-step process for designing D&I research. This process involves defining the problem, identifying viable implementation strategies, developing appropriate research questions, defining measurable outcomes, and utilizing a study design that suits the implementation. Investigators beginning this process should start by:

1. Defining a problem within the organization with situation analysis and knowledge synthesis to decide whether an intervention is needed before investing time and other resources.
2. Talking with key interdisciplinary stakeholders and future end-users to understand the problem at their level within the organization.
3. Identifying implementation strategies that make sense within the context of their organization.
4. Developing appropriate research questions and defining specific, measurable, actionable, relevant, and time-based outcomes.
5. Utilizing an experimental design that appropriate to the research question, which uses randomization where possible.

Potential pitfalls

1. Failure to utilize a situation analysis and knowledge synthesis may lead to significant time delays and resource costs.
2. Failure to involve key stakeholders and future end-users will impact the adoption of new programs by healthcare units.

References

1. US Investments in Medical and Health Research and Development: 2013–2018. ResearchAmerica. https://www.researchamerica.org/sites/default/files/Publications/InvestmentReport2019_Fnl.pdf.
2. Bauer MS, Kirchner J. Implementation science: what is it and why should I care? *Psychiatr Res.* 2020;283:112376. https://doi.org/10.1016/j.psychres.2019.04.025, 2020/01/01/.
3. Mosteller F. Innovation and evaluation. *Science.* 1981;211(4485):881. https://doi.org/10.1126/science.6781066.
4. Grant J, Green L, Mason B. Basic research and health: a reassessment of the scientific basis for the support of biomedical science. *Res Eval.* 2003;12(3):217–224. https://doi.org/10.3152/147154403781776618.
5. Morris ZS, Wooding S, Grant J. The answer is 17 years, what is the question: understanding time lags in translational research. *J R Soc Med.* 2011;104(12):510–520. https://doi.org/10.1258/jrsm.2011.110180, 2011/12/01.
6. Chalmers I, Glasziou P. Avoidable waste in the production and reporting of research evidence. *Lancet.* 2009;374(9683):86–89. https://doi.org/10.1016/S0140-6736(09)60329-9.
7. Institute-of-Medicine. *The State of Quality Improvement and Implementation Research: Expert Views: Workshop Summary.* National Academies Press; 2007.
8. Damanpour F. Organizational innovation: a meta-analysis of effects of determinants and moderators. *Acad Manag J.* 1991;34(3):555–590. https://doi.org/10.5465/256406, 1991/09/01.
9. Eccles MP, Mittman BS. Welcome to implementation science. *Implement Sci.* 2006;1(1):1. https://doi.org/10.1186/1748-5908-1-1, 2006/02/22.
10. Glasgow RE, Vogt TM, Boles SM. Evaluating the public health impact of health promotion interventions: the RE-AIM framework. *Am J Publ Health.* 1999;89(9):1322–1327. https://doi.org/10.2105/ajph.89.9.1322.
11. Frambach RT, Schillewaert N. Organizational innovation adoption: a multi-level framework of determinants and opportunities for future research. *J Bus Res.* 2002;55(2):163–176. https://doi.org/10.1016/S0148-2963(00)00152-1, 2002/02/01.
12. Aarons GA, Hurlburt M, Horwitz SM. Advancing a conceptual model of evidence-based practice implementation in public service sectors. *Admin Policy Mental health.* 2011;38(1):4–23. https://doi.org/10.1007/s10488-010-0327-7.
13. Brown CH, Curran G, Palinkas LA, et al. An overview of research and evaluation designs for dissemination and implementation. *Annu Rev Publ Health.* 2017;38:1–22. https://doi.org/10.1146/annurev-publhealth-031816-044215.
14. Marx MV, Sabri SS. Interventional radiology residency: steps to implementation. *J Am Coll Radiol.* 2015;12(8):854–859. https://doi.org/10.1016/j.jacr.2015.04.003, 2015/08/01.
15. Halsted MJ, Froehle CM. Design, implementation, and assessment of a radiology workflow management system. *Am J Roentgenol.* 2008;191(2):321–327. https://doi.org/10.2214/AJR.07.3122, 2008/08/01.
16. Bell LTO, James R, Rosa JA, Pollentine A, Pettet G, McCoubrie P. Reducing interruptions during duty radiology shifts, assessment of its benefits and review of factors affecting the radiology working environment. *Clin Radiol.* August 2018;73(8):759.e19–759.e25. https://doi.org/10.1016/j.crad.2018.04.007.
17. Othman K, Bashir O, Arabi M. Implementation of electronic medical record in vascular and interventional radiology. *J Radiol Nurs.* 2019;38(3):201–209. https://doi.org/10.1016/j.jradnu.2019.03.007, 2019/09/01/.
18. Trumm CG, Morhard D, Ertl-Wagner B, Glaser C, Reiser MF. Impact of RIS/PACS integrated speech recognition on report availability. *Radiol Manag.* Nov-Dec 2008;30(6):16–23. quiz 24-6.
19. Trumm CG, Glaser C, Paasche V, et al. Impact of a PACS/RIS-integrated speech recognition system on radiology reporting time and report availability. *Röfo.* April 2006;178(4):400–409. https://doi.org/10.1055/s-2006-926610. Einfluss eines PACS-/RIS-integrierten Spracherkennungssystems auf den Zeitaufwand der Erstellung und die Verfügbarkeit radiologischer Befunde.
20. Tung EL, Dibble EH, Jindal G, Movson JS, Swenson DW. Implementation and impact of a comprehensive radiology report categorization system on communication of important results. *J Am Coll Radiol.* 2021;18(2):248–256. https://doi.org/10.1016/j.jacr.2020.07.026, 2021/02/01/.
21. Morbi AHM, Hamady MS, Riga CV, et al. Reducing error and improving efficiency during vascular interventional radiology: implementation of a preprocedural team rehearsal. *Radiology.* 2012;264(2):473–483. https://doi.org/10.1148/radiol.12110530, 2012/08/01.
22. Desender LM, Van Herzeele I, Lachat ML, et al. Patient-specific rehearsal before EVAR: influence on technical and nontechnical operative performance. A randomized controlled trial. *Ann Surg.* 2016;264(5).
23. Desender L, Van Herzeele I, Lachat M, et al. A multicentre trial of patient specific rehearsal prior to EVAR: impact on procedural planning and team performance. *Eur J Vasc Endovasc Surg.* 2017;53(3):354–361. https://doi.org/10.1016/j.ejvs.2016.12.018, 2017/03/01/.
24. Raja AS, Ip IK, Prevedello LM, et al. Effect of computerized clinical decision support on the use and yield of CT pulmonary angiography in the emergency department. *Radiology.* 2012;262(2):468–474. https://doi.org/10.1148/radiol.11110951, 2012/02/01.
25. Gupta S, Rajiah P, Middlebrooks EH, et al. Systematic review of the literature: best practices. *Acad Radiol.* 2018;25(11):1481–1490. https://doi.org/10.1016/j.acra.2018.04.025.
26. Sucharew H, Macaluso M. Methods for research evidence synthesis: the scoping review approach. *J Hosp Med.* 2019;7:416–418.
27. Faltinsen EG, Storebø OJ, Jakobsen JC, Boesen K, Lange T, Gluud C. Network meta-analysis: the highest level of medical evidence? *BMJ Evid Based Med.* 2018;23(2):56. https://doi.org/10.1136/bmjebm-2017-110887.
28. Hartling L, Guise JM, Kato E, et al. *EPC Methods: An Exploration of Methods and Context for the Production of Rapid Reviews.* Agency for Healthcare Research and Quality (US); 2015.

29. JBI, Aromataris E, Z M, eds. *JBI Manual for Evidence Synthesis*. JBI; 2020.
30. Debray TPA, de Jong VMT, Moons KGM, Riley RD. Evidence synthesis in prognosis research. *Diagn Prognostic Res*. 2019;3(1):13. https://doi.org/10.1186/s41512-019-0059-4, 2019/07/11.
31. Shinkins B, Yang Y, Abel L, Fanshawe TR. Evidence synthesis to inform model-based cost-effectiveness evaluations of diagnostic tests: a methodological review of health technology assessments. *BMC Med Res Methodol*. 2017;17(1):56. https://doi.org/10.1186/s12874-017-0331-7, 2017/04/14.
32. Antman EM, Lau J, Kupelnick B, Mosteller F, Chalmers TC. A comparison of results of meta-analyses of randomized control trials and recommendations of clinical experts: treatments for myocardial infarction. *JAMA*. 1992;268(2):240–248. https://doi.org/10.1001/jama.1992.03490020088036.
33. Ioannidis JPA. Contradicted and initially stronger effects in highly cited clinical research. *JAMA*. 2005;294(2):218–228. https://doi.org/10.1001/jama.294.2.218.
34. Powell BJ, Waltz TJ, Chinman MJ, et al. A refined compilation of implementation strategies: results from the Expert Recommendations for Implementing Change (ERIC) project. *Implement Sci*. 2015;10(1):21. https://doi.org/10.1186/s13012-015-0209-1, 2015/02/12.
35. Landsverk J, Brown CH, Smith JD, et al. Design and analysis in dissemination and implementation research. In: Brownson RC, Colditz GA, Proctor EK, eds. *Dissemination and Implementation Research in Health: Translating Science to Practice*. Oxford Scholarship Online; 2017.
36. Walker TJ, Brandt HM, Wandersman A, et al. Development of a comprehensive measure of organizational readiness (motivation × capacity) for implementation: a study protocol. *Implement Sci Commun*. 2020;1(1):103. https://doi.org/10.1186/s43058-020-00088-4, 2020/11/11.
37. Teeters LA, Heerman WJ, Schlundt D, Harris D, Barkin SL. Community readiness assessment for obesity research: pilot implementation of the Healthier Families programme. *Health Res Pol Syst*. 2018;16(1):2. https://doi.org/10.1186/s12961-017-0262-0, 2018/01/15.
38. Moullin JC, Dickson KS, Stadnick NA, Rabin B, Aarons GA. Systematic review of the exploration, preparation, implementation, sustainment (EPIS) framework. *Implement Sci*. 2019;14(1):1. https://doi.org/10.1186/s13012-018-0842-6, 2019/01/05.
39. Aarons GA, Ehrhart MG, Farahnak LR, Sklar M. Aligning leadership across systems and organizations to Develop a strategic climate for evidence-based practice implementation. *Annu Rev Publ Health*. 2014;35(1):255–274. https://doi.org/10.1146/annurev-publhealth-032013-182447, 2014/03/18.
40. Proctor EK, Landsverk J, Aarons G, Chambers D, Glisson C, Mittman B. Implementation research in mental health services: an emerging science with conceptual, methodological, and training challenges. *Admin Policy Mental Health*. 2009;36(1):24–34. https://doi.org/10.1007/s10488-008-0197-4.
41. Proctor E, Silmere H, Raghavan R, et al. Outcomes for implementation research: conceptual distinctions, measurement challenges, and research agenda. *Admin Policy Mental Health*. 2011;38(2):65–76. https://doi.org/10.1007/s10488-010-0319-7.
42. Saldana L. The stages of implementation completion for evidence-based practice: protocol for a mixed methods study. *Implement Sci*. 2014;9(1):43. https://doi.org/10.1186/1748-5908-9-43, 2014/04/05.
43. Chamberlain P, Brown CH, Saldana L. Observational measure of implementation progress in community based settings: the Stages of implementation completion (SIC). *Implement Sci*. 2011;6(1):116. https://doi.org/10.1186/1748-5908-6-116, 2011/10/06.
44. Glasgow RE, Magid DJ, Beck A, Ritzwoller D, Estabrooks PA. Practical clinical trials for translating research to practice: design and measurement recommendations. *Med Care*. 2005;43(6):551–557. https://doi.org/10.1097/01.mlr.0000163645.41407.09.
45. Hwang S, Birken SA, Melvin CL, Rohweder CL, Smith JD. Designs and methods for implementation research: advancing the mission of the CTSA program. *J Clin Transl Sci*. 2020;4(3):159–167. https://doi.org/10.1017/cts.2020.16.
46. Ramalingam V, Kies DD, Dariushnia SR, Seale T, Kim HS, Yamada K. Evaluation of patient safety and complication rates after implementation of a pre-procedure gastrostomy tube clinic. *J Vasc Intervent Radiol*. 2014;25(3):S39–S40. https://doi.org/10.1016/j.jvir.2013.12.096.
47. Ewusie JE, Soobiah C, Blondal E, Beyene J, Thabane L, Hamid JS. Methods, applications and challenges in the analysis of interrupted time series data: a scoping review. *J Multidiscip Healthc*. 2020;13:411–423. https://doi.org/10.2147/JMDH.S241085.
48. Smith JD. Single-case experimental designs: a systematic review of published research and current standards. *Psychol Methods*. 2012;17(4):510–550. https://doi.org/10.1037/a0029312.
49. Reichardt CS. *Nonequivalent group design. Encyclopedia of Statistics in Behavioral Science*. 2005.
50. Ali MS, Prieto-Alhambra D, Lopes LC, et al. Propensity score methods in health technology assessment: principles, extended applications, and recent advances. Review. *Front Pharmacol*. 2019;10(973). https://doi.org/10.3389/fphar.2019.00973, 2019-September-18.
51. Bouttell J, Craig P, Lewsey J, Robinson M, Popham F. Synthetic control methodology as a tool for evaluating population-level health interventions. *J Epidemiol Commun*. 2018;72(8):673. https://doi.org/10.1136/jech-2017-210106.
52. Hilton Boon M, Craig P, Thomson H, Campbell M, Moore L. Regression discontinuity designs in health: a systematic review. *Epidemiology*. 2021;32(1).
53. Brown CH, Liao J. Principles for designing randomized preventive trials in mental health: an emerging developmental epidemiology paradigm. *Am J Community Psychol*. 1999;27(5):673–710. https://doi.org/10.1023/A:1022142021441, 1999/10/01.
54. Stuart EA, Cole SR, Bradshaw CP, Leaf PJ. The use of propensity scores to assess the generalizability of results from randomized trials. *J Roy Stat Soc*. 2001;174(2):369–386. https://doi.org/10.1111/j.1467-985X.2010.00673.x.
55. Stiell IG, Clement CM, Grimshaw J, et al. Implementation of the Canadian C-Spine Rule: prospective 12 centre cluster randomised trial. *BMJ*. 2009;339:b4146. https://doi.org/10.1136/bmj.b4146.
56. Palen TE, Sharpe RE, Shetterly SM, Steiner JF. Randomized clinical trial of a clinical decision support tool for improving the appropriateness scores for ordering imaging studies in primary and specialty care ambulatory clinics. *Am J Roentgenol*. 2019;213(5):1015–1020. https://doi.org/10.2214/AJR.19.21511, 2019/11/01.
57. van der Leest M, Cornel E, Israël B, et al. Head-to-head comparison of transrectal ultrasound-guided prostate biopsy versus multiparametric prostate resonance imaging with subsequent magnetic resonance-guided biopsy in biopsy-naïve men with elevated prostate-

specific antigen: a large prospective multicenter clinical study. *Eur Urol*. 2019;75(4):570–578. https://doi.org/10.1016/j.eururo.2018.11.023, 2019/04/01.

58. Collins LM, Baker TB, Mermelstein RJ, et al. The multiphase optimization strategy for engineering effective tobacco use interventions. *Ann Behav Med*. 2011;41(2):208–226. https://doi.org/10.1007/s12160-010-9253-x.
59. Collins LM, Nahum-Shani I, Almirall D. Optimization of behavioral dynamic treatment regimens based on the sequential, multiple assignment, randomized trial (SMART). *Clin Trials*. 2014;11(4):426–434. https://doi.org/10.1177/1740774514536795.
60. Lei H, Nahum-Shani I, Lynch K, Oslin D, Murphy SA. A "SMART" design for building individualized treatment sequences. *Annu Rev Clin Psychol*. 2012;8:21–48. https://doi.org/10.1146/annurev-clinpsy-032511-143152.
61. Ong M-S, Magrabi F, Post J, et al. Communication interventions to improve adherence to infection control precautions: a randomised crossover trial. *BMC Infect Dis*. 2013;13(1):72. https://doi.org/10.1186/1471-2334-13-72, 2013/02/06.
62. Brown CH, Wyman PA, Guo J, Peña J. Dynamic wait-listed designs for randomized trials: new designs for prevention of youth suicide. *Clin Trials*. 2006;3(3):259–271. https://doi.org/10.1191/1740774506cn152oa, 2006/06/01.
63. Brown CA, Lilford RJ. The stepped wedge trial design: a systematic review. *BMC Med Res Methodol*. 2006;6. https://doi.org/10.1186/1471-2288-6-54, 54-54.

Chapter 92

Mixed methods research

Ashwin Deshmukh[1] and Joshua Cornman-Homonoff[2]

[1]*Department of Radiology, Yale New Haven Health Bridgeport Hospital, Bridgeport, CT, United States;* [2]*Department of Radiology and Biomedical Imaging, Yale University School of Medicine, New Haven, CT, United States*

Key points

- Mixed methods research involves integrating qualitative and quantitative methodology into a single research study.
- It aims to combine the advantages of both quantitative and qualitative research and eliminate their individual drawbacks.
- It is a relatively new and underutilized methodology in interventional radiology research and may have interesting potential applications.

Introduction and definition

Mixed methods research (MMR) broadly involves combining or integrating qualitative and quantitative research approaches into a single research study. This contrasts with "multimethods research," which involves combining either multiple qualitative studies alone or multiple quantitative methods alone. Thus, the mixing of qualitative and quantitative research approaches aims to provide answers and multidimensional insights into complex research questions and enrich one approach with the other. The integration or "mixing" can occur at any or all phases of the research process.

Although a combination of qualitative and quantitative techniques has been employed since the early 20th century for social and anthropological studies, Campbell and Fiske first demonstrated using both methodologies for validation purposes.[1] The term triangulation was coined by Webb et al.,[2] while Denzin[3] defined triangulation for research methodology as "the combination of methodologies in the study of the same phenomenon" and described data triangulation, investigator triangulation, theory triangulation, and methodological triangulation. Over the years, many different names have been given to this third research methodology, for example, integrative research, multiple methods, triangulated studies, ethnographic residual analysis, and mixed research.[4] The definition of MMR has been evolving, and there have been several attempts to define what constitutes MMR. Johnson et al. reviewed various definitions by experts in the field of MMR and defined it as "the type of research in which a researcher or team of researchers combines elements of qualitative and quantitative research approaches (e.g., use of qualitative and quantitative viewpoints, data collection, analysis, inference techniques) for the broad purposes of breadth and depth of understanding and corroboration"[4]. MMR has found increasing applications in the fields of social sciences and healthcare-related research. Its applications in healthcare have been growing to address more complex research questions, especially in healthcare delivery and policy.

The research paradigms

Paradigm is defined as "the set of beliefs and practices that guided the field."[5] Traditionally, researchers have aligned to either the quantitative (positivist) or qualitative (constructivist) paradigms. The quantitative methodology involves collecting data that can be measured and subjected to statistical analysis. It is thought to be objective and unbiased. Historically, healthcare research has almost exclusively used the quantitative methodology. The qualitative methodology, on the other hand, involves understanding views and opinions of participants, asking questions that are broad in nature, and analyzing these responses for common themes. This method of inquiry is inherently subjective and biased. Some of the techniques of data collection include interviews, open-ended questionnaires, and focus groups. There has been considerable debate about the merits of quantitative versus qualitative methodology as well as the compatibility of combining these two paradigms. The pragmatist philosophy focuses on the outcomes rather than the process and advocates "a needs-based or contingency approach to research method and concept selection."[6] It lays a foundation for the

Translational Interventional Radiology. https://doi.org/10.1016/B978-0-12-823026-8.00078-X

combination of qualitative and quantitative approaches. MMR can provide a broader view of the study phenomena and reveal insights that individual methodologies may not be able to. It aims to combine the benefits of both quantitative and qualitative methods and eliminate potential biases from both methodologies.

Types of MMR

There have been various attempts at classifying MMR study designs. Leech and Onwuegbuzie described one classification based on timing, weighting, and level of mixing of quantitative and qualitative components.[7] Timing refers to whether the qualitative and quantitative components occurred simultaneously (concurrent study design) or one after another (sequential study design). Weighting refers to whether the study employed qualitative and quantitative methods equally or whether one of these components was given higher priority or more emphasis than the other. Level of mixing refers to whether the study was partially mixed or fully mixed. Thus, the combination of these three dimensions produces many combinations, as listed in the following.

1. Concurrent, equal status, partially mixed
2. Concurrent, equal status, fully mixed
3. Concurrent, dominant, partially mixed
4. Concurrent, dominant, fully mixed
5. Sequential, equal status, partially mixed
6. Sequential, equal status, fully mixed
7. Sequential, dominant, partially mixed
8. Sequential, dominant, fully mixed

The previous classification is not exhaustive, and other types include multiphase, embedded, conversion, and transformative designs. An important consideration is the stage of integration. Morse and Niehaus described two possible points of integration—the results point of integration and the analytical point of integration,[8] whereas Teddlie and Tashakkori described possible integration at any of the four different stages of investigation, viz. conceptualization, data collection, analysis, and inference.[9] Tables 92.1 and 92.2 summarize the previously described classification and some common types of study designs, respectively. It is important to note that many of the published studies utilizing MMR are complex and cannot be fully classified based on any single system. A detailed and thorough description of the study methodology bears more practical utility than classifying the design used in that study.

Advantages of MMR

Researchers have described numerous purposes of mixing in MMR. Greene et al. described five distinct purposes of mixing; whereas, Bryman elaborated many more.[10,11] Broadly speaking, the benefits of mixing are as follows:

1. It allows greater validity and/or credibility of a study by comparing and corroborating qualitative and quantitative data. It also provides a more complete picture from the study.
2. The limitations of either of the two approaches are minimized when they are utilized in a single study design.
3. Results from one approach can be used to design or implement the other approach.
4. MMR may be able to address a unique and broader range of questions that a quantitative or qualitative approach may be unable to answer when used individually.
5. MMR allows viewing a research subject from multiple angles, thus adding diversity of views and perspectives.
6. When unusual findings are generated using one approach, using the other approach may provide an explanation for these results.
7. One approach may increase the utility and/or applicability of results obtained from the other approach.

Potential drawbacks of MMR

1. Increases study complexity:

Researchers need to be familiar with both qualitative and quantitative approaches, as well as the appropriate technique of mixing these methodologies to suit the study design and research question.

TABLE 92.1 Classification of MMR.

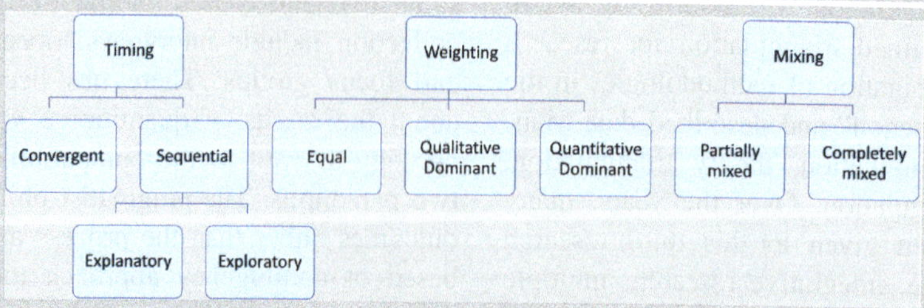

Based on classification proposed by Leech, NL, Onwuegbuzie, AJ. A typology of mixed methods research designs. *Qual Quant.* 2009;43:265-275. https://doi.org/10.1007/s11135-007-9105-3.

TABLE 92.2 Some of the commonly used study designs in mixed methods research.

Convergent	Both quantitative and qualitative data collection and analysis occur at the same time.
Sequential	*Explanatory*—quantitative data collection and analysis followed by qualitative data collection and analysis.
	Explanatory—qualitative data collection and analysis followed by quantitative data collection and analysis.
Embedded	Quantitative and qualitative data are collected and analyzed within a quantitative or a qualitative design.
Multiphase	Quantitative and qualitative studies are conducted separately and later analyzed together.

2. Greater resource requirements:

A single researcher may not be able to undertake both qualitative and quantitative research, especially when these approaches are used concurrently. Similarly, more resources are needed to employ both qualitative and quantitative study designs. It often translates to significant increases in the cost of conducting research.

Examples from published medical literature

Very few publications related to interventional radiology (IR) have used MMR as their study methodology. These three examples provide an insight into how MMR could cater to a unique and diverse range of research questions in the field of IR:

1. Munn et al.[12] employed MMR to study the use of and attitudes toward safety checklists in IR in multiple hospitals in Australia. They used quantitative methodology for assessing checklist compliance and for conducting a retrospective medical record audit. This provided measurable, quantitative data in terms of rates and percentages. Qualitative methodology was used in the form of focus group analysis, which involved discussion with radiologists practicing in those hospitals. From this qualitative analysis, several common themes emerged that identified problems and barriers in implementing safety checklists and strategies for improvement. Thus, qualitative methodology provided not only insight into understanding the reasons for low checklist compliance but also potential solutions.
2. England et al.[13] used MMR to develop and validate a questionnaire to assess health-related quality of life in patients with vascular malformations. They used qualitative methodology in the form of interviews with patients to develop a pilot questionnaire. The questionnaire consisted of Likert scale items as well as free-response questions. The validity of this questionnaire was studied using quantitative methodology. Thus, this was a complex study design in which qualitative methodology was used to design a questionnaire that utilized both quantitative and qualitative assessment, and the validity of the newly developed questionnaire was tested using a quantitative methodology.
3. Schonberg et al.[14] studied the psychological impact and experience of benign breast biopsy in older women. This was a prospective longitudinal study in which patients were interviewed at the time of biopsy as well as 6 months after biopsy. At the time of biopsy, they employed quantitative methodology in the form of a Psychological Consequence Questionnaire, which consisted of a score ranging from 0 to 36. Additional quantitative scoring systems were also utilized. At the follow-up assessment, the perception of information provided to them after the breast biopsy, as well as concerns about breast cancer, was assessed using measurable responses. Qualitative components included describing the role family and friends played in decision-making after their abnormal mammogram, as well as assessing other thoughts about their experience after their abnormal mammogram. Common themes were identified from these responses and a conceptual framework was developed that highlighted, for example, the factors associated with increased or decreased anxiety. Similarly, common factors were identified that improved or worsened patients' experience at biopsy. Thus, this study identified the degree of psychological impact (quantitative component) as well as the factors contributing to this impact (qualitative component).

Conclusion

MMR is a relatively new and rarely used methodology in IR research. Although more demanding in terms of complexity and resources, it could provide unique insights and the ability to answer research questions that may not be optimally answered using either a quantitative or a qualitative research methodology alone. The previously listed examples illustrate the immense and yet untapped potential of MMR in IR.

Potential pitfalls

- MMR can be expensive and utilizes more resources than either a qualitative or a quantitative design alone. Thus, the additional benefit of utilizing both types of approaches should justify the increased cost and resources utilized.
- While classification systems provide a reference for understanding and designing a mixed methods study, many study designs can be unique to that study and do not fully conform to a classification system.

Get started

- Browse PubMed for additional publications that have used MMR in medical research.
- Read the articles referenced later to gain an in-depth understanding on conducting MMR.
- Engage with not only the biostatisticians but also researchers from social sciences department at your university to gain unique perspectives about MMR.

References

1. Campbell DT, Fiske DW. Convergent and discriminant validation by the multitrait-multimethod matrix [cited 2021 Jul 13] *Psychol Bull.* 1959;56(2):81–105. Available from:/record/1960-00103-001.
2. Webb EJ, Campbell DT, Schwartz RD, Sechrest L. *Unobtrusive Measures: Nonreactive Research in the Social Sciences.* Rand McNally & Company; 1968.
3. Denzin NK. *The Research Act: A Theoretical Introduction to Sociological Methods.* McGraw-Hill. McGraw-Hill; 1979. cited 2021 Jul 13]. 370 p. Available from: https://books.google.com/books/about/The_Research_Act.html?id=gfS1AAAAIAAJ.
4. Johnson RB, Onwuegbuzie AJ. Toward a definition of mixed methods research. *J Mix Methods Res.* 2007;1(2):112–133.
5. Morgan DL. *Paradigms Lost and Pragmatism Regained: Methodological Implications of Combining Qualitative and Quantitative Methods*; 2016. https://doi.org/10.1177/2345678906292462 [cited 2021 Jul 24];1(1):48–76. Available from: https://journals.sagepub.com/doi/10.1177/2345678906292462.
6. Johnson RB, Onwuegbuzie AJ. *Mixed Methods Research: A Research Paradigm Whose Time Has Come*; 2016 https://doi.org/10.3102/0013189X033007014 [cited 2021 Jul 24];33(7):14–26. Available from: https://journals.sagepub.com/doi/10.3102/0013189X033007014.
7. Leech NL, Onwuegbuzie AJ. A typology of mixed methods research designs. *Qual Quant*; 2009 [cited 2021 Jul 18];43(2):265–75. Available from: https://link.springer.com/article/10.1007/s11135-007-9105-3.
8. Morse JM, Niehaus L. *Mixed Method Design: Principles and Procedures.* Left Coast Press; 2009.
9. Teddlie C, Tashakkori A. *Foundations of Mixed Methods Research: Integrating Quantitative and Qualitative Approaches in the Social and Behavioral Sciences*; 2009. Available from: https://www.worldcat.org/title/foundations-of-mixed-methods-research-integrating-quantitative-and-qualitative-approaches-in-the-social-and-behavioral-sciences/oclc/213480119.
10. Greene JC, Caracelli VJ, Graham WF. Toward a conceptual framework for mixed-method evaluation designs. *Educ Eval Policy Anal.* 1989;11(3):255–274 [cited 2021 Jul 21]. Available from: https://journals.sagepub.com/doi/10.3102/01623737011003255.
11. Bryman A. Integrating quantitative and qualitative research: how is it done? *Qual Res.* 2006;6(1):97–113 [cited 2021 Jul 21]. Available from: https://journals.sagepub.com/doi/10.1177/1468794106058877.
12. Munn Z, Giles K, Aromataris E, et al. Mixed methods study on the use of and attitudes towards safety checklists in interventional radiology. *J Med Imaging Radiat Oncol.* 2018;62(1):32–38.
13. England RW, Motaghi M, Kohler B, Hemmingson TE, Wu AW, Weiss CR. Development and preliminary validation of the patient-reported outcome measure for vascular malformation questionnaire: a prospective cohort study. *J Vasc Interv Radiol [Internet].* 2021;32(5):683–690. https://doi.org/10.1016/j.jvir.2021.01.263. e4. Available from:.
14. Schonberg MA, Silliman RA, Ngo LH, et al. Older women's experience with a benign breast biopsy—a mixed methods study. *J Gen Intern Med.* 2014;29(12):1631–1640 [cited 2021 Jul 22]. Available from: https://pubmed.ncbi.nlm.nih.gov/25138983/.

Further reading

Creswell JW, Plano Clark VL. *Designing and Conducting Mixed Methods Research* 2011.

The Journal of Mixed Methods Research. Sage Journals.

Chapter 93

Population- and setting-specific implementations

N. El Sehemawi[1], John T. Moon[2], S. Nourouzpour[3], Menelaos Konstantinidis[4], Li Ka Shing[5] and Nariman Nezami[6,7]

[1]*Faculty of Medicine, University of Montreal, Montreal, QC, Canada;* [2]*Division of Interventional Radiology and Image-Guided Medicine, Department of Radiology and Imaging Sciences, Emory University School of Medicine, Atlanta, GA, United States;* [3]*Toronto General Hospital, University Health Network, Toronto, ON, Canada;* [4]*Institute of Health Policy, Management and Evaluation, University of Toronto, Toronto, ON, Canada;* [5]*Knowledge Institute, St. Michael's Hospital, Unity Health Toronto, Toronto, ON, Canada;* [6]*Division of Vascular and Interventional Radiology, Department of Diagnostic Radiology and Nuclear Medicine, University of Maryland School of Medicine, Baltimore, MD, United States;* [7]*Experimental Therapeutics Program, University of Maryland Marlene and Stewart Greenebaum Comprehensive Cancer Center, Baltimore, MD, United States*

Introduction

Research in the medical field traditionally centers around testing the efficacy and effectiveness of interventions using scientific experiments such as transitional studies and randomized controlled trials to assess internal validity. However, results from such trials often lack the external validity needed to allow the implementation of a given intervention in real-life settings[1] and fail to achieve actual progress in healthcare practices.[2] Implementation research focuses on developing specific solutions to apply scientific discoveries into the daily medical practice. It can be defined as a "multidisciplinary specialty that seeks generalizable knowledge about the behavior of stakeholders, organizations, communities, and individuals in order to understand the scale of, reasons for, and strategies to close the gap between evidence and routine practice for health in real-world contexts."[3]

There are different theoretical frameworks for implementation research that expound on the topic, including (1) the Consolidated Framework for Implementation Research, (2) the National (US) Implementation Research Network's Active Implementation Frameworks, (3) the World Health Organization's Introductory implementation research guide, and (4) the Standards for Reporting Implementation Studies. The common goal of these strategies is to provide guidelines to improve the translation of scientific advancements into actual practices.

Interventional radiology (IR) is a growing field that presents many challenges regarding implementation research due in part to its rapidly evolving technologies and its heterogeneous practice. In this chapter, relevant settings and population groups in the context of IR will be compared and discussed to highlight the development of translational research in the field and shed light on the areas of possible improvement.

Setting and implementation

Effective dissemination and implementation of an intervention depend on various factors such as relevance, feasibility, and associated costs in a given setting.[4] Therefore, implementation strategies may depend on the specific setting in which an intervention is developed. Prostate artery embolization (PAE), for example, is not recognized as the primary treatment option by the American Urological Association; however, it is a widely accepted minimally invasive surgery by multiple radiology societies for the treatment of benign prostatic hyperplasia as well as the treatment of lower urinary tract symptoms, a common comorbidity.[5] Many advances in PAE research have been achieved in the private sector, and the evidence supporting PAE benefits continues to grow. In the United States, it is estimated that at least 60% of practicing interventional radiologists are in the private sector.[6] As such, it is not surprising to see instances in which translational research originates from and is implemented by interventional radiologists in private practice. However, despite such advancements, patient eligibility for this procedure, individualized care,[7,8] and

Translational Interventional Radiology. https://doi.org/10.1016/B978-0-12-823026-8.00011-0

generalizability of results in various settings[4] are of significant concerns. In fact, not all patients are eligible or meet the criteria for such procedures, and not all centers offer such a procedure.[9]

Further research should focus on establishing and evaluating the implementation and standardization of this procedure to offer additional options to patients who are ineligible for currently available surgical treatments.[10] Nonetheless, PAE's advancements illustrate the implementation of translational research originating from the private sector with growth that has since included academic partnerships. The development and implementation of PAE is a paradigm for many lesser-known therapeutics that are typically implemented in private practices, such as cosmetic sclerotherapy.

An intervention's characteristics may also significantly impact the appropriate setting in which the implementation research is conducted. For instance, as an evolving intervention, yttrium-90 (Y90) radioembolization may encounter the least resistance in academic settings.[11] Currently, only two microspheres are available for clinical use: the SIR-Spheres and TheraSpheres. Given the multidisciplinary nature of cancer treatment, the stakeholders involved in Y90 development have included oncologists, radiation oncologists, diagnostic radiologists, interventional radiologists, nuclear medicine specialists, physicists, hepatologists, transplant surgeons, and pathologists. Not surprisingly, their involvement in patient selection criteria and support of clinical trials are more readily available and supported by the infrastructure of academic centers. However, subsequent to refined methodology, use, and patient criteria within academic settings, Y90 radioembolization has been successfully implemented at the community level as well.[12] Thus, Y90 represents an excellent example of implementation research beginning within the academic setting and being disseminated to community practices. However, the iterative process of therapy improvement and refinement will continue in academic settings, as demonstrated by ongoing phases II and III trials conducted in university-affiliated hospitals.[11,13]

Finally, as resources are often scarce in community and rural settings, practices vary widely between these settings and urban hospitals. Acute pulmonary embolism (PE) represents a substantial health burden as it is associated with significant morbidity and mortality. Treatments for this condition have evolved to include many interventional therapies, such as catheter-directed thrombolysis or thrombectomy, surgical embolectomy, and anticoagulation therapies as the first-line treatment.[14] However, evident disparities exist in the prevalence of thrombolytic interventions between rural settings and metropolitan hospitals.[15] Management of PE must be efficient and address the many steps between triage and the appropriate treatment. The pulmonary embolism response team, a multidisciplinary team developed in urban centers, aims to standardize the treatment approach to PE. It has proven to increase accuracy in clinical risk stratification,[16] reduce in-hospital mortality,[17,18] and reduce the length of stay, among other advantages.[19] Yet, most institutions are not equipped or staffed for specialized PE care and require that patients be transferred from their local hospitals (mainly tertiary centers) to well-equipped high service centers.[15] Further implementation research must therefore be targeted at meeting the needs for management of interhospital patient transfers.[20]

Population and implementation

Implementation research must also consider the interventions' targeted population. The sex of the population limits some procedures it targets, such as uterine artery embolization (UAE). 10%–30% of women suffer from heavy menstrual bleeding, 40% of whom present with uterine fibroids.[21] Many controlled trials, as well as a Cochrane review, have already demonstrated the validity and safety of UAE in treating this medical condition.[22,23] However, despite the intervention's demonstrated benefit, its implementation is still limited. Solutions that fall within the implementation research framework have been suggested, such as option grids to help women make their choices regarding their treatment and help the physicians better guide their patients in this decision.[24]

Race also plays a significant role in the distribution of certain medical conditions. Studies have also demonstrated that postpartum hemorrhages disproportionately affect African American women, with reports of higher morbidity and mortality.[25,26] The reason for such disparity has yet to be elucidated and is an improvement opportunity for future research. Systematic reviews have shown that UAE is an effective embolization technique that minimizes the risks of complications and avoids the need for general anesthesia while preserving the uterus for future fertility.[27] However, such a procedure might not be readily available in centers where it is needed the most. As such, implementation research in IR concerning UAE should focus on helping targeted populations by ensuring that this procedure is widely available and implemented.[28]

Various interventional radiology procedures are inherently palliative in nature. They provide patients with significantly reduced postprocedural pain, reduce postoperative hospital stays, and eliminate the need for general anesthesia.[29] Many IR interventions aim at improving or relieving a patient's symptoms without the direct intention of curing disease or prolonging life. McCullough et al. have found that as much as 80% of IR referrals fall into the palliative care category.[30] Palliative care follows a distinct approach that prioritizes a patient's overall well-being and desires over the prolongation of his life. Measuring the validity of a procedure by its technical success becomes obsolete as it may not meet the patient's desires. As such, guidelines to translate IR advancements into this type of practice should focus on involving patients and their families in the decision-making process and measuring an intervention's success on outcome measures such as symptoms or quality-of-life assessments.[31] Palliative care patients also present a clear opportunity for interventional radiologists to incorporate palliative care skills into their practice to help further guide the patients and make use of the technology developed in this field.[32]

Conclusion

Interventional radiology is a rapidly growing field with clear benefits in many clinical settings and populations. However, further work is needed in (1) using evidence-based frameworks for developing recommendations of possible IR treatments for candidate patients who are otherwise ineligible for other treatments, (2) translating skills and practices from highly specialized centers (often in urban settings) to suburban or rural centers, or improving interhospital transfer to reduce the resource disparities, (3) ensuring that the practice of IR is ubiquitously available across the population (i.e., irrespective of race or other factors), and (4) increasing patient engagement at every step in the translational research pipeline.

In developing recommendations for potential IR patients, even in the absence of substantial, high-quality evidence, frameworks such as Grading of Recommendations, Assessment, and Evaluations Evidence to Decision should be used.[33] This will allow for a systematic process whereby healthcare providers at any level can establish evidence-based practices for IR solutions and would allow the opportunity to foster relationships between academic centers and private practice.

Additionally, providing top-of-the-line training to residents irrespective of the setting of their primary institutions and by engaging in cross-institutional collaboration will help ensure that (1) future IRs have the necessary skills to practice in rural or suburban settings, and (2) rural centers and clinicians in those centers have the skillset to provide high-quality care to their patients.

In conclusion, IR has employed both traditional frameworks of implementation research, including (1) those that start in academic settings and spread to community practices, as well as (2) those initiated by private practices in collaboration with academia. Implementation frameworks help more patients access an intervention of need and implicitly bridge disparities and inequities that exist in access to care.

Potential pitfalls

- Considerations regarding the generalizability of results from private sector settings to hospital settings for a given procedure should be addressed to ensure the value of IR treatment.
- Implementational research should focus not only on developing new and innovative technologies, but it should also investigate the operation settings' limitations in terms of staffing and equipment.
- Research in interventional radiology should continue to focus on addressing specific population concerns and ensure its availability for the population in question.

Examples

- **Intervention development setting:** Prostate artery embolization is a minimally invasive procedure which illustrates the successful translation of interventions from private sectors to academic partnership and Vis versa.
- **Implementation setting:** The distinct management of acute pulmonary embolism in rural and tertiary care centers highlights the disparities in access to interventional radiology in healthcare. In fact, most institutions are not equipped or staffed for specialized PE care and require that patients be transferred from their local hospitals (mainly tertiary centers) to well-equipped high service centers.
- **Population concerns:** As many interventional radiology procedures are palliative in nature, a procedure's success should be based on symptom reduction and quality-of-life assessment rather than by its technical success.

Resources/additional readings

- The (updated) Consolidated Framework for Implementation Research: https://implementationscience.biomedcentral.com/articles/10.1186/s13012-022-01245-0
- National (US) Implementation Research Network's Active Implementation Frameworks https://www.activeimplementation.org/wp-content/uploads/2021/04/AIRN-AFixsen-FrameworksAnalysis-2021.pdf
- World Health Organization's Introductory implementation research guide: https://apps.who.int/iris/bitstream/handle/10665/91758/9789241506212_eng.pdf
- The Standards for Reporting Implementation Studies: https://bmjopen.bmj.com/content/bmjopen/7/4/e013318.full.pdf

Get started

- Understand the various factors, such as relevance, feasibility, associated costs and an intervention's characteristics that can influence the setting in which an intervention is developed.
- Recognize the significant disparities between high services centers and community and rural hospitals.

- Reflect on the importance of reducing this gap by translating skills to different settings and finding efficient ways to provide more equitable access to care.
- Recognize the necessity of guidelines to translate IR advancements into daily medical practice and the role of the populations involved.

References

1. Damschroder L, Peikes D, Petersen D. *Using Implementation Research toGuide Adaptation, Implementation, and Dissemination of Patient-Centered Medical Home Models.* Rockville, MD: Agency for Healthcare Research and Quality; February 2013. AHRQ Publication No. 13-0027-EF.
2. Pinnock H, Barwick M, Carpenter CR, et al. Standards for reporting implementation studies (StaRI) statement. *BMJ.* 2017;356:i6795. https://doi.org/10.1136/bmj.i6795.
3. Odeny TA, Padian N, Doherty MC, et al. Definitions of implementation science in HIV/AIDS. *Lancet HIV.* 2015;2:e178–e180.
4. Peters DH, Tran NT. *Taghreed Adam. Implementation Research in Health: A Practical Guide. Alliance for Health Policy and Systems Research.* World Health Organization; 2013.
5. Dias Jr US, de Moura MRL, Viana PCC, et al. Prostatic artery embolization: indications, preparation, techniques, imaging evaluation, reporting, and complications. *Radiographics.* 2021 Sep-Oct;41(5):1509–1530. https://doi.org/10.1148/rg.2021200144. Epub 2021 Aug 20. PMID: 34415807.
6. Peck S, Van Erp J, Watabe J, Costantino M, Varma MK. Interventional radiology private practice patterns. *Semin Intervent Radiol.* March 2019;36(1):3–9. https://doi.org/10.1055/s-0039-1683356. Epub 2019 Mar 29. PMID: 30936608; PMCID: PMC6440908.
7. Young S, Golzarian J. Prostate artery embolization: state of the evidence and societal guidelines. *Tech Vasc Intervent Radiol.* September 2020;23(3):100695. https://doi.org/10.1016/j.tvir.2020.100695. Epub 2020 Oct 5. PMID: 33308532.
8. Young S. Prostate artery embolization for benign prostatic hyperplasia: the hunt for the ideal patient population. *Diagn Interv Imaging.* March 2021;102(3):119–120. https://doi.org/10.1016/j.diii.2021.01.005. Epub 2021 Jan 29. PMID: 33518450.
9. Ontario Health (Quality). Prostatic artery embolization for benign prostatic hyperplasia: a health technology assessment. *Ont Health Technol Assess Ser.* June 4, 2021;21(6):1–139. PMID: 34188733; PMCID: PMC8202600.
10. Picel AC, Hsieh TC, Shapiro RM, Vezeridis AM, Isaacson AJ. Prostatic artery embolization for benign prostatic hyperplasia: patient evaluation, anatomy, and technique for successful treatment. *Radiographics.* September–October, 2019;39(5):1526–1548. https://doi.org/10.1148/rg.2019180195. Epub 2019 Jul 26. PMID: 31348735.
11. Woerner AJ, Johnson GE. Advances in Y-90 radioembolization for the treatment of hepatocellular carcinoma. *Hepatoma Res.* 2022;8:2. https://doi.org/10.20517/2394-5079.2021.122.
12. De La Ossa J, Nguyen H, Mohan K, et al. Y90 radioembolization: a case series in a community-based setting. *Am J Gastroenterol.* October 2020;115:S1710–S1711. https://doi.org/10.14309/01.ajg.0000715148.67905.b0.
13. Tong AK, Kao YH, Too CW, Chin KF, Ng DC, Chow PK. Yttrium-90 hepatic radioembolization: clinical review and current techniques in interventional radiology and personalized dosimetry. *Br J Radiol.* June 2016;89(1062):20150943. https://doi.org/10.1259/bjr.20150943. Epub 2016 Mar 24. PMID: 26943239; PMCID: PMC5258157.
14. Giri J, Sista AK, Weinberg I, et al. On behalf of the American heart association council on peripheral vascular disease; council on arteriosclerosis, thrombosis and vascular biology; council on cardiopulmonary, critical care, perioperative and resuscitation; and council on cardiovascular surgery and anesthesia. Interventional therapies for acute pulmonary embolism: current status and principles for the development of novel evidence: a scientific statement from the American heart association. *Circulation.* 2019;140:e774–e801. https://doi.org/10.1161/CIR.0000000000000707.
15. Rali P, Sacher D, Rivera-Lebron B, et al. Interhospital transfer of patients with acute pulmonary embolism: challenges and opportunities. *Chest.* November 2021;160(5):1844–1852. https://doi.org/10.1016/j.chest.2021.07.013. Epub 2021 Jul 14. PMID: 34273391.
16. Brailovsky Y, Kunchakarra S, Lakhter V, et al. Pulmonary embolism response team implementation improves awareness and education among the house staff and faculty. *J Thromb Thrombolysis.* January 2020;49(1):54–58. https://doi.org/10.1007/s11239-019-01927-5. PMID: 31396791.
17. Annabathula R, Dugan A, Bhalla V, Davis GA, Smyth SS, Gupta VA. Value-based assessment of implementing a pulmonary embolism response team (PERT). *J Thromb Thrombolysis.* January 2021;51(1):217–225. https://doi.org/10.1007/s11239-020-02188-3. PMID: 32542527; PMCID: PMC8325945.
18. Myc LA, Solanki JN, Barros AJ, et al. Adoption of a dedicated multidisciplinary team is associated with improved survival in acute pulmonary embolism. *Respir Res.* June 22, 2020;21(1):159. https://doi.org/10.1186/s12931-020-01422-z. PMID: 32571318; PMCID: PMC7310489.
19. Wright C, Goldenberg I, Schleede S, et al. Effect of a multidisciplinary pulmonary embolism response team on patient mortality. *Am J Cardiol.* December 15, 2021;161:102–107. https://doi.org/10.1016/j.amjcard.2021.08.066. PMID: 34794606; PMCID: PMC8761489.
20. Rali P, Sacher D, Rivera-Lebron B, et al. Interhospital transfer of patients with acute pulmonary embolism: challenges and opportunities. *Chest.* November 2021;160(5):1844–1852. https://doi.org/10.1016/j.chest.2021.07.013. Epub 2021 Jul 14. PMID: 34273391.
21. Istre O, Qvigstad E. Current treatment options for abnormal uterine bleeding: an evidence-based approach. *Best Pract Res Clin Obstet Gynaecol.* 2007;21(6):905–913. https://doi.org/10.1016/j.bpobgyn.2007.03.021.
22. de Bruijn AM, Huisman J, Hehenkamp WJK, et al. Implementation of uterine artery embolization for symptomatic fibroids in The Netherlands: an inventory and preference study. *CVIR Endovasc.* June 3, 2019;2(1):18. https://doi.org/10.1186/s42155-019-0061-5. PMID: 32026034; PMCID: PMC6966393.
23. Gupta JK, Sinha A, Lumsden MA, Hickey M. Uterine artery embolization for symptomatic uterine fibroids. *Cochrane Database Syst Rev.* 2014;12:CD005073.

24. Riggan KA, Stewart EA, Balls-Berry JE, Venable S, Allyse MA. Patient recommendations for shared decision-making in uterine fibroid treatment decisions, 23743735211049655 *J Patient Exp*. October 20, 2021;8. https://doi.org/10.1177/23743735211049655. PMID: 34692992; PMCID: PMC8532210.
25. Sabato, Emily DO, Gobern JMD, Meske SMS, Buckley MMS. Racial disparities in postpartum hemorrhage in a community hospital: a retrospective cohort analysis [24C]. *Obstet Gynecol*. May 2020;135:35S−36S. https://doi.org/10.1097/01.AOG.0000663324.67045.80.
26. Caldwell CJ, McCullagh L. Recent data indicate that black women are at greater risk of severe morbidity and mortality from postpartum haemorrhage, both before and after adjusting for comorbidity. *Evid Base Nurs*. 2019;22:57.
27. Ruiz Labarta FJ, Pintado Recarte MP, Alvarez Luque A, et al. Outcomes of pelvic arterial embolization in the management of postpartum haemorrhage: a case series study and systematic review. *Eur J Obstet Gynecol Reprod Biol*. 2016;206:12−21. https://doi.org/10.1016/j.ejogrb.2016.07.510.
28. Gyamfi-Bannerman C, Srinivas SK, Wright JD, et al. Postpartum hemorrhage outcomes and race. *Am J Obstet Gynecol*. August 2018;219(2):185.e1−185.e10. https://doi.org/10.1016/j.ajog.2018.04.052. Epub 2018 May 9. PMID: 29752934.
29. Lintin L, Uberoi R. *Interventional Radiology in Palliative Care: Introduction to IR*. Springer. Springer Nature Switzerland AG; 2021. https://doi.org/10.1007/978-3-030-65463-4. corrected publication 2021 https://link.springer.com/content/pdf/10.1007/978-3-030-65463-4.pdf.
30. McCullough HK, Bain RM, Clark HP, Requarth JA. The radiologist as a palliative care subspecialist: providing symptom relief when cure is not possible. *AJR Am J Roentgenol*. February 2011;196(2):462−467. https://doi.org/10.2214/AJR.10.4672. PMID: 21257901.
31. Khemlani MH. *Interventional Radiology in Palliative Care: Clinical Applications of Outcome Measurement*. Springer. Springer Nature Switzerland AG; 2021. corrected publication 2021 https://link.springer.com/content/pdf/10.1007/978-3-030-65463-4.pdf.
32. Buss MK. The intersection of palliative care and interventional radiology: enhancing understanding and collaboration. *Semin Intervent Radiol*. June 2017;34(2):140−144. https://doi.org/10.1055/s-0037-1602756. Epub 2017 Jun 1. PMID: 28579682; PMCID: PMC5453779.
33. Alonso-Coello P, Schünemann HJ, Moberg J, et al, GRADE Working Group. GRADE Evidence to Decision (EtD) frameworks: a systematic and transparent approach to making well informed healthcare choices. 1: Introduction. *BMJ*. June 28, 2016;353:i2016. https://doi.org/10.1136/bmj.i2016. PMID: 27353417.

Part X

Public health

Chapter 94

Public health research

Siddharth Venkatraman[1] and Capt. Tej Ishaan Mehta[2,3]
[1]*The Johns Hopkins University School of Medicine, Baltimore, MD, United States;* [2]*Department of Interventional Radiology, The Johns Hopkins Hospital, Baltimore, MD, United States;* [3]*United States Air Force Medical Corps, Falls Church, VA, United States*

Key points

- Public health is concerned with the prevention of disease and disability
- Public health research draws on biostatistics, epidemiology, surveillance, social sciences, and biology to analyze population-level data
- Improvements to community health are made through the development and implementation of interventions that address a problem in a population at risk

What is public health?

Public health is often defined as the science and the art of preventing disease, prolonging life, and promoting physical health and efficiency. More specifically, public health aims to understand the influence of factors that determine population health and propose interventions and health policies based on scientific knowledge and evidence. The field is classically divided into three domains:[1]

(1) Health improvement (psychosocial aspects of health; impacts of social determinants of health)
(2) Health protection (communicable disease control; environmental, chemical, radiation, and nuclear threats; occupational health),
(3) Healthcare service delivery (healthcare systems, service quality, evidence-based practice, clinical effectiveness, and health economics).

Research and practice within these three domains draw upon the underlying core skills of epidemiology, biostatistics, law, and ethical practice.

Public health research

Public health researchers study biological, behavioral, social, or environmental determinants of health, which allows them to assess interventions to improve population health and prevent disease. Like translational research, public health research is a multidisciplinary process that draws on biostatistics, epidemiology, surveillance, social sciences, and biology to analyze data on population samples. Population-level data is collected via the following:

- Cohort studies: monitoring a group(s) of individuals with a common characteristic(s) or environment over time to determine the onset of a disease or health outcome.
- Case-control studies: recruiting participants with a characteristic and participant(s) without the characteristic (control group) to assess the effects of interventions or risk/protective factors.
- Cross-sectional studies: analyze various characteristics of a population at the time of recruitment (akin to taking a snapshot of a group at a certain time).
- Morbidity registries: collected by the CDC, outlining the prevalence of various diseases across the population.
- Medical administrative databases: created for nonresearch purposes to collect data on patient care, but very useful to evaluate the usage and efficacy of various interventions, medications, and procedures.

The impact of public health research and initiatives cannot be overstated. By using statistical techniques to make comparisons between large groups of individuals, important facts about the nature and causes of disease that might not come to light in any other way could be uncovered. It represents one of the only ways to study the subtle effects of environmental agents such as ionizing radiation, tobacco smoke, petroleum fumes, and healthcare inequity on human health (Fig. 94.1).

Translational Interventional Radiology. https://doi.org/10.1016/B978-0-12-823026-8.00041-9

Converting translational research into public health outcomes

The term "translational research" is often associated with the phrase "bench to bedside," but the expedited movement of biomedical advances from laboratory to clinical trials is only the first phase of the translational process. The second phase of translation takes innovations from the bedside and applies them to real-world practice. Though this is equally important, it receives far less attention. Recently, translational research has adopted a broader definition that recognizes the need to integrate public health. The field of translational research now encompasses laboratory studies, clinical demands, public health and health management, policies, and economics. Research is especially important in the setting of public policy decision-making as it raises the quality of the debate about health issues to include scholarly evidence as well as anecdotes and biases. By promoting evidence-based policymaking, we can use novel clinical advances to improve the lives of the entire population. For example, the addition of Angiotensin-converting enzyme inhibitors for treating heart failure in pay-for-performance physician payment reform models and the implementation of state mandates for human papillomavirus vaccination are two meaningful translations of basic and clinical science findings with health services and public policy.[2] At a global level, some stunning success stories can be attributed to this approach. For example, recognition of chronic Hepatitis B infection from mother to child via vertical viral transmission helped identify a leading cause of hepatocellular carcinoma in Southeast and East Asia. Large-scale Hepatitis B vaccination strategies were then implemented, leading to a dramatic drop in the incidence of this lethal malignancy.[3]

Applications of interventional radiology to public health

Epidemiological and public health research has been instrumental in helping interventional radiology (IR) become an independent field and optimizing various endovascular and minimally invasive procedures. A few examples are presented in the following:

- *Cost-effectiveness of interventional radiology*: Epidemiological studies determined that IR has been shown to be generally more cost-effective than alternative surgical treatments. As a result, there has been a tremendous increase in the number of IR procedures performed worldwide and the market cap for interventional radiology has increased significantly.[3]
- *Occupational safety for radiologists*: Before the development of modern radiation safety practices, radiation-induced skin cancer, breast cancer, and certain types of leukemias were increased among radiologists.[4] The development of radiation safety practices beginning in the 1950s has since markedly reduced the occupational exposure of radiologists to ionizing radiation, with a concomitant decrease in radiation-related sequelae.[4] Interventional radiologists have higher rates of exposure to ionizing radiation than noninterventional radiologists, with concomitant increased risks of lower extremity epilation and cataracts, but these risks have also been markedly reduced in modern practice.[4,5]

Public health research and interventions

Key principles of epidemiology

Epidemiology is defined as "the study of the distribution and determinants of disease frequency or health outcomes in human populations." Two kinds of frequency measures are commonly used in epidemiology: incidence rates and prevalence rates. **Incidence** refers to the rate of new cases in a defined population over a period. **Prevalence** is the total number of cases in a population at a particular time, often measured using a survey. **Death rates**, or mortality rates (the incidence of death), are the third measure of frequency, used for diseases that are usually fatal.[6]

Limitations of epidemiological studies

While epidemiologic studies have the advantage of studying populations of humans rather than experimental animals, certain limitations are associated with this as well. Each type of epidemiologic study has distinct strengths and weaknesses.[6]

FIGURE 94.1 The four stages of translational research, which ultimately realize their potential through public health outcomes. Bidirectional arrows emphasize that new knowledge and hypotheses are generated at each step.

Well-designed cohort studies partially eliminate issues relating to differences in exposure conditions. For example, individual human behaviors may vary greatly, but between large, well-selected exposed and nonexposed groups, these differences may be essentially randomly distributed. Nonetheless, **confounding variables** may exist and be nonrandomly distributed, affecting measures of differences between the groups of interest. Cohort studies are generally conducted prospectively, however, if adequate records are available, they may be conducted retrospectively.

Case-control studies generally involve identifying individuals with an outcome of interest (cases) and near-identical individuals without the outcome of interest (controls) and then retrospectively identifying common exposures. Case-control studies may be subject to significant bias from many fronts, most notably regarding accurate identification of well-matched case and control groups and in accurately identifying the exposure(s) of interest.

Most cross-sectional studies are designed to characterize the prevalence of given health outcomes in specific populations at a specific point in time. Cross-sectional studies are relatively easy and cost-effective to conduct compared to cohort and case-control studies and may be an effective way to determine the prevalence of specific health outcomes; however, cross-sectional studies are also often prone to significant biases and have significant inherent limitations. Cross-sectional studies are unable to make causal inferences or measure the incidence of disease as they lack a time-dependent variable. Moreover, accurately collecting data on a large enough sample with a health condition of interest to make accurate population-level inferences is often not feasible, limiting the generalizability of cross-sectional studies.

These difficulties do not mean that no valid conclusion can be drawn from epidemiologic studies, rather, they demonstrate the types of errors to which different kinds of studies may be prone and alert researchers about what to watch out for in choosing a study design and in interpreting the results.

Get started: designing public health interventions

Just as the problems facing researchers are multifaceted, our interventions can take on many forms to address them. It can be a program, a policy initiative in a hospital or a state, or a certain practice that becomes favorable. Public health interventions may be considered population-based if they consider all levels of practice:[7]

(1) Systems-focused: targeting the systems that affect health through changing organizations, policies, laws, and power structures.
(2) Community-focused: changing community norms, attitudes, awareness, practices, and behaviors
(3) Individual or family-focused: targeting individuals belonging to a population at risk by changing their knowledge, attitudes, beliefs, practices, and behaviors.

Interventions should be dynamic, iterative, creative, open to change, and forward-looking to be effective and adaptable to change during implementation. The major steps in developing a public health intervention are outlined in the following:[8]

(1) **Planning the development process:** identify the problem to target and understand the problem to determine its importance and determine what aspects of it can be changed by an intervention. A protocol and timeline of the project including a list of resources, funding, and collaborators would be generated at this stage as well.
(2) **Involve collaborators and stakeholders:** at this stage, it is important to involve those who will deliver and benefit from the intervention (i.e., patients, public, target population, service providers, and policymakers).
(3) **Create teams and establish decision-making processes:** bring together a team with relevant expertise who can determine how to engage stakeholders and make final decisions regarding the implementation of the intervention.
(4) **Review published research evidence and draw on existing theories/frameworks:** look for evidence of existing successful interventions or failed interventions and create an evidence base for each component of the theoretical intervention. These theories are usually based on psychological, sociological, organizational, or implementation theories and can be used to inform their intervention and determine what is important or feasible.
(5) **Articulate program theory:** a program theory describes how a specific intervention is expected to lead to its effects and under what conditions. It shows the causal pathway between the intervention, immediate outcomes, and long-term goals. This is especially important when trying to communicate with stakeholders and funding agencies.
(6) **Primary data collection:** surveys or qualitative interviews are often used to assess the acceptability, values, and priorities of those delivering and receiving the intervention. Pretest and posttest measures of the intermediate outcomes can be useful during the early stages of the implementation process to ensure that outcomes are aligned with long-term goals.
(7) **Understand context**: a successful population health Intervention requires a thorough understanding of the individual and population-level characteristics (i.e., geography, socioeconomic status, culture, politics, organization funding policies).

(8) **Develop a plan for future implementation**: many interventions that are designed do not actually lead to changes in policy or practice because they are not sustainable, so it is important to examine how the intervention can be "scaled up" to the entire target population while remaining cost-effective and aligned with goals.
(9) **Design and refine the intervention**: this refers to the creative process where ideas are generated about the content and delivery of the proposed intervention. This is then followed by the creation of a prototype or "proof of concept" that can be evaluated by stakeholders. After successive iterations of refinement, the intervention becomes optimized and ready for implementation by policymakers and service providers.

Though these are displayed as a sequence, they should be performed dynamically and in parallel for the greatest efficiency and success, as many of these steps overlap with each other.

Resources

1. Minnesota Department of Health (2001). Public health interventions: Applications for public health nursing practice.
2. Fleming ES, Perkins J, Easa D et al. The Role of Translational Research in Addressing Health Disparities: A Conceptual Framework. *Ethn Dis*. 2008; 18(2 Suppl 2):S2-155-160.

References

1. Ogilvie D, Craig P, Griffin S, Macintyre S, Wareham NJ. A translational framework for public health research. *BMC Publ Health*. 2009;9(1):1–10.
2. Mirvis DM. From research to public policy: an essential extension of the translation research agenda. *Clin Transl Sci*. 2009;2(5):379–381.
3. Messina LM, Schneider DB, Chuter TA, et al. Integrated fellowship in vascular surgery and intervention radiology: a new paradigm in vascular training. *Ann Surg*. 2002;236(4):408.
4. Parikh JR, Geise RA, Bluth EI, et al. Potential radiation-related effects on radiologists. *Am J Roentgenol*. 2017;208(3):595–602.
5. Wiper A, Katira R, Roberts D. Interventional cardiology: it'sa hairy business. *Heart*. 2005;91(11):1432.
6. Schneider M-J. *Introduction to Public Health*. Jones & Bartlett Learning; 2020.
7. Rosenstock IM, Hochbaum GM. Some principles of research design in public health. *1961. Am J Public Health*. 2010;100(10):1861–1863.
8. O'Cathain A, Croot L, Duncan E, et al. Guidance on how to develop complex interventions to improve health and healthcare. *BMJ Open*. 2019;9(8):e029954.

Chapter 95

Epidemiology

Alexander Martinek[1], Neal Patel[2] and Christopher Awad[3]
[1]Morehouse School of Medicine, Atlanta, GA, United States; [2]Nova Southeastern University Dr. Kiran C. Patel College of Osteopathic Medicine, Ft. Lauderdale, FL, United States; [3]Emory University School of Medicine, Atlanta, GA, United States

Key points

- An introduction to epidemiological research within the field of interventional radiology.
- Summary of the different types of epidemiological studies and questions available.
- Two real-world examples of epidemiological research from published literature within interventional radiology.

Objectives

1. Analyze published examples of epidemiological research in interventional radiology and use said publications as resources to guide new epidemiological research studies.
2. Clearly define key terms such as epidemiology, cohort study, case-control, selection bias, incidence, morbidity, and mortality among others.
3. Understand the similarities and differences within the unique types of epidemiological studies.
4. To gain a comprehensive understanding of how epidemiological research is conducted within the field of interventional radiology.
5. Analyze trends within epidemiological studies including incidence, morbidity, and mortality rates.

Introduction

Epidemiology is not just the study of disease but also encompasses markers of health, such as disability, injury, mortality, hospitalization, or quality of life measures.[1] The study of epidemiology has been perceived as not translational, meaning it lacks direct improvement of patient care and public health. When considering the role of epidemiology in translational research, Khourey et al. expand upon a framework called translational epidemiology. This framework describes how observational, experimental, and theoretical functions of epidemiology are required to connect laboratory and clinical aspects of research with population research.[2] In the field of interventional radiology, epidemiologic studies can be used to develop data, which identify limitations of knowledge, that allow further research and interventions to be developed. Because epidemiology is in the spectrum of biomedical and clinical research, physicians that utilize it can gain a better understanding of the distribution of a particular condition or disease of interest, as well as the other various statistics that surround a disease. This plays a vital role in shaping patient outcomes. Knowing these factors can allow for an increase in the detection of a disease or condition, a reduction in risk factors among patients, and play a defining factor in determining which treatment modalities allow for the best outcomes in certain populations. The traditional approach to epidemiology and the translational approach differ in assessing the consequences of disease and the efficacy of possible interventions. The findings from a study more suited toward the translational approach to epidemiology can be utilized to aid clinician decision-making, and it will often employ randomized, controlled, and cross-over study designs.[3]

Epidemiology is often introduced as a scientific discipline focused on the study of pathological conditions, but it is increasingly embodying the multidisciplinary aspect of clinical and translational research. Using epidemiological principles, effective interventions can be discovered, identified, and implemented to benefit individuals and public health. This chapter will introduce epidemiology as it is relevant to research in interventional radiology. It will then describe epidemiological questions, methods, concepts, and components that are effective in conducting this type of study. A discussion that includes modern published

Translational Interventional Radiology. https://doi.org/10.1016/B978-0-12-823026-8.00096-1

examples of epidemiology use in interventional radiology is included. By the end of the chapter, action items and additional resources to get started will be explained. This chapter is intended to demonstrate basic epidemiological principles and context that allow previous chapters to be understood. Many elements of epidemiology research have been described in detail in other chapters.

Chapter text

Epidemiology uses data in a systematic approach to answer questions concerned with the incidence, cause, control, and monitoring of the health status of populations.[1] Osterholm et al. 1 state that epidemiology involves the careful description of events within populations and the comparison of rates at which these events occur between groups within those populations.[4] These studies allow for many advancements to be made within the public health sphere in the fields of disease diagnosis, prevention, and treatment. Epidemiology can be broken down into two main approaches, descriptive and analytic.[1]

Descriptive epidemiology answers the who, when, and where questions that relate to a disease or condition. This data is imperative when conducting research in interventional radiology for several reasons. Data needs to be developed that can display detailed and illustrative depictions of the health of a population through variables such as weight, age, or gender. Data also allows for the identification of patterns associated with the public health problem in question, and specific groups or areas can be identified based on rates of a disease or condition, which is important for developing a testable hypothesis. Descriptive epidemiology answers questions that are useful for clinicians and public health.[3] Table 95.1 shows the five W's of descriptive epidemiology as described by the Center for Disease and Control (CDC).[1]

The data and interpretation resulting from descriptive epidemiology frequently serve as the starting point for analytic epidemiology. Analytical tools such as study designs and statistical tests are often utilized to further investigate and extrapolate descriptive data sets.[3] A summary of the study designs is provided in Table 95.2 from Rivas-Ruiz et al., which describes the different types of epidemiological designs used in research today.[5]

Understanding the differences in the types of designs used will aid in understanding prior literature published on epidemiology as well as guide future research being conducted. For example, a case-control study, a form of retrospective research, allows for analysis of prior studies that can shed light on unknown associations that may have been initially overlooked. In contrast, cohort studies follow subjects throughout time, assessing subjects at different intervals looking for changes from the previous cross-section or baseline.[1] Questions that have direct implications for decision-making are important, especially when considering the fact that individual preferences for the types of research questions and analytics approaches remain diverse (Table 95.3).[4]

Real-world examples epidemiology in interventional radiology

Example 1

A pulmonary embolism occurs when one or both of the pulmonary arteries in the lungs are blocked by a blood clot. The most common cause of pulmonary emboli is blood clots found in the deep veins of the lower extremities (deep vein thrombosis) that travel up to and lodge in the pulmonary arteries. It is the third most common cause of cardiovascular death and is associated with multiple inherited and acquired risk factors as well as advanced age.[6] The extent of pulmonary emboli can be directly correlated to the "degree of obstruction" and the pathophysiology of the clot. Furthermore, the past medical history of the patient is also imperative when diagnosing and treating patients with pulmonary emboli.

The objective of epidemiological studies is that physicians are able to gain an understanding of the cause and prognosis of the disease including demographics, incidence, prevalence, risk factors, treatment outcomes, and mortality/morbidity, which will allow the physician to accurately and carefully treat said disease. Patients can gain a better understanding of their prognosis and truly understand the severity that may increase patient compliance. The precise number of patients in the United States with pulmonary embolism is uncertain, it is suspected that about 9,00,000 patients are affected each year according to the

TABLE 95.1 The five W's of descriptive epidemiology.

What = health issue of concern
Who = person
Where = place
When = time
Why/how = causes, risk factors, modes of transmission

TABLE 95.2 Types of epidemiological designs.

Designs	Definition
I. Experimental studies	The investigator controls assignment and intervenes in the design. Best designs for generating hypotheses
I.A. Controlled clinical trials	Prospective analysis is made of the effect of an intervention in a group of patients randomly selected from a target population
I.B. Community-based intervention studies	Interventions are made in large community-based samples
I.C. Quasiexperimental studies	Group rather than individual randomized assignment; all subjects in each group receive or do not receive a given intervention
II. Observational studies	
II.A. Analytical	Alternative to experimental studies for generating hypotheses
II.A.1. Case-control studies	Studies of retrospective nature, since we start from the effect to study the antecedents of exposure in two groups of subjects called cases and controls, according to whether they have a disease or not
II.A.2 Follow-up (cohort) studies	Follow-up of cohorts over time (exposed and nonexposed) with the purpose of assessing the hypothesis of association between a given exposure and some effect
II.B. Descriptive	
II.B.1. Ecological studies	Studies using pooled information, that is, the analytical unit is the group or population, and the information on exposure is the average in each pooled study unit
II.B.2. Case reports or case series	Descriptions of clinical observations corresponding to isolated cases or groups of individuals with the same diagnosis
II.B.3. Cross-sectional (prevalence) studies	Determination of the proportion of individuals presenting a certain disease or risk factor at a given moment in time

TABLE 95.3 Goals of epidemiological research.

1.	Describe patterns of infection and disease occurrence in populations
2.	Identify outbreaks or unusual rates of disease occurrence.
3.	Facilitate laboratory-based efforts to identify infectious agents.
4.	Describe the occurrence of asymptomatic infection and the spectrum of disease associated with specific agents.
5.	Provide population-based descriptions of clinical illness to improve the specificity of diagnosis for individual diseases.
6.	Assist in the understanding of disease pathogenesis.
7.	Identify and characterize factors in the chain of infection that contribute to agent transmission and the development of disease.
8.	Develop and evaluate treatment protocols through clinical trials.
9.	Develop and evaluate primary, secondary, and tertiary prevention and control measures for individuals.
10.	Describe and assess the use of prevention measures on a community-wide basis.

US CDC.[7] The CDC also estimates that around 60,000–1,00,000 Americans succumb to the disease every year.[7] Venous thrombo-emboli affect the elderly more than they affect the youth, in particular, those above the age of 70 who are particularly susceptible. A multitude of risk factors contributes to the increase in pulmonary emboli including but not limited to hypercoagulability, stasis of blood, damage to endothelial cells, obesity, cigarette smoking, and finally, hypertension.

There are two types of pulmonary emboli that need to be addressed by interventional radiologists: the submassive and the massive subtypes. There are several associated symptoms present when a patient is dealing with a pulmonary embolism, such as, but not limited to, dyspnea, chest pain, cough, and syncope. There is a multitude of ways to treat different subtypes of pulmonary emboli ranging from anticoagulant therapy for low-risk individuals to systemic fibrinolytic therapy for more serious high-risk patients. The simplest and most widespread interventional radiology technique is the use of pigtail catheters to fragment the thrombus by continuous rotation of the catheter.[8] The proximal fragmentation of the thrombus leads to distal embolization of smaller thrombi; however, some authors have reported pulmonary hypertension with the use of this technique.[8] This is of epidemiological relevance to those interventional radiologists that may choose to use the pigtail method in patients with a prior history of pulmonary hypertension. While we do not want to exacerbate the symptoms, a different approach may be taken to fragment the clot. Many new devices are being developed that have shown some promising results and may be widely available in the future, such as mechanical aspiration systems as well as ultrasonic devices to lyse the clots.

Example 2

Renal cell carcinoma (RCC) is the most common type of kidney cancer seen in adults. RCC is defined as a type of kidney cancer that begins within the tubules of the kidneys. RCC is the 14th most prevalent cancer in all adults and is associated with high morbidity and mortality rates.[9] It is now the ninth most common neoplasm in the United States.[10] The prognosis of RCC varies greatly on the stage of the disease. Metastatic disease only has a 12% survival rate.[10] The earlier the renal cell carcinoma is found, and treated, the better are the survival outcomes. Many modifiable risk factors including smoking, obesity, hypertension, and recreational use of drugs have all been linked to an increase in renal cell carcinoma incidence.

The epidemiology of renal cell carcinoma is of particular importance as it allows physicians to remain informed about the current statistics about the disease. Knowing the risk factors that are modifiable is also of particular importance in that patients can be informed before acquiring the disease. The incidence in the United States between 2006 and 2010 is reported to be 15.3 per 100,000 people per year.[9] The total number of deaths from RCC was approximately 72,019 in 2008.[9] This number rose to 1,75,000 10 years later in 2018 as reported by GLOBOCAN. While the rate of mortality has remained near identical year after year in the United States, the increase in those patients with modifiable risk factors is of growing concern. A global age-standardized mortality rate of 2.2 per 100,000 people per year was noted.[11] Survival rates have been around 76% in the United States from the years 2009 to 15.[10] These statistics are beneficial to interventional radiologists and allow for better decision-making with regard to treatment options as well as informing the patient about the prognosis.

The traditional approach to treat renal cell carcinoma is performing a nephrectomy; this approach has been replaced by both laparoscopic as well as IR procedures. These new treatments have become mainline therapies due to the improvement in outcomes, preserved renal function, decreased complications, and faster recovery rates. Especially for patients who have severe comorbidities and advanced age, radiofrequency ablation (RFA) is appropriate for treating small renal masses,[12] as it has the most long-term follow-up data with regards to its safety, efficacy, and survival outcomes and is cost-effective. Other alternative therapies such as microwave ablation and cryoablation are just as effective. Percutaneous and laparoscopic RFA are both effective therapies against RCC however both have their individual advantages and disadvantages. The laparoscopic approach allows direct visualization of exophytic and parenchymal renal tumors, while the percutaneous approach is less invasive and tailored toward outpatient care.[12] Current epidemiological studies can help to decide which treatment method is best for each specific patient when evaluating incidence, risk factors, efficacy, comorbidities, and morbidity/mortality.

Potential pitfalls within epidemiological research

1. The first major pitfall regarding epidemiological research in interventional radiology would be to not include individual patient experiences in the day-to-day practice of interventional radiology. Synthesizing macroscopic epidemiologic trends with microscopic individual cases will allow for the holistic integration and advancement of epidemiological research.
2. Another pitfall frequently faced is ensuring epidemiological research is included when designing future clinical trials for new techniques or procedures in interventional radiology. Accurate analysis of incidence, risk factors, and comorbidities are all variables that need to be assessed when designing new studies.
3. A third potential pitfall occurs when all epidemiological research is not considered when evaluating a specific disease or interventional radiology procedure. Global as well as state and local epidemiological trends need to be considered when evaluating different disease prognosis.

4. One final pitfall commonly encountered includes epidemiological research's reliance on observational studies. There is a need to reduce different biases associated with observational studies along with all epidemiological research as a whole. Eliminating biases will allow for a more comprehensive and cohesive analysis of data in research.

Data sources

When conducting epidemiological research, data may be obtained from a plethora of sources such as direct data collection, previously published literature, and online registered databases to name a few. Epidemiological data is unique in that it must be analyzed through the lens of time instead of just objectively. According to the CDC Principles of Epidemiology in public health practice, changes in data trends over time such as an increase in disease incidence or mortality are crucial factors to recognize when conducting research.[13] Data that can be organized into a table, graph, or chart is crucial in being able to understand and recognize changes in trends over time. Maps are also particularly useful when trying to identify areas of the country or world where a particular disease has a higher incidence. Data can also be analyzed by numerous other factors such as age, race, gender, and severity. Knowledge of the most recent studies can help physicians to better suspect certain diseases and accurately diagnose and report cases to the proper entity ensuring that new statistics can be added to existing data sets. For example, it was reported by Rawla et al. that the number of incident cases of liver cancer in both sexes and all age groups is estimated to increase from 8,41,080 in 2018 to 1,361,836 in 2040 (an overall change of +61.9%). The estimated number of deaths from liver cancer in both sexes and all age groups is estimated to increase from 7,81,631 in 2018 to 12,84,252 in 2040 (an overall change of +64.3%).[14] This data clearly highlights the drastic hypothesized increase in hepatocellular carcinoma over the next 20 years. After analyzing this example, physicians should be able to recognize the increase in hepatocellular carcinoma among the at-risk population and emphasize to those patients to reduce their comorbidities and risk factors.

Get started

After reading this chapter, the reader should now have a better understanding of how epidemiological research is conducted in the field of interventional radiology. The reader should try to keep in mind the aforementioned pitfalls when conducting his or her own research studies as well as when searching through published literature. For more information on Epidemiology in the public health practice, you can refer to the CDC's textbook on Principles of Epidemiology at https://www.cdc.gov/csels/dsepd/ss1978/SS1978.pdf. The reader can also visit any of the additional references listed at the end of this chapter, listed under further reading, that are invaluable resources to gain a better understanding of epidemiological research.

References

1. Dicker R, Parrish RG, Koo D, Coronado F. *Principles of Epidemiology in Public Health Practice: An Introduction to Applied Epidemiology and Biostatistics*. Atlanta, GA: U.S. Department of Health and Human Services, Centers for Disease Control and Prevention (CDC), Office of Workforce and Career Development; 2012.
2. Khoury MJ, Gwinn M, Ioannidis JP. The emergence of translational epidemiology: from scientific discovery to population health impact. *Am J Epidemiol*. 2010;172(5):517–524. https://doi.org/10.1093/aje/kwq211.
3. Robertson D, Williams GH. *Clinical and Translational Science: Principles of Human Research*. Amsterdam: Elsevier/Academic Press; 2017.
4. Osterholm MT, Hedberg CW. *Epidemiologic Principles. Mandell, Douglas, and Bennett's Principles and Practice of Infectious Diseases*. 2015:146–157. https://doi.org/10.1016/B978-1-4557-4801-3.00013-8. e2.
5. Rivas-Ruiz F, Expósito-Ruiz M, Domínguez-Almendros S. In: Pérez-Fernández V, ed. *Research Designs in Clinical Epidemiology*. Elsevier; 27 Jan. 2012. www.elsevier.es/es-revista-allergologia-et-immunopathologia-105.
6. Turetz M, Sideris AT, Friedman OA, Triphathi N, Horowitz JM. Epidemiology, pathophysiology, and natural history of pulmonary embolism. *Semin Intervent Radiol*. 2018;35(2):92–98. https://doi.org/10.1055/s-0038-1642036.
7. Beckman MG, Hooper WC, Critchley SE, Ortel TL. Venous thromboembolism: a public health concern. *Am J Prev Med*. April 2010;38(4 Suppl):S495–S501.
8. De Gregorio MA, Guirola JA, Lahuerta C, Serrano C, Figueredo AL, Kuo WT. Interventional radiology treatment for pulmonary embolism. *World J Radiol*. 2017;9(7):295–303. https://doi.org/10.4329/wjr.v9.i7.295.
9. Ridge CA, Pua BB, Madoff DC. Epidemiology and staging of renal cell carcinoma. *Semin Intervent Radiol*. 2014;31(1):3–8. https://doi.org/10.1055/s-0033-1363837.
10. Padala SA, Barsouk A, Thandra KC, et al. Epidemiology of renal cell carcinoma. *World J Oncol*. 2020;11(3):79–87. https://doi.org/10.14740/wjon1279.
11. Ferlay J, Shin HR, Bray F, Forman D, Mathers C, Parkin DM. Estimates of worldwide burden of cancer in 2008: GLOBOCAN 2008. *Int J Cancer*. 2010;127(12):2893–2917.
12. Nguyen D, vanSonnenberg E, Kang P, Mueller PR. Urologic and interventional radiology treatment of renal cell carcinomas-similarities and differences. *Ann Transl Med*. 2019;7(Suppl 3):S113. https://doi.org/10.21037/atm.2019.05.24.
13. *"Lesson 5: Public Health Surveillance Section 5: Analyzing and Interpreting Data." Principles of Epidemiology in Public Health Practice, Third Edition an Introduction to Applied Epidemiology and Biostatistics*. U.S. Department of Health and Human Services; 2012.

14. Rawla P, Sunkara T, Muralidharan P, Raj JP. Update in global trends and aetiology of hepatocellular carcinoma. *Contemp Oncol.* 2018;22(3):141–150. https://doi.org/10.5114/wo.2018.78941.

Further reading

1. Jayachandran A, Shrestha R, Bridle KR, Crawford DHG. Association between hereditary hemochromatosis and hepatocellular carcinoma: a comprehensive review. *Hepatoma Res.* 2020;6(8). https://doi.org/10.20517/2394-5079.2019.35.
2. Bosetti C, Turati F, La Vecchia C. Hepatocellular carcinoma epidemiology. *Best Pract Res Clin Gastroenterol.* 2014;28(5):753–770. https://doi.org/10.1016/j.bpg.2014.08.007. ISSN 1521-6918.
3. Molla N, AlMenieir N, Simoneau E, et al. The role of interventional radiology in the management of hepatocellular carcinoma. *Curr Oncol.* 2014;21(3):e480–e492. https://doi.org/10.3747/co.21.1829.

Chapter 96

Good questions

Brian Covello and Roberto Fourzali
Department of Interventional Radiology, HCA Florida Aventura Hospital, Aventura, FL, United States

The birth of interventional radiology (IR) took place in the hands of cowboys. Seldinger and Dotter were innovators as much as they were physicians, and oftentimes the early paradigm was one of possibility before acceptability. As interventional radiology blossomed into a distinct medical specialty, interventional radiologists have been called upon to step into the age of evidence-based medicine. Along with this call, comes the need for large randomized controlled trials. Data has become the tool by which IR proves itself in the obstacle course of medical specialties, insurance companies, and indeed patients themselves. In this environment, public health research has become as important to the field of interventional radiology as other basic science and translational work. While many trainees understandably concentrate on gaining the scientific and statistical skills needed to perform research, they must also focus on the arguably more difficult skill of learning how to ask the right questions. Research questions are the fundamental core of any scientific study. Providing a focus, it helps to determine the methodology, analysis, and reporting style. This chapter explores what is needed to ask the right public health research questions by examining three paramount examples within recent literature.

The formation of public health research questions is similar to the formulation of any focused clinical question. The PICO model helps provide a general framework. The acronym PICO stands for patient/population, intervention/indicator, compare/control, and outcome. In public health questions, a population is typically studied. One should determine the relevant age, sex, and specific demographic information of the studied population. For intervention/indicator, the management strategy, diagnostic test, or group exposure is considered. The compare/control portion includes a comparison to an alternative management strategy or no intervention. The outcome includes the relevant public health consequences of a given intervention. This may include aspects such as quality of life or progression-free survival (Table 96.1). In addition to PICO, impactful public health research questions are those that are meaningful, measurable, and feasible (Table 96.2).

Clinical patient experience and a firm grounding in the field oftentimes provide the wisdom needed to decipher what constitutes a meaningful question. To those early in their careers, mentors are a resource to gain this foothold. The concept of a meaningful question extends to the population, the intervention, the control, and the outcome of the experiment. For the population, the researcher must ask which population is most likely to benefit from the research. A broader public health need should be determined. By choosing the patient population that stands to benefit, larger and more impactful comparisons to the wider demographic population may be made. A good research question should involve a meaningful intervention. Is there a need for a new type, a different type, or an alternative therapy than the one that currently exists? Choosing a meaningful comparison or control group is also beneficial to the impact that the research will have and serves to provide a statistical basis for the study. Most importantly, the outcome should have a meaningful impact on the population being studied.

The topic of uterine-artery embolization provides an example of meaningful public health questions. In the 1990s, uterine-artery embolization began to provide an alternative to surgical myomectomy or hysterectomy for women with uterine fibroids. Some of the earliest research focused not only on the scientific outcomes of uterine fibroid volume, size, and sonographic appearance but also the symptomatic control that women were able to achieve as assessed through questionnaires.[1] One of the first published reviews on this topic recognized the tendency of radiologists to quickly adopt new interventional procedures and stressed the importance of establishing randomized controlled trials.[2] The field of interventional radiology responded en masse, with further research focusing on safety, efficacy, and importantly, quality of life. From a public health perspective, monitoring women's quality of

Translational Interventional Radiology. https://doi.org/10.1016/B978-0-12-823026-8.00081-X

TABLE 96.1 The PICO acronym and its meaning are described. Examples are taken from recent public health research related to interventional radiology.

Acronym	Meaning	Public health examples
P (patient/ population)	Population to be studied	**1)** Patients with uterine fibroids **2)** Patients with colon cancer and liver metastasis **3)** atients with benign prostatic hypertrophy
I (intervention)	Interventional radiology procedure to be studied	**1)** Uterine artery embolization **2)** Yttrium-90 embolization **3)** Prostate artery embolization
C (compare/ control)	Baseline group that does not receive the intervention	**1)** Patients with uterine fibroids who undergo myomectomy **2)** Patients with colon cancer liver metastasis who undergo traditional treatment **3)** Patients with benign prostatic hypertrophy who undergo transurethral resection of the prostate
O (outcome)	The outcome to be tested	**1)** UFS-QOL improvement **2)** Progression-free survival **3)** IPSS improvement

life was paramount to determining the effectiveness of uterine artery embolization versus other more invasive interventions. The Society of Interventional Radiology quickly adopted the uterine fibroid symptom health-related quality of life questionnaire (UFS-QOL) to measure both objective pretreatment and posttreatment fibroid-related symptoms. An excerpt of questions from the UFS-QOL is provided in Table 96.3. To date, uterine embolization trials such as the Fibroids of the Uterus: Myomectomy versus Embolization (FUME) trial and the Uterine-Artery Embolization or Myomectomy for Uterine Fibroids (FEMME) trial have used quality-of-life measurements as the backbone of their clinical outcomes.[3,4] The landmark FEMME trial was a multicenter randomized controlled trial that investigated health outcomes between women wishing to avoid hysterectomy who were treated with either embolization or myomectomy. The primary outcome for this trial was the score on the related quality-of-life domain of the UFS-QOL questionnaire. The FEMME investigators recognized that the questionnaire had sufficient discriminatory validity to assess clinical outcomes.[4] UFS-QOL scores at 6 months and 1 year were included as a secondary

TABLE 96.2 An overview is provided for how the concepts of meaningful, measurable, and feasible relate to the PICO model.

	Meaningful	Measurable	Feasible
Patient/ Population	What population is most likely to benefit from the research question?	How will the population be measured?	Are there enough patients to power the study? Is there access to the population that stands to benefit?
Intervention	Is there a potential for the intervention to have a significant impact that affects general public health?	What interventional effect is being studied? How many types of interventions are being studied? How many interventions of any type will be performed?	Is the intervention feasible and safe to perform? Are there enough interventionalists available to perform the experiment? What type of equipment is needed? What are the required ancillary services?
Compare/ Control	Has a relatable and understandable comparison group been formed?	What type of measurement is being performed between the experimental and control groups? How are the baseline population characteristics being studied?	Is there an adequate control group to properly demonstrate an effect?
Outcome	Is the specified outcome meaningful to the general population?	How will the outcome be measured?	How well does the experiment prove or disprove an outcome?

TABLE 96.3 An excerpt of questions taken from the UFS-QOL is provided.

During the previous 3 months how distressed were you by	Not at all (1)	A little bit (2)	Somewhat (3)	A great deal (4)	A very great deal (5)
1. Heavy bleeding during menstrual period	☐	☐	☐	☐	☐
2. Passing blood clots during your menstrual period	☐	☐	☐	☐	☐
3. Feeling fatigued	☐	☐	☐	☐	☐
During the previous 3 months, how often have your symptoms related to uterine fibroids ...					
1. Made you feel anxious	☐	☐	☐	☐	☐
2. Made you feel as if you are not in control of your life	☐	☐	☐	☐	☐
3. Made you less productive	☐	☐	☐	☐	☐
During the previous 3 months, how often have your symptoms related to fibroids ...					
1. Caused you embarrassment	☐	☐	☐	☐	☐
2. Diminished your sexual desire	☐	☐	☐	☐	☐
3. Made you feel irritable	☐	☐	☐	☐	☐

outcome. The FEMME trial concluded superiority of myomectomy over uterine-artery embolization for health-related quality of life while also noting a similar incidence of perioperative and postoperative complication between the two groups.[4] Early IR researchers knew that meaningful public health questions addressed those topics most pertinent to the patient as a whole.

A good research question must be measurable. Measurability has many forms, which are dependent on the type of research conducted. The PICO acronym helps provide the general structure in which to form a research question. The measurability of the population should be considered. This may include an appropriate grouping of the population, as well as measurements that will be taken to determine differences between the population studied. This concept is also applied to the control group. A list of measurement baseline characteristics for the control group should be created. The intervention being studied, the number of interventions to take place, and interventional follow-up should also be considered. Lastly, the characteristics of how to measure the outcome should be determined. This may include the type of outcome studied, how the outcome will be determined, and the statistical analysis required to prove whether the outcome was achieved.

For interventional radiologists, the type of outcome being measured is critical. As consultants, interventionalists must produce research that is relatable to referring physicians. In interventional oncology, the measurement of progression-free survival is a heavily relied upon measurement. This measurement has broad implications for oncologists. Progression-free survival in malignancies is defined as the length of time after the treatment of cancer in which a patient lives with the malignancy without it becoming worse. In the early 1950s, researchers identified that approximately 80% of hepatic tumor blood supply was from the hepatic artery, providing a separation of blood supply from the normal hepatic parenchyma. By 1965, the first reports of using yttrium-90 radioactive isotope beads within the hepatic artery supply for hepatic tumors were established.[5] By the late 1970s, studies reported a decrease in tumor size when hepatic liver metastasis was treated with yttrium-90.[6] The late 1980 and 1990s brought phase I clinical trials. Finally, a phase 3 clinical trial measuring chemotherapy versus chemotherapy plus resin microspheres for liver metastasis demonstrated progression-free survival in the combination group. This ultimately led to resin microspheres gaining FDA premarket approval in the treatment of colorectal cancer liver metastasis. Yttrium-90 studies provide an example of the various types of measurability. In the early studies, more analytic outcomes such as tumor volume were recorded. As research progressed, broader public health questions such as progression-free survival provided the ultimate pathway for treatment approval. Understanding the distinction between types of measurements is critical for formulating the best research question possible.

A good research question can be reasonably answered by an experiment. PICO provides the framework by which to assess feasibility. First, the feasibility of the population to be studied is considered. The number of patients needed to appropriately power the study should be calculated. The power of a study is the probability that the study will detect a difference in measurement between the experimental and control group. The power depends on a preset alpha and the sample size. The sample size is therefore directly related to project feasibility. Similarly, there must be an adequate control group to properly demonstrate an effect. The feasibility of the intervention itself is also determined by analyzing the safety of the intervention, the ability of interventional radiologists to perform the procedure, and adequate interventional equipment. The final question to consider is how well the experiment can demonstrate a significantly different outcome if one exists.

Many factors can impact the feasibility of an intervention within a public health research experiment. This topic is embedded with such complexity that entire feasibility studies are conducted. Feasibility studies are studies that determine whether a given intervention is suitable for further investigation.[7] The history of prostate artery embolization (PAE) provides an example of the importance of feasibility beyond feasibility studies and into the public health domain. In 1980, researchers first investigated bilateral internal iliac artery embolization for prostatic hemorrhage in dogs.[8] By 2008, research groups had shown the technical feasibility and safety of PAE in animal models for benign prostatic hypertrophy.[9] The transition to human studies began around this time, with the first successful human PAE for benign prostatic hypertrophy (BPH) published in 2010, with follow-up research showing symptomatic improvement up to 30 months.[10] As further research was ignited, groups began publishing case series with increasing population size. As PAE research expanded beyond technical feasibility, other more massive questions awaited investigators. Researchers sought to increase the power of their studies by including more patients. Interventionalists shared techniques for a feasible approach and methodology. Experiments were designed around feasible outcomes, such as quality of life, prostatic volume, and patients' International Prostate Symptom Score (IPSS), a questionnaire scoring the severity of BPH symptoms. In 2014, researchers from China conducted the first randomized controlled trial of PAE versus transurethral resection of the prostate.[11] Today, PAE research is developing toward multicenter, prospective randomized controlled trials with long-term follow-up. At each stage, feasibility, as it relates to population, intervention, control, and outcome, has helped spur the rapid growth for this treatment option.

In conclusion, the future expansion of interventional radiology depends greatly on the quality of research produced. Public health research is especially important as interventionalists increase their clinical knowledge and expertise. Good research questions are the core of any impactful scientific study. While the PICO model has classically provided a framework by which to formulate a research question, questions must also be meaningful, measurable, and feasible. This chapter has reviewed each of these concepts with recent examples from the history of interventional radiology.

References

1. Goodwin SC, Vedantham S, McLucas B, Forno AE, Perrella R. Preliminary experience with uterine artery embolization for uterine fibroids. *J Vasc Intervent Radiol.* 1997;8(4):517–526. https://doi.org/10.1016/s1051-0443(97)70603-1.
2. Reidy JF, Bradley EA. Uterine artery embolization for fibroid disease. *Cardiovasc Intervent Radiol.* 1998;21(5):357–360. https://doi.org/10.1007/s002709900280.
3. Manyonda IT, Bratby M, Horst JS, Banu N, Gorti M, Belli AM. Uterine artery embolization versus myomectomy: impact on quality of life–results of the FUME (Fibroids of the Uterus: myomectomy versus Embolization) Trial. *Cardiovasc Intervent Radiol.* 2012;35(3):530–536. https://doi.org/10.1007/s00270-011-0228-5.
4. Manyonda I, Belli AM, Lumsden MA, et al. Uterine-artery embolization or myomectomy for uterine fibroids. *N Engl J Med.* 2020;383(5):440–451. https://doi.org/10.1056/NEJMoa1914735.
5. Ariel IM. Treatment of inoperable primary pancreatic and liver cancer by the intra-arterial administration of radioactive isotopes (Y90 radiating microspheres). *Ann Surg.* 1965;162:267–278. https://doi.org/10.1097/00000658-196508000-00018.
6. Grady ED. Internal radiation therapy of hepatic cancer. *Dis Colon Rectum.* 1979;22(6):371–375. https://doi.org/10.1007/BF02586901.
7. Bowen DJ, Kreuter M, Spring B, et al. How we design feasibility studies. *Am J Prev Med.* 2009;36(5):452–457. https://doi.org/10.1016/j.amepre.2009.02.002.

8. Darewicz J, Cylwik B, Musierowicz A, Boczon S. The effect of internal iliac artery embolization of the prostatic gland in the dog. *Int Urol Nephrol*. 1980;12(1):37–41. https://doi.org/10.1007/BF02085379.
9. Sun F, Sanchez FM, Crisostomo V, et al. Benign prostatic hyperplasia: transcatheter arterial embolization as potential treatment–preliminary study in pigs. *Radiology*. 2008;246(3):783–789. https://doi.org/10.1148/radiol.2463070647.
10. Carnevale FC, Antunes AA, da Motta Leal Filho JM, et al. Prostatic artery embolization as a primary treatment for benign prostatic hyperplasia: preliminary results in two patients. *Cardiovasc Intervent Radiol*. 2010;33(2):355–361. https://doi.org/10.1007/s00270-009-9727-z.
11. Gao Y, Huang Y, Zhang R, et al. Benign prostatic hyperplasia: prostatic arterial embolization versus transurethral resection of the prostate–a prospective, randomized, and controlled clinical trial. *Radiology*. 2013;270(3). https://doi.org/10.1148/radiol.13122803, 929-8.

Chapter 97

Translational interventional radiology: Population and environmental-specific considerations

Ahmed Al-Nowfal

University College Hospital, London, United Kingdom

A thorough understanding of the population and environment to which a research project will be applied allows the innovations to reach those who need it the most. A population consists of all the subjects or events of a certain type about which researchers seek knowledge or information. A population might be broad in scope (e.g., adults within a country) or narrow (e.g., patients undergoing lower limb angioplasty with nonhealing ulcers). Even with a relatively small population, it is often impractical or impossible to gather data about each member of the population. Instead, researchers select a subset of the population, called a *sample*, which is a manageable size for observation. From their observations about the sample, researchers generalize about the population from which the sample was chosen. The ability of researchers to generalize and apply them into translation depends on a well-selected sample.[1]

To draw a satisfactory sample is sometimes more challenging than analyzing the observations made on it. A well-chosen sample will need to contain most of the information about a particular population so that all inferences made will accurately represent the native population. Therefore, it is essential that researchers define the population before selecting the sample. This means considering a variety of factors which include age, sex, occupation, family history, exposure to risk, inoculation state, the country lived in or visited, and many other genetic or environmental factors. It is advisable to examine samples when drawn to see whether they are, on average, comparable in these respects.[2]

This is creating a new challenge in translational research. For example, globally, the spectrum of the vascular disease has been continuously changing with the aging population, changes in lifestyle, and the diabetic epidemic, meaning the sample might not accurately represent the population that needs to be treated.

The random process of selection is intended to make samples comparable and representative, but sometimes it can by chance lead to disparities. To guard against this possibility the sampling may be stratified. No matter how well selected a sample is, studies are often narrowly focused and have restrictive inclusion and exclusion criteria. These conditions are inherently more static than those encountered in the real world, so traditional clinical research trials are not necessarily an indication of real-world effectiveness. There may be drastic differences in rates between individuals of different races, gender, and socioeconomic status. However, even when stratifying for measurable variables it is important to consider the importance of environmental risk factors as a cause of disease burden.

Environmental health is a growing area of knowledge, continually increasing and updating the body of evidence linking the environment to human health. The Global Burden of Disease project states that there are 26 environmental and occupational risk factors that need to be considered.[3] They are defined as all the external physical, chemical, biological, and work-related factors that affect a person's health, excluding factors in natural environments that cannot reasonably be modified. For example, even though collectively, "African American" may exhibit high rates of some adverse health outcomes, there may be drastic differences in rates between migrant workers from the Caribbean or Central America and fifth-generation urban-dwelling African Americans. Establishing more accurate estimates of the incidence and prevalence of diseases in minority populations will require further study of sociocultural and environmental risk factors.

Translational Interventional Radiology. https://doi.org/10.1016/B978-0-12-823026-8.00103-6

In 2012, the World Health Organization (WHO) estimated that 12.6 million global deaths, representing 23% (95% CI: 13%–34%) of all deaths, were attributable to the environment. For example, it is estimated that 35% (52 million) of lower-respiratory diseases each year are secondary to air pollution and second-hand smoke. Noncommunicable diseases are also related to air pollution, chemicals, and second-hand smoke, and these are responsible for 119 million cardiovascular diseases each year, 49 million cancers, and 32 million chronic respiratory diseases. However, this is only the tip of the iceberg when looking at all evidence related to environmental exposures and human health.[4]

The interrelation between population and environment will also need to be considered. A change in one environmental variable will have a clear effect on others. For example, the removal of air pollution will lead to a drastic effect on population health and an increase in pollutants will cause the health of the population degrade simultaneously. This will affect the size of the population as the general population's health degrades and the population decreases because of the presence of diseases. Similarly, the size of the population will also contribute to the environmental changes, an increased population will lead to further industrialization and urbanization, with further pollution.

Therefore, when conducting translational research it is vital to look at the population that the new discoveries will be applied to, appreciating the nuances within each population and environment, and catering the study accordingly.

References

1. Fleming ES. The role of translational research in addressing health disparities: a conceptual framework. *Ethn Dis*. 2008;18(2 Suppl 2):S2, 155–60.
2. The BMJ, Populations and samples sourced from: https://www.bmj.com/about-bmj/resources-readers/publications/statistics-square-one/3-populations-and-samples Cited on. 31/May/2022.
3. Prüss-Ustün A, Neira M. *Preventing Disease through Healthy Environments: A Global Assessment of the Environmental Burden of Disease*Vol 259. Geneva, Switzerland: World Health Organization; 2016.
4. Stanaway JD, Afshin A, Gakidou E, et al. Global, regional, and national comparative risk assessment of 84 behavioural, environmental and occupational, and metabolic risks or clusters of risks for 195 countries and territories, 1990–2017: a systematic analysis for the global burden of disease study 2017. *Lancet*. 2018;392:1923–1994.

Chapter 98

Ethics, policy, and law

S. Nourouzpour[1], N. El Sehemawi[2], John T. Moon[3], E.A. Lalla[4], M. Veneranda[5], G. Lopez-Reyes[5], Nariman Nezami[6] and Menelaos Konstantinidis[7,8]

[1]*Toronto General Hospital, University Health Network, Toronto, ON, Canada;* [2]*Faculty of Medicine, University of Montreal, Montreal, QC, Canada;* [3]*Division of Interventional Radiology and Image-Guided Medicine, Department of Radiology and Imaging Sciences, Emory University School of Medicine, Atlanta, GA, United States;* [4]*Centre for Research in Earth and Space Science, Lassonde School of Engineering, York University, Toronto, ON, Canada;* [5]*Universidad de Valladolid, Valladolid, Spain;* [6]*Division of Vascular and Interventional Radiology, Department of Diagnostic Radiology and Nuclear Medicine, University of Maryland School of Medicine. Baltimore, MD, United States;* [7]*Division of Biostatistics, Dalla Lana School of Public Health, University of Toronto, Toronto, ON, Canada;* [8]*Child Health Evaluative Sciences, Hospital for Sick Children, Toronto, ON, Canada*

Key points

- Studies should involve animals sparingly, appealing to alternatives when possible, and where not possible, doing so humanely and with respect.
- Funding in IR can be challenging, and further work is required on the part to recognize IR as an established specialty. Conflict of interests (COI) should be disclosed for any study, irrespective of the type of intervention.
- Rigorous observational studies should be conducted, and level 1 evidence should make use of methodologies that can incorporate various types of evidence (e.g., observational studies and randomized controlled trials (RCTs)).
- Off-label procedures, while often necessary, should be conducted with care, especially when there is a reason for concern of liability.
- Patient data must be vigorously protected, especially when complex collaborations exist.

Introduction

Despite IR's emergent independence as a recognized primary specialty in medicine, ethical challenges remain a persistent problem within the field, affecting all members of the specialty, from researchers to clinicians. Key issues to consider surrounding research and practice in IR include animal testing, limitations in research funding and collaborations, as well as remuneration and conflicts of interest.

Given that research plays a significant role in knowledge acquisition, an unstable and tenuous foundation may have a domino effect on knowledge translation, and ultimately, on clinical practice. As a result of insufficient knowledge and unclear guidelines surrounding the laws and policies in IR, patient-centered care presents its own set of challenges, such as the use of automated devices in patient care, as well as security concerns among data sharing and privacy.

Basic and preclinical research

Stemming from the principles of evidence-based medicine, behind every clinical decision, there must exist a clear body of evidence borne out of rigorous scientific research. Such research, however, is not without ethical and legal challenges.

Animal models

Advancements in the medical field are largely dependent on systematic research, which are often accompanied by a host of ethical and moral implications. The use of animal models in experimental designs have widely been used in evidence-based medicine as they allow scientists to find repeatable observations while avoiding patient harm. Despite this, animal models present unequivocal advantages in medical research.[1] Transplantation research, for instance, has largely benefitted from the use of mouse and porcine animal models in preclinical research.[2] However, there are undeniable moral implications in the use of healthy animals for the sake of human benefits. Thus, the most commonly used guideline is that of the 3R's of animal research: (1) replacement (i.e., using alternative methods when possible), (2) reduction (i.e., limiting the number of animals necessary to reach the study validity), and (3) refinement (i.e., enhancing animal health and welfare through the use of appropriate anesthesia and

Translational Interventional Radiology. https://doi.org/10.1016/B978-0-12-823026-8.00080-8

enrichment of their environment as well as reducing harm and distress in animals).[1,2] Respect is a fourth widely promoted standard and supports the care of the animals used in experimental models.[1]

Funding

The allocation of funding is largely distributed to research areas that are the most "promising," often characterized as research areas that have demonstrated great scientific success and are clinically feasible and favorable in a clinical setting.[3] As a result, specialties with less marketable success struggle with funding contributions. However, despite the Society of Interventional Radiology Political Action Committee's (SIRPAC) efforts over the last 12 years, their last report in 2018 suggests that the SIRPAC has yet to reach the funding level of other medical specialties. In addition to the pervasive difficulties in obtaining funding, several key ethical issues exist in funding research and remuneration for investigators.[4]

In the absence of conventional avenues for funding, research is often funded by third parties, often from industry (e.g., pharmaceutical, or medical device companies). While such an engagement may be beneficial by providing funding to the investigating team, this may result in perverse incentives. For example, for fear of losing future funding, researchers may conduct themselves unethically in favor of the industry partner's products (e.g., specific stents or beads in IR), thereby biasing the estimated effect of interventions being studied. Furthermore, in developing innovative interventions, companies often require the domain expertise of practicing IR specialists. However, clinicians may again be perversely incentivized by taking on roles such as consultants or board/advisory members (with monetary or in-kind remuneration). Cumulatively, such dynamics speak to the need for policy and laws around disclosure and permissibility of conflicts of interest.

COIs are often attributed to pharmacological research, and thus, COIs are frequently and improperly overlooked when nonpharmacological studies are published. This is particularly important in the context of IR, in which device-based interventions are often studied. COIs can damage the quality of the evidence presented and can hinder the credibility of science.[5] Recently, guidelines such as the updated 2020 Cochrane policy concerning COIs have been developed to attain objectivity in research and ensure confidence in the evidence provided in the published research.[6] Additionally, the Sunshine Act was designed to increase transparency in financial relationships between manufacturers of healthcare products, and physicians and teaching hospitals.[7] This policy requires disclosing any financial support given from manufacturers to physicians.[8] It is, however, worth mentioning that overzealous policies in this direction may disincentivize innovation.

RCTs and guidelines

RCTs are considered the gold standard when it comes to medical research. However, due to the lack of standardized care in IR practice and its fast-paced growth, RCTs become difficult to develop, thus trailing behind new technological advancements and the clinical insights that can be derived from them in the field.[9] This, coupled with technical and ethical barriers in funding clinical research, makes it even more difficult for rigorous clinical trials to be conducted.

An important downstream consequence of this is the quality of the evidence upon which guideline development and the standard of care are based. For example, it is well known that guidelines often draw from systematic reviews and meta-analyses (the gold standard for which is to include only RCTs). However, if the elaboration of RCTs is inhibited, conventional meta-analyses will necessarily be of a lower quality or not be conducted at all, thereby limiting the quality of the evidence from which clinical guidelines can reliably draw. Considering such challenges, previous work has suggested that well-designed observational studies can still provide insights on clinical best practices and can complement RCTs in establishing promising therapies. Additionally, methodological developments have been made to allow for meta-analysis of different types of evidence (i.e., observational and RCTs),[10] thus allowing for high-quality evidence to be disseminated and used.

Clinical practice

In addition to the challenges posed by basic and preclinical research, IR faces many ethical and legal challenges.

Technology

Due to the innovative nature of the field, IR physicians often find themselves performing novel off-label procedures, sometimes with limited medical research, on cases referred to them because of their time-sensitive nature.[11] This fast-paced environment is not conducive to ethical deliberations as most of the time is devoted to innovation. As such, insufficient time is attributed to reflecting on the complications incurred by those innovative interventions; ethical discussions and research are scarce.[11] IR interventions are sometimes used as a last resort procedure, relying on newly developed imagery and surgical technology, and have been particularly coveted in the palliative care department or for critically ill patients.[12,13] Additionally, because IR is a nascent field, other specialties may not refer candidate patients to IR when patient-centered care would dictate they should. In such cases, blame is easily attributed to the physician in charge. Thus, future translational research should focus on improving patient safety while delivering transparent, quality care.[12]

The same is true, if not altogether exacerbated, when algorithms borne out of machine learning and artificial intelligence are calibrated to inform or augment clinical practice, given that such algorithms may lack internal or external validity, thus putting into question on whom the ethic onus lies, and may also be inequitable (e.g., not calibrated to minority group patients). Ethical dilemmas regarding addressing such complications with the patient must be discussed to ensure patients receive the appropriate support and ensure that the IR values are aligned with their physicians.[11,14]

Data sharing

In the digital era, sensitive patient information and highly confidential data are commonly stored on automated devices, rendering patient information vulnerable to data leaks, where sensitive material may be collected by an unknown third party.[15] This is particularly concerning as third-party players may take advantage of important patient identifiers, such as patient age, sex, or date of birth, that may be stored and utilized by external threats.[16] Similar problems arise in multicenter and multinational studies, where sharing data across institutions and borders may put patient data at risk, thereby requiring safeguards in data storage and guidelines around its use. This is especially problematic as the standard of protection among patient participants remains variable across location sites and inconsistencies in ethical guidelines around patient privacy and anonymity.

Conclusion

Although IR has accomplished significant strides in medicine, it is impossible to ignore many ethical barriers that are often overlooked. Awareness and recognition surrounding such concerns are the first step to improving the care and well-being of all parties involved within the field. Considering that clinical guidelines and practice are based on evidence-based information, adequate funding is necessary to support high-quality research, an integral component in securing a name for IR within the medical field. Furthermore, studies should involve animals sparingly, appealing to alternatives when possible, and when not possible, doing so humanely and with respect. Moreover, improved transparency and communication between healthcare professionals, third parties, and patients remain critical in hopes of providing a secure, ethical foundation that is driven to help improve patients' security, privacy, and overall well-being.

Get started

Understand the role of animal models, funding, clinical trials, and evidence syntheses in pre-clinical research.

Understand the role of technology and data sharing in clinical research.

Highlight measures put forth by the Sunshine Act to ensure this transparency and to reduce the reporting burden, include delegating the reporting requirements on manufacturers and organizations.

Reflect on the ethical challenges and mitigating strategies for translational research.

Potential pitfalls

- Notable considerations should be addressed for the limitation of translating animal studies to human subjects.
- Conflict of interests (COI) should be disclosed for any study, irrespective of the type of intervention, as it can create bias and hinder the research's validity.

Real-world examples

- **Animal Models:** The world's first vaccine was discovered by British doctor, Edward Jenner, who used living cows to demonstrate immunity against the cowpox virus. This discovery led to the immunization and eventual eradication of the devastating smallpox virus.
- **RCTs:** The most recent global pandemic, novel coronavirus disease (COVID-19) is an excellent example of a real-world use of a randomized controlled trial. Scientists on the hunt for a quick, effective, and safe vaccine to help combat COVID-19 ran multiple randomized controlled trials, consisting of both placebo and intervention arms, to help determine vaccine efficacy.
- **Off-Label Procedures:** Beta-blockers are drugs that are licensed for use in the management of high blood pressure, abnormal heart rhythms, and more. Despite this, beta-blockers have gained increased popularity in treating those with anxiety disorders as well, regardless of its intended use.

Resources

1. https://www.bmj.com/commercial-influence
2. https://medinform.jmir.org/2022/5/e36388
3. https://arxiv.org/abs/1808.00023

References

1. Raymond J, Salazkin I, Gevry G, Nguyen TN. Interventional neuroradiology: the role of experimental models in scientific progress. *AJNR Am J Neuroradiol*. 2007;28(3):401–405. PMID: 17353303; PMCID: PMC7977832.
2. Robinson NB, Krieger K, Khan FM, et al. The current state of animal models in research: a review. *Int J Surg*. 2019;72:9–13. https://doi.org/10.1016/j.ijsu.2019.10.015. Epub 2019 Oct 15. PMID: 31627013.

3. Mandal J, Ponnambath DK, Parija SC. Ethics of translational medical research. *Trop Parasitol.* 2017;7(2):62.
4. Keller EJ, Collins JD, Crowley-Matoka M, Chrisman HB, Milad MP, Vogelzang RL. Why vascular surgeons and interventional radiologists collaborate or compete: a look at endovascular stent placements. *Cardiovasc Intervent Radiol.* 2017;40(6):814–821. https://doi.org/10.1007/s00270-017-1570-z. Epub 2017 Jan 6. PMID: 28062898.
5. Clark AM, Choby A, Ainsworth K, Thompson DR. Addressing conflict of interest in non-pharmacological research. *Int J Clin Pract.* 2015;69(3):270–272. https://doi.org/10.1111/ijcp.12569. PMID: 25727801.
6. Macbeth F, Webster A, Foxlee R, et al. *"CoI Policy Cochrane Library." CoI Policy Cochrane Library | Cochrane Training.* Cochrane; 2020. v3.3 https://training.cochrane.org/sites/training.cochrane.org/files/public/uploads/Conflict%20of%20Interest%20policy_Jan%202022_v18_1.pdf, 2020.
7. Toroser D, Robinson M, Gegner J, et al. Systematic review of reports describing potential impact of the sunshine act on peer-reviewed medical publications. *Curr Med Res Opin.* 2016;32(3):547–553. https://doi.org/10.1185/03007995.2015.1131151. Epub 2015 Dec 28. PMID: 26652290.
8. Agrawal S, Brennan N, Budetti P. The sunshine Act – effects on physicians. *N Engl J Med.* 2013;368(22):2054. https://doi.org/10.1056/NEJMp1303523. PMID: 23718163.
9. Pereira PL, Helmberger T, Taieb J, et al. Scientific societies should start sponsoring observational studies. *Cardiovasc Intervent Radiol.* 2020;43:1865–1867. https://doi.org/10.1007/s00270-020-02649-5.
10. Efthimiou O, Mavridis D, Debray TPA, et al. Combining randomized and non-randomized evidence in network meta-analysis. *Stat Med.* 2017;36:1210–1226. https://doi.org/10.1002/sim.7223.
11. Keller EJ. Reflect and remember: the ethics of complications in interventional Radiology. *Semin Intervent Radiol.* 2019;36(2):104–107. https://doi.org/10.1055/s-0039-1688423. Epub 2019 May 22. PMID: 31123380; PMCID: PMC6531013.
12. Higgins MCSS, Herpy JP. Medical error, adverse events, and complications in interventional Radiology: liability or opportunity? *Radiology.* 2021;298(2):275–283. https://doi.org/10.1148/radiol.2020202341. Epub 2020 Dec 15. PMID: 33320064.
13. Rexha I, et al. Role of 3D quantitative tumor analysis for predicting overall survival after conventional chemoembolization of intrahepatic cholangiocarcinoma. *Sci. Rep.* 2021;11:9337. https://doi.org/10.1038/s41598-021-88426-x.
14. Konstantinidis M, Lalla EA. Clinical anisotropy: A case for shared decision making in the age of too much data and patient dis-integration. *J. Eval. Clin. Pract.* 2020;26(2):604–609. https://doi.org/10.1111/jep.13312.
15. Castelvecchi D. Can we open the black box of AI? *Nature.* 2016;538(7623):20–23. https://doi.org/10.1038/538020a. PMID: 27708329.
16. Geis JR, Brady AP, Wu CC, et al. Ethics of artificial intelligence in radiology: summary of the joint European and North American multisociety statement. *Radiology.* 2019;293(2):436–440. https://doi.org/10.1148/radiol.2019191586. Epub 2019 Oct 1. PMID: 31573399.

Chapter 99

Healthcare institutions and systems

Syamak Moattari[1] and Babak Mohit[2]

[1]*Health Sciences Department, Worcester State University, Worcester, MA, United States;* [2]*Sleep Disorders Center, Division of Pulmonary and Critical Care Medicine, Department of Medicine, University of Maryland School of Medicine, Baltimore, MD, United States*

What is the role of healthcare organizations?

Healthcare systems are so important to the overall mission of maintaining good health that the World Health Organization (WHO) dedicated its landmark 2000 World Health Report to health systems, and how to make these systems more effective and efficient around the globe.[1–3] The report suggests three main goals for every health system:

- Maintenance of good health
- Responsiveness to the population expectations
- Fairness of financial contribution

The report also highlights four vital functions of any health system, which include service provision, resource generation, financing, and stewardship. At its core, the WHO argues that all healthcare systems should:

1 Provide access to a wide range of services including, prevention, diagnosis, treatment, and rehabilitation.
2 Protect the population against the cost of sickness by introducing some type of insurance.
3 Improving the health of populations by good governance and stewardship.

What are the different types of healthcare organizations?

On the provider side, the organizations within a healthcare system include facilities for prevention, tests and diagnosis, treatment, rehabilitation, and therapy.[4] Hospitals of different types—general, specialty, teaching—are central to any healthcare system. Alternative Therapics Centers are another type of healthcare organization. It is estimated that over $9 billion per year is spent on such services.[5] Because of demographic transition and aging of the population, the long-term care facilities such as nursing homes, long-term care facilities, and rehabilitation services are also critical in the integration of healthcare services. Charity and voluntary agencies also provide services from bench to bedside, ranging from support in natural disasters to establishing cancer clinics for children.[5]

Who pays for healthcare?

Delving further into all three WHO guidelines, Reid (2010) examines four different healthcare finance models, which are used across the world. These models include the Beveridge model, the Bismarck model, National Health Insurance (NHI), and the Out-of-Pocket model.[6]

The Beveridge model[6]

Named after William Henry Beveridge (1879–1963), a British economist and politician who played a central role in designing Britain's National Health Service.[7] In this arrangement of the health system, health is considered an essential human right and funded by tax payments. Government is the sole payer for health and owns many hospitals and clinics, and the entire health system is organized around a national health service. The United Kingdom, New Zealand, most Scandinavian countries, and Cuba are examples of this model.

The Bismarck model[6]

Otto von Bismarck (1815–98) unified Germany in 1871 and served as its first chancellor until 1890. The Bismarck healthcare model uses an insurance system usually financed jointly by employers and employees through payroll deduction.[6] It means, the health system operates through national health insurance. This type of insurance plan is universal and tends to cover everybody. Today Germany has over 240 different funds, while tight regulation gives

Translational Interventional Radiology. https://doi.org/10.1016/B978-0-12-823026-8.00099-7

the government much of the cost-control power that the single-payer Beveridge model provides. The health systems of Germany, France, Belgium, the Netherlands, Japan, and Switzerland follow this model.

The National Health Insurance model[6]

This system has elements of both the Beveridge and the Bismarck model.[6] In this system, the private sector is quite active in healthcare. A government-run insurance program pays for the services and is able to negotiate with the private sector to get better prices. The classic NHI system was developed in Canada, but some newly industrialized countries such as Taiwan and South Korea, have also adopted the NHI model.[6]

The out-of-pocket (fee for service) model[6]

This is the world's oldest model of healthcare financing. In this model, patients are responsible for their doctor bills. Health is considered a personal good and a commodity in the market. Patients who have money can buy it, and those who cannot afford have no access. In rural regions of Africa, India, China, and South America, hundreds of millions of people live their whole lives without ever seeing a doctor. They may have access, though, to a village healer using home-brewed remedies that may or not be effective against the disease.

The special case of the United States

In the United States, various government and marketplace health insurance plans, as well as the fee-for-service model are available to meet different needs. Some marketplace plans restrict members to choose their providers from a specific network, some let you see a specialist without any primary doctor referral. Some do not let the members see specialists unless they have a referral from a primary doctor.

The government sector covers two different populations

Medicare

Medicare is a governmental health insurance program, designed to pay primarily for the healthcare needs of the retired US citizens over the age of 65. This program is less restrictive compared to marketplace plans. Medicare is funded and administered by the US Federal Government and operates homogeneously across the United States, even though local variation in the availability and quality of services may exist.

Medicaid

Medicaid is the healthcare part of the Federal safety net welfare program of the United States. Medicaid is funded to a large extent by the US Federal Government; however, each of the 50 states is largely independent in the way they chose to spend their Medicaid funds. Therefore, large differences in terms of eligibility, availability, and quality of services may exist.

Market place insurance for working population

The majority of the US insured population is insured through plans offered through their workplace. Some examples of plan types available through the marketplace include the following:

Preferred provider organizations

Preferred provider organizations (PPOs) are a type of health plan. A network of doctors, hospitals, and other providers who have agreed to charge less for plan members to work within the PPO. By using a PPO, usually, patients do not need a referral from their primary care provider when they want to see a specialist. With PPO coverage, patients may go outside the network and see a provider they wish to visit, with additional costs.[8,9]

Exclusive provider organization

In this type of plan, the members are restricted to use doctors, specialists, or hospitals in the plan's network. Networks in this type of plan usually are local. However, if the patient chooses to get care outside the plan's network, it usually will not be covered (except in an emergency). Exclusive provider organizations (EPOs) generally have lower monthly premiums. In this model, sometimes patients are required to have a primary care doctor (PCP), but not always, and may see a specialist without a referral.[9]

Health maintenance organization

A type of health insurance plan that usually limits coverage to care from doctors and hospitals who work for or contract with the health maintenance organization (HMO). It usually has lower monthly premiums than a PPO or an EPO health plan. It generally will not cover out-of-network care except in an emergency. An HMO may require patients to live or work in its service area to be eligible for coverage. In this plan, patients must choose a PCP to coordinate their healthcare. HMOs often provide integrated care and focus on prevention and wellness.[9]

Point of service[9]

This is a hybrid plan. As with HMOs, in network care is affordable and guided by a PCP. The point-of-service (POS) plans also give the patient the freedom of PPO plans to see providers outside the network. However, POS plans require patients to get a referral from their primary care doctor to see a specialist. POS participants pay lower premiums than those in a PPO, but more than those with HMO coverage.[9]

The role of the radiologist within the healthcare system

Knechtges and Carlos[10] argue that "the traditional view of the radiologist as a physician who adds value to the health care system solely by generating and interpreting diagnostic images is outdated."[10] They explore other roles such as economic gatekeeping, political advocacy, public health delivery, patient safety, quality of care improvement, and information technology as evolving roles that radiologists are increasingly involved with.[10] One of the current discourses around the role of radiology in translational research is around the use of artificial intelligence (AI) in this field. Important AI tools for diagnostic imaging include algorithms for disease detection and classification, image optimization, radiation reduction, and workflow enhancement.[11] The development of advanced AI technology is occurring rapidly, but the translation of these findings to clinical practices is relatively slow.[11] By facilitating more collaboration between radiologists and AI developers, we can see better use of AI in clinical practices.

References

1. Reid PPCW, Grossman JH, et al., eds. *National Academy of Engineering (US) and Institute of Medicine (US) Committee on Engineering and the Health Care System. Building a Better Delivery System: A New Engineering/Health Care Partnership. Connecting Patients, Providers, and Payers*. Washington, DC: National Academies Press (US); 2005. https://www.ncbi.nlm.nih.gov/books/NBK22827/.
2. Ritz D, Althauser C, Wilson K. 3. Introducing the key stakeholders: patients, providers, PAYORS, and policymakers (the four P'S). In: *Connecting Health Information Systems for Better Health: Leveraging Interoperability Standards to Link Patient, Provider, Payor, and Policymaker Data*. PATH and Joint Learning Network for Universal Health Coverage; 2014.
3. World Health Organization (WHO). *The World Health Report 2000: Health Systems: Improving Performance*. World Health Organization; 2000. https://www.who.int/whr/2000/en/whr00_en.pdf.
4. Brunot T. *What Are the Different Types of Healthcare Organizations?*; 2019. https://www.sapling.com/7394733/different-types-health-care-organizations.
5. Sultz H. *Health Care USA*. 6th ed. Jones & Bartlett Learning; 2008.
6. Reid TR. *The Healing of America: A Global Quest for Better, Cheaper, and Fairer Health Care*. Penguin Books Ltd.; 2010.
7. Midgley J, Beveridge LW. *Encyclopedia of Social Work*. 19 ed Vol 3. NASW Press; 1995:2574.
8. HealthCare.gov. *Preferred Provider Organization (PPO)*; 2021. https://www.healthcare.gov/glossary/preferred-provider-organization-ppo/.
9. Cigna Health and Life Insurance Company (CHLIC). *What's the Difference between an HMO, EPO, and PPO? Get Help Choosing between an HMO, EPO, and PPO Plan*; 2021. https://www.cigna.com/individuals-families/understanding-insurance/hmo-ppo-epo.
10. Knechtges PM, Carlos RC. The evolving role of radiologists within the health care system. *J Am Coll Radiol: JACR*. 2007;4(9):626–635.
11. Allen Jr B, Seltzer SE, Langlotz CP, et al. A road Map for translational research on artificial intelligence in Medical imaging: from the 2018 national institutes of health/RSNA/ACR/the academy workshop. *J Am Coll Radiol*. 2019,16(9 Pt A):1179–1189.

Chapter 100

Public health institutions and systems

Syamak Moattari[1] and Babak Mohit[2]
[1]Health Sciences Department, Worcester State University, Worcester, MA, United States; [2]Sleep Disorders Center, Division of Pulmonary and Critical Care Medicine, Department of Medicine, University of Maryland School of Medicine, Baltimore, MD, United States

Introduction

What is the definition of a public health institution? How are these institutions functioning? Who pays for these organizations? How are they related to the health system in general? Before discussing these particular organizations, we need to talk about the health systems.

The World Health Organization defines a health system as "comprising all the organizations, institutions and resources that are devoted to producing health actions. A health action is defined as any effort, whether in personal health care, public health services or through intersectoral initiatives, whose primary purpose is to improve health."[1]

On the other hand, health systems are composed of agencies that plan, fund, and regulate the healthcare system. Financial resources and some functions such as preventive, clinical, and rehabilitative services including educating health professionals are among health systems' activities.

Every system is composed of different parts, such that the system as a whole has a greater function than the sum of individual parts. Merriam-Webster dictionary defines a system as "... an organization forming a network especially for distributing something or serving a common purpose." Health systems, organizations, people, and money that shape the system each may be coming from different backgrounds. Some of these organizations are public (governmental) entities, while others may be private and for profit, non-for-profit, or nongovernmental organizations.

The public health system of the United States

Taking the health system of the United States as an example, the federal government acts in six main areas related to population health: (1) policymaking, (2) financing, (3) public health protection, (4) collecting and disseminating information about US health and health care delivery systems, (5) capacity building for population health, and (6) direct management of services.[2] At the center of public health services is the Department of Health and Human Services, which is a Cabinet-Level Department and besides internal offices and undersecretaries, includes the following agencies (in alphabetical order)[3]:

- Administration for Children and Families
- Agency for Healthcare Research and Quality
- Agency for Toxic Substances and Disease Registry
- Centers for Disease Control and Prevention
- Centers for Medicare and Medicaid Services
- Food and Drug Administration (FDA)
- Health Resources and Services Administration
- Indian Health Service
- National Institutes of Health (NIH)
- Substance Abuse and Mental Health Services Administration

Given that in the United States, public health governance is determined at the state level, the United States Department of Health and Human Services (DHHS) and above-named Federal agencies have an advisory and oversight role and their influence on policymaking may differ from state to state. State health departments also have varied different types of relationships with regional/local public health departments. Different models of public health governance can impact the delivery of health services, which are exampled as follows.

Types of governance health structures[4,5]:

- Centralized or largely centralized structure: employees of the state lead the local health units. For example, the Alabama Department of Public Health is a largely centralized model.
- Decentralized or largely decentralized structure: local health units are primarily led by employees of local governments. The California Department of Public Health is an example of a decentralized model.

Translational Interventional Radiology. https://doi.org/10.1016/B978-0-12-823026-8.00091-2

- Mixed structure: state employees lead some health units at the local level and local governments lead some of the health units. The Tennessee Department of Health had a mixed structure.
- Shared or largely shared structure: local health units might be led by employees of the state or by employees of local government. The Kentucky Department for Public acts as a shared structure.

At the local level, Local Health Departments (LHDs) are the front-line organizations for delivering basic public health services to most communities.[6] According to the National Association of County and City Health Officials, there are approximately 2800 LHDs in the United States.[7]

LHDs are different in size and nature of population they serve, urban and rural mix, governmental structure. These organizations with limited monetary and human resources serve diverse populations. Local Health Departments also are direct providers of a variety of clinical programs and services. These services may range from adult and child vaccinations to screening and treatment of both chronic and infectious diseases, as well as maternal and child health services. These departments also provide various population-based programs and services that include epidemiology and surveillance, primary, secondary, and tertiary prevention, forming regulations and enforcing them through inspection, or licensing, as well as environmental health services.[7]

Nongovernmental agencies, such as the Joint Commission (https://www.jointcommission.org/) and the Institute of Medicine, as well as various Medical (such as Radiological Society of North America) and Public Health such as American Public Health Association professional organizations also play an important role in shaping the public debate and modus operandi of public health authorities.

How can radiology help public health?

Radiology and public health have historically collaborated in different ways. Public Health can use radiology's imaging data for epidemiologic assessments and reaction to population health problems.[8] In disease surveillance systems for outbreak detection and bioterrorism, such as the Electronic Surveillance System for the Early Notification of Community-based Epidemics, the use of radiological data has been tested.[8] In recent years, the transition from analog to digital formats imaging created an opportunity for the public health system to use radiology data more broadly. Another interesting factor is the role of radiology in a vast majority of the medical field and surgical subspecialties' practice; therefore, it created a foundation for a uniform database of images and reports. It is possible to map radiological data in different geographical areas and recognize disease patterns in different settings.[8]

In the recent COVID-19 pandemic, radiology played an important role in helping public health through monitoring disease outbreaks. For example, in 2020 and in absence of enough COVID-19 diagnosis tests, chest radiography and CT-Scans became an important and lifesaving alternative to screen patients for COVID-19 pneumonia.[9-15]

How can public health help radiology?

The Institute of Medicine has defined the mission of public health as "fulfilling society's interest in assuring the conditions in which people can be healthy."[2,6]

Imagine, a radiologist is in process of discovering a new imaging drug or contrast agent. At the discovery stage, the National Institute of Biomedical Imaging and Bioengineering (NIBIB—https://www.nibib.nih.gov/) part of the NIH provides various grants for discoveries that may impact the health of the public.

Furthermore, at some point, before the invention is commercialized, it becomes the role of the FDA to supervise testing to assure that the drug or agent is safe and efficacious for the purpose it has been filed. Having access to nutritious and safe food, and effective medical products are essential to the health of the population, and the FDA's mission is aligned with this objective.[16] The FDA has a specialized program for this purpose. Its mission is to "protect the public from hazardous or unnecessary radiation exposure from radiation-emitting electronic products."[17] Most related to the field of Radiology, the FDA's Center for Devices and Radiological Health (CDRH) is closely aligned with the mission.[17] CDRH defines its mission as follows:

> *Regulating radiation-emitting electronic products used for both medical and non-medical applications, under the authorities of the Electronic Product Radiation Control provisions and the Medical Device Amendments of 1976 to the Food, Drug, and Cosmetic Act; Regulating facilities that perform mammography under the authority of the Mammography Quality Standards Act.*

CDRH works with a wide array of stakeholders including public agencies (from the federal, state, and local levels) as well as nongovernmental groups such as (professional organizations, academic institutions, industry manufacturers, and consumer groups).

Toward its mission, the CDRH is focused on several key objectives.[17]

"Maintains awareness of new and existing radiation-emitting products, their manner of use, and their manufacturers;

Studies the biological effects of radiation-emitting products and their potential risks to health;

Assesses radiation emission levels from products in various applications;

Provides direction and guidance to the general public and users of radiation-emitting products to minimize unnecessary radiation exposure; and

Oversees product manufacturer compliance with all applicable requirements and pursues regulatory and enforcement action to address public health problems."

Conclusion

As a branch of medicine, public health has a long history of collaborating with other fields of medicine toward not only improving the health and well-being of the individual but also communities and society as a whole. The field of radiology has and continues to work with public health, each contributing to the progress of the other. Public health institutions, in the public, private, and nongovernmental and professional society domains, have key roles in furthering progress toward evidence-based diagnosis, treatment, and policymaking.

References

1. World Health Organization (WHO). *The World Health Report 2000: Health Systems: Improving Performance*. World Health Organization; 2000. https://www.who.int/whr/2000/en/whr00_en.pdf.
2. Institute of Medicine (IOM). *The Future of the Public's Health in the 21st Century*. The National Academies Press; 2003.
3. U.S. Department of Health & Human Services (DHHS). *HHS Agencies & Offices*; 2015. https://www.hhs.gov/about/agencies/hhs-agencies-and-offices/index.html.
4. Association of State and Territorial Health Officials (ASTHO). *Profile of State Public Health*. Vol 5. Association of State and Territorial Health Officials (ASTHO); 2019.
5. Centers for Disease Control and Prevention (CDC). *Health Department Governance*; 2020. https://www.cdc.gov/publichealthgateway/sitesgovernance/index.html.
6. Institute of Medicine (IOM). *The Future of the Public's Health in the 21st Century: The Governmental Public Health Infrastructure*. The National Academies Press; 2002.
7. National Association of County and City Health Officials (NACCHO). *National Profile of Local Health Departments*. National Association of County and City Health Officials (NACCHO); 2019. https://www.naccho.org/uploads/downloadable-resources/Programs/Public-Health-Infrastructure/NACCHO_2019_Profile_final.pdf.
8. Mollura DJ, Carrino JA, Matuszak DL, et al. Bridging radiology and public health: the emerging field of radiologic public health informatics. *J Am Coll Radiol*. 2008;5(3):174−181.
9. Behzad S, Aghaghazvini L, Radmard AR, Gholamrezanezhad A. Extrapulmonary manifestations of COVID-19: radiologic and clinical overview. *Clin Imag*. 2020;66:35−41.
10. Demirjian NL, Fields BKK, Gholamrezanezhad A. Role of chest CT in resource-driven healthcare systems. *AJR Am J Roentgenol*. 2020;215(3):W36.
11. Demirjian NL, Fields BKK, Song C, et al. Impacts of the Coronavirus Disease 2019 (COVID-19) pandemic on healthcare workers: a nationwide survey of United States radiologists. *Clin Imag*. 2020;68:218−225.
12. Eslambolchi A, Maliglig A, Gupta A, Gholamrezanezhad A. COVID-19 or non-COVID viral pneumonia: how to differentiate based on the radiologic findings? *World J Radiol*. 2020;12(12):289−301.
13. Keshavarz P, Rafiee F, Kavandi H, Goudarzi S, Heidari F, Gholamrezanezhad A. Ischemic gastrointestinal complications of COVID-19: a systematic review on imaging presentation. *Clin Imag*. 2021;73:86−95.
14. Kooraki S, Hosseiny M, Myers L, Gholamrezanezhad A. Coronavirus (COVID-19) outbreak: what the department of radiology should know. *J Am Coll Radiol*. 2020;17(4):447−451.
15. Shin H, Abdelhalim A, Chau S, Shah S, Desai B, Gholamrezanezhad A. Responding to coronavirus disease 2019: LA County hospital experience. *Emerg Radiol*. 2020;27(6):785−790.
16. Hamburg MA, Sharfstein JM. The FDA as a public health agency. *N Engl J Med*. 2009;360(24):2493−2495.
17. Food and Drug Administration (FDA). *FDA Radiological Health Program*; 2018. https://www.fda.gov/radiation-emitting-products/fda-radiological-health-program.

Part XI

Practical resources

Chapter 101

Presenting data

Ragheed Al-Dulaimi
Department of Radiology and Imaging Sciences at the University of Utah- School of Medicine, Salt Lake City, UT, United States

Key points

- Data presentation is a crucial skill for researchers in interventional radiology.
- There are multiple ways of presenting research data.
- The choice between different graphical and tabular formats depending on data specification and preferences of the researcher.

Why it matters?

There are multiple ways to present research results in interventional radiology. You could incorporate the information within the text in the result section of the manuscript or choose to present it separately in a tabular or graphic format. We frequently hear the adage "a picture worth a 1000 words." Many epidemiologists and clinical investigators would agree that graphical illustration of the research results can be a very effective way to summarize findings and convey the research message as a visually appealing and easy-to-understand presentation.[1] Therefore, data presentation is a crucial skill for researchers in interventional radiology to master. This chapter will provide the basis for this but further reading and practice are essential.

Basic concepts

To select the best format to present your results, you have to understand the type of the data and the graphical options available. The data is composed of elements called variables.[2] These are usually gathered during the data collection phase to be analyzed by a statistician. In general, the variables are either numerical or categorical variables.[3]

Numerical variables that are collected as part of a continuous scale are called continuous variables.[3] Common examples include patient age (in years), procedure time (min), volume of fluid removed during paracentesis (mL), and survival time after Y90 (months). On the other hand, some numerical variables are discrete.[3] In other words, they only accept one of the distinct whole values and cannot be described with fractions like for example number of biopsy's attempts and number of sessions for varicose veins treatment.[4]

Categorical variables could be either nominal or ordinal.[3,4] Nominal variables came in categories without specific orders.[3] These could only have two categories and in this case called binary or dichotomous. For example, gender (male and female), Yes and No variables, and whether patient had unilateral or bilateral uterine artery embolization. However, it is not uncommon for nominal variables to have multiple categories such as the type of embolization material used (coils, gel foam, particles, ...). In contrast, categorical variables that have three or more categories with particular order are called ordinal variables.[3] For example, stages of cancer (stage I– IV), classification of adverse events and complications into mild, moderate, or severe and categorization of bleeding risk into low, moderate, and high.

Presentation of categorical variables

Using tables

Presentation of categorical variables in table format should be self-explanatory. The categorical variables are summarized as the number of observations in each category of the variable. This is also called count or frequency. Often, relative frequency to the total number of observations is also obtained. This is called percentage. Categorical variables can be presented as frequency and percentages, or percentages only. Authors may also choose to present one category of dichotomous variable and in this case, it will be implied that the other category will be the remaining percentage and frequency.

Using graphs

The two most commonly used graphs to describe categorical variables are pie diagram and bar chart:

1. Pie chart: A circular diagram in which each group of a categorical variable is shown in a segment (piece of the pie) that is proportional to their relative percentage of the total.[4] Limitation of pie charts includes difficulty

Translational Interventional Radiology. https://doi.org/10.1016/B978-0-12-823026-8.00094-8

comparing multiple variables. A waffle plot can be used as an alternative to the pie chart. It uses different icons and colors to describe various categories.

2. Bar chart: In this case, each group of the categorical variable will have a bar with the height of the bar based on their frequency or relative frequency (percentage).[4,5] A grouped bar chart allows the illustration of more than one categorical variable using different colors.[4] Another extension of the standard bar chart that could be used to look at two variables instead of one is a stacked bar chart.[1] In this case, there are subbars stacked on top of each other to represent the frequency of relative frequency of the second variable.[1] Several modifications of this could also be seen in the literature such as mosaic and spine plot but the description of those is beyond the scope of this chapter (Fig. 101.1).

Presentation of continuous variables

Using tables

Presentation of continuous variables in table format depends on their distribution. The decision of what descriptive numbers to be used in the table depends on the distribution of the variable. Distribution can easily be determined during statistical analysis. Variables could have a normal distribution in which the data is symmetrically distributed around the mean and the mean, median, and mode are equal. In this case, mean and standard deviation are usually used. Some researchers may choose to include the range (minimum and maximum) instead or in addition to the standard deviation if relevant. However, it is not uncommon in research to have variables with a skewed distribution where most of the observations took values near the minimum or the maximum of the variables. In this case, mean, median, and mode will have different values, and mean will be easily affected by extreme values. In this case, it is a wiser choice to use the median instead of the mean in the table. Some researchers may choose to include the interquartile range. This represents the difference between the upper and lower quartile.

In a given table that summarizes several continuous variables with normal or skewed distribution, it is not uncommon to choose to describe all continuous variables using the median to be consistent and make the table easy to comprehend.

FIGURE 101.1 Common types of graphs used for categorical variables. (A) Grouped bar chart. (B) Pie diagram.

Using graphs

Several types of the graph can be used to present continuous variables. The most common ones are as follows:

1. Histogram and or frequency polygon: Histogram is a diagram made of rectangles that show the distribution of the measured value for the continuous variable.[4] On the other hand, a frequency polygon is a linear chart or curve that describes the distribution of the data.[4] Both the frequency polygon and the histogram can be used to compare the distribution of multiple continuous variables in one graph.[4]
2. Dot plot: This is a display of individual data points with one more summary statistics such as the mean or median.[1] This could be used when you have a small number of observations.[1,5]
3. Box plot: In this case, the data is summarized in a box that shows the first and third quartiles and the median as a central line.[1,2,4,5] A common variation of this is called a box-and-whisker plot, referred to as a box plot that has two whiskers to show the upper and lower extreme values.[1] Another variation of the box plot is the violin plot in which the box is replaced by an estimated outline of the distribution.[1,5] This is usually used for variables with a more complex distribution (Fig. 101.2).

General consideration for table design

Researchers frequently use tabular format to present their data. Commonly tables include a summary of different variables (categorical, continuous, or both). The table should be clear and self-explanatory.[3] The variables should be arranged in a meaningful way. Some researchers may choose to group variables together. For example, they may start with demographics where they put the summary of their patient cohort age, gender, and race. This will be followed by other categories as patients' relevant clinical characteristics (diabetes, heart disease, ...), procedure-related variables (time of procedure, type of embolization, ...), and finally, outcome-related variables.

Do and Don't

- Do include enough details in the title or the table footage. This will help the reader to understand the content without the need to refer to the accompanying text.[3]
- Do avoid large tables as you can. In other words, do not include too much information. Tables with multiple

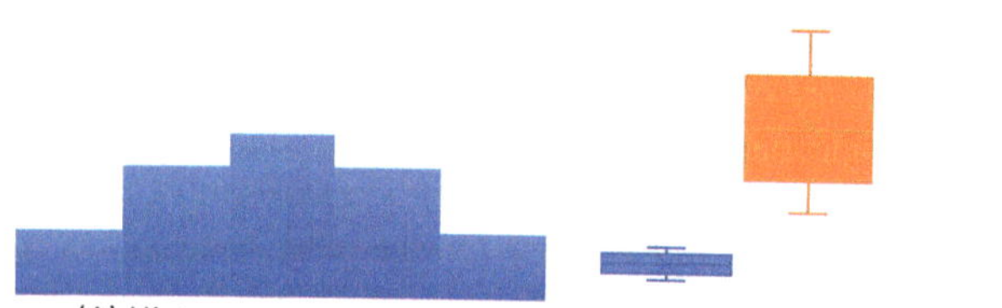

FIGURE 101.2 Common Types of graphs used for continuous variables. (A) Histogram. (B) Box and whisker plot.

comparisons and too many columns are harder to read and understand. For example, some authors may choose not to include a separate column for *P*-value if most of the variables are within one category (statistically significant or nonsignificant) and choose to provide that information as a footnote. Similarly, a column for missing may not be needed and that information could be in a footnote.

- Do consider using greater than (>) or less than (<) symbols if it makes your table easier to use while at the same time conveys the message. For example, you could use <0.05 or <0.001 when stating the *P*-value that indicates statistical significance versus reporting the actual value of *P*-value.
- Do use percentage when describing categorical variables in addition to or instead of frequency and be consistent. For example, whether you decide to use row or column percentage, use it for all variables in the table.
- Do avoid long multidigit numbers. You can do this by rounding fractions and expressing large numbers in thousands or millions when appropriate such as a number of platelets or white blood cells.
- Do avoid long footnotes. The footnote is to make the table easy to read and understand but not a replacement for the manuscript.

General consideration for graphical presentation

Researchers frequently use figures as needed to comprehend information that is too complex to be described efficiently in a paragraph. Graphs have to be visually appealing with a careful selection of colors.

Do and Do not

- Do present data clearly. This includes using clear title and labels.[1] If you use symbols or different colors, do include a legend.
- Do make the graph with the appropriate size. It should not be too large that it fills the whole page nor too small that makes it hard to read.
- Do not use three-dimension instead of two.
- Do not change the scale, intervals, and unit to emphasize certain sections. This may mislead the reader to the wrong conclusion.
- For multicolor graphs, be cautious that some readers may be color-blind and do use color-blind friendly palette or consider sticking to black and white.[5]
- For multicolor graphs, do use different endpoint colors such as one warm and one cold or one dark and one bright ...

Other types of graphs

There are many other types of graphs that are used in clinical research including translational research of interventional radiology. Some of the common ones are listed in the table[1] but describing them in detail is beyond the scope of this chapter.

Figure type	Common use	How it is look like
Donut chart	Graphical modification of stacked bar chart.[1] Used to summarize categorical variables. A common modification of donut chart is a sunburst chart that used to illustrate hierarchal data.	
Scatter plot	Graphical evaluation for relationship between two quantitative variables. Each observation is represented by a dot with each of the coordinates corresponding to one of the variables.[4]	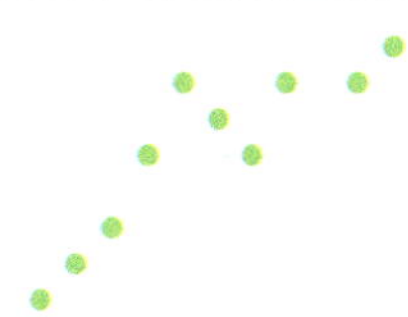
Forest plot	Graphical comparison of results from different studies in meta-analysis.[4]	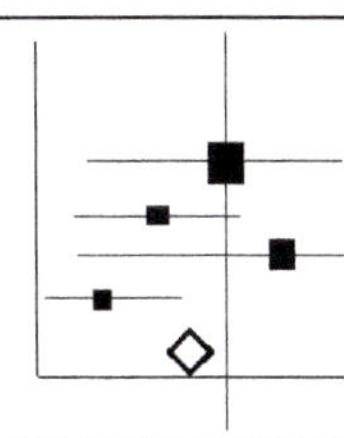
Line graph	Commonly used to track changes over time. Time on the x-axis and value of the variable on the y-axis. Each time point will be presented as a dot then a line will connect between the different dots (time point) to show changes over time.[4]	

Get started

1. Take a look at example articles of published literature in interventional radiology such as the one below and review their tables and graph.
2. If you involve in research, do discuss the choice of a particular graph or table format with your statistician or senior author.
3. Take a look at further information about data presentation. Some of those are listed below.
4. The best way to get better at data presentation is through practice. Get involved with drafting the results section and designing tables and graphs and with time you will be better.

Real-world examples

1. Ho K, Rao S, Honeybul S et al. *A Multicenter Trial of Vena Cava Filters in Severely Injured Patients.* New England Journal of Medicine, 2019.[6]
2. Higashigaito K, Schmid T, Puippe G, Morsbach F et al. *CT Angiography of the Aorta: Prospective Evaluation of Individualized Low-Volume Contrast Media Protocols.* Radiology, 2016.[7]
3. Hinde K, Maingard J, Hirsch J, Phan K et al. *Mortality Outcomes of Vertebral Augmentation (Vertebroplasty and/or Balloon Kyphoplasty) for Osteoporotic Vertebral Compression Fractures: A Systematic Review and Meta-Analysis.* Radiology, 2020.[8]

References

1. Hertel J. A picture tells 1000 words (but most results graphs do not): 21 Alternatives to Simple Bar and Line Graphs. *Clinics in Sports Medicine*. 2018:441–462. https://doi.org/10.1016/j.csm.2018.04.001.
2. Spriestersbach A, Röhrig B, du Prel J-B, Gerhold-Ay A, Blettner M. Descriptive Statistics The Specification of Statistical Measures and Their Presentation in Tables and Graphs Part 7 of a Series on Evaluation of Scientific Publications. *Deutsches Ärzteblatt International*. 2009;106(36):578–583. https://doi.org/10.3238/arztebl.2009.0578.
3. Duquia R, Bastos J, Bonamgio R, González-Chica D, Martínez-Mesa J. Presenting data in tables and charts. *An Bras Dermatol*. 2014;89(2): 280–285. https://doi.org/10.1590/abd1806-4841.20143388.
4. Mishra P, Pandey CM, Singh U, Gupta A. Scales of measurement and presentation of statistical data. *Ann Cardiac Anaesthes*. 2018;21(4): 419–422. https://doi.org/10.4103/aca.ACA_131_18.
5. Weissgerber T, Winham S, Heinzen E, et al. Reveal, don't conceal: transforming data visualization to improve transparency. *Circulation*. 2019;140(18):1506–1518. https://doi.org/10.1161/CIRCULATIONAHA.118.037777.
6. Ho K, Rao S, Honeybul S, Zellweger R, Wibrow B, Lipman J, et al. A multicenter trial of Vena Cava filters in severely injured patients. *N Engl J Med*. 2019;381(4). https://doi.org/10.1056/NEJMoa1806515.
7. Higashigaito K, Schmid T, Puippe G, Morsbach F, Lachat M, SEifert B, et al. CT angiography of the aorta: prospective evaluation of individualized low-volume contrast media protocols. *Radiology*. 2016;280(3):960–968. https://doi.org/10.1148/radiol.2016151982.
8. Hinde K, Maingard J, Hirsch J, Phan K, Asadi H, Chandra R. Mortality outcomes of vertebral augmentation (Vertebroplasty and/or Balloon Kyphoplasty) for osteoporotic vertebral compression fractures: a systematic review and meta-analysis. *Radiology*. 2020;295(1):96–103. https://doi.org/10.1148/radiol.2020191294.

Chapter 102

Manuscript preparation

Christopher Awad[1] and Alexander Martinek[2]

[1]*Emory University School of Medicine, Atlanta, GA, United States;* [2]*Morehouse School of Medicine, Atlanta, GA, United States*

See Table 102.1 for a worksheet to take you through the major lessons of this chapter

- Introduction—outline of questions or topics to tackle throughout the chapter, and description of how the chapter will deal with them.
  - While arguably the most critical skill for any scientist is the ability to formulate a hypothesis and test that hypothesis in a repeatable and scalable fashion, manuscript writing proves to be equally as important for scientists in order for them to share the process, knowledge gained from an experience, and the ways that they believe their investigation moved a conversation forward productively. Writing manuscripts can be technically challenging

TABLE 102.1 The ABC of manuscript writing worksheet.

Areas	Description
#1 Outcomes	Identity research questions that are Relevant Demonstrable Specific
#2 Materials methods	What materials were used? Who were the subjects of our study? How was our study designed? How did our design influence our outcomes? What procedure did we follow? A note on verb tense: The verbs are written in the simple pat tense and are passive
#3 Results	What are our results? What are our most significant results what leads us to their significance? What are out most interesting results why? Be clear, concise, and objective in describing your Results **Notes on verb tenses:** **A** For a general report on results, the simple past tense and passive sentence tends to be used, for example, the difference between the X and Y groups was significant **B.** When referring to or highlighting results in tables, figures, or charts, the simple present tense is most commonly used, for example, "the most interesting aspects of this graph is ...," "table 1 shows the summary statistics for ...," "looking at Figure 3, it is apparent that ..."
#4 Introduction	Why is your research meaningful? What is commonly understood/known about the topic? What are your other research objectives/questions? **How to share a substantive introduction establish the field** **A** Show that the general research area is important, central, interesting, and problematic in some way: **B.** Introduce and review terms of previous interest/exploration in the area

Translational Interventional Radiology. https://doi.org/10.1016/B978-0-12-823026-8.00102-4

and provides authors with the opportunity to share their work in organized and comprehensible ways.[1] This chapter describes the classical steps that guide all authors to systematically record and share quality knowledge of the following components; *materials and methods, outcomes, introduction, references, discussion, abstract, and title.* Radiological manuscript writing has evolved and a wealth of experience in the community has suggested this be the order for radiological manuscripts that advance readership.[2]

- Topic 1—Outcomes: *What ended up happening?*
  - Selecting appropriate results from any investigation helps reflexively define the way scientists share their question, while also providing the key incentive for readers of any manuscript. As demonstrated in Fig. 102.1, the outcomes help refine the manuscript process in several ways:
    - narrows the literature review of relevant topics to help your readers understand your investigation more foundationally
    - structures the conversation on terms that can be supported by your investigation and can be corroborated by the field
    - supports the results that you share in your manuscript from the plethora of understanding your team gained during their scientific exploration
    - champions the methodology and rationale for your methodology within the field
- Topic 2—Materials and Methods: *How did it happen?*
  - This section provides particularly high value for your work as it tends to be the most highly scrutinized or the most highly praised; clear and detailed descriptions of the actions taken to achieve the results to be shared are critical to providing others the opportunity to repeat the experience or trust the results you have achieved. Structure your manuscript in the simple past passive tense so that your language would read *"The images were taken before the patient visit every week."* A general outline for these sections includes the following:
    - Study design and topics
    - Imaging/procedural/study design techniques
    - Techniques and methods of Analysis with or without justification
    - Statistical analysis of the data
- Topic 3—Introduction: *Why you did what you did*
  - Providing readers with a rationale at the beginning of your manuscript drives home the answers to two critical questions for readers: "Why should I care about what you did?" and "Why does it matter?" Introductions are meant to answer those two questions while being aimed at a specific audience; when manuscripts are clear about who their intended beneficiary is, there is less room for confusion around the outcomes and there are more clearly defined stakeholders. Introductions require three things: (1) explanation of the area of interest and relevance to readers; (2) demonstration of the gaps within the field and the need for further exploration; (3) purpose of your work. Sharing the relevance of the issue, and its significance to the population or discipline, naturally feeds into a summary of the need for the work done by the manuscript writers. That foundational work is necessary to justify the purpose of your work and the hypothesis to be tested (Fig. 102.2). A narrowed and focused rationale should rely on an introduction to the work that is steeped confidently in an understanding of the field and coupled with a demonstrable need.[4]

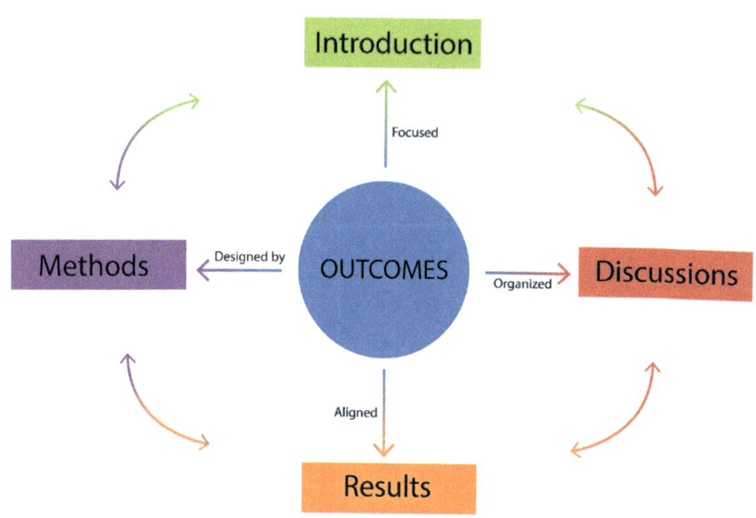

FIGURE 102.1 The cycle of manuscript refinement.

Table 1: The ABC of manuscript writing worksheet

Areas	Description
#1 Outcomes	Identify research questions that are: • Relevant • Demonstrable • Specific
#2 Materials & Methods	• What materials were used? Who were the subjects of our study? • How was our study designed? • How did our design influence our outcomes? • What procedure did we follow? **A note on verb tenses: the verbs are written in the simple past tense and are passive*
#3 Results	• What are our results? • What are our most significant results & what leads us to their significance? • What are our most interesting results & why? ***Be clear, concise, and objective in describing your Results* ******Notes on Verb Tenses:*** a. For a general report on results, the simple past tense and passive sentence tends to be used e.g., "*The difference between the X and Y groups was significant.*" b. When referring to or highlighting results in tables, figures or charts, the simple present tense is most commonly used, e.g., "*The most interesting aspect of this graph is …*", "*Table 1 shows the summary statistics for …*", "*Looking at Figure 3, it is apparent that …*"
#4 Introduction	• Why is your research meaningfulful? • What is commonly understood/known about the topic? • What are your other research objectives/questions? **How to share a substantive Introduction** **Establish the field** a. Show that the general research area is important, central, interesting, and problematic in some way; b. Introducc and review items of previous interest/exploration in the area.

FIGURE 102.2 The ABC of manuscript writing: a worksheet.

Establish the gap

Indicate how your work fills a gap in the previous research, or extends previous knowledge in some way.

Share your purpose

a. Outline purposes or state the nature of the present research
b. List research questions
c. Announce principle findings;
d. State the value of the present research;

#5 Discussion

1. What are our major findings?
2. What is the significance/implication of the findings?
3. How do the findings advance the field?

The way to discuss:

a. Summarise the result that we want to discuss;
b. Share a brief comparison of individual findings with findings from other studies (*making sure to highlight some similarities and differences*);
c. Provide your hypothesis and postulated understanding that explains your findings while providing evidence from your investigation and the literature that supports your conclusion
e. Plainly state the limitations, weaknesses, and assumptions of your study
f. Indicate the importance of your work by sharing or demonstrating applications, recommendations, and further/future implications

---

#6 Discussion

a. Summarise and bring together the main areas covered in the composition
b. Share a concluding/final comment or judgement on what the manuscript presents as a whole and offer suggestions for improvement and speculation on future directions

---

#7 Abstract & Title

Write concise, self-contained and good abstract that reflect the outlook of your manuscript with summary, concision, and word count as guiding factors.

FIGURE 102.2 cont'd

- Topic 4—References: *Where to go next if you want to know more*
  - This evaluation of the field for critically relevant work gives readers the confidence that your analysis is steeped in, and in complement to, published and validated work. Citing recent references highlights the relevance of your study. Authors should be familiar with the avenues they would like to explore for submission and publication. Format the citations and bibliography according to the Journal's instructions. Do not exceed the maximum allowed number of citations.
- Topic 5—Discussion: *What this means to you and others*
  - Providing readers with a meaningful interpretation of the data that is both specific to them and novel for their consideration leads most discussions to be constructed with five key points:
    - Summary of the most relevant ideas: Starting off with a reiteration of the purpose of the study, manuscript writers should briefly share an accurate and summative version of the evidence that supports the conclusion. Here it is appropriate to state the answer, as the writer sees it, to the initial research purpose or hypothesis of the investigation
    - Interpretation: Here, the reader is expecting a synthesized version of the meaning of the results. Emphasis should be placed on the following factors:
    - Significance or originality of each finding
    - Advancements that are provided by these results, while avoiding self-promoting language
    - Relationship of your results to the field and to previous studies
    - Statement of the following:[3]
    - Relevant positive or negative data
    - Discrepancies in methodology or underlying rationale
    - Argument for the primacy of used rational/methodology/etc.
    - Limitations of the study
    - Summary of the most relevant details and themes from your manuscript
    - Statement of the primary conclusion(s) that the manuscript aims to communicate with readers
    - Potential for future studies, investigative changes, or collaboration
- Topic 6—Abstract: This short statement previews for the reader what ahead.[4] Other than the title, this section of a manuscript serves as its billboard: readers use the title as the first step to reading the abstract, and the abstract as the step before investigating the rest of the paper. The elements that readers are commonly looking for in the abstract include, but are not limited to:
  - Importance of the subject of the manuscript and the summarized identification of the knowledge gap
  - Aim(s) of the manuscript's focus
  - Indication/Summary of the methods used
  - Summary of the key finding(s)
  - Implications of the findings and/or value of the current study
- Topic 7—Title: The importance of the Title will not be understated by this resource: search engines commonly use words in manuscript titles to categorize, organize by relevancy, and display articles en mass to the curious reader.[5,6] Titles should be limited to 15 words or less, should be declarative in nature, and should offer an immediate understanding to your readers of what they will find inside if they investigate further.[7] More than anything else, titles should be composed to be informative, relevant, and brief. The previous sentence contained more words than your title should. Answering the question of what readers can expect by clicking on the title would be a good place to start your naming process and should come at the end of your manuscript preparation. Do not forget about the keywords! These are great ways to enhance the algorithm and help readers focus on the scope of the manuscript's presentation.

Get started

Looking to get started on manuscript writing? Here are some things you can do right away (before even reading this chapter)!

1. Get a handle on your topic and materials: collect, organize, and edit your starting materials, tables, and references
2. Look at the guidelines for authorship at a publication venue: reviewing authorship guides or even articles published in your target publication venue can help jump start thinking and planning
3. Scrap your first outline: Plan on creating a draft, outline, or structured display of your thoughts that you will not have any pressure around.
4. Start with others: Creating accountability with deadlines, group work, early drafts and comments can help create a foundation that will quickly provide form and momentum to manuscript teams.

Potential pitfalls

While many seasoned authors will inevitably have their own wizardry associated with the steps involved in manuscript writing, editing, submitting, and revising, there are a few key points, summarized below, that are good to keep in mind. Regardless of your topic or discipline, the following ideas will be helpful to revisit as another 'checklist' throughout your manuscript creation.

1. Losing the Point
   a. Authors often find creative and new ways of describing the outcomes of their work through the manuscript writing process. Driving narrowly attentive, consistent language without repetition will help keep your work in focus and unburden the steps taken after drafting.
2. Don't play catch-up
   a. The details in this chapter are a succinct summary of many approaches to manuscript writing. It emphasizes the need to start with the most concrete parts of your investigation and end with the most 'abstract' (ie We suggest you craft the title last). Many manuscripts would benefit from this stepwise approach in order to assure the correct consideration of audience, novelty, clarity of purpose, copy-editing, publication venue and guidelines, editorial nuances, and the ever-evolving 'review process.'
3. Over- and Under-tones
   a. To this point, pitfalls have been concerned with rhetorical nuance: communicating clearly and taking the correct overall steps. This pitfall involves the use of language to communicate something in the text in a subversive or even overt way that is not the main point of your manuscript. Whether those are imitative undertones of other published work, or blatant overtones of misappropriated texts, language should be marshaled to communicate the evidence-based idea under investigation. The ways that teams avoided confirmation bias should be on display throughout the investigations' construction and execution, transparently and brilliantly reflected in the writing of the manuscript as well.
4. Citations
   a. While many other common pitfalls can be stratified, the last one will be summarized around the use of citations. It is precisely because all manuscript teams have great intentions to add to the growing body of knowledge that citations are to be taken so seriously. Citing critically and appropriately contributes to the readers' understanding of your work and your contribution while also bolstering the argument at hand. Citation should be cross-referenced by co-authors to assure the highest rigor in citation use. The ways that manuscript teams carve out their citations for a work help orient audiences to the complex relationships that they want to add to. Proper citation use also helps manuscript authors avoid the first pitfall and keep the message of the manuscript clear.

Real-world examples

- Examples in literature—*These three examples focus on various components of the manuscript and offer reasons for their exemplary nature*:
  1. de Oliveira et al.[8] have done a great job keeping the title concise, using detailed and relevant keywords, and providing appropriate citations for the work. While a case report is a more straightforward example, the key elements are followed, and the reader is satisfied with the description of a patient diagnosed with coronavirus with bleeding from varicose veins in an emergency with worsening of his lung function.
  2. Plascencia et al.[9] have engaged readers with an intimate review of a decent population for the topic of the manuscript with detailed visuals, helpful graphics, and powerful analytics and conversation. The discussion is specific and focuses on the key message that the authors want to deliver: the efficacy of embolization for Infantile Hemangiomas.
  3. E Ray Jr et al. have demonstrated what an efficient cross-sectional literature review of 893 papers over two distinct time periods to share striking conclusions on the shifts of metrics like authorship and academic focus.[10] By introducing readers to the point of the review early with exacting language, the authors masterfully summarize changes in the landscape of the radiological literature statistically and contextually, grounding the research by taking manuscript length to address the potential biases of the author and to discuss the contextualizing and appreciable factors that might explain the "globalizing" phenomenon they point to (Table 102.1).

Resources

For other resources, see the following:

- Katz M. *From Research to Manuscript: A Guide to Scientific Writing*. Springer Science & Business Media; 2009.
- Makin, TR, de Xivry JJO. Ten common statistical mistakes to watch out for when writing or reviewing a manuscript. *Elife*. 2019;8.

- Trail GT, James JD. Seven deadly sins of manuscript writing: reflections of two experienced reviewers. *J Glob Sport Manag*. 2016;1(3–4), 142–156.

References

1. Swales JM, Feak CB. *Academic Writing for Graduate Students*. 3rd ed. University of Michigan Press; 2012. https://doi.org/10.3998/mpub.2173936.
2. Bannas P, Reeder SB. How to write an original radiological research manuscript. *Eur Radiol*. 2017;27:4455–4460. https://doi.org/10.1007/s00330-017-4879-8.
3. Matosin N, Frank E, Engel M, Lum JS, Newell KA. Negativity towards negative results: a discussion of the disconnect between scientific worth and scientific culture. *Dis Model Mech*. 2014;7:171–173 [PubMed: 24713271].
4. Morley J. *Academic Phrasebank*. University of Manchester; 2017.
5. Rosenfeldt FL, Dowling JT, Pepe S, Fullerton MJ. How to write a paper for publication. *Heart Lung Circ*. 2000;9:82–87 [PubMed: 16352000].
6. Wells WA. Me write pretty one day: how to write a good scientific paper. *J Cell Biol*. 2004;165:757–758 [PubMed: 15210725].
7. Sheiman R. *Writing an Original Research Manuscript: The Do's and Don'ts, Chicago*. 2009.
8. de Oliveira DS, Rímoli JR, Valle LG, Benini BB, de Brito Siqueira LT. Transjugular intrahepatic portosystemic shunt in coronavirus patient: a case report. *Am J Interv Radiol*. 2021;5:6.
9. Plasencia AR, O'Higgins TV. Large parotid and cheek hemangiomas refractory to medical treatment: is there a role for embolization? *Am J Interv Radiol*. 2020;4:21.
10. Ray Jr CE, Gupta R, Blackwell J. Changes in the American interventional radiology literature: comparison over a 10-year time period. *Cardiovasc Intervent Radiol*. 2006;29(4):599–604. https://doi.org/10.1007/s00270-005-0209-7. PMID: 16565796.

Chapter 103

Research team: Standard personnel and roles

Shelly Bhanot[1] and Merve Ozen[2]
[1]*Rush University, Chicago, IL, United States;* [2]*University of Kentucky, Lexington, KY, United States*

Key points

- Identifying the specific roles of the research team members is crucial to create a capable team.
- Agreement among the team members on the process of obtaining and sharing data, establishing credit and authorship early in a project is imperative for success.
- Creating expectations in a team to work toward a shared null or alternative hypothesis optimizes project success.

Introduction

Imagine that you are in the trauma bay, and a new patient is rolled in. You know that the patient was in a motor vehicle accident with a loss of consciousness. With any trauma, there are strict protocols, but they cannot be completed without a cohesive team to carry them out. Similarly, when conducting research, the development of a team allows for the means to complete a project with efficiency and outcomes. In the literature, research teams are referred to as the "fundamental social unit of science."[1] Behind the research that discovers the future of science is a well-constructed research team. Every research project is unique and has different components. This chapter is dedicated to ensuring that a research team consists of the necessary members and provides strategies that promote synergy and success.

Chapter

The effect of collaborative team efforts on scientific discovery are well known. In one study analyzing approximately 20 million papers and 2.1 million patents over 5 decades, it was shown that the number of coauthored articles has grown in contrast to the number of single-authored papers. This study demonstrated that in the last 45 years, the average number of authors on a paper for science and engineering has increased from 1.9 to 3.5 authors. It does not come as a surprise that teams tend to produce high-impact research relative to standalone authors. Some theories of thesis shift include the fact that research has increased in scale, complexity, and cost over the recent decades.[2] There has been some skepticism that, despite the more excellent collective knowledge that comes from a team, productive losses occur with lack of coordination among the team members; ultimately leading to low impact research, as Scott Fitzgerald once said, "No grand idea was ever born in a conference."[3] To fully take advantage of the potential impact of a research team, the foremost important aspect is to establish a shared vision. With scientific research, a shared vision comes in the form of a hypothesis. Two types of hypotheses remain dominant throughout the development of research: null and alternative hypotheses (please refer to Chapter 60 for more detailed definitions). The first step is to propose a null hypothesis. Next, the researchers will perform experimentation or observation, which will determine if the null hypothesis is rejected or not. If the null hypothesis is rejected, based on parameters such as *P*-value or confidence intervals, then the project shifts perspectives and begins to explore the alternative hypothesis.[4] Research often follows this pathway and having a unified hypothesis often serves as a strong starting point before forming a plan and research team to inspect that hypothesis.

Research teams often have core members to help in the investigation of a hypothesis. Typically, the principal investigator (PI) and their coinvestigator (Co-I) will be the creator of the fundamental theory and most responsible for the research being conducted. These members oversee the project as a whole and guide the additional research members such as trainees helping with the project. These members also aid in ensuring that projects comply with institutional and federal policies. An institutional review board (IRB) often requires that a PI and Co-I are explicitly acknowledged.

Translational Interventional Radiology. https://doi.org/10.1016/B978-0-12-823026-8.00097-3

A project's significance is determined by the impact of its data. Therefore, a project's data coordinators and biostatisticians become imperative members of a team. Their responsibilities consist of the collection and interpretation of data. The conversation of data into meaning statistics requires thorough knowledge of statistical principles detailed in the "Statistical Principles" section. It is the use of statistical data that ultimately dictates the acceptance or rejection of a project hypothesis.

In addition to the interpretation of the data, the management of an extensive clinical project relies on research coordinators (RCs). RCs are integral professional figures of significant research. As clinical research increases in complexity, RCs are vital to preparing, coordinating, and directing clinical trials under the supervision of the PI. This role is typically recognized as critical to large-scale projects. One study out of Italy even discussed how it is essential to have specific employment contracts for this role as clinical trial units continue to grow across the country.[5]

Lastly, it is essential to stress the significance of synergy in a research team. Having the appropriate hypothesis and team members increases a project's chance of success; however, a well-integrated team is what allows for collaboration, integration, and flexibility within an organization. Being clear about defining roles, responsibilities, and authorship expectations, in the beginning, is critical. It is only when each member understands their position and significance in a project that a joint mission can be created. Sharing recognition and credit within a work team fosters trust. Trust is the real building block for collaboration that allowed a project to grow and flourish.[6]

Get started

Below are the best steps to create a highly effective and integrative team to optimize a research endeavor (Fig. 103.1):

- Have a clear hypothesis (null vs. alternative) that each team member clearly understands.
- Team members should be delegated specific roles and responsibilities.
- Each member of the team should understand their contribution in terms of expectations and authorship toward the final research product.
- Anticipate challenges and promote a synergistic culture that allows for feedback and collaboration among members.

Potential pitfalls

There are several potential pitfalls when developing or working in a research team. Here, we will discuss three major ones: distraction, conflict, and time scale.

A classic analogy for a major con of a research team can be seen with a group of studying medical students. In a group study session, when there are 3–5 participants, there can be productive conversation and exchange of ideas. Difficult concepts can be elucidated through group teaching. However, unrelated distractions and discussion can insidiously sidetrack a group session. Similarly, in a research team, by nature of a group of people working together, it can be difficult to stay on task. Clear agenda and pre-meeting planning are ways to control the variables of distraction.

In a similar vein, when coordinating a larger group of team members, it is important to navigate each other's conflicting opinions or communication styles. The principal investigator becomes important in these situations as leadership skills are integral to team success. Here lies another potential pitfall: weak or uninterested management of team members could be a major deterrent to exploring a hypothesis. Strong leadership from not only the principal investigator but also every member of a team is important to fulfill designated tasks and weave each other's work into a cohesive project.

Finally, time scale and its limitations are a major pitfall of a research team. Imagine a scenario where the members of the team are performing research *in addition to* their normal expected duties. Coordinating schedules and carving time among many members can elongate the time it takes to complete a project. Understanding the elongation of time in a research team is important and is often good. Taking time to produce clear goals, hypotheses, and methods often leads to excellent results.

Real-world examples

Real world examples of research teams are ever-present in any scientific community. Whether in medicine or another field which engages in research, a cohesive team is necessary to make meaningful advancements. The impact of a research team can be seen in large scale multi-center trials or single center improvement projects.

One small, but salient, example includes a recent quality improvement project performed at Rush University Medical center in Chicago, Illinois. Radiology resident's noticed high phone call volumes to the on-call telephone. Anyone in the hospital who had a radiology-associated question (e.g. "Where is the neuroradiology reading room?", "Can you take a look at a stat chest radiograph for me?") would call the same particular extension – the on-call telephone. These phone calls were tedious to handle and disruptive to the workflow expected of the radiology resident holding that on-call telephone. Two co-investigators, under the supervision of a supporting attending physician (the principal investigator), organized a project to create an automated answering service that could re-direct callers to designated workrooms. The hypothesis was that if there was an organized answering service, then the hospital providers could directly access the reading radiology physicians that could best answer their questions; this would, in turn, improve resident workflow. Two data coordinators gathered data based on the current call volumes and which reading rooms were most commonly requested (e.g. the neuroradiology and chest radiology reading rooms). Using this data, the co-investigators assembled interdisciplinary meetings with reading room representatives to decide designated phone extensions for each reading room.

The creation of the automated answering service also lended itself to post-creation data analysis. After the answering service was instated, a biostatistician performed a two-tailed analysis of the call volumes both before and after the automated answering service. Ultimately, with the help of many members of a research team sharing a common vision, the automated service was found to statistically reduce call volumes thus improving radiology resident workflow. This is a single-center example of how organized research teams can lead to substantial impact.

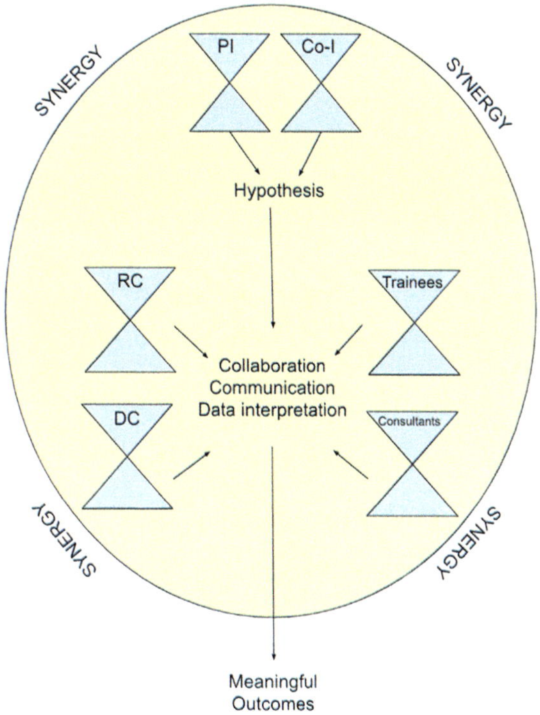

FIGURE 103.1 Research team algorithm.

Resources

Resources involved in a research team are broad. When performing research, information is often sorted into two major categories: primary and secondary. Primary resources are usually first-hand accounts or original works. Secondary resources typically include evaluation and analysis of primary sources. Blending these types of resources is useful towards creating directives and hypotheses. Minds of a multi-person research team can synthesize many different resources into clean ideas and data.

Resources of a research team are also defined by their respective institutions. Based on time and leadership constraints, assembling a team is often defined by availability of resources from each institution.

References

1. Milojević S. Principles of scientific research team formation and evolution. *Proc Natl Acad Sci Unit States Am.* 2014;111(11):3984. https://doi.org/10.1073/pnas.1309723111.
2. Wuchty S, Jones BF, Uzzi B. The increasing dominance of teams in production of knowledge. *Science.* 2007;316(5827):1036–1039. https://doi.org/10.1126/science.1136099.
3. Donaldson S. A Fitzgerald autobiography. *The F Scott Fitzgerald Review.* 2002;1(1):143–157. https://doi.org/10.1111/j.1755-6333.2002.tb00063.x, 2002/01/01.
4. Ludbrook J, Dudley H. Issues in biomedical statistics: statistical inference. *Aust N Z J Surg.* 1994;64(9):630–636. https://doi.org/10.1111/j.1445-2197.1994.tb02308.x, 1994/09/01.
5. Cagnazzo C, Guarrera A, Cenna R, et al. [Clinical research: enough players to get out there?]. *Recenti Prog Med.* 2019;110(6):285–291. https://doi.org/10.1701/3182.31609. Ricerca clinica: abbastanza giocatori per vincere la partita?
6. Stokols D, Hall KL, Taylor BK, Moser RP. The science of team science: Overview of the field and introduction to the supplement. *Am J Prev Med.* 2008;35(2):S77–S89. https://doi.org/10.1016/j.amepre.2008.05.002.

Chapter 104

Patent basics

Tushar Garg[1] and Apurva Shrigiriwar[2]
[1]*Division of Vascular and Interventional Radiology, Russell H. Morgan Department of Radiology and Radiological Science, The Johns Hopkins University School of Medicine, Baltimore, MD, United States;* [2]*Division of Gastroenterology and Hepatology, The Johns Hopkins University School of Medicine, Baltimore, MD, United States*

Key points

- A patent is the grant of property rights that is issued by the United States Patent and Trademark Office to the inventor of an invention.
- An invention can be only be patented if it fulfills the criteria of being useful, novel, and nonobvious.
- Medical patents help to improve the quality and delivery of medical care to patients all over the world, and they are essential for the development of new treatment and investigation modalities.

Why it matters?

Any original creation of the human intellect, like artistic, literary, technical, or scientific creation is referred to as an intellectual property (IP) of the individual, and intellectual property rights (IPR) are the legal rights given to an inventor or a creator to protect their invention or creation for a specific time after it is registered.

To protect the IPR all over the world, an organization called World Intellectual Property Organization (WIPO) was formed in 1967 in which was later adapted into an agency of the United Nations in 1974. WIPO outlines the framework for IP and IPR, and it also regulates the various policies concerned with IPR throughout the world. In the United States, the United States Patent and Trademark Office (USPTO or Office), an agency of the US Department of Commerce, is responsible for protecting the IP by granting patents for the protection of inventions and registering trademarks.

There are various types of intellectual properties (see Fig. 104.1), but not all of them are recognized in the United States. The United States recognizes four categories of intellectual properties: copyrights, trademarks, trade secrets, and patents. A brief description of these properties is provided below with a special focus on patents and the patent obtaining process.

Copyrights

Copyright is a form of protection that is granted by the law for original works of authorship fixed in a tangible medium of expression. The provisions for granting copyright are mentioned in the US Constitution. It is a form of intellectual property law that protects original works, including literary, dramatic, musical, and artistic works. It does not protect facts, ideas, systems, or methods of operation even though it can protect the way these things are expressed.

Trademarks

A trademark is a word, phrase, symbol, or design that identifies and distinguishes the source of the good of one party from those of others. Trademarks are different from patents and copyrights, in the sense that they do not expire after a set term of years, and the trademark rights duration is based on the actual use of the trademark by the concerned party.

Trade secrets

A trade secret is any practice or process of a company that is generally not known outside of the company. To

FIGURE 104.1 Different types of intellectual properties.

Translational Interventional Radiology. https://doi.org/10.1016/B978-0-12-823026-8.00104-8

be considered a trade secret, the information should provide a company an economic advantage over its competitors and should be a product of internal research and development.

Patent

A patent is an intellectual property right granted to an inventor by the concerned government office for a novel invention. Here the term invention refers to a solution to a problem in the form of the development of a product or process. By obtaining a patent, the owner does not gain a right to practice an invention; instead, it confers them the right to exclude others from making, using, or selling the claimed invention. The patent owner's right to exclude is available for the life of the patient, which is fixed by statute in most cases to 20 years from its earliest relevant filing with the concerned government authority.

To obtain a US patent, the investor must file the patent application, which contains one or more "claims" with the USPTO. Two types of patent applications are recognized by the USPTO, provisional, and nonprovisional. At the time of filing a patent, the inventor must disclose all relevant information in a descriptive way to the patent office in the application, and this is available to the public even after the patent is granted for exclusive rights of an invention to an inventor. The process of obtaining a patent from the first step of coming up with an idea of obtaining a patent is visualized in Fig. 104.2.

For an invention to be suitable to gain a patent, it needs to fulfill the following criteria:

- Usefulness: the invention must have some real-life applications, the invention can either be used for industrial purposes or can be used for a practice purpose
- Novelty: the invention must be new technology that has not been published or available in the country of filing or elsewhere in the world before the date of patent filing.
- Nonobviousness: any invention that can be developed or performed by any ordinary skilled person is considered obvious and cannot be patented.

The different types of patents that are granted by USPTO are discussed in Fig. 104.3.

Types of medical patents

The main categories of medical patents are medical devices, chemical products, and pharmaceutical drugs, healthcare information technology, medical and surgical methods, and regenerative medicine.

Medical device patents include physical devices used by physicians, hospitals, and other providers and include items such as diagnostic instruments, stents, implantable devices (prostheses), surgical tools, and many others.

How to get started: the process of obtaining a patent

Step 1: Search if the invention has already been patented on http://patft.uspto.gov/.

If the invention has not been patented then.

Step 2: Decide the type of patent application most applicable to your invention? Design Patent, Plant Patent, or Utility Patent (most common).

Step 3: Determine your patent filling strategy, that is, if you want to file the patent globally for international protection or you want to file it only in the United States.

The most common types of patents filed are utility patents, so the process of filing a utility patent is described in the next steps.

Step 4: Decide which type of utility patent you want to file? Provisional Patent or Nonprovisional Patent.

Step 5: Decide if you want an expedited examination? these include prioritized examination, accelerated examination program, first action interview, and patent prosecution highway.

Step 6: Decide who will file the patent? Yourself (Pro Se) or a registered attorney or agent (preferred)

Step 7: Now prepare an electronic filing by paying the application processing fees and applying for a customer number and digital certificate.

Step 8: Apply for a patent by using the Electronic Filing System as a registered eFiler and submit your application.

Step 9: The application is now examined by USPTO.

If the application is accepted proceed to step 10a otherwise see step 10b.

Step 10a: Pay the issue and publication fees.

Step 10b: File replies, requests for reconsideration and appeals and then if accepted follow step 10a.

Step 11: USPTO Grants Patent.

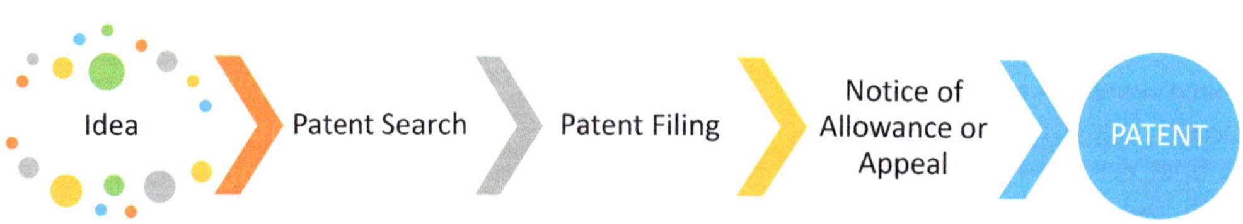

FIGURE 104.2 The patent process from start to finish.

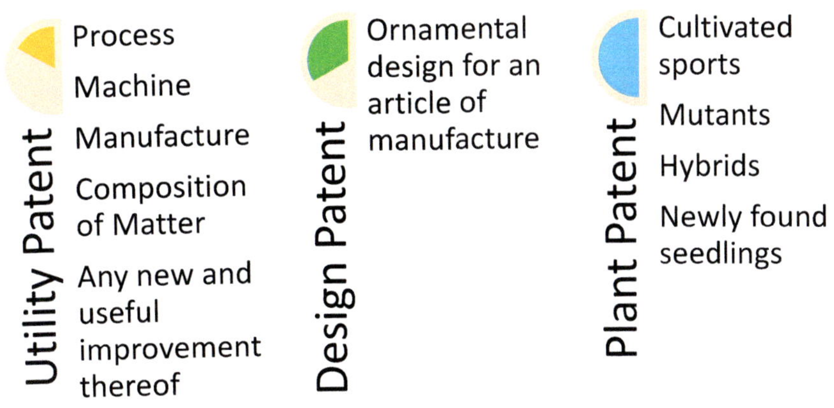

FIGURE 104.3 Different types of patents granted by USPTO.

Step 12: Pay maintenance fees due at 3 1/2, 7 1/2, and 11 1/2 years after the patent is granted.

References

1. Murphy A, Stramiello M, Lewis S, Irving T. Introduction to intellectual property: a U.S. perspective. *Cold Spring Harbor Perspect Biol.* 2015;7(8). https://doi.org/10.1101/cshperspect.a020776.
2. Mayfield DL. Medical patents and how new instruments or medications might Be patented. *Mo Med.* 2016;113(6):456–462. http://www.ncbi.nlm.nih.gov/pubmed/30228529. Accessed June 26, 2020.
3. Process for Obtaining a Utility Patent I USPTO. Accessed June 26, 2020. https://www.uspto.gov/patents-getting-started/patent-basics/types-patent-applications/utility-patent/process-obtaining.
4. General information concerning patents I USPTO. Accessed June 26, 2020. https://www.uspto.gov/patents-getting-started/general-information-concerning-patents.

Chapter 105

Venture pathways

William Borror
Department of Diagnostic and Interventional Imaging, University of Texas, Houston, TX, United States

Bullet-point summary of the chapter

- Creating a business plan is essential as it serves as a roadmap for the business and a marketing tool to present to potential investors
- Funding for new ventures comes in many different forms and the founders of a new venture need to have an understanding of the different types
- All new ventures must have an exit strategy as new investors will be hesitant to become involved with the business if there is no plan to recoup their investment

Prior chapters have focused on identifying a need in the market, developing a solution, and subsequently testing that solution. This chapter will focus on taking that solution from the lab bench to the marketplace by starting a company. Entire books have been written on starting a company as well as the individual topics that will be discussed in this chapter. This chapter will serve as a brief overview and provide a broad foundation in the basics of building a company. Putting together a business plan, funding the business, and exit strategies will be discussed.

Before starting a business, it is important to reemphasize that the product must solve an unmet need and in addition to this, there must be sufficient demand in the market that customers are willing to pay for that solution. Furthermore, the product must provide value to the customer. Value is commonly defined as the monetary worth of the benefits that the product provides to the customer. It is important to note that value and the price paid by the customer are not the same. For example, let us imagine a hypothetical product, The Widget, which is being sold for $50. ABC company decides to purchase The Widget because the product can increase their yearly sales and thus, their profit. By buying The Widget, ABC company can increase its yearly profit by $100. To ABC Company, the widget has a value of about $100. In this example, the value was equivalent to the additional profit derived from purchasing The Widget, but this is not always the case as the value can be subjective and may vary between customers. The value that a product provides to the customer should be greater than the price paid by the customer.

After determining that a product meets a market need and has the potential to provide value to its intended customers, a business plan should be assembled. Tables 105.1 and 105.2 summarize the different sections of a business plan and provide a brief description of each section. These sections provide a general starting point; however, there can be variability in the structure depending on the industry and personal preference. The business plan will serve as a roadmap as well as a medium for conveying your business idea to potential financiers, investors, and partners. A good business plan tells the audience a story and includes the people involved with the business, the opportunity, the context, and the balance of the risk to the reward. The people involved in the business venture are important to potential investors as without a strong management team and execution skills, even an excellent product can fail. Investors will want to see that management has experience in the industry, strong ties within the industry, and is potentially well known. Opportunities must exist for the industry that the business is trying to enter as investors will want to know whether the industry is large, rapidly growing, or a combination of the two. It is much easier to gain a share of a market that is large or is rapidly growing than to compete with established companies in a market that is either shrinking or has limited growth expectations. Context is also key as investors will want to know the regulatory environment of the industry and how the current macroeconomic environment will either create challenges or opportunities. Details of how management will surmount the challenges or embrace the opportunity should be elucidated.

Translational Interventional Radiology. https://doi.org/10.1016/B978-0-12-823026-8.00084-5

TABLE 105.1 Anatomy of a business plan.

Section	Description
Cover page and table of contents	The cover page is the first thing your audience will see and should be clean and professional in appearance. This page should include the business name, contact information, and company logo or emblem. The table of contents should provide structure and organization to the plan, allowing the audience to easily locate the information they might be interested in.
Executive summary	Just like the abstract of a journal article, this should capture the reader's attention and give a brief overview of the business. This may be the only section a reader browses before making a decision.
Business description	This section should state the purpose of your business, business objectives, and why the business will succeed. It should include a history of the business and a description of the product.
Business environment analysis	Readers should be convinced in this section of your business's potential, the industry in which the business will compete, and the market where your product will be sold. Competitors and similar products that they sell should be detailed here as well.
Industry background	Describes the range of products offered, barriers to entry, the overall profit of the industry, and predicted growth rate of profits and revenue. Whether the industry is made up of many smaller companies or is dominated by a few large competitors is detailed here.
Competitive analysis	Identify which companies solve the same problems as your product. For each of these companies, detail their strengths/weaknesses, market share, and why they have been successful. Explain what differentiates your product from your competitors. Detail the threats your company will face from competitors. In other words, how will competitors react when you try to enter the market?

TABLE 105.2 Anatomy of a business plan cont.

Section	Description
Market analysis	This section will focus on who will be purchasing your product and your target customers. Target customers and their characteristics are defined. It will answer if there is room for another product in the market if that market can expand to accommodate your product and how the demand for your product will grow.
Marketing plan	How will you bring your product to the market, how will your product be priced, and how will you motivate customers to buy your product? Identify marketing objectives, sales milestones, and how long you estimate it will take to reach them. Determine how much it will cost to acquire a customer, analyze buying behavior, and figure out how you will build loyalty for your product.
Operations plan	Provide an overview of the processes that will be used to turn ideas and raw materials into the end product. The cost to produce and sell the product along with the number of products that need to be sold to cover these costs should be explained. Outline where materials will be sourced from and the manufacturing process.
Management summary	Provide an overview of each manager's resume, their commitment to the business, and what they bring to the table. Address any weakness and answer whether anyone has successfully worked together previously.
Financial plan	How much money will be required to start your business and how will the money be used? Include projected revenue, operating margin, operating profit, net income, and net margin over a 3–5 year period. Provide documented sources for your projections such as literature or expert opinion.
Attachments and milestones	List major events in the development of the business. Provide more detail on other aspects of the business plan that you did not have a chance to go into more depth on in a prior section.

Lastly, an overview of the risk that potential investors would be taking and how this risk will be managed. In the end, investors are looking for reward and the reward must be sizable enough to compensate for the risk. In other words, risk and rewards are directly proportional to investors. As risk increases, an increase in the potential reward is also expected. For example, an article by Manigarta et al. noted that venture capitalists investing in early-stage companies expect a yearly return on investment of between 36% and 45% for early-stage investments and 26%–30% for a later stage, more established company. These expected returns can vary depending on the current macro environment as investors weigh the opportunity cost for a given investment. For example, the S&P 500 has returned an average of 8% per year from 1957 through 2018. When weighing a potential investment, investors may compare the returns to an S&P 500 index fund that is diversified and carries significantly less risk relative to the venture they are looking to invest in. Investors will want a significant spread between the estimated return of the venture and the expected forward returns of the index fund. This spread is needed to compensate for the increased risk.

Coming up with a business idea and a plan for that business often requires very little monetary investment. However, taking the idea from paper to the marketplace often requires a significant investment of capital. The entrepreneurs can invest their own capital in the beginning, but that initial investment is often insufficient, and outside investors are often needed. Common sources of funding include angel investors, venture capital, small business loans, private placements, proceeds from business competitions, and startup incubators (Fig. 105.1). Angel investors are high net worth individuals that may be a friend or family member of a partner within the business. They are present in the earliest stages of the business and give the first infusion of capital in exchange for a portion of ownership in the company to help get the business off the ground. These investors are looking for a significant return on their investment, but they are often more invested in helping the company get a good start rather than utilizing more predatory tactics of other types of investors.

Small business loans, otherwise known as debt financing, is another option that can be obtained from a bank. Rather than purchasing a portion of the equity in the business, the bank loans the business money and expects repayment of the original amount plus interest. These loans can either be secured or unsecured. Secured being that the business owner puts up some form of collateral such as the deed to a house to back the loan if the business would fail. This reduces the risk to the bank and results in a lower interest rate for the borrower. Alternatively, an unsecured loan does not have collateral and is only based on the borrower's creditworthiness as measured by their credit score among other factors. Similar to a loan, private placements are another option. Instead of borrowing money from a bank, the bank creates a financial product called a bond and sells those bonds to a preselected group of individuals or institutions. The business then repays these individuals or institutions the amount that was borrowed plus interest.

Business incubators can indirectly be a source of funding for early-stage businesses; however, their expectations and the services they provide can be very heterogeneous. Commonly, business incubators provide a place for entrepreneurs to work on their business and a network of contacts to help them succeed. This can reduce the new business's overhead; however, in exchange for this, the incubator may require a portion of the equity in the

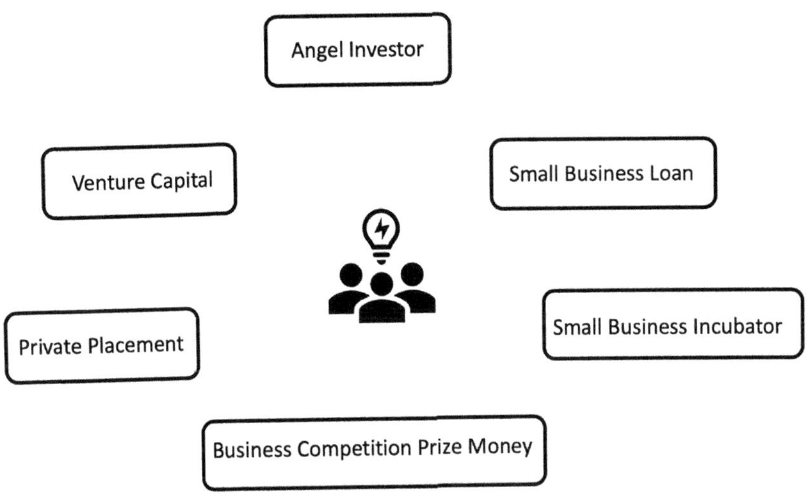

FIGURE 105.1 Sources of funding.

business. Incubators serve as a springboard for the launch of the business and provide opportunities for the business to participate in business competitions that provide prize money and exposure to potential investors. Potential contacts developed through the incubator could help foster partnerships between the new businesses and established players in the industry.

The last type of funding is venture capital, which is one of the most talked-about forms of funding. Venture capital (VC) is a form of private equity investing that focuses on companies that are in their earliest stages and have high growth potential. VC firms commonly raise money for a fund from investors such as public or private pension funds, banks, or high net worth individuals. This pool of money is then invested in many early-stage companies with high growth potential. In the end, the goal of the VC fund is to reach a liquidity event with the portfolio companies in which they invested and extract their initial investment plus a profit on that initial investment.

The last topic that will be discussed is exit strategies. At a certain point, investors or the founders of the company will want to sell a portion or all of their equity in the company, which results in a liquidity event. The most common exit strategies include an initial public offering (IPO), acquisition, and licensing. After the company reaches a level of maturity where it is believed to be able to meet all the requirements put forth by the US Securities and Exchange Commission, it can begin the process of filing for an IPO. The company usually works with an investment bank that will perform due diligence on pricing the shares for the IPO and lining up potential large purchasers of the shares when they become available on the public markets. By selling some or all of their shares, the initial founders or employees who have equity in the company can receive cash for their share of ownership. Similarly, investors can realize their investment gains at this point.

Besides doing an IPO, undergoing an acquisition is another commonly used exit strategy. The new company may be acquired by a competitor, a private equity fund, or a company in which its product would complement its existing product line. Startups commonly form relations with established companies early in their development. The rationale for this is varied, but the established companies often provide some form of support or guidance. After the startup reaches a mature stage, the established company may decide to purchase the startup as the product may fit well within their business model. The acquirer at this point already has a strong relationship with the startup and its founders, making acquisition and integration easier. Besides acquisition by an established company, startups are occasionally acquired by a private equity (PE) fund. PE is similar to VC in that both invest in companies that are not listed on the public marketplaces. However, PE typically invests in mature companies that have already demonstrated profitability. PE funds acquire mature companies that they believe can be transformed into more profitable and thus more valuable companies that later be sold again at a profit.

Finally, for companies that do not undergo an IPO or are not acquired, licensing is another option. In licensing, a startup might allow another company to sell the product or service that they have developed in exchange for a fee. This can be beneficial for both the startup and the company licensing the product or service. Often, the cost of actually producing a product in the volumes needed to meet customer demand can require large initial investments of capital. In this case, the startup can license its product to an established company that already has the equipment and necessary materials to manufacture the product. This can be mutually beneficial in that the startup avoids the high capital costs associated with setting up manufacturing of their product and while the established company has a desirable product to sell without the associated expenses of the initial development of that product such as research and development costs. The fees generated from licensing can be split based upon the percentage of ownership between employees of the company and founders as well as the outside investors.

Get started

- Now that you have a business idea, retest your hypothesis that your solution meets a market meet and that customers are willing you purchase your product
- Assemble a diverse team of advisors outside of your area of expertise to help along the way as you start your business. Key advisors might include an accountant, lawyer, marketing professional, and insiders within the industry your company plans to enter.
- With the help of your advisors, assemble the business plan and assess the amount of funding that will be required and what sources will be utilized for that funding.
- Come up with a list of potential exit strategies for the business

Further reading

1. Ganti A. *Angel Investor*. Investopedia; 2020. https://www.investopedia.com/terms/a/angelinvestor.asp.
2. Anderson JC, Narus JA. *Business Marketing: Understand what Customers Value. Harvard Business Review*; 1998 (November-December 1998). https://hbr.org/1998/11/business-marketing-understand-what-customers-value. Accessed June 20, 2020
3. Press HB. *Creating a Business Plan: Expert Solutions to Everyday Challenges*. Harvard Business Review Press; 2007.

4. Press HBS. *Creating a Business Plan: Expert Solutions to Everyday Challenges*. Harvard Business School Publishing; 2007.
5. Manigart S, De Waele K, Wright M, et al. Determinants of required return in venture capital investments: a five-country study. *J Bus Ventur*. 2002;17(4):291–312. https://doi.org/10.1016/S0883-9026(00)00067-7.
6. Maverick JB. *What is the Average Annual Return for the S&P 500?* Investopedia; 2020. https://www.investopedia.com/ask/answers/042415/what-average-annual-return-sp-500.asp.

Chapter 106

Small business innovation research/small business technology transfer (SBIR/STTR)

Brandon Ewing
Department of Radiology, University at Buffalo Jacobs School of Medicine and Biomedical Sciences, Buffalo, NY, United States

Introduction

Interventional radiology (IR) is a rapidly developing field with significant opportunities for research and development. Conducting impactful research is a time-consuming and expensive endeavor. The NIH, through its SBIR/STTR initiatives, can provide IR physicians with funding for projects that have the ultimate goal of producing commercialized, marketable products.

About the SBIR/STTR initiatives

Developed in the late twentieth century, the SBIR and the STTR are initiatives designed to provide monetary funding for small businesses seeking to develop innovative research across a wide domain of interests, including energy, defense, and healthcare.[1] Currently, there are 11 federal agencies associated with the SBIR and five in the STTR. Each agency associated administers its own program within the context of the overarching objectives of the SBIR/STTR initiatives.[1] For physician-scientists, the NIH is the recognized distributor of SBIR/STTR funds, specifically for those looking to develop and distribute new technologies that improve health and save lives.[2] This chapter will examine the difference between the SBIR/STTR programs and will explain the application process so that individuals can submit grants successfully.

Distinguishing between SBIR/STTR

Before discussing the application process, it is important to distinguish between the SBIR and STTR. The SBIR focuses on using small businesses to stimulate the development of new technological innovations that can be commercialized into products.[1] In contrast, the STTR focuses on the transfer of newly-developed technologies by a nonprofit research institution through a small, for-profit business.[1] In other words, to be eligible for STTR grants, small businesses *must* be teamed with a nonprofit research institution. Regardless of the particular program, small businesses that are successful in receiving a grant still maintain complete ownership of their project and ideas.

Both of the initiatives require the following basic principles to be met[1]:

- The research must be conducted with the specific goal of developing a novel invention that can be commercialized
- The research must involve a component of technical risk
- The product must not have already been developed

Each program is organized into three phases as follows[3]: Phase I serves to establish the feasibility of proposed research developments. Project funding during this phase is capped at $150,000 for 6 months (SBIR) or 12 months (STTR). Phase II allows research efforts to expand and provides up to $1,000,000 of additional funding for 2 years. Phase III focuses on the commercialization of projects developed during phase I/II and is not funded by the SBIR/STTR programs.

Basic eligibility requirements

To be able to participate, an applicant must be a for-profit small business.[1] For eligibility purposes, a small business

Translational Interventional Radiology. https://doi.org/10.1016/B978-0-12-823026-8.00079-1

is defined as having less than 500 employees, regardless if said employees are full time or part time.[1] Additionally, 51% of the business must be owned by US citizens and/or permanent residents.[1]

Each prospective applicant must designate a principle investigator (PI) who will possess overall responsibility for the project.[1] The PI must have an appropriate educational background to conduct the project.[1] For SBIR grants, the PI must be primarily employed at the small business.[1] In contrast, for STIR grants, the PI can either be employed at a research institution or at the small business.[1] The NIH grants also allow for coinvestigators to be named, as long as the investigators justify it in writing.[1] Table 106.1 provides a brief overview regarding the basic eligibility requirements.

Application components

Once a small business has an appropriate idea and has reviewed the eligibility requirements, it can start gathering the required materials. In general, submissions are considered on a triannual basis, with application cycles beginning in January, April, and September.[3] Before submission, applicants must complete all five of the following registrations available at their respective websites: (1) Data Universal Number System, (2) System for Award Management, (3) Grants.gov, (4) eRA commons, and (5) Small Business Adminstration (SBA) Company Registry.[2] It is important to note that the bulk of the grant application and associated forms are managed through the eRA commons. Table 106.2 provides an overview of the purpose and relevance of these registrations to the application process.

Application review

Once all of the appropriate materials have been received, the application status can be tracked in the eRA Commons. It is important to note that applications must be periodically checked for errors and/or warnings. Errors and warnings differ in that an error will prevent an application from moving forward unless addressed, while warnings do not halt the evaluation process. Applicants are able to correct warnings at their discretion.[3]

If all errors have been addressed, the application will then proceed for review. The NIH utilizes a "dual peer review system" to ensure that grant funds are awarded to the best-fit applicants.[3] In the first level of review, a referral officer from the NIH Center for Scientific Review will assign the application to a particular small business study section and to the appropriate review center.[3] A SRO will then organize the appropriate peer reviewers to examine the application.[3] Postpeer review, a final impact score is generated based on the overall scientific merit of the research.[3] A second review of the application is performed by an advisory board consisting of relevant scientists from the community and public representatives.[3] Based on their input, the director of the advisory board will make the final decision on whether an application is accepted.[3]

Application acceptance

Application decisions are made available through the eRA commons. Successful applicants will receive an official NoA.[3] The NoA is a legal document that must be presented to the designated disbursement agency upon award redemption.[3] Unsuccessful applicants are able to reach out to a representative and discuss reasons for rejection as well as what alterations are needed for the project to be accepted in the future.[3] Applicants who wish to be reconsidered can resubmit applications as early as the following cycle and no later than 37 months following rejection.[3]

Currently accepted IR projects/ real world examples

There have been several ongoing IR projects that have successfully received grant funding. Table 106.3 provides a description of a few of these projects so that prospective readers can get a better idea of what constitutes a successful project.

Potential pitfalls

The application and acceptance process for SBIR/STTR grants is undoubtedly complex. One of the most common pitfalls when applying is not being an appropriate candidate.[1] As previously stated, these grants only apply for research that is currently in development.[1] Small businesses with already developed products cannot use grant money retroactively and hence cannot participate.[1] In addition, pitfalls are often centered on the proposed content itself. Research proposals that receive SBIR/STTR grant funding must clearly address why a problem is important, how it can be solved with a commercialized product and how it relates to the goals of the NIH.[8] It is recommended that prospective applicants first consult with an NIH program officer to discuss the general feasibility of their project before beginning the application process. Lastly, another notable pitfall is failing to allot sufficient time for completing the required applications.[3] It is recommended that applications are started well in advance of the submission deadline for the cycle.[3] For example, registration for System for Award Management can take days to weeks to complete.[3] Hence, it is important to make sure all requirements are completed in a timely manner for proposal consideration.

TABLE 106.1 Basic eligibility requirements.[1]

	SBIR	STTR
Primary applicant	For-profit small business	For-profit small business
Principle investigator	**Required**; primarily employed with business	**Required**; employed either with business or research institution
Partnership	**None required**; but encouraged	**Required**; must be nonprofit research institution
Involvement	**Phase I**: Research partners can complete up to 30% of work **Phase II**: Research partners can complete up to 50% of work	**Small business** performs at least 40% of the work. **Research institution** performs at least 30% of work

TABLE 106.2 Required registration and processing time.[3,4]

Registration	Purpose	Processing time
Dun and Bradstreet Universal Numbering System	Provides a unique universal identification number for the applicant's business (and businesses in general)	2 days
System for Award Management	Allows applicants (businesses) to be awarded government contracts	Several weeks
Grants.gov	Assists business with finding and applying for grant funding. Forwards applicants to the involved agency	1 day
eRA Commons	NIH-managed system that allows applicants to view information about their grant as well as track its status. The bulk of the application is created and processed here	Up to 10 days
Small Business Adminstration (SBA) Company Registration	Helps agencies (i.e., NIH) administer SBIR/STTR programs and registration by providing an additional layer of assurance that all the necessary steps are met	2 days

TABLE 106.3 Examples of successfully submitted projects.[5–7]

Award year	Business	Project name	Description
2014	Chemofilter, INC	*ChemoFilter: A Novel Catheter Device to Enable High-Dose Chemotherapy Treatment*	Project seeks to modulate the intravenous dosing of drugs through the development of a catheter filtration device that prevents drug escape from the target organ
2014	Triple Ring Technologies, INC	*A low dose fluoroscope for interventional radiology procedures*	Project seeks to reduce the harmful effects of radiation during IR procedures via creating a "low-dose fluoroscopic system" that widens the field of view to reduce entrance X-ray exposure
2017	Puracath Medical, INC	*A device to reduce catheter-related bloodstream infection rates in patients with peripherally inserted central catheters PICCs*	Project seeks to reduce catheter-related bloodstream infections via developing an FDA approved catheter hub disinfection system

Get started

- Determine whether you fit the basic eligibility criteria for SBIR/STTR grants at the NIH
- View ongoing IR projects and examine the similarities and differences between yours and theirs
- Find an approved NIH officer to discuss your project feasibility (see resource section)

Resources

1. An online course that provides a basic overview of the SBIR/STTR programs. https://www.sbir.gov/tutorials/program-basics/
2. A step-by-step application guide on submitting the SBIR/STTR grant paperwork. https://grants.nih.gov/grants/how-to-apply-application-guide/forms-f/sbir-sttr-forms-f.pdf
3. A journal article that provides tips on writing a strong project summary and abstract. https://files.eric.ed.gov/fulltext/EJ888519.pdf
4. A list containing links to the full application of several successfully submitted proposals. https://www.niaid.nih.gov/grants-contracts/sample-applications
5. A list of NIH officers that can provide insight regarding project feasibility. https://sbir.nih.gov/engage/ic-contacts

References

1. Program Basics. Small Business Administration. https://www.sbir.gov/tutorials/program-basics. Accessed February 28, 2021.
2. *Committee on Capitalizing on Science, Technology, and Innovation: An Assessment of the Small Business Innovation Research Program—phase II; Board on Science, Technology, and Economic Policy; Policy and Global Affairs; National Academies of Sciences, Engineering, and Medicine. SBIR/STTR at the National Institutes of Health*. Washington (DC): National Academies Press (US); November 12, 2015.
3. SBIR/STTR Application Process Infographic. National Institute of Health. https://sbir.nih.gov/infographic. Accessed February 26, 2021.
4. Stein E, Lee D, Gooneratne N. SBIR/STTR grants: application guidance. *Acad Entrepreneurship Med Health Scient*. September 30, 2019;1(2):1–12.
5. Patel A. ChemoFilter: A Novel Catheter Device to Enable High-Dose Chemotherapy Treatment. SBIR/STR. https://www.sbir.gov/sbirsearch/detail/706912. Accessed March 13, 2021.
6. Funk T. A low-dose fluoroscope for interventional radiology procedures. SBIR/STR. https://www.sbir.gov/sbirsearch/detail/678716. Accessed March 13, 2021.
7. Rasooly J. A device to reduce catheter related bloodstream infection rates in patients with peripherally inserted central catheters PICCs. SBIR/STR. https://www.sbir.gov/sbirsearch/detail/1304965. Accessed March 13, 2021.
8. Greenwood J. Lessons learned: common mistakes made in phase I & II proposals. In: *National SBIR Conference: JW Marriott Austin*. December 1, 2015 (Austin, Texas).

Chapter 107

Sample forms and templates

Aakash N. Gupta[1], Robert J. Lewandowski[1,2,3] and Andrew C. Gordon[1]
[1]Department of Radiology, Section of Interventional Radiology, Northwestern University, Feinberg School of Medicine, Chicago, IL, United States;
[2]Department of Medicine, Division of Hematology and Oncology, Northwestern University, Feinberg School of Medicine, Chicago, IL, United States;
[3]Department of Surgery, Division of Transplant Surgery, Northwestern University, Feinberg School of Medicine, Chicago, IL, United States

Key points

- Grant proposals, institutional review board (IRB) protocols, and research manuscripts are important documents generated at various stages of a research project.
- Conducting research may require creating budgets, implementing Consolidated Standards of Reporting Trials (CONSORT) diagrams, and utilizing other practical tools for study design.
- Writing a grant proposal or manuscript can be intimidating, but starting early and going through many iterations with mentors are key to success.

Why it matters?

Grant proposals, institutional review board (IRB) protocols, budgets, Consolidated Standards of Reporting Trials (CONSORT) diagram, manuscript formatting, and practical tools for study design are all essential elements of research study design. They are key steps to leading a successful research project, and mastery of these components can improve the quality of your research and the validity of your findings. Herein, we discuss each component, offer suggestions, and provide real-world examples relevant to IR.

Grant proposals

A grant proposal is a comprehensive document detailing a proposed research plan and is submitted to an organization/institution to request support or funding. While the exact structure and components of each grant proposal will vary, here we review the core components. Be sure to adhere to the specific guidelines provided by the organization or institution that the grant proposal will be submitted to.

Short overview

The short overview (also called "Abstract," "Executive Summary," or "Specific Aims Page") provides a succinct overview of the proposed research plan. This can vary in length from a paragraph to two pages in length depending on the guidelines, but it can often be the most important part of the grant proposal. A strong overview section takes an "hourglass" shape—starts with a broad introduction, boils down to proposal specifics, and culminates with a strong base. The introduction highlights the larger problem, discusses current knowledge, and identifies a specific gap in knowledge that the grant proposal seeks to fill. A brief description of the proposed goals, hypothesis, rationale, and objectives should follow. Next, specific aims to test the hypothesis are listed. Each aim should contain a testable hypothesis, relevant experiments, and impact. Finally, the conclusion paragraph underscores the significance, innovation, expected outcomes, and positive impact of your proposal.

The overview should draw from all important aspects of the grant proposal. It is usually advised to write this section last. As this is often what a reviewer will read first, careful attention should be given such that the overview is clear, direct, and succinct.

Research strategy

The research strategy (also called "Study Design," "Research Plan," or "Scientific Methods") details the specifics of the proposed experiments in an organized, methodical fashion. Often, this begins with the background and significance, which consists of an evidence-based introduction to the field, current knowledge about the identified problem, key statistics and outcomes, relevant prior studies, and ongoing challenges. A central hypothesis should be presented with additional hypotheses for each topic introduced. Data from preliminary studies are helpful

Translational Interventional Radiology. https://doi.org/10.1016/B978-0-12-823026-8.00015-8

to demonstrate the feasibility of the proposed experiments and also highlight the investigators' experience. A description of the team structure includes the key members, departments, and collaborations. This is especially important for grant proposals with an emphasis on multidisciplinary collaboration. The timeline lays out when each component of each aim will be carried out. The aims should be numbered and consist of the title, rationale, subsections for each key component, measurements and analysis, and statistical methods. Finally, a description of potential pitfalls and potential solutions, expected outcomes, and future directions can be helpful to convey an expert understanding of the relevant topic areas.

This section should be consistently organized with an outline or numbering system with bolded headers. An organization system in combination with key figures should act as a map to guide readers through your proposal. Upon finishing reading the research strategy, a reader should feel confident that the proposed project is important to the advancement of the field and that your team is well designed and capable of carrying out the proposed project. An example of a Society of Interventional Radiology (SIR) Foundation-funded research plan from a translational IR project is provided in Appendix 1.

Institutional review board protocols

Before initiating a research study involving human subjects, a research protocol must be submitted and approved by an IRB. The key sections of an IRB protocol are: (1) review category, (2) research design, (3) research procedures, (4) study population, (5) informed consent, (6) potential risks and benefits, and (7) data security and patient confidentiality. The review categories include exempt review, expedited review, and full review. If a grant proposal was written, the research strategy section can be used to fill the research design, procedures, and study population sections of the IRB protocol. Potential risks cover physical, social, psychological, and other risks; potential benefits cover benefits to individual participants and/or society. Finally, patient confidentiality and data security is critical and may include secure servers, anonymization of data, physically locked cabinets, and a list of team members with data access. Depending on your institution or department, dedicated resources and research staff may be available to assist with drafting and submitting IRB protocols. An IRB protocol template is included in Appendix 2.

Budgets

Budgets are typically geared toward funding research materials and studies, support for personnel, equipment, or a combination thereof. An example is provided in Table 107.1.

Consort diagram

The CONSORT statement is a set of recommendations to standardize reporting of randomized controlled trials with the goal of improving understanding of the trial design, analysis, and interpretation (1). The CONSORT 2010 statement consists of a 25-item checklist and flow diagram, both of which are freely available online (2). Fig. 107.1 shows an example of a CONSORT flow diagram for a randomized controlled trial in Interventional Oncology comparing yttrium-90 radioembolization (Y90) with conventional transarterial chemoembolization (cTACE) for the treatment of hepatocellular carcinoma (3).

Manuscript formatting

Original research manuscripts are generally structured into an introduction, materials and methods, results, discussion, and conclusion. The introduction starts generally and funnels down to specifics over a series of two to four paragraphs that each introduce a topic in the field, discuss what is currently known, and highlight a current gap in knowledge that the manuscripts will address. A hypothesis should be included at the end of the introduction. The methods and results parallel each other with subsections that organize each component of the project, ranging from the patient cohort, data collection, and outcome variables to statistical methods. The discussion section begins with a summary of the key findings and delves deeper into (1) how the findings address knowledge gaps, (2) how the results should be interpreted in the context of the broader literature, and (3) what limitations are present in the current study. The conclusion presents the key message of the paper and must address the stated hypothesis in the introduction.

There is no one correct method to write a manuscript, but an excellent overview is provided by Albert A. Nemcek (4). The most important aspect of writing a successful manuscript is choosing an important and relevant topic. This happens long before the manuscript is written and the key to this is having a strong research team and invested mentors. Each journal has a specific audience and knowing your audience is absolutely critical to tailoring your writing and presentation of the data (tables/figures). Performing a literature review and drafting the methods/results are often the most helpful first steps to begin preparing a manuscript. Presenting results to coauthors and mentors can be valuable to distill the key findings and anticipate limitations. The introduction and discussion both frame the results and assist the reader in understanding how the manuscript fits into the existing literature. Finally, figures and tables are vital to concisely present results, emphasize key findings, illustrate data analysis methodology, and draw reader's interest. The abstract is how most reviewers and readers will interact with your work and usually the first thing they

TABLE 107.1 Example of a budget from a translational IR research grant proposal to support the investigation of image-guided transcatheter thermoradiotherapy of liver tumors in the VX2 rabbit model (see Appendix 1 for more details).

Category	Item	Quan	Unit	Unit cost	Subtotal
Equipment and services	MRI	72	1 h/$460	$230/rabbit	$16,560
	ICP-MS	30	1 h	$35	$1050
Animal	Rabbit	36 HCC+	$10_{hindlimb}$	$228.35	$10,504.10
	Procedure + delivery fee	3		$100	$300
	Housing	1269	1 cage/day	$3.81	$4834.89
Supplies	Sutures	12	Box of 12	$22	$264
	Cotton applicator				
	Needles, blades, *etc*	3		$50	$150
	Ketamine	6	10*ml*	$30	$180
	Meloxicam	2	10*ml*	$30	$60
	Catheters	15	Catheter	$85	$1275
	YAS microspheres	1	25 *g*	$1666	$1666
Pathology	H & E staining/*immunohisto*	164	Slide	$15	$2460
Travel	SIR 2014	1		$230	$230
	SIR 2015	1		$230	$230
Total					$39,763.99

HCC, hepatocellular carcinoma; *ICP-MS*, inductively coupled plasma mass spectrometry; *MRI*, magnetic resonance imaging; *SIR*, Society of Interventional Radiology; *YAS*, yttrium aluminosilicate.

read. Therefore, the abstract must be superb, without typos, and focus on the essential key findings. This is the elevator pitch of the manuscript and many positive or negative reviews are largely influenced by the quality of the abstract. Initial rejections are common and inescapable. Use them to refine and improve your work.

Practical tools

Here we provide some practical tools to aid in study design and manuscript preparation:

(1) **G*Power:** G*Power is a freely available tool developed by the Universität Düsseldorf and used to compute statistical power across a variety of tests (5).
(2) **Reference manager:** A reference manager can help organize a literature search, keep track of papers, and automatically manage the reference list in a grant or manuscript. EndNote (Clarivate, Philadelphia, PA), Zotero (George Mason University, Fairfax, VA), Mendeley (Elsevier, Amsterdam, Netherlands), Papers (ReadCube, Cambridge, MA), and RefWorks (Ex Libris Group, Jerusalem, Israel) are widely used. University and local libraries often have workshops on specific software.
(3) **Statistical packages:** Commonly used statistical packages include Stata (StataCorp, College Station, TX), SPSS (IBM, Armonk, NY), SAS (SAS Institute, Cary, NC), JMP (SAS Institute, Cary, NC), and R (R Core Team, Vienna, Austria). Most of these packages are pre-loaded with statistical tests, options for importing and manipulating data, data visualization tools, and online documentation.
(4) **Graphic design:** Figure design and layout can be performed in a variety of powerful graphic design tools, ranging from Microsoft PowerPoint (Microsoft, Redmond, WA) to Adobe Photoshop (Adobe, San Jose, CA).

Get started

- Start reading and writing! The best way to write a successful grant or manuscript is to go through many iterations with feedback from mentors. Getting a first draft on paper is often the hardest step.

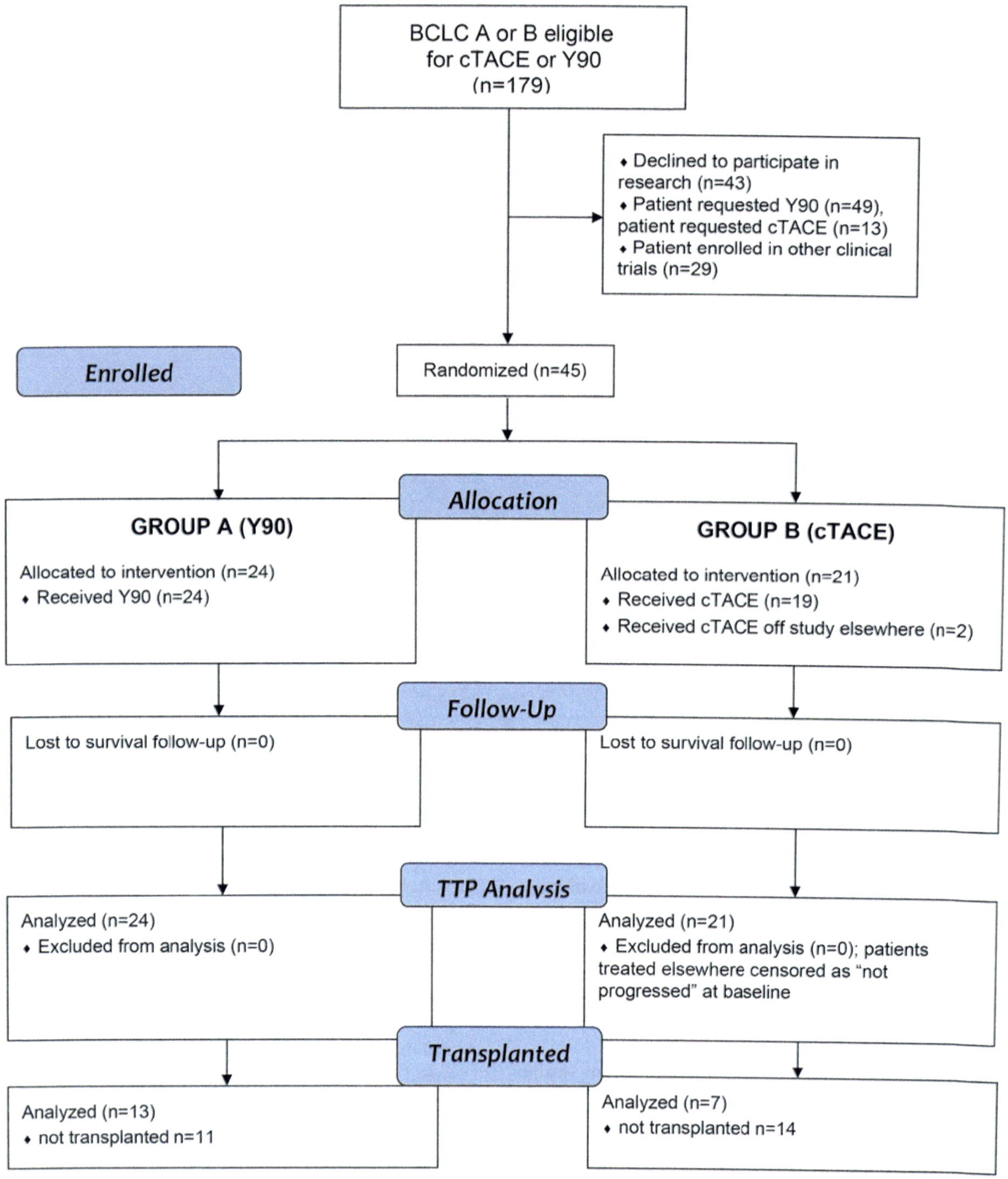

FIGURE 107.1 Example of a CONSORT diagram. BCLC, Barcelona Clinic Liver Cancer; cTACE, conventional transarterial chemoembolization; Y90, yttrium-90 radioembolization; TTP, time to progression. *Reprinted from Salem R, Gordon AC, Mouli S, et al. Y90 Radioembolization Significantly Prolongs Time to Progression Compared with Chemoembolization in Patients with Hepatocellular Carcinoma.* Gastroenterology. *2016; 151(6):1155–1163.e1152, with permission from Elsevier. (Supplementary Figure 1 in original).*

- Explore your institution's resources for opportunities such as writing workshops, research design consultation, and presubmission peer reviews.
- Submit a grant application for departmental or institutional funding. You might be surprised who you meet and what you learn in the process!

Potential pitfalls

1) When generating grant proposals and manuscripts, it is imperative to involve key collaborators throughout the process. This is particularly important for multi-disciplinary research which involves specialty expertise and impacts stakeholders in other fields. Such collaboration elevates the impact and quality of research.
2) From grant proposal to manuscript, the course of a research project inevitably changes. As such, rigid proposals, specific aims, and budgets can hinder research progress. A successful research team will anticipate such change by building in flexibility and planning to iterate as new obstacles and new data arise.

Real-world examples

See Appendices 1 and 2 for real-world examples of a grant proposal for a translational project in interventional radiology and of an IRB protocol for research. See Table 107.1 and Fig. 107.1 for examples of a budget and CONSORT diagram, respectively.

Resources

Further resources are identified in the Practical Tools section and include specific programs/software for statistical analyses, reference managers, and graphic design. Often, these resources are readily available at the research institution.

Appendix 1: example of a grant proposal research plan

Specific aims

This proposal will address the following Specific Aims in the rabbit VX2 hepatocellular carcinoma (HCC) model:

Aim 1: Determine whether concomitant hyperthermia enhances the efficacy of intra-arterial brachytherapy in liver tumors through a tumor-specific increase in apoptosis and reduction in angiogenesis. To study the role of hyperthermia as a possible enhancer of radiosensitivity in liver tumors, we will use IO-YAS MPs to compare lone hyperthermia, lone radiation, and combined hyperthermia/radiation under equivalent delivery conditions. Cell culture experiments will define radiation and hyperthermia dose relationships. Next, we will assess tumor volume and early immunohistochemistry markers as indicators of therapeutic efficacy and therapy response in the VX2 rabbit liver tumor model. **Hypothesis:** We expect concomitant hyperthermia ≥43°C to significantly reduce tumor volumes and increase active apoptosis while reducing tumor angiogenesis.

Aim 2: Develop MR imaging techniques for in vivo **tracking and alternating magnetic field (AMF) treatment planning.** To harness unique MR imaging capabilities for therapeutic planning, we will use MRI to map IO-YAS distributions and use specific absorption rate (SAR) measurements in W/g to evaluate thermal performance in vivo. We expect MRI R2* relaxometry measurements to allow accurate mapping of IO-YAS MP distributions, allowing for prediction of SARs under defined AMF protocols. MRI-based IO-YAS mapping will be performed and select AMF protocols will be applied at 430, 450, 470A coil currents; direct temperature measurements upon AMF application will be used to determine spatial SARs for a given AMF protocol and this data will be correlated with R2 and R2* maps. In this way, we will be able to link MRI R2* relaxometry measurements to distributions for rigorous AMF treatment planning through SAR estimation. **Hypothesis:** We expect R2* maps to accurately predict heterogeneous spatial SAR values within tumors targeted with intra-arterial therapy.

Background and significance

Liver cancer is a significant world health problem and the second and sixth leading cause of cancer deaths in men and women, respectively.[1] Although resection and transplantation offer the best 5-year survival rates, 70%−85% of patients are not candidates for resection and a severely limited number of organs are available for transplantation.[2–6] New therapies that offer a survival advantage either independently or as a bridge to resection or transplantation for these patients are desperately needed. Yttrium-90 radioembolization is a minimally invasive, nonsurgical procedure that delivers localized radiation therapy for the treatment of unresectable liver cancer. Our long-term goal is to improve the therapeutic efficacy of radioembolization in liver cancer patients while minimizing adverse toxicity.

Hyperthermia is known to enhance radiotherapy in many cancers; however, thermoradiotherapy has not yet become a standardized practice in the treatment of unresectable hepatic tumors. Current thermoradiotherapy methods used to treat tumors outside the liver generally involve the use of separate therapeutic vehicles, time separation between serial treatments, and invasive delivery, all of which may reduce efficacy.[1,7–22] Alternatively, we advocate the use of a single delivery system to maximize spatiotemporal overlap of radiation and hyperthermia to enhance radiosensitivity in magnetically targeted tumors. Our proposal utilizes iron oxide (IO) based nano agents for noninvasive electromagnetic (EM) hyperthermia with IO labeled yttrium aluminosilicate microparticles (IO-YAS MPs). In prior work, I have designed such IO-YAS MPs and demonstrated both highly specific delivery to hepatic tumors and induction of localized intra-tumoral hyperthermia in vivo upon exposure to an AMF. My immediate goal is to understand the interactions and efficacy of radioembolization with concomitant hyperthermia and to develop new MRI-based imaging techniques for monitoring both intra-tumoral delivery and AMF-induced hyperthermia. We hypothesize that (a) selective internal thermoradiotherapy will improve therapeutic efficacy in comparison to radioembolization alone through increased apoptosis and inhibition of angiogenesis and that (b) the addition of quantitative MR imaging will permit accurate tracking of intra-hepatic MP delivery and prediction of spatial SAR values.

Hyperthermia in combination cancer radiotherapies

In many cancers, hyperthermia can serve to synergistically augment radiosensitivity and even improve survival; however, the underlying mechanism is unknown and time separation between therapies may greatly diminish efficacy (maximal cell kill achieved when limiting time separation between radiation and heat exposure).[2–8,11,23–27] The many pathways implicated in tumor cell kill during hyperthermia strongly suggest that this treatment modality should be broadly applicable.[8,11,12,28] This is evidenced by clinical trials for thermoradiotherapy in tumors of the chest wall, cervix, rectum, bladder, melanoma, head, and neck that have shown significantly improved locoregional control over radiation alone despite the great variability in delivery and dosing strategies.[14] Furthermore, one preclinical study has shown that external beam radiation sequentially applied with regional hyperthermia of 40–42°C in a rodent model of HCC significantly reduced tumor progression when compared to lone radiotherapy.[29] In contrast, we are interested in a targeted endovascular approach for simultaneous thermoradiotherapy because the normal liver is highly radiosensitive. It is well known that cells under conditions of impaired nutrient supply, low pH, or hypoxic conditions are extremely sensitive to the combination of radiation and hyperthermia that suggests hyperthermia as a modulator of tumor microenvironment for radiation enhancement.[8,11] Given the implicit importance of the microenvironment, we theorize that tight spatiotemporal overlap delivered via a synchronous thermoradiotherapy is optimal. Until now, no studies have been equipped to use the same vehicle for hyperthermia and brachytherapy to interrogate these relationships. We are uniquely prepared to accomplish this goal.

AMF hyperthermia

While clinical trials have shown success with combination treatments, tumors in such studies have generally been superficial and therefore accessible. Deep tumors may be treated with the use of iron oxide (IO) localizing agents when exposed to an AMF. These agents generate heat due to energy released in magnetic hysteresis losses, Brownian, or Néel relaxation in special cases[10,13,17,19,30,31] and the energy transfer is a function of field strength, frequency, and intrinsic particle and tissue properties.[7,10,23–27,31–33] A multitude of prior studies have explored the use of IO nanoparticles (IO NPs) to treat tumors with EM hyperthermia.[7,8,23–27] Heating of magnetic NPs taken up by a tumor with an AMF to temperatures >43°C may independently induce cell death in cancerous tissues.[8] The thermal effects of magnetic NPs exposed to external EM fields have been explored in vivo using animal models of HCC,[34] human breast cancer,[35,36] glioma,[37] renal cancer,[38] and prostate cancer.[39–41] Our own promising development of dual-mode brachytherapy MPs to increase radiosensitivity with EM hyperthermia builds on this work. Recently, a form of thermoradiotherapy combining magnetic hyperthermia of ferrofluids and external beam radiation was approved for clinical treatment of glioblastoma patients in the European Union[20–22] and is also being investigated for the treatment of prostate cancer with promising initial results.[42,43] Selective delivery is critically important for these hyperthermia methods; for liver tumors, we advocate a highly selective transcatheter approach.

Thermal dosimetry

Induced hyperthermia is potentially incomplete or heterogeneous based on numerous data from human clinical trials where patients may not achieve their defined target therapy (i.e., 42.5 or 43°C for 30 or 60 min),[15] there is great variability in the temperatures and thermal doses achieved,[16,18,44] and these temperatures fluctuate considerably as a function of space and time[42]—focal "hot spots" with high heating rates can even result in burns from hyperthermia treatments.[45] Moreover, incomplete heating may not only result in reduced tumor cell kill but also thermal tolerance to subsequent hyperthermic therapies.[46] The typical temperature distributions achieved in living tissues are highly heterogeneous due to heat-sink effects produced by large blood vessels, limitations of current technologies in the delivery of thermal therapies, and thermoregulatory mechanisms that influence temperatures over time. Presently, such heterogeneous distributions are not adequately predicted using bioheat transfer models alone because the approximations within these models are challenged by irregularities of tissue thermal conductivities, complex anatomical geometries, and dynamic blood flow effects. The present gold standard in monitoring thermal dosimetry in hyperthermic oncology is invasive temperature measurement with poor spatial resolution (although it can be increased with an increasing # of probes). We propose use of advanced MR imaging techniques to noninvasively predict spatial SAR distributions and guide AMF treatment planning.

Radioembolization for unresectable hepatic tumors

Yttrium-90 radioembolization is a minimally invasive brachytherapy for inoperable liver cancer. The transcatheter procedure relies on the fact that the hepatic artery preferentially supplies blood to tumors within the liver. ^{90}Y microspheres localize in tumor blood vessels and emit radiation inducing cancerous cell death. These high-energy β emissions have low penetration and a short half-life (64.1 h), resulting in minimal damage to surrounding normal tissues and low toxicity.[47–52] Clinical microspheres are 20–40 μm in diameter while having densities of 3.29–3.34 g/cc; these remain in embolized vessels for the remainder of patients' lives, however, rapid decay minimizes microsphere activity to 6% in 11 days. We designed our IO-YAS MPs to mimic clinical microsphere profiles with 20–64 μm diameters and a density of 3.59 g/cc. IO-YAS MP activities and energy spectra remain to be characterized. ^{90}Y selective internal radiation therapy has a 42% response rate (by WHO criteria)[53] but improvements in patient survival over transcatheter chemoembolization remain to be seen. We believe hyperthermia will improve radiosensitivity of hepatic tumors for increased survival either directly or as a bridge to resection or transplantation.

Preliminary studies

In preliminary studies, I have developed IO-YAS MPs (Fig. 107A.1) for synchronous thermoradiotherapy. These MPs transduce EM energy to thermal energy under AMFs produced by a radiofrequency generator (EASYHEAT 2.4, Ameritherm, Scottsville, NY) while maintaining a high yttrium content. The AMF was produced via a 2.4 kW radiofrequency (RF) generator yielding a controlled current through the coil (specified by the operator) alternating at a specific frequency in the 150–400 kHz range. This RF generator has been thoroughly tested and certified to produce consistent, reproducible currents through the specified coil. Current through the coil was chosen based on the desired field strength. Temperatures exceeding 43°C in 2 mL samples are produced in less than 4 min. Furthermore, results showed that thermal response was proportional to both MP concentrations (Fig. 107A.1B) and field strengths (Fig. 107A.1C) applied, allowing for tunable hyperthermia treatments. These initial studies show IO-YAS microparticles can achieve therapeutic temperatures upon exposure to an AMF. The proposed experiments will be the first to investigate the interactions of hyperthermia and radiation emitted from the same source.

This preliminary study was selected for oral presentation at the 2010 SIR Annual Scientific Meeting in Tampa, Florida.

To test our ability to generate hyperthermia in vivo and the potential heat-sink effects of hepatic perfusion, I experimented with AMF hyperthermia in the VX2 rabbit model of HCC and demonstrated highly localized hyperthermia. Tumor induction in 21 rats was accomplished with MRI scans confirming tumor growth at 1 week. Treatment rats (N = 7) received intratumoral injection of IO-YAS MPs under direct visualization following laparotomy. A neodymium magnetic provided magnetic localization and fiber optic temperature probes were placed in the tumor, NHP of the same lobe (within 2 cm of tumor periphery), and rectum. A rapid and sustained thermal response ≥43°C was observed in the treated tumors with minimal heating of adjacent tissue (Fig. 107A.2B). Importantly, AMF control rats (N = 7) receiving no IO-YAS MPs demonstrated a mild thermal response of 1–2°C with no individual temperatures in excess of 39°C (Fig. 107.2C).

This study was selected for oral presentation and was one of 15 featured abstracts at the 2011 SIR Annual Scientific Meeting in Chicago, Illinois.

I have previously shown that direct injection of MPs with magnetic localization in treatment rats (N = 7) provided selective delivery as quantified by inductively coupled plasma optical emission spectrometry (ICP-OES; Fig. 107A.3). We expect to accomplish highly selective transcatheter tumor delivery of IO-YAS MPs in the VX2 rabbit model of HCC.

The IO-YAS MPs we have developed are highly visible under MRI (Fig. 107A.4) and promote strong signal losses at the site of the tumor on T2w imaging. Intratumoral deposition was confirmed in each treatment

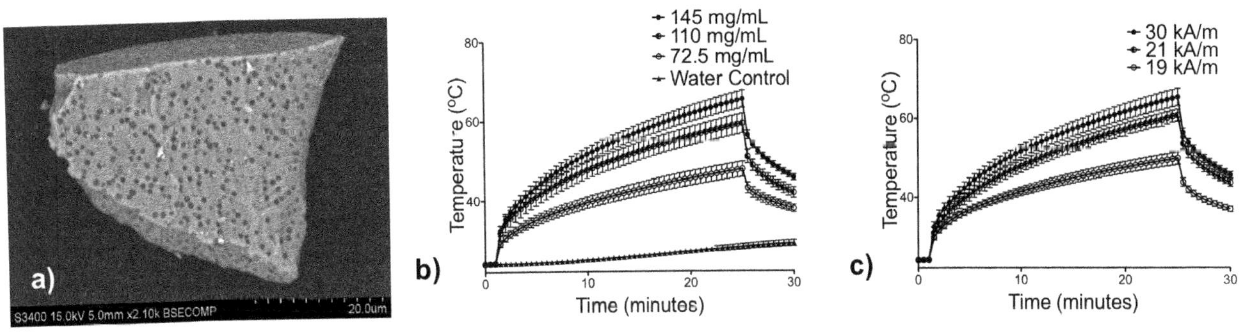

FIGURE 107A.1 IO-YAS MPs. (A) SEM image of a single microparticle (22–64 μm, 3.59 g/mL) doped with electron dense IO for applications in thermoradiotherapy. (B) Varied concentrations and water control exposed to a field strength of 30 kA/m alternating at 193 kHz. (C) 145 mg/mL exposed to varied field strengths at 193 kHz. Curves represent five independent experiments. Bars = standard deviation.

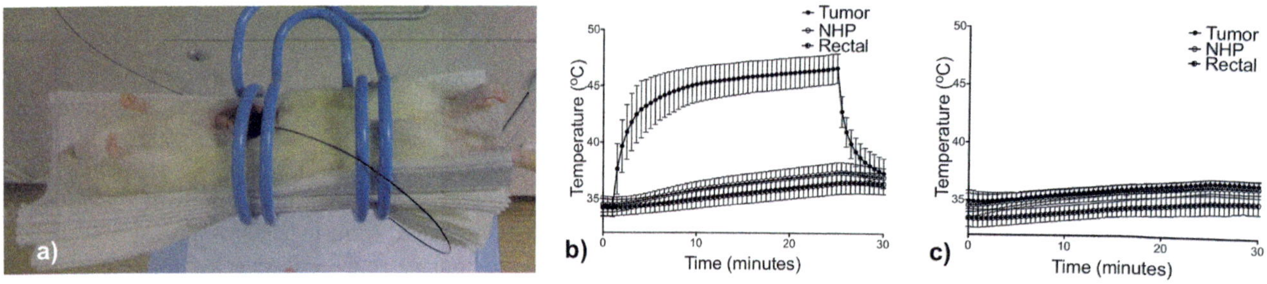

FIGURE 107A.2 in vivo AMF hyperthermia. (A) Each rat, while still anesthetized, was placed in the water-cooled Helmholtz-style coil and baseline temperatures were recorded with three fiber optic probes for 1 min after which time the AMF (30 kA/m, 193 kHz) was applied for 24 min. (B) Treatment group temperature measurements in the tumor, normal hepatic parenchyma (NHP), and rectum (N = 7). (C) AMF control group temperature measurements (N = 7). Bars = standard deviation. This study was approved by our institutional animal care and use committee and was performed in accordance with institutional guidelines.

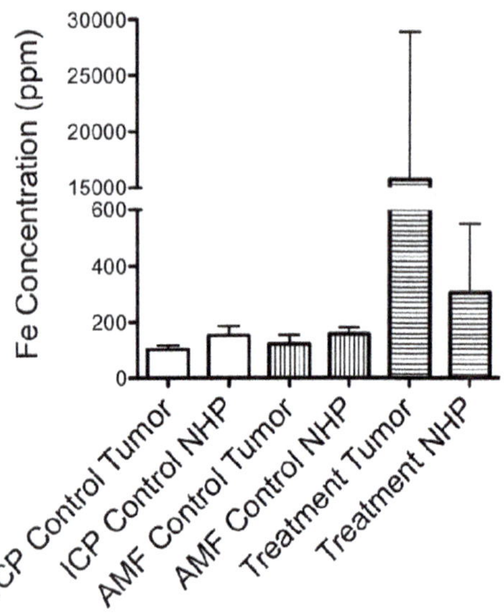

FIGURE 107A.3 Localized intratumoral delivery. Isolated AMF and procedural exposure did not significantly alter tumor and NHP [Fe]s nor the tumor:NHP ratio as measured by ICP-OES. Percutaneous delivery of IO-YAS MPs with magnetic targeting under direct vision yielded a mean tumor:NHP Fe ratio of 75.2 versus 0.77 in the AMF control group ($P < .05$).

rat with histopathology (Fig. 107A.5). Artifacts associated with Fe due to the high Fe_3O_4 content of the IO-YAS MPs are quick markers for identification of particle distributions but may preclude measurement of proton resonance frequency shifts for MR thermometry due to substantially shortened T2*.[54–58] Alternative MRI techniques like ultra-short echo-time 59 or gradient echo sampling of the spin echo (GESSE)[60] methods may permit quantitative MRI measurements of IO-YAS MP distributions. In the proposed study, we theorize GESSE will provide accurate IO-YAS MP mapping and that this information will allow prediction of spatial SAR values for a given AMF protocol.

All preliminary data, unless otherwise noted, were generated by the applicant.

Research design

Aim 1: Determine whether concomitant hyperthermia enhances the efficacy of intra-arterial brachytherapy in liver tumors through a tumor-specific increase in apoptosis and reduction in angiogenesis.

Rationale

Our therapeutic approach uses a single agent for magnetic tumor targeting, imaging, generation of hyperthermia, and delivery of radiotherapy. This may provide several advantages such as magnetic targeting for localized delivery, optimizing spatial and temporal overlap of hyperthermia and radiation, unique imaging capabilities, and the potential for subsequent serial treatments. The immediate spatial overlap is expected to be particularly beneficial due to the usual heterogeneity of microsphere deposition that would limit alternative approaches.[32,33] Importantly, the localized nature of this therapy enables high local doses of radiation and hyperthermia with minimal systemic toxicity. This approach contrasts greatly with the sequential dosing strategies used in prior thermoradiotherapies that likely limited therapeutic potential.[61] In addition, embolization due to the size of the MPs may prevent extravasation and uptake by the reticuloendothelial system and allow for a thermal distribution specific to the level of the vascular network.[28] Hyperthermia is expected to be efficacious when induced with microspheres; intra-arterial ferromagnetic embolization hyperthermia with AMF exposure has been readily accomplished in the VX2 rabbit model[34] and rigorous theoretical modeling data suggest that therapeutic tumor core temperatures (≥43°C) may be achieved even when microsphere distributions are highly heterogeneous.[28] If ^{90}Y radioembolization can be combined with AMF hyperthermia to deliver a more effective treatment, then it may be possible to treat patients with less radiation while maintaining safe and thorough treatment of targeted tumors. Ideally, this new therapy would take advantage of a hyperthermia radiosensitizing phenomenon observed in

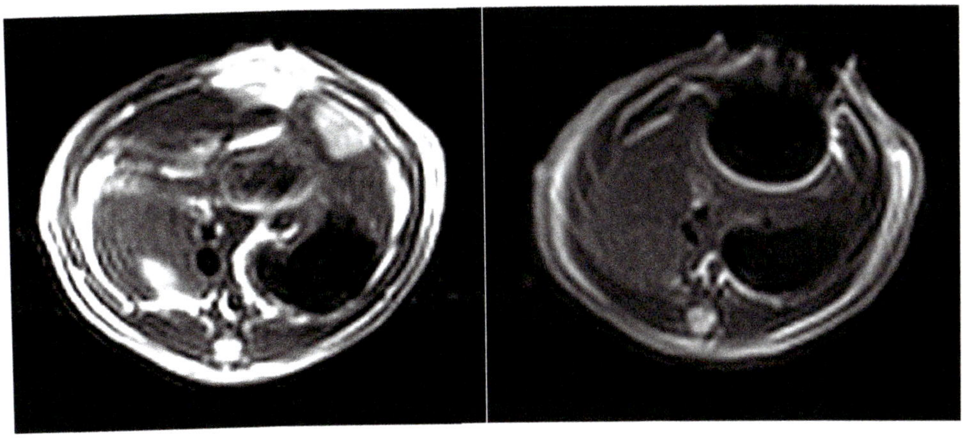

FIGURE 107A.4 T2w TSE imaging (transverse view) for treatment rat showing pre- (left) and postinjection (right) scans. Administration of IO-YAS microparticles resulted in ring artifact and strong losses in signal intensity that suggested intratumoral deposition. Intratumoral depostion was confirmed by histopathology.

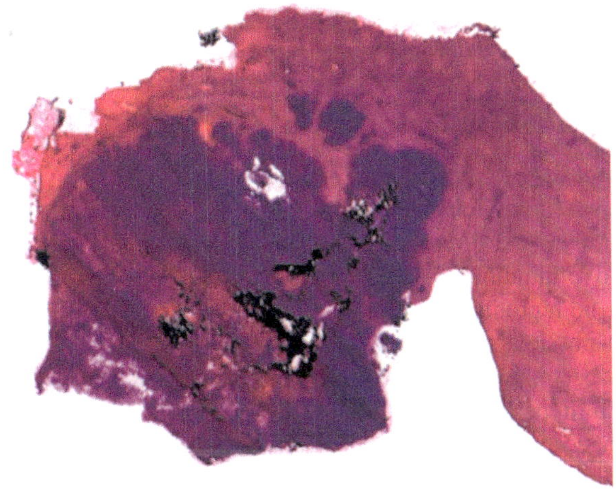

FIGURE 107A.5 H&E staining confirming intratumoral deposition of IO-YAS. In each treatment rat (N = 7), histopathology corroborated MR imaging findings with dense intratumoral IO-YAS deposition.

many other cancers to improve survival in patients with hepatic tumors. In preliminary studies, I have shown highly localized hyperthermia in the setting of solid tumor in vivo in the N1–S1 rat model of HCC (please see preliminary data above).

Experimental design

Thermal dose–response relationships with brachytherapy in vitro

Thermal dose–response relationships with ^{90}Y radioembolization are presently unknown. Thus, we will first compare relative cytotoxicity of IO-YAS MPs in 2 cell lines HepG2 (human liver HCC cell line; ATCC, HB-8065, Manassas, VA) and normal hepatocytes exposed to thermoradiotherapy, lone radiation, or lone hyperthermia at 24-, 48-, 72-, 96-, and 120-h posttherapy. The effects of modulating, radiation dose, temperature, duration of therapy, and time separation between therapies on survival fractions will be compared in both HepG2 cells (ATCC, HB-8065, Manassas, VA) and normal hepatocytes. These cells will be analyzed using a FACScalibur flow cytometer (BD Biosciences, San Jose, CA), yielding percentages of apoptotic (annexin V positive, PI negative) and necrotic cells. We will fit the experimental surviving fraction data to a linear-quadratic function and incorporate a target theory model. These data will allow the thermal enhancement ratio (TER = radiation dose/radiation dose with hyperthermia for isosurvival) to be plotted at various temperatures in HepG2 cells and normal hepatocytes. The ratios of the TERs for a given temperature will yield a plot for the therapeutic gain ratio ($TGR = TER_{HepG2}/TER_{Normal}$), an indication of preferential toxicity for malignant cells. Cellular S values (defined in Gy/Bq s for ^{90}Y given size of nucleus and cytoplasm size in μm) will determine appropriate doses in each case. We expect hyperthermia will confer a TER>2 and a selective radiosensitization of HepG2 cells over normal hepatocytes for a TGR>1.3 that is maximal when hyperthermia is added to brachytherapy within 2–4 h.

Thermal doses will be calculated by converting time-dependent temperature recordings into a therapeutically equivalent time at 43°C (t_{43}):

$$t_{43} = \sum_{n=0}^{n=final} (\Delta t_n) R^{(43 - T_{ave}(\Delta t_n))}$$

where t_{ave} (Δt_n) is the average temperature over time interval Δt_n, R = 0.5 when $t_{ave} > 43°C$ and $R = 0.25$ when $t_{ave} < 43°C$. These R values are based on the observation by Sapareto and Dewey that an increase of 1°C above 43°C, halved the treatment time while a decrease of 1°C below 43°C increased the treatment time fourfold for a given therapeutic outcome and radiation dose in select cell lines.[62,63] Although these R values are used to guide therapies in human patients, they are derived from the cells of rats and mice. It is conceivable that these values may differ in humans and depend on fundamental differences in specific types of cancer. Using the data obtained from cell studies in Aim 1, we will define R values to build an empirically driven dosing formula for therapeutically equivalent times at 43°C that is specific to HepG2 cells representative of human HCC.

Intra-arterial brachytherapy ± hyperthermia versus lone intra-arterial hyperthermia in vivo

The VX2 tumor model will be used because the VX2 tumor blood supply is almost entirely derived from the hepatic artery, similar to that of human hepatocellular carcinoma, and because rabbit anatomy is sufficiently large to facilitate catheterization procedures that closely parallel the clinical paradigm in human transcatheter liver-directed therapies.[64] The rabbit model will permit the measurement of temperatures and efficacy within liver tumors and normal tissues for the proposed treatment. This model is absolutely essential to allow for rigorous histopathology and characterization of thermal and radiation dosing strategies that would not yet be possible in a clinical setting. Radioembolization with magnetic targeting of ^{90}Y microspheres in the VX2 rabbit model of HCC has also been described.[65] Our lab has extensive experience with the VX2 rabbit model and its use for transfemoral hepatic arterial catheterization in liver-directed therapies including yttrium microsphere delivery.[66–70] To assess the effects of intra-arterial brachytherapy with concomitant hyperthermia in hepatic tumors, we will compare the combination therapy (Group HT + BT) to lone hyperthermia (HT) and lone brachytherapy (BT) in our rabbit model of HCC. After sufficient tumor growth is confirmed via MRI, rabbits will undergo anesthesia[71] with intra-arterial infusion and neodymium magnet localization of IO-YAS MPs or radioactively functionalized IO-[72] Y MPs (25 mg for 5.6 MBq in 5 mL saline) for generation of the three groups (HT + BT, HT, BT). Fiber optic temperature probes will allow procedural monitoring at 30-s intervals (rectal temp. not to exceed 41°C). Tumor volumes will be assessed by liver explantation and caliper measurement at 8-day intervals out to 32 days. Body weight will also be recorded at 5-day intervals as a surrogate measure of toxicity. We will explant the liver tissue and perform histopathology and ICP for the detection of Fe concentrations ([Fe]s). We will use the Biostatistics Core Facility of the Robert H. Lurie Comprehensive Cancer Center for all statistical analyses. Comparison of $[Fe]_{Tumor}$ and $[Fe]_{NHP}$ in each group with the paired t-test will be an internal control. Plots of mean tumor diameters over time will provide a readout for therapeutic efficacy. Histopathology will include H&E and TUNEL staining for quantification of tumor apoptosis versus necrosis. We will also compare immunohistochemical measures in HT + BT versus BT groups at 1, 3, and 7-day intervals after therapy for pathology assessment of early treatment response. We consulted with a biostatistician for sample sizes. We expect requiring 24 + 12 = 36 rabbits (2 per time point per group). Explanted tissue from these additional groups will be stained for CD34 as a marker of tumor angiogenesis with sinusoidal capillarization,[73,74] for caspase-3 as a marker of active apoptosis[75] and H&E for overall cell morphology and nuclear structure. We expect hyperthermia to increase active apoptosis while reducing tumor angiogenesis. Analysis will be performed with one-way ANOVA and Tukey post hoc correction to compare the percentage of tumor apoptosis/necrosis, and weight differences between rabbits in the separate groups. We expect hyperthermia ≥43°C to synergistically reduce tumor burden and disease progression with concomitant increases in caspase-3 and decreases in CD34 expression, and this synergy will be maximal when hyperthermia and radiation are delivered synchronously.

Limitations and alternatives

Feasibility studies I have performed in recent years have shown highly localized intratumoral hyperthermia of 45°C; however, the direct-injection approach enabled highly selective intratumoral deposition of IO-YAS MPs that might not be reproduced with transcatheter infusion. Even if it is not possible to generate hyperthermia exceeding 43°C as has been the practice historically, clinical trials have shown mild hyperthermia of 40–42°C to remain a significant enhancer of radiotherapy.[13,17,19] Another alternative is to explore therapeutic efficacy with direct injection. While this does not immediately model clinical radioembolization, such experiments would still provide essential information regarding the role of hyperthermia for the enhancement of brachytherapy in hepatic tumors. Additionally, we have the ability to label our YAS MPs with many different IO agents. Should the present compositions have insufficient SARs, I can further optimize our compositions with other IO agents such as EMG series ferrofluid (FerroTec, San Jose, CA) and MPIOs (microsize particles of iron oxide, Bangs Laboratories, Fishers, IN) using a synthesis process I have already developed and implemented (through collaboration with the Mo-Sci Corporation). I can test new compositions for their elemental content and SARs under AMF heating. I expect these compositions to provide high SARs ≥30 W/g (at ≤30 kA/m) allowing for 45°C intra-arterial hyperthermia. While we have chosen an anticipated effective dosing scheme based on the literature, this may need to be adjusted with a dose escalation protocol given initial findings and yttrium to IO ratios may need to be further optimized. Future preclinical studies building on these essential first steps will be necessary to determine tumor volumes that can be appropriately treated with such dosing schemes. Another possibility is that hyperthermia does not enhance brachytherapy in HepG2 cells or the VX2 rabbit model. It is possible to also explore other cell lines such as Huh-7 (HCC cells) and even without validation in liver cancer, the development and investigation of the described therapeutic and imaging technologies would prove useful in many other settings for embolization thermoradiotherapy or magnetic hyperthermia with IOs.

Aim 2: Develop MR imaging techniques for in vivo tracking and AMF treatment planning.

Rationale

In Aim 1, we characterize the potential therapeutic benefits of hyperthermia. Clinical practice has shown ^{90}Y radioembolization results in heterogeneous microsphere deposition.[32,33] Therefore, patient-specific imaging for prediction of IO-YAS distributions is an essential step for visualization of delivery and estimation of both thermal and radiation dosing. Accurate prediction of spatial SAR values may inform the correct AMF treatment parameters to minimize unnecessary exposure to particular EM field strengths and frequencies that may initiate eddy currents in subcutaneous fat and increase the risk for nonspecific heating and burns. Even the most advanced models for noninvasive thermometry have not yet proven reliable enough to eliminate the need for invasive temperature monitoring in the clinical setting. Thus, our approach is primarily intended to facilitate the estimation of SAR values rather than predict temperatures over the entire course of therapy sessions. We expect linear, SAR-dependent thermal responses under early AMF exposure with high SAR IO-YAS MPs that minimize early perfusion influences on heat transfer. Future studies may build on these essential first steps by implementing bioheat transfer models for noninvasive thermometry during AMF therapy sessions.[10,76–78]

Our lab has described the implement-ation of IOs as contrast agents in MR imaging and techniques for the tracking of IO-YAS MPs with relaxometry for [IO] estimation (Fig. 107A.6).[68] Arterial infusion of superparamagentic IOs with MRI tracking has been accomplished previously in HCC patients and the VX2 rabbit model.[68,79,80] Additionally, my previous studies show visualization of IO-YAS MPs utilized in the hyperthermia experiments described earlier with in vivo 3T MRI (Magnetom Trio clinical scanner, Siemens Medical Solutions, Washing-ton, D.C.). In those studies, T2-weighted turbo spin echo (T2w TSE) imaging confirmed tumor growth at 1 week with a signal intensity of 1256 ± 344 (mean $\pm$ SD; $N = 7$) arbitrary units (a.u.). Administration of IO-YAS MPs resulted in a signal loss to 130 ± 51 a.u. ($N = 7$) with histology confirming intratumoral MP deposition and sparse distributions in the tumor periphery and NHP.

We believe noninvasive imaging of MP distributions in vivo will inform treatment planning and additionally allow for quantitative assessment of spatial dosimetry. Quantitative MRI of IO-YAS will be used for in vivo estimation of the SAR distributions for three specific AMF parameters. We will implement T2w TSE approaches for localization and apply gradient echo sampling of the spin echo (GESSE) sequences for R2* and R2 measurements. We expect GESSE to allow for R2* and R2 measurements in the presence of field inhomogeneities induced by the IO-YAS MPs to be implemented in this proposal.[60] We will test this technique in vivo and expect accurate prediction of SARs in superior and inferior tumor quadrants ($n = 8$) for a given AMF protocol.

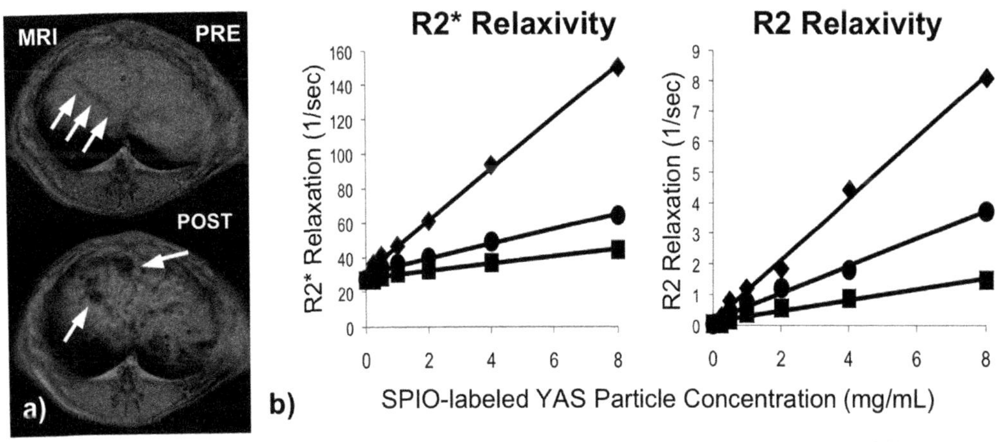

FIGURE 107A.6 MRI and relaxometry. (A) in vivo T2*-weighted MRI before and after transcatheter infusion of IO-YAS MPs in the VX2 rabbit (triple arrow denotes tumor margin) with dense deposition in tumor periphery (single arrows). (B) Plots of R2* and R2 relaxation times versus IO-YAS concentration in phantoms demonstrated strong linear relationships (all $r^2 > 0.98$) with increasing relaxivities for particles with increased IO content (diamonds = 5.0%, circles = 1.0%, boxes = 0.5% by mass IO). *Modified images courtesy of former lab member Twinkle Gupta, PhD.*

Experimental design

MR imaging protocols for estimation of IO-YAS MP concentrations

Purpose	Sequences	Scan parameters
Anatomic imaging	T2-weighted TSE	TR/TE = 2000/45 ms, turbo-factor = 28, 256 × 180 matrix, FOV = 50 × 35 mm, 100 coronal slices, 0.2 mm
R2, R2* measurements	GESSE	TR = 2000 ms, GRE ETL = 21, ES = 2.0 ms, 50 mm FOV, 0.2 mm, 256 matrix, 600 Hz/px BW

We will perform quantitative MRI voxel-wise R2*, and R2 measurements for agarose gel phantoms of known IO-YAS MP concentrations. These data will establish a predicted relationship between in vivo MRI R2* measurements and IO-YAS MP concentrations. All MRI studies will be performed using a 7T Bruker MRI scanner (Siemens Medical Solutions) within the NWU Olson Center for Translational Imaging. Coronal and transverse T2w TSE images of the entire abdomen will be acquired for localization. Our in vivo R2* and R2 measurements will be compared to the actual concentrations as determined with histopathology and ICP analysis. We will perform voxel-wise calculation to produce MRI-based MP concentration measurements $[IO]_{MRI}$ (based upon the R2* relaxivities) for comparison to reference ICP measurements, $[IO]_{ICP}$ within each volume of interest (superior and inferior tumor quadrants). We will evaluate the relationship between MRI-based and ICP-based measurements of MP concentrations using Lin's concordance coefficient and Pearson correlation coefficients. Sample size of 36 has 80% power to detect correlation of 0.52 or greater, assuming two-tailed test and $\alpha = 0.05$.

Spatial SAR estimation and validation

We believe the high SAR values of these IO-YAS MPs should allow accurate prediction of early thermal response (i.e., increase in temperature due to AMF application) in tissues based upon our MRI [IO-YAS MP] distribution measurements, with negligible early influences of perfusion. Rabbits from *Aim 1* (derivation set, N = 36) will provide data from percutaneous fiber-optic temperature probes inserted in superior and inferior tumor quadrants during AMF treatment sessions (Fig. 107A.7). Spatial and temporal temperature measurements will be used to calculate SARs at given 430, 450, and 470 Amp AMF protocols. Each scan will provide a voxel-wise map of R2* for each volume of interest (n = 8) and these mappings will be compared to the corresponding spatial distribution of SARs under rapid heating. Thus, actual relationships between SARs and MRI measurements of [IO-YAS MP] will be determined. We will then test these methods as they would be used clinically. For a given R2* map generated after trans-catheter infusion of IO-YAS MPs, we will use the above imaging methods for estimation of spatial SARs under specific AMF protocols in *each* rabbit, apply the protocol, and measure temperatures over time as explained previously for validation of this image-guided approach (validation set, N = 6, 2 per AMF protocol). The relationships between measured and MRI-based SAR estimates will be compared using Lin's concordance coefficient and Pearson correlation coefficients. A sample

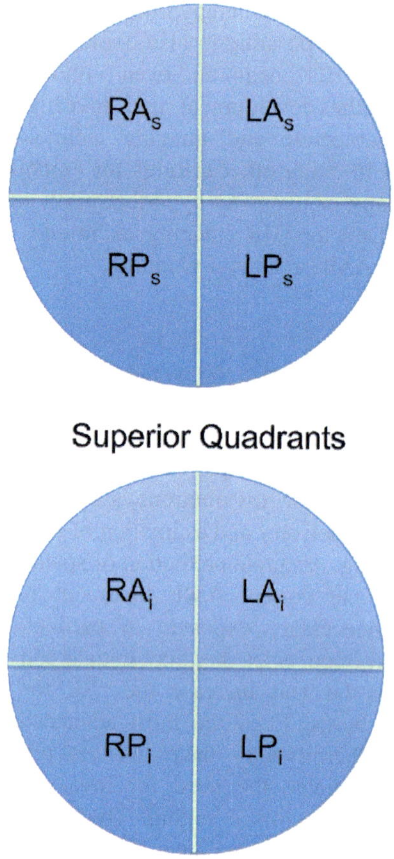

FIGURE 107A.7 Temperature probe locations during AMF treatment.

size 36 has 80% power to detect a correlation of 0.6 or greater, assuming a two-tailed test and $\alpha = 0.05$. The sample size for validation set animals may require adjustment pending the results of Aim 1.

AMF protocol	Field strength	AMF parameters
1	High	Coil current: 470 A, frequency: 190–200 kHz, ramping: None, pulses: None
2	Medium	Coil current: 450 A, frequency: 190–200 kHz, ramping: None, pulses: None
3	Low	Coil current: 430 A, frequency: 190–200 kHz, ramping: None, pulses: None

A, anterior; *i*, inferior; *L*, left; *P*, posterior; *R*, right; *s*, superior.

Limitations and alternatives

The use of holmium microspheres (^{166}Ho) with MRI tracking and γ-scintographic imaging was performed previously with a sequence similar to GESSE. This study aimed to use the paramagnetic nature of the ^{166}Ho radioisotope to accomplish dual-mode imaging but in vivo MRI measurements increasingly underestimated ex vivo microsphere concentrations as the infused concentration increased.[81] We believe our choice of IO compositions, use of phantom data, and larger sample size will yield a superior outcome. Should quantitative imaging strategies poorly predict [IO]s, the validation and refinement of MR imaging strategies remains a worthy pursuit that would allow for intraprocedural monitoring and direct visualization of intrahepatic IO-YAS MP delivery for thermoradiotherapy providing early detection of shunting to extrahepatic fields. Finally, it is possible that early SAR measurements do not correlate with MR relaxometry measurements in vivo due to differences in thermoregulatory responses in individual rabbits. Based on human data, heating to 40°C results in an immediate but gradual increase in cardiac output and stroke volume with decreases in splanchnic, renal, and muscle blood flow in normal tissues.[82] These effects occur on the order of minutes whereas we will take early SAR measurements at 30 s. It should also be noted that thermoregulatory mechanisms in humans and rabbits differ. Furthermore, we believe the intrinsically aberrant tumor vasculature will insulate the tumor from early thermoregulatory effects following AMF hyperthermia but we have no data to support such a theory at this time.

Timeline

PhD-1: Completion of nine required courses. Aim 1.1 will be performed concurrently with already synthesized and characterized IO-YAS MPs. **PhD-2:** Aim 1.2 to determine therapeutic efficacy of thermoradiotherapy versus lone radioembolization and Aims 2.1,2.2 studies to correlate MP concentrations, SAR and MR measurements. **PhD-3:** Aims 1.2, 2.1, 2.2 continued. Manuscript submissions/revisions. **PhD-4:** Aim 2.2 validation set initiated. Thesis preparation. Defense. Manuscript submissions/revisions. **MD-3:** Continued clinical clerkships. Manuscript submissions/revisions. **MD-4:** Electives and subinternships. Manuscript submissions/revisions. Residency interviews.

Appendix 2: IRB protocol template for biomedical research

Instructions

- Depending on the nature of what you are doing, some sections may not be applicable to your research. If a section is not applicable, delete. You may delete subsections that are not applicable.
- Add the STU# from eIRB + to the header of this document.
- Add the Version Date to the footer of this document.
- When you save and upload this document, add the protocol version date to the title.
- Keep an electronic copy of this document for your files. You will need to modify this copy when making changes.
- As you are creating this protocol, remove all instructions in italics and/or parentheses so that they are not contained in the final version of your protocol.

Protocol title: (include the full protocol title.)

PRINCIPAL INVESTIGATOR:
  Name
  P.I. Department sponsoring/supporting the study
  Telephone Number
  Email Address
  VERSION DATE:
  (Include the version date here and in the footer.)

Study summary

Investigational Agent(s) (Drugs or devices)

IND/IDE/HDE #	
Indicate Special Population(s)	Children Children who are wards of the state Adults unable to consent Cognitively impaired adults Neonates of uncertain viability Pregnant women Prisoners (or other detained/ paroled individuals) Students/Employees
Sample size	
Funding source	
Indicate the type of consent to be obtained	Written Verbal/Waiver of documentation of informed consent Waiver of HIPAA authorization Waiver/Alteration of consent process
Site	Lead site (for a multiple site research study) Data coordinating center (DCC)
Research related radiation exposure	Yes No
DSMB/DMC/IDMC	Yes No

Objectives

(Describe the purpose, specific aims, or objectives. State the hypotheses to be tested.)

Background

(Describe the relevant prior experience and gaps in current knowledge.)

(Describe any relevant preliminary data.)

(Provide the scientific or scholarly background for, rationale for, and significance of the research based on the existing literature and how will it add to existing knowledge.)

Study endpoints

(Describe the primary and secondary study endpoints.)

(Describe any primary or secondary safety endpoints.)

Study intervention(s)/investigational agent(s)

(Describe the study intervention and/or investigational agent (e.g., drug, device) that is being evaluated.)

(*Drug/Device Handling: If the research involves drugs or devices, describe your plans to store, handle, and administer those drugs or devices so that they will be used only on participants and be used only by authorized investigators.*)

- *If the control of the drugs or devices used in this protocol will be accomplished by following an established, approved organizational Standard Operating Procedures (SOP)* (e.g., *Research Pharmacy SOP for the Control of Investigational Drugs, etc.), please reference that SOP in this section.*

If the drug is has an Investigational New Drug (IND) or the device has an Investigational Device Exemption (IDE) or a claim of abbreviated IDE (nonsignificant risk device), include the following information:

- *Identify the holder of the IND/IDE/Abbreviated IDE.*
- *Explain procedures followed to comply with sponsor requirements for FDA-regulated research.*

(Is this an NSR (Abbreviated IDE) Device?)

- *If yes, provide rationale for your determination.*
- *Identify where research procedures will be performed.*

Procedures involved

(Describe and explain the study design.)

(Provide a description of all research procedures being performed and when they are performed, including procedures being performed to monitor participants for safety or minimize risks.)

Describe:

- *Procedures performed to lessen the probability or magnitude of risks.*
- *Delineate which procedures are considered standard of care and which are considered research-related. (For example, if the frequency of CT scans is within standard of care, this should be indicated)*
- *All drugs and devices used in the research and the purpose of their use, and their regulatory approval status.*
- *The source records, including medical or educational records that will be used to collect data about participants. (Attach all surveys, scripts, and data collection forms.)*

What data will be collected during the study and how will that data be obtained?

- *If there are plans for long-term follow-up (once all research related procedures are complete), what data will be collected during this period.*

(For Humanitarian Use Device (HUD): provide a description of the device, a summary of how you propose to use the device, including a description of any screening procedures, the HUD procedure, and any patient follow-up visits, tests or procedures.)

(If your study uses research-related radiation, attach a Radiation Dosimetry Form in the Supporting Documents section of the application in eIRB+ and explain how much and how often radiation will be used.)

Audio/Video Recording/Photography: If applicable, describe:

- the type of recording/photography being utilized
- why the type of recording is necessary to the research
- how the recordings/photograph(s) will be utilized in the research (e.g., data analysis only)
- how and where the recordings/photograph(s) are stored, who has access to them, and if/when they will be destroyed.

Data and specimen banking

If data or specimens will be banked for future use, describe

- *where the specimens will be stored,*
- *how long they will be stored,*
- *how the specimens will be accessed,*
- *who will have access to the specimens, and*
- *the data to be stored or associated with each specimen.*

(Describe the procedures to release data or specimens, including: the process to request a release, approvals required for release, who can obtain data or specimens, and the data to be provided with specimens.)

(If collecting and sharing genetic data, review the https://irb.northwestern.edu/sites/irb/files/files/HRP-442%20-%20CHECKLIST%20-%20Biobanking.docxchecklist and include the information addressed in the questions.)

Sharing results with participants

Describe if study results or individual participant results [such as results of investigational diagnostic tests, genetic tests, or incidental findings] will be shared with participants or anyone else (e.g., the participant's primary care physician).

(Describe how the results will be shared.)

Study timelines

Describe:

- the duration of an individual's participation in the study,
- approximately how long it will take to enroll all study participants, and
- the estimated date for the investigators to complete this study's primary analyses.

Inclusion and exclusion criteria

Describe:

- how individuals will be screened for eligibility,
- the criteria that define who will be included or excluded in your final study sample,
- specify if you will include or exclude each of the following special populations (members of the populations below may not be included in your research unless you indicate this in your inclusion criteria):
- Adults unable to consent
- Individuals who are not yet adults (infants, children, teenagers)
- Pregnant women
- Prisoners
- Vulnerable populations
- If this study excludes certain populations, explain the rationale for the exclusion in detail.

Vulnerable populations

(If the research involves individuals who are vulnerable to coercion or undue influence, describe additional safeguards included to protect their rights and welfare. Refer to the checklists below for specifics.)

"CHECKLIST: Pregnant Women (HRP-412)"

"CHECKLIST: Neonates (HRP-413)"

"CHECKLIST: Neonates of Uncertain Viability (HRP-414)"

"CHECKLIST: Prisoners (HRP-415)"

"CHECKLIST: Children (HRP-416)"

"CHECKLIST: Cognitively Impaired Adults (HRP-417)"

Participant population(s)

Accrual number:	Category/group: (adults/children special/vulnerable populations)	Consented: maximum number to be consented or reviewed/collected/screened	Enrolled: number to complete the study or needed to address the research question

Local			
Study-wide			
Total:			

Recruitment methods

(Describe when, where, and how potential participants will be recruited.)

(Describe the source of participants.)

(Describe the methods that will be used to identify potential participants.)

Describe materials that will be used to recruit participants. (Attach copies of these documents with the eIRB + application. For advertisements, attach the final copy of printed advertisements.)

(When advertisements are taped for broadcast, attach the final audio/video file. You may submit the wording of the advertisement before taping to preclude retaping because of inappropriate wording, provided the IRB reviews the final audio/video file.)

Compensation for participation in research activities

Describe the amount, timing, and method of any payments to participants. (e.g., gift card, ClinCard, check.)

(If payment is by check, you must request name, address and Social Security Number to issue a check for participation. Study payments are considered taxable income and are reportable to the IRS.)

(If USDOD, then please be aware that compensation will need to check with their supervisor before they can accept any payment for participation in research.)

(If the investigator believes that the biologic specimens obtained could be part of or lead to the development of a commercial product, indicate if the participant will have any right to compensation or ownership interest related to such development.)

(Describe when and how participants will be informed of the results of the research.)

Withdrawal of participants

Describe:

- *any anticipated circumstances under which participants will be withdrawn from the research without their consent,*
- *any procedures for orderly termination,*
- *procedures that will be followed when participants withdraw from the research, including partial withdrawal from procedures with continued data collection.*

Risks to participants

(List the reasonably foreseeable risks, discomforts, hazards, or inconveniences to the participant related to participation in the research. Include, for the IRB's consideration, a description of the probability, magnitude, duration, and reversibility of the risks.)

(Consider physical, psychological, social, legal, and economic risks.)

If applicable, indicate:

- *which procedures may have risks to the participants that are currently unforeseeable.*
- *which procedures may have risks to an embryo or fetus should the participant be or become pregnant.*
- *risks to others who are not participants.*

Potential benefits to participants

(Describe the potential benefits that individual participants may experience from taking part in the research. Include, for the IRB's consideration, the probability, magnitude, and duration of the potential benefits.)

(Indicate if there is no direct benefit. Do not include benefits to society or others.)

Data management and confidentiality

(Describe the data analysis plan, including any statistical procedures, power analysis, and a justification for your target enrollment number.)

(Describe the steps that will be taken to secure the data (e.g., training, authorization of access, password protection, encryption, physical controls, certificates of confidentiality, and separation of identifiers and data) during storage, use, and transmission.)

If applicable: refer to the HRP-335 GDPR (General Data Protection Regulations (GDPR) Data Protection Worksheet for additional requirements if you are collecting data from participants in the European Economic Area).

(Describe any procedures that will be used for quality control of collected data.)

Describe how data or specimens will be handled study-wide:

- What information will be included in that data or associated with the specimens?
- Where and how data or specimens will be stored?
- How long the data or specimens will be stored?
- Who will have access to the data or specimens?

- Who is responsible for receipt or transmission of the data or specimens?
- How data or specimens will be transported?

Provisions to monitor the data to ensure the safety of participants

(This section is required when research involves more than Minimal Risk to participants.)

Describe:

- *The plan to periodically evaluate the data collected regarding both harms and benefits to determine whether participants remain safe. The plan might include establishing a data monitoring committee (DSMB/DMC/IDMC) and a plan for reporting data monitoring committee findings to the IRB and the sponsor.*
- *The frequency of DSMB Meeting.*
- *What data are reviewed, including safety data, untoward events, and efficacy data.*
- *How the safety information will be collected (*e.g., *with case report forms, at study visits, by telephone calls with participants).*
- *The frequency of data collection, including when safety data collection starts.*
- *Who will review the data.*
- *The frequency or periodicity of review of cumulative data.*
- *The statistical tests for analyzing the safety data to determine whether harm is occurring.*
- *Any conditions that trigger an immediate suspension of the research.*

Provisions to protect the privacy interests of participants

(Describe the steps that will be taken to protect participants' privacy interests. "Privacy interest" refers to a person's desire to place limits on whom they interact or whom they provide personal information.)

(Describe what steps you will take to make the participants feel at ease with the research situation in terms of the questions being asked and the procedures being performed. "At ease" does not refer to physical discomfort, but the sense of intrusiveness a participant might experience in response to questions, examinations, and procedures.)

(Indicate how the research team is permitted to access any sources of information about the participants.)

Compensation for research-related injury

(If the research involves more than Minimal Risk to participants, describe the available compensation in the event of research-related injury.)

(Provide a copy of contract language, if any, relevant to compensation for research-related injury.)

Economic burden to participants

Describe any costs that participants may be responsible for because of participation in the research.

Consent process

Indicate if you will you be obtaining consent; and if so, describe:

- *Where the consent process takes place.*
- *Any waiting period available between informing the prospective participant and obtaining the consent.*
- *Any process to ensure ongoing consent.*
- *The role of the individuals listed in the application as being involved in the consent process.*
- *The language used by those obtaining consent and the language understood by the prospective participant or the legally authorized representative.*
- *The time that will be devoted to the consent discussion.*
- *Steps that will be taken to minimize the possibility of coercion or undue influence.*
- *Steps that will be taken to ensure the participants' understanding.)*
- *If applicable: If you will be obtaining consent from individuals in the European Economic Area refer to the GDPR Guidance for participant rights and for the two step process of consent including the use of the GDPR Compliance Consent Document template (HRP-590).*

Non-english speaking participants

(Indicate what language(s) other than English are understood by prospective participants or representatives.)

(If participants who do not speak English will be enrolled, describe the process to ensure that the oral and written information provided to those participants will be in that language. Indicate the language that will be used by those obtaining consent.)

Waiver or alteration of consent process

(consent will not be obtained, required information will not be disclosed, or the research involves deception).

(Review the "(Waiver or Alteration of Consent Process)" to ensure you have provided sufficient information for the IRB to make these determinations.)

(If the research involves a waiver the consent process for planned emergency research, review the "Waiver of Consent for Emergency Research" to ensure you have

provided sufficient information for the IRB to make these determinations.)

- ***Participants who are not yet adults (infants, children, teenagers)***
- *Describe the criteria that will be used to determine whether a prospective participant has not attained the legal age for consent to treatments or procedures involved in the research under the applicable law of the jurisdiction in which the research will be conducted. (E.g., individuals under the age of 18 years.)*
- *For research conducted in the state, review "SOP: Legally Authorized Representatives, Children, and Guardians (HRP-013)" to be aware of which individuals in the state meet the definition of "children."*
- *For research conducted outside of the state, provide information that describes which persons have not attained the legal age for consent to treatments or procedures involved the research, under the applicable law of the jurisdiction in which research will be conducted. One method of obtaining this information is to have a legal counsel or authority review your protocol along the definition of "children" in "SOP: Legally Authorized Representatives, Children, and Guardians (HRP-013)."*
- *Describe whether parental permission will be obtained from:*
- *Both parents unless one parent is deceased, unknown, incompetent, or not reasonably available, or when only one parent has legal responsibility for the care and custody of the child.*
- *One parent even if the other parent is alive, known, competent, reasonably available, and shares legal responsibility for the care and custody of the child.*
- *Describe whether permission will be obtained from individuals other than parents, and if so, who will be allowed to provide permission. Describe the process used to determine these individuals' authority to consent to each child's general medical care.*
- *Indicate whether assent will be obtained from all, some, or none of the children. If assent will be obtained from some children, indicate which children will be required to assent.*
- *When assent of children is obtained describe whether and how it will be documented.*
- ***Cognitively Impaired Adults***
- *Describe the process to determine whether an individual is capable of consent. The IRB allows the person obtaining assent to document assent on the consent document and does not routinely require assent documents and does not routinely require cognitively impaired adults to sign assent documents.*
- Adults Unable to Consent
- *List the individuals from whom permission will be obtained in order of priority. (e.g., durable power of attorney for health care, court appointed guardian for health care decisions, spouse, and adult child.)*
- *For research conducted in the state, review "SOP: Legally Authorized Representatives, Children, and Guardians (HRP-013)" to be aware of which individuals in the state meet the definition of "legally authorized representative."*
- *For research conducted outside of the state, provide information that describes which individuals are authorized under applicable law to consent on behalf of a prospective participant to their participation in the procedure(s) involved in this research. One method of obtaining this information is to have a legal counsel or authority review your protocol along the definition of "legally authorized representative" in "SOP: Legally Authorized Representatives, Children, and Guardians (HRP-013)."*
- *Describe the process for assent of the adult participants. Indicate whether:*
- *Assent will be required of all, some, or none of the participants. If some, indicated, which participants will be required to assent and which will not.*
- *If assent will not be obtained from some or all participants, an explanation of why not.*
- *Describe whether assent of the participants will be documented and the process to document assent. The IRB allows the person obtaining assent to document assent on the consent document and does not routinely require assent documents and does not routinely require participants to sign assent documents.*
- ***Humanitarian Use Devices***
- *For HUD uses, provide a description of how the patient will be informed of the potential risks and benefits of the HUD and any procedures associated with its use.*
- ***Process to Document Consent in Writing***
- *Describe how consent of the participant will be documented in writing.*
- *If your research presents no more than minimal risk of harm to participants and involves no procedures for which written documentation of consent is normally required outside of the research context, the IRB will generally waive the requirement to obtain written documentation of consent.*
- *(If you will document consent in writing, please add this when prompted in your EIRB + Application) If you will obtain consent, but not document consent in writing, attach a consent script. Review "CHECKLIST: Waiver of Written Documentation of Consent (HRP-411)" to ensure that you have provided sufficient information. You may use "TEMPLATE CONSENT DOCUMENT (HRP-502)"to create the consent document or script.)*

- *Setting*
- *Describe the sites or locations where your research team will conduct the research.*
- *If a Lead Coordinating Center or Data Coordinating center please see describe the role that NU has in this research study (for definitions please see below)*
- *Describe the composition and involvement of any community advisory board.*
- *For research conducted outside of the organization and its affiliates describe:*
- *Site-specific regulations or customs affecting the research for research outside the organization.*
- *Local scientific and ethical review structure outside the organization.*
- *Lead Coordinating Center:*
- *A lead coordinating center is defined as a site that provides the administrative, clinical, and technical expertise and leadership in the design and coordination of the multisite collaborative research for a multicenter trial.*
- *The principal investigator will be responsible for all site monitoring and for the coordination of participant recruitment, screening, enrollment and retention, data and safety monitoring, data collection and analysis, adherence to the protocol-directed procedures and guidelines, and the prompt review, reporting and resolution of adverse events.*
- *Data Coordinating Center:*
- *A data coordinating center is defined as a site that is responsible only for the collection and storage of data collected from all sites involved in a multisite trial.*

Protected health information (PHI and HIPAA)

(HIPAA applies to Protected Health Information (PHI). PHI is individually identifiable health information that is created or maintained by a covered entity (health care providers, hospitals, physician offices, health care clearing houses, health care plans), or their business associate(s).)

(If your research does not involve the use of medical record information maintained by a covered entity and if the information generated from research will not be placed into the medical record, then HIPAA does not apply.)

Indicate the following:

- *Does the study involve the creation, use, or disclosure of Protected Personal Health Information?*
- *Will a HIPAA Authorization be obtained from for all or some participants?*
- *If HIPAA Authorization will not be obtained, indicate what alternatives will be used.*

- Names
- Geographic Subdivisions: All geographic subdivisions smaller than a state, including street address, city, county, precinct, ZIP code and their equivalent geographical codes, except for the initial three digits of a ZIP code if, according to the current publicly available data from the Bureau of the Census: (a) The geographic unit formed by combining all ZIP codes with the same three initial digits contains more than 20,000 people. (b) The initial three digits of a ZIP code for all geographic units containing 20,000 or fewer people are changes to 000.
- Dates and Age: All elements of dates (except year) for dates directly related to an individual, including birth date, admission date, discharge date, date of death; and all ages of 89 and all elements of dates (including year) indicative of such age, except that such ages and elements may be aggregated into a single category of age 90 or older.
  - Telephone numbers
  - FAX numbers
  - Social Security Numbers
  - Medical Record Numbers
  - Health Plan Beneficiary Numbers
  - Account Numbers
  - Certificate/License Numbers
  - Vehicle Identifiers and Serial Numbers, including License Plate Numbers
  - Device Identifiers and Serial Numbers
  - Web Universal Resource Locators
  - Internet Protocol (IP) Address Numbers
  - Biometric Identifiers, including Fingerprints and Voiceprints
  - Full-Face Photographic Images and any Comparable Images
  - Any other Unique Identifying Number, Characteristic, or Code, unless otherwise permitted by the Privacy Rule for reidentification

(Review Waiver of HIPAA Authorization and provide protocol specific justifications if HIPAA applies and you will not be obtaining a signed HIPAA Authorization.)

Qualifications to conduct research and resources available

Describe the resources available to conduct the research: For example, as appropriate:

- *Justify the feasibility of recruiting the required number of suitable participants within the agreed recruitment period. For example, how many potential participants do you have access to? What percentage of those potential participants do you need to recruit?*

- *Describe the time that you will devote to conducting and completing the research.*
- *Describe your facilities.*
- *Describe the availability of medical or psychological resources that participants might need as a result of an anticipated consequences of the human research.*
- *Describe your process to ensure that all persons assisting with the research are adequately informed about the protocol, the research procedures, and their duties and functions.*
- *Multisite Research (Complete if NU is the Lead Site and/or Data Coordinating Center)*
- *Studywide Number of Participants*
- *If this is a multicenter study, indicate the total number of participants to be accrued across all sites.*

Multisite research

(If this is a multicenter study and participants will be recruited by methods not under the control of the local site (e.g., call centers, national advertisements) describe those methods. Local recruitment methods are described later in the protocol.)

Describe:

- *when, where, and how potential participants will be recruited.*
- *the methods that will be used to identify potential participants.*
- *materials that will be used to recruit participants. (Attach copies of these documents with the application. For advertisements, attach the final copy of printed advertisements. When advertisements are taped for broadcast, attach the final audio/video tape. You may submit the wording of the advertisement before taping to preclude re-taping because of inappropriate wording, provided the IRB reviews the final audio/video tape.)*

(If this is a multisite study where you are the lead investigator, describe the processes to ensure communication among sites. See "WORKSHEET: Communication and Responsibilities (HRP-830)." All sites have the most current version of the protocol, consent document, and HIPAA authorization.)

All required approvals (initial, continuing review and modifications) have been obtained at each site (including approval by the site's IRB of record).

(All modifications have been communicated to sites, and approved (including approval by the site's IRB of record) before the modification is implemented.)

(If this is a multi-site study where you are the lead investigator, describe plan of how information will be managed to ensure protection of participants. All engaged participating sites will safeguard data, including secure transmission of data, as required by local information security policies.)

(All local site investigators conduct the study in accordance with applicable federal regulations and local laws.)

Indicate any location-specific regulations or customs affecting your project, including any local scientific and ethical review structure including if applicable, considerations of the GDPR for data collected in the European Economic Area applies (see GDPR guidance).

(All noncompliance with the study protocol or applicable requirements will be reported in accordance with local policy.)

Describe the method for communicating to engaged participating sites (see "WORKSHEET: Communication and Responsibilities (HRP-830)"):

- Problems (inclusive of reportable events).
- Interim results.
- The closure of a study)

If this is a multicenter study where you are a participating site/investigator, describe the local procedures for maintenance of confidentiality. (See "WORKSHEET: Communication and Responsibilities (HRP-830).")

- *Where and how data or specimens will be stored locally?*
- *How long the data or specimens will be stored locally?*
- *Who will have access to the data or specimens locally?*
- *Who is responsible for receipt or transmission of the data or specimens locally?*
- How data and specimens will be transported locally?.

References

1. Jemal A, et al. Global cancer statistics. *CA Canc J Clin.* 2011;61:69–90.
2. Mazzaferro V, et al. Liver transplantation for the treatment of small hepatocellular carcinomas in patients with cirrhosis. *N Engl J Med.* 1996;334:693–699.
3. Llovet JM, Fuster J, Bruix J. Intention-to-treat analysis of surgical treatment for early hepatocellular carcinoma: resection versus transplantation. *Hepatology.* 1999;30:1434–1440.

4. Llovet J, Schwartz M, Mazzaferro V. Resection and liver transplantation for hepatocellular carcinoma. *Semin Liver Dis*. 2005;25:181–200.
5. Llovet JM, Burroughs A, Bruix J. Hepatocellular carcinoma. *Lancet*. 2003;362:1907–1917.
6. Sherman M. Hepatocellular carcinoma: epidemiology, surveillance, and diagnosis. *Semin Liver Dis*. 2010;30:3–16.
7. Eggeman AS, Majetich SA, Farrell D, Pankhurst QA. Size and concentration effects on high frequency hysteresis of iron oxide nanoparticles. *Ieee T Magn*. 2007;43:2451–2453.
8. Overgaard J. Effect of hyperthermia on malignant cells in vivo. A review and a hypothesis. *Cancer*. 1977;39:2637–2646.
9. Wust P, et al. Hyperthermia in combined treatment of cancer. *Lancet Oncol*. 2002;3:487–497.
10. Lv Y-G, Deng Z-S, Liu J. 3-D numerical study on the induced heating effects of embedded micro/nanoparticles on human body subject to external medical electromagnetic field. *Ieee T Nanobiosci*. 2005;4:284–294.
11. Hildebrandt B, et al. The cellular and molecular basis of hyperthermia. *Crit Rev Oncol Hematol*. 2002;43:33–56.
12. Horsman MR, Overgaard JH. A potent enhancer of radiotherapy. *Clin Oncol-Uk*. 2007;19:418–426.
13. Valdagni R, Amichetti M. Report of long-term follow-up in a randomized trial comparing radiation therapy and radiation therapy plus hyperthermia to metastatic lymph nodes in stage IV head and neck patients. *Int J Radiat Oncol Biol Phys*. 1994;28:163–169.
14. Overgaard J, et al. Randomized trial of hyperthermia as adjuvant to radiotherapy for recurrent or metastatic malignant-melanoma. *Lancet*. 1995;345:540–543.
15. Vernon C, et al. Radiotherapy with or without hyperthermia in the treatment of superficial localized breast cancer: results from five randomized controlled trials. *Int J Radiat Oncol Biol Phys*. 1996;35:731–744.
16. Sneed PK, et al. Survival benefit of hyperthermia in a prospective randomized trial of brachytherapy boost +/- hyperthermia for glioblastoma multiforme. *Int J Radiat Oncol Biol Phys*. 1998;40:287–295.
17. van der Zee J, et al. Comparison of radiotherapy alone with radiotherapy plus hyperthermia in locally advanced pelvic tumours: a prospective, randomised, multicentre trial. Dutch Deep Hyperthermia Group. *Lancet*. 2000;355:1119–1125.
18. Jones EL, et al. Randomized trial of hyperthermia and radiation for superficial tumors. *J Clin Oncol*. 2005;23:3079–3085.
19. Franckena M, et al. Long-term improvement in treatment outcome after radiotherapy and hyperthermia in locoregionally advanced cervix cancer: an update of the Dutch Deep Hyperthermia Trial. *Int J Radiat Oncol Biol Phys*. 2008;70:1176–1182.
20. Maier-Hauff K, et al. Intracranial thermotherapy using magnetic nanoparticles combined with external beam radiotherapy: results of a feasibility study on patients with glioblastoma multiforme. *J Neuro Oncol*. 2007;81:53–60.
21. van Landeghem FKH, et al. Post-mortem studies in glioblastoma patients treated with thermotherapy using magnetic nanoparticles. *Biomaterials*. 2009;30:52–57.
22. Maier-Hauff K, et al. Efficacy and safety of intratumoral thermotherapy using magnetic iron-oxide nanoparticles combined with external beam radiotherapy on patients with recurrent glioblastoma multiforme. *J Neuro Oncol*. 2011;103:317–324.
23. Kalambur VS, Longmire EK, Bischof JC. Cellular level loading and heating of superparamagnetic iron oxide nanoparticles. *Langmuir*. 2007;23:12329–12336.
24. Hergt R, et al. Maghemite nanoparticles with very high AC-losses for application in RF-magnetic hyperthermia. *J Magn Magn Mater*. 2004;270:345–357.
25. Fortin J-P, Gazeau F, Wilhelm C. Intracellular heating of living cells through Néel relaxation of magnetic nanoparticles. *Eur Biophys J*. 2008;37:223–228.
26. Gruettner C, et al. Synthesis and antibody conjugation of magnetic nanoparticles with improved specific power absorption rates for alternating magnetic field cancer therapy. *J Magn Magn Mater*. 2007;311:181–186.
27. Kim D-H, Nikles DE, Johnson DT, Brazel CS. Heat generation of aqueously dispersed CoFe2O4 nanoparticles as heating agents for magnetically activated drug delivery and hyperthermia. *J Magn Magn Mater*. 2008;320:2390–2396.
28. Tsafnat N, Tsafnat G, Lambert T, Jones S. Modelling heating of liver tumours with heterogeneous magnetic microsphere deposition. *Phys Med Biol*. 2005;50:2937–2953.
29. Hehr T, et al. Experimental thermoradiotherapy in malignant hepatocellular carcinoma. *Int J Radiat Oncol Biol Phys*. 2003;55:1374–1380.
30. Hergt R, Dutz S, Zeisberger M. Validity limits of the Neel relaxation model of magnetic nanoparticles for hyperthermia. *Nanotechnology*. 2010;21:015706.
31. Purushotham S, Ramanujan RV. Modeling the performance of magnetic nanoparticles in multimodal cancer therapy. *J Appl Phys*. 2010;107:114701.
32. Campbell AM, Bailey IH, Burton MA. Analysis of the distribution of intra-arterial microspheres in human liver following hepatic yttrium-90 microsphere therapy. *Phys Med Biol*. 2000;45:1023–1033.
33. Campbell A, Bailey I, Burton M. Tumour dosimetry in human liver following hepatic yttrium-90 microsphere therapy. *Phys Med Biol*. 2001;46:487–498.
34. Moroz P, Jones SK, Winter J, Gray BN. Targeting liver tumors with hyperthermia: ferromagnetic embolization in a rabbit liver tumor model. *J Surg Oncol*. 2001;78(22–9). discussion 30–1.
35. DeNardo S, et al. Development of tumor targeting bioprobes (In-111-chimeric L6 monoclonal antibody nanoparticles) for alternating magnetic field cancer therapy. *Clin Cancer Res*. 2005;11:7087S–7092S.
36. Hilger I, et al. Thermal ablation of tumors using magnetic nanoparticles: an in vivo feasibility study. *Invest Radiol*. 2002;37:580–586.
37. Rabias I, et al. Rapid magnetic heating treatment by highly charged maghemite nanoparticles on Wistar rats exocranial glioma tumors at microliter volume. *Biomicrofluidics*. 2010;4:024111.
38. Bruners P, et al. Thermoablation of malignant kidney tumors using magnetic nanoparticles: an in vivo feasibility study in a rabbit model. *Cardiovasc Inter Rad*. 2010;33:127–134.

39. Johannsen M, et al. Thermotherapy using magnetic nanoparticles combined with external radiation in an orthotopic rat model of prostate cancer. *Prostate*. 2006;66:97–104.
40. Johannsen M, et al. Magnetic fluid hyperthermia (MFH)reduces prostate cancer growth in the orthotopic Dunning R3327 rat model. *Prostate*. 2005;64:283–292.
41. Johannsen M, et al. Evaluation of magnetic fluid hyperthermia in a standard rat model of prostate cancer. *J Endourol*. 2004;18:495–500.
42. Johannsen M, et al. Clinical hyperthermia of prostate cancer using magnetic nanoparticles: presentation of a new interstitial technique. *Int J Hyperther*. 2005;21:637–647.
43. Johannsen M, et al. Thermotherapy of prostate cancer using magnetic nanoparticles: feasibility, imaging, and three-dimensional temperature distribution. *Eur Urol*. 2007;52:1653–1661.
44. Stauffer PR, et al. Conformal microwave array (CMA) applicators for hyperthermia of diffuse chest wall recurrence. *Int J Hyperther*. 2010;26:686–698.
45. van der Zee J, et al. Reirradiation combined with hyperthermia in breast cancer recurrences: overview of experience in Erasmus MC. *Int J Hyperther*. 2010;26:638–648.
46. Nielsen OS, Overgaard J. Importance of preheating temperature and time for the induction of thermotolerance in a solid tumour in vivo. *Br J Cancer*. 1982;46:894–903.
47. Salem R, et al. Treatment of unresectable hepatocellular carcinoma with use of 90Y microspheres (TheraSphere): safety, tumor response, and survival. *J Vasc Intervent Radiol*. 2005;16:1627–1639.
48. Riaz A, et al. Complications following radioembolization with yttrium-90 microspheres: a comprehensive literature review. *J Vasc Intervent Radiol*. 2009;20:1121–1130. quiz 1131.
49. Atassi B, et al. Biliary sequelae following radioembolization with Yttrium-90 microspheres. *J Vasc Intervent Radiol*. 2008;19:691–697.
50. Sangro B, et al. Liver disease induced by radioembolization of liver tumors: description and possible risk factors. *Cancer*. 2008;112:1538–1546.
51. Young JY, et al. Radiation dose limits and liver toxicities resulting from multiple yttrium-90 radioembolization treatments for hepatocellular carcinoma. *J Vasc Intervent Radiol*. 2007;18:1375–1382.
52. Kennedy AS, et al. Treatment parameters and outcome in 680 treatments of internal radiation with resin 90Y-microspheres for unresectable hepatic tumors. *Int J Radiat Oncol Biol Phys*. 2009;74:1494–1500.
53. Salem R, et al. Radioembolization for hepatocellular carcinoma using Yttrium-90 microspheres: a comprehensive report of long-term outcomes. *Gastroenterology*. 2010;138:52–64.
54. Hey S, et al. Simultaneous T1 measurements and proton resonance frequency shift based thermometry using variable flip angles. *Magn Reson Med*. 2012;67:457–463.
55. de Senneville BD, Ries M, Maclair G, Moonen C. MR-guided thermotherapy of abdominal organs using a robust PCA-based motion descriptor. *IEEE Trans Med Imag*. 2011;30:1987–1995.
56. Fite BZ, et al. Magnetic resonance thermometry at 7T for real-time monitoring and correction of ultrasound induced mild hyperthermia. *PLoS One*. 2012;7:e35509.
57. Puls R, et al. Laser ablation of liver metastases from colorectal cancer with MR thermometry: 5-year survival. *J Vasc Intervent Radiol*. 2009;20:225–234.
58. Wlodarczyk W, et al. Comparison of four magnetic resonance methods for mapping small temperature changes. *Phys Med Biol*. 1999;44:607–624.
59. Robson MD, Gatehouse PD, Bydder M, Bydder GM. Magnetic resonance: an introduction to ultrashort TE (UTE) imaging. *J Comput Assist Tomogr*. 2003;27:825–846.
60. Yablonskiy DA, Haacke EM. An MRI method for measuring T2 in the presence of static and RF magnetic field inhomogeneities. *Magn Reson Med*. 1997;37:872–876.
61. Jordan A, et al. Presentation of a new magnetic field therapy system for the treatment of human solid tumors with magnetic fluid hyperthermia. *J Magn Magn Mater*. 2001;225:118–126.
62. Sapareto SA, Dewey WC. Thermal dose determination in cancer therapy. *Int J Radiat Oncol Biol Phys*. 1984;10:787–800.
63. Dewhirst MW, Sim DA, Sapareto S, Connor WG. Importance of minimum tumor temperature in determining early and long-term responses of spontaneous canine and feline tumors to heat and radiation. *Cancer Res*. 1984;44:43–50.
64. Geschwind JF, et al. Chemoembolization of liver tumor in a rabbit model: assessment of tumor cell death with diffusion-weighted MR imaging and histologic analysis. *J Vasc Intervent Radiol*. 2000;11:1245–1255.
65. Kobeiter H, et al. Targeted transarterial therapy of Vx-2 rabbit liver tumor with Yttrium-90 labeled ferromagnetic particles using an external magnetic field. *Anticancer Res*. 2007;27:755–760.
66. Virmani S, et al. Comparison of two different methods for inoculating VX2 tumors in rabbit livers and hind limbs. *J Vasc Intervent Radiol*. 2008;19:931–936.
67. Guo Y, et al. Electroporation-mediated transcatheter arterial chemoembolization in the rabbit VX2 liver tumor model. *Invest Radiol*. 2012;47:116–120.
68. Gupta T, et al. MR tracking of iron-labeled glass radioembolization microspheres during transcatheter delivery to rabbit VX2 liver tumors: feasibility study. *Radiology*. 2008;249:845–854.
69. Deng J, et al. Diffusion-weighted PROPELLER MRI for quantitative assessment of liver tumor necrotic fraction and viable tumor volume in VX2 rabbits. *J Magn Reson Imag*. 2008;27:1069–1076.
70. Deng J, et al. In vivo diffusion-weighted imaging of liver tumor necrosis in the VX2 rabbit model at 1.5 Tesla. *Invest Radiol*. 2006;41:410–414.
71. Davis JA. Mouse and rat anesthesia and analgesia. *Curr Protoc Neurosci Appendix*. 2008;4. Appendix 4B.
72. Bilbao R, et al. A blood-tumor barrier limits gene transfer to experimental liver cancer: the effect of vasoactive compounds. *Gene Ther*. 2000;7:1824–1832.
73. Varma M, Jasani B. Diagnostic utility of immunohistochemistry in morphologically difficult prostate cancer: review of current literature. *Histopathology*. 2005;47:1–16.
74. Fox SB, Harris AL. Histological quantitation of tumour angiogenesis. *APMIS*. 2004;112:413–430.
75. Najimi M, Smets F, Sokal E. Hepatocyte apoptosis. *Methods Mol Biol*. 2009;481:59–74.
76. Arunachalam K, Maccarini PF, Craciunescu OI, Schlorff JL, Stauffer PR. Thermal characteristics of thermobrachytherapy surface applicators for treating chest wall recurrence. *Phys Med Biol*. 2010;55:1949–1969.

77. Rabin Y. Is intracellular hyperthermia superior to extracellular hyperthermia in the thermal sense? *Int J Hyperther.* 2002;18:194–202.
78. Atkinson W, Brezovich I, Chakraborty D. Useable frequencies in hyperthermia with thermal seeds. *Ieee T Bio-Med Eng.* 1984;31:70–75.
79. Dudeck O, et al. Local arterial infusion of superparamagnetic iron oxide particles in hepatocellular carcinoma: a feasibility and 3.0 T MRI study. *Invest Radiol.* 2006;41:527–535.
80. Lee K-H, et al. Distribution of iron oxide-containing Embosphere particles after transcatheter arterial embolization in an animal model of liver cancer: evaluation with MR Imaging and implication for therapy. *J Vasc Intervent Radiol.* 2008;19:1490–1496.
81. Seppenwoolde J, et al. Internal radiation therapy of liver tumors: qualitative and quantitative magnetic resonance imaging of the biodistribution of holmium-loaded microspheres in animal models. *Magn Reson Med.* 2005;53:76–84.
82. Rowell LB. Human cardiovascular adjustments to exercise and thermal stress. *Physiol Rev.* 1974;54:75–159.

Further reading

1. Begg C, Cho M, Eastwood S, et al. Improving the quality of reporting of randomized controlled trials. The CONSORT statement. *JAMA.* 1996;276(8):637–639.
2. Schulz KF, Altman DG, Moher D, Group C. CONSORT 2010 statement: updated guidelines for reporting parallel group randomised trials. *BMJ.* 2010;340:c332.
3. Salem R, Gordon AC, Mouli S, et al. Y90 radioembolization significantly Prolongs time to progression compared with chemoembolization in patients with hepatocellular carcinoma. *Gastroenterology.* 2016;151(6):1155–1163. e1152.
4. Nemcek A. *How to Get Your Paper Published. Lecture Presented at: SIR 38th Annual Scientific Meeting.* 2013 (New Orleans, LA).
5. Faul F, Erdfelder E, Lang AG, Buchner AG. Power 3: a flexible statistical power analysis program for the social, behavioral, and biomedical sciences. *Behav Res Meth.* 2007;39(2):175–191.

Glossary terms

3R's

- Replacement: the use of alternative methods to traditional animal-based research.
- Reduction: limiting the number of animals used in research.
- Refinement: when it is necessary to use animals in research, their health and welfare should be considered. Some examples include the use of appropriate anaesthesia, reducing the distress and reducing the harm in animals.

510(K) Notification Premarket notification showing the FDA that the product is safe and effective.

Abbreviated new drug application A request to approve a pharmaceutically equivalent version of an already FDA-approved drug.

Adherence Health-related behavior that adheres to the recommendations of a physician, healthcare provider, or investigator in a research project. The word *adherence* aims to avoid the authoritarian connotations of *compliance*, formerly used to describe this behavior.

Adaptive randomization Manner by which randomization itself occurs is adjusted after monitoring study results over time. For instance, patients who are subsequently enrolled in a trial are more likely to be placed into the treatment group(s) that were deemed more effective among subjects earlier on during the course of the study.

Adaptive study design In terms of a definition, the Pharmaceutical Research Manufacturer Association Working Group on Adaptive Design defined it as "a clinical trial design that uses accumulating data to decide how to modify aspects of the study as it continues, without undermining the validity and integrity of the trial."

Adverse event (AE) An unexpected medical problem that happens during treatment with a therapy or a known side effect of the treatment given. Adverse events may be mild, moderate, or severe and may be caused by something other than the drug or therapy being given. This can also be called an adverse effect (National Cancer Institute definition).

Agency for Healthcare Research and Quality (AHRQ) A United States federal agency focused on improving the safety and quality of America's health care system. AHRQ aims to develop the knowledge, tools, and data needed to improve the healthcare system and support patients, healthcare professionals, and policy makers in making informed health decisions.

Allergenics Substance that causes an allergic reaction.

Alpha value The probability of rejecting a true, null hypothesis, and used as the significance level within a study. Generally, set at a value of 0.05 within medical research studies.

Alternative hypothesis The statement being tested against the null hypothesis in statistical hypothesis testing.

Analysis A step within the scientific method. The process of interpreting and attempting to identify patterns or trends from collected data.

Analysis of variance (ANOVA) A collection of statistical models and their associated estimation procedures used to analyze the differences among means.

Antibiotic A medicine that inhibits the growth or destroys microorganisms.

ARIA system Active postmarket risk identification and analysis (ARIA) System is part of the Sentinel System, and has programs that allow the FDA to analyze healthcare data processed by the Sentinel System.

Artifactual nonadherence Deviations from the study protocol that are intentional and sometimes caused by "professional subjects," or subjects that enroll in studies for primary or secondary gain.

Artificial intelligence (AI) The perceiving, synthesizing and inferring of information demonstrated by machines.

Artificial neural networks (ANN) A computational framework inspired by the organization and function of the brain.

Attrition Reduction in the number of participants in a study as it progresses. Losses may be due to withdrawal, dropouts, or protocol deviations.

Autonomy (1) The capacity for self-governance, that is. the ability to make reasonable decisions. (2) A moral principle barring interference with autonomous decision-making.

BCLC Barcelona Clinic Liver Cancer staging system is commonly used to stratify hepatocellular carcinoma.

Belmont report A report issued by the US National Commission for the Protection of Human Subjects in Biomedical and Behavioral Research in 1979, which has had a significant influence over human subjects' research ethics, regulation, and policy. The report provided a conceptual foundation for the Common Rule and articulated three principles of ethics: respect for persons, beneficence, and justice.

Beneficence The ethical obligation to do good and avoid causing harm.

Beveridge model A health care system in which the government provides health care for all its citizens through income tax payments.

Bias Systemic derivation of results or inferences from the truth due to various causes including errors in study design, measurement, or statistical interpretation.

Biologic Any pharmaceutical drug product manufactured in, extracted from, or semi-synthesized from biological sources.

Biological products Biological products are regulated by the Food and Drug Administration (FDA) and are used to diagnose, prevent, treat, and cure diseases and medical conditions. Biological products are a diverse category of products and are generally large, complex molecules. These products may be produced through biotechnology in a living system, such as a microorganism, plant cell, or animal cell, and are often more difficult to characterize than small molecule drugs. There are many types of biological products approved for use in the United States, including therapeutic proteins (such as filgrastim), monoclonal antibodies (such as adalimumab), and vaccines (such as those for influenza and tetanus).

Biologicals A therapeutic substance that is derived from a biological source.

Bioprinting "3-D printing" to an industry standard term such as "additive manufacturing".

Bismarck model Also referred as "Social Health Insurance Model" is a limited health care system, in which people pay a fee to a fund that in turn pays health care activities, that can be provided by State-owned institutions, other Government body-owned institutions, or a private institution.

Blinding Concealment of group allocation from the patient, team of investigators, and/or the committee monitoring the response variables.

Block randomization Block randomization is a form of randomization that tries to keep a steady balance of study participants in different treatment arms throughout the period of enrollment, with the goal of allocating participants into treatment arms in a fixed ratio (e.g., 1:1 for a two-arm design) within small blocks.

Bond A loan between an investor and a borrower which is packaged as a tradable security.

Bonferroni's method A post hoc analysis that uses thresholds based on the T-distribution to compare groups.

Botox Injectable botulinum toxin is used in medicine for a variety of purposes. Specifically in IR, botox has seen utility in the treatment of musculoskeletal or neuromuscular disorders and been used for cosmesis among other utilities.

BRUSO Brief, relevant, unambiguous, specific, and objective.

Case series Type of medical research that investigates a group of patients who have undergone the same type of procedure or have the same medical conditions over a specified time period.

Case-control study A retrospective study in which two groups with different outcomes, one with a certain outcome and one with no outcome (control) are examined to define some attributes that occurred that may have led to the different outcomes.

Case-control study Type of medical research that investigates a group of patients who have a known medical condition(s), disease state, and exposure(s) or have undergone a certain type of procedure and a control group of patients who lack the topic of interest.

CDRH Center for Devices and Radiological Health.

Center for Drug Evaluation and Research (CDER) Division of the FDA responsible for monitoring most drugs outlined in the Food, Drug, and Cosmetic Act.

Centers for Medicare and Medicaid Services (CMS) A United States federal agency that collaborates with state governments to manage the Medicare program. Additionally, CMS offers resources for researchers who require existing stored data or to fund new investigations.

Champion PHOENIX trial Cangrelor versus Standard Therapy to Achieve Optimal Management of Platelet Inhibition PHOENIX trial, a double-blinded study involving patients undergoing urgent or elective percutaneous coronary intervention for coronary insufficiency, randomly assigned to get cangrelor (IV antiplatelet agent) or clopidogrel (oral antiplatelet agent).

Chemistry, manufacturing, and controls A critical part of the IND application geared toward the characterization of the drug product and assurance of the safety and consistent manufacturing of the drug product.

Class I devices Devices for which the general controls of the Act are sufficient to provide reasonable assurance of their safety and effectiveness. They typically present minimal potential for harm to the user and the person being tested. They are subject to general controls, which include registration and listing, labeling, and adverse event reporting requirements (Section 513(a) (1) (A) of the Act). Most Class I devices are exempt from premarket notification, subject to certain limitations found in Section 510(l) of the Act and in 21 CFR § 862.9, 864.9, and 866.9. Some are also exempt from the "Quality Systems Regulation" found in 21 CFR Part 820.

Class II devices Devices for which general controls alone are insufficient to provide reasonable assurance of their safety and effectiveness and for which establishment of special controls can provide such assurances. Special controls may include special labeling, mandatory performance standards, risk mitigation measures identified in guidance, and postmarket surveillance (Section 513(a) (1) (B) of the Act). Some Class II devices are exempt from premarket notification, subject to limitation in 21 CFR § § 862.6, 864.9, and 866.9.

Class III devices Devices for which insufficient information exists to provide reasonable assurance of safety and effectiveness through general or special controls. Class III devices are usually those that support or sustain human life, are of substantial importance in preventing impairment of human health, or which present a potential, unreasonable risk of illness or injury (Section 513(a) (1) (C) of the Act). Most Class III devices require premarket approval. A limited number of Class III devices that are equivalent to devices legally marketed before enactment of the Medical Device Amendments of 1976 may be marketed through the premarket notification (510(k)) process, until FDA has published a requirement for manufacturers of that generic type of device to submit PMA data.

Class medical devices Higher associated risk and require additional testing to assure safety and effectiveness, including IDE approval before undergoing clinical trials.

Clinical equipoise A state of genuine uncertainty in regard to the therapeutic merits of each arm in a clinical trial.

Clinical question A specific research question related to a particular population, intervention, comparison, and outcome measure.

Clinical significance The meaningful effect that an intervention offers patients, magnitude of a treatment's effect as measured through the absolute or relative change of clinical outcomes.

Clinical strategy Defines the overall clinical pathway for the development of an innovation and helps define and prioritize each study's endpoints.

Clinically meaningful endpoint An outcome that directly measures how an individual feels, functions, or survives.

Cluster sampling Target population is divided into multiple groups (or clusters).

Code of Federal Regulations (CFR) FDA's interpretation of *The Federal Food, Drug, and Cosmetic Act* and related statutes and document the Federally required actions for drug sponsors.

Coefficient of determination The coefficient of determination is a percentage that can be used to explain how much the variance of the dependent variable is explained by the independent variable.

Cohort study A longitudinal study in which participants are grouped together based on some defining variable and data are collected at various cross-sections of time.

Co-investigator (Co-Is) From an Institutional Review Board (IRB) perspective, are delegated with particular tasks towards the project. Co-Is must have the necessary training and experience for participating in the project. Their involvement needs to be explicitly acknowledged in the IRB review process.

Collagen injections Used in the augmentation of soft tissue for cosmetic enhancements. Can be used for skin depression, scarring, or wrinkle correction. Involved in injecting collagen into the skin.

Combination products Therapeutic and diagnostic products that combine drugs, devices, and/or biological products.

Commerce The act of buying and selling.

Commercial IND Application submitted by a corporate entity for the investigation of a drug intended to be commercialized through a marketing application.

Commercialization Introduction of a device or product into a marketplace, with associated financial gain.

Compliance The act of meeting set rules or standards.

Complication Any deviation from the normal posttreatment course. This can be an injury or negative physical or mental effect caused by medical or surgical or interventional management (rather than the disease process itself) that results in a prolonged hospital stay, disability at discharge, or death.

Composite endpoints An outcome that is reached whenever a subject experiences any one of several distinct but related endpoints.

Confidence interval A range of estimates for an unknown parameter computed at a given confidence level.

Confidence level The long run proportion of confidence intervals that theoretically contain the true value of the unknown parameter.

Conflict of Interests (COI) Circumstances in which a responsible individual(s) may compromise the integrity of the research as a result of personal interests and/or gain.

Confounding variable An extraneous factor related to the independent and dependent variable that influences the cause-and-effect relationship between the variables of interest and may affect the outcome of the experiment.

CONSORT Consolidated Standards of Reporting Trials statement is a set of reporting standards for randomized trials.

Consecutive All patients meeting case series inclusion criteria are included.

Consumer The intended recipient of the product. May be a physician, patient population, or anyone else in the clinical setting.

Contingency table Also known as an RxC table and defined by contained rows and columns, a contingency table is a type of table that displays frequencies of different variables.

Convergent study A type of mixed method research in which both quantitative and qualitative data collection and analysis occur at the same time.

Copyright It is a form of intellectual property law that protects original works, including literary, dramatic, musical, and artistic works.

Correlation coefficient A numerical measure of correlation.

Correlation Any statistical relationship between two random variables or bivariate data.

Cosmetics in IR Cosmetic-based medical procedures done by Interventional radiologists or utilizing interventional radiological techniques/equipment.

Cox PH model Cox proportional hazard model.

CPMP (Committee for Proprietary Medicinal Products) European Medical Agency's committee responsible for expressing the Agency's opinions on issues regarding medicinal products for use in humans.

Crossover bias Bias arising when the control group is mistakenly exposed to the proposed intervention or procedure.

Cross-sectional study A type of observational study that analyzes data from a population or a representative subset of a population at a specific point in time.

cTACE Conventional transarterial chemoembolization is a catheter-directed delivery of chemotherapy and simultaneous embolization.

Customer requirements Customer requirements are specifications that originate with customers as opposed to internal stakeholders and are used to inform the design process.

Data and safety monitoring board See data monitoring committee.

Data coordinators Data coordinators, or *biostatisticians*, are integral members of the team in that their contribution produces the ability for the study to make an impact. While these members are not always individually unique from PIs or Co-Is, the use of numerical input to create meaningful statistics requires a specific understanding of data production. Please refer to the "Statistical Principles" section for additional information.

Data extraction The process of extracting relevant data from the selected studies for inclusion in a systematic review or meta-analysis.

Data leaks The exposure of confidential information to unauthorized parties.

Data monitoring committee An independent data review entity, comprised of a panel of experts, that provides advice on trial continuation based on unmasked data available only to the committee Data and safety monitoring board: See data monitoring committee.

Data presentation The format(s) by which salient data points are summarized and represented.

Decentralized Control is maintained through smaller governing bodies as opposed to a larger solitary authority.

Department of Health and Human Services (HHS) A department of the executive branch of the US government entrusted with the goal of ensuring the health of all Americans and providing essential human services.

Design for assembly Design for assembly (DFA) is the process of designing products for ease of assembly, often with the goal of reducing total part count.

Design for manufacturing Design for manufacturing (DFM) is the process of designing parts with the goal of simplifying manufacturing techniques, such as reducing mold complexity, using commonly available materials, and relaxing dimensional tolerances.

Descriptive A type of statistics that includes measures of central tendency (mean, median, mode), variation range, standard deviation, variance, and percentile/quartile.

Determinants of health The range of personal, social, economic, and environmental factors that determine the health status of individuals or populations.

Devices An instrument, apparatus, implement, machine, contrivance, implant, in vitro reagent, or other similar or related articles, including any component, part, or accessory, which is intended for use in the diagnosis of disease or other conditions, or in the cure, mitigation, treatment, or prevention of disease, in man or other animals, or which does not achieve its primary intended purposes through chemical action within or on the body of man or other animals and which is not dependent upon being metabolized for the achievement of its primary intended purposes.

Directive An authoritative instruction or guidance.

Dissemination phase The phase in which investigators share results via conference presentations, publications, traditional media, social media, and implement strategies for translating results into practice.

Drug product A required component of the CMC application listing the components used in the manufacture of investigational drug products including alternatives used in the manufacturing.

Drug substance A required component of the CMC application that provides a description of the physical and chemical properties of the drug product, the manufacturer, the method of preparation, methods used to ensure identity, purity, and quality of the product, and information supporting the stability of the drug product during toxicological and clinical studies.

Drug-eluting stents (DES) A stent coated with a time release medication with the intent to block cell proliferation and reduce the risk of stenosis or fibrosis post angioplasty.

Drugs A substance intended for use in the diagnosis, cure, mitigation, treatment, or prevention of disease. A substance (other than food) intended to affect the structure or any function of the body.

Dunnett method A post hoc analysis based on the T-distribution to compare experimental groups to control groups.

Efficacy The ability to produce the intended or desired result.

Electronic cigarettes An electronic device that is shaped like a cigarette and contains a nicotine-based liquid that is vaporized and inhaled.

Electronic health record Digital medical record which contains information from multiple providers, ideally providing a more complete view of a patient's medical history.

Embedded design A type of design in which one type of data is secondary to the other and is used to support the primary data's evidence.

Embedded study A type of mixed method research in which quantitative and qualitative data are collected and analyzed within a quantitative or a qualitative design.

Embolization To form a natural or artificial embolus. IR deals in forming artificial emboli through coils, particles, gel foam, etc.

EndoAVF Endovascular arterial venous fistula. A hemodialysis fistula which may be formed through a variety of endovascular techniques.

Endpoint The clinical study objective that is being measured.

Epidemiology A branch of medical science that deals with the incidence, distribution and possible control of diseases and other factors relating to health.

Equivalence margin(s) Predetermined quantity/range within which the novel intervention must not differ from the control with regards to primary clinical endpoint in a trial to be deemed noninferior/equivalent.

Equivalence trial Trial to evaluate whether novel intervention differs from standard intervention by greater than a predetermined range with regards to the primary clinical endpoint.

Evidence-based medicine A practice of medicine that utilizes the best evidence present to treat patients and disease.

Ex vivo Referring to procedures taking place outside of a living organism.

Executive planning committee The committee is responsible for oversight of the multicenter clinical trial and all of its subcommittees.

Experiment A step within the scientific method. A specific plan implemented to accurately identify an answer to a generated research question.

Explanatory clinical trial Randomized clinical trial designed to assess the efficacy of an intervention in ideal circumstances.

Exploratory sequential study A type of mixed method research in which qualitative data collection and analysis is followed by quantitative data collection and analysis.

Exploratory survey A type of survey used to gain general insight into a specific topic that is not clearly understood.

External pilot study Small scale test runs of the larger study that are designed to troubleshoot the mechanics of the study protocol and design.

External validity The validity of applying conclusions from a scientific study to contexts outside of that study.

Factor An independent variable or a variable of interest in the context of a factorial trial.

Factorial design Alternative clinical trial design allowing investigators to simultaneously evaluate two or more variables in an investigation or evaluate interactions between interventions.

Failure Mode and Effects Analysis (FMEA) The process of early analysis of all possible routes of device failure, with the goal to proactively design protections against failure early on in the design process.

Feasibility studies Performed by manufacturers to obtain preliminary data on safe and effectiveness, typically with a small number of subjects, which subsequently guides the design for a pivotal study.

Feasibility The possibility, capability or likelihood of something being done or accomplished.

Federal health organizations Governmental organizations that with objective of maintaining and enhancing public health in the United States under the direction of the Department of Health and Human Services including Administration for Children and Families (ACF), Agency for Healthcare Research and Quality (AHRQ), Agency for Toxic Substances and Disease Registry (ATSDR), Centers for Disease Control and Prevention (CDC), Centers for Medicare and Medicaid Services (CMS), Food and Drug Administration (FDA), Health Resources and Services Administration (HRSA), Indian Health Service (IHS), National Institutes of Health (NIH), and Substance Abuse and Mental Health Services Administration (SAMHSA).

Fee for service A payment method in which health care providers are paid by the patient for each service performed.

Food and Drug Administration (FDA) The United States Food and Drug Administration possesses several functions, among which include ensuring that foods, drugs, vaccines, biological products, and medical devices that are to be used/consumed by people are safe and efficacious.

***F*-test** Any statistical test in which these test statistic has an F distribution under the null hypothesis.

Futility The inability of a trial to achieve its objectives.

Gaussian distribution A theoretical continuous probability distribution in the shape of a bell curve that can be used to infer the normal distribution of data in research studies and statistical tests.

General controls The most basic of regulatory controls including device registration and listing, manufacturing requirements, and prohibitions on adulteration or misbranding, which apply to all medical devices and are deemed sufficient to ensure safety and effectiveness in Class I devices.

Generic (drug) A drug that contains the same chemical substance as a drug that was previously protected by chemical patents.

Geopolitical Politics linked to that of geographical location.

Globalization The process by which various entities interact and integrate on a worldwide scale.

Glossary terms and definitions Equivalence trial: Trial to evaluate whether novel intervention differs from standard intervention by greater than a predetermined range with regards to the primary clinical endpoint.

Glossary

- Progression-free survival: time from random assignment in a trial to disease progression or death.
- Objective response rate: proportion of patients with a complete response or partial response to treatment generally defined using the RECIST criteria or a derivation thereof.
- Time to progression: length of time from initial date of diagnosis (or start of treatment for a particular disease) until disease progression. Time to progression may be utilized to assess the efficacy of a given treatment.

Grant proposal A comprehensive document containing specifics for proposed experiments and submitted to request funding or support.

Graphics processing units (GPUs) A type of circuit originally designed to manipulate and alter memory to accelerate the creation of images in a frame buffer intended for output to a display device. They may additionally be used to process large blocks of data in parallel.

GxP Good x Practices where x may stand for D-distribution, M-manufacturing, S- storage—a set of regulations set forth by the FDA to ensure compliance as well as safety of the drug products throughout their distribution, manufacturing, and storage.

HCC Hepatocellular carcinoma.

HDE Humanitarian device exemption.

Health governance The set of relationships and responsibilities established by a health service between its patients, providers, payers, and policymakers as well as other stakeholders.

Health information technologies Health IT is the processing, storage, and exchange of health information in an electronic environment.

Health monitoring Includes blood pressure, heart rate, respiratory rate, and other basic measures taken during clinical research studies.

Health system The organization of people, institutions, and resources that deliver health care services to meet the health needs of target populations.

Healthcare access bias Bias arising when patients analyzed at a certain hospital are not representative of cases in the overall community population.

HUD Humanitarian use device.

Human subjects Term refers to patients or participants in a clinical research study.

Hydrogel-coated embolization coils Hydrogel-coated embolization coils are a drug/device combination product using a platinum coiling wire surrounded by an expandable hydrogel polymer.

Hypothesis A step within the scientific method. A logical, well-reasoned estimation of the expected answer to a generated research question.

ICP-MS Inductively coupled plasma mass spectrometry, a type of mass spectrometry that uses inductively coupled plasma to ionize a given sample decomposing the sample into basc atoms and small polyatomic ions which may then be further analyzed using mass spectrometry techniques.

IDE Investigational device exemption.

IDEAL framework Innovation, Development, Exploration, Assessment, and Long-Term (IDEAL) framework designed to promote high-quality surgical and interventional procedural innovation.

Implementation research Implementation research focuses on developing specific solutions to apply scientific discoveries into the daily medical practice.

In vitro (of a biological process) made to occur in a laboratory vessel or other controlled experimental environment rather than within a living organism or natural setting.

In vivo (of a biological process) occurring or made to occur within a living organism or natural setting.

Incidence The occurrence, rate, or frequency of a disease in the context of medicine or public health.

Inclusion and exclusion criteria Criteria used to select studies for inclusion or exclusion in the context of a systematic review or meta analysis.

Independent samples *t*-test Statistical test used to compare means from two independent samples.

Informed consent The process of making a free and informed decision (such as to participate in research). Individuals who provide informed consent must be legally competent and have enough decision-making capacity to consent to research. Research regulations specify the types of information that must be disclosed to the subject.

INHANCE trial Indacaterol to Help Achieve New COPD Treatment Excellence trial, a randomized clinical trial of inhaled indacaterol (long-acting beta 2 agonist bronchodilator) for the treatment of chronic COPD.

Institutional review board (IRB) a committee responsible for reviewing and overseeing human subjects research. An IRB may also be called a research ethics committee (REC) or research ethics board (REB). IRBs usually include members from different backgrounds and disciplines, with institutional and outside members, scientists, and non-scientists.

Institutional Review Board A committee whose main focus is to ensure that research studies remain ethical and do not unnecessarily bring harm to human participants.

Intellectual property It is a property that results from a person's original creative thoughts. These include patents, copyrights, trademarks, and trade secrets.

Intention-to-treat analysis A fundamental way to analyze a randomized control trial in which all subjects allocated to each arm of the trial are analyzed "as intended" upon randomization, whether or not they received the exposure allocated or completed treatment.

Interaction A situation in which the effect of one causal variable on an outcome depends on the state of any additional N causal variables i.e. The effect of two causal variables on one outcome do not create an additive effect.

Internal pilot study Studies are designed to help adjust sample size and statistical power calculations using realistic recruitment rates and to help set realistic expectations for how long the study will take to complete.

Internal validity A measure of reliability and soundness of research. Internal validity may express whether the research was conducted appropriately.

Interventionalists Broad term to describe physicians who perform interventions.

Intervention In the context of clinical practice, interventions are generally considered minimally invasive medical procedures usually achieved through some percutaneous means.

Invention It is a novel, useful process, improvement, or design which did not exist beforehand that was developed as a product of some unique intuition or work.

Investigational device exemption (IDE) Application-based regulatory process to evaluate preclinical safety and efficacy.

Investigational use in vitro diagnostic (IVD) product An IVD product being shipped or delivered for product testing prior to full commercial marketing (e.g., for use on specimens derived from humans to compare the usefulness of the product with other products or procedures in current use or recognized as useful). These products must be labeled according to 21 CFR 812.5 for nonsignificant risk or significant risk devices and according to 21 CFR 809.10(c) (2) (ii) for devices that are exempt under 812.2(c).

Investigator Manages and reports events associated with the product during its study according to the given protocols.

Investment bank A financial services company that serves an advisor for financial transactions and creates financial products such as the underwriting of debt and equity securities.

Investigator bias Bias arising when an investigator's unconscious conflict of interest which may influence the general attitude toward a research project.

Investigator IND Application submitted by a physician who wishes to initiate and conduct a study involving an unapproved investigational agent or an approved investigational agent for off-label indications or population.

IRB protocol A protocol for the research study of human subjects submitted to the institutional review board before the initiation of the study.

ITT (intention to treat) analysis Analysis of results based on initial treatment assignment regardless of eventual treatment received.

IVC filter Device placed in the IVC to reduce the risk of pulmonary embolism, generally in patients with contraindications to anticoagulation.

Justice (1) Treating people fairly. (2) An ethical principle that obligates one to treat people fairly. Distributive justice refers to allocating benefits and harms fairly; procedural justice refers to using fair processes to make decisions that affect people; formal justice refers to treating similar cases in the same way.

K-award NIH's career development awards, given to support research career development for promising clinicians.

KL2 program A research training and development grant program sponsored by NIH found at partnering institutions that provides supportive funding to candidates with a completed M.D, D.O., or Ph.D. degree to participate in a curriculum to gain additional training in research in clinical and translational sciences (1–5 years of total training), with scholars eventually obtaining a master's or doctoral degree in clinical and translational science while as they are given dedicated time for laboratory or clinical research.

Kruskal–Wallis test A nonparametric test to compare ordinal variables between three or more populations from independent samples.

Lasagna's law of research The incidence of patient availability sharply decreases when a clinical trial begins and returns to its original level near trial completion.

Legal authority Power designated by the law to oversee and enforce rules and regulations.

Level Number of divisions of one factor in a factorial trial.

Linear regression analysis If the dependent variable is continuous and normally distributed, the linear regression equation ($Y = a + b \times X$) can be used to estimate the dependent variable for different values of the independent variable.

Liquidity event This occurs when the founders of a business or early investors in a business receive cash for their portion of equity.

Livestock Domesticated animals raised in an agricultural setting to produce labor and commodities such as meat, eggs, milk, fur, leather, and wool.

LM/GLM Linear models/generalized linear models.

Local health department Local government organization within each local jurisdiction (county province, city) with the objective of maintaining and enhancing public health in the jurisdiction.

Logistic regression analysis A statistical model that models the probability of an event occurring by having the log odds of the event be a linear combination of one or more independent variables.

Machine learning (ML) A subset of AI devoted to algorithms that can improve at performing tasks and making predictions through exposure to data.

Macro environment Uncontrollable set of external conditions that can affect the development of a business which commonly include inflation, interest rates, social conditions, government policies, technological developments, and climate changes.

Main effect Relationship between the factor and clinical outcome.

Maximum tolerable dose (MTD) Maximum tolerable dose (MTD) is the highest dose of a drug or treatment that does not cause unacceptable side effects. The maximum tolerated dose is determined in clinical trials by testing increasing doses on different groups of people until the highest dose with acceptable side effects is found.

Medicinal A substance or plant that has healing properties.

Meta-analysis A statistical method for combining the results of multiple studies to estimate the overall effect of an intervention or exposure on a specific outcome.

Microbial limits testing A required test in the Drug Product Section of the CMC that is used to determine if drug products are within acceptable microbial standards.

Minimization A method of adaptive stratified sampling used in clinical trials to minimize the imbalance between patients in various treatment/control groups over a number of preselected factors.

Mixed-method research A type of research methodology that involves combining or integrating qualitative and quantitative research approaches into a single research study.

Morbidity A measure related to the disease burden and symptoms from a disease.

Mortality A measure related to the number of deaths caused by the health event under investigation.

Multicenter clinical trial A collaborative study in human subjects involving multiple, independent care centers that investigates the risks and benefits of a novel clinical intervention with the goal of maximizing the generalizability of the results.

Multimethod research A type of research methodology that involves combining either multiple qualitative studies alone or multiple quantitative methods alone.

Multiphase study A type of mixed-method research in which quantitative and qualitative studies are conducted separately and later analyzed together.

Municipal Relating to a city or town or its governing body.

Narrative review A type of evidence synthesis that provides an account of what is believed to be the state of the art and consequent conclusions, based on the opinions of an expert or group of experts.

National Institutes of Health (NIH) An agency within HHS which provides grant funding for the vast majority of medical research.

National Institute of Health US governmental agency involved in conducting research in public health and biomedical sciences

National Science Foundation (NSF) An independent government agency which provides critical research funding to non-medical scientific fields including engineering.

NIH RePORT NIH's Research Portfolio Online Reporting Tools catalog which allows users to search catalog research projects which have received funding.

NN Neural network.

Noncomparative Term for any study without a comparison or control group.

Nonconsecutive Some patients meeting case series inclusion criteria are included.

Nongovernmental organizations A nonprofit civil society group that functions independently of any government and can be organized on a community, national, and international levels to serve a social or political goal such as humanitarian causes.

Noninferiority trial Trial to evaluate whether novel intervention is less than the standard intervention by greater than a predetermined quantity with regards to the primary clinical endpoint.

Nonrandomized controlled trial A study in which subjects are assigned into two groups deliberately: control and treatment groups. In some cases, researchers make participant assignments, while in others, participants effectively assign themselves.

Nonsignificant risk (NSR) device A device that does not meet the definition of significant risk (SR) device (see definition below). An IDE is considered approved for a NSR investigational device study once sponsors meet the abbreviated requirements found in the "Investigational Device Exemptions" regulation at 21 CFR 812.2(b). The risk determination for an investigational device study should be based on the proposed use of the device in the investigation in addition to the characteristics of the device.

Nonsignificant risk No potential for serious risk to health, safety, or welfare of a subject and only require Institutional Review Board approval as sufficient for meeting IDE requisites.

Nonstratified randomization For this randomization strategy, random chance essentially determines the assignment of study participants.

Nontechnical error A human error associated with team working and/or general cognitive skills.

Not classified device Postamendment device types that have yet to be classified.

Notice of award (NoA) Official notice of grant award.

Null hypothesis The statement or position that there is no relationship between two analyzed variables.

Observation study A study where no intervention is made, providing estimates and associations of events in the natural setting, hence the investigator does not alter the exposure status.

Observation A step within the scientific method. The act of paying attention to the surrounding world to recognize important details and generate questions.

Observational study Also called epidemiological studies, observational studies are those where the investigator is not acting upon study participants, but instead observing natural relationships between factors and outcomes.

Odds ratio The ratio of odds. A measure of the likelihood of a particular outcome, generally defined as probability of an event occurring divided by one minus the probability of an event occurring.

Office-based labs (OBLs) Outpatient care offices opened by IR physicians, which allow for a delivery-focused care model in a nonhospital setting that is patient-centered. Also known as Office Interventional Suites.

Off-label procedures Unapproved use of a drug or treatment for a specific disease, illness, or health condition.

Oncologic Relating to the study and treatment of tumors.

Opportunity cost This represents the missed benefits that result from an investor or business choosing one investment over the other.

Other adverse event Any unwanted or unintended side effect associated with recent medical treatment that may or may not be directly related to the treatment.

Outcome measures The primary data that address the trial's central research question.

Overall survival The percent of subjects in a given group that remain alive after a given period of time from the designated start point, usually from a time of initial diagnosis or initiation of a select treatment.

Overhead An expense related to running a business that is not directly attributed to the production of a product or services. A common example would be rent on an office building.

PACS Picture archiving and communication system.

Paired samples *t*-test Statistical test used to compare means from two related samples.

Palliative care Palliative care is a subspeciality of medicine that focuses on the improvement of quality of life of patients and their families who are facing challenges associated with life-threatening illness. It involves a range of services delivered by a range of professionals including physicians, nursing, support workers, paramedics, pharmacists, physiotherapists, and volunteers.

Partial or complete hold An order for the delay or suspension due to deficiency in a clinical investigation.

Patent It is exclusive, which is given to an inventor by the government for a certain number of years to make, use, or sell an invention.

Patient-Centered Outcomes Research (PCOR) Research that compares the impact of two or more preventive, diagnostic, treatment, or healthcare delivery approaches on health outcomes, including those that are meaningful to patients.

Patient-Centered Outcomes Research Institute (PCORI) A nonprofit institute created through the 2010 Patient Protection and Affordable Care Act that financially supports comparative effectiveness research. It assists consumers, clinicians, purchasers, and policy makers in making informed decisions intended to improve healthcare at both the individual and population levels.

Patient-reported outcome measure (PROM) A tool (e.g., questionnaire) that can record the consequences of disease and/or a treatment received as reported by the patient, for example, the patient's perceptions of their health or physical/mental well-being, experienced symptoms, physical functioning, and treatment satisfaction.

Pearson's chi-square test A frequently used test used to determine if a difference exists between the observed and expected frequencies of categorical variables.

Pearson's correlation coefficient A measure of linear correlation between two sets of data.

Per protocol analysis Analysis which involves the describes only the outcomes of study participants who adhered to the research protocol, excluding those who deviated. Also referred to as *explanatory analysis* or *analysis by treatment administered.*

Pharmacodynamics (PD) Pharmacodynamics (PD) is the study of a drug's molecular, biochemical, and physiologic effects or actions. It comes from the Greek words "pharmakon," meaning "drug," and "dynamikos," meaning "power."

Pharmacokinetics (PK) Pharmacokinetics (PK) is the study of how the body interacts with administered substances for the entire duration of exposure.

Pharmacology The study of uses, effects, and modes of action of drugs.

Phase I clinical trial Early clinical trial where researchers test a new intervention within a small group of people with safety being the primary outcome of interest.

Phase II clinical trial Clinical trial where researchers study a novel intervention within a moderate size group of subjects with safety and dosage being the outcomes of interest.

Phase III clinical trial Large research study comparing a novel intervention to other accepted treatments with efficacy and adverse events as the primary outcome measures.

Phase IV clinical trial Postmarketing clinical trial study to investigate the risk, benefits, and ideal patient population for a new intervention.

Piloting A small-scale preliminary studying before implementing the full intended study.

Pioneer drug Brand name or a patented version of the drug.

Pivotal studies Performed by manufacturers to support a marketing application with a purpose to obtain data on safety and effectiveness for a specified intended use, usually with a statistically justified number of subjects.

Placebo An inert treatment or substance that is designed to have no therapeutic effect.

Planning phase The phase in which investigators define the research question, systematically review the literature, identify outcome measures, conduct pilot studies to ensure feasibility, and estimate power.

PMA Premarket approval, a rigorous premarket submission process often involving clinical trials and other requirements, which apply to Class III devices, befitting their high-risk designation.

PMN Premarket notification.

Policymaker A person responsible for or contributing the formulation of ideas and plans to be implemented, typically at the government or business level.

Postamendment device Device types that were not in existence prior to the Medical Device Amendments of 1976 to the Federal Food, Drug, and Cosmetic Act, thus requiring premarket notification or approval depending on the device classification.

Postmarket device safety monitoring The close monitoring of a medical device once it has been approved and released into the market for usage, to ensure that a particular device continues to function as clinically intended, as well as to closely monitor patients for any possible adverse effects/outcomes arising from the usage of a particular device. Any adverse effects/outcomes should be immediately documented and reported to the appropriate agencies to ensure that this is made known and closely monitored. There are multiple types of monitoring, for instance, active or passive monitoring, which may be utilized.

Power The probability of correctly rejecting a true null hypothesis in the context of statistics.

Pragmatic clinical trial Randomized clinical trials designed to assess the effectiveness of an intervention in real-world, routine clinical practice.

Preamendment device Device types that were in existence prior to the Medical Device Amendments of 1976 to the Federal Food, Drug, and Cosmetic Act and thus did not require premarket notification or approval.

Preclinical research Research done before human testing and clinical trials. Typically includes basic sciences and testing in animal models.

Premarket approval application (PMA) The application for approval required prior to the marketing of most Class III medical devices (Section 515 of the Act, 21 USC 360e). (See definitions of Class I, II, and III devices above.) PMA approval is based on a determination by FDA that the applicant's submission provides sufficient valid scientific evidence to provide reasonable assurance that the device is safe and effective for its intended use(s). The PMA regulation is 21 CFR Part 814.

Premarket approval An application with results from a pivotal study that must show the device is either safe and/or more effective in it's desired outcome than the current standard of care.

Prevalence Total number of individuals in a population who have a disease or health condition at a specific period of time.

Primary clinical endpoint The main clinical result which is being investigated and for which the subjects are randomized and for which the trial is powered.

Primary endpoint A precisely defined variable that reflects the clinical outcome most important for a study to observe and thus dictates how the study is powered.

Principal investigator (PI) Is responsible for all activities associated with conducting research and complying with institutional and federal policies. This member is ultimately the overseer of the project as a whole. Students, trainees, and fellows function under the advisor-ship of the PI.

Principle investigator Primary individual responsible for conducting and managing a particular research project.

Private equity (PE) An alternative investment class that includes private companies that are not listed on public stock exchanges.

Professional subject A person who may enroll in studies in a serially monogamous manner without a washout period, or more than one study at a time with no intention of adhering to the protocol. This is typically done as a way to make money or gain attention.

Product A device, procedure, technique, or other implementation of evidence-based medicine that is meant to satisfy the need(s) of a designated consumer.

Project development phase The phase in which investigators identify collaborators, develop the protocol and research operations manual, prepare grant applications, obtain ethical approval, execute subside contracts, register the study in a clinical trial registry, form a manuscript committee, and conduct feasibility testing and data validation at each site.

Prospective A study or investigation that looks forwards in time.

Protocol Detailed plan for conducting a review or meta-analysis.

Provisional patent It means the license appliable only for the time being and needs to be converted into a permanent one.

Public health The science of protecting, promoting, and improving the health of entire communities of people.

Publication bias Bias arising when media publishes studies showing positive or significant findings with exaggerated headlines or far-reaching implications of study implications, which may falsely lead to increased perceived efficacy of an intervention.

***p*-Value** The probability of obtaining test results at least as extreme as the result actually observed under the assumption that the null hypothesis is correct.

Qualitative interaction An effect from one factor that causes the main effect of another factor to reverse its effect.

Qualitative methods of research The use of nonnumerical data to evaluate behaviors, perceptions, and experiences.

Quality appraisal The process of evaluating the quality and relevance of studies included in a systematic review or meta-analysis, using established tools and criteria.

Quality assurance (QA) The product-oriented concept of design and development to meet specified criteria.

Quality control (QC) The process-oriented concept and action of analyzing and verifying a product's suitability for use according to specified criteria.

Quality management (QM) The combination of QC and QA, a concept meant to encompass all aspects required for a product to meet specified criteria.

Quality management system (QMS) A comprehensive and systemic approach to study design, which allows QM to be integrated at every step of the protocol.

Quality of life A broad range of human experiences related to one's overall well-being that are likely to be defined by subjective experiences and perceptions.

Quality team (QT) The individual(s) delegated by the sponsor to monitor the study to ensure the resources, standard operating procedure (SOP), and information follow the quality management criteria, and alert the sponsor to any deviations.

Quantitative interaction An effect from one factor that causes minor changes to the main effect of another factor but does not reverse its effect.

Quasi-experimental design A type of experimental study design where subjects are assigned to control and treatment groups in a nonrandom fashion. An archetypical example is nonrandomized controlled trials.

Question A step within the scientific method. The act of specifically defining what a researcher is trying to determine through their research.

Radiation Emission or transmission of energy in the form of waves or particles through a medium.

Randomization Formal process by which participants are randomly assigned to a treatment group at a predetermined ratio, eliminating allocation bias and generating comparable groups.

Randomized clinical trial Type of blinded prospective study design in which participants are randomized and assigned to either a control group or treatment group(s). This type of study design has several merits, as well as drawbacks.

Randomized controlled trials (RCTs) A type of study that randomly assigns participants into an experimental group or a control group. As the study is conducted, the only expected difference between the control and experimental groups is the outcome variable being studied.

Real-world nonadherence Unintentional deviations from the study protocol that occur due to lack of patient education, cognitive function, or resources, or unexpected crossover that occurs due to chance.

Recall bias Bias occurring when patients are more or less likely to recall and disseminate information about their exposure(s) depending on their outcomes.

Reduction to practice The process of developing an invention past the conception stage. Can be *actual*, creating a working device, or *constructive*, creating a patent application with appropriately detailed descriptions.

Registry An organized system of data collection used to evaluate specified outcomes for a population defined by a particular disease, condition, or exposure, and that serves a predetermined scientific, clinical, or policy purpose.

Regulator The highest position in the hierarchy of quality control. Regulators establish the overall national (or international) guidelines and standards for quality systems.

Reliability A characteristic that describes how well a measure will demonstrate the same outcome when repeated in similar clinical circumstances.

Reporting bias Combination of publication bias and citation bias.

Research (noncommercial) IND Application submitted by an individual investigator.

Research coordinators (RCs) Prepare, direct, and coordinate clinical trials under the supervision of a PI.

Research operations manual A step-by-step guide to policies and standard operation procedures associated with the multicenter clinical trial that includes a breakdown of study team members, study flow, inclusion and exclusion criteria, study methods, training, and details regarding data collection and sharing.

Research protocol A document that details the background, rationale, objectives, methods, and organization of a clinical study along with statistical and ethical considerations.

Respondent fatigue Decline in the engagement of survey respondents due to becoming tired or bored of the task.

Responsiveness A characteristic that describes how sensitive an outcome measure is toward detecting changes in the risk/benefit an intervention offers.

Retrospective A study or investigation that looks back-ward in time.

RF Random forest.

Risk/benefit analysis A process for determining an acceptable level of risk, given the potential benefits of an activity or technology. See also Risk Management, Precautionary Principle.

RO1 Also known as research project grant, which is the most commonly used grant program within NIH.
RO3 NIH's small grant program, which offers funding for short-term and smaller projects.
Safety Based on valid scientific evidence, that the probable benefits to health from the use of the device for its intended uses and conditions of use, when accompanied by adequate directions and warnings against unsafe use, outweigh any probable risks.
Sample frame A list used to define a researcher's population focused on the specific interest the researcher plans to pursue for his or her study. It defines a set of elements that can aid the researcher in selecting a sample from the target population.
Sample representativeness A particular set of parameters utilized in approximating the sample size as well as sample composition.
Sample size The sample size refers to the number of individuals that are included within a particular study.
Scheffe's method An exploratory post hoc analysis based on the *F*-distribution that analyzes all pairwise combinations to generate a hypothesis.
Scientific method A systematic approach to scientific discovery. Most often consists of the following steps A systematic approach to scientific discovery. Most often consists of the following stepsObservation, Question, Hypothesis, Experiment, Analysis, and Conclusion.
Scientific review officer (SRO) National Institute of Health (NIH)-employed individual that manages the peer review process for grant applicants.
Secondary endpoint Additional clinical outcomes of interest to investigators that exercise less bearing on the research protocol than primary endpoints.
Selection bias A form a bias introduced by nonrandomized selection of individuals or groups for analysis.
Self selection bias Bias arising when individuals who volunteer to participate in a study have different relevant clinical characteristics than those who elect not to join the study.
Sentinel initiative Created by the Food and Drug Administration (FDA) to determine new and improved methods by which to assess the safety of medical products approved within the United States of America, for instance—medications, vaccines, and medical devices.
Serious adverse event Death or life-threatening illness warranting hospitalization after medical intervention.
Shelf-life The length of time for which an item remains useable, fit for consumption, or saleable.
Significant risk (SR) device An investigational device that presents a potential for serious risk to the health, safety, or welfare of a subject and (1) is intended as an implant; (2) is purported or represented to be for use in supporting or sustaining human life; (3) is for a use of substantial importance in diagnosing, curing, mitigating, or treating disease, or otherwise preventing impairment of human health; or (4) otherwise presents a potential for serious risk to the health, safety, or welfare of a subject. The risk determination for an investigational device study should be based on the proposed use of the device in the investigation in addition to the device characteristics. Sponsors of significant risk device studies must apply to FDA for an Investigational Device Exemption (IDE) (see definition above). (21 CFR 812.3(a), 812.3(m); 812.20).
Significant risk Potential for serious risk to health, safety, or welfare of a subject.
Simulation The imitative representation of the functioning of one system or process by means of the functioning of another.
SIR Society of interventional radiology is an organization focused on advancement of minimally invasive, image-guided therapeutic interventions for disease management.
Small business Business with less than 500 employees predominately owned in the United States.
Small Business Innovation Research (SBIR)/Small Business Technology Transfer Program (STTR) Governmental programs that provide monetary funding to small businesses seeking to develop commercialized products that coincide with government interests.
SMART A commonly used objective setting framework used in project management that focuses on creating goals that are specific, measurable, achievable, realistic, and time sensitive.
Spearman's correlation coefficient A nonparametric test that can be utilized in nonnormal distributions, which calculates the correlation coefficient from continuous data ranks.
Special controls Controls, often device-specific, that include the meeting of certain performance standards, postmarket surveillance, and special labeling and data requirements, to which Class II devices are subject.
SPIRIT Standard Protocol Items: Recommendations for Interventional Trials.
Sponsor Develops the study protocol and performance criteria according to the needs of the consumer.
Sponsor–investigator The investigator conducting the proposed clinical investigation who also submits the IND application.
Stability testing Description of tests used to confirm stability of the drug product during toxicology and clinical studies.
Standard of care Research done before human testing and clinical trials. Typically includes basic sciences and testing in animal models.
State health department State government organization within each state with the objective of maintaining and enhancing public health in the state's jurisdiction.
Statistical assumptions Fundamental inferences made by statistical tests to base calculations and conclusions.
Statistical power See Power.
Statistical significance The probability that an observed difference is attributable to the intervention.
Stratified randomization A sampling method whereby the whole study population is first stratified into subgroups with similar attributes or characteristics, then the stratified groups undergo simple random sampling.
Stratified sampling Total target population is divided into smaller subgroups.
Study conduct phase Part of the patient screening process; specifically implementing inclusion and exclusion criteria.
Study design Refers to the method and format in which the study tests and gathers results.
Study execution phase The phase in which investigators recruit and enroll patients, perform quality assurance measures, and analyze data.

Study follow-up period Helps in identifying patients that are compliant vs. noncompliant. This can affect the power of the study.

Study protocol phase Trial planning phase where researchers categorize their trial based on efficacy and effectiveness.

Subject recruitment Aids in identifying crucial strategies that benefit subject inclusion.

Surrogate endpoints Responsive and easily measured laboratory results and physical exam signs that reflect a biological response to therapy to predict future clinical benefit.

Survey instrument The questionnaire that serves as primary source of data collection.

Survey questionnaire Research tools (often consisting of a set of open-ended and/closed-ended questions) to collect data from study participants; also referred to as survey instrument.

Survey research The collection of information from a sample of individuals through their responses to questions.

Swine Any of various stout-bodied short-legged omnivorous artiodactyl mammals (family Suidae) with a thick bristly skin and a long flexible snout.

Synthesis The process of combining the results of multiple studies to estimate the overall effect of an intervention or exposure on a specific outcome.

Systematic review A systematic and transparent process for identifying, selecting, and synthesizing the results of all relevant studies on a specific research question.

Systemic random sampling Individuals are selected based on fixed intervals in a numbered population.

TACE Transcatheter arterial chemoembolization—a minimally invasive procedure performed by an interventional radiologist that uses embolic particles coated with chemotherapeutic drugs with the goal of restricting a tumor's blood supply.

TAGGS HHS' Tracking Accountability in Government Grant Systems allowing users to search HHS-funded grants via a variety of criteria.

Technical benchmarking Technical benchmarking is the process of quantitatively assessing or testing the capabilities of competitor devices, to generate functional baselines for the design of a new device.

Technical error A human error associated with knowledge, technical skill, or expertise.

Technical risk The likelihood that a project will not succeed due to challenges related to product construction.

Tensor processing units (TPUs) Is an AI accelerator application-specific integrated circuit (ASIC) developed by Google for neural network machine learning, using Google's own TensorFlow software.

TL1 program A research training and development grant program sponsored by NIH found at partnering institutions that support medical students, predoctoral students, and postdoctoral students with funding to participate in a curriculum that trains students to gain research skills and experience in clinical and translational sciences through coursework, didactic lectures, and dedicated time to participate in laboratory or clinical research.

Toxicology The measurement and analysis of potential toxins, intoxicating or banned substances, and prescription medications present in a person's body.

Trade secret It is a secret process, technique, or method which is used by an organization or an individual to provide them a competitive advantage over others.

Trademark It is a word, phrase, symbol, or design which identifies and distinguishes the source of the good of one party from those of others.

Transarterial chemoembolization See TACE.

Transarterial radioembolization A transarterial catheter directed therapy for the treatment of various cancers, generally using Yttrium-90 or Holmium-166 doped microparticles.

Transcatheter Generally referring to minimally invasive non-surgical techniques or therapies, typically from a percutaneous approach, using devices such as wires and catheters.

Translational science The process of transforming laboratory observations into interventions that benefit the health of the public.

Triangulation A method that uses different but complementary information on the same topic.

True experimental design A type of experimental study design where subjects are assigned to control and treatment groups randomly. An archetypical example is randomized controlled trials.

TTP Time-to-progression is the time from disease diagnosis or treatment to the time of progression of disease.

Tukey's method A post hoc analysis that compares every possible pair to determine if there is a difference between the means.

Type 2 statistical error Failure to reject a false null hypothesis.

Unclassified device Preamendment device types that did not get classified during the amendments process started in 1976 and are still unclassified to this time.

Validity A characteristic that describes the degree to which an outcome is able to measure what it intends to.

Value Inflection An event that results in a significant change in the value of a company or process; in an engineering process, it centers around the concept of minimizing risk.

Value proposition Conveys why a novel intervention is unique and useful from the perspective of *specific* group of patients or other stakeholders.

Value The monetary worth of the benefits that the product provides to the customer.

Varicose veins Nonuniform, twisted, and engorged veins are usually in the lower extremity. Can cause pain but is also a cosmetic concern for patients when superficial.

Venture capital (VC) It is a form of private equity that includes private companies not listed on public stock exchanges. The companies which take venture capital are much earlier stage companies than those of traditional private equity and they use that investment to finance their startup.

VIRTEX A data centralized registry released in 2020 offered by the Society of Interventional Radiology (SIR).

Voluntariness The ability to make a free (un-coerced) choice.

Woodchucks A grizzled thickset marmot (Marmota monax) chiefly of Alaska, Canada, and the north-eastern United States; synonym: groundhog.

YAS Yttrium aluminosilicate is a glass microparticle containing yttrium.

Yttrium-90 A radioactive isotope of Yttrium currently used as a treatment for certain cancers including hepatocellular carcinoma and metastatic colon cancer.

Index

Note: 'Page numbers followed by "f" indicate figures and "t" indicate tables.'

D

Q

R

X

Y

Z

9780128230268